Edited by
Armando Córdova

Catalytic Asymmetric Conjugate Reactions

Related Titles

Dai, L.-X., Hou, X.-L. (eds.)

Chiral Ferrocenes in Asymmetric Catalysis

Synthesis and Applications

2010
ISBN: 978-3-527-32280-0

Kollár, L (ed.)

Modern Carbonylation Methods

2008
ISBN: 978-3-527-31896-4

Börner, A. (ed.)

Phosphorus Ligands in Asymmetric Catalysis

Synthesis and Applications

3 Vol
2008
ISBN: 978-3-527-31746-2

Yamamoto, H., Ishihara, K. (eds.)

Acid Catalysis in Modern Organic Synthesis

2 Vol
2008
ISBN: 978-3-527-31724-0

Maruoka, K. (ed.)

Asymmetric Phase Transfer Catalysis

2008
ISBN: 978-3-527-31842-1

Stepnicka, P. (ed.)

Ferrocenes

Ligands, Materials and Biomolecules

2008
ISBN: 978-0-470-03585-6

Christmann, M., Bräse, S. (eds.)

Asymmetric Synthesis – The Essentials

2008
ISBN: 978-3-527-32093-6

Hudlicky, T., Reed, J. W.

The Way of Synthesis

Evolution of Design and Methods for Natural Products

2007
ISBN: 978-3-527-32077-6

Hiersemann, M., Nubbemeyer, U. (eds.)

The Claisen Rearrangement

Methods and Applications

2007
ISBN: 978-3-527-30825-5

Cornils, B., Herrmann, W. A., Muhler, M., Wong, C.-H. (eds.)

Catalysis from A to Z

A Concise Encyclopedia

2007
ISBN: 978-3-527-31438-6

Edited by Armando Córdova

Catalytic Asymmetric Conjugate Reactions

WILEY-VCH

WILEY-VCH Verlag GmbH & Co. KGaA

The Editor

Prof. Armando Córdova
Department of Organic Chemistry
Stockholm University
Arrhenius Laboratory
106 91 Stockholm
Schweden

All books published by **Wiley-VCH** are carefully produced. Nevertheless, authors, editors, and publisher do not warrant the information contained in these books, including this book, to be free of errors. Readers are advised to keep in mind that statements, data, illustrations, procedural details or other items may inadvertently be inaccurate.

Library of Congress Card No.: applied for

British Library Cataloguing-in-Publication Data
A catalogue record for this book is available from the British Library.

Bibliographic information published by the Deutsche Nationalbibliothek
The Deutsche Nationalbibliothek lists this publication in the Deutsche Nationalbibliografie; detailed bibliographic data are available on the Internet at <http://dnb.d-nb.de>.

© 2010 WILEY-VCH Verlag GmbH & Co. KGaA, Boschstr. 12, 69469 Weinheim, Germany

All rights reserved (including those of translation into other languages). No part of this book may be reproduced in any form – by photoprinting, microfilm, or any other means – nor transmitted or translated into a machine language without written permission from the publishers. Registered names, trademarks, etc. used in this book, even when not specifically marked as such, are not to be considered unprotected by law.

Cover Design Grafik Design Schulz, Fußgönheim
Typesetting Laserwords Private Limited, Chennai, India
Printing and Binding Strauss GmbH, Mörlenbach

Printed in the Federal Republic of Germany
Printed on acid-free paper

ISBN: 978-3-527-32411-8

Contents

Preface *XIII*
List of Contributors *XV*

1	**Rhodium- and Palladium-Catalyzed Asymmetric Conjugate Additions** *1*	
	Guillaume Berthon and Tamio Hayashi	
1.1	Introduction *1*	
1.2	Rh-Catalyzed ECA of Organoboron Reagents *1*	
1.2.1	α,β-Unsaturated Ketones *2*	
1.2.1.1	A Short History *2*	
1.2.1.2	Catalytic Cycle *4*	
1.2.1.3	Model for Enantioselection *7*	
1.2.1.4	Organoboron Sources Other than Boronic Acids *9*	
1.2.1.5	Rh Precatalysts *10*	
1.2.1.6	Ligand Systems *11*	
1.2.1.7	Bidentate Phosphorus Ligand *11*	
1.2.1.8	Monodentate Ligand *14*	
1.2.1.9	Diene Ligands *16*	
1.2.1.10	Bis-Sulfoxide *18*	
1.2.1.11	Mixed Donor Ligands *19*	
1.2.1.12	Trapping of Boron Enolates *19*	
1.2.1.13	α,β-Unsaturated Aldehydes *20*	
1.2.2	α,β-Unsaturated Esters and Amides *22*	
1.2.2.1	Diastereoselective Conjugate Addition *26*	
1.2.2.2	Fumarate and Maleimides *26*	
1.2.2.3	Synthetically Useful Acceptors for Rh-Catalyzed ECAs *29*	
1.2.3	Other Alkenes *30*	
1.2.3.1	Alkenylphosphonates *31*	
1.2.3.2	Nitroalkenes *31*	
1.2.3.3	Sulfones *33*	
1.2.3.4	1,4-Addition/Enantioselective Protonation *34*	
1.3	Rh-Catalyzed ECA Organotitanium and Organozinc Reagents *38*	
1.4	Rh-Catalyzed ECA of Organosilicon Reagents *41*	
1.5	Rh-Catalyzed ECA with Other Organometallic Reagents *44*	

Catalytic Asymmetric Conjugate Reactions. Edited by Armando Córdova
Copyright © 2010 WILEY-VCH Verlag GmbH & Co. KGaA, Weinheim
ISBN: 978-3-527-32411-8

1.6	Rh-Catalyzed ECA of Alkynes	44
1.7	Rh-Catalyzed Tandem Processes	47
1.7.1	Tandem ECA/Aldol Reaction	47
1.7.2	Tandem Conjugate Addition/1,2-Addition	48
1.7.3	Tandem ECA/Mannich Reaction	49
1.7.4	Tandem Conjugate Addition/Michael Cyclization	50
1.7.5	Tandem Carborhodation/Conjugate Addition	50
1.8	1,6-Conjugate Additions	54
1.9	Pd-Catalyzed ECA	56
1.9.1	Catalytic Cycle	62
1.10	Conclusions	63
	References	64

2 Cu- and Ni-Catalyzed Conjugated Additions of Organozincs and Organoaluminums to α,β-Unsaturated Carbonyl Compounds 71

Martin Kotora and Robert Betík

2.1	Introduction	71
2.2	General Aspects	72
2.2.1	Properties of Organozinc and Organoaluminum Compounds	72
2.2.1.1	Reaction Mechanisms of Cu-Catalyzed Conjugated Addition	72
2.2.1.2	Reaction Mechanisms of Ni-Catalyzed Conjugated Addition	73
2.2.2	Preparation of Organozinc Compounds	74
2.2.2.1	Preparation by a Direct Insertion into the Carbon–Halide Bond	74
2.2.2.2	Preparation by Lithium–Zinc Transmetallation	75
2.2.2.3	Preparation by an Iodine–Zinc Exchange Reaction	76
2.2.2.4	Preparation by a Boron–Zinc Exchange	76
2.2.2.5	Preparation by Other Metal–Zinc Exchange	77
2.2.2.6	Commercial Availability	77
2.2.3	Preparation of Organoaluminum Compounds	77
2.3	Conjugated Additions	78
2.3.1	Cu-Catalyzed Conjugated Addition	78
2.3.2	Ni-Catalyzed Conjugated Addition	79
2.4	Ligands for Cu-Catalyzed Enantioselective Conjugated Additions	81
2.4.1	Phosphoramidites	81
2.4.2	Phosphines Bearing an Amino Acid Moiety	93
2.4.3	Phosphines	100
2.4.4	Phosphites	101
2.4.5	NHC-Compounds	107
2.4.6	Various Ligands (Ligands with Mixed Functionalities)	109
2.4.7	Selected Experimental Procedures	116
	Kinetic resolution of racemic 5-methylcyclohex-2-enone by Cu-catalyzed enantioselective conjugated addition of Et_2Zn in the Presence of L2a (100 mmol scale) [136]	116
	Cu-catalyzed enantioselective addition of Et_2Zn to *N*-[4-chlorophenyl (toluene-4-sulfonyl)methyl]formamide in the presence of L2c [78]	117

Cu-catalyzed asymmetric conjugated addition of Me_3Al to
5-methylpent-3-en-2-one in the presence of L2d [74]   *118*
Cu-catalyzed conjugated addition of Et_2Zn to
(E)-1-[(3-Phenyl)acryloyl]pyrrolidin-2-one in the presence
of L2d [62]   *118*
Cu-catalyzed conjugated addition of (*E*)-(2-phenyl-1-propen-1-yl)
dimethylaluminum to cyclohexenone in the
presence of L2d [74]   *119*
Cu-catalyzed asymmetric conjugated addition of Et_2Zn to
cycloheptenone in the presence of L8 [70]   *119*
Cu-catalyzed conjugated addition of *i*-Pr_2Zn to 1-acetylcyclopentene in
the presence of L12 [80]   *120*
Cu-catalyzed conjugated addition of Et_2Zn to
[3-[6-(*tert*-butyldimethylsilanyloxy)hex-2-enoyl]oxazolidin-2-one
in the presence of L14 [83]   *120*
Cu-catalyzed conjugated addition of Et_2Zn to
4,4-(dimethyl)cyclohexenone in the presence of L25 [92]   *121*
Cu-catalyzed conjugated addition of diethylzinc to
3,3-dimethoxy-1-nitro-1-propene in the presence of L5a [67]   *121*
Cu-catalyzed conjugated addition of Ph_2Zn to 3-methylcyclohexenone
in the presence L38b [116]   *122*
Cu-catalyzed conjugated addition of $[AcO(CH_2)_4]_2Zn$ to
tert-butyl 2-methyl-6-oxocyclohex-1-enecarboxylate in the
presence L55a [136]   *123*

2.5	Ligands for Ni-Catalyzed Enantioselective Conjugated Additions	*124*
2.5.1	Amino Alcohols and Aminothiolates *124*	
2.5.2	Aminothiolates and Thioethers *126*	
2.5.3	Pyridino-Alcohols *127*	
2.5.4	Diamines *127*	
2.5.5	Aminoamides *128*	
2.5.6	Sulfoximines *128*	
2.5.7	Cyanobisoxazolines *128*	
2.6	Application of Conjugated Additions in the Synthesis of Natural Compounds *129*	
2.6.1	Application of Non-Asymmetric Conjugated Additions *129*	
2.6.1.1	Synthesis of (±)-β-Cuparenone *129*	
2.6.1.2	Synthesis of a Guanacastepene Intermediate *130*	
2.6.1.3	Synthesis of Prostaglandins *131*	
2.6.1.4	Synthesis of (±)-Scopadulcic Acid *131*	
2.6.2	Asymmetric Conjugated Additions *132*	
2.6.2.1	Synthesis of Prostaglandins *132*	
2.6.2.2	Formal Synthesis of Clavukerins *132*	
2.6.2.3	Synthesis of Muscone *133*	
2.6.2.4	Synthesis of (+)-Ibuprofen *133*	
2.6.2.5	Synthesis of Erogorgiaene *134*	

2.6.2.6	Synthesis of (−)-Pumiliotoxin C	134
2.6.2.7	Synthesis of Phthiocerol	135
2.6.2.8	Synthesis of Leaf Miner Pheromones	135
2.6.2.9	Synthesis of β-Amino Acids	136
2.6.2.10	Synthesis of Clavularin B	136
2.6.2.11	Formal Synthesis of Axanes	137
2.7	Conclusions	137
	References	138

3 ECAs of Organolithium Reagents, Grignard Reagents, and Examples of Cu-Catalyzed ECAs 145
Gui-Ling Zhao and Armando Córdova

3.1 Introduction 145
3.2 Enantioselective Conjugate Addition of Lithium Reagents 145
3.3 Catalytic Enantioselective Conjugate Addition of Grignard Reagents 152
3.4 Cu-Complexes as Catalysts for Enantioselective Conjugate Additions 159
3.5 Conclusions 164
References 164

4 Asymmetric Bifunctional Catalysis Using Heterobimetallic and Multimetallic Systems in Enantioselective Conjugate Additions 169
Armando Córdova

4.1 Introduction 169
4.2 Dinuclear Zn-Complexes in Catalytic ECAs 171
4.3 Heterobimetallic Rare-Earth–Alkali Metal-Binol Complexes in ECAs 175
4.4 Heterobimetallic Rare-earth–Alkali Metal-Binol Complexes in ECAs of Heteroatom Nucleophiles 179
4.5 Miscellaneous 184
4.6 Conclusion 188
References 189

5 Enamines in Catalytic Enantioselective Conjugate Additions 191
Ramon Rios and Albert Moyano

5.1 Introduction and Background 191
5.2 Mechanistic Considerations 193
5.3 Ketone Conjugate Additions 195
5.3.1 Ketone Conjugate Additions to Nitroolefins 195
5.3.1.1 Secondary Amines 196
5.3.1.2 Primary Amines 202
5.3.2 Amine–Thiourea Catalysts 202
5.3.3 Ketone Conjugate Additions to α,β-Unsaturated Carbonyl Compounds 204

5.3.4	Other Reactions	204
5.4	Aldehyde Conjugate Additions	205
5.4.1	Aldehyde Conjugate Additions to Nitroolefins	206
5.4.2	Aldehyde Conjugate Additions to Vinyl Sulfones	208
5.4.3	Other Reactions	209
5.5	Tandem or Cascade Reactions	212
5.6	Conclusions	214
5.7	Experimental	215
	Ketone Addition to Nitrostyrenes (as reported by List [12])	215
	Asymmetric Michael Reaction of Aldehydes and Nitroalkenes (as reported by Hayashi [49])	215
	Catalytic Conjugate Addition of Aldehydes to Vinyl Sulfones (as reported by Palomo [57])	215
	General Procedure for the Synthesis of Cyclohexene Derivatives (as reported by Enders [67])	215
	References	216

6 Iminium Activation in Catalytic Enantioselective Conjugate Additions 219

Jose L. Vicario, Efraim Reyes, Dolores Badía, and Luisa Carrillo

6.1	Introduction	219
6.2	Mechanistic Aspects of the Iminium Activation Concept, and Factors Influencing Stereocontrol in Michael Additions	220
6.3	Michael Reactions	224
6.3.1	1,3-Dicarbonyl Compounds as Nucleophiles	224
6.3.2	Nitroalkanes as Nucleophiles	230
6.3.3	Other Acidic Carbonyl Compounds as Nucleophiles	235
6.3.4	Silyl Enol Ethers and Enamides as Nucleophiles	238
6.3.5	Aldehydes as Nucleophiles	240
6.4	Conjugate Friedel-Crafts Alkylations	241
6.5	Conjugate Hydrogen-Transfer Reactions	245
6.6	Conjugate Additions of Heteronucleophiles	249
6.6.1	*N*-Nucleophiles	250
6.6.2	*P*-Nucleophiles	255
6.6.3	*O*-Nucleophiles	256
6.6.4	*S*-Nucleophiles	258
6.7	Cascade Reactions	259
6.7.1	Michael/Aldol Cascade Reactions	260
6.7.2	Michael/Knoevenagel Cascade Reactions	266
6.7.3	Cascade Michael/*N*-Acyliminium Cyclization Reaction	267
6.7.4	Michael/Michael Cascade Reactions	268
6.7.5	Michael/Morita–Baylis–Hilman Cascade Reactions	269
6.7.6	Michael/Michael/Aldol Triple Cascade Reactions	270
6.7.7	Michael/α-Alkylation Cascade Reactions	274

6.7.8	Cascade Processes Initiated by Conjugate Friedel–Crafts Reaction *277*	
6.7.9	Cascade Processes Initiated by Conjugate Hydrogen-Transfer Reaction *280*	
6.7.10	Cascade Processes Initiated by Hetero-Michael Reactions *282*	
6.8	Concluding Remarks and Outlook *287*	
	References *287*	
7	**Organocatalytic Enantioselective Conjugate Additions of Heteroatoms to α,β-Unsaturated Carbonyl Compounds** *295*	
	Shilei Zhang and Wei Wang	
7.1	Introduction *295*	
7.2	''N'' as Nucleophiles *295*	
7.2.1	Intermolecular aza–Michael Reactions *295*	
7.2.2	Intramolecular aza–Michael Reactions *299*	
7.2.3	Nitrogen Initiated aza–Michael Cascade Reactions *300*	
7.3	''O'' as Nucleophiles *304*	
7.3.1	Intermolecular oxa–Michael Reactions *304*	
7.3.2	Intramolecular oxa–Michael Reactions *306*	
7.3.3	Oxygen-Initiated oxa–Michael Cascade Reactions *307*	
7.4	''S'' as Nucleophiles *310*	
7.4.1	Intermolecular thia–Michael Reactions *310*	
7.4.2	Sulfur-Initiated thia–Michael Cascade Reactions *314*	
7.5	''P'' as Nucleophiles *316*	
7.6	Concluding Remarks *317*	
	References *317*	
8	**Domino Reactions Involving Catalytic Enantioselective Conjugate Additions** *321*	
	Lutz F. Tietze and Alexander Düfert	
8.1	Introduction *321*	
8.2	Metal-Mediated Domino Michael/Aldol Reactions *322*	
8.3	Metal-Mediated Domino Michael Reaction/Electrophile Trapping with Noncarbonyl Compounds *332*	
8.4	Organocatalytic Domino Michael Reactions/Electrophilic Trapping *335*	
8.5	1,4-Conjugate Additions Followed by a Cycloaddition, Hydrogenation, Rearrangement, or Other Reactions *344*	
8.6	Conclusion *347*	
	References *347*	
9	**Asymmetric Epoxidations of α,β-Unsaturated Carbonyl Compounds** *351*	
	Alessandra Lattanzi	
9.1	Introduction *351*	

9.2	Metal-Catalyzed Epoxidations *352*	
9.2.1	Epoxidation of α,β-Unsaturated Ketones Mediated by Chirally Modified Zn- and Mg-Alkyl Peroxides *352*	
9.2.2	Epoxidation of α,β-Unsaturated Ketones, Amides, and Esters Mediated by Lanthanide–BINOL Systems *357*	
9.3	Organocatalyzed Epoxidations *362*	
9.3.1	Phase-Transfer Catalysts *362*	
9.3.2	Polyamino Acids *369*	
9.3.3	Optically Pure Alkyl Hydroperoxides or Ligands *373*	
9.3.4	Guanidine-Based Catalysts *376*	
9.3.5	Pyrrolidine-Based Catalysts *377*	
9.3.6	Imidazolidinone Salt Catalysts *381*	
9.3.7	Primary Amines *382*	
9.4	Conclusions *385*	
9.5	Experimental *386*	
	Typical Procedure for Asymmetric Epoxidation of α,β-Unsaturated Esters [39] *386*	
	Asymmetric Epoxidation of *trans*-Chalcone Catalyzed by Diaryl Prolinol 90/TBHP System [83] *387*	
	References *387*	
10	**Catalytic Asymmetric Baylis–Hillman Reactions and Surroundings** *393*	
	Gui-Ling Zhao	
10.1	Introduction *393*	
10.2	The Reaction Mechanism *393*	
10.3	Asymmetric Intermolecular Baylis–Hillman Reaction *396*	
10.3.1	Diastereoselective Baylis–Hillman Reaction *396*	
10.3.2	Enantioselective Baylis–Hillman Reaction *402*	
10.3.2.1	Chiral Tertiary Amine Catalysts *402*	
10.3.2.2	Chiral Tertiary Phosphines *412*	
10.3.2.3	Chiral Sulfides *420*	
10.3.2.4	Chiral Acids *421*	
10.3.3	Chiral Reaction Media *429*	
10.4	Asymmetric Intramolecular Morita–Baylis–Hillman Reaction *431*	
10.5	Conclusions *433*	
	References *435*	

Index *439*

Preface

The stereoselective conjugate reaction is one of the most important transformations in organic synthesis to achieve asymmetric carbon–carbon and heteroatom-carbon bond-forming reactions. There is today no such book that is focused on this topic. In particular, there is need for a book that covers catalytic asymmetric methods. The last book on the general topic of conjugate reactions in organic synthesis was published in 1992 and the discussion of stereoselective reactions is a small part. This book covers catalytic asymmetric methods based on conjugate additions, which are catalyzed by organometallic complexes or small organic molecules. Significant efforts have been made in asymmetric catalysis during this decade and pioneers of this field were awarded the Nobel prize in 2001. Thus, a book that focuses on modern methods on catalytic stereoselective conjugate addition reactions is highly desirable for the chemistry community.

I would like to thank all the distinguished scientists and their coauthors for their rewarding and timely contributions. I gratefully acknowledge the Wiley-VCH editorial staff, in particular to Dr. Elke Maase for proposing to me this excellent topic and to Dr. Waltraud Wuest who was of precious help for the development of this project.

Stockholm, April 2010 *Armando Córdova*

List of Contributors

Dolores Badía
University of the Basque Country
Faculty of Science and
Technology
Department of Organic
Chemistry II
P.O. Box 644, 48080 Bilbao
Spain

Guillaume Berthon
Syngenta Crop Protection
Münchwilen AG
Schaffhauserstrasse
Postfach 4332, Stein
Switzerland

Robert Betík
Department of Organic and
Nuclear Chemistry
Faculty of Science
Charles University in Prague
Hlavova 8, 128 43 Praha 2
Czech Republic

Luisa Carrillo
University of the Basque Country
Faculty of Science and
Technology
Department of Organic
Chemistry II
P.O. Box 644, 48080 Bilbao
Spain

Armando Córdova
Stockholm University
Department of Organic
Chemistry
Arrhenius Laboratory
Svante Arrhenius väg 16 C
plan 6, 10691 Stockholm
Sweden

Alexander Düfert
Universität Göttingen
Institut für Organische und
Biomolekulare Chemie
Tammannstr. 2
37077 Göttingen
Germany

Tamio Hayashi
Kyoto University
Graduate School of Science
Department of Chemistry
Rigakubu-ichi-goukan
Kyoto 606-8502
Japan

Martin Kotora
Department of Organic and
Nuclear Chemistry
Faculty of Science
Charles University in Prague
Hlavova 8, 128 43 Praha 2
Czech Republic

and

Institute of Organic
Chemistry and Biochemistry
Academy of Sciences of the
Czech Republic
Flemingovo n. 2, 166 10 Praha 6
Czech Republic

Alessandra Lattanzi
Università di Salerno
Dipartimento di Chimica
Via Ponte don Melillo
84084 Fisciano
Italy

Albert Moyano
University of Barcelona
Department of Organic
Chemistry
Martí i Franquès, 1-11
08028 Barcelona
Spain

Efraim Reyes
University of the Basque Country
Faculty of Science and
Technology
Department of Organic
Chemistry II
P.O. Box 644, 48080 Bilbao
Spain

Ramon Rios
University of Barcelona
Department of Organic
Chemistry
Martí i Franquès, 1-11
08028 Barcelona
Spain

Lutz F. Tietze
Universität Göttingen
Institut für Organische und
Biomolekulare Chemie
Tammannstr. 2, 37077 Göttingen
Germany

Jose L. Vicario
University of the Basque Country
Faculty of Science and
Technology
Department of Organic
Chemistry II
P.O. Box 644, 48080 Bilbao
Spain

Wei Wang
University of New Mexico
Department of Chemistry and
Chemical Biology
Clark Hall B-56
Albuquerque, NM 87131-0001
USA

Shilei Zhang
University of New Mexico
Department of Chemistry and
Chemical Biology
Clark Hall B-56
Albuquerque, NM 87131-0001
USA

Gui-Ling Zhao
Stockholm University
Department of Organic
Chemistry
Arrhenius Laboratory
Svante Arrhenius väg 16 C
plan 6, 10691 Stockholm
Sweden

1
Rhodium- and Palladium-Catalyzed Asymmetric Conjugate Additions

Guillaume Berthon and Tamio Hayashi

1.1
Introduction

Since the seminal report by Uemura [1] in 1995 for palladium, and by Miyaura in 1997 for rhodium [2], the late transition metal-catalyzed conjugate addition of organoboron reagents to activated alkenes has emerged as one of the most functional group-tolerant and reliable carbon–carbon bond-forming processes. The maturity of this methodology is such that it has become an ideal testing ground for new ligand concepts and design, as will be illustrated throughout this chapter. A true statement to the robustness of this process is the application of Rh-catalyzed enantioselective conjugate addition (ECA) on a kilogram-scale for the manufacture of advanced pharmaceutical ingredients, and its use as a key step in the synthesis of complex natural products [3–5].

In this chapter, an overview will be provided – spanning from 2003 to mid-2009 – of the developments in the field of rhodium- and palladium-catalyzed ECA of organometallic reagents (B, Si, Zn, and Ti) to activated alkenes. The chapter is not intended to be comprehensive, and will include only selected examples of this powerful methodology. For more in-depth and comprehensive accounts, the reader should consult a number of excellent reviews that are available on this subject [6–16].

1.2
Rh-Catalyzed ECA of Organoboron Reagents

This section will include details of the state of the art for the rhodium-catalyzed ECA of organoboron reagents to activated olefins. Special emphasis will be placed on α,β-unsaturated ketones, as this substrate class has attracted the most attention and undergone thorough investigation with a plethora of different ligand systems. Many of the findings described for α,β-unsaturated ketones are also applicable to other olefin classes and other nucleophilic organometallic reagents, unless otherwise specified.

Catalytic Asymmetric Conjugate Reactions. Edited by Armando Córdova
Copyright © 2010 WILEY-VCH Verlag GmbH & Co. KGaA, Weinheim
ISBN: 978-3-527-32411-8

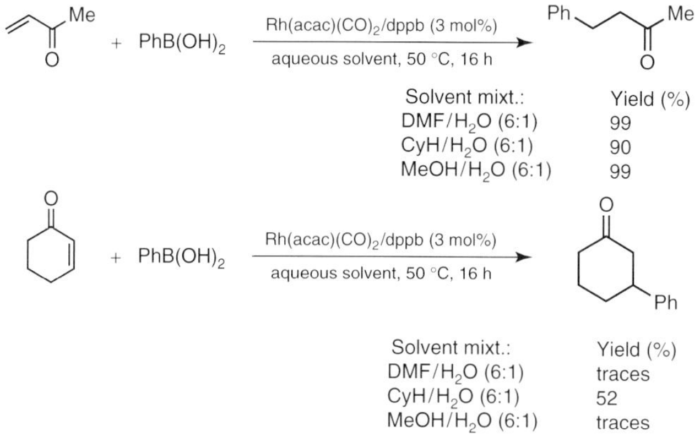

Scheme 1.1 Seminal report of Rh-catalyzed conjugate addition of organoboronic acids.

1.2.1
α,β-Unsaturated Ketones

1.2.1.1 A Short History

The first example of conjugate addition of an arylboronic acid to an enone catalyzed by transition metal complexes can be traced back to a report from 1995 by Uemura and coworkers [1]. This reaction was carried out ligand-less and with a high catalyst loading (10 mol%). The interest in this reaction remained limited until 1997, when Miyaura reported that the [Rh(acac)(CO)$_2$]/dppb (acac = acetylacetonato; dppb = 1,4-bis(diphenylphosphino)butane) system would efficiently catalyze the conjugate addition of a wide range of aryl- and alkenylboronic acids to methyl vinyl ketone (MVK) in high yields, and also to β-substituted enones including 2-cyclohexenone, albeit in lower yields (Scheme 1.1) [2].

The hallmarks of this reaction are: (i) no competitive uncatalyzed reaction of the organoboronic acids onto the enone; (ii) no 1,2-addition of the organoboron reagent; and (iii) a large functional group tolerance which is in contrast to organolithium and Grignard reagents.

A real breakthrough in this methodology came in 1998, when Hayashi and Miyaura described the first example of a rhodium-catalyzed ECA [17]. For the first time, a wide range of aryl and alkenyl fragments could be added in high yields and with exquisite enantioselectivity to an α,β-unsaturated ketone using (S)-binap ((S)-L1) as the chiral diphosphine ligand (Scheme 1.2) [17]. Since this initial report, great progress has also been made in the copper-catalyzed ECA using Grignard and organozinc reagents (this subject is treated in detail in Chapter 3) [18–21]. In order to achieve such high enantioselectivities in the Rh-catalyzed ECA, several factors had to be modified from the original conditions: namely, the rhodium precursor was changed from [Rh(acac)(CO)$_2$] to [Rh(acac)(C$_2$H$_4$)$_2$]; the solvent system was changed to 1,4-dioxane/H$_2$O (10:1); the temperature was increased to 100 °C; and the reaction time was shortened to 5 h. The scope of the reaction was very broad, and a wide range of arylboronic acids (2)

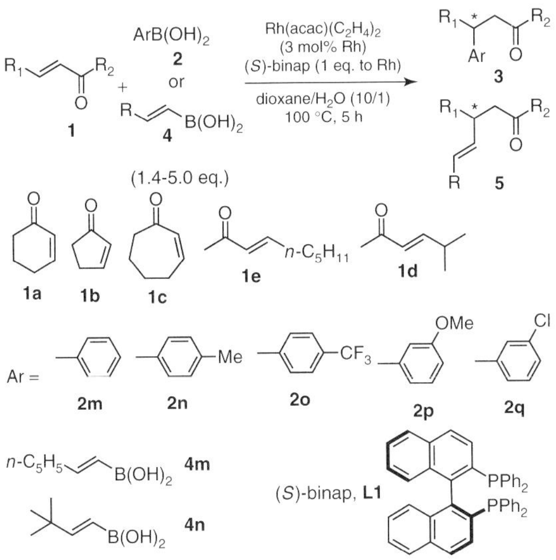

1.2 Rh-Catalyzed ECA of Organoboron Reagents

Entry	Enone 1	Boronic acid 2 or 4 (eq to 1)	Yield (%) Ketone 3 or 5		ee (%)
1	1a	2m (5.0)	3am	>99	97 (S)
2	1a	2n (5.0)	3an	>99	97
3[a]	1a	2o (2.5)	3ao	70	99
4	1a	2p (5.0)	3ap	97	96
5	1a	2q (5.0)	3aq	94	96
6	1b	2m (1.4)	3bm	93	97 (S)
7	1c	2m (1.4)	3cm	51	93
8	1d	2m (5.0)	3dm	82	97
9	1e	2m (2.5)	3em	88	92
10	1a	4m (5.0)	5am	88	94
11	1a	4n (5.0)	5an	76	91
12	1b	4m (2.5)	5bm	64	96

[a]In 1-propanol/H_2O (10:1).

Scheme 1.2 ECA of organoboronic acids to α,β-unsaturated ketones catalyzed by [Rh(acac)(C_2H_4)]/(S)-binap.

substituted with electron-donating or -withdrawing groups could be added to 2-cyclohexenone (**1a**) with high enantioselectivities (Scheme 1.2, entries 2–5). The ECA of alkenylboronic acids **4m** and **4n** to 2-cyclohexenone and 2-cyclopentenone was also very selective (Scheme 1.2, entries 11 and 12). Linear enones having a *trans* geometry also gave high enantioselectivities (Scheme 1.2, entries 8 and 9).

These key seminal findings set the stage for an intense research activity in the area of rhodium-catalyzed ECAs, and related processes, which goes unabated in

2009. In the following sections, the mechanism will be discuss and a working model postulated for the observed enantioselectivity. Using the insights brought by the mechanistic studies, the importance of the Rh precatalyst, as well of the organoboron derivatives that can be used, will also be discussed. An overview will then be presented, by ligand class, of all ligands systems reported until mid-2009, after which the different substrate classes and other competent organometallic reagents in Rh-catalyzed ECA will be reviewed. Following a section devoted to tandem processes involving a conjugate addition step, the chapter will conclude with details of the palladium catalysts for ECAs.

1.2.1.2 Catalytic Cycle

In 2002, Hayashi and coworkers established the detailed mechanistic cycle for the rhodium-catalyzed ECA. An example of this catalytic cycle for the ECA of phenylboronic acid onto 2-cyclohexenone (**1a**) is given in Scheme 1.3 [22]. The cycle goes through three identifiable intermediates: the hydroxyrhodium **A**; the phenylrhodium **B**; and the oxa-π-allylrhodium (rhodium-enolate) **C** complexes. These intermediates are related to the cycle as follows. The reaction is initiated through the transmetallation of a phenyl group from boron to hydroxyrhodium **A** to generate the phenylrhodium **B**. The 2-cyclohexenone will subsequently insert into Rh–Ph bond of **B** to form the oxa-π-allylrhodium **C**. The rhodium enolate **C** is unstable under protic condition, and will be readily hydrolyzed to regenerate **A** and liberate the ECA product. It is important to note that, throughout the catalytic cycle, rhodium remains at a constant oxidation state of +I.

This cycle was validated through the observation of **A**, **B**, **C** in stoichiometric nuclear magnetic resonance (NMR) experiments [22] in which the [Rh(OH)((S)-binap))]$_2$ dimer was used as the precursor to monomeric species **A**, and extraneous triphenylphosphine was added to stabilize coordinatively unsaturated complex **B**. Remarkably, in these NMR experiments all of the elementary steps readily occurred at 25 °C. This was in contrast to the reaction catalyzed by [Rh(acac)(binap)] (**6**), which required a temperature of 100 °C in order for it to occur. This apparent discrepancy in activity was pinpointed to the presence of acac-H in the reaction mixture when [Rh(acac)(binap)] was used. Acac-H readily reacts with [Rh(OH)((S)-binap)] at 25 °C to regenerate [Rh(acac)((S)-binap)], which

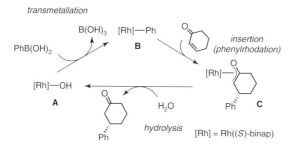

Scheme 1.3 Catalytic cycle for an ECA catalyzed by a hydroxyrhodium complex.

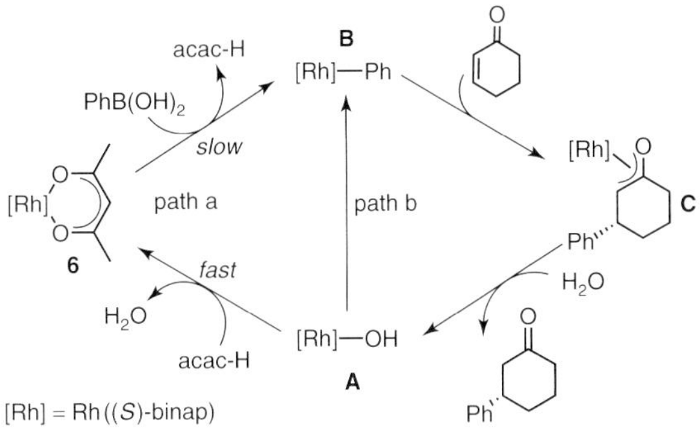

Scheme 1.4 Catalytic cycle for an ECA catalyzed by a rhodium acac complex.

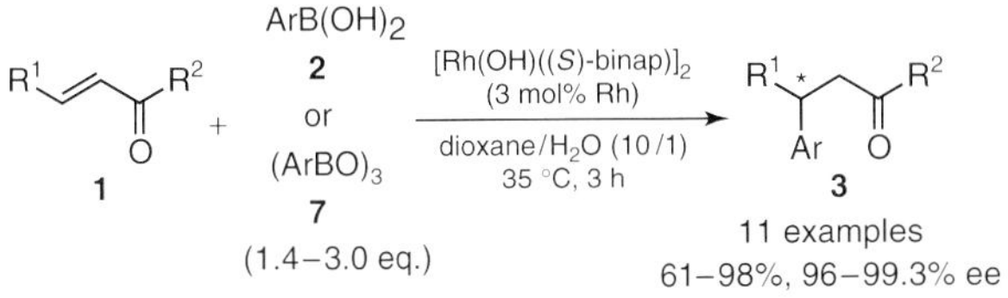

Scheme 1.5 ECA with [Rh(OH)(binap)]$_2$.

in turn transmetallates very slowly with PhB(OH)$_2$ at the same temperature (Scheme 1.4). Therefore, the acetylacetonato ligand inhibits the reaction, and should be avoided if a high catalytic activity is required.

Based on the information gleaned from the mechanistic studies, the [Rh(OH)((S)-binap))]$_2$ dimer (an acac-free rhodium source) was evaluated as a catalyst (Scheme 1.5). This increased the reaction rate drastically and enabled the reaction to be completed in 3 h at 35 °C. In turn, lowering the temperature increased the enantioselectivity, the yield, and also limited the protodeboration of the arylboronic acid. Thus, only 1.4 equiv. of ArB(OH)$_2$ were necessary relative to the acceptor. In this improved protocol, arylboroxines (**7**) were found to be good precursors to boronic acids (cf. Section 1.4).

Following this initial mechanistic study, Hayashi and coworkers performed a detailed kinetic study of the catalytic cycle of Rh-catalyzed ECA using the reaction calorimetry methodology and analysis developed by Blackmond [23, 24]. For this study, the reaction between phenylboronic acid and MVK in the presence of boric acid (B(OH)$_3$) was used (Schemes 1.6 and 1.7). These conditions were chosen to achieve a high reproducibility and sufficiently exothermic reactions for an accurate calorimetric analysis. The reaction network was modeled as follows: The catalytically inactive [Rh(OH)((R)-binap)]$_2$ dimer (**9**) is in equilibrium with the weakly solvated hydroxyrhodium monomer **A** that transmetallates with phenylboronic acid at a rate k_1. The subsequent steps – the insertion of MVK and hydrolysis to yield **8**, (which are kinetically indistinguishable) – are combined in one rate constant, k_2. The analysis, under catalytically relevant conditions, revealed that $K_{dimer} = 8 \times 10^2$ M^{-1} and $k_1 = 0.5$ M^{-1} s^{-1}; however k_2 was too large to be obtained with

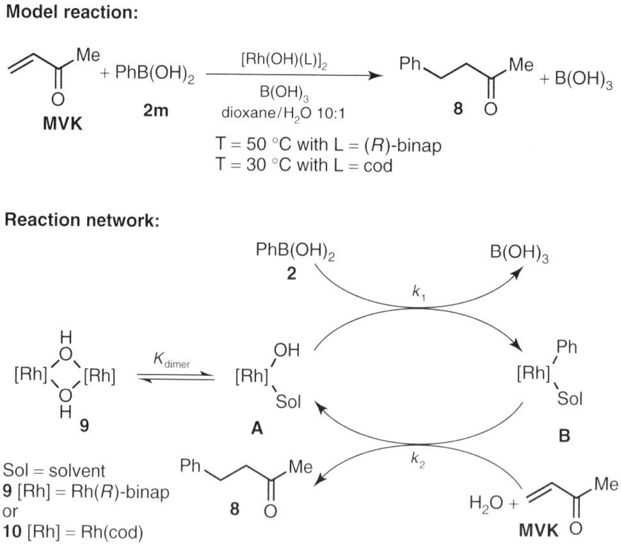

Model reaction:

[Scheme showing MVK + PhB(OH)$_2$ (2m) → product 8 with [Rh(OH)(L)]$_2$, B(OH)$_3$, dioxane/H$_2$O 10:1, T = 50 °C with L = (R)-binap, T = 30 °C with L = cod]

Reaction network:

[Catalytic cycle showing dimer 9 [Rh]$_2$(OH)$_2$ ⇌ K_{dimer} [Rh]–OH (A) → [Rh]–Ph (B) with PhB(OH)$_2$/B(OH)$_3$ (k_1) and MVK/product 8 (k_2)]

Sol = solvent
9 [Rh] = Rh(R)-binap
or
10 [Rh] = Rh(cod)

Scheme 1.6 Model reaction and reaction network used for conjugate addition of PhB(OH)$_2$ to MVK in the presence of Rh/binap and Rh/cod catalysts.

a statistically meaningful value. Thus, the rate-determining step of the catalytic cycle was the transmetallation from boron to rhodium. Furthermore, the large value for K_{dimer} indicated that most of the rhodium lay in the catalytically dormant [Rh(OH)((R)-binap)]$_2$ (**9**) dimer.

Using the same methodology, Hayashi and coworkers performed the kinetic analysis of the reaction catalyzed by [Rh(OH)(cod)]$_2$ (**10**) (cf. Section 1.5) [25]. Under identical conditions, the rate of the reaction with [Rh(OH)(cod)]$_2$ (**10**) was 20-fold faster than with [Rh(OH)((R)-binap)]$_2$ (**9**). For the quantitative analysis, the reaction temperature was set to 30 °C, which led to $K_{dimer} = 3.2 \times 10^2$ M^{-1} and $k_1 = 1.3$ M^{-1} s^{-1}, and k_2 was estimated to be 16 M^{-1} s^{-1}. On the assumption that K_{dimer} did not vary much between 30 and 50 °C, k_1 was calculated as 6.7 M^{-1} s^{-1}. Thus, the remarkably large catalytic activity of rhodium–diene complexes can be attributed to both a lower K_{dimer} and the higher rate of transmetallation of [Rh(OH)(cod)] (**10**) versus [Rh(OH)(binap)] (**9**). There is, to date, no rational why π-accepting diene ligands accelerate the rate-determining transmetallation by an order of magnitude relative to diphosphine ligands. Other π-accepting ligands, such as phosphoramidites, also have a beneficial effect on the reaction rate [26].

The transmetallation step is arguably one of the least understood process in homogeneous catalysis. In the case transmetallation from boron to rhodium or palladium under the conditions of conjugate addition, the mechanism is thought to occur through a metal hydroxy complex **A** which can coordinate to highly oxophilic arylboronic acid to give intermediate **D**; the latter can then deliver the aryl fragment to rhodium in an intramolecular fashion to furnish the aryl-rhodium species **B** (Scheme 1.7) [16, 27].

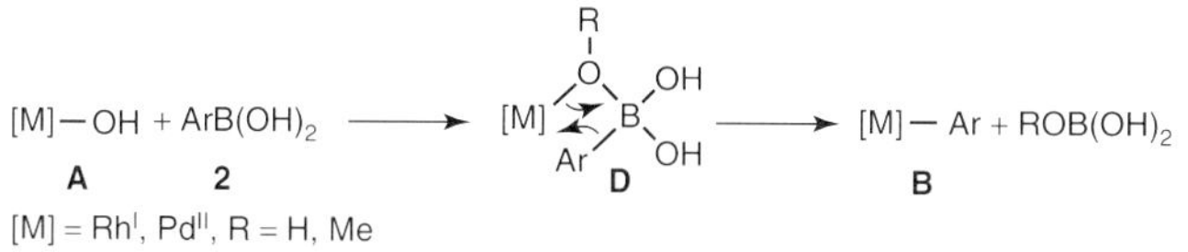

$$[M]-OH + ArB(OH)_2 \longrightarrow [M]\underset{Ar}{\overset{R}{\underset{D}{\bigotimes}}}\underset{OH}{\overset{O}{\underset{OH}{\bigotimes}}}B \longrightarrow [M]-Ar + ROB(OH)_2$$

A   **2**   **D**   **B**

[M] = Rh^I, Pd^{II}, R = H, Me

Scheme 1.7 Proposed mechanism for the transmetallation of organoboronic acids.

Direct evidence for this mechanism was provided by Hartwig and coworkers, who showed that a boronic acid would react cleanly with the hydroxy dimer **11**, **12**, or the Rh-enolate **13** to give complex **14** which could be isolated [28]. Upon heating, **14** would rearrange to form a Rh-aryl bond and extrude an insoluble boroxine oligomer (Scheme 1.8).

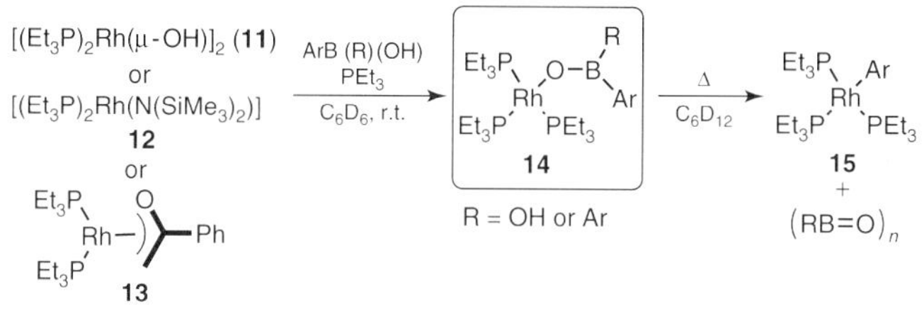

Scheme 1.8 Direct observation of the transmetallation from boron to rhodium.

Although this process occurs under neutral conditions, it is greatly accelerated by the presence of stoichiometric amounts of base. This is rationalized by the quaternization of the arylboronic acid, which facilitates rupture of the B–C$_{sp2}$ bond [16, 29].

1.2.1.3 Model for Enantioselection

Scheme 1.10 shows the proposed stereochemical pathway in the reaction catalyzed by rhodium complex coordinated with (S)-binap (**L1**) [17]. According to the highly skewed structure known for transition metal complexes coordinated with a binap ligand [30], the (S)-binap rhodium intermediate **E** should have an open space at the lower part of the vacant coordination site, the upper part being blocked by one of the phenyl rings of the binap ligand. The olefinic double bond of 2-cyclohexenone (**1a**) coordinates to rhodium, with its α*Si* face forming **F** rather than with its α*Re* face, which undergoes migratory insertion to form a stereogenic carbon center in **G**, the absolute configuration of which is *S*.

This stereocontrol model can be applied to predict the absolute configurations of all 1,4-addition products with the (S)-binap–rhodium intermediate. This intermediate attacks the α*si* face of α,β-unsaturated ketones, both cyclic and linear forms, as well as other electron-deficient olefins such as α,β-unsaturated esters and amides, nitroalkenes, and alkenylphosphonates. For the ECA of linear alkenes, the geometry of the double bond plays a determining role in the stereochemical outcome (Schemes 1.9 and 1.10). This is again explained

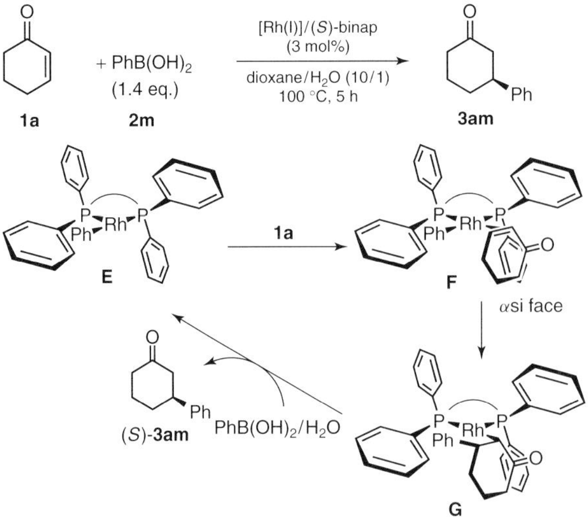

1 Rhodium- and Palladium-Catalyzed Asymmetric Conjugate Additions

Scheme 1.9 Proposed model for the enantioselection with [Rh((S)-binap)] complex.

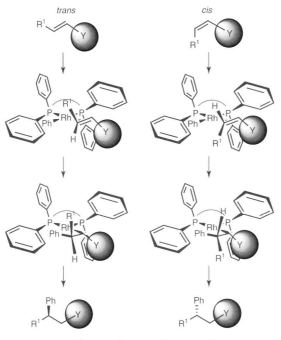

Y = C(O)R², C(O)OR², C(O)N(R²)₂, P(O)(OR²)₂, NO₂

Scheme 1.10 Stereocontrol model for the ECA of *trans* and *cis* olefins.

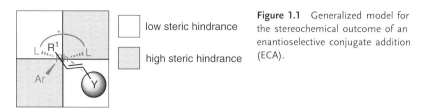

Figure 1.1 Generalized model for the stereochemical outcome of an enantioselective conjugate addition (ECA).

by which enantiotopic face the electron-deficient olefins can coordinate to the [Rh(Ph)((S)-binap)] intermediate.

This stereochemical model can be extended to a wide range of C_2 symmetric bidentate ligands by considering how the ligand, when it is coordinated to rhodium, is capable of bisecting the space around the rhodium into a quadrant, and which enantiotopic face of the alkene will minimize steric interaction upon coordination to Rh (Figure 1.1) [22].

1.2.1.4 Organoboron Sources Other than Boronic Acids

Although organoboronic acids (**2**) are the most practical and widespread source of organoboron reagents, other sources have proven to be equally effective (Figure 1.2). Boronate esters from catechol (**16**), ethylene glycol (**17**), and pinacol (**18**) can be used in Rh-catalyzed ECAs [31, 32]. The rate of the ECA reaction of these boronate esters is directly related to the ease of their hydrolysis back to the corresponding boronic acid. While catechol boronates will react quickly, pinacol boronic esters react slowly in Rh-catalyzed ECAs [31]. This feature can be used advantageously in the one-pot alkyne hydroboration with catecholborane, followed by Rh-catalyzed ECA [32].

Boronic acids are in equilibrium with oligomeric species of various degrees of hydration. The complete dehydration of organoboronic acid leads to the well-defined cyclic organoboroxine (**7**). Boroxines are readily hydrolyzed back to the corresponding boronic acid with 1 equiv. of water relative to boron under basic aqueous conditions [33]. Organoboroxines have become the preferred reagents for Rh-ECAs,

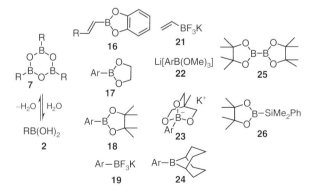

Figure 1.2 Organoboron reagent competent for Rh-catalyzed ECAs.

because they enable addition of the organoboron reagent with a precise stoichiometry, and are also more stable towards protodeboration than boronic acids, especially at an elevated temperature (ca. 100 °C).

Potassium aryltrifluoroborate salts (**19**) have become a very popular source of organoboron reagents [34–36], because they are more stable than the corresponding boronic acids while still being reactive in Rh-cat. ECAs [37]. One particularly useful reagent is potassium vinyltrifluoroborate (**21**), which enables the introduction of a vinyl group in excellent yields with high enantioselectivities [38]. The corresponding vinylboronic acid cannot be used because it is unstable and readily undergoes protodeboration and polymerization [39]. It is important to note, that potassium organotrifluoroborates do not transmetallate directly to rhodium(I), but rather that a monohydroxyborate (**20**) is probably the boron species that effects the transmetallation step, with a mechanism akin to that depicted in Scheme 1.8. This monohydroxyborate (**20**) has been observed to be in equilibrium with the corresponding potassium organotrifluoroborates (**19**) under basic aqueous conditions for Rh-catalyzed ECAs (Eq. (1.1)) [40–44].

$$\text{Ar}-\text{BF}_3\text{K} + \text{Base} + \text{H}_2\text{O} \rightleftharpoons \boxed{\text{Ar}-\text{BF}_2(\text{OH})\text{K}} + \text{Base}\cdot\text{HF}$$
$$\text{19} \qquad\qquad\qquad\qquad\quad \text{20}$$

Species involved in transmetallation (1.1)

Lithium trimethylarylborate (**22**) is also a very active reagent for the ECA, but is relatively unstable and is best formed *in situ* by lithium/halogen exchange on an aryl bromide, followed by addition of trimethoxyborane [31, 45]. Cyclic aryl triolborates (**23**) are also convenient and reactive reagents for Rh-catalyzed ECAs [46, 47]; these reagents have the advantage of being very stable in air and water, and more soluble in organic solvents than related potassium organotrifluoroborates. The reactive ArB(9-BBN) (**24**) derivatives can be used in ECA reactions in aprotic solvents, and also in the absence of base to yield a stable boron enolate [48]. The chiral boron enolate can be further reacted with an electrophile to yield a ketone with a high diastereoselectivity (see Scheme 1.18). Reagents such as bis(pinacolato)diboron (**25**) [49] and dimethylphenylsilylpinacolatoboron (**26**) [50, 51] have been used to introduce a boron and silyl moiety in Rh-catalyzed addition reactions (cf. Section 1.6; see Scheme 1.59).

1.2.1.5 Rh Precatalysts

As noted above, the nature of the Rh precatalyst used in the ECA reaction is of crucial importance. In the first step, the Rh precursor must enable a rapid exchange of ligands in order to form quantitatively the enantioselective catalytic species. Thus, when [Rh(acac)(C_2H_4)$_2$] is used in conjunction with a chiral bidentate ligand, high enantioselectivities are observed, whereas when [Rh(acac)(CO)$_2$] is used lower selectivities are obtained because of the strong binding of CO [17]. As discussed the acac ligand has an inhibiting effect on the reaction; therefore, acac-free [Rh(Cl)(C_2H_4)$_2$]$_2$ is the precatalyst of choice because it enables rapid and irreversible ligand exchange. Other cod-containing rhodium precursors, such a [Rh(Cl)(cod)]$_2$ and [Rh(μ-OH)(cod)]$_2$, should be avoided because of their higher catalytic activity than the Rh–phosphine complex [29, 52]. Thus, only trace amounts of these complexes (due to incomplete ligand exchange) will suffice to significantly lower the observed enantioselectivity of an ECA.

Chiral cationic rhodium complexes, formed *in situ* by the reaction of [Rh(cod)$_2$]$^+$BF$_4^-$, [Rh(cod)(MeCN)$_2$]$^+$BF$_4^-$ [53–60], or [Rh(nbd)$_2$]$^+$BF$_4^-$ [29, 61] with a chiral ligand, are also active rhodium precatalysts [62]. Under the basic aqueous conditions used for Rh-catalyzed ECAs, the cationic precursors are presumably converted *in situ* into the neutral Rh–OH species bearing the chiral ligands. Thus, the [Rh(nbd)$_2$]$^+$BF$_4^-$ complex should be preferred because the corresponding [Rh(μ-OH)(nbd)]$_2$ complex is a poor catalyst for the reaction. One practical advantage of using cationic rhodium precursors is that they enable more robust reaction conditions, and lead more consistently to higher enantioselectivities when the catalyst is generated *in situ* with a chiral ligand than with the corresponding neutral Rh precursor [63]. This is presumably due to the faster exchange of the diene for the chiral ligands on the cationic rhodium relative to the neutral precursors [63]. Furthermore, with cationic rhodium precursors, Et$_3$N can be used instead of KOH as the activator, thus making the ECA protocol more functional group-tolerant [64].

Hayashi and coworkers found that traces of phenol (0.05 ± 0.02 mol%) present in commercial phenylboronic acid can significantly deactivate chiral diene rhodium catalysts [65]. This deactivation pathway becomes prevalent under low-catalyst loading conditions (below 0.05 mol%). The phenol impurity can be removed by dehydration of the boronic acid to the boroxine (**7**), followed by washing with hexanes. These findings are probably more applicable to other Rh systems for ECAs, which are used at low catalyst loadings.

1.2.1.6 Ligand Systems

In this section, an overview will be presented of the different ligand designs and concepts that have been applied to the rhodium-catalyzed ECAs of organoboronic acids to α,β-unsaturated ketones. Very early on, the ECA of phenylboronic acid onto 2-cyclohexenone was chosen as a model reaction for Rh-catalyzed ECAs. The wealth of reports using this model reaction enables the direct comparison of a wide gamut of ligand structures. To facilitate the comparison, the ligands have been grouped by families (i.e., phosphorus-based bidentate, monodentate, mixed ligands, and others). When a family of ligands was prepared, only the best-performing ligand will be discussed. It should be borne in mind that this is only a comparison on a fixed model reaction, and some ligands might be better suited for specific substrates; when possible, this will be highlighted.

1.2.1.7 Bidentate Phosphorus Ligand

Following the initial breakthrough for Rh-catalyzed ECA, using binap as a ligand, a flurry of reports has emerged employing bidentate phosphorus ligands. These results are summarized in Schemes 1.11 and 1.12. Diop (**L2**) [66], and chiraphos

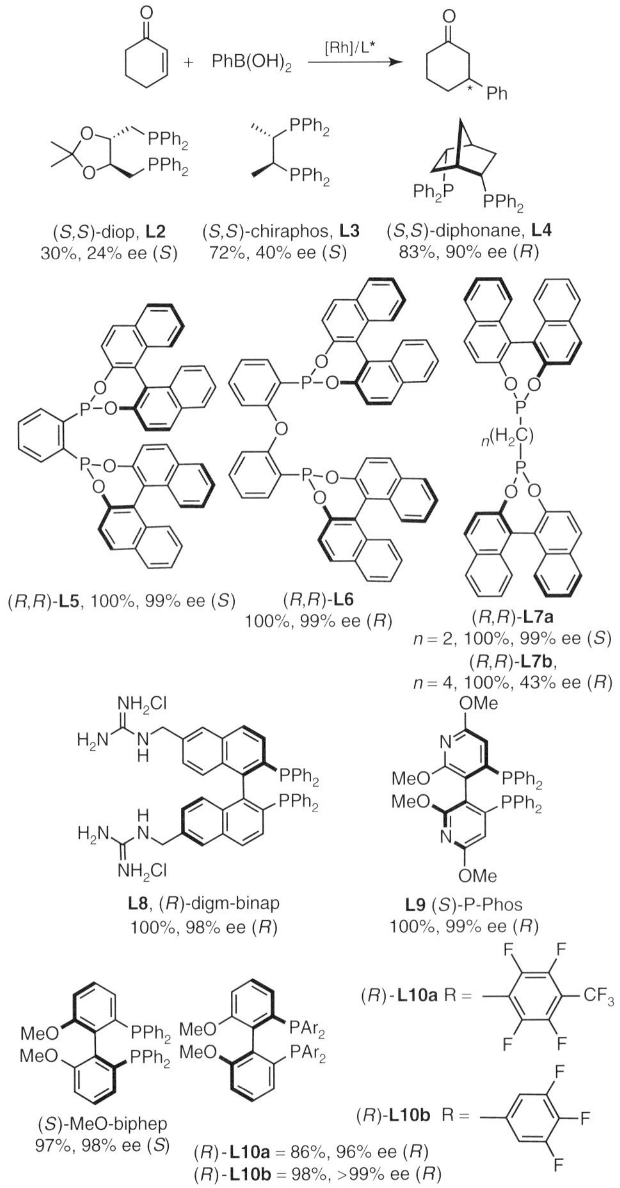

Scheme 1.11 ECA catalyzed by C_2-symmetric bidentate ligand rhodium complexes.

(**L3**) [66] gave low enantioselectivities. The diphonane ligand **L4** bears an interesting backbone and gives good selectivities [67]. Binol-based bisphosphonites **L5** and **L6** performed well in Rh-catalyzed ECA and interestingly, depending on the carbon chain length separating the phosphonites in **L7a** and **L7b**, the enantioselectivity is reversed [68]. Similarly, binol-based bisphosphoramidite **L11** linked together in the

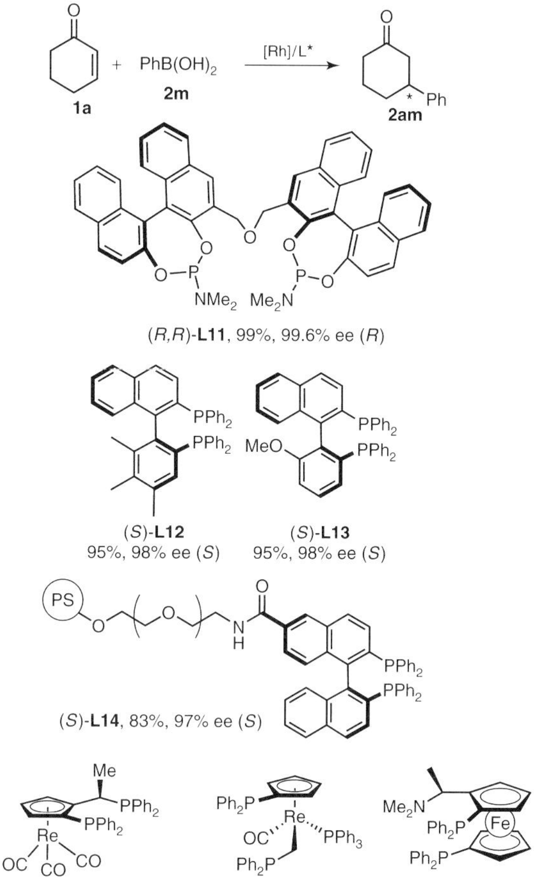

(*R*,*R*)-**L11**, 99%, 99.6% ee (*R*)

(*S*)-**L12** 95%, 98% ee (*S*)

(*S*)-**L13** 95%, 98% ee (*S*)

(*S*)-**L14**, 83%, 97% ee (*S*)

(R,S_p)-**L15** 93%, 95% ee (*S*)

(*R*)-**L16** 30%, 40% ee (*R*)

(*S*,*R*)-bppfa **L17** 5%, 3% ee (*S*)

Scheme 1.12 ECA catalyzed by bidentate ligand rhodium complexes.

3-position gave excellent results [64, 69]. The water-soluble binap-based ligand **L8** catalyzed the ECA in aqueous media with a turnover number (TON) of 13 200 [70]. The observation that π-accepting ligands accelerate the rate-determining transmetallation in Rh-catalyzed ECAs was confirmed by the use of π-accepting ligands **L10a** and **L10b**, which displayed a higher catalytic activity than the corresponding MeO-biphep [71]. In general, axially chiral ligands such as **L9** [72], (*S*)-MeO-biphep [71] **L10** [71], **L12** [73], **L13** [73], and polystyrene-supported binap **L14** [74] all produce excellent enantioselectivities on par with binap. Substitution in remote positions (not 3 and 3′) of the binaphthyl backbone of the binap ligands has little influence on the stereochemical outcome [75, 76]. The diphosphine ligand **L15** [77], **L16** [78], and **L17** [66] bearing planar chirality were investigated, although only Re-based **L15** gave high enantiomeric excess (ee) values.

1.2.1.8 Monodentate Ligand

Although monodentate chiral ligands (P-stereogenic) were the first class of ligands to be used in Rh-catalyzed asymmetric homogeneous catalysis, they have since been replaced by rigid bidentate ligands [79]. Early attempts to use the monodentate (R)-MeO-mop (**L18**) as a ligand for Rh-catalyzed ECA proved to be disappointing, with low conversion and enantioselectivities [66]. The discovery by de Vries and Feringa that binol-based phosphoramidites where highly active and enantioselective in Rh-catalyzed hydrogenation [80] has spurred a renewed interest in monodentate ligands in asymmetric catalysis. A great impetus for this growing trend in homogeneous catalysis is the cheap and rapid synthesis of monodentate ligands over bidentate ones. As for asymmetric hydrogenation, phosphoramidite ligands performed very well in Rh-catalyzed ECAs due to their strong π-accepting properties, which renders them more reactive than phosphines (cf. Section 1.2) [26, 38, 81–86]. For the addition phenylboronic acid onto 2-cyclohexenone, H_8-binol-based phosphoramidite (S)-**L19** proved to be the most efficient [82]. Phosphoramidite ligands have also been used for the addition of potassium organotrifluoroborates [38]. In an early and elegant report of a chiral monodentate N-heterocyclic carbene (NHC), Andrus and coworkers showed that the cyclophane-based NHC (**L20**) was very effective for Rh-catalyzed ECA [87, 88]. Importantly, only 1 equiv. of chiral NHC relative to rhodium was necessary to obtain high selectivity, this being in stark contrast to other monodentate ligands, which require 2 equiv. per rhodium (Scheme 1.13).

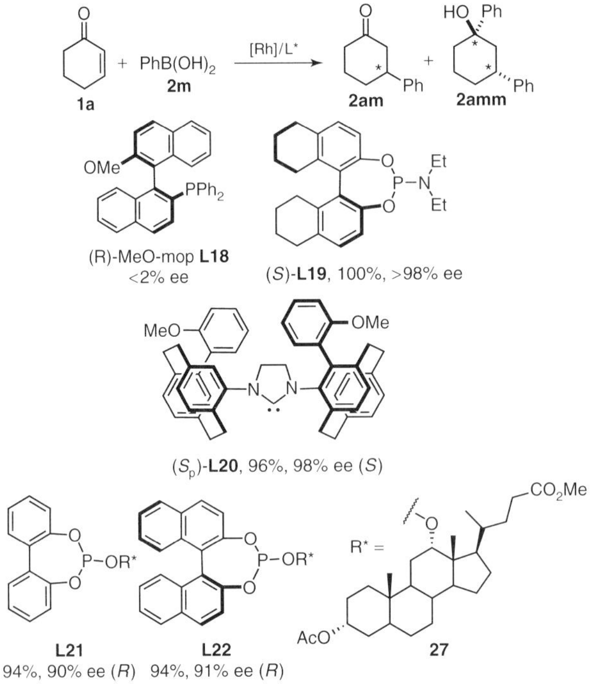

Scheme 1.13 ECA catalyzed by monodentate ligand rhodium complexes.

The use of inexpensive methyl deoxycholic ester **27** as the source of chirality in phosphite **L21** proved efficient [88, 89]. The deoxycholic moiety induces only one conformation in the *tropos* biphenyl backbone (Eq. (1.2)).

$$\text{(aR)-L*} \rightleftharpoons \text{(aS)-L*}$$
R = H, *t*-Bu, Me (1.2)

This approach alleviates the need to use binol as the source of chirality. When the deoxycholic moiety is paired with each enantiomer of binol, only one the diastereoisomers (**L22**) was catalytically active, demonstrating the value of having a flexible *tropos* backbone. In addition, depending on the molar ratio of phosphite **L21** relative to Rh, the reactivity and selectivity of the addition could be modulated. With only 1 equiv. of phosphite per Rh, the major product was **2am**, while with 2 equiv. of phosphite **L21**, **2am** underwent a diastereoselective 1,2-addition to furnish the bisphenylated **2amm** product as a single diastereoisomer [89].

The use of a monodentate ligand offered the tantalizing possibility to mix different monodentates together to quickly generate combinatorial libraries of complexes. This fascinating approach goes beyond the traditional parallel preparation of modular ligands. Thus, mixtures of monodentate ligands L^a and L^b can, upon exposure to a transition metal [M], form not only the two homocombinations $[M(L^a)_2]$ and $[M(L^a)_2]$ but also the heterocombination $[M(L^a)(L^b)]$. For example, a 1:1:1 mixture of L^a, L^b, and metal precursor is used (and no other interaction exists) a statistical mixture of $[M(L^a)_2]$, $[M(L^a)_2]$ and $[M(L^a)(L^b)]$ in a 1:1:2 ratio will be obtained (Eq. (1.3)). If the heterocombination is more reactive and selective than the homocombinations, an improved catalyst system is formed without the need to synthesize new ligands. This approach – dubbed *combinatorial transition-metal catalysis* – has been reviewed in an excellent article by Reetz [90].

$$L^a + L^b + [M] \longrightarrow [M(L^a)_2] + [M(L^b)_2] + [M(L^a)(L^b)]$$
(1:1:1)   Homo-L^a Homo-L^b Hetero
Statistical distribution: 1:1:2 (1.3)

The first example of this approach applied to Rh-catalyzed ECA was described by Feringa with mixtures of phosphoramidites [84]. Each homo combination of **L23** and **L24** performed significantly less well than the hetero combination generated *in situ* (Scheme 1.14) [84]. An interesting extension of this approach is to use mixtures of *tropos* monodentate phosphoramidites ligand. In this example, the combinatorial mixing of ligand led also to the identification of hetero combination **L25** and **L26** that performed significantly better than both homocombinations of ligands (Scheme 1.14) [91].

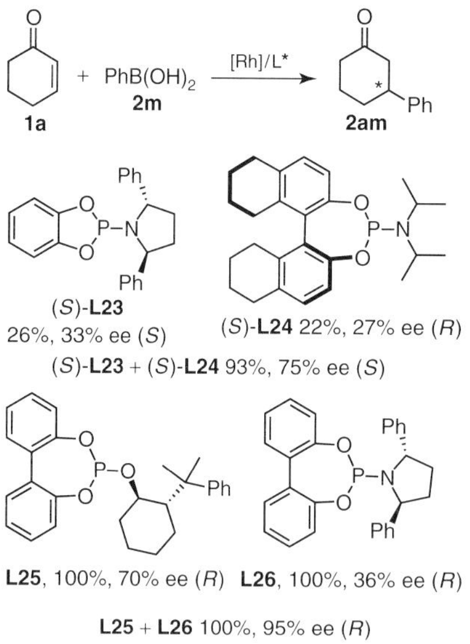

Scheme 1.14 Rh-catalyzed ECA using mixtures of chiral monodentate ligands.

(S)-L23
26%, 33% ee (S)

(S)-L24 22%, 27% ee (R)

(S)-L23 + (S)-L24 93%, 75% ee (S)

L25, 100%, 70% ee (R)   L26, 100%, 36% ee (R)

L25 + L26 100%, 95% ee (R)

1.2.1.9 Diene Ligands

The seminal observation by Miyaura and coworkers that [Rh(μ-OH)(cod)$_2$] is the most active catalyst (with a TON of up to 375 000) for rhodium-catalyzed conjugate addition [29, 92] prompted the investigation of optically active diene as ligand for this transformation. Since the first application of chiral (S,S)-2,5-dibenzylbicyclo[2.2.1]heptadiene ((S,S)-L27b) in the addition of phenylboronic acid to cyclohex-2-enone by Hayashi and coworkers [93], a variety of bicyclic diene scaffolds [94–104] has been successfully applied (Scheme 1.15). Independently, Carriera and coworkers reported the application of a chiral diene for Ir-catalyzed allylic substitution [105]. These independent discoveries have spurred intense research efforts in homogeneous catalysis which were compiled in 2008 in an excellent review by Carreira and Grützmacher [106].

An examination of the substituent-effect in chiral bicyclo[2 : 2 : 1]heptadiene scaffold revealed that just two methyl groups in ligand **L27a** are sufficient to impart a high enantioselectivity (95% ee) in the ECA reaction [99]. Moreover, variation of the alkyl group to Bn (**L27b**), Cy (**L27d**), allyl (**L27f**), and *i*-Bu (**L27e**) does not significantly influence the selectivity when compared to **L27a** [107]; only when moving to a phenyl substituent is a higher ee-value obtained. A systematic exploration of different bicyclic scaffolds for chiral diene, revealed that the 2,5-disubstituted bicyclo[2.2.1]heptadiene (**L27**) [93, 108], bicyclo[2.2.2]octadiene (**L28, L29, L30, L31, L32**) [95, 105, 109–115], and bicyclo[3.3.0]octadiene (**L33**) [100, 116] gave high enantiomeric excesses over a wide range of substrates.

On the other hand, the first-generation 2,6-disubstitued bicyclo[3.3.1]nonadiene (**L34**) [117, 118] and bicyclo[3.3.1]decadiene (**L35**) [118] gave inferior results. A re-examination of the substitution pattern on these dienes revealed that 3,7-disubstituted bicyclo[3.3.1]nonadienes (**L36**) gave excellent results and greatly surpassed previous chiral dienes as ligands for the ECA of alkenyl boronic acid

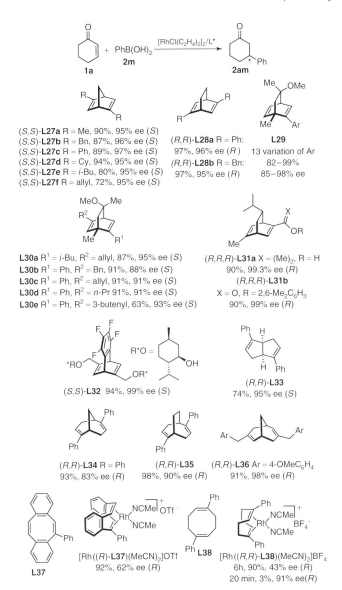

Scheme 1.15 Chiral diene ligands in the Rh-catalyzed addition of phenylboronic acid to 2-cyclohexenone.

onto β-silyl α,β-unsaturated ketones [104]. Using the scaffold developed by Carreria (**L30**), it was found that just the mono-substitution of the bicyclo[2.2.2]octadiene framework with an aryl moiety (**L29**) was necessary to achieve high enantioselectivities [44, 103]. In this first systematic investigation of the steric effects of *mono*-substituted chiral dienes, the *ortho* positions on the aromatic moiety (i.e., 2,6-Me$_2$ C$_6$ H$_3$) were found to be important to achieve highest selectivities.

The lengthy synthesis of **L27** and the need to resolve **L28**, **L34**, **L35**, and **L36** by preparative high-performance liquid chromatography (HPLC) has impeded the widespread investigation of chiral dienes. To overcome this limitation, several readily accessible scaffolds have been synthesized. Ligand **L30** is conveniently synthesized in seven steps from readily available (−)-carvone [105]. Interestingly, the allyl-functionalized **L30c** gave significantly higher ee-values than the *n*-propyl-substituted **L30d**. A reinvestigation of the synthesis of the 2,5-disubstituted bicyclo[2.2.1]heptadiene family (**L27**) led to a much shorter and efficient five-step synthesis from inexpensive bicyclo[2:2:1]heptadiene, using an iron-catalyzed cross-coupling as the key step [99]. A very concise two-step entry into the chiral bicyclo[2.2.2]octadiene framework (**L31**) was reported by Hayashi and Rawal, using as the key step a Diels–Alder reaction between (*R*)-α-phelandrene and propiolate esters, followed by the addition of MeLi for **L31a** or a simple trans-esterification for **L31b** [102, 119]. Dienes **L31a** proved to be highly effective ligands for the Rh-catalyzed ECA on a range of cyclic and linear α,β-unsaturated ketones, giving near-perfect enantioselectivity. Strikingly, the α,β-unsaturated ester moiety in diene **L31b** does not insert in the Rh-aryl bond during the catalytic cycle, most likely because the alkene in the rigid bicyclic framework cannot be co-planar to the coordination plane of Rh, which is necessary for insertion into the Rh-aryl bond. Furthermore, the use of strongly π-accepting α,β-unsaturated ester significantly accelerates the reaction rate, presumably by facilitating the transmetallation process to form the Rh-aryl bond *trans* to the enoate moiety of **L31b**. This *trans* influence on the transmetallation process is highly beneficial because it forces the substrate to coordinate *cis* to the bulky ester moiety, leading to a more effective enantioselection. This *trans* effect could also explain the high enantioselectivities obtained with mono-substituted chiral diene **L29**.

Chiral tetrafluorobenzobarralene **L32** was obtained directly from the [4+2] cycloaddition of 1,4-bis((−)-menthoxymethyl)benzene with *in situ*-generated tetrafluorobenzyne, albeit with a low (8%) yield. The resulting diastereoisomers were separated by column chromatography to afford the desired chiral diene [101].

The Ph-dbcot (**L37**) [98] and 1,5-Ph-cod (**L38**) [97] are achiral dienes, but when coordinated to Rh they become conformationally locked, which leads to a pair of enantiomers that can be subsequently resolved such that, ultimately, enantiomerically pure cationic rhodium complexes [Rh((*R*)-**L37**)(MeCN)$_2$](OTf) and [Rh((*R,R*)-**L38**)(MeCN)$_2$](BF$_4$) can be obtained. Complex [Rh((*R*)-**L37**)(MeCN)$_2$](OTf) leads to a moderate 62% ee [97], while complexes [Rh((*R,R*)-**L38**)(MeCN)$_2$](BF$_4$) gave high ee-values at low conversion. However, the enantioselectivity was eroded at higher conversions due to the conformational instability of ligand **L38** of the complex [98].

1.2.1.10 Bis-Sulfoxide

Bis-sulfoxides are an emerging class of ligands in homogeneous catalysis [120, 121]. Dorta and coworkers reported the first chiral bis-sulfoxides ligand **L39** [122] and **L40** [123], and found them to be exceptional ligands for the Rh-catalyzed ECA of arylboronic acids, giving near-perfect enantioselectivities over a wide range of cyclic α,β-unsaturated ketones (Scheme 1.16). The biphenyl bis-sulfoxide **L40** was found to be more active and selective than binaphthyl derivative **L39** [123]. A comparison of the X-ray crystal structures of [RhCl((*R*)-binap)]$_2$, [RhCl((*S,S*)-**L28a**)]$_2$ [124], and [RhCl((*R,R*)-**L39**)]$_2$ [123] indicated that the ligating properties of *bis*-sulfoxides might lie somewhere between that of diene and *bis*-arylphosphine ligand.

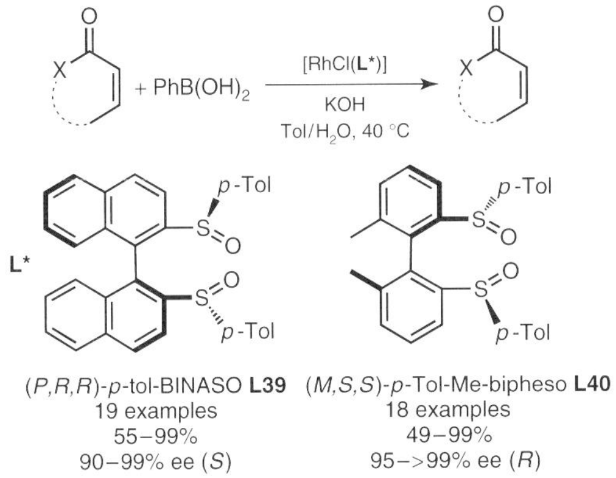

Scheme 1.16 Chiral *bis*-sulfoxide ligands for Rh-cat ECA.

(*P,R,R*)-*p*-tol-BINASO **L39**
19 examples
55–99%
90–99% ee (*S*)

(*M,S,S*)-*p*-Tol-Me-bipheso **L40**
18 examples
49–99%
95–>99% ee (*R*)

1.2.1.11 Mixed Donor Ligands

A range of chiral ligands bearing a phosphorus center and another coordinating functionality have been investigated in the Rh-catalyzed ECA of boronic acid onto α,β-unsaturated carbonyls (Scheme 1.17). Interestingly, the ligand (*S*)-ip-phox family performed poorly in this transformation [66], while the L-proline-derived amido phosphines (*S*)-**L42** was found to be very active and selective [125, 126], and has been applied in diastereoselective ECA processes [127]. The combination of NHC and a phosphine moiety **L42** was successful [128]. Based on the observation that phosphines coordinate more strongly to late transition metals than do alkenes, but that the latter provide an effective chiral environment, Shintani and Hayashi synthesized the chiral phosphines-alkene ligand **L43**, and this proved to be highly effective for both the ECA onto enones and maleimides [129]. The kinetic study of the Rh/**L43** system in ECA, revealed that the catalytic activity with **L43** to be intermediate to that of diphosphines and dienes [130]. Good activities and enantioselectivities were obtained with ligands **L44** [131], amidophosphine-alkene **L45** [132] (first synthesized by Carreira [133]) and with the chiral phosphine-olefin **L46** [134].

1.2.1.12 Trapping of Boron Enolates

In the rhodium-catalyzed ECA of organoboron reagents to electron-deficient alkenes, the protic solvent is essential for hydrolyzing oxo-π-allylrhodium to regenerate the active hydroxorhodium species and liberate the 1,4-addition product (cf. Scheme 1.3). However, this process does not take full advantage of the chiral rhodium-enolate that is transiently formed, and which could further react either

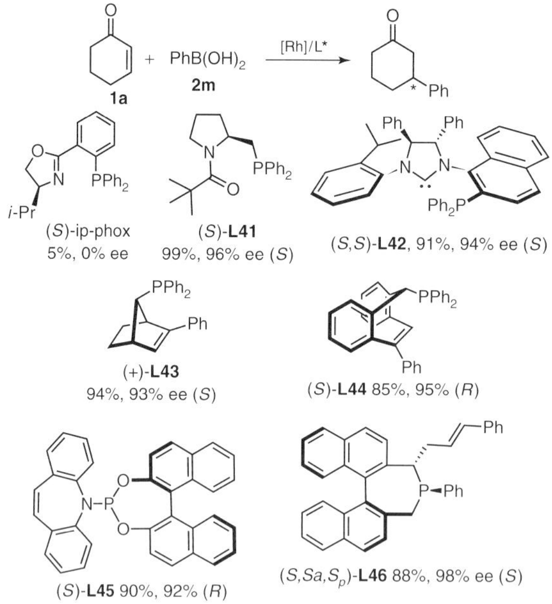

Scheme 1.17 ECA catalyzed by bidentate ligand rhodium complexes.

inter- or intramolecularly with an electrophile. To tap into this reactivity, Hayashi and coworkers found that, under aprotic conditions, the use of ArB(9-*BBN*) (**24**) was sufficiently reactive to form quantitatively – and with exquisite selectivity – the chiral boron enolates (see Scheme 1.18) [48, 135]. As a typical example, the reaction of 2-cyclohexenone (**1a**) with 1.1 equiv. of PhB(9-*BBN*) (**24a**) in the presence of a rhodium catalyst generated from [Rh(OMe)((S)-binap)]$_2$ and in toluene gave a high yield of the boron enolate (S)-**28** with 98% ee. The boron enolate formation was not observed in the reaction with phenylboronate esters, phenylboroxine, or tetraphenylborate. Unfortunately, the scope of this reaction is limited and a chiral boron-enolate was isolated only for 2-cyclohexenone (**1a**) and 2-cycloheptenone (**1c**). One of the most useful reactions of the resulting boronates is the aldol reaction with aldehydes (such as propanal), which is known to proceed through a well-organized transition state to give the *anti*-aldol product as a single diastereoisomer (**29**). Treatment of the boron enolate with *n*-butyllithium at −78 °C, followed by reaction with allyl bromide, gave 2-allylcyclohexanone **30** as a single diastereoisomer.

1.2.1.13 α,β-Unsaturated Aldehydes

Aldehydes are among the most versatile functional groups in organic chemistry, thus an ECA protocol to generate chiral 3,3-diarylpropanals (Scheme 1.19; **32**) is highly desirable and can be used in the synthesis of biologically active substances. However, enals represent an especially challenging class of substrates in Rh-catalyzed ECAs [112, 109, 136]. This can be attributed to the high reactivity of

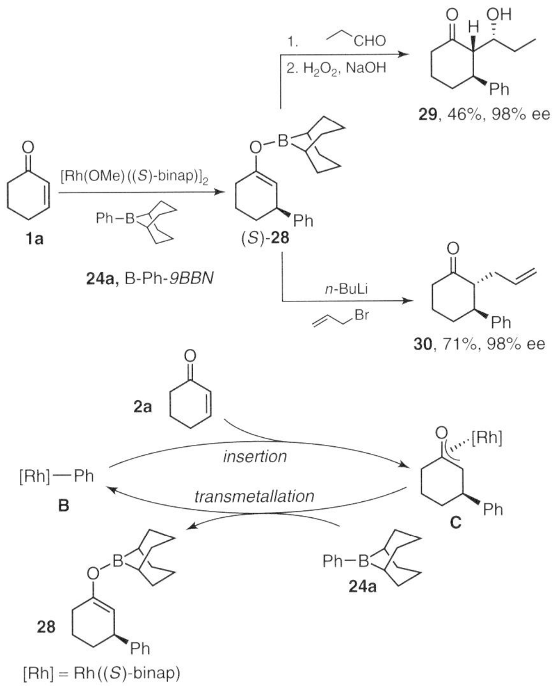

Scheme 1.18 Tandem Rh-catalyzed ECA with PhB(9-*BBN*) onto 2-cyclohexenone.

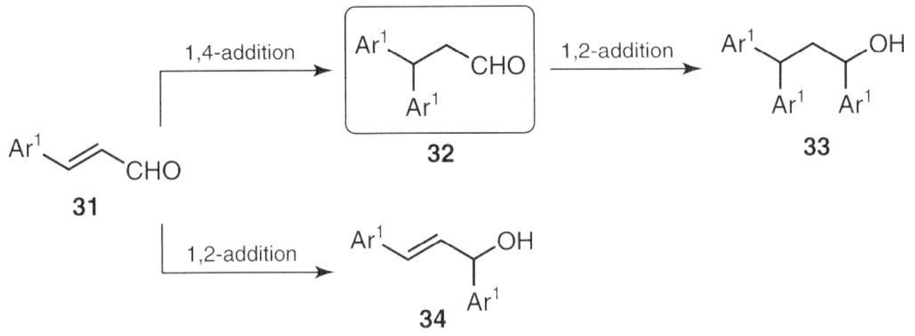

Scheme 1.19 Reaction pathways in the Rh-catalyzed conjugate addition of arylboronic acids to enals.

aldehydes, which can undergo competitive 1,2-addition either in competition with (**34**) or after the 1,4-addition (**33**) (Scheme 1.19).

The influence of the ligand on the selectivity of the transformation is shown in Scheme 1.20 [137]. Whereas, the use of phosphines ligands in the addition of phenylboronic acid to cinnamaldehyde **31a** led selectively to the allyl alcohol **34am**, the Rh/diene-catalyzed process resulted in the formation of the desired 1,4-adduct **32am**.

The use of chiral dienes (**L30f**, **L27b**, and **L28b**) was optimal for the formation of a wide range of enantiomerically enriched 3,3-diarylpropanals and 3-arylalkanals.

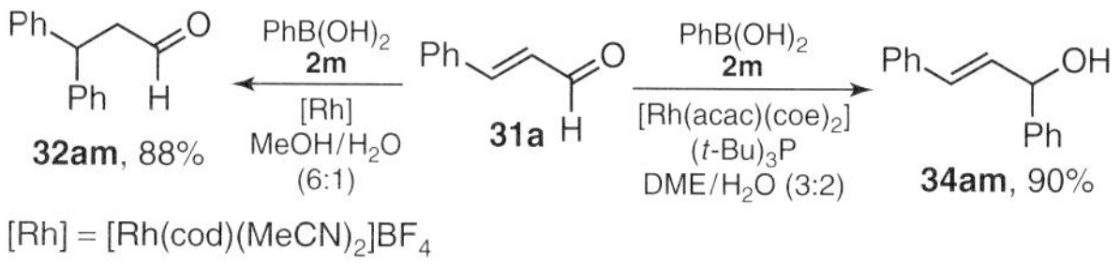

1 Rhodium- and Palladium-Catalyzed Asymmetric Conjugate Additions

Scheme 1.20 Ligand control of the selectivity for 1,2- or 1,4-additon to enal **31a**.

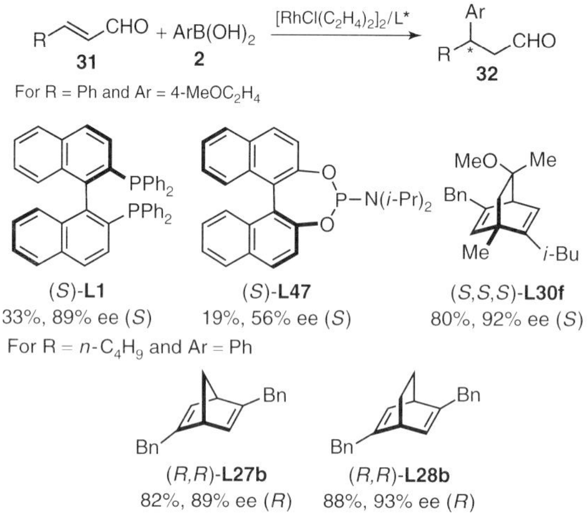

Scheme 1.21 Rh-catalyzed ECA of arylboronic acids to enals.

However, poorer results were obtained with conventional ligands such as (R)-binap (**L1**) or phosphoramidite **L47** (Scheme 1.21).

1.2.2
α,β-Unsaturated Esters and Amides

Following the initial discovery of rhodium-catalyzed ECA organoboron reagents onto α,β-unsaturated ketones, the methodology was rapidly expanded to other classes of α,β-unsaturated carbonyl compounds. Cyclic α,β-unsaturated esters (**35**) react well with arylboronic acids in the ECA, catalyzed by the Rh/(S)-binap system [31, 62]. However, for linear enoates (**37**) the more reactive LiArB(OMe)$_3$ (**22**, generated *in situ*) organoboron reagent is necessary to obtain acceptable yields (Scheme 1.22). In addition, the bulkier ester function (R = t − Bu) has a positive effect on the enantioselectivity on the process, but lowers the yield of the ECA.

The Rh-catalyzed ECA on α,β-unsaturated esters has been applied as the key step in the synthesis of a series of biologically active compounds such as baclofen (**41**) and tolterodine (**44**) (Scheme 1.23) [4, 138, 139]. Especially noteworthy, is the first reported application of a Rh-catalyzed ECA on a multi-kilogram scale on **45** (25 kg) [4]. A key result of the multi-kilogram scale-up of this reaction is the unexpected discovery that the use of a minimal quantity of 2-propanol (1 equiv.), rather than water as the co-solvent, reduces the extent of rhodium-mediated protodeboration

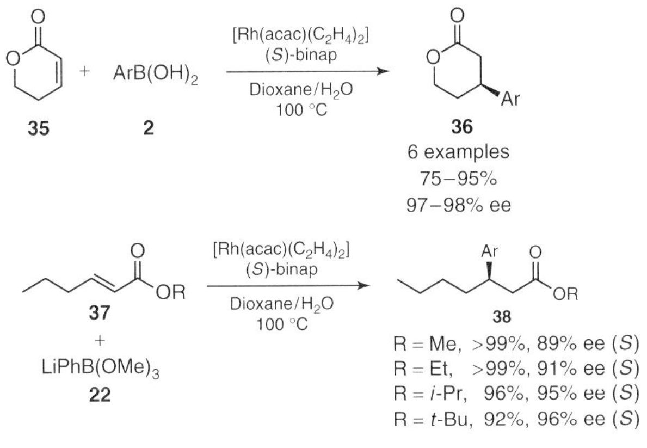

Scheme 1.22 [Rh]/(S)-binap-catalyzed ECA of organoboron reagent to cyclic and linear α,β-unsaturated esters.

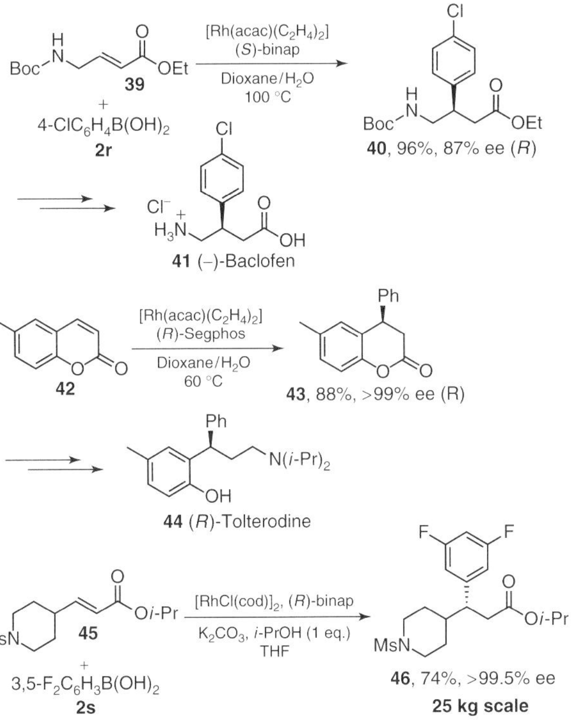

Scheme 1.23 Applications of the Rh-catalyzed ECA of ester in the synthesis of active pharmaceutical ingredients.

of the boron species. In addition, potassium carbonate (K_2CO_3) was found to be a useful additive.

A cationic rhodium(I)–chiraphos (**L3**) system was developed by Miyaura for the enantioselective preparation of β-diaryl carbonyl compounds (**48**) via the ECA of arylboronic acids (**2**) to β-aryl-α,β-unsaturated ketones or esters (**47**) (Scheme 1.24) [61]. As with other systems, the increase in the size of the ester group to CO_2t-Bu, led to a slight increase in enantioselectivity. The chiraphos ligand (**L3**) was quite effective for these substrates, but was a poor ligand for the Rh-catalyzed

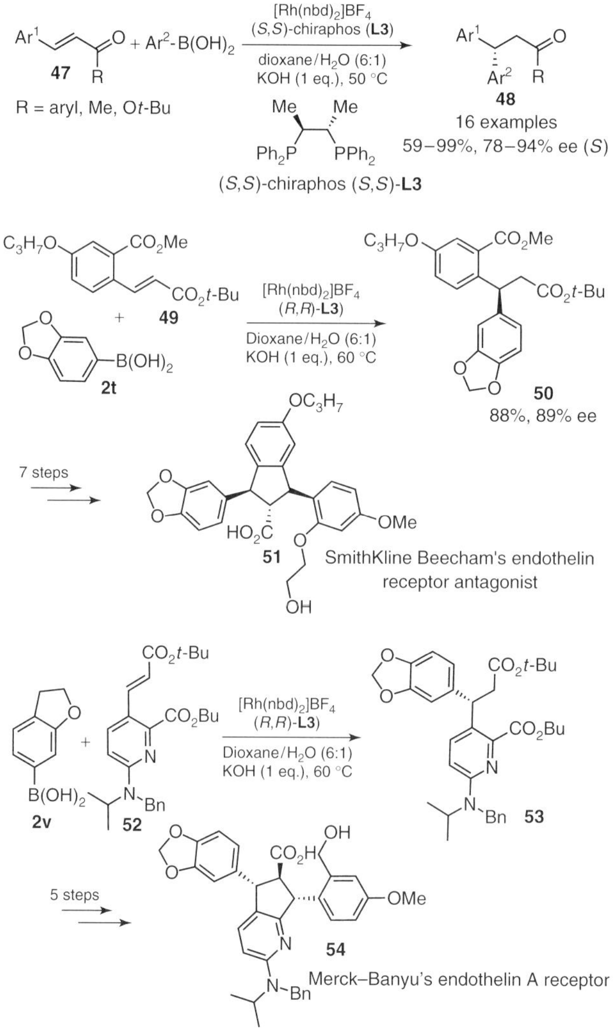

Scheme 1.24 Rh-catalyzed ECA of arylboronic acid on β-aryl α,β-unsaturated esters and applications to the synthesis of active pharmaceutical ingredients.

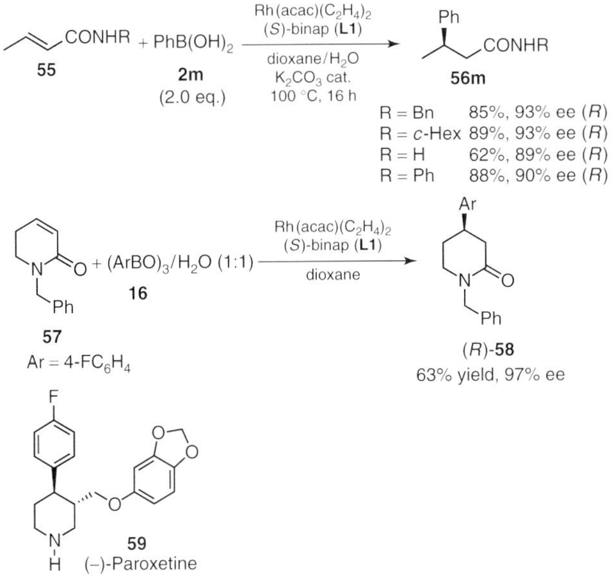

Scheme 1.25 [Rh]/(S)-binap catalyzed ECA of organoboron reagent to linear and cyclic α,β-unsaturated amides.

ECA of cyclic enones (cf. Scheme 1.11). This system proved quite versatile and functional group-tolerant, and was successfully applied as the key enantioselective step in the synthesis of two endothelin receptor antagonists **51** [140] and **54** [141].

Although less reactive than enones and enoates, linear α,β-unsaturated amides **55** perform similarly well under standard Rh-catalyzed ECA conditions [62]. The use of K_2CO_3 as an activator significantly increased the overall reaction yield. Comparable results were obtained for the addition of 4-fluorophenylboronic acid to 5,6-dihydro-2(1H)-pyridinones **57** to yield **58**, a key intermediate in the synthesis of paroxetine [33]. The asymmetric addition onto α,β-unsaturated amides has also been applied in total synthesis [142] (Scheme 1.25).

The use of boroxines with the Rh/phosphoramidite (**L19**) catalytic system enabled the synthesis of 2-aryl-4-piperidones **60**, which serve as a useful framework and are found in a number of active pharmaceutical ingredients (APIs) (Scheme 1.26) [85]. The use of aryl boroxines (**7**) and the slow addition of water were necessary to

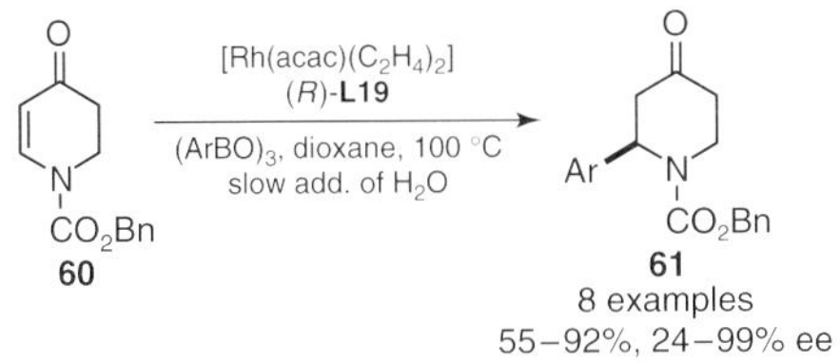

Scheme 1.26 Rh-catalyzed ECA on 2,3-dihydro-4-pyridones.

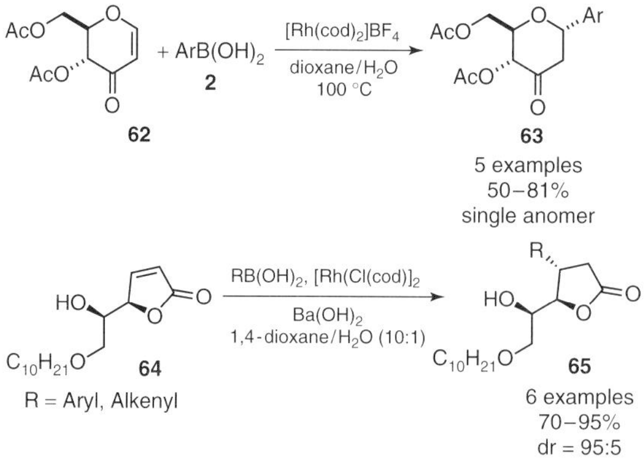

Scheme 1.27 Examples of a diastereoselective Rh-catalyzed conjugate addition.

minimize protodeboration. For a variant of this transformation with organozinc, see Scheme 1.48 [143].

1.2.2.1 Diastereoselective Conjugate Addition

The presence of a chiral center in close proximity to the β position of an α,β-unsaturated carbonyl compound can be used to control the diastereoselectivity of a Rh-catalyzed conjugate addition, without the need for extraneous chiral ligand on Rh. The first example of such a diastereoselective reaction was reported for the synthesis C-glycosides **63** from the enantiomerically pure cyclic esters **62** [144]. The reaction was efficiently catalyzed by a cationic Rh source to give the C-glycosides **63** as single anomers. Another example is the Rh-catalyzed addition of aryl- and alkenylboronic acids onto butenolide **64**, leading to the products **65** with high diastereoselectivities (Scheme 1.27) [145]. In this regard, the presence of an unprotected hydroxyl group may also provide an enhancement of the diastereocontrol at the addition.

A similar approach was used for the asymmetric synthesis of functionalized pyrrolizidinones **67** (Scheme 1.28) [146]. In this case, chiral diene (S,S,S)-**L30f** was used to enhance the diastereoselectivity of the process leading to (S,R)-**67**. On the other hand, the use of the other enantiomer of the ligand (R,R,R)-**L30f**, reversed the diastereoselectivity of the ECA to afford (S,S)-**67** (Scheme 1.28). Therefore, the process is under ligand control.

1.2.2.2 Fumarate and Maleimides

Although the ECA products of fumarates and maleimides are synthetically useful 2-substituted 1,4-dicarbonyl compounds, they represent a difficult class of substrates because they are relative unreactive. In the ECA of phenylboronic acid onto di-*tert*-butyl fumarates (**68**), traditional diphosphine ligands (**L1**) and phosphoramidite ligands (**L19**) gave poor yields and enantioselectivities, while the bulky chiral diene **L27g** gave higher yields and synthetically useful enantiomeric excesses (Scheme 1.29) [108].

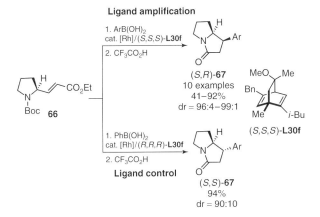

Scheme 1.28 Asymmetric synthesis of functionalized pyrrolizidinones using [Rh]/diene catalytic system.

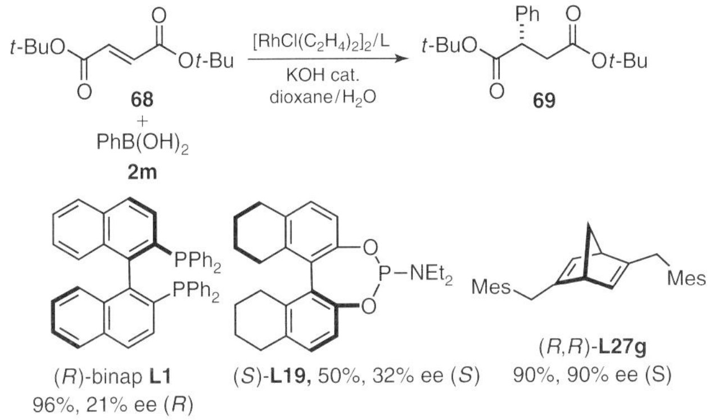

Scheme 1.29 Rh-catalyzed ECA of phenylboronic acid onto di-tert-butyl fumarate.

Similarly, phosphine-based ligands such as (R)-binap (**L1**) lead only to moderate enantioselectivity for the ECA of phenylboronic acid (**2m**) onto benzyl maleimide [129]. First-generation chiral dienes such as **L27g** showed increased reactivity [129], but the enantioselectivity remained low. A breakthrough was achieved with the use of phosphorus–olefin hybrid ligands **L43** [129, 130], and **L44** [131] that gave excellent yields and enantioselectivities.

The efficient diastereoselective synthesis of axially chiral N-arylsuccinimides **73** has been achieved by using chiral diene **L28a** (Scheme 1.30) [108, 147]. Diphosphine ligands gave lower diastereoselectivities and enantioselectivities (Scheme 1.31).

An elegant application of maleimide **73** is the use of the axial chirality as a stereochemical relay for subsequent transformations, such as an alkylation of **73m** to generate a quaternary chiral center with diastereoselectivity and near-perfect

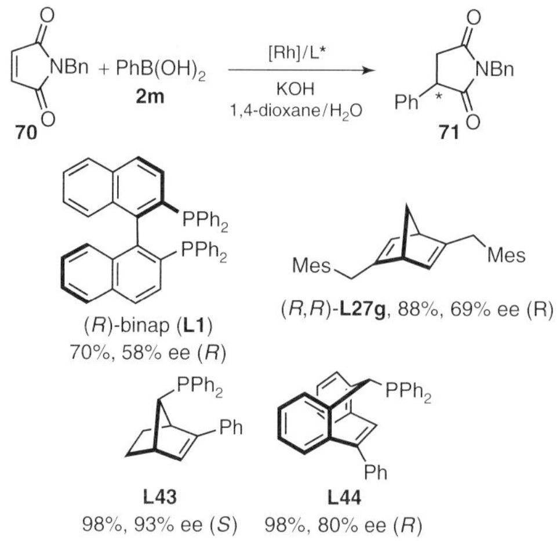
70
NBn
+ PhB(OH)2
2m
[Rh]/L*
KOH
1,4-dioxane/H2O
Ph
NBn
71
PPh2
PPh2
(R)-binap (L1)
70%, 58% ee (R)
Mes
Mes
(R,R)-L27g, 88%, 69% ee (R)
PPh2
Ph
L43
98%, 93% ee (S)
PPh2
Ph
L44
98%, 80% ee (R)

Scheme 1.30 Activity and selectivity of different ligand in the Rh-catalyzed ECA of $PhB(OH)_2$ onto maleimide **70**.

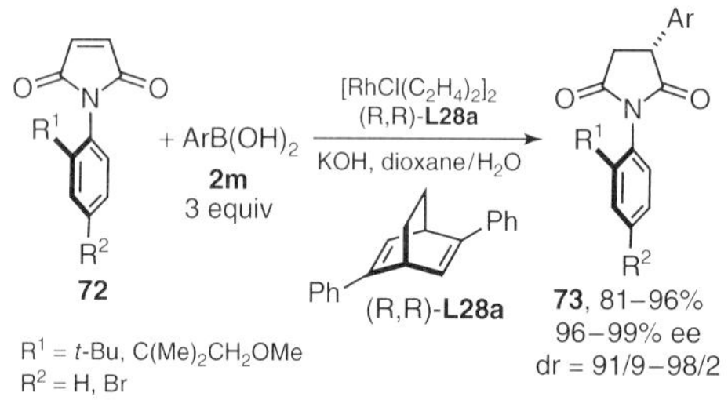
72
+ ArB(OH)2
2m
3 equiv
[RhCl(C2H4)2]2
(R,R)-L28a
KOH, dioxane/H2O
Ph
Ph
(R,R)-L28a
Ar
73, 81–96%
96–99% ee
dr = 91/9–98/2
R1 = t-Bu, C(Me)2CH2OMe
R2 = H, Br

Scheme 1.31 Diastereoselective synthesis of axially chiral arylsuccinimides.

enantioselectivity. The maleimide **73m** can also be oxidized to dienophile **75** which undergoes a smooth stereoselective Diels–Alder to yield bicycle **76** (Scheme 1.32) [147].

The effective construction of chiral quaternary carbons is arguably one of the biggest challenges in asymmetric catalysis [148–150]. There are very few examples

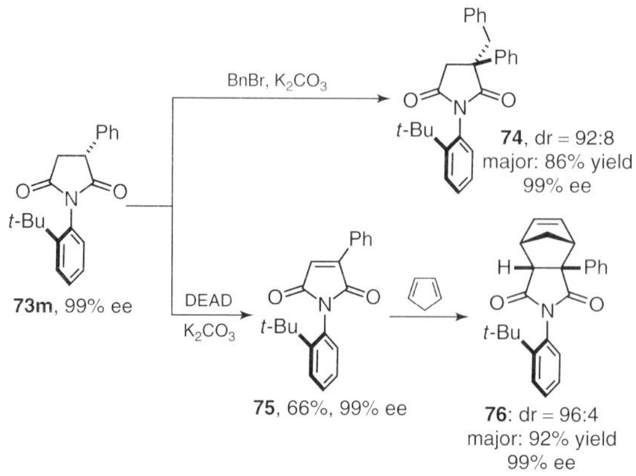
BnBr, K2CO3
74, dr = 92:8
major: 86% yield
99% ee
73m, 99% ee
DEAD
K2CO3
75, 66%, 99% ee
76: dr = 96:4
major: 92% yield
99% ee

Scheme 1.32 Transfer of the axial chirality of **73m**.

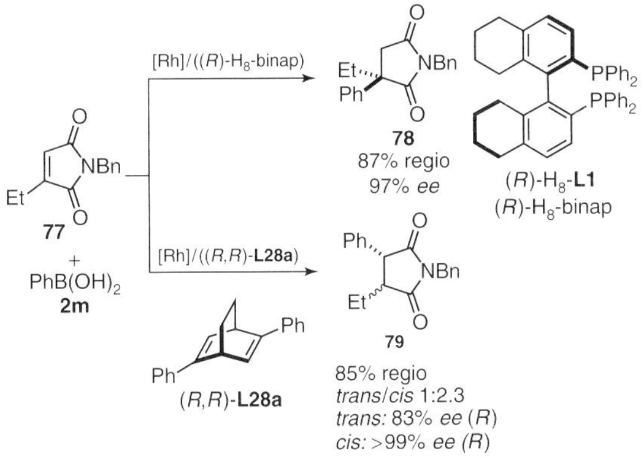

Scheme 1.33 Influence of the ligand on the regioselectivity of Rh-catalyzed ECA onto maleimides.

of ECA to β,β-disubstituted α,β-unsaturated carbonyl compounds [149]. During an examination of the use of substituted maleimides, Hayashi and coworkers discovered that the regioselectivity of the addition is a function of the ligand employed (Scheme 1.33) [151]. Whereas, Rh/(H$_8$-binap)-catalyzed processes give rise preferably to 1,4-adducts **78** with a quaternary stereogenic center, the Rh/((R,R)-**L28a**) catalyst leads to *cis/trans* mixtures of **79**.

1.2.2.3 Synthetically Useful Acceptors for Rh-Catalyzed ECAs

Today, a range of synthetically useful acceptors is available that can be used in the Rh-catalyzed ECA (see Figure 1.3). For example, the use of β-silyl-substituted α,β-unsaturated carbonyl compounds **80** as acceptors is of special interest as these compounds can be transformed to β-hydroxyketones by Tamao–Fleming oxidation [111]. In addition, the introduction of an alkenyl group onto the β-silyl enone **80** leads to a chiral allylsilane that can further react with the ketone moiety in an intramolecular Sakurai [152] reaction [153]. The 3,7-disubstituted bicyclo[3.3.1]octadiene **L36** performs particularly well with these substrates [104]. Another family of acceptors that enable the straightforward modification of the resulting adducts are α,β-unsaturated esters **47** [95], and α,β-unsaturated Weinreb

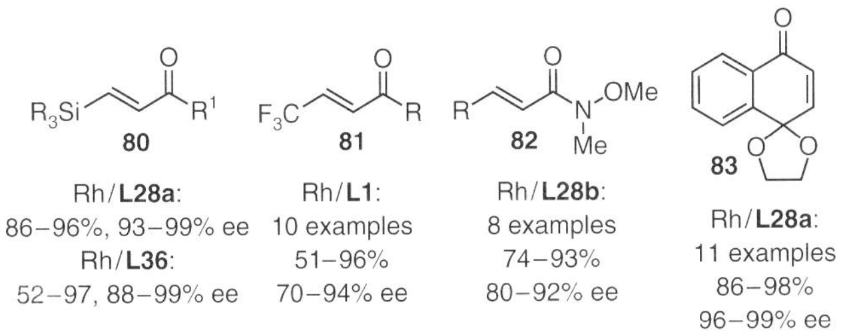

Figure 1.3 Synthetically useful acceptors in Rh-catalyzed ECAs.

amides **82** [154]. The ECA product using β-trifluoromethyl-α,β-unsaturated ketones **81** are of particular interest in the medicinal, pharmaceutical, and agricultural fields [155]. Lastly, α-arylated tetralones can be accessed in high yields and stereoselectivity by Rh/diene-catalyzed ECA of organoboron reagents to quinone monoketal **83** [114].

The Rh-catalyzed ECA of alkenyltrifluoroborate (**86**) onto quinone monoketals **85** was used by Corey and coworkers as a key step in the synthesis of platensimycin, a potent anti-cancer agent [3, 156] (Scheme 1.34).

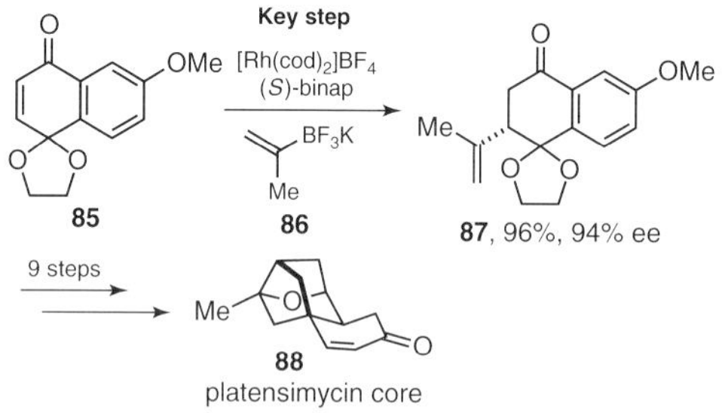

Scheme 1.34 Application of Rh-catalyzed ECA as a key step in the total synthesis of platensimycin.

To summarize the general trends in Rh-catalyzed ECAs of organoboron reagents onto α,β-unsaturated carbonyl compounds, the rate of the conjugate addition decreases with the reactivity of the acceptor and its steric bulk. Thus, acceptors can be classified into the following order of decreasing reactivities: enals > enone > enoate > enamides > fumarates > maleimides. Because effective coordination of the acceptor to rhodium is crucial for the reaction, acceptors that bear steric bulk in proximity to the unsaturation will be less reactive than unhindered ones. In addition, the activity of Rh/ligand catalytic systems increases with increasing π-accepting properties of the ligand.

1.2.3
Other Alkenes

The enantioselective construction of stereogenic carbon centers substituted with two aryl groups and one alkyl group is a subject of importance, because this structural motif is often found in pharmaceuticals and natural products. As discussed previously in Scheme 1.21 and Figure 1.3, their asymmetric synthesis has been reported by the chiral diene/rhodium-catalyzed asymmetric 1,4-addition of arylboronic acids to β-aryl-α,β-unsaturated aldehydes (**31**) and esters (**47**). An alternative approach to such chiral building blocks has been demonstrated by Hayashi and coworkers with the use of arylmethylene cyanoacetates **89** as substrates (Scheme 1.35) [115]. The ECA product **90** can be easily decarboxylated to give the corresponding enantiopure β, β'-diaryl nitrile. The best results were obtained with the chiral diene (*R,R*)-**L28a** which enabled consistently high enantioselectivities to be achieved. As shown in the table in Scheme 1.35, the presence of both

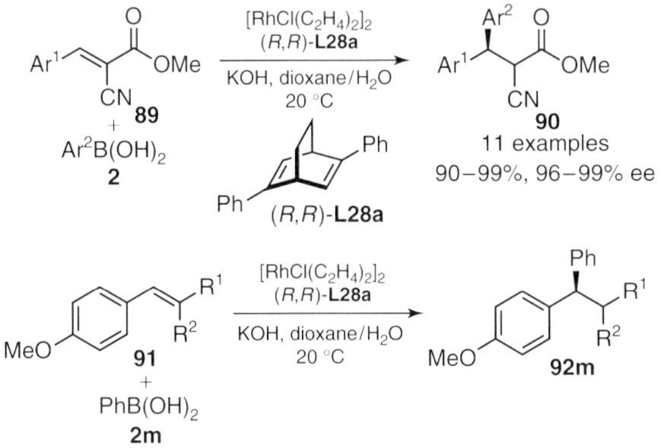

Entry	Substrate	R¹	R²	Product	Yield (%)	ee (%)
1	91a	CO$_2$Me	CN	92am	99	99 (R)
2	91b	CN	CN	92bm	9	n.d.
3	91c	CO$_2$Me	CO$_2$Me	92cm	11	n.d.
4	91d	CN	H	92dm	74	52 (R)
5	91e	CO$_2$Me	H	92em	99	57 (R)

Scheme 1.35 Rh-catalyzed ECA of arylboronic acids to arylmethylene cyanoacetates.

cyano and ester groups at the α-position of the substrates is essential for the high reactivity and enantioselectivity in the present reaction. Other combinations gave poor results.

1.2.3.1 Alkenylphosphonates

The Rh-catalyzed ECA onto alkenylphosphonates was first reported by Hayashi and coworkers. This class of substrate is less reactive than α,β-unsaturated carbonyl compounds, and the use of arylboroxine (**7**) instead of arylboronic acids (**2**), in the presence of 1 equiv. of water, was necessary to obtain high yields (Scheme 1.36) [157]. The enantioselectivities and chemical yields were slightly higher in the reaction catalyzed by the rhodium complex coordinated with unsymmetrically substituted binap ligand, (S)-u-binap, which has diphenylphosphino and *bis*(3,5-dimethyl-4-methoxyphenyl)- phosphino groups at the 2 and 2′ positions on the 1,1′-binaphtyl skeleton (Scheme 1.36). In agreement with the stereochemical model depicted in Scheme 1.10, the *trans* and *cis* geometries of alkenylphosphonate **93** give rises to opposite enantiomers.

1.2.3.2 Nitroalkenes

Nitroalkenes are good substrates for the rhodium-catalyzed ECA of organoboronic acids [158]. Hayashi reported that the reaction of 1-nitrocyclohexene (**96**) with phenylboronic acid (**2m**) in the presence of the rhodium/(S)-binap catalyst gave an 89% yield of the 2-phenyl-1-nitrocyclohexane (**97m**) (Scheme 1.37). The main phenylation product **97m** is a *cis* isomer (cis/trans = 87/13), and both the *cis*

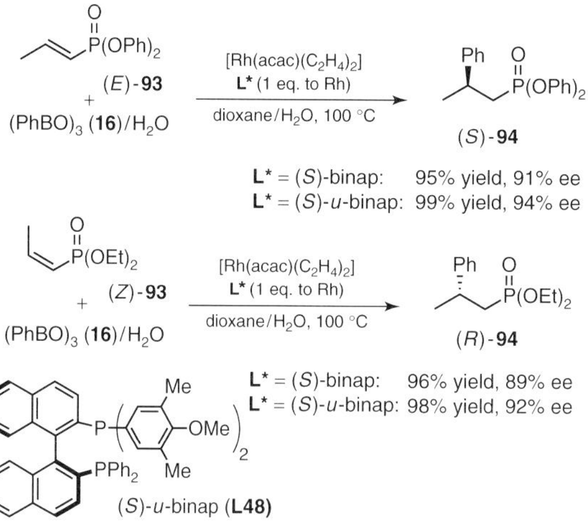

Scheme 1.36 Rh-catalyzed ECA of arylboronic onto alkenylphosphonates.

and *trans* isomers were 98% enantiomerically pure. Treatment of the *cis*-rich mixture with sodium bicarbonate in refluxing ethanol caused *cis–trans* equilibration, giving thermodynamically more stable *trans* isomer (*trans/cis* = 97/3). It should be noted, that this rhodium-catalyzed asymmetric phenylation produced thermodynamically less stable *cis* isomer of high enantiomeric purity, and it can be isomerized, if so desired, into the *trans* isomer without any loss of its enantiomeric purity. The preferential formation of *cis*-**97** in the catalytic arylation may

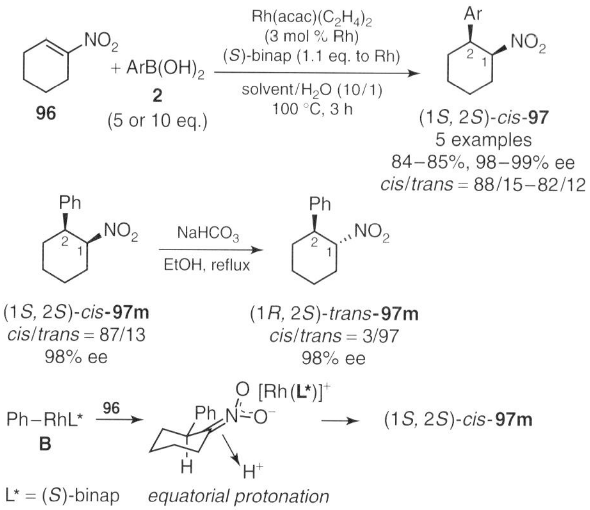

Scheme 1.37 Rh-catalyzed ECA of arylboronic onto nitroalkenes.

indicate the protonation of a rhodium nitronate intermediate in the catalytic cycle (Scheme 1.38).

The optically active nitroalkanes obtained by the present method are useful chiral building blocks, which can be readily converted into a wide variety of optically active compounds by taking advantage of the versatile reactivity of nitro compounds. An example of the usefulness of this transformation was reported by the research group at Merck, who performed a Rh-catalyzed ECA of difluorophenylboronic acid onto nitroalkene **98** as the key step in the synthesis of nitroalkane **99** on a 2 kg scale, this being a precursor to a migraine headache treatment [5]. Bicarbonate was found to be a useful activator in this reaction.

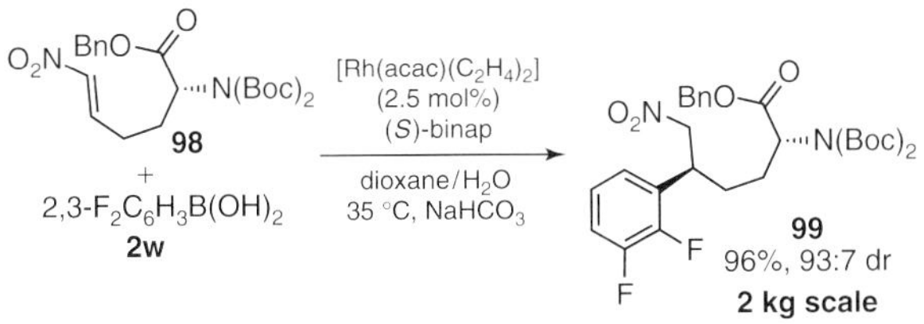

Scheme 1.38 Application of Rh-catalyzed ECA of a nitroalkene on a kilogram-scale.

1.2.3.3 Sulfones

Hayashi disclosed that α,β-unsaturated phenyl sulfones do not react with organoboron reagents under the usual rhodium-catalyzed conditions. When nucleophilic aryltitanium reagents were employed, an elimination of the sulfonyl group occurred after the conjugate addition, leading to desulfonylated alkenes as final products (cf. Scheme 1.48) [159]. The key to this synthetic challenge was found through the use of a rhodium-coordinating α,β-unsaturated 2-pyridylsulfone (**100**). With these pyridyl sulfones, it was possible to obtain a general methodology for providing β-substituted sulfones in high yields and enantioselectivities ranging from 76 to 92% ee with (*S*,*S*)-chiraphos (**L3**) as the chiral ligand (Scheme 1.39) [160, 161]. The corresponding 4-pyridyl sulfone analogs displayed no reactivity, demonstrating the necessity of the 2-pyridyl sulfone moiety to stabilize **H**. The *cis*-alkenylsulfone (*cis*-**100**) gave the opposite enantiomer to *trans*-alkenylsulfone (*trans*-**100**), and this can be rationalized by the general stereochemical model depicted in Scheme 1.12 [160, 161]. The chiral β-substituted sulfones **101** readily participates in a Julia–Kociensky olefination [162] to provide a novel approach to the enantioselective synthesis of allylic substituted alkenes *trans*-**102**. In addition, the chiral sulfones **101** can be alkylated and the sulfonyl removed by Zn reduction [161]. This approach was extended to the addition of alkenylboronic acids to β-aryl-β-methyl-α,β-unsaturated pyridylsulfones, which enabled the efficient stereoselective formation of quaternary centers with up to 99% ee [163].

A novel and original approach for the synthesis of enantiomerically enriched 2-arylbut-3-enols **105** makes use of *cis*-allyldiol **103** and arylboroxines (**7**)

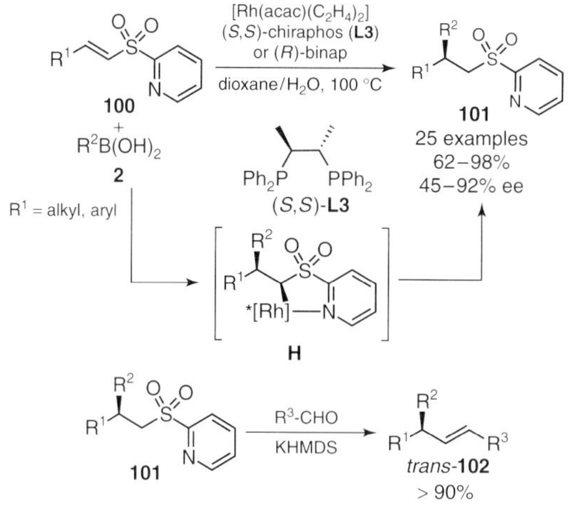

Scheme 1.39 Rh-catalyzed ECA of organoboronic acid with alkenyl sulfones.

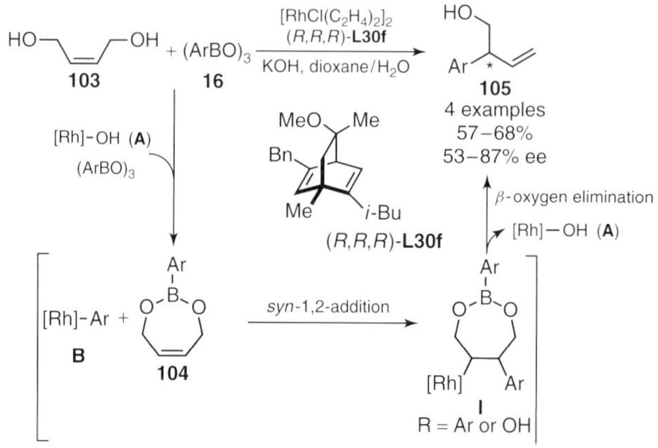

Scheme 1.40 Rh-catalyzed substitutive arylation of a *cis*-allylic diol with arylboroxines.

(Scheme 1.40) [164]. Under the reaction conditions, *cis*-diol **103** readily forms the cyclic arylboronic ester **104** which serves as an acceptor for a *syn*-1,2-carborhodation by [Rh(μ-OH)(L30f)] to give intermediate **I**. A subsequent β-oxygen elimination regenerates the active rhodium-hydroxide species **A** and releases the optically active alcohols **105**.

1.2.3.4 1,4-Addition/Enantioselective Protonation

So far, the Rh-catalyzed ECA process with β-substituted α,β-unsaturated carbonyl compounds has been reviewed. With this family of substrates, the enantio-determining step is the 1,2-insertion of the acceptor into the Rh–C$_{sp2}$ bond. However, when α-substituted α,β-unsaturated carbonyl compounds are employed, the enantio-determining step is the protonation of the oxo-π-allylrhodium species (see Scheme 1.10). The asymmetric variant of this mechanistic manifold was first exploited by Reetz and coworkers, in the ECA of α-acetamidoacrylic ester

106 by phenylboronic acid (**2m**), giving the phenylalanine derivative (**107m**) in up to 77% ee (Scheme 1.41) [68]. A similar asymmetric transformation has also been reported by Frost and coworkers [165]. This reaction represents a convenient alternative to the synthesis of unnatural phenylalanine derivates (**107**), which are commonly accessed through Rh-catalyzed asymmetric hydrogenation of β-aryl-α-acetamidoacrylic esters [166].

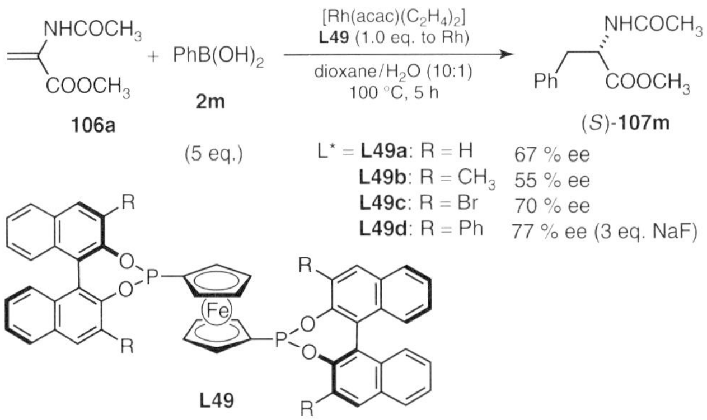

Scheme 1.41 Rh-catalyzed conjugate addition/enantioselective protonation.

In recognizing the importance of the proton source in this process, Genêt and Darses investigated a wide range of phenols instead of water as the proton donor [37, 167]. Guaiacol (**108**) proved to be the best proton donor and enabled a satisfactory enantioselectivity with a range of potassium organotrifluoroborates (Scheme 1.42). The ratio of metal to ligand had a notable effect on the enantioselectivity of the reaction, with 2.2 equiv. of chiral ligand relative to metal being optimal. Higher temperatures and the use of toluene or dioxane increased both conversion and enantioselectivity. Importantly, the presence of water accelerated the reaction 10-fold relative to guaiacol, but drastically decreased the enantioselectivity to 16% ee. The use of organoboronic acids also leads to lower yields and enantioselectivities, presumably due to residual traces of water. As with the ECA of enoate (see Scheme 1.22), increasing the size of the ester substituent in **106** leads to a slight increase in enantiomeric excesses. The use of more π-acidic (S)-difluorophos (**L50**) yields faster and more selective transformation relative to binap (**L1**). In-depth mechanistic studies and density functional theory (DFT) calculations into this process suggested that the actual mechanism goes through a sequential conjugate addition and β-hydride elimination to form an imine and a Rh-hydride species. The Rh-hydride subsequently reinserts into the imine enantioselectively, and the Rh-amino bond is then hydrolyzed to generate the reaction product [37].

The scope of substrates amenable to the enantioselective protonation was expanded to include dimethyl itaconate (**109**) [168], α-benzyl acrylates (**112**) [169], and α-aminomethyl acrylates (**114**) [170] (Scheme 1.43) [165]. In all of

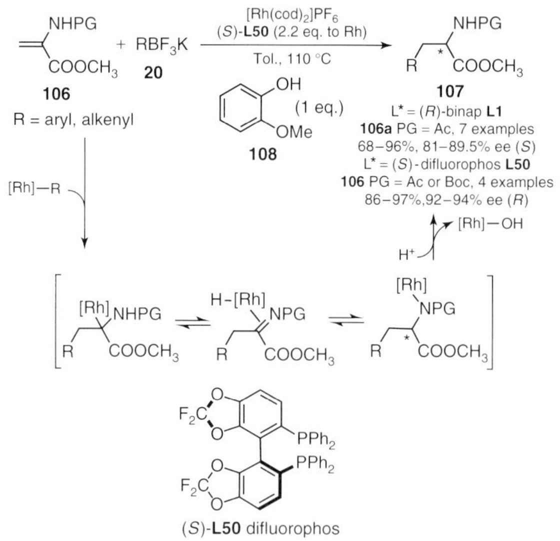

Scheme 1.42 Rh-catalyzed conjugate addition/enantioselective protonation using guaiacol of the sole proton source.

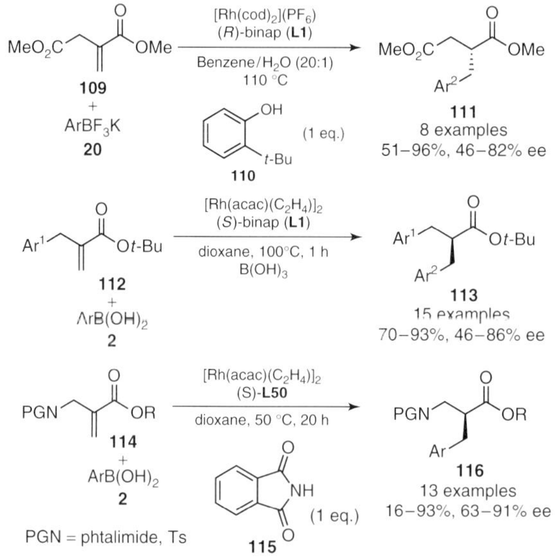

Scheme 1.43 Applications of Rh-catalyzed addition/enantioselective protonation.

these examples the choice of the additional proton source (phenol **110**, boric acid (B(OH)$_3$), or phthalimide **115**) was critical to obtain high yields and good selectivities.

An interesting application of the Rh-catalyzed enantioselective protonation is the peptide modification via site-selective residue interconversion, through an elimination and conjugate addition sequence (Scheme 1.44). Thus, a serine or cysteine (**117**) can be selectively eliminated from a peptide chain to form a dehydroalanine fragment (**118**) that can subsequently undergo a Rh-catalyzed ECA to yield peptide with modified Ph-alanine fragment (**119**). Although the diastereoisomeric excesses are modest, this transformation can be applied on a range of di- and tripeptides [171].

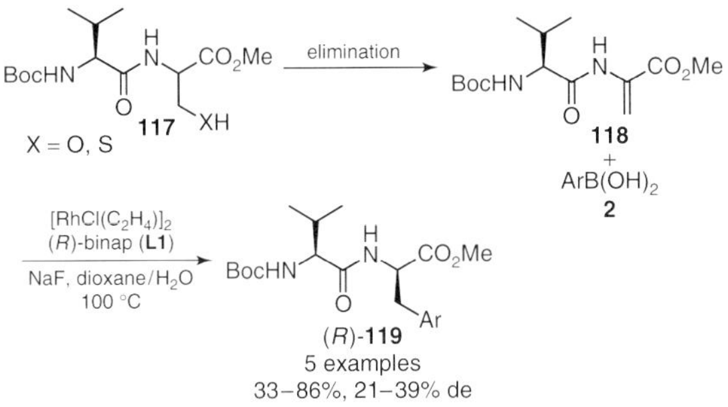

Scheme 1.44 Application of the addition/enantioselective protonation to the synthesis unnatural peptides through site interconversion.

The enantioselective protonation was also applied to the hydroarylation of diphenylphosphinylallenes [172]. Unlike the oxo-π-allylrhodium species in previous examples, the π-allylrhodium intermediate **K** can be protonated α or γ to the phosphorus center to give intermediates **121a** and **121b**, respectively. Tetrahydrofuran (THF) was found to minimize the amount of **121b** formed. Furthermore, bulkier R groups (*t*-Bu) in allene **120** led to **121b** because protonation at the least-hindered position was favored. Under these conditions, the boronic acid would act as the proton source (Scheme 1.45).

An excellent overview of enantioselective protonation has been recently published [173].

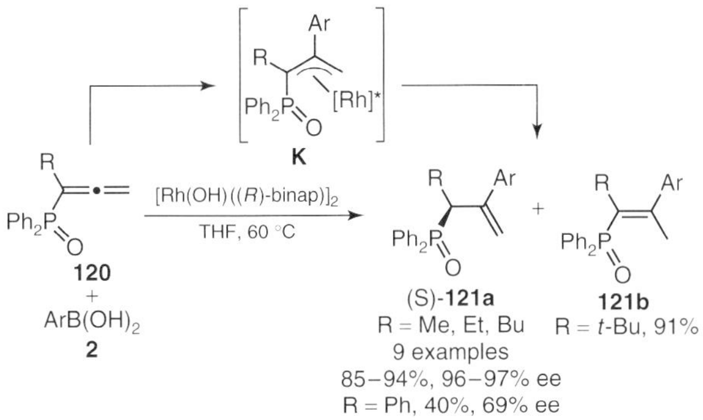

Scheme 1.45 Rh-catalyzed enantioselective addition to diphenylphosphinylallenes.

1.3
Rh-Catalyzed ECA Organotitanium and Organozinc Reagents

Hayashi and coworkers expanded the scope of nucleophilic reagents amenable to Rh-catalyzed ECA transformation to organotitanium [159, 174, 175] and organozinc [97, 136, 143, 176] reagents. These reagents are far more nucleophilic than the corresponding organoboron derivatives, and can be added to α,β-unsaturated compounds at room temperature. Furthermore, the reactions are run under aprotic conditions which enables trapping of the stable titanium and zinc enolates formed *in situ* with a range of electrophiles. The aryltitanium and arylzinc reagents are conveniently generated *in situ* through the addition of the corresponding aryllithium to ClTi(O*i*-Pr)$_3$ [174] and ZnCl$_2$, respectively [143].

The titanium-enolate (**123am**) obtained from 2-cyclohexenone (**1a**) was treated with lithium isopropoxide, to generate a more reactive titanate, and with ClSiMe$_3$ to yield a silyl enol ether (**124**). Treatment of the titanium enolate with propanal resulted in formation of the *exo-E*-enone (**125**) by aldol addition and elimination (Scheme 1.46), but no elimination occurred with boron enolate (cf. Scheme 1.20). Allyl bromide could also be added to **123am** to give (2*R*,3*S*)-*trans*-3-phenyl-2-allylcyclohexanone (**126**) in 82% yield as a single diastereomer, the enantiomeric purity of which was more than 99% (Scheme 1.46).

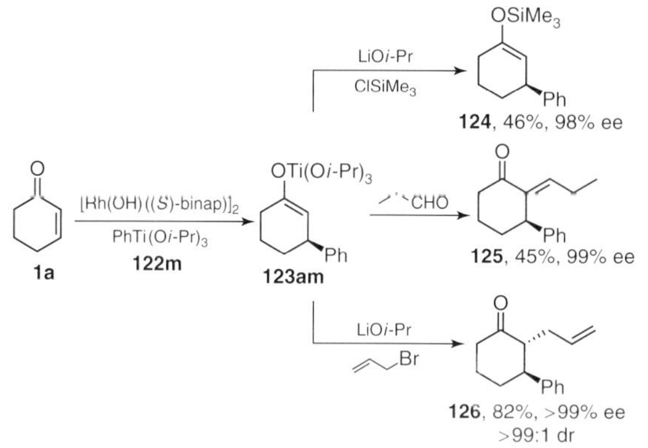

Scheme 1.46 Rh-catalyzed ECA of phenyltitanium reagent onto 2-cyclohexenone.

α,β-Unsaturated sulfones, bearing no coordinating function, are resilient to Rh-catalyzed ECA with organoboron reagents. However, the Rh-catalyzed addition of aryl titanium reagents (**122**) on α,β-unsaturated sulfones revealed an interesting mechanistic pathway, in which the *cine*-substitution product **128** was formed (Scheme 1.47) [159]. This reaction was successfully exploited to form enantiopure allylic arenes (**128**). The reaction is thought to proceed through the stereoselective aryl-rhodium into the alkenyl sulfone (**L**), followed by β-hydrogen elimination

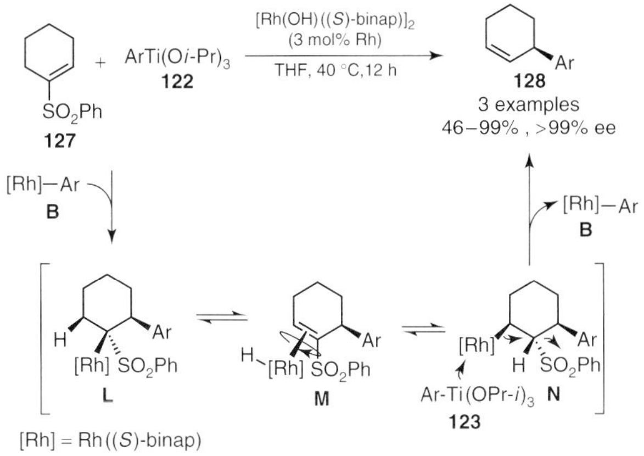

1.3 Rh-Catalyzed ECA Organotitanium and Organozinc Reagents | 39

Scheme 1.47 Rh-catalyzed asymmetric cine-substitution of alkenyl sulfones.

(**M**)/hydride insertion sequence to form the alkyl-rhodium intermediate **N**. Subsequent *anti*-elimination of a sulfinyl produces the allylic arenes **128** and regenerates the aryl-rhodium species (Scheme 1.47).

It was found that a combination of lithium aryltitanates (LiArTi(O*i*-Pr)$_4$) and chlorotrimethylsilane constitutes an effective arylating reagent, producing high yields of silyl enol ethers as 1,4-addition products [177]. Such titanates are generated *in situ* from the addition of an aryllithium onto Ti(O*i*-Pr)$_4$. A mechanistic study of this system revealed that the reaction passed through similar intermediates as with organoboron reagents. The ClSiMe$_3$ was necessary to promote the addition of organotitanium reagents.

The stereoselective synthesis of chiral β-substituted piperidones through a Rh-catalyzed ECA is challenging because of the low reactivity of vinologous amides **60**. To overcome this, the more reactive arylzinc reagents (**129**), in conjunction with a Rh/binap catalyst, were employed (Scheme 1.48) [143]. Importantly, only a small excess of arylzinc relative to the enone was necessary to obtain a full conversion, while at least 3 equiv. of phenylboronic acid were required (Scheme 1.48).

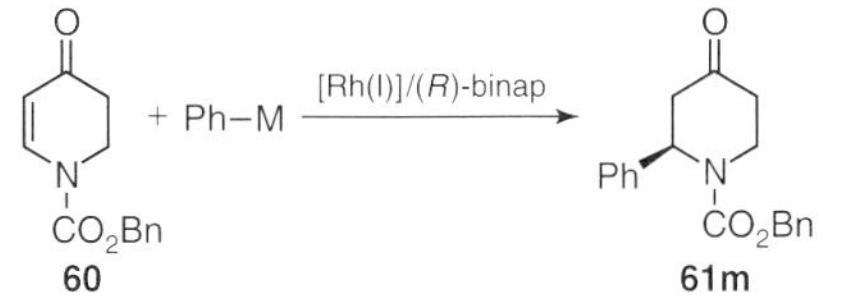

Entry	Ph-M (eq.)	Yield (%)	ee (%)
1	PhB(OH)$_2$ (3)	78	98
2	PhB(9-*BBN*) (3)	33	97
3	PhTi(O*i*-Pr)$_3$ (1.5)	70	>99.5
4	PhZnCl (**129m**) (1.5)	95	>99.5

Scheme 1.48 Rh/(*R*)-binap-catalyzed ECA of organometallic reagents.

These conditions enabled the synthesis of a range of β-arylpiperidones **61** with near-perfect enantioselectivities.

In addition, the stable zinc-enolate formed during the reaction could be subsequently trapped with electrophiles such as pivaloyl chloride and allyl bromide to yield **131** and **132** (single diastereoisomer), respectively [143] (Scheme 1.49).

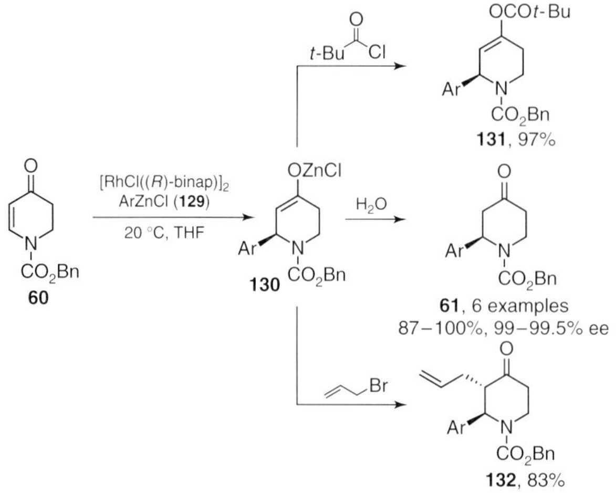

Scheme 1.49 Rh-catalyzed ECA with arylzinc onto piperidones.

The combination of arylzinc reagent (**129**) and ClSiMe$_3$ was found to be well suited to the Rh-catalyzed ECA onto enals, a challenging class of substrates (cf. Section 1.8) (Scheme 1.50). Only a limited amount of 1,2-addition product was observed (ca. <20%). Similar to the addition of organotitanates, ClSiMe$_3$ is necessary to activate the arylzinc reagents, the product being obtained as the corresponding silyl enol ether (**133**). For this transformation, arylzinc reagents were superior to organoboron and organotitanium reagents.

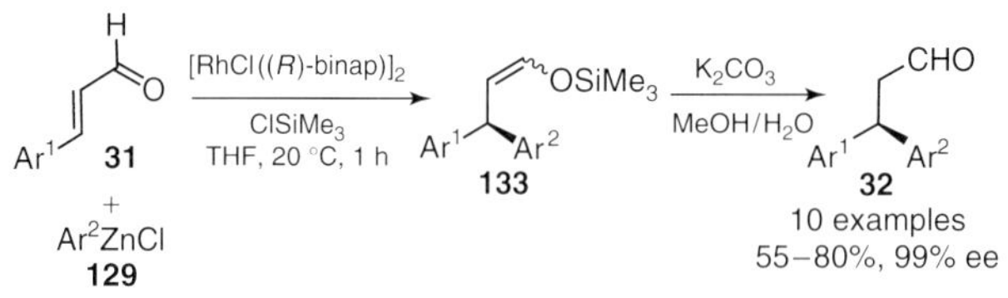

Scheme 1.50 Rh-catalyzed ECA of arylzinc/ClSiMe$_3$ onto enals.

The combination of heteroarylzinc reagent (**134**) and ClSiMe$_3$ chloride has also been applied to the ECA of heteroaromatic compounds to α,β-unsaturated ketones and esters (Scheme 1.51) [178]. Interestingly, (*R*,*R*)-Me-duphos (**L51**) proved to be the best chiral ligand for this transformation, but it performed poorly for the ECA of organoboron reagents (cf. Scheme 1.11).

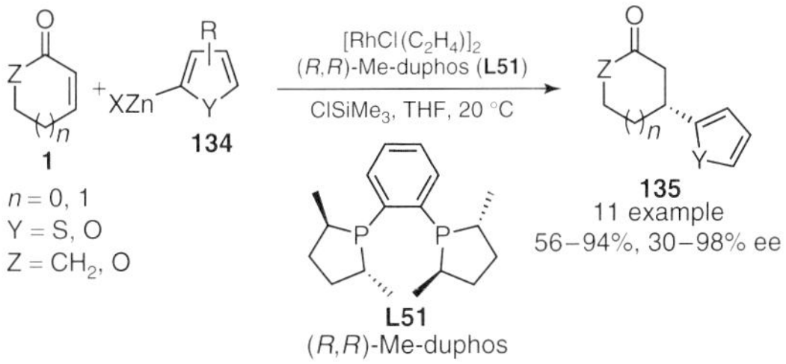

Scheme 1.51 Rh-catalyzed ECA of heteroaromatic zinc reagents to α,β-unsaturated ketones and esters.

1.4
Rh-Catalyzed ECA of Organosilicon Reagents

Today, organosilicon reagents play a growing role in organic synthesis due to their low cost, low toxicity, ease of handling, tolerance to a variety of functional groups, and simplicity of byproduct removal [179–181]. Unfortunately, organosilicon reagents are far less reactive than their boron, tin, titanium, and zinc counterparts, and this has led to a relatively slow development as nucleophiles in Rh-catalyzed ECA reaction. The first Rh-catalyzed addition of organosilicon reagent to α,β-unsaturated carbonyls was independently reported in 2001 by Mori and Li [182, 183]. By analogy with organoboronic acids, Mori made use of silanediols 136 as the nucleophiles and [Rh(OH)(cod)]$_2$ as the catalyst (Scheme 1.52). Depending on whether the reaction conditions where anhydrous or not, either the Mizoroki–Heck-type product 137 or the addition product 3 were observed. The Mizoroki–Heck-type product is due to the β-hydride elimination from the rhodium-enolate O, followed by conjugate reduction of an acrylate and subsequent transmetallation of the resulting Rh-enolate P (Scheme 1.52). Thus, as had been

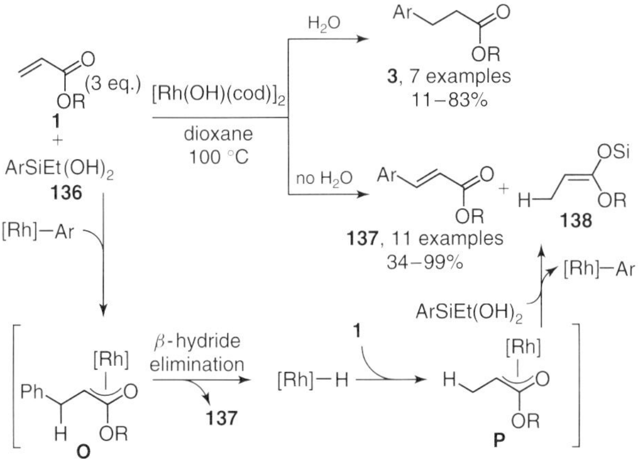

Scheme 1.52 Rh-catalyzed addition of arylsilanol to acrylates.

observed for organoboron reagent, water proved to be essential in these reactions to hydrolyze the Rh-enolate and regenerate the active hydroxyrhodium species. Using similar conditions, poly(phenylmethylsiloxane) (arylated silicone oil) proved to be a competent organosilicon derivative in Rh-catalyzed conjugate addition [184].

Li and coworkers found that a combination of dichlorodiarylsilanes and NaF, using water as the solvent and a cationic rhodium source ([Rh(cod)$_2$]BF$_4$), was optimal for the addition onto a wide range of acceptors, including challenging maleates and β-aminoacrylate [183]. It is likely that, under these conditions, silanols are formed.

Interestingly, Oi and coworkers found that organosiloxanes generated from alkoxysilanes (RSi(OR′)$_3$) **139** were a good alternative to hydrolytically unstable chloro- and fluoro-organosilicon derivatives, and [Rh(cod)(MeCN)$_2$]BF$_4$ proved to be the best catalyst for this transformation (Scheme 1.53) [56, 185]. With cationic rhodium complexes, organosiloxanes do not require external activator to transmetallate to rhodium. These conditions were applied successfully to the asymmetric variant of this reaction, using (S)-binap as the chiral ligand (Scheme 1.53) [58]. The enantioselectivities observed were very close to those obtained with organoboronic acid, indicating a similar reaction pathway [17, 60]. In a one-pot procedure, using the same cationic [Rh((S)-binap)(MeCN)$_2$]BF$_4$ complex, alkenylsilanes can be generated *in situ* by hydrosilylation of the corresponding alkyne, and then subjected to ECA [186, 187]. Organosiloxanes were also used in the Rh-catalyzed asymmetric addition to α-substituted acrylic esters (for the reaction with organoboronic acids; cf. Scheme 1.43) [188].

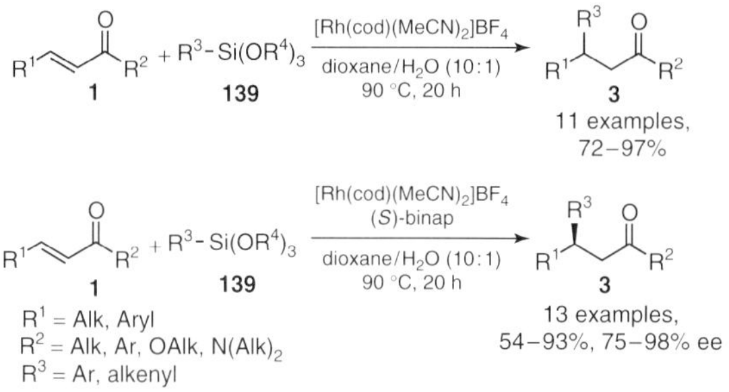

Scheme 1.53 Rh-catalyzed ECA of organosiloxanes onto α,β-unsaturated carbonyls.

Based on the seminal studies of Hudrilk [189, 190], Hiyama and Nakao introduced the stable and reusable tetraorganosilicon reagents, alkenyl-, aryl-, and silyl[2-(hydroxymethyl)phenyl]dimethylsilanes **140**, which undergo 1,4-addition reactions to α,β-unsaturated carbonyl acceptors under mild rhodium catalysis (Scheme 1.54) [153, 191–193]. The only byproduct in this case was the volatile cyclic siloxane **141**. These reagents were the first examples of stable tetraorganosilicon compounds that could efficiently transmetallate from Si to Pd and Rh under mild, fluoride-free conditions. The key to the success of the tetraorganosilicon **140** is the internal hydroxy nucleophilic trigger, which increases the polarization of the

1.4 Rh-Catalyzed ECA of Organosilicon Reagents

Ar–Si bond upon deprotonation. This reagent enables the introduction of a large gamut of aryl and alkenyl groups onto all α,β-unsaturated ketones, ester, amides, nitriles, vinylogous amides, fumarates, and maleimides (Scheme 1.54) [192]. Chiral diene ligands **L28** are extremely effective for this transformation, using the tetraorganosilicon reagents **140**.

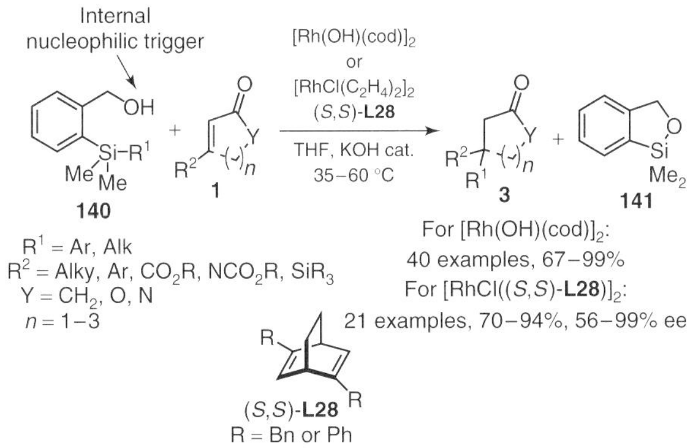

Scheme 1.54 Rh-catalyzed ECA of tetraorganosilicon reagent **141** to α,β-unsaturated carbonyls.

The aryl derivative of reagent **140a** can be easily regenerated by addition of the corresponding Grignard reagent onto **141**, whereas the alkenyl functionalized **140b** is obtained by reduction of **141** by LiAlH$_4$ followed by protection and subsequent regioselective alkyne hydrosilylation and deprotection (Scheme 1.55) [193]. The silyl variant (**142**) of reagent **140** can also function as a silyl transfer, albeit with only moderate yield due to competitive loss of silane [192].

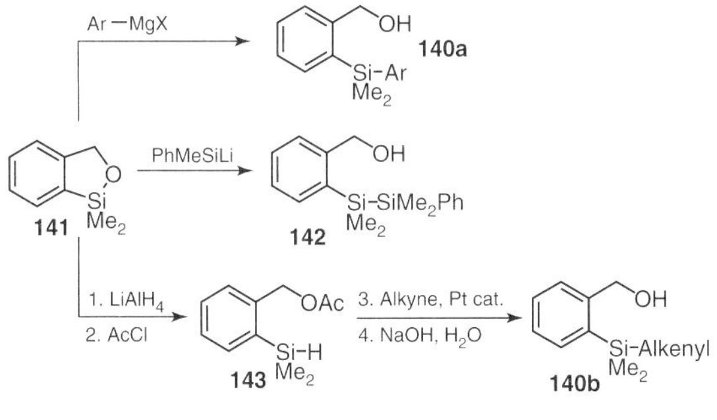

Scheme 1.55 Regeneration of the tetraorganosilicon reagents **141**.

In addition to aryl and alkenyl groups, bis(pinacolato)diboron (**25**) and silyl boronic ester **26** can be used in Rh-cat conjugate additions to transfer boron [49] and silyl groups respectively. Thus, the SiMe$_2$Ph group can be introduced onto linear and cyclic α,β-unsaturated carbonyl compounds in moderate yields and with high enantioselectivity (Scheme 1.56) [50, 51].

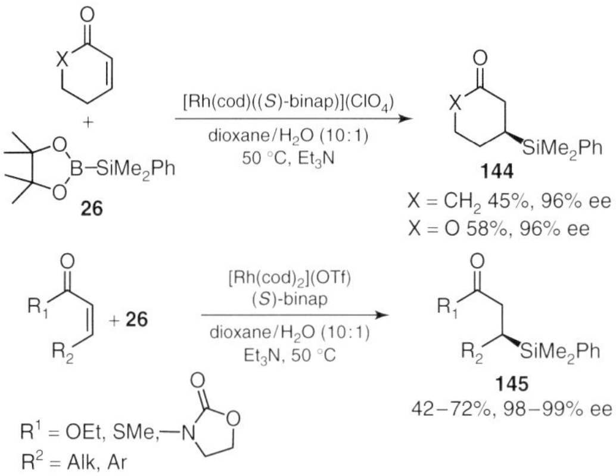

Scheme 1.56 Rh-catalyzed ECA with (pin)B-SiMe$_2$Ph.

1.5
Rh-Catalyzed ECA with Other Organometallic Reagents

Apart from B, Si, Ti, and Zn, several other organometallic reagents have been used successfully in Rh-catalyzed ECA. Alkenyl zirconium species, derived from the hydrozirconation [194] of primary alkynes by Schwartz reagent ([Cp$_2$Zr(H)(Cl)]), are useful reagents in Rh-catalyzed ECAs [59, 195, 196]. Aryl-lead [197], -tin [54, 57, 198], and triarylbismuth [199] compounds are also competent aryl-transfer agents in Rh-catalyzed conjugate additions, even in the presence of air and water [200]. Diaryl indium hydroxides represent an interesting class of organometallic reagents in rhodium-catalyzed ECAs. These are readily obtained from the reaction of the corresponding aryl lithium or aryl Grignard onto indium trichloride followed by hydrolysis; more importantly, they are nontoxic and stable under aqueous conditions [201]. As with organoboronic acids, the presence of an hydroxide on indium is necessary for smooth transmetallation, and a similar transmetallation step can be surmised (cf. Scheme 1.8) [201].

It thus appears, that the main criteria for effectiveness of an organometallic reagent are that: (i) it can smoothly and efficiently transmetallate to the rhodium center; and (ii) it is stable under the conditions necessary to hydrolyze the rhodium-enolate, or that the reagent can transmetallate directly with the Rh-enolate.

1.6
Rh-Catalyzed ECA of Alkynes

Alkyne as a nucleophile for Rh-catalyzed conjugate addition has been studied minimally, with the earliest reports employing only MVK as the acceptor and having long reaction times [202, 203]. The major hurdle with Rh-catalyzed alkynylation is that an primary alkyne (**146**) is generally more reactive than an enone towards Rh-alkynylides (**147**), and this results in the preferential formation of alkyne dimer

(**149**) rather than the conjugate addition product **148** (Scheme 1.57). The state of the art for asymmetric alkylation methodologies was reviewed in 2009 [204].

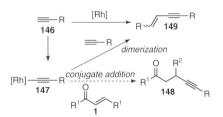

Scheme 1.57 Competitive reaction pathways for ECA of alkynes onto α,β-unsaturated compounds.

Hayashi and coworkers developed three methods to overcome competitive alkyne dimerization in rhodium-catalyzed ECAs of alkynes to enones [205, 206]. The first method consisted of the rearrangement of racemic alkynyl-alkenyl alcohols of type **150** (obtained from the 1,2-addition of a silylated alkynyl lithium onto an enone) to the conjugate addition product **153** (Scheme 1.58). This rearrangement was proposed to occur by the formation of rhodium-alkoxide **151**, followed by a β-alkyne elimination to form a chiral alkynyl-rhodium complex **152**, from which the ECA occurred. The key to this reaction is that the enone is in the coordination sphere of the rhodium center as the Rh-alkynyl is generated, and that no free alkyne is present during the reaction, thus shutting down the alkyne dimerization pathway. Interestingly, for indenone derivative **154**, the diene ligand (*S*,*S*)-**L28a** gave higher yields and enantioselectivity than (*R*)-binap (Scheme 1.58).

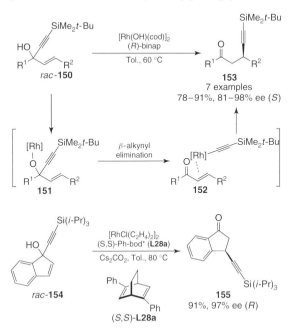

Scheme 1.58 Rh-catalyzed rearrangement of *rac*-**150** into chiral β-alkynyl ketones.

46 | *1 Rhodium- and Palladium-Catalyzed Asymmetric Conjugate Additions*

The second method for Rh-catalyzed conjugate alkynylation is the bimolecular addition of an alkyne onto an enone. The success of this method relies on the careful tuning of the catalyst and substrates (Scheme 1.59). Hence, to prevent competitive alkyne dimerization, the size of the alkyne was increased by using (triisopropylsilyl)acetylene (**156**). The use of the bulky (*R*)-DTBM-Segphos (**L52**) was also instrumental in accessing high enantioselectivities. Interestingly, cyclic enones were also alkynylated with excellent enantioselectivities, albeit in lower yields (55% for indenone). Although this method is high yielding and enantioselective, the scope of ketone amenable to this transformations is rather limited.

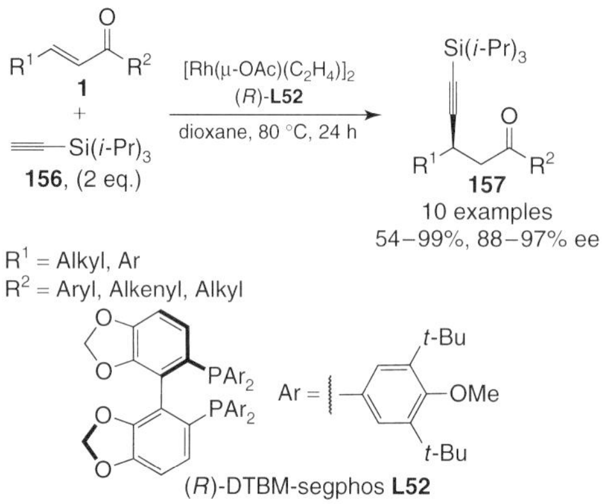

Scheme 1.59 Direct Rh-catalyzed ECA of alkyne onto α,β-unsaturated ketones.

The final method improved on these results by using bulky internal bis-silylated alkyne **158** [207]. The silanol in alkyne **158** enables a smooth transmetallation of the alkyne from Si to Rh while maintaining **158** bulky enough to prevent undesired alkyne dimerization. By using the alkynylsilanols **158**, the reaction time

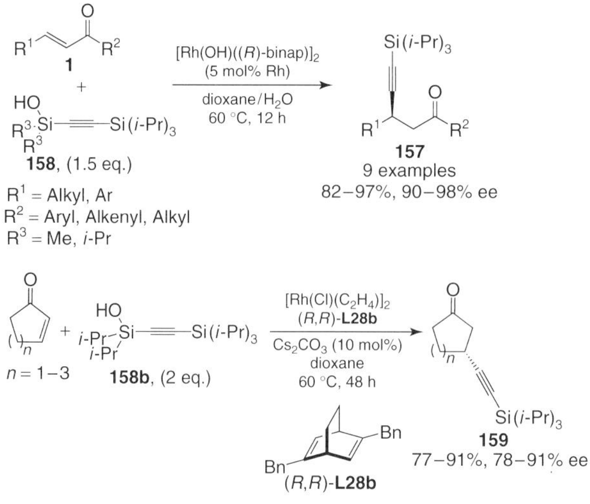

Scheme 1.60 Rh-catalyzed ECA of alkynylsilanols onto α,β-unsaturated ketones.

could be reduced to 12 h and the temperature lowered to 60 °C. Importantly, this methodology enabled the ECA of alkynes onto cyclic ketones, which were previously too unreactive to compete against alkyne dimerization, to form **159** (Scheme 1.60). In this case a catalytic amount of base (Cs_2CO_3) was necessary to accelerate the rate of the reaction.

1.7
Rh-Catalyzed Tandem Processes

In the previous sections, the rhodium-catalyzed ECAs of organometallic reagents on a wide range of acceptors has been described. In this case, the more electron-deficient the unsaturated substrate is, the more readily it will react with the organorhodium species generated by transmetallation from boron to rhodium. Thus, it is possible to program carborhodation cascades by assembling inter- or intramolecularly acceptors of differing reactivity. The cascade is initiated by carborhodation onto the most reactive acceptors, which subsequently react with the second most reactive acceptors, and so forth, until the sequence is terminated by protonolysis of the organorhodium intermediate. Such cascade sequences, which consist of multiple carbometallation steps, provide powerful methods for the construction of structurally complex molecules in an efficient and atom-economic manner [208]. These transformations have been the object of several reviews [209–211].

1.7.1
Tandem ECA/Aldol Reaction

An elegant three-component Rh-catalyzed tandem ECA/aldol reaction was developed by Hayashi and coworkers [135]. The reaction of ArB(9-*BBN*) (**24**), vinyl ketone **160**, and propanal catalyzed by [Rh(OH)((*S*)-binap)]$_2$ as a catalyst gave optically active products, *syn*-(4*S*,5*R*)-**162** in 41% ee and *anti*-(4*R*,5*R*)-**162** in 94% ee, though the *syn/anti* selectivity was only 0.8/1 (Scheme 1.61). Formation of the enantiomerically enriched products demonstrated that the reaction proceeded

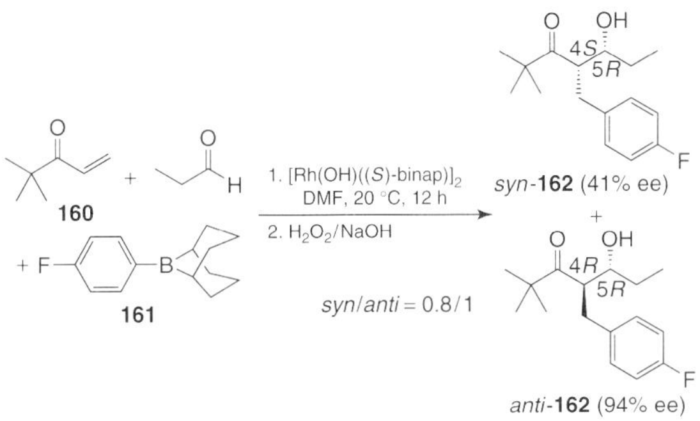

Scheme 1.61 Three-component Rh-catalyzed tandem ECA/aldol reaction.

through the (oxa-π-allyl)rhodium complex **C** coordinated with (*S*)-binap ligand, and underwent an aldol-type reaction, with aldehyde forming rhodium aldolate **Q** (Scheme 1.62). The aldolate further reacted with **24** to form **163** and regenerate **B**. The boron enolate as an intermediate was ruled out, which would lead to a racemic aldol product. An analogous zirconium-based three-component coupling reaction was reported by Nicolaou and coworkers [196].

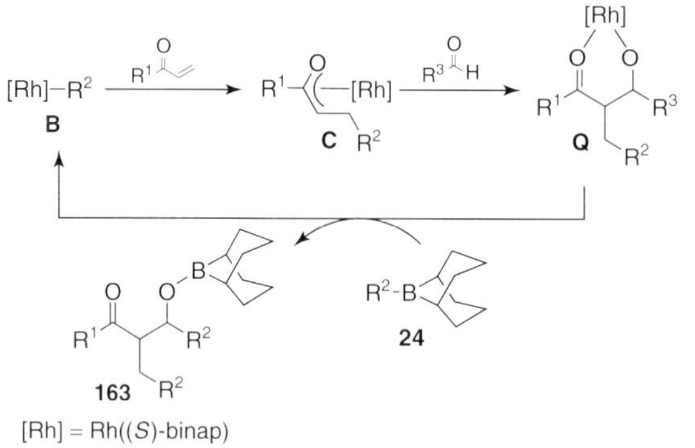

Scheme 1.62 Rh-catalyzed tandem ECA/aldol reaction.

Krische and coworkers reported the intramolecular trapping of a chiral Rh-enolate by an electrophile [212]. The ECA of phenylboronic acid onto an α,β-unsaturated ketone having a pendant ketone (**164**) results in the formation of an (oxa-π-allyl)rhodium species, which subsequently undergoes an intramolecular aldol addition onto the ketone fragment. Because of the close proximity of the electrophile, the reaction is kinetically favored and protic conditions can be used. This sequence produces cyclic aldol adduct **165** with a near-perfect control of relative and absolute stereochemistries (Scheme 1.63) [212]. For example, the reaction of keto-enone **164a** (n = 1) with phenylboronic acid in the presence of a rhodium/(*R*)-binap catalyst gave the five-membered ring aldol product **165am** exclusively, in 88% chemical yield and 94% ee. This method was elegantly applied to the desymmetrization of diketoenones **166**, which resulted in the stereoselective formation of bicycle **167** with four contiguous stereogenic centers, including a quaternary center (Scheme 1.63) [213].

1.7.2
Tandem Conjugate Addition/1,2-Addition

Cyano groups can also serve as electrophiles for Rh-catalyzed tandem cyclization triggered by the conjugate addition of arylboron reagents to form five- and six-membered β-enamino esters **169** from the cyano enoate **168** (Scheme 1.64) [214]. An (oxa-π-allyl)rhodium intermediate, generated by the initial conjugate addition of an arylrhodium species, undergoes a facile intramolecular addition to the cyano group followed by sequential transmetallation with Ar-B(9-*BBN*) (**24**).

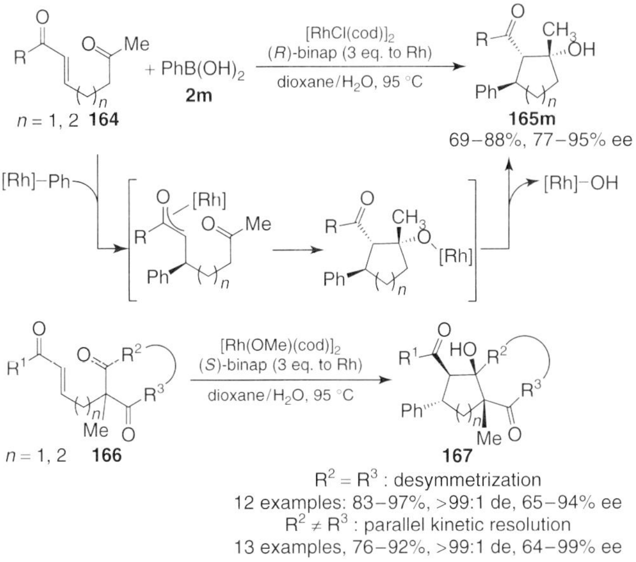

Scheme 1.63 Rh-catalyzed intramolecular tandem ECA/aldol reaction.

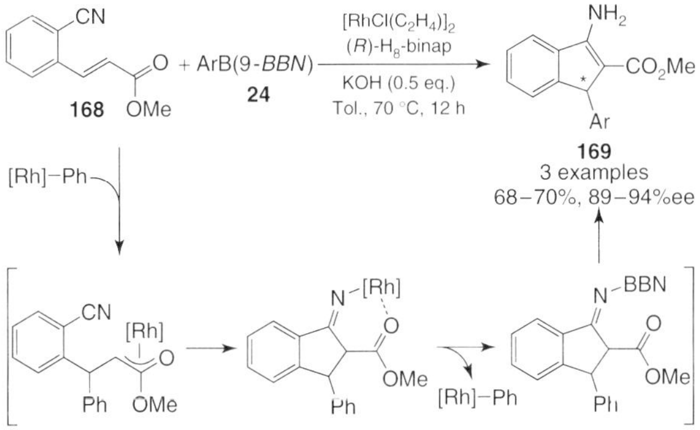

Scheme 1.64 Tandem Rh-catalyzed ECA/1,2-addition onto a cyano group.

The use of (R)-H$_8$-binap for the reaction of **168** gave the ECA/1,2-addition product **169** with up to 95% ee.

1.7.3
Tandem ECA/Mannich Reaction

A Rh-catalyzed tandem conjugate addition/Mannich cyclization reaction of imine-substituted electron-deficient alkenes **170** with arylboronic acids has been achieved (Scheme 1.65) [215]. This sequence provides a versatile entry into highly functionalized tetrahydroquinolines **171** with acceptable diastereoselectivities.

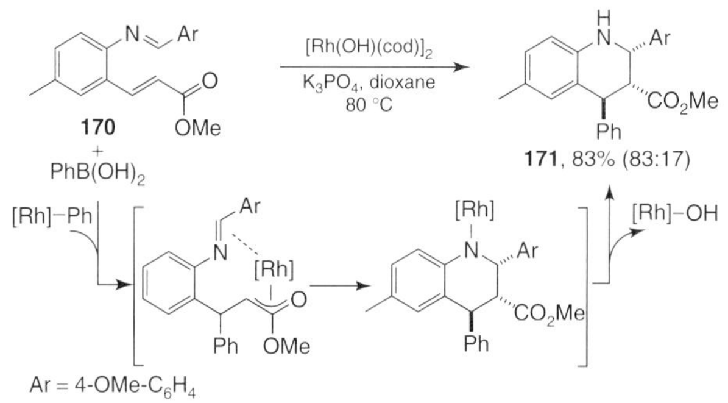

Scheme 1.65 Tandem Rh-catalyzed conjugate addition/1,2 addition onto an imine function.

1.7.4
Tandem Conjugate Addition/Michael Cyclization

A stereoselective Rh-catalyzed tandem annulation reaction triggered by the conjugate addition of arylboronic acids to enones **172**, followed by an intramolecular Michael reaction leading to **173**, has been reported (Scheme 1.66) [216]. This sequence afforded 1,2,3-trisubstituted indanes **173** in a highly regio- and diastereoselective fashion. On the other hand, the corresponding reactions with α,β-unsaturated esters gave rise to a complex mixture of reaction products.

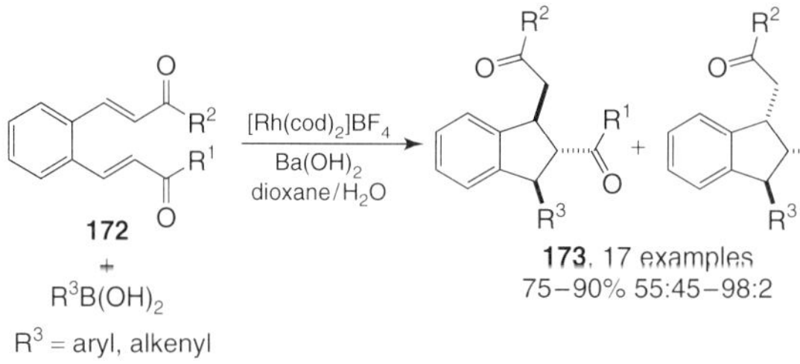

Scheme 1.66 Tandem Rh-catalyzed conjugate addition/Michael cyclization.

1.7.5
Tandem Carborhodation/Conjugate Addition

In 2000, Miura disclosed that strained alkenes, such as 2-norbornene, could efficiently undergo sequential carborhodation processes [217]. Lautens and coworkers exploited this observation for the rapid assembly of highly functionalized polycyclic systems **176** in a diastereoselective fashion (Scheme 1.67) [218, 219]. This sequence relies on the subtle coexistence, within **174**, of an arylboronate esters and an α,β-unsaturated carbonyl function which cannot react intramolecularly or dimerize under Rh-catalyzed reaction conditions, but which can undergo a carborhodation with **175** followed by an intramolecular conjugate addition. The reaction is performed in aqueous media with water-soluble electron-rich phosphines as ligands. Under these conditions, the boron pinacol ester is slowly hydrolyzed to generate a boronic acid, which will subsequently transmetallate with rhodium to generate intermediate **R** (Scheme 1.67). The arylrhodium species generated adds onto the strained alkene through the *exo* face (Scheme 1.67, **S**). In the resulting alkylrhodium species, rhodium is presumably coordinated to the internal pendant olefin. Subsequent conjugate addition occurs by a 5-*exo*-trig process to generate a rhodium-enolate **T**. Rapid hydrolysis of the Rh-enolate **T**, releases the fused indane product and regenerates the hydroxorhodium catalyst. The scope of this sequence has been expanded to bifunctional boronates esters/Michael acceptor building blocks bearing heteroaromatic rings (i.e., indole, benzofuran, benzothiophene, and thiophene) [220].

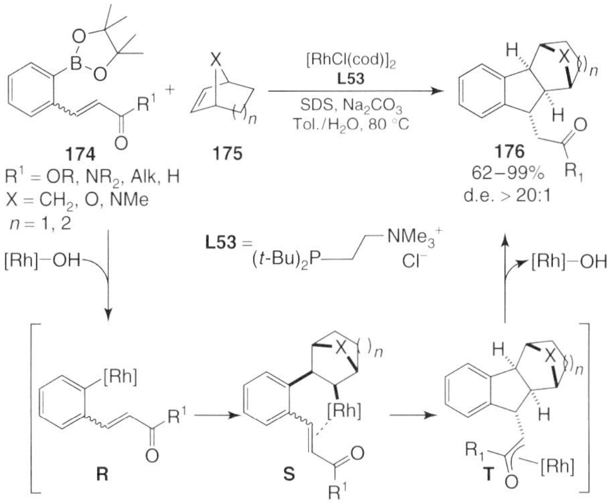

Scheme 1.67 Tandem Rh-catalyzed carborhodation/conjugate addition.

Activated alkynes such a **178** can react with **177** under similar conditions, to afford 1*H*-indenes **179** (Scheme 1.68) [221].

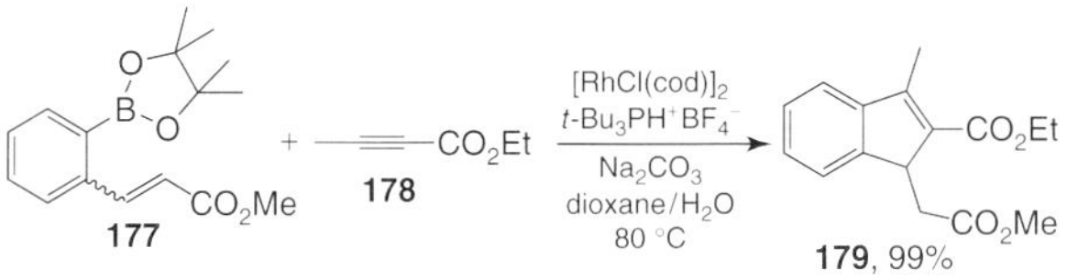

Scheme 1.68 Tandem Rh-catalyzed carborhodation/conjugate addition.

A closely related Rh-catalyzed tandem cyclization reaction has been achieved by using a boronate ester **180** in which the aryl function has been replaced by a *cis*-alkene (Scheme 1.69) [222, 223]. With this substrate class, after the initial carborhodation with **175** (intermediate **U**), a vinylcyclopropanation reaction occurs through a rare selective 1,6-addition of an alkylrhodium species in preference to

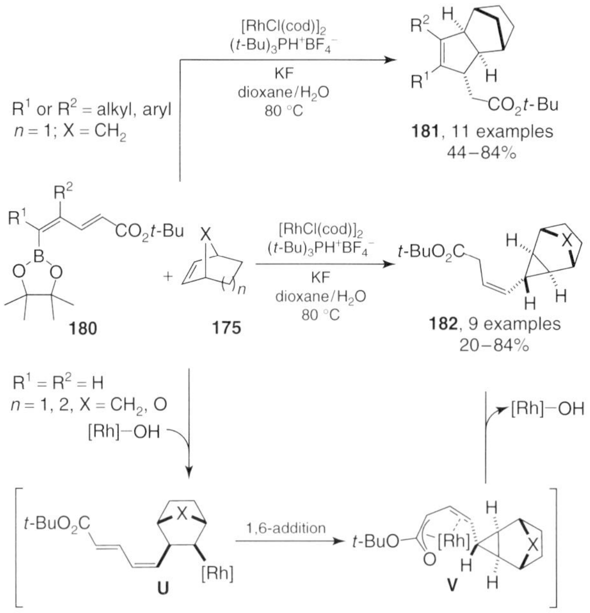

Scheme 1.69 Tandem Rh-catalyzed carborhodation/conjugate addition.

1,4-addition to form exclusively cyclopropane **182**. The exclusive formation of a *cis*-alkene was thought to be favored due to internal coordination of the carbonyl group, which locks the (oxa-π-allyl)rhodium **V** into one conformation. Depending on the substitution pattern of the dienylboronate esters, the reaction pathway can switch from an intramolecular 1,6-addition leading to **182**, to a 1,4-addition mechanism affording carbocycles **181** containing cyclopentene moiety instead of a vinylcyclopropane. Moreover, low yields are observed if the dienylboronate esters **180** are substituted α or γ to the ester function [223].

Interestingly, a Rh/diene (**L28b**) catalyst is far more chemo- and enantioselective at promoting arylative cyclization of alkyne-tethered electron-deficient olefin **183** to cyclopentene **184** than a Rh/diphosphine catalyst (Scheme 1.70) [224]. The observed chemoselectivity, which involves an initial carborhodation of the triple bond of **183** instead of the conjugated double bond, is in accord with the observation that the Rh/diene catalyst displays a higher activity in the arylation of alkynes than in the 1,4-addition to α,β-unsaturated esters, whereas a Rh/diphosphine catalyst behaves in the opposite manner. This behavior could be due to the more electrophilic nature of a Rh/diene complex relative to a Rh/phosphine center. A related addition/cyclization, involving the initial carborhodation of an alkyne followed by an enantioselective 1,2-addition onto a tethered aldehyde, was reported by Hayashi and coworkers [225]. For this reaction, chiral dienes **L27b** and **L28b** also proved instrumental in accessing high activity and enantioselectivity.

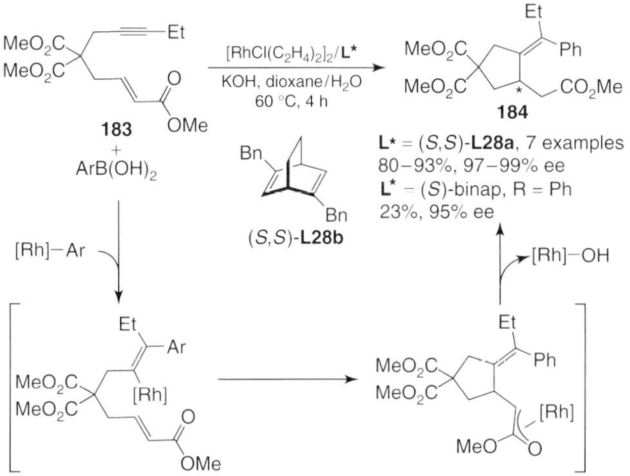

Scheme 1.70 Rh/diene-catalyzed arylative cyclization of alkyne-tethered electron-deficient olefins.

Syn-1,2-addition is the most common pathway for the insertion of an alkyne onto a Rh–carbon bond. However, 1,1-carborhodation pathway was recently observed with **185**, in which an *endo*-olefin cyclic product **186** is formed (Scheme 1.71) [226]. This novel addition/cyclization reaction occurred through an alkylidenerhodium-mediated 1,1-carborhodation process. Following the formation of vinylidenerhodium **W**, α-migration of the R^2 group from the Rh center to the vinylidene ligand provides alkenylrhodium intermediate **X**, which then undergoes addition to the pendant enone to give rhodium enolate **Y**. Finally, protonation of **Y** produces cyclopentene derivatives and regenerates the methoxyrhodium species.

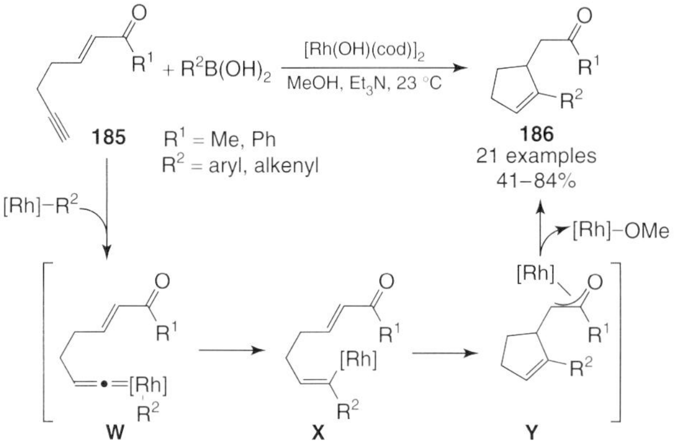

Scheme 1.71 Tandem 1,1-carborhodation/conjugate addition.

1.8
1,6-Conjugate Additions

From this review, it is apparent that Rh-catalyzed ECA additions are now a well-understood process. Arguably, the new frontier in Rh-catalyzed ECA methodology is now the control of the regio- and enantioselectivity in 1,6-additions processes. The 1,6-additions onto $\alpha,\beta,\gamma,\delta$-di-unsaturated carbonyls are particularly challenging because of the multitude of possible reaction pathways which are under substrate control. This is clearly illustrated in Scheme 1.72 [227] where, depending on the substitution pattern of the $\alpha,\beta,\gamma,\delta$-di-unsaturated esters **187** and the nature of the organoboron reagent, three competitive reaction pathways coexist. For unhindered dienoates **187** (R^1 = H or Me) the 1,6-addition product **189** is favored. However, when R^1 and R^2 are aromatic the 1,4-addition (**188**) becomes predominant, and when R^2 is an alkenyl group the fully conjugated Heck-type product **190** is observed.

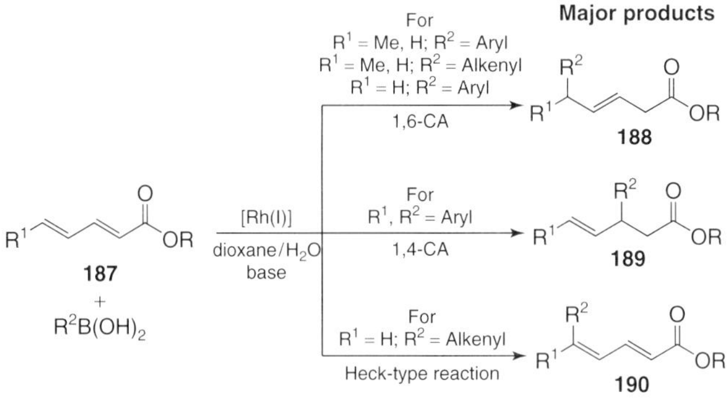

Scheme 1.72 Rh-catalyzed 1,6-addition of organoboronic acid onto $\alpha,\beta,\gamma,\delta$-diunsaturated esters.

Hayashi and coworkers were able to control the regioselectivity of the 1,6-addition β-substituted dienoates by using reactive arylzinc reagent (**129**) in conjunction with ClSiMe$_3$ (Scheme 1.73) [176]. Lower yields were observed with anionic aryltitanium reagents, and no reaction occurred with arylboronic acids. The β-disubstitution in **191** is essential to prevent competitive 1,4-ECA. The 1,6-addition products were obtained as the corresponding silyl-enol ether as a mixture of geometrical isomers which, after hydrolysis, gave (*R*)-**192** in near-quantitative yields and good to excellent enantioselectivities [176]. A related transformation, involving the enantioselective 1,6-addition of aryltitanate to 3-alkynyl-2-en-1-ones (**193**) in the presence of chlorotrimethylsilane enabled the formation of enantiomerically enriched allenes **194** (Scheme 1.73) when (*R*)-Segphos was used as a chiral ligand.

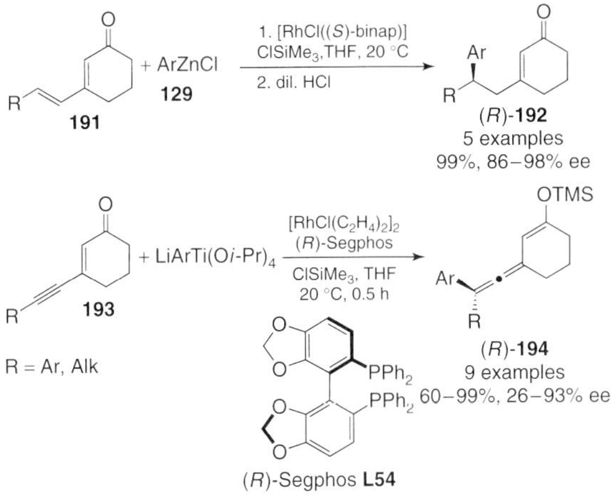

1.8 1,6-Conjugate Additions

Scheme 1.73 Rh-catalyzed asymmetric 1,6-addition of arylzinc reagents onto α,β,γ,δ-diunsaturated ketones.

In 2006, Hayashi and coworkers reported a breakthrough in transition metal-catalyzed 1,6-addition methodology with organoboronic acids, through the use of [Ir(μ-OH)(cod)]$_2$ the catalyst (Scheme 1.74) [228]. Importantly, this was the first report of an iridium-catalyzed addition of organoboronic acid onto an electron-deficient alkene or diene. The 1,6-addition products **191** were obtained preferentially as the *cis* isomer, and compound **196** was hydrogenated to **197** to facilitate analysis. The high 1,6-selectivity obtained with the iridium catalyst was in stark contrast to that observed with the parent [Rh(μ-OH)(cod)]$_2$ complex as catalyst, under the same reaction conditions (Scheme 1.74). Thus, the rhodium-catalyzed reaction gave the 1,4-adduct as the main isomer (55% yield) and a minor amount (34% yield) of 1,6-adducts **196** (as a mixture of geometric isomers).

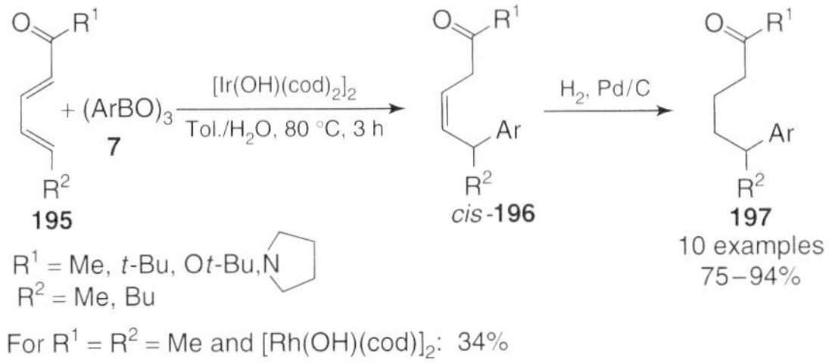

Scheme 1.74 Ir-catalyzed 1,6-conjugate addition with arylboroxines.

Competition experiments revealed that the iridium catalyst had a much stronger reactivity toward the dienone than the enone, while the opposite reactivity was observed for [Rh(μ-OH)(cod)]$_2$. On the basis of the high reactivity towards the diene moiety and the high *cis* selectivity in the 1,6-addition product, the catalytic cycle that was surmised is depicted in Scheme 1.75. Here, transmetallation of an aryl group

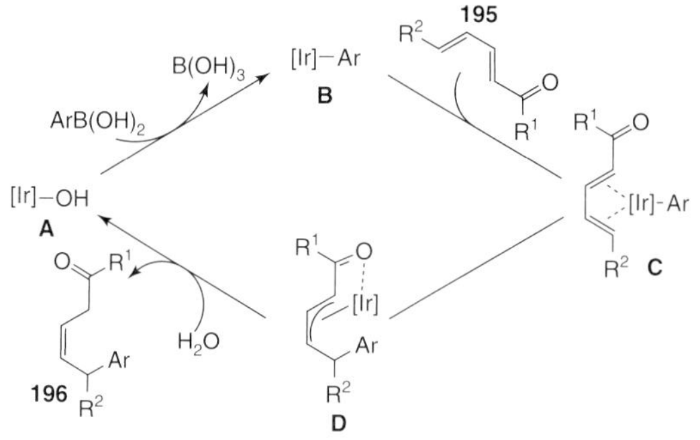

Scheme 1.75 Proposed catalytic cycle to the Ir-catalyzed 1,6-addition with arylboronic acids.

from the boron to iridium–hydroxide **A** forms an aryl–iridium species **B** [229]; the coordination of the dienone to the phenyl–iridium complex with a cisoid diene moiety results in the formation of a (η^4-diene)–iridium complex **C**. Insertion of the diene into the phenyl–iridium bond then occurs, forming the π-allyl–iridium moiety **D**, and this is followed by a selective hydrolysis of **D** at the α position to the carbonyl function with the assistance of phenylboronic acid or boric acid to give the 1,6-addition product cis-**196** and regenerate the hydroxo–iridium species **A**.

1.9
Pd-Catalyzed ECA

As described above, the rhodium-catalyzed ECA is now a relatively mature and well-understood technology, providing a very robust and flexible approach to the introduction of aryl and alkenyl groups in high yields, with excellent enantioselectivities and chemoselectivities. However, from an economic point of view, rhodium remains the most onerous metal, its price having been the subject of intense speculation over the past decade.[1] Such high, and fluctuating, prices hampers the use of Rh-catalyzed ECA on a large industrial scale, and creates a need for alternative methodologies based on cheaper transition metals, such as palladium. Several reviews have been produced on the subject of Pd-catalyzed ECAs [15, 230–234].

The first use of palladium as a catalyst for conjugate addition onto α,β-unsaturated ketones can be traced back to the studies of Cacchi and coworkers, who employed organotin [235] and organomercury [236] reagents in an acidic two-phase system. In 1995, Uemura and coworkers reported [1] that organoboronic could be used in the Pd-catalyzed addition onto enones. This reaction was performed under acidic

---

1) At its peak in 2008, the price of Rh (US$ 10 000 per ounce; US$ 300 per gram) was 23-fold higher than that of Pd (US$471 per ounce; US$15 per gram).

conditions and with a high catalyst loading (Scheme 1.76). Additionally, NaBPh$_4$ in conjunction with SbCl$_3$ could be used to deliver the phenyl fragment to the enone.

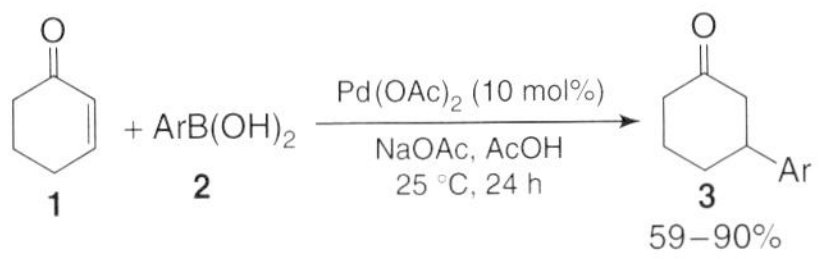

Scheme 1.76 First report of a Pd-catalyzed conjugate addition of organoboronic acids.

Although these early reports showed the potential of the palladium-catalyzed conjugate addition, the methodology lagged behind the rhodium-catalyzed reaction. The most likely reason for the underdevelopment of Pd-catalyzed conjugate addition is the propensity of neutral Pd-enolates, which are carbon-centered rather than oxygen-centered (i.e., Rh-enolates), to undergo competitive β-hydride elimination rather than hydrolysis (Scheme 1.77). This generates Heck-type coupling with concomitant Pd-black formation.

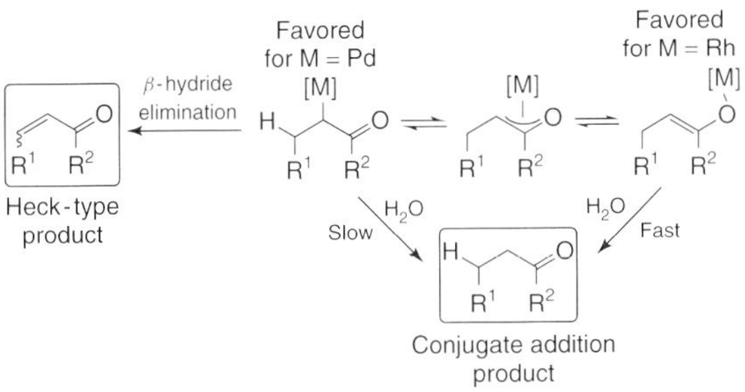

Scheme 1.77 Difference in reactivity between neutral Pd- and Rh-enolates.

Based on the findings that cationic palladium-enolates are much more susceptible to hydrolytic Pd–carbon bond cleavage than are neutral species [237], and that cationic-palladium species are much more reactive towards the addition to alkenes, Miyaura and coworkers developed a highly efficient Pd-catalyzed conjugate addition using dicationic [Pd(dppe)(MeCN)$_2$](SbF$_6$)$_2$ complex as the catalyst (Scheme 1.78) [238, 239]. The use of a cationic palladium source in the presence of water causes the β-hydride elimination from the Pd-enolate to be completely shut down by increasing the hydrolysis rate. The cationic Pd source is very active, promoting conjugate addition at room temperature. The use of a bidentate phosphine ligands with a two-carbon spacer, such as 1,2-*bis*(diphenylphosphino)ethane (dppe), proved essential to obtain reactivity. Diphosphine ligands with larger bite-angles [240] were ineffective [diphenylphosphinopropane (dppp), 2,2′-bis(diphenylphosphino)-1,1′-binaphthyl (binap), and Ph$_3$P gave no catalytic activity]. Although the presence of a base such as K$_2$CO$_3$ accelerates the reaction rate, it also promotes β-hydride elimination. Variation of the non-coordinating anion (SbF$_6^-$, BF$_4^-$, PF$_6^-$) had no significant influence on the reactivity. Under these conditions, addition to β-arylenals **31** proved completely chemoselective, with only the 1,4-addition product being observed. Enoates were sluggish substrates, while the α,β-unsaturated amides were unreactive.

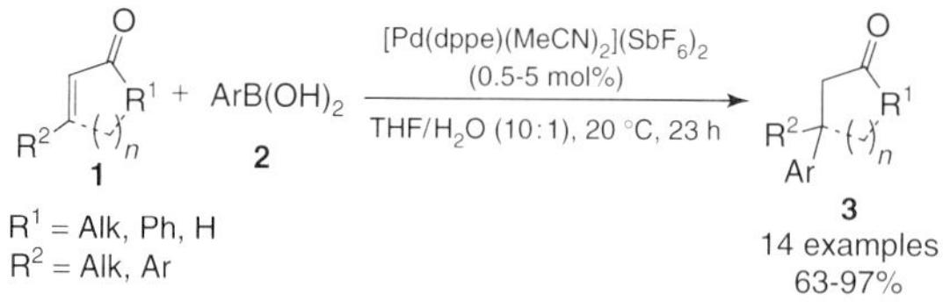

Scheme 1.78 Conjugate addition of organoboronic acid catalyzed by cationic Pd/dppe complex.

Initially, it was considered that the nitrile ligands (acetonitrile or benzonitrile) on the palladium were necessary to stabilize the catalyst and to obtain catalytic activity. However, later studies revealed that nitrile-free complexes generated by the *in situ* oxidation of Pd(dba)$_2$ with Cu(BF$_4$)$_2$ in the presence of phosphorus ligands were much more active catalysts. This enabled an expansion of the scope to use arylsiloxanes [239, 241] as the nucleophilic component.

In 2004, Miyaura and coworkers reported the first Pd-catalyzed ECA onto α,β-unsaturated ketones using triarylbismuth reagents [242]. In this seminal study, the combination of [Pd(MeCN)$_2$](SbF$_6$)$_2$ with (S,S)-dipamp (**L55**) or (S,S)-chiraphos (**L3**), as the chiral ligands, with Cu(BF$_4$)$_2$ as a substoichiometric additive in a MeO/H$_2$O (6:1) mixture, proved to be the most reactive and enantioselective system (Scheme 1.79). The organobismuth reagents were especially reactive and gave full conversion at just −5 °C. Using the same conditions, the scope of the reaction could be expanded to potassium aryltrifluoroborates and aryltrifluorosilanes [243]. The yields were high with both the triarylbismuth and ArBF$_3$K reagents, but slightly lower with ArSiF$_3$. Most importantly, the enantioselectivities observed did not vary with the nature of the nucleophilic reagent (BiAr$_3$, ArSiF$_3$, or ArBF$_3$K).

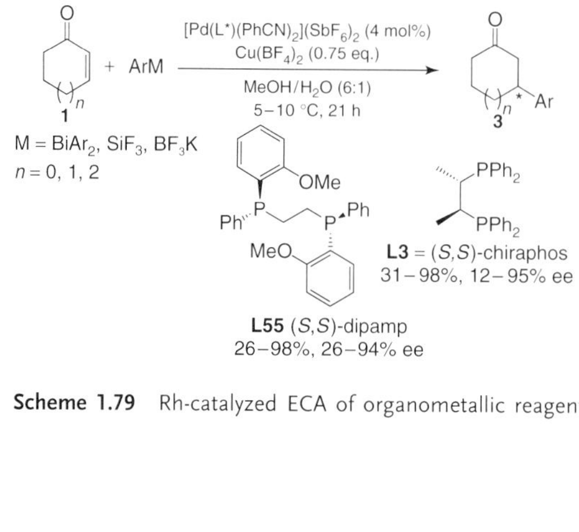

Scheme 1.79 Rh-catalyzed ECA of organometallic reagents onto α,β-unsaturated ketones.

Miyaura and coworkers further developed the methodology to realize the use of more readily available arylboronic acids. Under optimized conditions, acetone was found to be a more suitable solvent than MeOH, while the addition of 5–10 mol% AgBF$_4$ or AgSbF$_6$ was found to improve the catalytic activity and stability of the dicationic Pd(II) catalyst, presumably by oxidizing the inactive Pd(0) complexes back to the dicationic active species [239]. With these additives, the catalyst loading could be lowered to 0.01 mol% [244], under which conditions the β-arylenones underwent ECA smoothly with organoboronic acid at 0 or 25 °C, and with a low catalyst loading (Scheme 1.80). One interesting application of this methodology was the efficient synthesis of optically active chromenes **199** through a ECA/dehydration sequence of β-arylenones **198** (Scheme 1.80) [244].

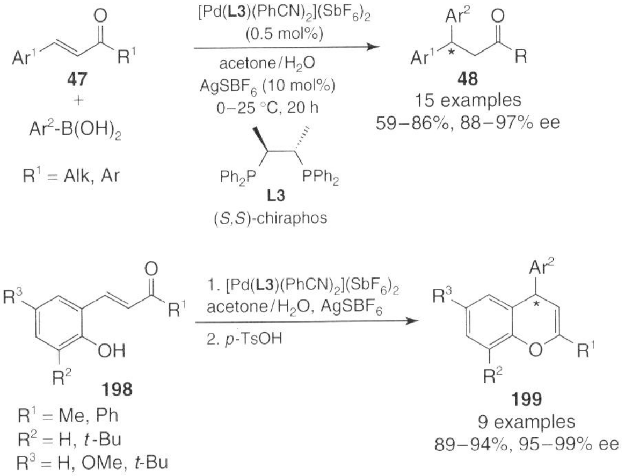

Scheme 1.80 Pd-catalyzed ECA of arylboronic acids to β-arylenones.

This process was ingeniously used for the synthesis of enantiopure 1-aryl-1 H-indenes **201** via a tandem Pd-catalyzed ECA/aldol condensation of substrate **200**, as shown in Scheme 1.81 (for the related Rh-catalyzed transformation, see Scheme 1.63.) [245].

The Pd-catalyzed ECA with linear α,β-unsaturated esters and amides has proved problematic due to the competing formation of Heck-type products. An alternative entry into this substrate class is to use the α,β-unsaturated aryl ester **202** [246] and α,β-unsaturated N-benzoylamides **204** [247], which afford the ECA product

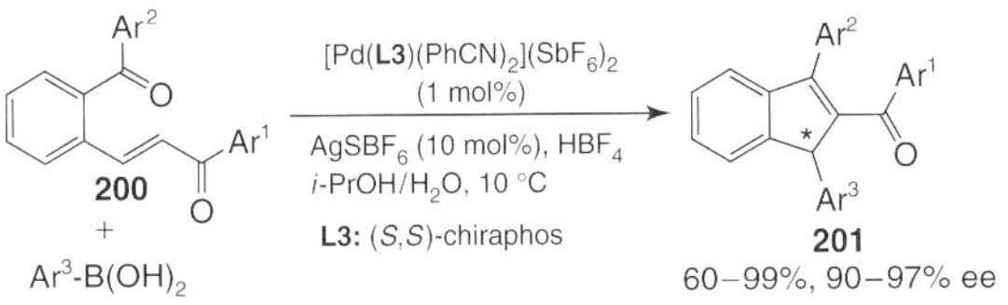

Scheme 1.81 Tandem Pd-catalyzed ECA/aldol condensation.

203 and 205 in high yields and enantioselectivities, without the concomitant formation of Heck-type products (Scheme 1.82). The effectiveness of amide 204 for these substrates resides in the coordinating of the two carbonyls, which shifts the C-centered Pd-enolate to an O-centered one. Similarly, maleimides 206 [247] proved to be competent substrates in Pd-catalyzed ECA, although the enantioselectivity was seen to depend heavily on the N-substituent (Scheme 1.82).

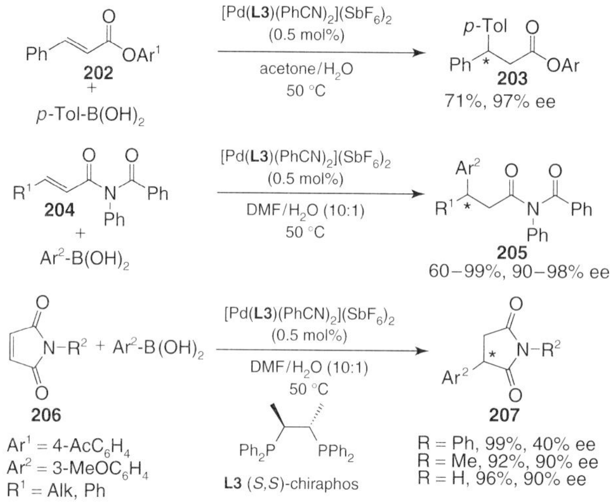

Scheme 1.82 Pd-catalyzed ECA of organoboronic acids to α,β-unsaturated esters and amides.

When the methodology was extended to β-aryl enals (31), it was again found that the addition of HBF$_4$ and AgBF$_4$ caused a dramatic increase in the reaction rate. It was proposed that an additional role of HBF$_4$ was to accelerate the rate of exchange between aldehyde and their corresponding hydrates, which is their favored form in aqueous solvents (Scheme 1.83) [248].

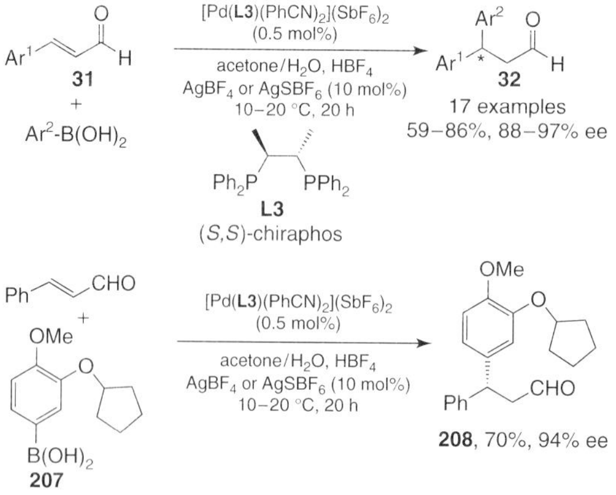

Scheme 1.83 Pd-catalyzed ECA of organoboronic acids onto β-aryl enals.

1.9 Pd-Catalyzed ECA

Minnaard and coworkers reported that the Pd-catalyzed ECA is efficiently catalyzed by Pd(OCOCF$_3$)$_2$ as a catalyst precursor [249]. In this case, an active cationic Pd species generated from Pd(OCOCF$_3$)$_2$ and (R,R)-Me-duphos (**L51**) proved to be the most effective ligand. The reaction did not take place if Pd(OAc)$_2$ was used as the precatalyst, as it required additional activation with triflic acid (CF$_3$SO$_3$H), though the yields were variable. The yields and enantioselectivities were high for cyclic α,β-unsaturated ketone and esters, but the ee-value was much lower for linear enals (**209**) and enones (**210**). For linear enoate **211**, formation of the Heck-type product became predominant (Scheme 1.84). A variation of this system could be used for the ECA of organosiloxanes, although ZnF$_2$ was required as an activator.

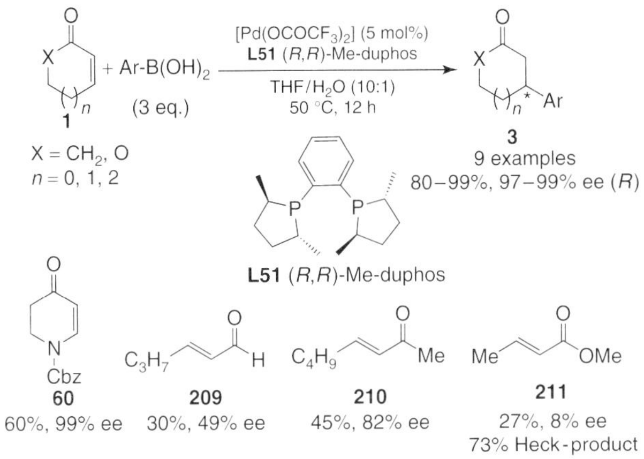

Scheme 1.84 Pd-catalyzed ECA of arylboronic acids onto α,β-unsaturated carbonyls.

Cationic palladium(II) complexes (R)-**212** bearing a chelating chiral bidentate NHC was reported to be active in the asymmetric conjugate addition of arylboronic acids to cyclic enones (Scheme 1.85) [250].

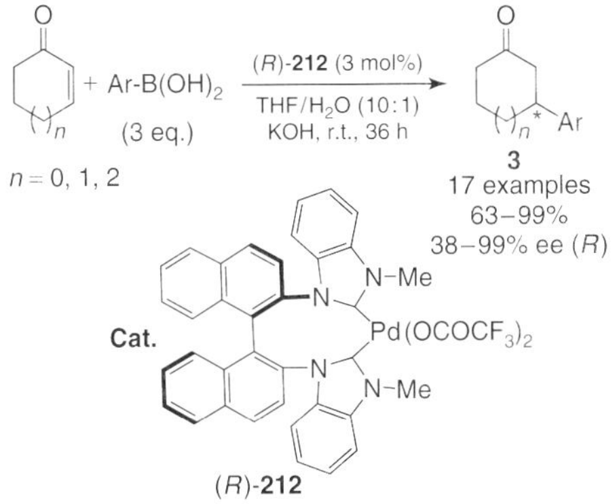

Scheme 1.85 Pd-catalyzed ECA of arylboronic acids onto α,β-unsaturated carbonyls.

It was also reported that a Pd(OAc)$_2$/2,2'-bipyridine catalytic system is highly effective for the addition of arylboronic acids to enones, enals, nitroalkenes and, very interestingly, also to cinnamates and acrylates (although the latter substrates have posed longstanding problems) [251]. The reaction can also be run in water in the presence of anionic surfactants [252].

In 2007, Hu and coworkers showed that palladacycle **213** is a highly efficient catalyst for the addition of organoboronic acids to enones (Scheme 1.86) [253]. The reaction is performed in toluene at room temperature, using K$_3$PO$_4$ as the base to activate the boronic acid. In contrast to dicationic Pd catalysts, a Lewis acidic activator such AgSbF$_6$ or HBF$_4$ is not necessary. Notably, palladacycle **213** can also promote the 1,2-addition to α-ketoesters [253], while palladacycle **214** derived from the inexpensive and π-acidic *tris*(2,4-di-*tert*-butylphenyl)phosphite (a plasticizer) was also found to promote the conjugate addition (Scheme 1.86) [254]. The enantiopure palladacycle **215**, generated *in situ* from optically active ferrocenyl phosphine ligand and Pd(dba)$_2$, was applied to the conjugate addition of arylboronic acids to cyclohexenone; the reaction afforded good yields and a promising 71% ee (Scheme 1.86) [255]. Considering that palladacycles are extremely stable and robust species, and that no onerous activator is needed, these type of catalyst will surely be developed further.

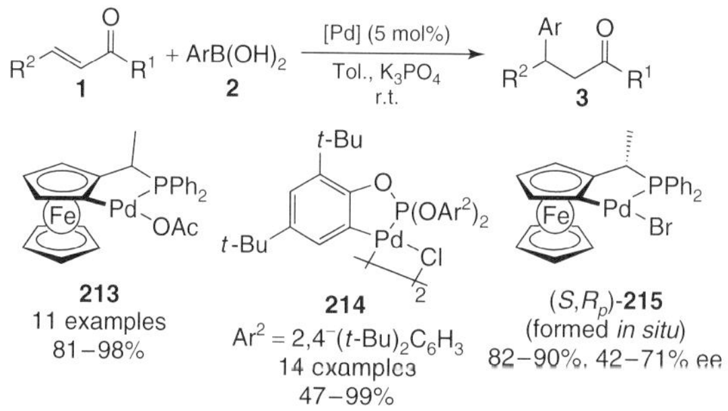

Scheme 1.86 Palladacycles as catalysts for with the conjugate addition of organoboronic acids.

In analogy to the tandem carborhodation/ECA processes developed for rhodium (cf. Section 1.5), Lu and coworkers developed a Pd-catalyzed variation of this transformation using **216** and internal alkynes (**217**) (Scheme 1.87) [256]. Interestingly, ligand (*S*)-**L55** proved to be the best chiral ligand – a striking outcome considering that these types of ligand performed poorly in Pd-catalyzed ECA.

1.9.1
Catalytic Cycle

The proposed mechanism for the Pd-catalyzed conjugate addition is depicted in Scheme 1.88, and is closely related to that shown in Scheme 1.3 for rhodium. The main difference is that cationic palladium intermediates are involved instead

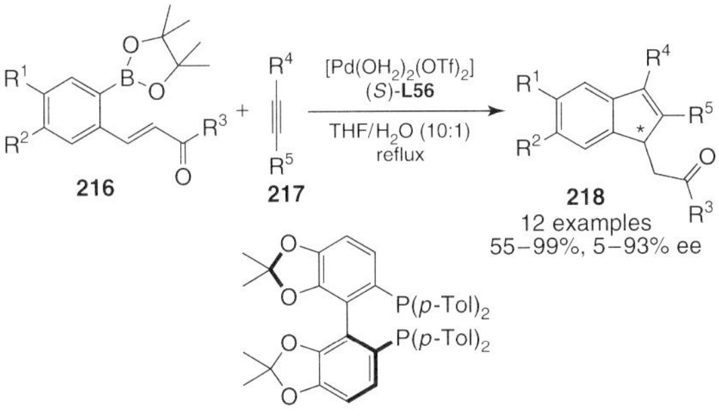

Scheme 1.87 Pd-catalyzed carborhodation/ECA.

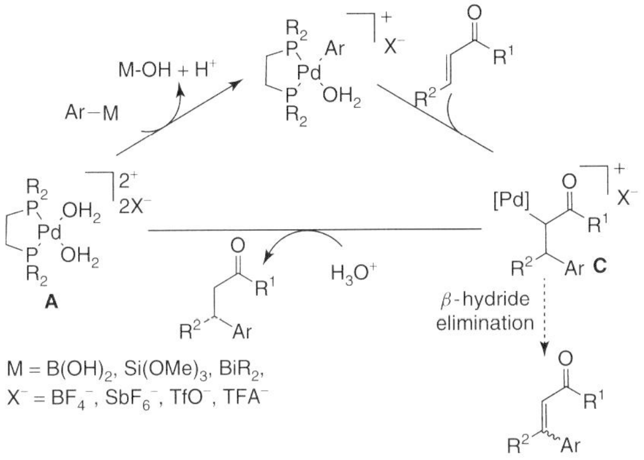

Scheme 1.88 Proposed catalytic cycle for Pd-catalyzed conjugate addition.

of neutral rhodium species. The easy transmetallation of the ArM reagent occurs from the dicationic precursor **A** to generate the cationic Pd–Ar species **B**; this subsequently adds to the α,β-unsaturated carbonyl to produce the C-centered Pd-enolate **C**, the hydrolysis of which yields the addition product and **A**. The slow hydrolysis of **C** results in a competitive β-hydride elimination.

The mode of enantioselection in the Pd-catalyzed ECA is thought to occur in a similar fashion as for Rh-catalyzed ECA (cf. Scheme 1.6).

1.10
Conclusions

Major progress has been achieved in the field of Rh-catalyzed ECAs since the initial reports during the late 1990s. Today, synthetic organic chemists have at their disposal a large toolbox of conditions and ligands from which to select, such that they can confidently tackle ECA in a complex setting, and on a large scale. Nonetheless, there remain some unsolved issues with rhodium-catalyzed ECA, such as the addition of an sp^3 carbon onto an activated alkene. In this respect, the Rh- and Pd-based methodologies are complementary to the copper-catalyzed processes that are well suited for the ECA of an alkyl Grignard, zinc, or aluminum reagent to an α,β-unsaturated carbonyl.

References

1. Cho, C.S., Motofusa, S., Ohe, K., and Uemura, S. (1995) *J. Org. Chem.*, **60**, 883–888.
2. Sakai, M., Hayashi, T., and Miyaura, N. (1997) *Organometallics*, **16**, 4229–4231.
3. Lalic, G. and Corey, E.J. (2007) *Org. Lett.*, **9**, 4921–4923.
4. Brock, S., Hose, D.R.J., Moseley, J.D., Parker, A.J., Patel, I., and Williams, A.J. (2008) *Org. Process Res. Dev.*, **12**, 496–502.
5. Burgey, C.S., Paone, D.V., Shaw, A.W., Deng, J.Z., Nguyen, D.N., Potteiger, C.M., Graham, S.L., Vacca, J.P., and Williams, T.M. (2008) *Org. Lett.*, **10**, 3235–3238.
6. Hayashi, T. (2001) *Synlett*, 879–887.
7. Hayashi, T. (2003) *Russ. Chem. Bull.*, **52**, 2595–2605.
8. Hayashi, T. and Yamasaki, K. (2003) *Chem. Rev.*, **103**, 2829–2844.
9. Lautens, M., Fagnou, K., and Hiebert, S. (2003) *Acc. Chem. Res.*, **36**, 48–58.
10. Fagnou, K. and Lautens, M. (2003) *Chem. Rev.*, **103**, 169–196.
11. Hayashi, T. (2004) *Bull. Chem. Soc. Jpn*, **77**, 13–21.
12. Hayashi, T. (2004) *Pure Appl. Chem.*, **76**, 465–475.
13. Yoshida, K. and Hayashi, T. (2005) in *Boronic Acids* (ed. D.G. Hall) Wiley-VCH Verlag GmbH, Weinheim, pp. 171–204.
14. Yoshida, K. and Hayashi, T. (2005) in *Modern Rhodium-Catalyzed Organic Reactions* (ed. P.A. Evans) Wiley-VCH Verlag GmbH, Weinheim, pp. 55–78.
15. Yamamoto, Y., Nishikata, T., and Miyaura, N. (2008) *Pure Appl. Chem.*, **80**, 807–817.
16. Miyaura, N. (2008) *Bull. Chem. Soc. Jpn*, **81**, 1535–1553.
17. Takaya, Y., Ogasawara, M., Hayashi, T., Sakai, M., and Miyaura, N. (1998) *J. Am. Chem. Soc.*, **120**, 5579–5580.
18. Krause, N. (1998) *Angew. Chem., Int. Ed. Engl.*, **37**, 283–285.
19. Alexakis, A. (2002) *Pure Appl. Chem.*, **74**, 37–42.
20. Lopez, F., Minnaard, A.J., and Feringa, B.L. (2007) *Acc. Chem. Res.*, **40**, 179–188.
21. Jerphagnon, T., Pizzuti, M.G., Minnaard, A.J., and Feringa, B.L. (2009) *Chem. Soc. Rev.*, **38**, 1039–1075.
22. Hayashi, T., Takahashi, M., Takaya, Y., and Ogasawara, M. (2002) *J. Am. Chem. Soc.*, **124**, 5052–5058.
23. Blackmond, D.G. (2005) *Angew. Chem., Int. Ed.*, **44**, 4302–4320.
24. Kina, A., Iwamura, H., and Hayashi, T. (2006) *J. Am. Chem. Soc.*, **128**, 3904–3905.
25. Kina, A., Yasuhara, Y., Nishimura, T., Iwamura, H., and Hayashi, T. (2006) *Chem. Asian J.*, **1**, 707–711.
26. Martina, S.L.X., Minnaard, A.J., Hessen, B., and Feringa, B.L. (2005) *Tetrahedron Lett.*, **46**, 7159–7163.
27. Matos, K. and Soderquist, J.A. (1998) *J. Org. Chem.*, **63**, 461–470.
28. Zhao, P.J., Incarvito, C.D., and Hartwig, J.F. (2007) *J. Am. Chem. Soc.*, **129**, 1876–1877.
29. Itooka, R., Iguchi, Y., and Miyaura, N. (2003) *J. Org. Chem.*, **68**, 6000–6004.
30. Ozawa, F., Kubo, A., Matsumoto, Y., Hayashi, T., Nishioka, E., Yanagi, K., and Moriguchi, K. (1993) *Organometallics*, **12**, 4188–4196.
31. Takaya, Y., Senda, T., Kurushima, H., Ogasawara, M., and Hayashi, T. (1999) *Tetrahedron: Asymmetry*, **10**, 4047–4056.
32. Takaya, Y., Ogasawara, M., and Hayashi, T. (1998) *Tetrahedron Lett.*, **39**, 8479–8482.

33. Senda, T., Ogasawara, M., and Hayashi, T. (2001) *J. Org. Chem.*, **66**, 6852–6856.
34. Darses, S. and Genêt, J.-P. (2003) *Eur. J. Org. Chem.*, 4313–4327.
35. Molander, G.A. and Figueroa, R. (2005) *Aldrichimica Acta*, **38**, 49–56.
36. Darses, S. and Genêt, J.P. (2008) *Chem. Rev.*, **108**, 288–325.
37. Navarre, L., Martinez, R., Genêt, J.P., and Darses, S. (2008) *J. Am. Chem. Soc.*, **130**, 6159–6169.
38. Duursma, A., Boiteau, J.G., Lefort, L., Boogers, J.A., de Vries, A.H., De Vries, J., Minnaard, A.J., and Feringa, B.L. (2004) *J. Org. Chem.*, **69**, 8045–8052.
39. Morrill, C. and Grubbs, R.H. (2003) *J. Org. Chem.*, **68**, 6031–6034.
40. Batey, R.A. and Quach, T.D. (2001) *Tetrahedron Lett.*, **42**, 9099–9103.
41. Molander, G.A., and Biolatto, B. (2002) *Org. Lett.*, **11**, 1867–1870.
42. Molander, G.A. and Biolatto, B. (2003) *J. Org. Chem.*, **68**, 4302–4314.
43. Yuen, A.K.L. and Hutton, C.A. (2005) *Tetrahedron Lett.*, **46**, 7899–7903.
44. Gendrineau, T., Genêt, J.P., and Darses, S. (2009) *Org. Lett.*, **11**, 3486–3489.
45. Takaya, Y., Ogasawara, M., and Hayashi, T. (1999) *Tetrahedron Lett.*, **40**, 6957–6961.
46. Yamamoto, Y., Takizawa, M., Yu, X., and Miyaura, N. (2008) *Angew. Chem., Int. Ed.*, **47**, 928–931.
47. Yu, X., Yamamoto, Y., and Miyaura, N. (2009) *Synlett*, 994–998.
48. Yoshida, K., Ogasawara, M., and Hayashi, T. (2003) *J. Org. Chem.*, **68**, 1901–1905.
49. Kabalka, G.W., Das, B.C., and Das, S. (2002) *Tetrahedron Lett.*, **43**, 2323–2325.
50. Walter, C., Auer, G., and Oestreich, M. (2006) *Angew. Chem., Int. Ed.*, **45**, 5675–5677.
51. Walter, C. and Oestreich, M. (2008) *Angew. Chem., Int. Ed.*, **47**, 3818–3820.
52. Sakuma, S. and Miyaura, N. (2001) *J. Org. Chem.*, **66**, 8944–8946.
53. Oi, S., Moro, M., and Inoue, Y. (1997) *Chem. Commun.*, 1621–1622.
54. Oi, S., Moro, M., Ono, S., and Inoue, Y. (1998) *Chem. Lett.*, 83–84.
55. Oi, S., Moro, M., and Inoue, Y. (2001) *Organometallics*, **20**, 1036–1037.
56. Oi, S., Honma, Y., and Inoue, Y. (2002) *Org. Lett.*, **4**, 667–669.
57. Oi, S., Moro, M., Ito, H., Honma, Y., Miyano, S., and Inoue, Y. (2002) *Tetrahedron*, **58**, 91–97.
58. Oi, S., Taira, A., Honma, Y., and Inoue, Y. (2003) *Org. Lett.*, **5**, 97–99.
59. Oi, S., Sato, T., and Inoue, Y. (2004) *Tetrahedron Lett.*, **45**, 5051–5055.
60. Oi, S., Taira, A., Honma, Y., Sato, T., and Inoue, Y. (2006) *Tetrahedron: Asymmetry*, **17**, 598–602.
61. Itoh, T., Mase, T., Nishikata, T., Iyama, T., Tachikawa, H., Kobayashi, Y., Yamamoto, Y., and Miyaura, N. (2006) *Tetrahedron*, **62**, 9610–9621.
62. Sakuma, S., Sakai, M., Itooka, R., and Miyaura, N. (2000) *J. Org. Chem.*, **65**, 5951–5955.
63. Lukin, K., Zhang, Q.Y., and Leanna, M.R. (2009) *J. Org. Chem.*, **74**, 929–931.
64. Yamamoto, Y., Kurihara, K., Sugishita, N., Oshita, K., Piao, D.G., and Miyaura, N. (2005) *Chem. Lett.*, **34**, 1224–1225.
65. Chen, F.-X., Kina, A., and Hayashi, T. (2006) *Org. Lett.*, **8**, 341–344.
66. Takaya, Y., Ogasawara, M., and Hayashi, T. (2000) *Chirality*, **12**, 469–471.
67. Vandyck, K., Matthys, B., Willen, M., Robeyns, K., Van Meervelt, L., and Van der Eycken, J. (2006) *Org. Lett.*, **8**, 363–366.
68. Reetz, M.T., Moulin, D., and Gosberg, A. (2001) *Org. Lett.*, **3**, 4083–4085.
69. Kurihara, K., Sugishita, N., Oshita, K., Piao, D., Yamamoto, Y., and Miyaura, N. (2007) *J. Organomet. Chem.*, **692**, 428–435.
70. Amengual, R., Michelet, V., and Genêt, J.P. (2002) *Synlett*, 1791–1794.
71. Korenaga, T., Osaki, K., Maenishi, R., and Sakai, T. (2009) *Org. Lett.*, **11**, 2325–2328.
72. Shi, Q., Xu, L.J., Li, X.S., Jia, X., Wang, R.H., Au-Yeung, T.T.L.,

Chan, A.S.C., Hayashi, T., Cao, R., and Hong, M.C. (2003) *Tetrahedron Lett.*, **44**, 6505–6508.
73. Madec, J., Michaud, G., Genêt, J.P., and Marinetti, A. (2004) *Tetrahedron: Asymmetry*, **15**, 2253–2261.
74. Otomaru, Y., Senda, T., and Hayashi, T. (2004) *Org. Lett.*, **6**, 3357–3359.
75. Shimada, T., Suda, M., Nagano, T., and Kakiuchi, K. (2005) *J. Org. Chem.*, **70**, 10178–10181.
76. Yuan, W., Cun, L., Mi, A., Jiang, Y., and Gong, L. (2009) *Tetrahedron*, **65**, 4130–4141.
77. Stemmler, R.T. and Bolm, C. (2005) *J. Org. Chem.*, **70**, 9925–9931.
78. Kromm, K., Eichenseher, S., Prommesberger, M., Hampel, F., and Gladysz, J.A. (2005) *Eur. J. Org. Chem.*, 2983–2998.
79. Lagasse, F. and Kagan, H.B. (2000) *Chem. Pharm. Bull.*, **48**, 315–324.
80. van den Berg, M., Minnaard, A.J., Schudde, E.P., van Esch, J., de Vries, A.H.M., de Vries, J.G., and Feringa, B.L. (2000) *J. Am. Chem. Soc.*, **122**, 11539–11540.
81. Boiteau, J.G., Imbos, F., Minnaard, A.J., and Feringa, B.L. (2003) *Org. Lett.*, **5**, 681–684.
82. Boiteau, J.G., Minnaard, A.J., and Feringa, B.L. (2003) *J. Org. Chem.*, **68**, 9481–9484.
83. Duursma, A., Hoen, R., Schuppan, J., Hulst, R., Minnaard, A.J., and Feringa, B.L. (2003) *Org. Lett.*, **5**, 3111–3113.
84. Duursma, A., Pena, D., Minnaard, A.J., and Feringa, B.L. (2005) *Tetrahedron: Asymmetry*, **16**, 1901–1904.
85. Jagt, R.B.C., de Vries, J.G., Feringa, B.L., and Minnaard, A.J. (2005) *Org. Lett.*, **7**, 2433–2435.
86. Jagt, R.B., Toullec, P., Schudde, E.P., De Vries, J., Feringa, B.L., and Minnaard, A.J. (2007) *J. Comb. Chem.*, **9**, 407–414.
87. Ma, Y.D., Song, C., Ma, C.Q., Sun, Z.J., Chai, Q., and Andrus, M.B. (2003) *Angew. Chem., Int. Ed.*, **42**, 5871–5874.
88. Facchetti, S., Cavallini, I., Funaioli, T., Marchetti, F., and Iuliano, A. (2009) *Organometallics*, **28**, 4150–4158.
89. Iuliano, A., Facchetti, S., and Funaioli, T. (2009) *Chem. Commun.*, 457–459.
90. Reetz, M.T. (2008) *Angew. Chem., Int. Ed.*, **47**, 2556–2588.
91. Monti, C., Gennari, C., and Piarulli, U. (2007) *Chem. Eur. J.*, **13**, 1547–1558.
92. Itooka, R., Iguchi, Y., and Miyaura, N. (2001) *Chem. Lett.*, 722–723.
93. Hayashi, T., Ueyama, K., Tokunaga, N., and Yoshida, H. (2003) *J. Am. Chem. Soc.*, **125**, 11508–11509.
94. Defieber, C., Paquin, J., Serna, S., and Carreira, E.M. (2004) *Org. Lett.*, **6**, 3873–3876.
95. Paquin, J., Stephenson, C.R.J., Defieber, C., and Carreira, E.M. (2005) *Org. Lett.*, **7**, 3821–3824.
96. Otomaru, Y., Okamoto, K., Shintani, R., and Hayashi, T. (2005) *J. Org. Chem.*, **70**, 2503–2508.
97. Kina, A., Ueyama, K., and Hayashi, T. (2005) *Org. Lett.*, **7**, 5889–5892.
98. Lang, F., Breher, F., Stein, D., and Grutzmacher, H. (2005) *Organometallics*, **24**, 2997–3007.
99. Berthon-Gelloz, G. and Hayashi, T. (2006) *J. Org. Chem.*, **71**, 8957–8960.
100. Helbig, S., Sauer, S., Cramer, N., Laschat, S., Baro, A., and Frey, W. (2007) *Adv. Synth. Catal.*, **349**, 2331–2237.
101. Nishimura, T., Nagaosa, M., and Hayashi, T. (2008) *Chem. Lett.*, **37**, 860–861.
102. Okamoto, K., Hayashi, T., and Rawal, V.H. (2008) *Org. Lett.*, **10**, 4387–4389.
103. Gendrineau, T., Chuzel, O., Eijsberg, H., Genêt, J.P., and Darses, S. (2008) *Angew. Chem., Int. Ed.*, **47**, 7669–7672.
104. Shintani, R., Ichikawa, Y., Takatsu, K., Chen, F.X., and Hayashi, T. (2009) *J. Org. Chem.*, **74**, 869–873.
105. Fischer, C., Defieber, C., Takeyuki, S., and Carreira, E.M. (2004) *J. Am. Chem. Soc*, **126**, 1628–1629.
106. Defieber, C., Grützmacher, H., and Carreira, E.M. (2008) *Angew. Chem., Int. Ed.*, **47**, 4482–4502.
107. Noël, T., Vandyck, K., and Van der Eycken, J. (2007) *Tetrahedron*, **63**, 12961–12967.
108. Shintani, R., Ueyama, K., Yamada, I., and Hayashi, T. (2004) *Org. Lett.*, **6**, 3425–3427.

109. Hayashi, T., Tokunaga, N., Okamoto, K., and Shintani, R. (2005) *Chem. Lett.*, **34**, 1480–1481.
110. Shintani, R., Kimura, T., and Hayashi, T. (2005) *Chem. Commun.*, 3213.
111. Shintani, R., Okamoto, K., and Hayashi, T. (2005) *Org. Lett.*, **7**, 4757–4759.
112. Paquin, J.F., Defieber, C., Stephenson, C.R.J., and Carreira, E.M. (2005) *J. Am. Chem. Soc.*, **127**, 10850–10851.
113. Chen, F., Kina, A., and Hayashi, T. (2006) *Org. Lett.*, **8**, 341–344.
114. Tokunaga, N. and Hayashi, T. (2007) *Adv. Synth. Catal.*, **349**, 513–516.
115. Soergel, S., Tokunaga, N., Sasaki, K., Okamoto, K., and Hayashi, T. (2008) *Org. Lett.*, **10**, 589–592.
116. Wang, Z.Q., Feng, C.G., Xu, M.H., and Lin, G.Q. (2007) *J. Am. Chem. Soc.*, **129**, 5336–5337.
117. Otomaru, Y., Tokunaga, N., Shintani, R., and Hayashi, T. (2005) *Org. Lett.*, **7**, 307–310.
118. Otomaru, Y., Kina, A., Shintani, R., and Hayashi, T. (2005) *Tetrahedron: Asymmetry*, **16**, 1673–1679.
119. Okamoto, K., Hayashi, T., and Rawal, V.H. (2009) *Chem. Commun.*, 4815–4817.
120. Chen, M.S., Prabagaran, N., Labenz, N.A., and White, C.M. (2005) *J. Am. Chem. Soc.*, **127**, 6970–6971.
121. Chen, M.S. and White, C.M. (2004) *J. Am. Chem. Soc.*, **126**, 1346–1347.
122. Mariz, R., Luan, X., Gatti, M., Linden, A., and Dorta, R. (2008) *J. Am. Chem. Soc.*, **130**, 2172–2173.
123. Bürgi, J.J., Mariz, R., Gatti, M., Drinkel, E., Luan, X., Blumentritt, S., Linden, A., and Dorta, R. (2009) *Angew. Chem., Int. Ed.*, **48**, 2768–2771.
124. Tokunaga, N., Otomaru, Y., Okamoto, K., Ueyama, K., Shintani, R., and Hayashi, T. (2004) *J. Am. Chem. Soc.*, **126**, 13584–13585.
125. Kuriyama, M. and Tomioka, K. (2001) *Tetrahedron Lett.*, **42**, 921–923.
126. Kuriyama, M., Nagai, K., Yamada, K., Miwa, Y., Taga, T., and Tomioka, K. (2002) *J. Am. Chem. Soc.*, **124**, 8932–8939.
127. Chen, Q., Soeta, T., Kuriyama, M., Yamada, K.I., and Tomioka, K. (2006) *Adv. Synth. Catal.*, **348**, 2604–2608.
128. Becht, J.M., Bappert, E., and Helmchen, G. (2005) *Adv. Synth. Catal.*, **347**, 1495–1498.
129. Shintani, R., Duan, W.L., Nagano, T., Okada, A., and Hayashi, T. (2005) *Angew. Chem., Int. Ed.*, **44**, 4611–4614.
130. Duan, W.L., Iwamura, H., Shintani, R., and Hayashi, T. (2007) *J. Am. Chem. Soc.*, **129**, 2130–2138.
131. Piras, E., Lang, F., Ruegger, H., Stein, D., Worle, M., and Grützmacher, H. (2006) *Chem. Eur. J.*, **12**, 5849–5858.
132. Mariz, R., Briceno, A., and Dorta, R. (2008) *Organometallics*, **27**, 6605–6613.
133. Defieber, C., Ariger, M.A., Moriel, P., and Carreira, E.M. (2007) *Angew. Chem., Int. Ed.*, **46**, 3139–3143.
134. Kasak, P., Arion, V.B., and Widhalm, M. (2006) *Tetrahedron: Asymmetry*, **17**, 3084–3090.
135. Yoshida, K., Ogasawara, M., and Hayashi, T. (2002) *J. Am. Chem. Soc.*, **124**, 10984–10985.
136. Tokunaga, N. and Hayashi, T. (2006) *Tetrahedron: Asymmetry*, **17**, 607–613.
137. Ueda, M. and Miyaura, N. (2000) *J. Org. Chem.*, **65**, 4450–4452.
138. Meyer, O., Becht, J.M., and Helmchen, G. (2003) *Synlett*, 1539–1541.
139. Chen, G., Tokunaga, N., and Hayashi, T. (2005) *Org. Lett.*, **7**, 2285–2288.
140. Eliott, J.D., Lago, M.A., Cousins, R.D., Gao, A.M.G., Leber, J.D., Erhard, K.F., Nambi, P., Elshourbagy, N.A., Kumar, C., Lee, J.A., Bean, J.W., DeBrosse, C.W., Eggleston, D.S., Brooks, D.P., Feuerstein, G., Ruffolo, R.R., Weinstock, J., Gleason, J.G., Peishoff, C.E., and Ohlstein, E.H. (1994) *J. Med. Chem.*, **37**, 1553–1557.
141. Song, Z.G.J., Zhao, M.Z., Desmond, R., Devine, P., Tschaen, D.M., Tillyer, R., Frey, L., Heid, R., Xu, F., Foster, B., Li, J., Reamer, R., Volante, R., Grabowski, E.J.J., Dolling, U.H., Reider, P.J., Okada, S., Kato, Y., and Mano, E. (1999) *J. Org. Chem.*, **64**, 9658–9667.
142. Frost, C.G., Penrose, S.D., and Gleave, R. (2008) *Org. Biomol. Chem.*, **6**, 4340–4347.

143. Shintani, R., Tokunaga, N., Doi, H., and Hayashi, T. (2004) *J. Am. Chem. Soc.*, **126**, 6240–6241.
144. Ramnauth, J., Poulin, O., Bratovanov, S.S., Rakhit, S., and Maddaford, S.P. (2001) *Org. Lett.*, **3**, 2571–2573.
145. Navarro, C., Moreno, A., and Csaky, A.G. (2009) *J. Org. Chem.*, **74**, 466–469.
146. Zoute, L., Kociok-Köhn, G., and Frost, C.G. (2009) *Org. Lett.*, **11**, 2491–2494.
147. Duan, W.L., Imazaki, Y., Shintani, R., and Hayashi, T. (2007) *Tetrahedron*, **63**, 8529–8536.
148. Trost, B.M. and Jiang, C. (2006) *Synthesis*, 369–396.
149. Riant, O. and Hannedouche, J. (2007) *Org. Biomol. Chem.*, **5**, 873–888.
150. Cozzi, P.G., Hilgraf, R., and Zimmermann, N. (2007) *Eur. J. Org. Chem.*, 5969–5994.
151. Shintani, R., Duan, W.L., and Hayashi, T. (2006) *J. Am. Chem. Soc.*, **128**, 5628–5629.
152. Jacques, T., Markó, I.E., and Pospísil, J. (2005) in *Multicomponent Reactions* (eds J. Zhu and H. Bienaymé), Wiley-VCH Verlag GmbH, Weinheim, pp. 398–452.
153. Shintani, R., Ichikawa, Y., Hayashi, T., Chen, J., Nakao, Y., and Hiyama, T. (2007) *Org. Lett.*, **9**, 4643–4645.
154. Shintani, R., Kimura, T., and Hayashi, T. (2005) *Chem. Commun.*, 3213–3214.
155. Konno, T., Tanaka, T., Miyabe, T., Morigaki, A., and Ishihara, T. (2008) *Tetrahedron Lett.*, **49**, 2106–2110.
156. Lalic, G. and Corey, E.J. (2008) *Tetrahedron Lett.*, **49**, 4894–4896.
157. Hayashi, T., Senda, T., Takaya, Y., and Ogasawara, M. (1999) *J. Am. Chem. Soc.*, **121**, 11591–11592.
158. Hayashi, T., Senda, T., and Ogasawara, M. (2000) *J. Am. Chem. Soc.*, **122**, 10716–10717.
159. Yoshida, K. and Hayashi, T. (2003) *J. Am. Chem. Soc.*, **125**, 2872–2873.
160. Mauleon, P. and Carretero, J.C. (2004) *Org. Lett.*, **6**, 3195–3198.
161. Mauleon, P., Alonso, I., Rivero, M.R., and Carretero, J.C. (2007) *J. Org. Chem.*, **72**, 9924–9935.
162. Plesniak, K., Zarecki, A., and Wicha, J. (2007) *Top. Curr. Chem.*, **275**, 163–250.
163. Mauleón, P. and Carretero, J. (2005) *Chem. Commun.*, 4961–4963.
164. Miura, T., Takahashi, Y., and Murakami, M. (2007) *Chem. Commun.*, 595–597.
165. Chapman, C.J., Wadsworth, K.J., and Frost, C.G. (2003) *J. Organomet. Chem.*, **680**, 206–211.
166. Au-Yeung, T.T.-L., Chan, S.-S., and Chan, A.S.C. (2004) in *Transition Metals for Organic Synthesis*, vol. **2** (eds M. Beller and C. Bolm), Wiley-VCH Verlag GmbH, Weinheim, pp. 14–28.
167. Navarre, L., Darses, S., and Genêt, J.P. (2004) *Angew. Chem., Int. Ed.*, **43**, 719–723.
168. Moss, R.J., Wadsworth, K.J., Chapman, C.J., and Frost, C.G. (2004) *Chem. Commun.*, 1984–1985.
169. Frost, C.G., Penrose, S.D., Lambshead, K., Raithby, P.R., Warren, J.E., and Gleave, R. (2007) *Org. Lett.*, **9**, 2119–2122.
170. Sibi, M.P., Tatamidani, H., and Patil, K. (2005) *Org. Lett.*, **7**, 2571–2573.
171. Chapman, C.J., Hargrave, J.D., Bish, G., and Frost, C.G. (2008) *Tetrahedron*, **64**, 9528–9539.
172. Nishimura, T., Hirabayashi, S., Yasuhara, Y., and Hayashi, T. (2006) *J. Am. Chem. Soc.*, **128**, 2556–2557.
173. Mohr, J., Hong, A., and Stoltz, B. (2009) *Nat. Chem.*, **1**, 359–369.
174. Hayashi, T., Tokunaga, N., Yoshida, K., and Han, J.W. (2002) *J. Am. Chem. Soc.*, **124**, 12102–12103.
175. Hayashi, T., Kawai, M., and Tokunaga, N. (2004) *Angew. Chem., Int. Ed.*, **43**, 6125–6128.
176. Hayashi, T., Yamamoto, S., and Tokunaga, N. (2005) *Angew. Chem., Int. Ed.*, **44**, 4224–4227.
177. Tokunaga, N., Yoshida, K., and Hayashi, T. (2004) *Proc. Natl Acad. Sci. USA*, **101**, 5445–5449.
178. Le Notre, J., Allen, J.C., and Frost, C.G. (2008) *Chem. Commun.*, 3795–3797.
179. Langkopf, E. and Schinzer, D. (1995) *Chem. Rev.*, **95**, 1375–1408.
180. Ojima, I., Li, Z.Y., and Zhu, J.W. (1998) in *Chemistry of Organic Silicon Compounds* (eds Z. Rappoport and Y. Apeloig), John Wiley & Sons, Ltd, England, pp. 1687–1792.

181. Rappoport, Z. and Apeloig, Y. (2001) *The Chemistry of Organosilicon Compounds*, 3rd edn, John Wiley & Sons, Ltd, Chichester.
182. Mori, A., Danda, Y., Fujii, T., Hirabayashi, S., and Osakada, K. (2001) *J. Am. Chem. Soc.*, **123**, 10774–10775.
183. Huang, T. and Li, C. (2001) *Chem. Commun.*, 2348–2349.
184. Koike, T., Du, X.L., Mori, A., and Osakada, K. (2002) *Synlett*, 301–303.
185. Murata, M., Shimazaki, R., Ishikura, M., Watanabe, S., and Masuda, Y. (2002) *Synthesis*, 717–719.
186. Otomaru, Y. and Hayashi, T. (2004) *Tetrahedron: Asymmetry*, **15**, 2647–2651.
187. Sanada, T., Kato, T., Mitani, M., and Mori, A. (2006) *Adv. Synth. Catal.*, **348**, 51–54.
188. Hargrave, J.D., Herbert, J., Bish, G., and Frost, C.G. (2006) *Org. Biomol. Chem.*, **4**, 3235–3241.
189. Hudrlik, P.F., Abdallah, Y.M., and Hudrlik, A.M. (1992) *Tetrahedron Lett.*, **33**, 6747–6750.
190. Hijji, Y.M., Hudrlik, P.F., and Hudrlik, A.M. (1998) *Chem. Commun.*, 1213–1214.
191. Nakao, Y., Imanaka, H., Sahoo, A.K., Yada, A., and Hiyama, T. (2005) *J. Am. Chem. Soc.*, **127**, 6952–6953.
192. Nakao, Y., Chen, J., Imanaka, H., Hiyama, T., Ichikawa, Y., Duan, W.L., Shintani, R., and Hayashi, T. (2007) *J. Am. Chem. Soc.*, **129**, 9137–9143.
193. Nakao, Y., Imanaka, H., Chen, J., Yada, A., and Hiyama, T. (2007) *J. Organomet. Chem.*, **692**, 585–603.
194. Wipf, P. (2004) *Top. Organomet. Chem.*, **8**, 1–25.
195. Kakuuchi, A., Taguchi, T., and Hanzawa, Y. (2004) *Tetrahedron*, **60**, 1293–1299.
196. Nicolaou, K.C., Tang, W.J., Dagneau, P., and Faraoni, R. (2005) *Angew. Chem., Int. Ed.*, **44**, 3874–3879.
197. Ding, R., Chen, Y.J., Wang, D., and Li, C.J. (2001) *Synlett*, 1470–1472.
198. Venkatraman, S., Meng, Y., and Li, C.J. (2001) *Tetrahedron Lett.*, **42**, 4459–4462.
199. Venkatraman, S. and Li, C.J. (2001) *Tetrahedron Lett.*, **42**, 781–784.
200. Huang, T.S., Venkatraman, S., Meng, Y., Nguyen, T.V., Kort, D., Wang, D., Ding, R., and Li, C.J. (2001) *Pure Appl. Chem.*, **73**, 1315–1318.
201. Miura, T. and Murakami, M. (2005) *Chem. Commun.*, 5676–5677.
202. Nikishin, G.I. and Kovalev, I.P. (1990) *Tetrahedron Lett.*, **31**, 7063–7064.
203. Lerum, R.V. and Chisholm, J.D. (2004) *Tetrahedron Lett.*, **45**, 6591–6594.
204. Trost, B. and Weiss, A. (2009) *Adv. Synth. Catal.*, **351**, 963–983.
205. Nishimura, T., Katoh, T., Takatsu, K., Shintani, R., and Hayashi, T. (2007) *J. Am. Chem. Soc.*, **129**, 14158–14159.
206. Nishimura, T., Guo, X., Uchiyama, N., Katoh, T., and Hayashi, T. (2008) *J. Am. Chem. Soc.*, **130**, 1576–1577.
207. Nishimura, T., Tokuji, S., Sawano, T., and Hayashi, T. (2009) *Org. Lett.*, **11**, 3222–3225.
208. Trost, B.M. (1991) *Science*, **254**, 1471–1477.
209. Guo, H.C. and Ma, J.A. (2006) *Angew. Chem., Int. Ed.*, **45**, 354–366.
210. Miura, T. and Murakami, M. (2007) *Chem. Commun.*, 217–224.
211. Youn, S. (2009) *Eur. J. Org. Chem.*, **2009**, 2597–2605.
212. Cauble, D.F., Gipson, J.D., and Krische, M.J. (2003) *J. Am. Chem. Soc.*, **125**, 1110–1111.
213. Bocknack, B.M., Wang, L.C., and Krische, M.J. (2004) *Proc. Natl Acad. Sci. USA*, **101**, 5421–5424.
214. Miura, T., Harumashi, T., and Murakami, M. (2007) *Org. Lett.*, **9**, 741–743.
215. Youn, S., Song, J., and Jung, D. (2008) *J. Org. Chem.*, **73**, 5658–5661.
216. Navarro, C. and Csákÿ, A.G. (2008) *Org. Lett.*, **10**, 217–219.
217. Oguma, K., Miura, M., Satoh, T., and Nomura, M. (2000) *J. Am. Chem. Soc.*, **122**, 10464–10465.
218. Lautens, M., Dockendorff, C., Fagnou, K., and Malicki, A. (2002) *Org. Lett.*, **4**, 1311–1314.
219. Lautens, M. and Mancuso, J. (2004) *J. Org. Chem.*, **69**, 3478–3487.
220. Tseng, N.W. and Lautens, M. (2009) *J. Org. Chem.*, **74**, 1809–1811.
221. Lautens, M. and Marquardt, T. (2004) *J. Org. Chem.*, **69**, 4607–4614.

222. Tseng, N.W., Mancuso, J., and Lautens, M. (2006) *J. Am. Chem. Soc.*, **128**, 5338–5339.
223. Tseng, N.W. and Lautens, M. (2009) *J. Org. Chem.*, **74**, 2521–2526.
224. Shintani, R., Tsurusaki, A., Okamoto, K., and Hayashi, T. (2005) *Angew. Chem., Int. Ed.*, **44**, 3909–3912.
225. Shintani, R., Okamoto, K., Otomaru, Y., Ueyama, K., and Hayashi, T. (2005) *J. Am. Chem. Soc.*, **127**, 54–55.
226. Chen, Y. and Lee, C. (2006) *J. Am. Chem. Soc.*, **128**, 15598–15599.
227. de la Herrán, G., Murcia, C., and Csákÿ, A.G. (2005) *Org. Lett.*, **7**, 5625–5632.
228. Nishimura, T., Yasuhara, Y., and Hayashi, T. (2006) *Angew. Chem., Int. Ed.*, **45**, 5164–5166.
229. Koike, T., Du, X.L., Sanada, T., Danda, Y., and Mori, A. (2003) *Angew. Chem., Int. Ed.*, **42**, 89–92.
230. Yamamoto, Y., Nishikata, T., and Miyaura, N. (2006) *J. Synth. Org. Chem. Jpn*, **64**, 1112–1121.
231. Gutnov, A. (2008) *Eur. J. Org. Chem.*, 4547–4554.
232. Kobayashi, K., Nishikata, T., Yamamoto, Y., and Miyaura, N. (2008) *Bull. Chem. Soc. Jpn*, **81**, 1019–1025.
233. Yamamoto, Y. and Miyaura, N. (2008) *J. Synth. Org. Chem. Jpn*, **66**, 194–204.
234. Miyaura, N. (2009) *Synlett*, **2009**, 2039–2050.
235. Cacchi, S., Misiti, D., and Palmieri, G. (1981) *Tetrahedron*, **37**, 2941–2946.
236. Cacchi, S., Latorre, F.F., and Misiti, D. (1979) *Tetrahedron Lett.*, 4591–4594.
237. Albeniz, A.C., Catalina, N.M., Espinet, P., and Redon, R. (1999) *Organometallics*, **18**, 5571–5576.
238. Nishikata, T., Yamamoto, Y., and Miyaura, N. (2003) *Angew. Chem., Int. Ed.*, **42**, 2768–2770.
239. Nishikata, T., Yamamoto, Y., and Miyaura, N. (2004) *Organometallics*, **23**, 4317–4324.
240. Kamer, P.C.J., van Leeuwen, P.W.N., and Reek, J.N.H. (2001) *Acc. Chem. Res.*, **34**, 895–904.
241. Nishikata, T., Yamamoto, Y., and Miyaura, N. (2003) *Chem. Lett.*, **32**, 752–753.
242. Nishikata, T., Yamamoto, Y., and Miyaura, N. (2004) *Chem. Commun.*, 1822–1823.
243. Nishikata, T., Yamamoto, Y., Gridnev, I.D., and Miyaura, N. (2005) *Organometallics*, **24**, 5025–5032.
244. Nishikata, T., Yamamoto, Y., and Miyaura, N. (2007) *Adv. Synth. Catal.*, **349**, 1759–1764.
245. Nishikata, T., Kobayashi, Y., Kobayshi, K., Yamamoto, Y., and Miyaura, N. (2007) *Synlett*, 3055–3057.
246. Nishikata, T., Kiyomura, S., Yamamoto, Y., and Miyaura, N. (2008) *Synlett*, 2487–2490.
247. Nishikata, T., Yamamoto, Y., and Miyaura, N. (2007) *Chem. Lett.*, **36**, 1442–1443.
248. Nishikata, T., Yamamoto, Y., and Miyaura, N. (2007) *Tetrahedron Lett.*, **48**, 4007–4010.
249. Gini, F., Hessen, B., and Minnaard, A.J. (2005) *Org. Lett.*, **7**, 5309–5312.
250. Zhang, T. and Shi, M. (2008) *Chem. Eur. J.*, **14**, 3759–3764.
251. Lu, X.Y. and Lin, S.H. (2005) *J. Org. Chem.*, **70**, 9651–9653.
252. Lin, S. and Lu, X. (2006) *Tetrahedron Lett.*, **47**, 7167–7170.
253. He, P., Lu, Y., Dong, C., and Hu, Q. (2007) *Org. Lett.*, **9**, 343–346.
254. Bedford, R., Betham, M., Charmant, J.P.H., Haddow, M.F., Orpen, A., Pilarski, L.T., Coles, S., and Hursthouse, M. (2007) *Organometallics*, **26**, 6346–6353.
255. Suzuma, Y., Yamamoto, T., Ohta, T., and Ito, Y. (2007) *Chem. Lett.*, **36**, 470–471.
256. Zhou, F., Yang, M., and Lu, X. (2009) *Org. Lett.*, **11**, 1405–1408.

2
Cu- and Ni-Catalyzed Conjugated Additions of Organozincs and Organoaluminums to α,β-Unsaturated Carbonyl Compounds

Martin Kotora and Robert Betík

2.1
Introduction

Since their discovery during the early 1880s, conjugated addition (1,4-addition) reactions have been an indispensable part of organic synthesis. The first example of conjugated addition, reported by Komnenos in 1883, described the addition of diethylsodiomalonate to diethyl ethylidenemalonate [1], and this was soon followed by pioneering works of Michael [2] and Claisen [3] in 1887.

Since those early days, the scope of the conjugated reaction has expanded beyond imaginable limits and today it has become a standard synthetic tool for the construction of new C–C bonds. Given the range of potential nucleophiles (C-nucleophiles) that can react with a number of electrophiles (an unsaturated system in conjugation with an activating group, usually an electron-withdrawing group), its synthetic application is virtually unlimited. The popularity of this reaction is nicely demonstrated by never-ending interest from the synthetic chemical community in its application [4, 5].

In this chapter, the conjugate addition of a class of mild C-nucleofiles such as organozinc and organoaluminum compounds under catalytic conditions conveyed by transition metal compounds will be described. Special attention will be paid to asymmetric induction – that is, enantioselective conjugated addition. Details will also be provided of their application in the syntheses of natural products.

Parts of this topic have been recently reviewed by several authors, and these deserve mention here as a source of detailed information summarizing various aspects of the conjugated addition of organozinc and aluminum compounds. Among these must be included the reviews of Knochel *et al.*, detailing the synthesis of organozinc compounds and their reactions [6], of Togni *et al.* [7], Lemaire *et al.* [8], and of Kwong *et al.* [9], which summarize the use of N-ligands in asymmetric catalysis, by Carreira *et al.* [10] and Trost *et al.* [11], which deal with conjugated additions of metallated alkynes, by von Zezschwitz *et al.* describing aspects of additions to carbonyl compounds [12], and finally by Alexakis *et al.* [13], Feringa

---

Catalytic Asymmetric Conjugate Reactions. Edited by Armando Córdova
Copyright © 2010 WILEY-VCH Verlag GmbH & Co. KGaA, Weinheim
ISBN: 978-3-527-32411-8

et al. [14], and Knochel [15], who have reviewed recent advances in asymmetric conjugated additions. Other information may also be found elsewhere [16].

2.2
General Aspects

2.2.1
Properties of Organozinc and Organoaluminum Compounds

Organozinc compounds (R_2Zn and $RZnX$) belong to one of the first organometallic compounds ever prepared. Diethylzinc was prepared by Frankland in 1849 by heating ethyl iodide with zinc [17, 18]. Due to a rather large difference in the electronegativities of aluminum (1.61) and zinc (1.65) in comparison with carbon (2.55), the corresponding Zn–C and Al–C bonds are highly polarized and hence reactive. Both organometallics – that is, organozincs and organoaluminums – exhibit moderate nucleophilicity and basicity that allows them only to react with reactive electrophiles. Importantly, they usually do not attack the carbonyl group under ambient conditions (unlike organomagnesium compounds), which makes them convenient and selective reagents for a number of synthetic transformations. Moreover, their lower reactivity can often be finely controlled (or tuned) by the presence of appropriate transition metal compounds. Further information on properties of organozincs and organoaluminums can be found elsewhere [19, 20].

Although there are similarities between organozinc and organoaluminums, considerable differences also exist. In this regard, the characteristic property of organoaluminum compounds derives mainly from Lewis acidity, which is related to the aluminum atom tendency to complete the electron octet. Due to strong bonds of aluminum with electron-negative atoms such as oxygen or the halogens (e.g., Al–O 138 kcal mol^{-1}), nearly all organoaluminum compound are particularly reactive with oxygen. This property – which is commonly described as *oxygenophilicity* – is of great value when designing selective synthetic reactions. Conjugated additions of organoaluminums to α,β-unsaturated carbonyl compounds may serve as typical examples. The organoaluminum compound is not only the source of the transferable group, but also activates the conjugated system by Lewis acid/base interaction between the carbonyl group (the donor of the lone electron pair) and aluminum atom (the acceptor of the lone electron pair).

2.2.1.1 Reaction Mechanisms of Cu-Catalyzed Conjugated Addition
It is generally accepted that the copper-catalyzed conjugated addition of organozincs and organoaluminums resembles that of organocuprates. The main difference consists in the nature of reactive species, which is reasonable to understand because of a large excess of an organometal with respect to the catalytic amount of a copper catalyst. There is no doubt that the reactive species are multimetal center clusters with different stoichiometry. It has been shown clearly that the appropriate choice of an anion in copper salts also plays an essential role. In addition, the effect of a chiral

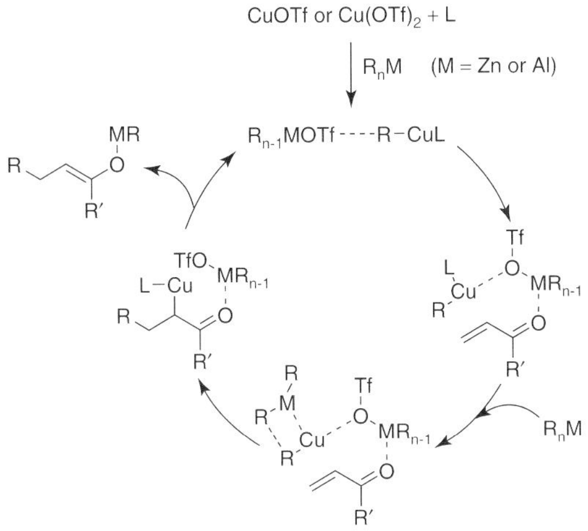

Scheme 2.1 Simplified reaction mechanism of copper-catalyzed conjugated addition.

ligand on the association of the cluster should be taken into account. Despite all of these factors that may affect the course and mechanism of the conjugated addition, it can be generalized and described in the following (much simplified) terms (Scheme 2.1). The first step is based on transmetallation between an organometal (M = zinc or aluminum) and copper(I) or (II) salts (in the case of copper(II) these are reduced to Cu(I) species) to produce the corresponding organocopper cluster species. A coordination of the metal (M = zinc or aluminum) to oxygen atom of the carbonyl group (Lewis acid–base pairing) and π-complexation of the copper to the double bond is then assumed. A reaction of the formed intermediate with another molecule of the organometal generates a highly reactive metal–zinc cluster that is capable of alkyl transfer to the double bond, forming a copper enolate. Finally, transmetallation of copper from the enolate by the reaction with the organometal releases the copper species back into the catalytic cycle. However, it should be noted that the actual course of the reaction might be much more complicated and may involve other species. The relevant issues regarding various aspects of the reaction mechanism have been discussed in reviews by Alexakis [13] and Feringa [14]. Some aspects of the reaction mechanism, such as a precatalyst formation [21], the double bond geometry of the Michael acceptor [22], and the "styrene effect" [23] have also been discussed.

2.2.1.2 Reaction Mechanisms of Ni-Catalyzed Conjugated Addition

In contrast to the copper-catalyzed conjugated addition, less attention has been paid to the reaction mechanism of the nickel-catalyzed process. The early proposed reaction mechanism for Ni-catalyzed conjugated addition of organoaluminum compounds to α,β-unsaturated ketones was reported by Bagnell *et al.* as early as in 1975 (Scheme 2.2) [24]. These authors claimed that, in the first step, a Ni(II) compound, Ni(acac)$_2$ or other Ni(II) salts, are reduced to Ni(0) species. There

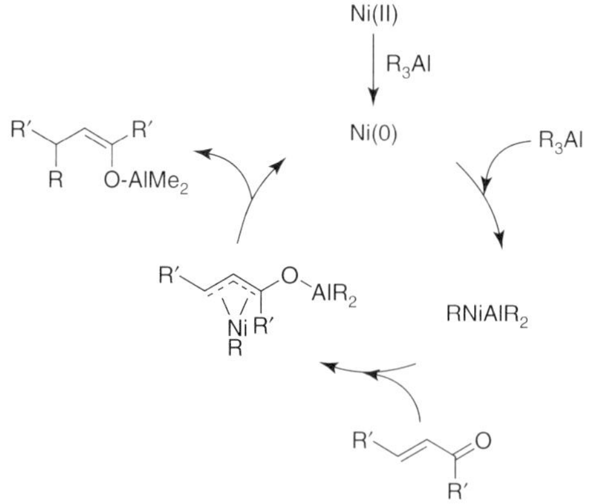

Scheme 2.2 Simplified reaction mechanism of nickel-catalyzed conjugated addition.

should then ensue an oxidative addition of the organoaluminum compound to give RNiAlR$_2$ species that coordinate to enone, forming π-allylic nickel species.

2.2.2
Preparation of Organozinc Compounds

Organozinc compounds can be prepared by several methods, and since this area has been extensively reviewed only the basic concepts will be noted at this point. (This topic was excellently reviewed by Knochel in 1993 [6], and more recently in 2005 [25]; additional information can be found elsewhere [19].) The usual procedures are based either on oxidative addition of zinc into the carbon–halide bond, or by exchange (transmetallation) reactions of zinc halides with other organometallics.

2.2.2.1 Preparation by a Direct Insertion into the Carbon–Halide Bond

This is historically the oldest procedure for the preparation of organozinc compounds. The rate of zinc insertion depends heavily on the nature of organic moiety, the halide, the reaction conditions, and the activation of zinc (of the zinc surface). The usual candidates for zinc insertion are organyl iodides, although chlorides and bromides or other organyl derivatives such as mesylates, tosylates, and so on may also be used. Although the common solvent most often used for the insertion is tetrahydrofuran (THF), also dimethylformamide (DMF) or mixture of benzene with DMA or DMF can also be used.

Zinc is normally used in the form of powder that must be activated, and this is most easily done by treating it successively with 1,2-dibromoethane and chlorotrimethylsilane prior to the organyl halide addition. The reaction usually proceeds under ambient conditions within a couple of hours. The mild and neutral reaction conditions can tolerate most functionalities such as ester, ketone, halide (chloride), amino- or substituted amino, amide, alkoxysilyl, sulfoxide, sulfide, sulfone, thioester, boronic ester, enone, and phosphate groups. On the other hand, nitro, azide (inhibit the zinc insertion), and hydroxyl groups (deprotonation) cannot

Table 2.1 Preparation of several organozinc compounds [6].

Alkyl iodide	Temperature (°C)	Time (h)	Solvent
n-BuI	45–50	3–4	THF
c-HexI	25–30	2	THF
NC(CH$_2$)$_4$I	20	2	THF
(EtO)$_2$(O)P(CH$_2$)$_2$Br	30	10	THF
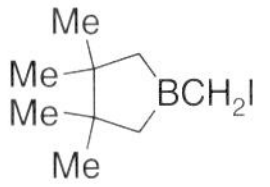 Me-C(Me)(Me)(Me)-BCH$_2$I	20	0.1	THF
PhSCH$_2$Cl	25	1	THF
(E)-1-Iodohexene	70	14	DMF
3-Iodocyclohexenone	25–50	0.5	THF
Benzyl bromide	0–5	2	THF
Allyl bromide	10	4–5	THF

be present in the molecule. Highly reactive zinc powder can be also prepared by the reduction of zinc salts with lithium metal or lithium naphthalenide. Recently, a highly reactive suspension of zinc powder in THF (known as *Rieke zinc* ®) has become commercially available [26, 27].

With regards to the structure of the organyl halide, allylic, and benzylic halides are the most reactive (even bromides and chlorides can be used as the starting material). The presence of polar functional groups (cyano, phosphate, thio, enone groups, etc.) in α or β position also greatly enhances the insertion rate. A similar approach can be also used for other types of organyl halide. Some typical examples of the preparation of organozinc compounds are displayed in Table 2.1.

2.2.2.2 Preparation by Lithium–Zinc Transmetallation

Although organolithiums are considered to be too reactive to tolerate most functional groups, they are excellent starting compounds for the preparation of organozincs. These are easily formed upon the addition of a dry zinc halide into a solution of the corresponding organolithium. Owing to the lower reactivity of alkenyl- and aryllithiums, these can be generated from starting materials bearing certain functional groups (ester, epoxy, cyano, nitro, or even halides such as Cl or Br). Thus, the prepared sp^2-hybridized organolithiums can be easily converted into the corresponding alkenyl- and arylzincs (Scheme 2.3).

Lithium to zinc transmetallation is also the underlying strategy of a one-pot procedure for the synthesis of diorganyl zincs which is based on the sonification of

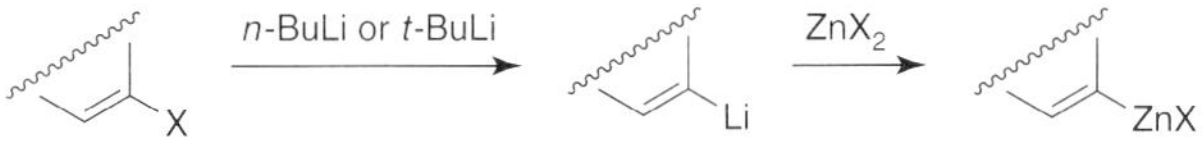

Scheme 2.3 Formation of organozincs via transmetallation of organolithiums.

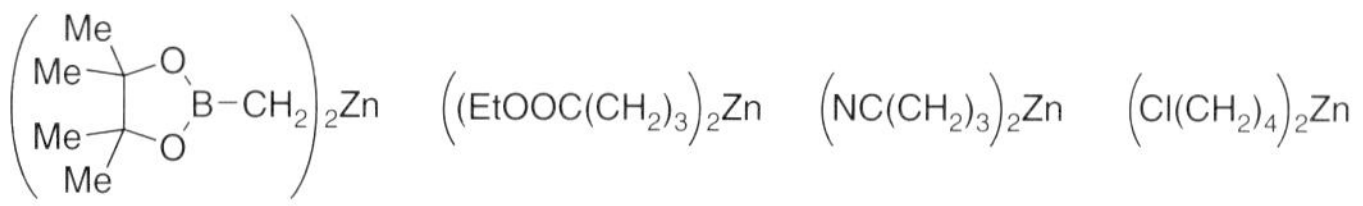

$\left(\begin{array}{c}\text{Me}\\\text{Me}\\\text{Me}\\\text{Me}\end{array}\rightarrow\begin{array}{c}\text{O}\\\text{B-CH}_2\\\text{O}\end{array}\right)_2\text{Zn}$ $\left(\text{(EtOOC(CH}_2)_3\right)_2\text{Zn}$ $\left(\text{NC(CH}_2)_3\right)_2\text{Zn}$ $\left(\text{Cl(CH}_2)_4\right)_2\text{Zn}$

Figure 2.1 Some examples of functionalized dialkylzinc compounds.

a mixture of an organic halide, lithium, and a zinc halide. The method is suitable for the preparation of dialkyl, dialkenyl, and diaryl zincs [28].

2.2.2.3 Preparation by an Iodine–Zinc Exchange Reaction

Diorganozinc compounds can be also prepared by a simple reaction of organyl halides with zinc, although for this purpose a different approach had to be developed. The most useful method with regard to the reaction conditions, as well as the availability of the starting material, is that based on the zinc–iodine exchange process, for which the first report appeared in 1966 [29]. The reaction can be applied to the synthesis of a number of diorganozincs, and involves the reaction of the corresponding organyl iodide with Et$_2$Zn in the presence of a catalytic amount of transition metal salts, usually CuI. The mild reaction conditions enable the preparation of diorganozincs bearing various other functional groups, such boronate esters, ester, nitriles, and halides (Figure 2.1). As far as the course of the reaction is concerned, it proceeds via radical mechanism.

Later, it was shown that the same process could be catalyzed even more effectively by Pd and Ni compounds [30], although in this case alkylzinc halides are obtained instead of diorganozincs. A typical example is the conversion of *n*-octyl iodide to the corresponding dioctylzinc (Cu catalysis) or octylzinc iodide (Pd catalysis) (Scheme 2.4).

2.2.2.4 Preparation by a Boron–Zinc Exchange

The versatility of the hydroboration of unsaturated C–C bonds with respect to a hydroboration agent as well as unsaturated substrates makes this reaction a popular tool for the synthesis of functionalized alkyl and alkenyl compounds. It is, therefore, not surprising that organoboranes can be converted into organozincs. One such method is based on the reaction of triorganoboranes with Me$_2$Zn, which yields the corresponding diorganozincs (Scheme 2.5). The driving force here is the formation of volatile BMe$_3$. This transmetallation has been used for the preparation of diallyl-, dibenzyl-, and dialkenylzincs [31–33]. A related method utilizes an exchange reaction between 2-alkyl-1,3-dithia-2-borolanes or diethylorganylboranes with Et$_2$Zn.

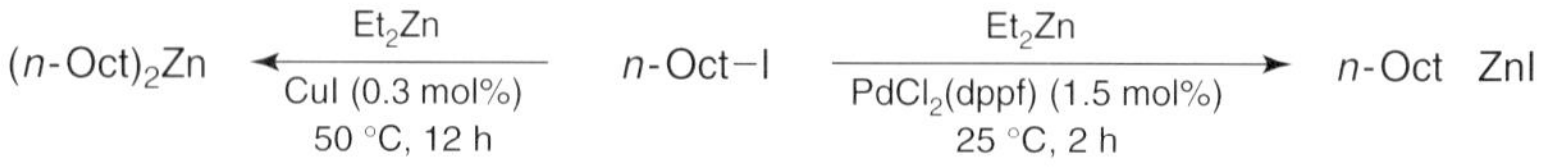

$(n\text{-Oct})_2\text{Zn}$ $\xleftarrow[\text{CuI (0.3 mol\%)}]{\text{Et}_2\text{Zn}}$ $n\text{-Oct-I}$ $\xrightarrow[\text{PdCl}_2\text{(dppf) (1.5 mol\%)}]{\text{Et}_2\text{Zn}}$ $n\text{-Oct ZnI}$
50 °C, 12 h   25 °C, 2 h

Scheme 2.4 Pd- or Cu-catalyzed reaction of alkyl iodides with diethylzinc.

$$3R^1_2Zn + 2R^2_3B \rightleftharpoons 3R^2_2Zn + 2R^1_3B$$

Scheme 2.5 Exchange of alkyl groups between zinc and boron.

2.2.2.5 Preparation by Other Metal–Zinc Exchange

In an analogical manner to the aforementioned methods, transmetallation between diorganomercury compounds and metallic zinc can also be used for the synthesis of various diorganozincs. The reaction is usually carried out in boiling toluene; however, in the presence of a catalytic amount of ZnX_2, it can be performed in THF at 60 °C within a short period of time [34]. Nevertheless, the development of less hazardous and more environmentally friendly methods for the synthesis of organozincs, avoiding the use of poisonous organomercury compounds, makes this approach rather obsolete. Organozincs can also be prepared by the transmetallation of alkenylzirconium compounds, prepared by the hydrozirconation of alkynes, with Et_2Zn or Me_2Zn to give mixed alkenyl(alkyl)zincs [35].

2.2.2.6 Commercial Availability

A comparatively easy access to organozinc halides and diorganozincs, and their good stability, are the underlying reasons why several of these compounds are available from commercial sources. Regarding the organozinc halides, their structural diversity spans from alkyl, cycloalkyl, alkenyl, aryl, to heteroaryl derivatives. Owing to the lower polarization of the C–Zn bond, and hence a lower basicity and nucleophilicity of the organozinc halides, a number of their derivatives bear various functional groups such as –CN, –COOR, halides such F, Cl, Br, and even I (in case of arylzincs), MeO, acetals. As far as diorganozincs are concerned, at present their offer is limited to a few dialkyl- (methyl, ethyl, i-propyl, and n-butyl) and diaryl zincs (phenyl, pentafluorophenyl).

2.2.3
Preparation of Organoaluminum Compounds

Organoaluminum compounds are not commonly prepared directly from the corresponding alkyl halides and metallic aluminum at the laboratory level. Usually, commercially available organoaluminums are used directly or as precursors for the preparation of other organoaluminums.

In this regard, hydroalumination constitutes probably the most general and convenient method for the synthesis of alkenylaluminums [36]. The reaction is usually carried out by the reaction of $i\text{-}Bu_2AlH$ with a terminal or internal alkyne. Alkynylation of the dialkylaluminum halides with lithium or magnesium alkynides is a method of choice for the preparation of alkynylaluminums, for which further information can be found elsewhere [20].

In contrast to organozincs, the commercial availability of organoaluminums is less broad, largely on the basis of their higher reactivity and instability. On the other hand, several trialkylaluminums (methyl, ethyl, and i-butyl) can be obtained from commercial sources, as well methyl- and ethylchloroaluminums and DIBAL-H (diisobutylaluminum hydride) that may serve as precursors for the synthesis of other organoaluminums.

2.3
Conjugated Additions

Conjugated additions (1,4-additions) of organometallic compounds of zinc and aluminum have been usually catalyzed by Cu(I) or Ni(0), or Ni(II) compounds. In all of these cases, the reactive species as well as the course of the reactions differ. They involve the participation of structurally different species and also, importantly, different roles for the transition metal compounds.

Probably the most-often used method for conjugated addition is based on the use of organocopper compounds that usually have been prepared from the corresponding organolithium or organomagnesium organometallics and copper(I) salts. It has been shown that varying the ratio of the reactants as well as the nature of the anion in the CuX salts, along with an appropriate choice of a solvent and additives, has a crucial influence on the formation and structure of the reactive species able to undergo conjugated addition. Although, the detailed studies usually involved well-defined organocopper compounds, the data acquired thus far have proven very valuable also for copper-salt-catalyzed conjugated reactions of organomagnesium compounds [37]. Analogously, a similar behavior has been expected also for those compounds generated from organozinc organometallics and copper salts.

2.3.1
Cu-Catalyzed Conjugated Addition

As organozinc compounds can be prepared with high functional group flexibility, they constitute ideal reactants for conjugated addition to α,β-unsaturated carbonyl compounds, allowing the synthesis of highly functionalized products. The conjugated addition is often carried out in the presence of a catalytic or stoichiometric amount of copper salts. In this regard, early experiments clearly showed that the use of THF-soluble copper salts (CuCN•2LiX) is mandatory to achieve high yields of products [6, 38]. It is assumed that, upon the reaction with the organozinc compound, a Cu–Zn species capable of reaction with the conjugated system is formed (Scheme 2.6). This can be explained by a faster transmetallation from zinc to copper.

Thus, the conjugated addition of the formed organozinc–copper species proceeded with a variety of β-monosubstituted α,β-unsaturated ketones or esters, furnishing products in high yields. On the other hand, in the case of β-disubstituted α,β-unsaturated carbonyl compounds, the reaction must be carried out either in

$$R-ZnX\ +\ 2LiCl\cdot CuCN\ \longrightarrow\ R-Cu(CN)ZnX$$

Scheme 2.6 Formation of cuprates by the reaction of organozincs with CuCN·2LiCl.

the presence of polar solvents such hexamethylphosphoramide (HMPA) [39], or in the presence of a Lewis acid (BF$_3$·Et$_2$O) [40] or Me$_3$SiCl [41] in order to activate the carbonyl group.

With regards to catalyst precursors, the above-mentioned soluble system for nonchiral conjugated additions has normally been used. Numerous examples can be found in the review of Knochel [6], although other adducts of CuX with various donor molecules such CuI·Me$_2$S [35] have also recently been used. The course of the reaction (usually the reaction rate) of the conjugated addition of organozincs to enones catalyzed by copper salt can be also influenced by the presence of suitable compounds capable of coordination to the copper atom. In this regard, an increase of the reaction rate of the addition of Et$_2$Zn to cycloenones in the presence of sulfonamide or *N*-heterocyclic carbene (NHC)-compound was observed by both Noyori [42] and Woodward [43].

In the case of enantioselective conjugated additions, copper salts that are soluble in various organic solvents, such as Cu(OTf)$_2$ and Cu(MeCN)$_4$X$_n$ (X = PF$_6$, BF$_4$), have often been used as a catalyst precursor. It has been found that the appropriate choice of anion may exert considerable influence on the asymmetric induction. Copper salts have also been used as catalysts in the conjugated addition of Me$_3$Al to linear and cyclic enones [44], as well as to α,β-unsaturated aldehydes [45], where a good selectivity in favor of 1,4-addition was observed.

2.3.2
Ni-Catalyzed Conjugated Addition

The conjugated addition of organoaluminum compounds to α,β-unsaturated carbonyl compounds has been recognized since the mid-1970s. However, in comparison with the copper-catalyzed process it has neither been extensively studied, nor used in organic synthesis. Nonetheless, the catalytic conjugated addition of organoaluminums to α,β-unsaturated carbonyl compounds has a considerable synthetic potential.

Mole *et al.* demonstrated that the conjugated addition of Me$_3$Al to various unsaturated ketones was greatly enhanced by the presence of a catalytic amount of Ni(acac)$_2$ [24]. The scope of this reaction ranged from simple enones to sterically hindered steroid substrates (Scheme 2.7). Analogously, the conjugated addition was carried out with cyclopropyl ketones [46], while similar results were independently reported also by Ashby *et al.* [47]; in both cases, the best results were obtained in diethyl ether. Until recently, the Ni-catalyzed addition of Me$_3$Al to sterically hindered ketones has not attracted any further attention. Importantly, it was

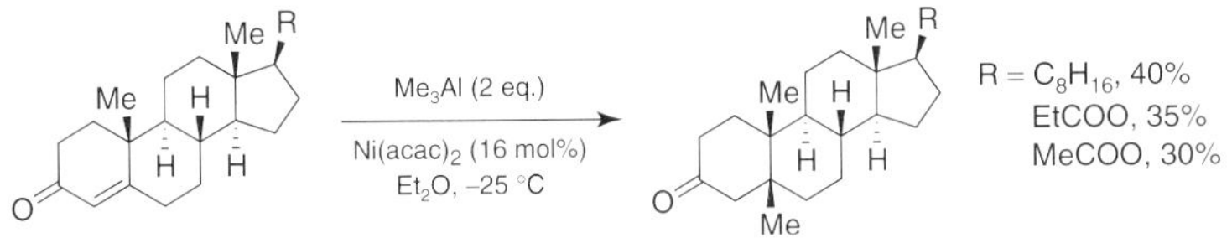

Scheme 2.7 Conjugated addition to steroid substrates.

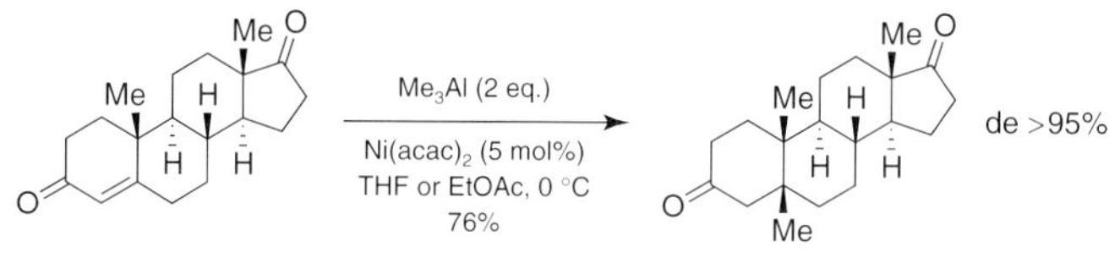

Scheme 2.8 Conjugated addition of Me$_3$Al to 3,17-diketo-$\Delta^{4,5}$-steroid.

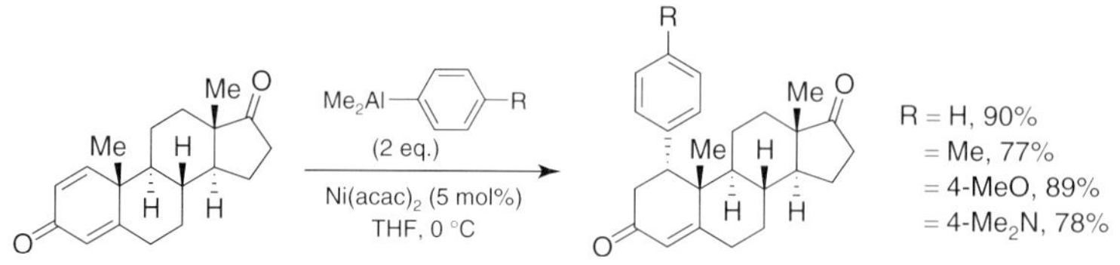

Scheme 2.9 Conjugated addition of arylaluminums to unsaturated steroids.

shown that the correct choice of reaction conditions, particularly of the solvent (THF, EtOAc) and reaction temperature, could lead to improved results [48]. The conjugated addition to 3,17-diketo-$\Delta^{4,5}$-steroid (Scheme 2.8), which proceeded in high yield (76%) and stereoselectivity (>95%), may serve as a typical example. As far as the scope of the reaction is concerned with respect to other alkylaluminums, the reaction is restricted only to reactants without β-hydrogens; that is, the addition is restricted to the transfer of methyl groups. The regioselective conjugated addition of Me$_3$Al to androstane-1,4-diene-3,17-dione was also studied [49].

The conjugated addition of other organoaluminums was carried out also with aryl- and vinylaluminum compounds. In the case of the former, a successful transfer of the aryl group was achieved by Westermann et al. [50], and the use of this method utility was demonstrated in the synthesis of 1α-arylsteroids bearing variously substituted aryl moieties (Scheme 2.9). Examples of the conjugated addition of vinylaluminums are even more scarce. A study demonstrating the preparation, characterization, and reactivity of the divinylalane [(CH$_2$ = CH)$_2$Al(μ–OCH$_2$CH$_2$NMe$_2$)] showed that it could efficiently transfer the vinyl group in Ni-catalyzed conjugated addition to 1,3-diphenyl-2-propene-1-one (*trans*-chalcone) (Scheme 2.10).

Organoindium compounds, which are more stable and less prone to hydrolysis, are closely related to organoaluminums. Interestingly, there are very few reports of their use in conjugated additions; the such first report, which appeared in 1998, involved the conjugated additions of triorganoindiums (*n*-butyl, *t*-butyl, and phenyl) to vinylmethylketone, acrylate, and acrylonitrile catalyzed by a nickel complex [51]. More recently, Murakami et al. reported a rhodium complex-catalyzed conjugated

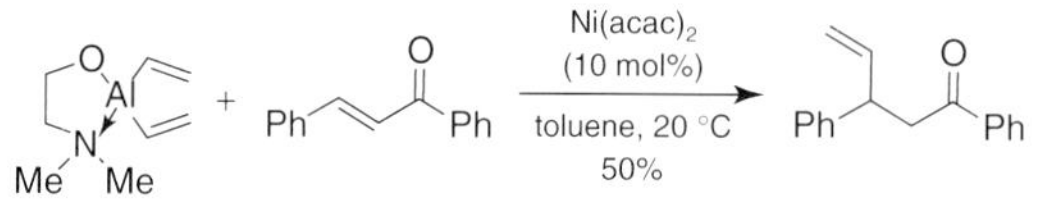

Scheme 2.10 Conjugated addition of vinylaluminum compounds to chalcone.

addition of diarylindium hydroxides to various α,β-unsaturated ketones, esters, and nitriles [52].

2.4
Ligands for Cu-Catalyzed Enantioselective Conjugated Additions

Copper-catalyzed conjugated addition has been intensively studied over past two decades, with various chiral ligands having been designed and tested in the enantioselective conjugated addition of organozinc or organoaluminum compounds. The group of chiral ligands comprised several classes, such as phosphoramidites, phosphines bearing amino acid (peptidic) moiety, phosphines, phosphites, NHC-compounds, and various other ligands (with mixed functionalities). With regards to the structure of the organometals, diorganylzincs and triorganylaluminum have normally been used.

2.4.1
Phosphoramidites

Phosphoramidite ligands constitute the largest group of chiral ligands used in enantioselective conjugated additions. Their popularity could be attributed to two main facts: (i) the modular nature of their preparation allows the rapid synthesis of large libraries of ligands (usually by the reaction of a chlorophosphite with an amine); and (ii) they exhibit exceptional selectivity and versatility with respect to a conjugated addition substrate and a nucleophile.

The phosphoramidite ligands may be classified into several subgroups, the first of which is constituted by those bearing chiral binaphthyl backbone, such as MonoPhos-type phosphoramidites **L1**, phosphoramidites bearing an amine moiety with one center of chirality **L2**, phosphoramidites bearing ferrocene rings **L3**, and phosphoramidites with a bis(tetrahydronaphthalene) moiety **L4** (Figure 2.2). A second group is based on phosphoramidites with a chiral **L5** or racemic biphenyl backbone **L6** (Figure 2.3), while the characteristic feature of the third group of ligands **L7** and **L8** is a chiral spirocyclic framework (Figure 2.4). The fourth group derives from Taddol **L9** (Figure 2.5) and phosphoramidites with a chiral 1,2-aminocyclohexane backbone **L10**.

The initial breakthrough in the enantioselective copper-catalyzed conjugated addition of organozincs to cyclic and linear enones was reported by Feringa *et al.* in 1996, and was achieved by using phosphoramidite ligands MonoPhos **L1a** [53]. As substrates for the conjugated addition of Et$_2$Zn, cyclohexenone and chalcone were tested and the corresponding products were obtained in 35% and 47% ee, respectively. Changing the methyl substituent for *iso*-propyl in the amine moiety (**L1b**) raised the enantioselectivity to 60% and 83% ee, respectively (Scheme 2.11).

The introduction of additional centers of chirality into the amine moiety, for example, by using (*R*,*R*)-bis(phenylethyl)amine, generated some **L2a** ligands capable of a higher asymmetric induction [54]. When the conjugated addition of

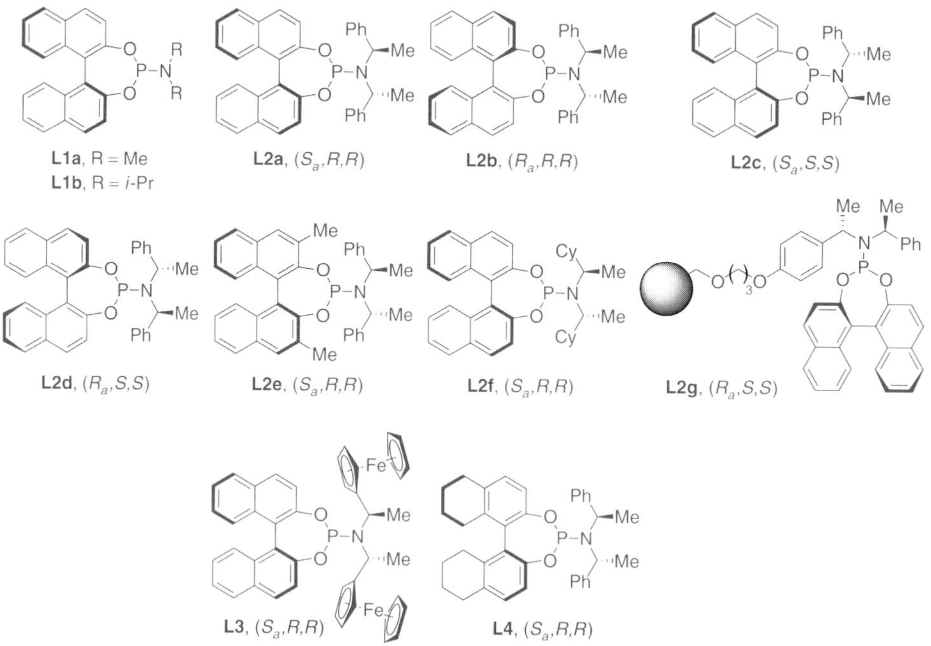

Figure 2.2 Ligands with chiral binaphthyl backbone.

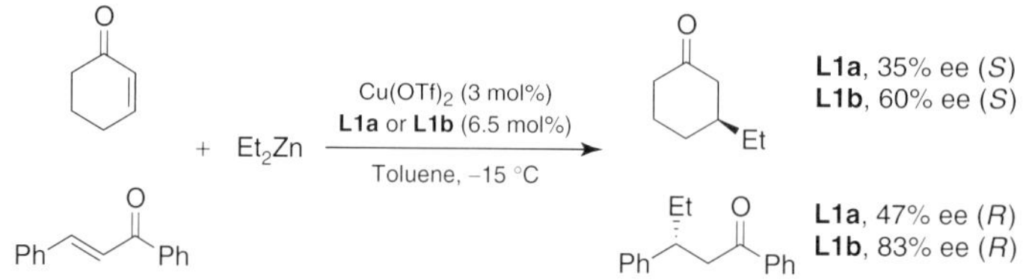

Scheme 2.11 Conjugated addition of Et$_2$Zn to cyclohexenone and chalcone in the presence of **L1a** and **L1b**.

various alkyl- and functionalized alkylzincs to cycloenones with five, six, and seven-membered rings was carried out, the highest enantioselectivity was observed for cyclohexenones with enantioselectivity in the range of 93–98% ee (Table 2.2). Interestingly, the addition to cyclopentenone or cycloheptenone proceeded with low selectivity of 10% and 53% ee, respectively. However, because of the interaction of the two centers of chirality in the amine moiety and the chiral binaphthyl backbone, mismatched and matched effects were observed. In this regard, the synthesis of a chiral phosphoramidite ligand bound to Merrifield resin **L2g**, and its application in the heterogeneous copper-catalyzed conjugated addition of Et$_2$Zn to cyclohexenone should be noted [55] as well. However, enantioselectivity was in the range of 65–84% ee, which was at least 10% lower compared to the use of **L2a** or its derivatives.

L5a, (S_a,R,R)

L5b, (S_a,R,R)

L6a

L6b

L6c

L6d

L6e

L6f

L6g

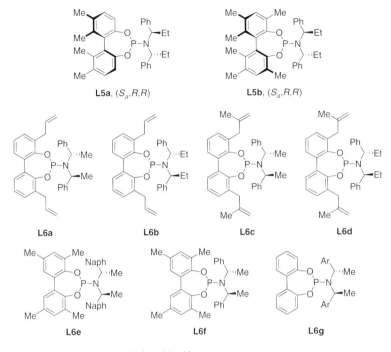

Figure 2.3 Ligands with biphenyl backbone.

L7

L8

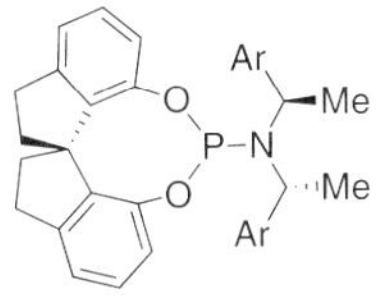

Figure 2.4 Ligands with spirocyclic backbone.

L9

L9a

L10

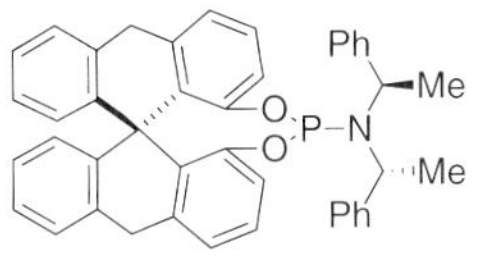

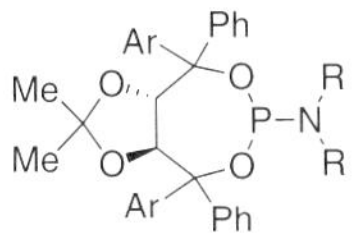

Figure 2.5 Taddol- and (R,R)-diaminocyclohexane-derived phosphoramidite ligand.

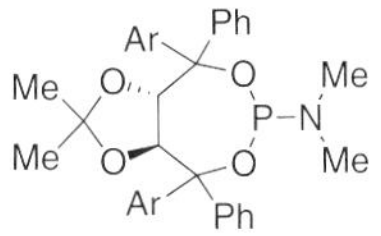

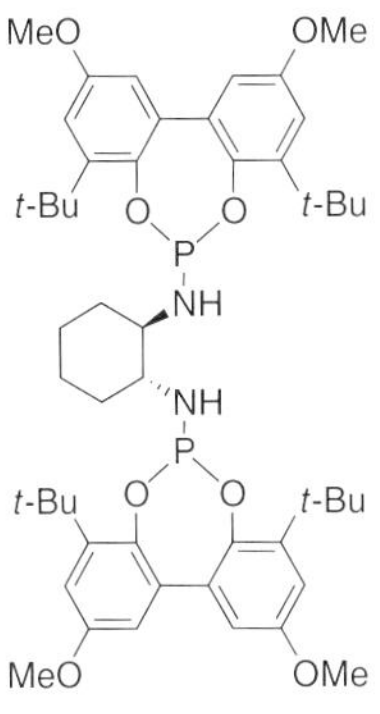

Table 2.2 Conjugated addition of organozincs to cycloenomes in the presence of **L2a** and **L3**.

Enone	R$_2$Zn	Ligand	Product	Yield (%)[a]	ee (%)	Reference
cyclopentenone	Et$_2$Zn	L2a	3-R-cyclopentanone	75	10 (R)	[54]
cyclohexenone	Et$_2$Zn	L3	3-R-cyclohexanone	91	32 (S)	[56]
	Et$_2$Zn	L2a		94	98 (S)	[54]
	Me$_2$Zn	L2a		72	98	[54]
	Hep$_2$Zn	L2a		95	95	[54]
	i-Pr$_2$Zn	L2a		95	94	[54]
	[Ph(CH$_2$)$_3$]$_2$Zn	L2a		53	95	[54]
	[AcOCH$_2$)$_5$]$_2$Zn	L2a		77	95	[54]
	[(EtO)$_2$CH(CH$_2$)$_3$]$_2$Zn	L2a		91	97	[54]
	[PivO(CH$_2$)$_6$]$_2$Zn	L2a		87	93	[54]
	Et$_2$Zn[b]	L3		100	96 (S)	[56]
4,4-dimethylcyclohexenone	Et$_2$Zn	L2a		74	>98	[54]
	Me$_2$Zn	L2a		68	>98	[54]
4,4-diphenylcyclohexenone	Et$_2$Zn	L2a		93	>98	[54]
cycloheptenone	Et$_2$Zn	L2a		83	53	[54]
	Et$_2$Zn	L3		96	98 (S)	[56]

[a] Isolated yields.
[b] 0 °C.

Recently, ferrocene-substituted phosphoramidite **L3** was also synthesized by Zheng et al. [56] and tested as a chiral ligand in the copper-catalyzed conjugated addition of Et$_2$Zn to cycloenones and nitroalkenes. In most cases, the enantioselectivity was >90% ee.

The chiral ligand **L2a** was also efficiently used to affect the kinetic resolution during the addition of Et$_2$Zn to racemic 5-substituted cyclohexenones [57]. The reaction proceeded at high enantioselectivity, with methyl, iso-propyl, phenyl, and trimethylsilyl derivatives (Table 2.3), but a somewhat lower enantioselectivity (83% ee) was observed for the kinetic resolution of 4-methylcyclohex-2-enones

Table 2.3 Kinetic resolution of 5-substituted cyclohexenones in the presence of **L2a**.

R¹ =	R₂Zn	Time (min)	Conversion (%)	ee (%)[a]
Me	Et₂Zn	20	53	99 (R)
	i-Pr₂Zn	60	55	84 (R)
	n-Bu₂Zn	30	54	99 (R)
	Me₂Zn	20	50	99 (R)
i-Pr	Et₂Zn	10	54	96
	n-Bu₂Zn	90	53	99
	Me₂Zn	25	48	85 (R)
Ph	Et₂Zn	–	55	89 (R)
Me₃Si	Et₂Zn	5	56	86
	n-Bu₂Zn	45	52	99
	Me₂Zn	15	50	99

[a] Configuration of the unconverted unsaturated ketone.

[57]. The enantioselective conjugated addition of organozincs in the presence of the same ligand to 6-substituted cyclohexenones to produce 2,5-substituted cyclohexanones was studied by Krause et al. [58]; the ensuing epimerization gave rise to enantiomerically and diastereoselectively pure products.

An interesting and synthetically attractive approach to the synthesis of enantiopure substituted cyclohexenones was proposed by Feringa et al. [59], where the underlying strategy was based on two consecutive conjugated additions using two enantiomerically opposite ligands. Thus, twofold enantioselective conjugated additions of Et₂Zn dienones in the presence of chiral ligands with different absolute configurations were carried out. Following the first enantioselective conjugated addition in the presence of **L2a**, a second addition was achieved either in the presence of the same ligand to yield the *trans*-substituted diastereoisomer, or in the presence of its enantiomer **L2b** to yield the *cis*-substituted diastereoisomer. In all the cases, the reaction proceeded with high stereoselectivity. Additions to 4,4-dimethoxycyclohexadienone (Scheme 2.12) and 4-methoxy-4-methylcyclohexadienone (Scheme 2.13) may serve as typical examples of this reaction.

The same ligand, **L2a**, was used in enantioselective conjugated addition to 5-(1-arylalkylidene) Meldrum's acids to create compounds with chiral quaternary carbon centers, with high enantioselectivity [60a]. The addition proceeded at high enantioselectivity (up to 95% ee) with a broad range of *para*-substituted substrates (Table 2.4), whereas in the case of *meta*-substituted substrates the enantioselectivity was between 78% and 85% ee. Similar results were obtained also for the addition

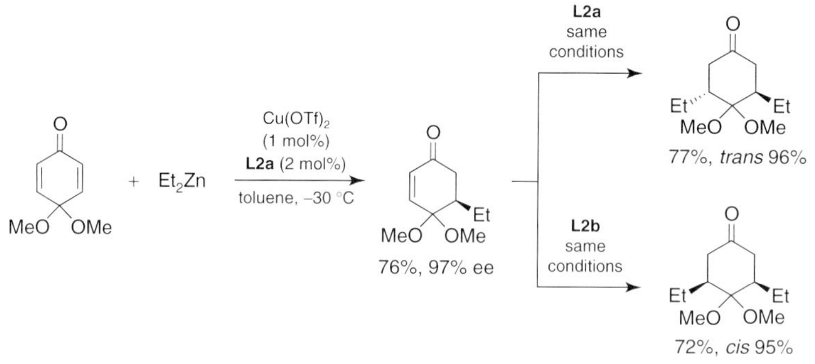

Scheme 2.12 Synthesis of enantiopure cyclohexanones.

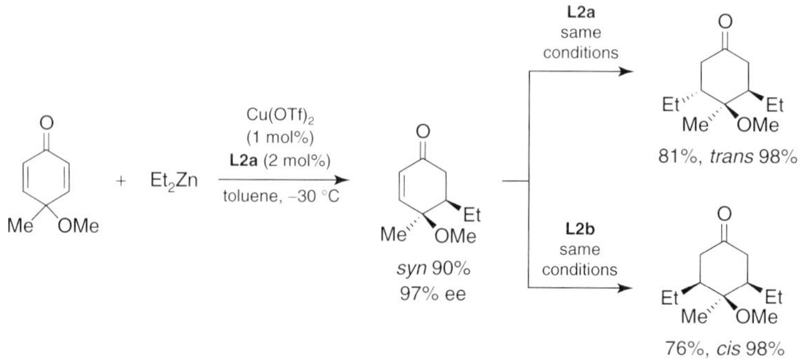

Scheme 2.13 Synthesis of enantiopure cyclohexanones.

of Me$_2$Zn to an aryl(ethyl)idene derivative [60b]. In a similar manner, the addition of Et$_2$Zn to alkylidene and arylidene Meldrum's acids was studied by Carreira et al. [61]. In this instance, when a chiral ligand with cyclohexyl rings **L2f** was used, the enantioselectivity for the alkylidene derivatives was >90% ee, but that for the arylidene derivatives did not exceed 88% ee (Scheme 2.14).

The enantioselective addition of Et$_2$Zn to unsaturated N-acyllactams (as masked derivatives of carboxylic acids) in the presence of **L2d** [62] is also worthy of mention. The size of the lactam ring has a dramatic effect on enantioselectivity, which was best when N-acylpyrrolidinones were used. As with the β-substituents, ee-values greater than 90% were observed in most cases (Table 2.5). For substituted cinnamic acid derivatives, the enantioselectivity was high regardless of the electron-donating or electron-accepting substituents at the *para* position. α,β-Unsaturated α'-oxyenones can be used in the presence of **L27** as another class of α,β-unsaturated carboxylic acid or aldehyde surrogates, and the conjugated addition of various dialkylzincs (Me, *i*-Pr, *n*-Bu) proceeded with up to 98% ee [63]. Likewise, the conjugated addition of Et$_2$Zn to α,β-unsaturated imines derived from α-amino acids in the presence of **L9a** was utilized [64].

Whilst the catalytic conjugated addition to α,β-unsaturated ketimines is a rather challenging process, a successful enantioselective copper-catalyzed conjugated

2.4 Ligands for Cu-Catalyzed Enantioselective Conjugated Additions

Table 2.4 Enantioselective conjugated addition to alkylidene Meldrum's acids.

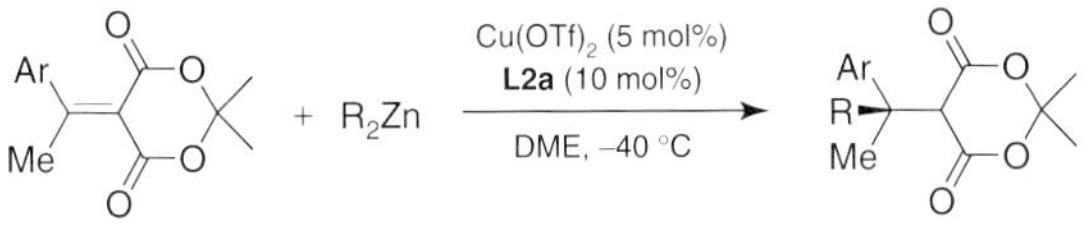

Ar =	R_2Zn	Yield (%)[a]	ee (%)	Ar	R_2Zn	Yield (%)[a]	ee (%)
Ph	Et_2Zn	95	84	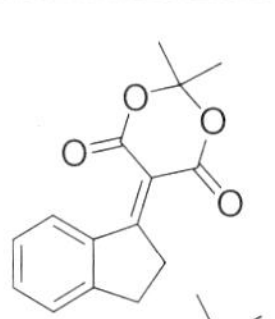	Et_2Zn	96	96
2-Naphthyl	Et_2Zn	66	95				
2-Furyl	Et_2Zn	97	91				
4-MeC_6H_4	Et_2Zn	82	89				
4-PhC_6H_4	Et_2Zn	76	95				
4-ClC_6H_4	Et_2Zn	88	95	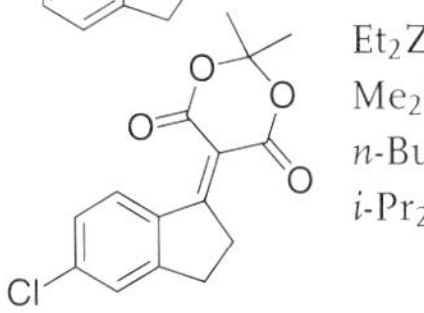	Et_2Zn	94	99
4-BrC_6H_4	Et_2Zn	84	92		Me_2Zn	98	99
4-FC_6H_4	Et_2Zn	83	92		$n\text{-Bu}_2Zn$	97	97
$4\text{-CF}_3C_6H_4$	Et_2Zn	87	92		$i\text{-Pr}_2Zn$	99	57
4-BnOC_6H_4	Et_2Zn	75	93				
4-ClC_6H_4	$i\text{-Pr}_2Zn$	99	65				
4-ClC_6H_4	$n\text{-Bu}_2Zn$	87	87				

[a]Isolated yields.

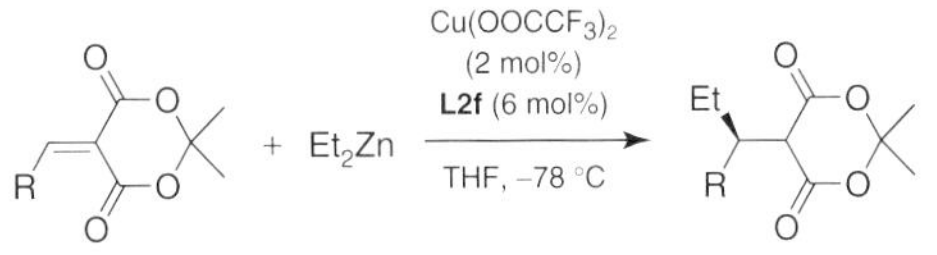

R =	Yield (%)	ee (%)
i-Pr	87	92 (R)
c-Hex	84	94 (R)
Ph	94	40 (R)
$p\text{-MeOC}_6H_4$	71	87 (R)
$p\text{-Me}_2NC_6H_4$	89	88
$p\text{-ClC}_6H_4$	66	44
$p\text{-BrC}_6H_4$	61	45
$p\text{-MeOC}_6H_4$	78	47
$p\text{-CF}_3OC_6H_4$	72	62
2-naphthyl	83	67
2-furanyl	64	80
2-thiophenyl	89	82

Scheme 2.14 Enantioselective conjugated addition of Et_2Zn to alkylidene Meldrum's acid derivatives in the presence of **L2f**.

addition of Me_2Zn was achieved to (2-pyridylsulfonyl)imines of chalcones [65]. The presence of a metal (copper)-coordinating 2-pyridylsulphonyl group was essential for this reaction to proceed, but once again the highest enantioselectivity (up to 80% ee) was achieved in the presence of **L2b** (Table 2.6).

The use of phosphoramidite with a tetrahydronaphthalene backbone **L4** in enantioselective addition did not show any high asymmetric induction [66]. When this route was tested by the addition of Et_2Zn to nitrostyrenes, the highest ee-value (37%) was obtained for 4-(trifluoromethyl)nitrostyrene (Table 2.7). The same reaction was also studied in the presence of phosphoramidites with a chiral

Table 2.5 Conjugated addition of alkylzincs N-acylpyrrolidinones in the presence of **L2d**.

R^1 =	R_2Zn	Yield (%)[a]	ee (%)
Me	Et_2Zn	75	87 (R)
	$i\text{-}Pr_2Zn$	78	60
Pr	Et_2Zn	80	84
	Me_3Al	70	36
i-Pr	Et_2Zn	75	95
Ph	Et_2Zn	78	98 (S)
	Bu_2Zn	62	85
	$i\text{-}Pr_2Zn$	75	81
$p\text{-}CF_3C_6H_4$	Et_2Zn	84	99
$p\text{-}MeOC_6H_4$	Et_2Zn	74	94
$m\text{-}CF_3C_6H_4$	Et_2Zn	55	94
$p\text{-}CF_3C_6H_4$	2	100	37
	1	100	77
(E)-MeCH=CH	1	16	96

[a] Isolated yields.

Table 2.6 Conjugated addition of alkylzincs to ketimines in the presence of **L2b**.

Ar^1 =	Ar^2 =	Yield (%)[a]	ee (%)
Ph	Ph	90	80
$p\text{-}MeOC_6H_4$	Ph	88	77
$p\text{-}ClC_6H_4$	Ph	91	70
2-Naph	Ph	90	71
Ph	$p\text{-}MeOC_6H_4$	72	76
Ph	$p\text{-}ClC_6H_4$	89	77
Ph	2-Naph l	86	74

[a] Isolated yields.

2.4 Ligands for Cu-Catalyzed Enantioselective Conjugated Additions

Table 2.7 Enantioselective conjugated addition of Et_2Zn to nitroalkenes in the presence of L4 and L5.

$$R\diagup\!\!\!\diagdown NO_2 + Et_2Zn \xrightarrow[\text{toluene}]{\text{Cu(OTf)}_2 \text{ (x mol\%)} \atop \text{L4 or L5 (2x mol\%)}} R\underset{*}{\diagup\!\!\!\diagdown}\overset{Et}{\!\!\!\diagup}NO_2$$

R =	Catalyst (mol%)	Ligand	Conversion (%)	ee (%)	Reference
p-MeC$_6$H$_4$	2	L4	100	35	[66]
	1	L5b	100	98	[67]
p-MeOC$_6$H$_4$	2	L4	100	27	[66]
	1	L5b	100	99	[67]
m-MeOC$_6$H$_4$	1	L5b	100	84	[67]
o-MeOC$_6$H$_4$	1	L5b	100	67	[67]
p-FC$_6$H$_4$	1	L5b	100	91	[67]
m-FC$_6$H$_4$	1	L5b	100	88	[67]
o-FC$_6$H$_4$	1	L5b	100	74	[67]
p-CF$_3$C$_6$H$_4$	2	L4	100	37	[66]
	1	L5b	100	77	[67]
o-CF$_3$C$_6$H$_4$	1	L5b	99	88	[67]
FurylC$_6$H$_4$	1	L5b	100	92	[67]
ThienylC$_6$H$_4$	1	L5b	100	96	[67]
(MeO)$_2$CH	1	L5a	100	97	[67]
	1	L5b	100	96	[67]
	2	L2a	100	90	[66]

biphenyl backbone **L5** [67], the best ligand proved to be **L5b** that, in most cases, resulted in an asymmetric induction of more than 90% ee.

A further modification was introduced by Alexakis et al., who utilized a combination of racemic or dynamic biphenyl moieties with chiral amines to synthesize ligands **L6**. The axial chirality of the biaryl moiety of the ligand is controlled by the chiral amine. The best results for the enantioselective conjugated addition of Et_2Zn to various cyclic or linear enones were obtained with ligands **L6a–L6d** [68a]. Likewise, the addition to nitroalkenes proceeded with good enantioselectivity, and a further improvement in the enantioselectivity of the addition reaction was observed in the case of cyclohexenone and cycloheptenone (>99 and 98% ee) when ligand **L6e** bearing a chiral naphthylethylamino group was used [68b]. Selected examples of this reaction are listed in Table 2.8.

Spirophosphoramidite ligands **L7** were tested in the CuX-catalyzed conjugated addition of Et_2Zn cyclohexenone and linear enones. The highest asymmetric induction (97% ee, Cu(OTf)$_2$, toluene, −20 °C) was obtained with **L7** bearing a chiral amine moiety, although when Cu(OAc)$_2$H$_2$O was used as the catalyst the enantioselectivity rose to 98% ee. The use of ligands having simple alkyl groups gave inferior results (34–70% ee), while the addition to linear α,β-unsaturated ketones proceeded with a rather mediocre selectivity (up to only 76% ee) [69].

Table 2.8 Conjugated addition of Et_2Zn to cycloenones in the presence of **L6**.

Enone	Ligand	Product	Yield (%)[a]	ee (%)	Reference
cyclohexenone	L6a	3-ethylcyclohexanone	99	96 (S)	[68a]
	L6b		99	96 (S)	[68a]
	L6c		99	89 (S)	[68a]
	L6d		99	87 (S)	[68a]
	L6e		99	99 (R)	[68b]
Me-CH=CH-C(O)-Me (with Me)	L6a	Me-CH(Et)-CH2-C(O)-Me (with Me)	99	88 (S)	[68a]
	L6b		80	84 (S)	[68a]
	L6c		99	95 (S)	[68a]
	L6d		99	92 (S)	[68a]
	L6e		99	41 (R)	[68b]
cycloheptenone	L6a	3-ethylcycloheptanone	97	73 (S)	[68a]
	L6b		99	81 (S)	[68a]
	L6c		99	63 (S)	[68a]
	L6d		99	42 (S)	[68a]
	L6e		99	98 (R)	[68b]
Ph-CH=CH-C(O)-Me	L6a	Ph-CH(Et)-CH2-C(O)-Me	99	93 (R)	[68a]
	L6b		30	80 (R)	[68a]
	L6c		71	89 (R)	[68a]
	L6d		16	51 (R)	[68a]
	L6e		99	82 (S)	[68b]
Ph-CH=CH-NO_2	L6a	Ph-CH(Et)-CH2-NO_2	99	92 (R)	[68a]
	L6b		99	95 (R)	[68a]
	L6c		99	90 (R)	[68a]
	L6d		99	95 (R)	[68a]
	L6e		99	72 (R)	[68b]

Reaction conditions: CuTC (2 mol%), **L6** (4 mol%), Et_2O, −30 °C.

[a] Isolated yields or conversions.

Considerably better results were obtained with the spirophosphoramidite ligand **L8**, as prepared by Zhang et al. [70], which gave a high enantioselectivity after the addition of Et_2Zn not only to cyclohexenone (99% ee), but also to cycloheptenone (98% ee). In contrast, the reaction with cyclopentenone gave the product in low yield and at low enantioselectivity (66%) (Scheme 2.15). The use of simple alkyl substituents on the nitrogen atom of the ligand resulted in a considerably reduced enantioselectivity (up to only 57% ee).

Taddol-based phosphoramidite ligands **L9** were prepared by Alexakis et al. [71]. However, their use in the conjugated addition to cyclic and linear enones gave products with only low asymmetric induction (up to 49% ee). Another class of

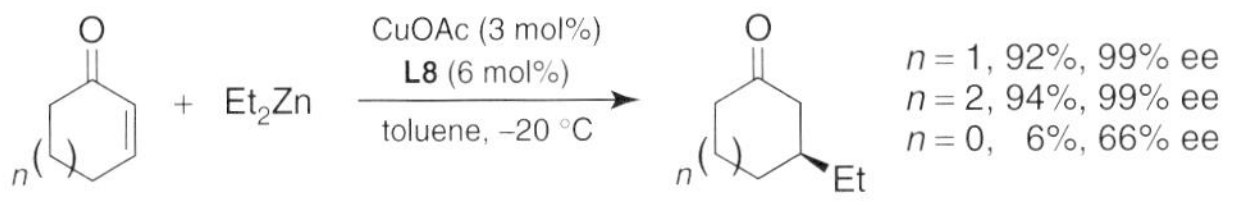

2.4 Ligands for Cu-Catalyzed Enantioselective Conjugated Additions

[Scheme 2.15 reaction: cycloenone + Et₂Zn → CuOAc (3 mol%), L8 (6 mol%), toluene, −20 °C → product with Et group]

n = 1, 92%, 99% ee
n = 2, 94%, 99% ee
n = 0, 6%, 66% ee

Scheme 2.15 Enantioselective conjugated addition to cycloenones in the presence of ligand **L8**.

phosphoramidites was based on the (*R*,*R*)-1,2-diaminocyclohexane framework **L10** [72]. In this case, the highest ee-value (70%) was obtained for the addition of Et₂Zn to cyclohexenone in the presence of Cu(OAc)₂•H₂O.

The copper-catalyzed enantioselective conjugated addition of organoaluminums has also been recently studied and, once again, the use of chiral phosphoramidite ligands proved to be the essential factor to achieve high enantioselectivity. One of the first reported examples was the enantioselective addition of tri-alkylaluminums to α,β-unsaturated lactams by Pineschi *et al.* (Scheme 2.16) [73]. However, the reaction proceeded with a mediocre enantioselectivity of 68%. Later, Alexakis *et al.* showed that a mixture of **L2d** with CuTC was a good catalytic system for the conjugated addition of Me₃Al and Et₃Al to linear and cyclic enones with high enantioselectivity (86–98%) (Table 2.9) [74]. This method was also used to carry out the first enantioselective conjugated addition of a trisubstituted alkenylalane, prepared by the Negishi carbometallation of phenylacetylene, to a cyclic enone with a good enantioselectivity of 77% (Scheme 2.17).

Owing to the enhanced Lewis acidity of aluminum (and its compounds) in comparison to zinc, the reaction also proceeded well with trisubstituted cyclic enones [75a]. In this case, the phosphoramidite ligands **L6f** and **L6e** were used, and the protocol allowed the synthesis of products with quaternary stereogenic centers with excellent enantioselectivities and yields (Table 2.9). This method was later expanded to a wide range of various trisubstituted cyclic enones [75b,c]. The addition of an alkenylalane, prepared by the hydroalumination of a terminal alkyne, to cyclohexenone was also attempted. The reaction required a high catalyst load and proceeded in a reasonable yield, but with a moderate enantioselectivity of 73% ee (Scheme 2.18) [75b].

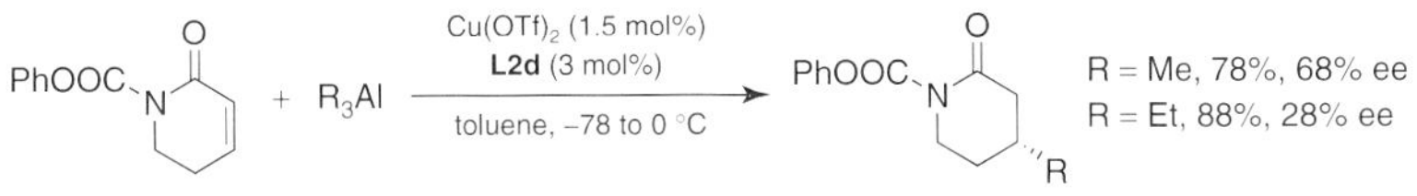

[Scheme 2.16 reaction: PhOOC-N lactam + R₃Al → Cu(OTf)₂ (1.5 mol%), L2d (3 mol%), toluene, −78 to 0 °C → product]

R = Me, 78%, 68% ee
R = Et, 88%, 28% ee

Scheme 2.16 Enantioselective conjugated addition of alkylaluminums to α,β-unsaturated lactams.

[Scheme 2.17 reaction: cycloenone + Ph/Me-AlMe₂ alkene → CuTC (1 mol%), L2d (4 mol%), Et₂O, −30 °C → product with Me and Ph groups]

n = 1, 54%, 77% ee
n = 2, 48%, 72% ee

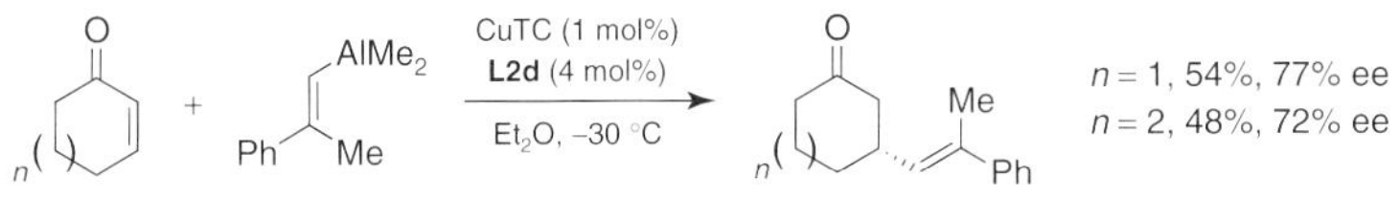

Scheme 2.17 Enantioselective conjugated addition of alkenylalane to cycloenones.

Table 2.9 Enantioselective conjugated addition of trialkylaluminums to trisubstituted cycloenones.

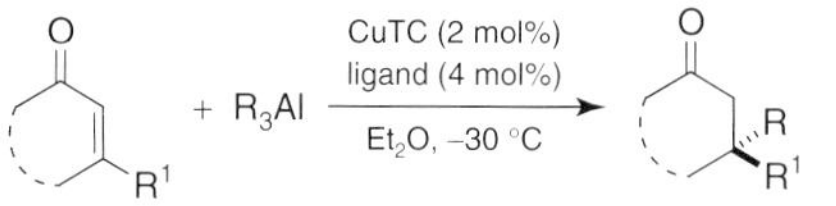

Enone	R^1 =	R$_3$Al	Ligand	Product	Yield (%)	ee (%)	Reference
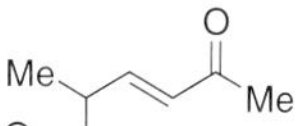 (acyclic Me-enone)	—	Et$_3$Al	L2b	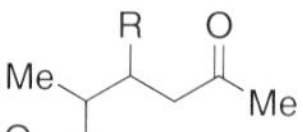	80	90 (R)	[74]
		Me$_3$Al	L2b		88	96 (R)	[74]
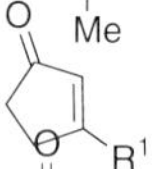 (furanone)	Me	Et$_3$Al	L6e	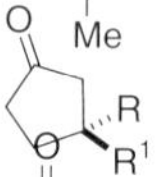	95	80 (R)	[75b]
	Me	Et$_3$Al	L6e		95	93 (R)^a	[75c]
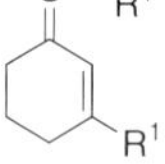 (cyclohexenone)	H	Et$_3$Al	L2b	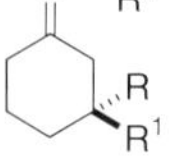	30	95 (R)	[74]
	H	Me$_3$Al	L2b		80	90 (R)	[74]
	Me	Et$_3$Al	L6f		77	94 (R)	[75a]
	Me	Et$_3$Al	L6e		95	97 (R)	[75a]
	Et	Me$_3$Al	L6f		95	94 (S)	[75a]
	Et	Me$_3$Al	L6e		84	96 (S)	[75a]
	i-Bu	Me$_3$Al	L6e		42	93 (R)	[75a]
	(CH$_2$)$_2$CH=CH$_2$	Me$_3$Al	L6e		95	95 (R)	[75a]
	(CH$_2$)$_3$CH=CH$_2$	Me$_3$Al	L6e		95	95 (R)	[75a]
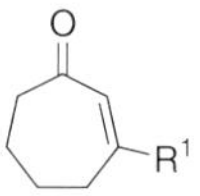 (cycloheptenone)	H	Et$_3$Al	L2b	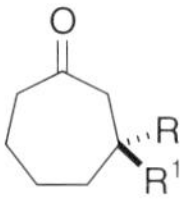	85	98 (R)	[74]
	H	Me$_3$Al	L2b		84	92 (R)	[74]
	Me	Et$_3$Al	L2d		58	95 (R)	[75b]

^{a} An inverse addition of reactants was applied.

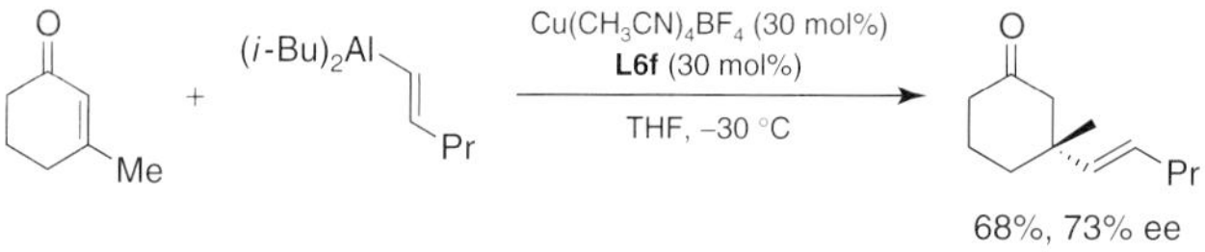

Scheme 2.18 Enantioselective conjugated addition of alkenylalane to 3-methylcyclohexenone.

A somewhat lower enantioselectivity was obtained by the use of phosphoramidite ligands with a more flexible biphenyl backbone [76]. The addition of Et$_3$Al to 3-methylcyclohexenone proceeded with up to 93% ee for **L6g**. Good enantioselectivity values (up to 93%) were also obtained for the addition of Me$_3$Al to nitroalkenes [77].

Recently, it was reported by Feringa that catalytic enantioselective conjugated additions can be carried out also to formyl imines (a hetero-α,β-unsaturated system) [78], with the corresponding formyl imines being generated *in situ* from aromatic and aliphatic α-amidosulfones. In general, the addition of alkylzincs or

2.4 Ligands for Cu-Catalyzed Enantioselective Conjugated Additions

Table 2.10 Conjugated addition of alkylzincs to ketimines in the presence of **L2c**.

R¹	RₙM	Yield (%)ᵃ	ee (%)
Ph	Et₂Zn	99	96 (R)
	i-Pr₂Zn	97	91 (R)
	n-Bu₂Zn	88	88 (R)
	Et₃Al	70	86 (S)
p-BrC₆H₄	Et₂Zn	94	99 (R)
p-MeOC₆H₄	Et₂Zn	99	97 (R)
p-MeC₆H₄	Et₂Zn	90	96 (R)
m-MeOC₆H₄	Et₂Zn	96	95
o-MeOC₆H₄	Et₂Zn	99	47
2-Naphthyl	Et₂Zn	94	80
PhCH₂CH₂	Et₂Zn	81	66
c-Hexyl	Et₂Zn	99	45
n-Hexyl	Et₂Zn	99	70

ᵃIsolated yields.

alkylaluminums proceeded with a high enantioselectivity of up to 99% ee in the presence of **L2c** (Table 2.10).

2.4.2
Phosphines Bearing an Amino Acid Moiety

The use of amino acid-based phosphine ligands is associated mainly with Hoveyda *et al.* who synthesized a library of bidentate ligands that were used in the conjugated addition of alkylzincs to various unsaturated compounds (Figure 2.6). The reaction in the presence of a catalytic amount of Cu(OTf)₂·C₆H₆ and dipeptidic ligand **L11** with various five-, six-, and seven-membered cycloenones proceeded with high ee-values (>98%), regardless of the ring size [79]. On the other hand, a similar level of the asymmetric induction (98% ee) for the conjugated addition to trisubstituted cyclic enones was obtained with the ligand **L12** [80]. Excellent asymmetric induction (95% ee) was also achieved in the conjugated addition to aliphatic enones with the modified dipeptidic ligand **L13** [81]. Some typical examples are shown in Table 2.11. Peptidophosphines found application as chiral ligands in the synthesis of several natural products; for example, peptidophosphine **L11** was used in the synthesis of clavularin B [79], and the ligand **L13** as the chiral ligand for two separate asymmetric conjugated additions of Me₂Zn to alicyclic enones during the synthesis of erogorgiaene [82].

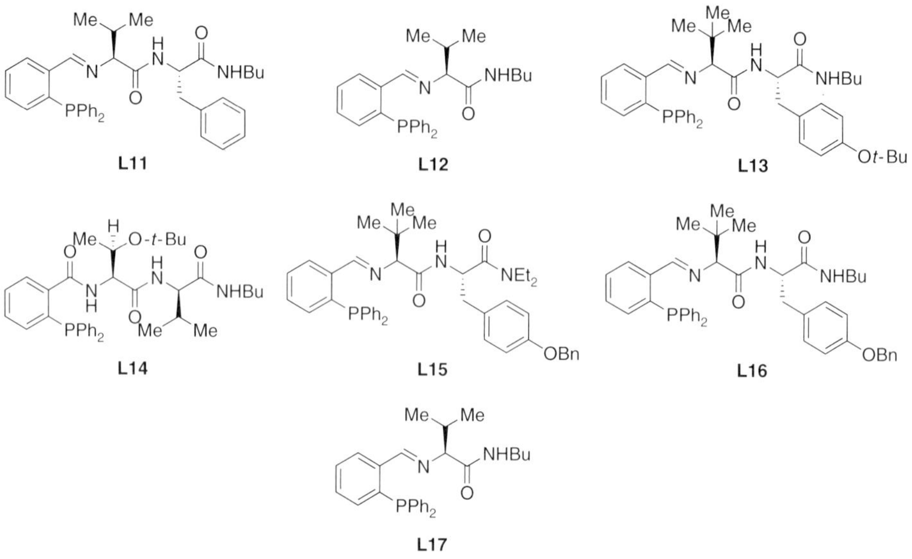

Figure 2.6 Hoveyda's amino acid-based phosphine ligands.

Similar results were also obtained in the addition of organozincs to N-acyloxazolidines, yielding products with an enantioselectivity of up to 98% in the presence of ligand **L14** (Scheme 2.19) [83]. Interestingly, the addition did not take place in the case of phenyl-substituted N-acyloxazolidines (R^1 = Ph).

Other substrates for conjugated addition are the nitroalkenes. In this case, the nitro group constitutes a synthetically interesting substituent, which can easily be transformed into other functional groups. In this regard, chiral peptidophosphines have also proved to be good ligands for enantioselective addition to nitroalkenes; among several chiral ligands tested, the best asymmetric induction was achieved with **L15**. The catalytic addition of various organozincs to aliphatic nitroalkenes proceeded with excellent enantioselectivity (up to 95% ee) and reasonable yields (Scheme 2.20) [84a]. However, the level of asymmetric induction depended on the

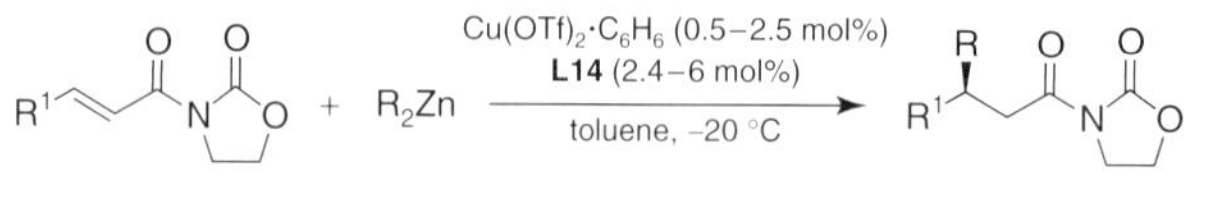

		Yield (%)	ee (%)
R^1 = Me	R = Et	95	95
R^1 = Me	R = i-Pr(CH$_2$)$_2$	61	93
R^1 = Me	R = i-Pr	95	76
R^1 = n-Pr	R = Et	86	94
R^1 = n-Pr	R = i-Pr(CH$_2$)$_2$	89	95
R^1 = n-Pr	R = Me	68	97
R^1 = (CH$_2$)$_3$OTBS	R = Et	95	98
R^1 = i-Pr	R = Et	88	92

Scheme 2.19 Conjugated addition to N-acyloxazolidines in the presence of **L14**.

2.4 Ligands for Cu-Catalyzed Enantioselective Conjugated Additions | 95

Table 2.11 Conjugated addition to enones in the presence of amino acid based phosphines.

Enone	R$_2$Zn	Ligand	Product	Yield (%)[a]	ee (%)
cyclopentenone	Me$_2$Zn	L11		78	97
	Bu$_2$Zn			92	98
	i-Pr$_2$Zn			90	79
	[AcOCH$_2$)$_4$]$_2$Zn			56	98
4,4-dimethylcyclopentenone	Et$_2$Zn	L11		56	97
cyclohexenone	Me$_2$Zn	L11		71	98
	Et$_2$Zn			98	98
	Bu$_2$Zn			93	95
	i-Pr$_2$Zn			98	72
	[AcOCH$_2$)$_4$]$_2$Zn			76	95
cycloheptenone	Me$_2$Zn	L11		80	98
	Et$_2$Zn			98	98
	Bu$_2$Zn			81	95
	i-Pr$_2$Zn			78	62
1-acetylcyclopentene	Me$_2$Zn	L12		42	97[b]
	Et$_2$Zn			66	97[c]
	i-Pr$_2$Zn			69	98[c]
	[AcOCH$_2$)$_4$]$_2$Zn			47	94[c]
1-acetylindene	Et$_2$Zn	L12		86	97[b]
1-acetylcycloheptene	Et$_2$Zn	L12		85	97[d]
Ph-CH=CH-C(O)Me	Et$_2$Zn	L13		90	93
n-Pent-CH=CH-C(O)Me	Me$_2$Zn	L13		71	94
	Et$_2$Zn			85	95

Table 2.11 (Continued)

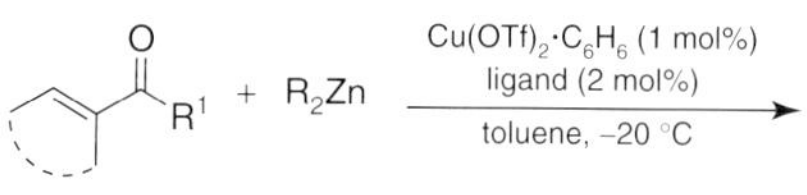

Enone	R$_2$Zn	Ligand	Product	Yield (%)[a]	ee (%)
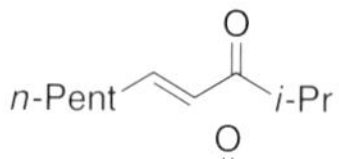 n-Pent ~~~ C(O)-i-Pr	Et$_2$Zn	L13	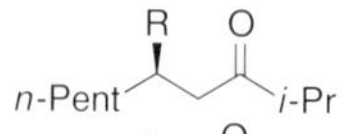 n-Pent-CH(R)-CH$_2$-C(O)-i-Pr	75	90
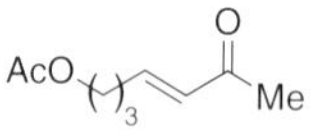 AcO(CH$_2$)$_3$-CH=CH-C(O)Me	Et$_2$Zn	L13	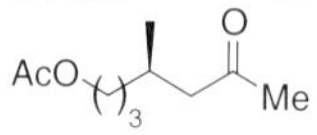 AcO(CH$_2$)$_3$-CH(R)-CH$_2$-C(O)Me	88	89

Reaction conditions: Cu(OTf)$_2$·C$_6$H$_6$ (1 mol%), ligand (2 mol%), toluene, −20 °C.

[a] Isolated yields.
[b] 16/1 anti/syn ratio after treatment with a base.
[c] >25/1 anti/syn ratio after treatment with a base.
[d] 4/1 anti/syn ratio after treatment with a base.

R^1−CH=CH−NO$_2$ + R$_2$Zn $\xrightarrow[\text{Toluene, 22 °C}]{\text{Cu(OTf)}_2\cdot\text{C}_6\text{H}_6\text{ (1 mol\%)}\atop\text{L15 (2 mol\%)}}$ R^1−CH(R)−CH$_2$−NO$_2$

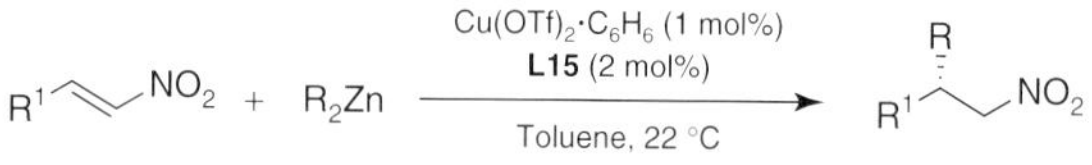

		Yield (%)	ee (%)
R^1 = Ph	R = Et	79	95
R^1 = Ph	R = Me	78	92
R^1 = Ph	R = Me$_2$CH(CH$_2$)$_3$	70	95
R^1 = Ph	R = AcO(CH$_2$)$_4$	60	89
R^1 = 4-MeOC$_6$H$_4$	R = Et	72	95
R^1 = 4-ClC$_6$H$_4$	R = Me	70	94
R^1 = 4-CF$_3$C$_6$H$_4$	R = Me	70	95
R^1 = 2-MeC$_6$H$_4$	R = Et	65	94
R^1 = 2-BrC$_6$H$_4$	R = Et	68	81
R^1 = 2-furyl	R = Et	78	95
R^1 = n-C$_5$H$_7$	R = Et	72	93
R^1 = c-C$_6$H$_{11}$	R = Et	52	95
R^1 = (MeO)$_2$CH	R = Et	64	84

Scheme 2.20 Conjugated addition to nitroalkenes in the presence of **L15**.

substituent attached to the double bond, as well as on the structure of the transferred group from the organozinc compound. Similar results were obtained also for the addition to cyclic nitroalkenes with the related ligand **L16**. The addition proceeded to give the 2-substituted nitrocycloalkanes with high enantioselectivity (up to 95% ee) and good *syn*-diastereoselectivity (~6 : 1) [84b]. The *syn*-stereoisomers could easily be converted into *anti*-stereoisomers by treatment with DBU. Especially attractive was the conjugated addition to trisubstituted nitroalkenes, resulting in the formation of nitroalkanes bearing chiral quaternary carbon centers [85] (Scheme 2.21). The corresponding nitroalkanes were obtained with excellent enantioselectivities (up

2.4 Ligands for Cu-Catalyzed Enantioselective Conjugated Additions

Scheme 2.21 Conjugated addition to trisubstituted nitroalkenes in the presence of **L15**.

		Yield (%)	ee (%)
R^1 = Ph, R^2 = Me	R = Bu	55	95
R^1 = 4-ClC$_6$H$_4$, R^2 = Me	R = Et	76	98
R^1 = 2-naphthyl, R^2 = Me	R = Et	79	95
R^1 = Ph, R^2 = n-Pr	R = Et	40	75
R^1 = Ph, R^2 = i-Pr	R = Me	85	73

to 98% ee). Nonetheless, the asymmetric induction depended on the substituents attached to the double bond, and in this case the best results were also obtained in the presence of ligand **L15**. Oxidation of the obtained nitroalkanes furnished chiral carboxylic acids.

A simple valine-based ligand **L17** proved to be the ligand of choice for the conjugated additions of dialkyl- and diarylzincs to β-silyl-α,β-unsaturated ketones [86]. Whilst the addition of alkylzincs proceeded with high enantioselectivity (up to 96% ee) in toluene, the addition of arylzincs had to be carried out in dimethoxyethane (DME) so as to achieve a similar level of asymmetric induction (up to 94% ee). The silylketones formed were transformed into substituted chiral allylsilanes, demonstrating the synthetic versatility of this reaction, some representatives of which are listed in Table 2.12.

Table 2.12 Enantioselective conjugated addition to β-silyl-α,β-unsaturated ketones in the presence of **L17**.

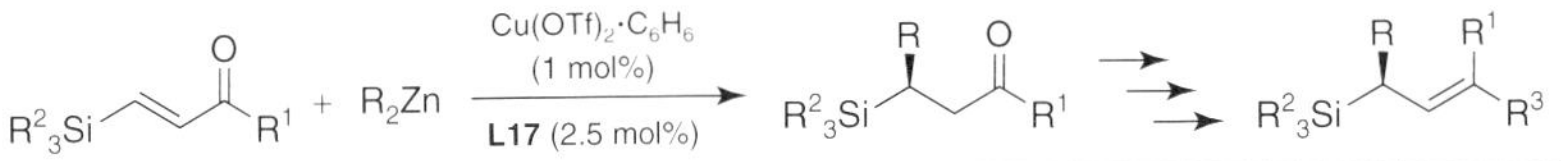

Enone	R$_2$Zn	Conditions	Product	Yield (%)[a]	ee (%)
Me$_3$Si—CH=CH—C(O)—Me	Me$_2$Zn	Toluene, 22 °C	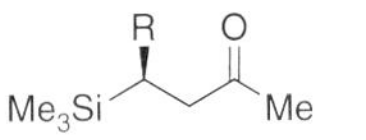 Me$_3$Si—CH(R)—CH$_2$—C(O)—Me	77	96
	Et$_2$Zn			59	95
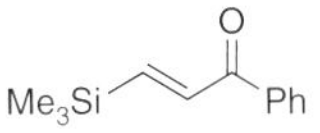 Me$_3$Si—CH=CH—C(O)—Ph	Me$_2$Zn	Toluene, 22 °C	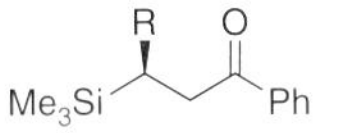 Me$_3$Si—CH(R)—CH$_2$—C(O)—Ph	91	89
	Et$_2$Zn			77	93
Me$_2$PhSi—CH=CH—C(O)—Me	Ph	DME, 0 °C	Me$_2$PhSi—CH(R)—CH$_2$—C(O)—Me	79	94
	4-(MeOC$_6$H$_4$)$_2$Zn			82	87
	4-(CF$_3$C$_6$H$_4$)$_2$Zn			66	87

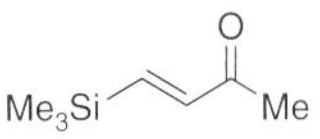

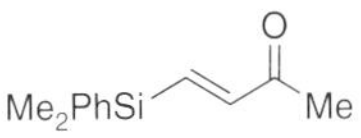

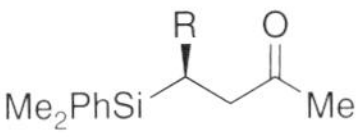

[a] Isolated yields.

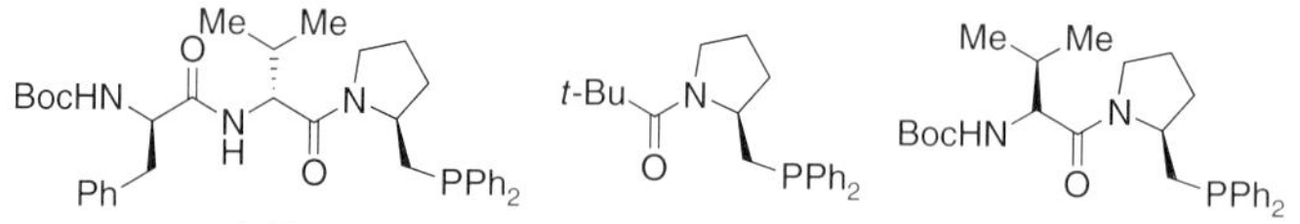

Figure 2.7 Peptidophosphines prepared by Tomioka.

A different set of amino acid-based phosphines was developed by Tomioka et al. (Figure 2.7), where the conjugated addition of organozincs in the presence of Cu(MeCN)$_4$BF$_4$ and the chiral peptidophosphine ligand **L18** to cyclic enones proceeded with high enantioselectivities (up to 98% ee) under mild reaction conditions [87a]. Various dialkylzincs were usually added to a substrate with high asymmetric induction (Table 2.13). Interestingly, in the case of Me$_2$Zn a sharp drop to 45% ee was observed. The same ligand was used in the kinetic resolution of racemic 5-substituted cyclohexenones by means of

Table 2.13 Enantioselective conjugated addition of alkylzincs to β-silyl-α,β-unsaturated ketones in the presence of **L18**.

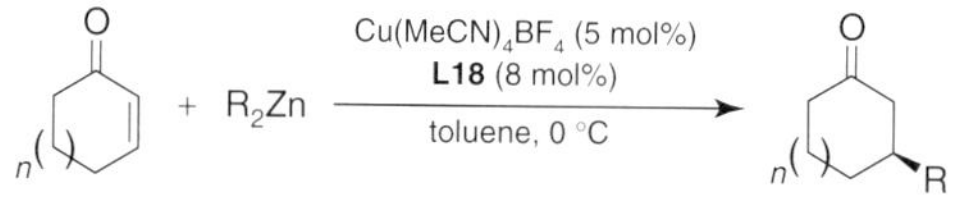

Enone	R$_2$Zn	Product	Yield (%)[a]	ee (%)
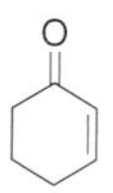 cyclohexenone	Et$_2$Zn	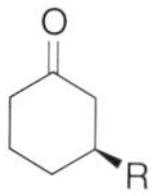	87	98
	i-Pr$_2$Zn		97	94
	[Ph(CH$_2$)$_3$]$_2$Zn		48	94
	[AcO(CH$_2$)$_5$]$_2$Zn		84	75
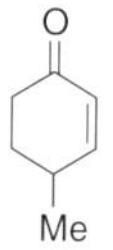 4-Me-cyclohexenone	Et$_2$Zn	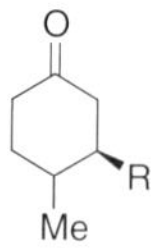	66	75
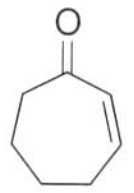 cycloheptenone	Et$_2$Zn	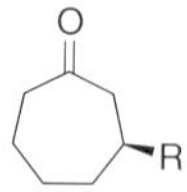	78	86
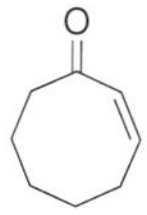 cyclooctenone	Et$_2$Zn	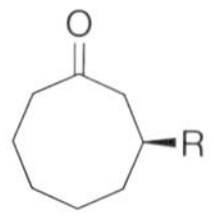	83	61

[a] Isolated yields.

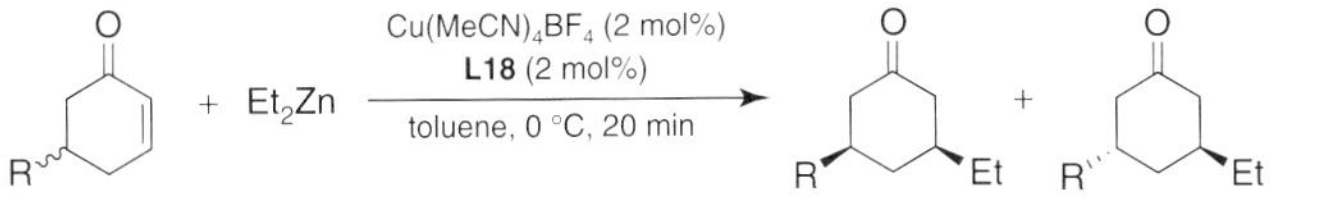

Scheme 2.22 Kinetic resolution of racemic 5-substituted cyclohexenones.

R	Yield (%)	ee (%)	Yield (%)	ee (%)
R = TMS	38	97	57	81
R = Me	38	98	58	87
R = Ph	41	88	53	87
R = 2-naphthyl	33	95	64	90

an enantioselective conjugated addition [87b]. The reaction, which was carried out at 0 °C for 20 min, furnished *trans*-3-alkylated-5-substituted cyclohexanones (*trans* selectivity 87–99%) with 81–90% ee, along with the recovered enantioenriched starting 5-substituted cyclohexenones with 88–98% ee (Scheme 2.22). The amidophosphine **L19** was studied as a chiral ligand in the conjugated addition of dialkylzincs to nitroalkenes [88]. The addition to aliphatic nitroalkenes yielded products with moderate enantioselectivity (54–68% ee), and a similar picture was observed for the addition to nitrocycloalkenes (62–80% ee). Interesting results were achieved in the catalytic conjugated addition of Et$_2$Zn to α,β-unsaturated N-2,4,6-triisopropylphenylsulfonylimines in the presence of amidophosphine **L20**, with the reaction proceeding from moderate to good enantioselectivity (up to 91%) and good yields [89] (Scheme 2.23). In this case, the corresponding chiral alcohols were obtained after hydrolysis and reduction of the obtained sulfonylimines.

A different set of peptidic phosphines based on valinol and prolinol was introduced by Kočovský *et al.* [90]. The ability of these materials to affect the enantioselective conjugated addition was tested by the reaction of Et$_2$Zn to six- and

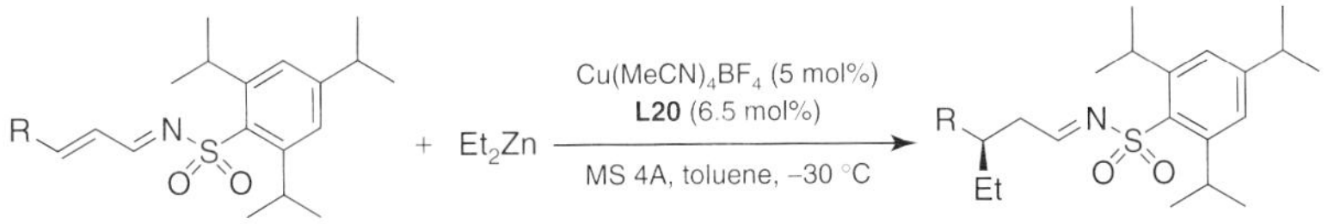

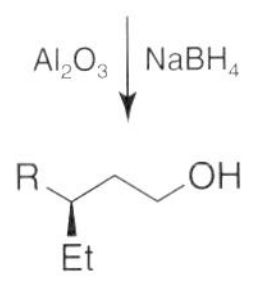

R	Yield (%)	ee (%)
R = Ph	80	80
R = 2-MeC$_6$H$_4$	84	83
R = 3-MeC$_6$H$_4$	80	81
R = 2-MeOC$_6$H$_4$	75	75
R = 4-MeOC$_6$H$_4$	78	79
R = 2-ClC$_6$H$_4$	79	82
R = 4-ClC$_6$H$_4$	78	78
R = 1-naphthyl	82	91
R = 2-naphthyl	85	75
R = 2-furyl	72	67

Scheme 2.23 Enantioselective conjugated addition of Et$_2$Zn to α,β-unsaturated N-2,4,6-triisopropylphenylsulfonylimines in the presence of **L20**.

		Yield (%)	ee (%)
n = 1	L21	92	83
n = 2	L21	90	81
n = 1	L22	86	86
n = 2	L22	81	87

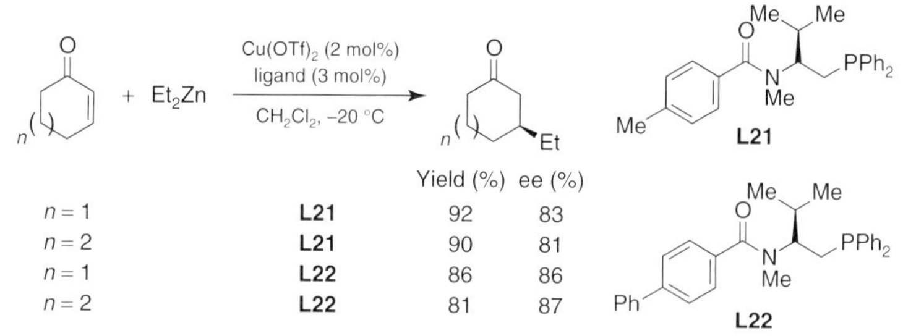

Scheme 2.24 Enantioselective conjugated addition of Et_2Zn to cyclic enones in the presence of **L21** and **L22**.

seven-membered ring cycloenones. The highest enantioselectivities (up to 87% ee) were obtained with ligands **L21** and **L22** (Scheme 2.24).

2.4.3
Phosphines

In the past, phosphines have been used only rarely for the enantioselective copper-catalyzed conjugated addition of organozincs, albeit with high enantioselectivity (Figure 2.8). One such example is the addition of Et_2Zn to chalcone in the presence of *N,P*-ferrocenyl ligands having central and planar chirality **L23** and **L24** [91]. The reaction with chalcone proceeded only with a modest enantioselectivity of 41% ee, whereas the addition to *ortho*-substituted chalcones resulted in a dramatic improvement in asymmetric induction. Thus, the highest enantioselectivities (up to 92% ee) were obtained with chalcones bearing an *ortho*-halophenyl group attached to the double bond and phenyl, or ferrocenyl, group to the carbonyl group (Scheme 2.25).

One excellent ligand proved to be aminophosphine **L25**, which was easily prepared from the corresponding amino alcohol by Leighton *et al.* [92]. The

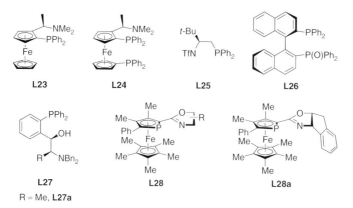

Figure 2.8 Chiral phosphine-based ligands for Cu-catalyzed conjugated addition.

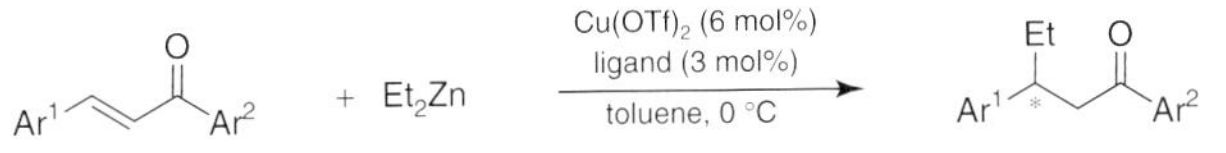

2.4 Ligands for Cu-Catalyzed Enantioselective Conjugated Additions

Ar¹	Ar²	Ligand	Yield (%)	ee (%)
Ar¹ = Ph	Ar² = Ph	L23	60	41
Ar¹ = Ph	Ar² = Ph	L24	30	15
Ar¹ = 2-ClC$_6$H$_4$	Ar² = Ph	L23	95	91
Ar¹ = 2-ClC$_6$H$_4$	Ar² = 4-ClC$_6$H$_4$	L23	96	85
Ar¹ = 2-ClC$_6$H$_4$	Ar² = 4-MeOC$_6$H$_4$	L23	90	81
Ar¹ = 2-MeOC$_6$H$_4$	Ar² = Fc	L23	89	92
Ar¹ = 2-BrC$_6$H$_4$	Ar² = Fc	L23	80	91

Scheme 2.25 Enantioselective conjugated addition of Et$_2$Zn in the presence of **L23** and **L24**.

conjugated addition of Et$_2$Zn to cyclohexenone proceeded with up to 97% ee, and similarly high enantioselectivities (usually >90% ee) were observed for additions to other cycloenones and also for the additions of other organozincs. It should also be noted that the addition of Et$_2$Zn to cyclohexenone was run on a large scale (10 g) with a catalyst loading as low as 0.1 mol% (0.25 mol% of the ligand), without any decrease in either enantioselectivity or yield (97% ee, 86% isolated yield). Some typical examples of this are listed in Table 2.14.

A reasonably high enantioselectivity was also observed for 2-diphenylphosphinyl-2′-diphenylphosphino-1,1′-binaphthalene **L26** [93]. The conjugated addition of Me$_3$Al to cyclohexenone and cycloheptenone proceeded with 89% ee, while that of Et$_3$Al proceeded with somewhat lower asymmetric induction, of 86% and 75% ee (Table 2.14).

Another example is a triphenylphosphine with the amino alcohol functionality in the side chain **L27** [94]. Here, the best results were obtained with **L27a**: the addition of Et$_2$Zn or Me$_2$Zn to acyclic α,β-unsaturated ketones proceeded with very high enantioselectivity of 98% ee, in most cases, as well as in the case of trisubstituted ketones and α,β-unsaturated amides. The only exception was the addition to cyclohexenone (82% ee). Among phosphine-based ligands, phosphaferrocene-oxazoline ligands **L28** [95] may also be included, with the asymmetric induction in the Cu(OTf)$_2$-catalyzed addition of Et$_2$Zn to various chalcones at up to 91% ee in the presence of **L28a**. It should be also noted that Carreira et al. [10] have utilized chiral phosphines as a ligand for the Cu-catalyzed enantioselective addition of terminal alkynes to Meldrum's acid derivatives.

Recently, conjugated additions of heteroarylzinc chlorides to cyclopentenones, catalyzed by rhodium chiral phosphine complexes (BINAP, DIOP, etc.), were attempted [96]. The highest asymmetric induction observed was as low as 18%, although in most cases racemic products were obtained.

2.4.4
Phosphites

One of the earliest examples of enantioselective conjugated additions catalyzed by chiral phosphate–copper complexes was reported by Alexakis et al. in 1997. In this

Table 2.14 Enantioselective conjugated additions in the presence of **L25**, **L26** and **L27a**.

Enone	R_nM	Catalysts	mol (%)	Ligand	mol (%)	Conditions	Product	Yield (%)[a]	ee (%)	Reference
Ph–CH=CH–C(O)Me	Et_2Zn	$Cu(OTf)_2$	3	**L27a**	3.6	CH_2Cl_2, 0 °C	Ph–CH(R)–CH$_2$–C(O)Me	90	98 (R)	[94]
	Et_2Zn	$Cu(OTf)_2 \cdot C_6H_6$	2.5	**L27a**	6	Toluene, 0 °C		82	87 (S)	[95]
n-Pent–CH=CH–C(O)Me	Et_2Zn	$Cu(OTf)_2$	3	**L27a**	3.6	CH_2Cl_2, 0 °C	n-C$_5$H$_{11}$–CH(R)–CH$_2$–C(O)Me	90	98 (S)	[94]
	Me_2Zn	$Cu(OTf)_2$	3	**L27a**	3.6	CH_2Cl_2, 0 °C		91	98 (S)	[94]
cyclopentenone	Et_2Zn	$Cu(OTf)_2$	2	**L25**	5	CH_2Cl_2, 23 °C	3-R-cyclopentanone	64	86	[92]
cyclopentenone	$i\text{-}Pr_2Zn$	$Cu(OTf)_2$	2	**L25**	5	CH_2Cl_2, 23 °C	3-R-cyclopentanone	58	84	[92]
cyclohexenone	Et_2Zn	$Cu(OTf)_2$	2	**L25**	5	CH_2Cl_2, 23 °C	3-R-cyclohexanone	83	97	[92]
	Bu_2Zn	$Cu(OTf)_2$	2	**L25**	5	CH_2Cl_2, 23 °C		52	94	[92]
	$i\text{-}Pr_2Zn$	$Cu(OTf)_2$	2	**L25**	5	CH_2Cl_2, 23 °C		89	96	[92]
	Me_3Al	$[Cu(MeCN)_4]PF_6$	2	**L26**	4	Et_2O, −45 °C		61	89 (R)	[93]
	Et_3Al	$[Cu(MeCN)_4]PF_6$	2	**L26**	4	Et_2O, −45 °C			86 (R)	[93]
	Et_2Zn	$Cu(OTf)_2$	3	**L27a**	3.6	CH_2Cl_2, 0 °C		92	82 (S)	[94]

2.4 Ligands for Cu-Catalyzed Enantioselective Conjugated Additions | 103

Substrate	Reagent	Cu source		Ligand		Solvent, T	Product	Yield	ee	Ref
cyclohexenone with Me/OMe	Et$_2$Zn	Cu(OTf)$_2$	2	L25	5	CH$_2$Cl$_2$, 23 °C		90	97	[92]
cycloheptenone with R	Et$_2$Zn	Cu(OTf)$_2$	2	L25	5	CH$_2$Cl$_2$, 23 °C		67	97	[92]
	i-Pr$_2$Zn	Cu(OTf)$_2$	2	L25	5	CH$_2$Cl$_2$, 23 °C		76	88	[92]
	Me$_3$Al	[Cu(MeCN)$_4$]PF$_6$	2	L26	4	Et$_2$O, −45 °C		57	89 (R)	[93]
	Et$_3$Al	[Cu(MeCN)$_4$]PF$_6$	2	L26	4	Et$_2$O, −45 °C		66	75 (R)	[93]
cyclopentenone-Me	Et$_2$Zn	Cu(OTf)$_2$	3	L27a	3.6	CH$_2$Cl$_2$, 0 °C		86	98[b]	[94]
Me-enone pyrrolidinone	Et$_2$Zn	Cu(OTf)$_2$	3	L27a	3.6	CH$_2$Cl$_2$, 0 °C		91	99 (R)	[94]

[a] Isolated yields.
[b] dr = 62/38.

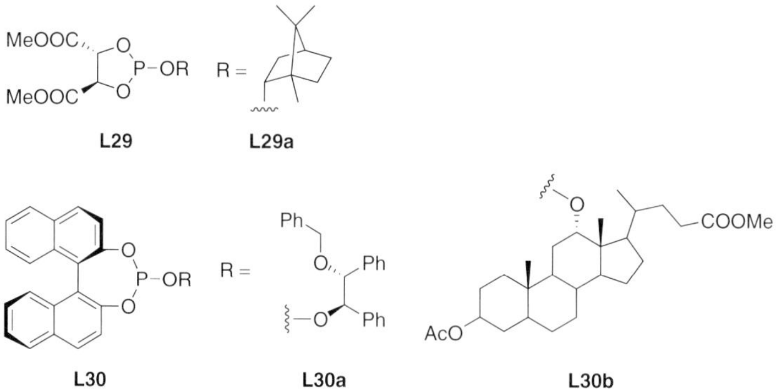

Figure 2.9 Phosphite **L29–L30**.

case, a number of phosphites derived from tartaric acid **L29** and various alcohols were tested in the addition of Et_2Zn to cycloenones [97]. The highest ee-value (65%) was obtained for the addition to benzalacetone in the presence of **L29a** (Figure 2.9; Table 2.15).

Later, phosphites derived from enantiopure binaphthols and chiral alcohols **L30** were used in the enantioselective additions of Et_2Zn to enones (Table 2.15). The highest ee-value (48%) was achieved for the addition to chalcone in the presence of **L30a**, whilst the addition to cyclohexenone proceeded with a low asymmetric induction (19%). A somewhat better result was obtained for the addition of Me_2Zn to (E)-cyclopentadec-2-enone, which yielded (−)-(R)-muscone with 78% ee [98]. Similar results were obtained with ligands bearing steroidal alcohols (deoxycholic acid derivatives) such as **L30b** (chalcone: 76% ee, cyclohexenone: 50% ee; (E)-cyclopentadec-2-enone: 63% ee) [99, 100]. Analogous results were obtained also for other deoxycholic acid-derived phosphites [101].

In comparison with monodentate ligands, bidentate phosphite ligands provided a much better enantioselectivity in the addition of organozincs to cycloenones. The bidentate phosphines with the chiral binaphthol scaffold **L31** were developed by Chan et al. (Figure 2.10; Table 2.15). The ligand **L31a** gave 90% ee in the conjugated addition of Et_2Zn to cyclohexenone [102], while in the case of 5,6-dihydro-2H-pyran-2-one the ee-value was up to 92% in the presence of **L31b** [103]. These conditions were also suitable for enantioselective additions to various unsaturated lactones [104]. Interestingly, asymmetric induction in the presence of diphosphites with pyranoside backbones of glucose and galactose of **L31c** and **L31d** did not exceed 88% [105]. Similar or inferior results were obtained also for other bis-phosphites with the glucosamine [106] and mannitol backbone [107].

Even higher enantioselectivities were obtained with ligands having bis(tetrahydronaphthol) moieties **L32** [108]. The copper-catalyzed conjugated addition of Et_2Zn to cyclopentenone, cyclohexenone, and cycloheptenone in the presence of **L32a** proceeded in the range of 96–98% ee (Table 2.15), with the enantioselectivity depending more dramatically on the ring size of the cycloenone

2.4 Ligands for Cu-Catalyzed Enantioselective Conjugated Additions

Table 2.15 Enantioselective conjugated addition to enones in the presence of **L29–L32**.

Enone	R_XM	Catalyst	mol (%)	Ligand	mol (%)	Conditions	Product	Yield (%)[a]	ee (%)	Reference
Ph⌢⌢Me (O)	Et_2Zn	$Cu(OTf)_2$	0.5	L29a	1	Et_2O, $-20\,°C$	Ph-CH(R)-CH$_2$-C(O)Me	77	65 (S)	[97]
Ph⌢⌢Ph (O)	Et_2Zn	$Cu(OTf)_2$	3	L30a	6	Toluene, $-40\,°C$	Ph-CH(R)-CH$_2$-C(O)Ph	66	48 (S)	[98]
	Et_2Zn	$Cu(OTf)_2$	2.5	L30b	3	Toluene, $-70\,°C$		86	76 (S)	[100]
cyclopentenone	Et_2Zn	$Cu(OTf)_2 \cdot C_6H_6$	1	L32a	2	Toluene, $-30\,°C$	3-R-cyclopentanone	50	90 (S)	[108]
	Et_3Al	$Cu(OTf)_2$	1	L31a	2	Toluene, $-30\,°C$		95	75	[104]
	Et_2Zn	$Cu(OTf)_2$	1	L31a	2	Toluene, $-30\,°C$		100	77 (S)	[102]
cyclohexenone	Et_2Zn	$Cu(OTf)_2$	3	L30a	6	Toluene, $-40\,°C$	3-R-cyclohexanone	98	19 (S)	[98]
	Et_2Zn	$Cu(OTf)_2$	2.5	L30b	3	Toluene, $-40\,°C$		89	50 (R)	[99]
	Et_2Zn	$Cu(OTf)_2$	1	L31a	2	Toluene, $0\,°C$		100	90 (S)	[102]
	Et_2Zn	$Cu(OTf)_2$	1	L31b	2	Toluene, $0\,°C$		100	92	[103]
	Et_2Zn	$Cu(OTf)_2$	1	L31a	2	Et_2O, $0\,°C$		99	98 (R)	[104]
	Et_2Zn	$Cu(OTf)_2 \cdot C_6H_6$	1	L31c	2	THF, $-78\,°C$		58	88 (S)	[105]
	Et_2Zn	$Cu(OTf)_2 \cdot C_6H_6$	1	L32a	2	Et_2O, $-30\,°C$		99	97 (S)	[108]
	Me_2Zn	$Cu(OTf)_2$	1	L32b	2	Et_2O, $-30\,°C$		99	93	[108]
	Me_3Al	$[Cu(MeCN)_4]BF_4$	1	L32b	2	CH_2Cl_2, $0\,°C$		81	96	[109a]
	Et_3Al	$Cu(OTf)_2$	1	L31a	2	Et_2O, $-30\,°C$		97	67	[104]

(continued overleaf)

Table 2.15 (Continued)

Enone	R$_X$M	Catalyst	mol(%)	Ligand	mol(%)	Conditions	Product	Yield (%)[a]	ee (%)	Reference
(cyclohexenone)	Me$_2$Zn	Cu(OTf)$_2$	1	L32b	2	Et$_2$O, −30 °C	(3-R-cyclohexanone)	85	99	[108]
	Et$_2$Zn	Cu(OTf)$_2$	1	L32b	2	Et$_2$O, −30 °C		99	97	[108]
(butenolide)	Et$_2$Zn	Cu(OTf)$_2$	1	L31a	2	Et$_2$O, −30 °C	(β-R-γ-butyrolactone)	70	87	[108]
	Et$_2$Zn	Cu(OTf)$_2$	1	L31b	2	Toluene, 0 °C		77	56	[103]
(dihydropyranone)	Et$_2$Zn	Cu(OTf)$_2$	1	L31a	2	Et$_2$O, −50 °C	(β-R-δ-valerolactone)	95	94	[104]
	Et$_2$Zn	Cu(OTf)$_2$	1	L31b	2	Toluene, 0 °C		100	92	[103]
	Me$_2$Zn	Cu(OTf)$_2$	1	L31a	2	Toluene, −50 °C		72	81	[104]
	Me$_3$Al	Cu(OTf)$_2$	1	L31a	2	Et$_2$O, −50 °C		60	85	[104]

[a] Isolated yields or conversions.

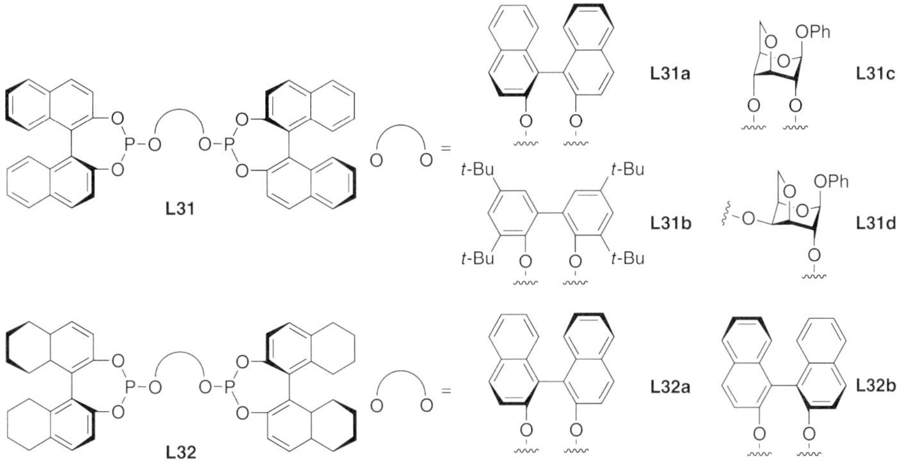

Figure 2.10 Phosphites with binaphthol framework **L31** and **L32**.

used in the addition of Me$_2$Zn. The highest ee-value (99%) was obtained in the case of cycloheptenone, but similar enantioselectivities were obtained also for the addition of Me$_3$Al and Et$_3$Al to cycloenones [109].

2.4.5
NHC-Compounds

The use of chiral *N*-heterocyclic carbenes as ligands for a copper-catalyzed conjugated addition was reported by Alexakis and coworkers [110] and Roland *et al.* [111] in 2001 (Figure 2.11). In this case, Alexakis *et al.* studied the addition of Et$_2$Zn to cyclic and acyclic ketones in the presence of various chiral ligands **L33**

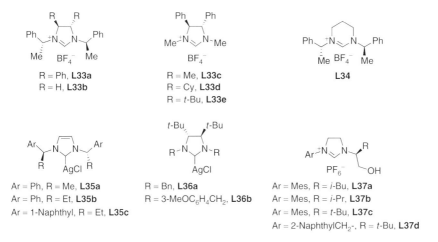

Figure 2.11 Chiral NHC-ligands.

Table 2.16 Enantioselective conjugated addition of Et₂Zn to enones and nitrostyrene in the presence of **L35–L37**.

Enone	Catalysts	mol %	Ligand	mol %	Conditions	Product	Yield (%)ᵃ	ee (%)	Reference
Me-CH=CH-C(O)-Me (Me substituted)	CuTC	4	L35a	4	Et₂O, −78 °C	Me-CH(Et)-CH₂-C(O)-Me (Me subst.)	44	49	[112]
Ph-CH=CH-C(O)-Me	CuT	4	L35a	4	Et₂O, −78 °C	Ph-CH(Et)-CH₂-C(O)-Me	60	42 (S)	[112]
cyclohexenone	CuTC	4	L35a	4	Et₂O, −78 °C	3-Et-cyclohexanone	99	62 (S)	[112]
	CuTC)₂	4	L36b	4	Et₂O, −78 °C		99	69 (S)	[112]
	Cu(OTf)₂	2	L37a	3	Et₂O, r.t.		99	86 (R)	[114]
	Cu(OTf)₂	2	L37b	3	Et₂O, r.t.		99	85 (R)	[114]
	Cu(OTf)₂	2	L37c	3	Et₂O, r.t.		99	87 (R)	[114]
	Cu(OTf)₂	2	L37d	3	Et₂O, r.t.		99	90 (R)	[114]
cycloheptenone	Cu(OAc)₂	4	L35a	4	Et₂O, −78 °C	3-Et-cycloheptanone	99	88 (R)	[112]
	Cu(OAc)₂	4	L35c	4	Et₂O, −78 °C		95	93 (S)	[112]
	CuTC	4	L36b	4	Et₂O, −78 °C		100	88 (S)	[112]
	Cu(OTf)₂	2	L37a	3	Et₂O, r.t.		99	90 (R)	[114]
	Cu(OTf)₂	2	L37d	3	Et₂O, r.t.		99	90 (R)	[114]
PhCH=CH-NO₂	Cu(OAc)₂	4	L35a	4	Et₂O, −78 °C	Ph-CH(Et)-CH₂-NO₂	100	75 (R)	[112]

ᵃIsolated yield or conversion.
r.t. = room temperature.

and **L34**. However, the asymmetric induction was rather low and did not exceed 27% ee for cyclohexenone (**L33a**) or 51% ee for cycloheptenone (**L33a**). Likewise, Roland and coworkers studied the conjugated additions to cyclohexenone in the presence of **L33e**, and this also proceeded with a low enantioselectivity (23% ee). The subsequent tuning of the ligand framework and the use of silver carbene complexes of **L35** and **L36** as ligand precursors led to a considerable improvement in enantioselectivity [112]. The conjugated addition proceeded to cyclohexenone with 69% ee (**L36b**) and to cycloheptenone with 93% ee (**L35c**). The addition to nitrostyrene also proceeded with a moderate enantioselectivity of 75% ee in the presence of **L35a**. For selected examples, see Table 2.16.

A further modification of the heterocyclic framework, by adding a side chain with a coordinating functionality (hydroxy group), resulted in the formation of bidentate ligands **L37** [113, 114]. These showed increased enantioselectivities (Table 2.16) in the CuTC-catalyzed addition of Et₂Zn to enones [113], with up to 87% ee for cyclohexenone (**L37c**) and 90% ee for cycloheptenone (**L37a**). Further elaboration of the N-substituents led to the synthesis of **L37d** [115], which showed a slightly higher asymmetric induction; for example, the addition of Et₂Zn

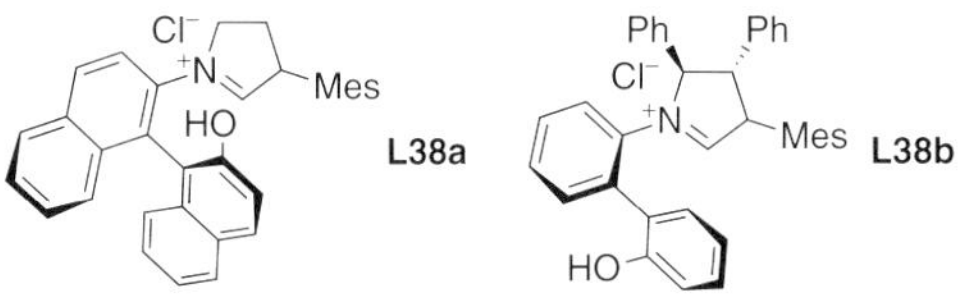

2.4 Ligands for Cu-Catalyzed Enantioselective Conjugated Additions | 109

Figure 2.12 Chiral NHC-ligands with binaphthyl and biphenyl backbones.

to cyclohexenone or cycloheptenone proceeded with 90 or 90% ee, respectively. Another library of chiral heterocyclic ligands was synthesized by Hoveyda *et al.*, and used for synthesis of the corresponding polynuclear Cu or Ag complexes with NHC ligands having a binaphthyl or biphenyl backbone **L38** (see Figure 2.12) [116]. The conjugated additions were carried out with various organozincs to trisubstituted cycloenones, to yield products with chiral quaternary centers and good to excellent enantioselectivities (74–96% ee) (Table 2.17).

2.4.6
Various Ligands (Ligands with Mixed Functionalities)

Except for the previously mentioned classes of ligands, there is a motley group of compounds which bear different ligating functionalities within their molecular framework. Typically, these bear two functional groups having P, N, O, or S moieties (Figure 2.13).

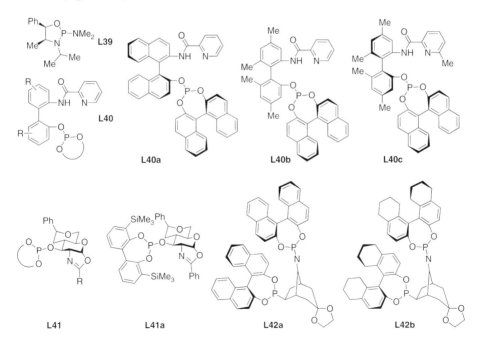

Figure 2.13 Various phosphite ligands bearing other ligating functionalities.

Table 2.17 Enantioselective conjugated addition of R_2Zn to enones in the presence of **L38b**.[a]

Enone	R_2Zn	Cu(OTf)$_2$·C$_6$H$_6$ (mol%)[b]	Product	Yield (%)[b]	ee (%)
3-methyl-2-cyclohexenone	Et$_2$Zn	2.5	3-methyl-3-R-cyclohexanone	92	93
	(n-Bu)$_2$Zn	2.5		83	86
	Ph$_2$Zn	2.5		95	97
	(4-MeOPh)$_2$Zn	5		89	90
3-allyl-2-cyclohexenone	Et$_2$Zn	5	3-allyl-3-R-cyclohexanone	92	84
	Ph$_2$Zn	5		88	89
3-(phenylethynyl)-2-cyclohexenone	Et$_2$Zn	2.5	3-(phenylethynyl)-3-R-cyclohexanone	78	74
3-phenyl-2-cyclohexenone	Et$_2$Zn	10	3-phenyl-3-R-cyclohexanone	85	90
3-methyl-2-cycloheptenone	Et$_2$Zn	2.5	3-methyl-3-R-cycloheptanone	83	85
	(n-Bu)$_2$Zn	5		67	77
	Ph$_2$Zn	5		88	96

[a] Reactions were run at −30 to −15 °C, Et$_2$O, 24–48 h.
[b] Cu salt/ligand = 1/1.
[c] Isolated yields.

One of the first ligands used in the copper-catalyzed enantioselective conjugated addition of organometallics to enones was prepared by Alexakis et al. [117]. This was based on chiral phosphorus compounds **L39** prepared from amino alcohols, and their use as a chiral ligand in the copper-catalyzed conjugated addition of Et$_2$Zn to cyclohexenone resulted in a mediocre enantioselectivity of 32% ee.

Another distinct group consisted of ligands bearing phosphite and amido groups **L40**, for which the best results were obtained with the ligand bearing two chiral binaphthyl backbones **L40a**. The conjugated addition of Et$_2$Zn to chalcone or cyclohexenone, catalyzed by [Cu(MeCN)$_4$]BF$_4$ in toluene, yielded the corresponding products in 97% and 56% ee, respectively [118]. A high asymmetric induction was observed with a related ligand **L40b**, which was a mixture of two diastereoisomers, bearing racemic biphenyl and chiral binaphthyl frameworks [119]. The conjugated addition of Et$_2$Zn to chalcone under the same conditions proceeded with up to 94% ee. The synthesis of ligands bearing also a chiral biphenyl backbone such as **L40c** enabled the enantioselectivity to be increased to 97% [120].

Another group of ligands is that of the phosphite-oxazolines, connected by a sugar moiety **L41** [106], although asymmetric induction in their presence was much lower. The conjugated addition of Et$_3$Al to cyclohexenone, catalyzed by Cu(OAc)$_2$ in the presence of **L41a**, proceeded with 64% ee. With regards to the addition to acyclic unsaturated ketones, the highest ee-value (78%) was obtained in the case of (*E*)-3-nonen-2-one under catalysis of [Cu(MeCN)$_4$]BF$_4$.

Mixed phosphate-phosphoramidites with the tropane skeleton were prepared by Laschat *et al.* and tested in the conjugated addition [121]. The best asymmetric induction (90% ee) was obtained with **L42a** for the addition of Et$_2$Zn to cyclohexenone, catalyzed by Cu(OAc)$_2$. The enantioselective addition to acyclic ketones proceeded with a low asymmetric induction that did not exceed 67% (for **L42b**). Recently, the synthesis of a pyridine-phosphite ligand was reported [122], in the presence of which the enantioselective conjugated addition of Et$_2$Zn to chalcones proceeded with up to 90% ee.

Ligands combining phosphine and amine moieties were also prepared (Figure 2.14), typical examples of which were compounds with phosphino-amide functionalities **L43** as prepared by Zhang *et al.* [123]. The conjugated addition of Et$_2$Zn to cyclohexenone catalyzed by copper salts ([Cu(MeCN)$_4$]BF$_4$) or Cu(OTf)$_2$C$_6$H$_6$) in the presence of ligands with the chiral binaphthyl backbone **L43a** and **L43b** resulted in a similar asymmetric induction of 91% and 92% ee, respectively; the addition to chalcone also proceeded with high enantioselectivity (96% ee). Analogous ligands, prepared by Hu *et al.*, with the chiral biphenyl backbone **L43c** and **L43d** gave similar results in the case of addition to chalcone [124].

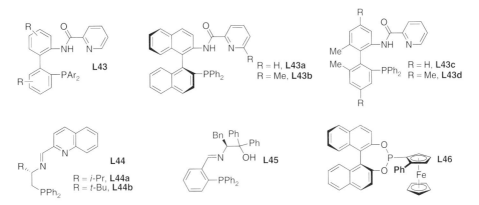

Figure 2.14 Phosphino-amine ligands.

Chiral amino alcohols served as the basic scaffold for the synthesis of another type of tridentate phosphino-amine ligand, **L144** and **L45**. Hayashi *et al.* synthesized two derivatives bearing *i*-Pr **L44a** and *t*-Bu groups **L44b** [125], whilst a high asymmetric induction (up to 99% ee) of the conjugated addition of Me$_2$Zn and Et$_2$Zn to various five-, six-, and seven-membered ring cycloenones was achieved in the presence of **L44b**. Interestingly, asymmetric induction in the addition of *n*-Bu$_2$Zn did not exceed 36% ee. Another tridentate ligand **L45** was prepared by Gau *et al.* [126], in the presence of which the conjugated addition of Et$_2$Zn to chalcones proceeded with asymmetric induction up to 97% ee.

A ligand bearing a chiral ferrocene and binaphthol framework **L46** was also synthesized by Woodward *et al.* [127]. In this case, the conjugated addition of Et$_3$Al to cyclohexenone proceeded with 92% ee, which was comparable with other ligands. However, its use in the enantioselective addition of (*E*)-(2-phenyl-1-propenyl)dimethylaluminum to cyclohexenone proceeded with a hitherto unmatched enantioselectivity of 85% ee, which was greater than those reported by Alexakis *et al.* [74] with **L2d** (Scheme 2.17).

Sulfur-containing ligands were also synthesized (Figure 2.15). In particular, binaphthyl-thiophosphoramidite ligands **L47** proved to promote the highly enantioselective addition of organozincs to a variety of enones [128]., with an enantioselectivity up to 98% being observed for **L47a** in the addition of Et$_2$Zn to acylic as well as cyclic enones, regardless of the ring size (Table 2.18). With regards to the addition of Me$_2$Zn and Ph$_2$Zn, the ee-values did not exceed 89%. In contrast to the thiophosphoramidite moiety, the presence of a selenophosphoramidite functionality **L47b** led to a decrease in enantioselectivity. The use of ligands based on chiral 1,2-diaminocyclohexane scaffold, such as iminothiophosphoramides ligands **L48** or sulfoamide-thiophosphoramide **L49**, on the enantioselective conjugated addition of Et$_2$Zn to enones exhibited an inferior enantioselectivity, with asymmetric induction in the addition to cyclohexenone not exceeding 75% ee (for **L48a**) [129] and 90% ee (for **L49a** in the presence of

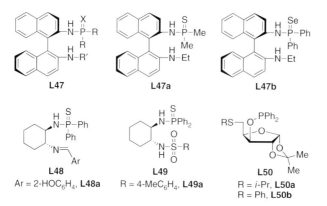

Figure 2.15 Thiophosphoramidite and thioether ligands.

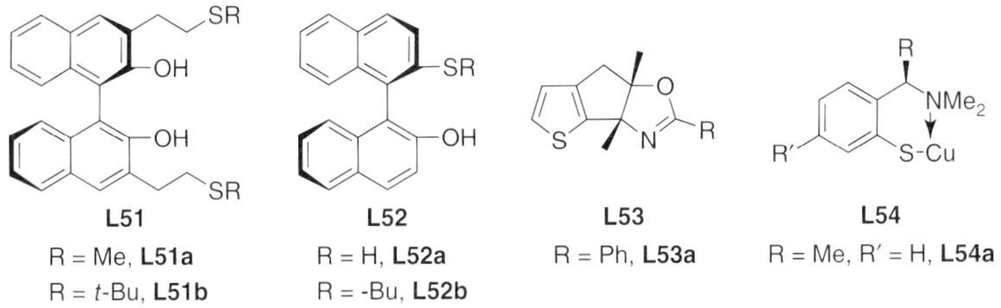

Figure 2.16 Various chiral ligands for conjugated addition.

LiCl) [130]. Thioether-phosphinite ligands with a xylose backbone **L50** [131] can be also assigned to sulfur-containing ligands. In this case, the conjugated addition of Et_2Zn to cyclohexenone proceeded with a moderate enantioselectivity (64% ee) with **L50a**, whilst following the addition of Et_3Al the best result was obtained with **L50b** (48% ee).

Among other ligands used for enantioselective conjugated addition may be included binaphthols with a thioether moiety **L51** [132], monothiolbinaphthol S-derivatives **L52** [133], oxazolines **L53** [134], and aminoarenethiolates **L54** [135] (Figure 2.16). The conjugated addition of Et_2Zn to cyclohexenone in the presence of **L51a** proceeded with 85% ee in MTBE, whereas for chalcone a 96% ee was obtained with **L51b**. When the addition to nitroalkenes was also tested, the simple monothiolbinaphthol **L52a** did not exhibit a reasonable enantioselectivity, generally being below 36% [133a]. In the case of the S-butyl derivative **L52b**, asymmetric induction in the addition of Me_3Al to enones proceeded with up to 93% ee (for 5,5-dimethylhex-3-en-2-one enone) [133c]. Rigid three-ring oxazolines **L53** were tested as chiral ligands in the conjugated addition of Et_2Zn to chalcone, the highest enantioselectivity (79% ee) being obtained with the phenyl derivative **L53a**. Finally, copper complexes of aminoarenethiolates also showed moderate enantioselectivity in the conjugated addition to cyclohexenone (83% ee, **L54a**).

Also worthy of mention with respect to asymmetric induction are peptide-based ligands **L55** (Figure 2.17), as introduced by Hoveyda *et al.* [136], that are related to the phosphine-peptide ligands **L13**. Fine-tuning of the peptide backbone resulted in the development of **L55a**, which was used in the asymmetric conjugated addition of organozincs to tetrasubstituted activated cycloenones (Table 2.19); in this case, the asymmetric induction was up to 95% ee.

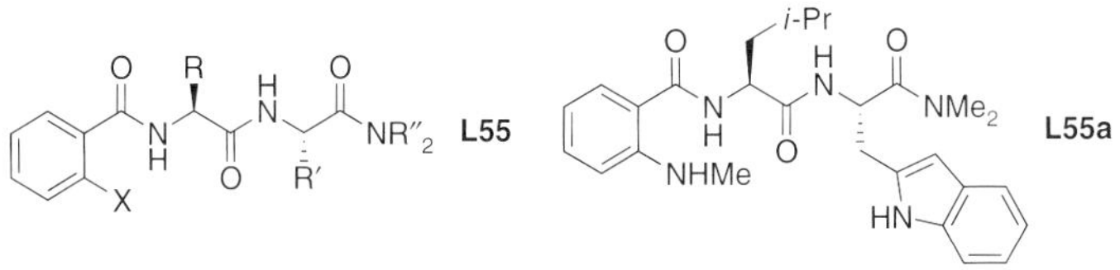

Figure 2.17 Peptide-based ligands.

Table 2.18 Enantioselective conjugated addition of R_2Zn to enones and cycloenones in the presence of ligands **L39–L41**, **L43**, **L44**, **L47**, **L51**, and **L53**.

Enone	R_xM	Catalyst	mol %	Ligand	mol %	Conditions	Product	Yield (%)[a]	ee (%)	Reference
Ph-CH=CH-C(O)- (chalcone)	Et_2Zn	$[Cu(MeCN)_4]BF_4$	1	L40a	2.5	Toluene, −10 °C	Ph-CH(R)-CH$_2$-C(O)-Ph	82	97 (S)	[118]
		$[Cu(MeCN)_4]BF_4$	1	L40b	2.5	Toluene, −20 °C		71	94 (S)	[119]
		$[Cu(MeCN)_4]BF_4$	1	L40c	2.5	Toluene, −20 °C		83	97 (S)	[120]
		$Cu(OTf)_2 \cdot C_6H_6$	1	L43b	5	Toluene/$(CH_2Cl)_2$, −20 °C		85	96 (S)	[123]
		$[Cu(MeCN)_4]BF_4$	1	L43c	2.5	Toluene/$(CH_2Cl)_2$, −10 °C		62	94 (R)	[124]
		$[Cu(MeCN)_4]BF_4$	1	L43d	2.5	Toluene, −10 °C		84	92 (R)	[124]
		$[Cu(MeCN)_4]BF_4$	3	L47a	6	Toluene, 20 °C		97	87 (S)	[128]
		$Cu(OAc)_2$	2	L51b	2.4	MTBE, −10 °C		91	96 (S)	[132]
		$Cu(OTf)_2$	2.1	L53a	4.2	Et_2O, −20 °C		49	79 (R)	[134]
n-C$_5$H$_{11}$-CH=CH-C(O)-Me	Me_3Al	$[Cu(MeCN)_4]BF_4$	1	L41a	4	Et_2O, −30 °C	n-C$_5$H$_{11}$-CH(R)-CH$_2$-C(O)-Me	94	78 (R)	[106]
	Me_3Al	$[Cu(MeCN)_4]BF_4$	5	L53a	5	THF, −46 °C		82	86 (R)	[132c]
	Et_2Zn	$[Cu(MeCN)_4]BF_4$	1	L40a	2.5	Toluene, −10 °C		90	53 (S)	[118]
	Et_2Zn	$[Cu(MeCN)_4]BF_4$	3	L47a	6	Toluene, 0 °C		75	98 (R)	[128]
cyclopentenone	Me_2Zn	$Cu(OTf)_2$	0.1	L44a	0.25	CH_2Cl_2, −20 °C	3-R-cyclopentanone	74	58 (S)	[125]
	Me_2Zn	$Cu(OTf)_2$	0.1	L44b	0.25	CH_2Cl_2, −20 °C		88	92 (S)	[125]
	Et_2Zn	$Cu(OTf)_2$	0.1	L44a	0.25	CH_2Cl_2, −40 °C		70	44 (S)	[125]
	Et_2Zn	$Cu(OTf)_2$	0.1	L44b	0.25	CH_2Cl_2, −40 °C		85	93 (S)	[125]

2.4 Ligands for Cu-Catalyzed Enantioselective Conjugated Additions

Cyclohex-2-enone (→ 3-R-cyclohexanone)

R$_2$Zn/R$_3$Al	Cu source	mol%	Ligand	t (h)	Solvent, T	Yield[a]	ee (config.)	Ref.
Et$_2$Zn	CuI	10	L39	20	THF, 20 °C	70	32 (S)	[117]
Et$_2$Zn	[Cu(MeCN)$_4$]BF$_4$	1	L40a	2.5	Toluene, −10 °C	90	31 (S)	[118]
Et$_3$Al	Cu(OAc)$_2$	1	L41a	4	Et$_2$O, −30 °C	50	64 (S)	[106]
Et$_2$Zn	Cu(OTf)$_2$·C$_6$H$_6$	1	L43a	5	Toluene, −20 °C	76	91 (S)	[123]
Et$_2$Zn	Cu(OTf)$_2$	1	L43b	5	Toluene/(CH$_2$Cl$_2$)$_2$, −20 °C	98	92 (S)	[123]
Et$_2$Zn	[Cu(MeCN)$_4$]BF$_4$	3	L47a	6	Toluene, 0 °C	95	97 (R)	[128]
Me$_2$Zn	[Cu(MeCN)$_4$]BF$_4$	3	L47a	6	Toluene, −20 °C	71	87 (R)	[128]
Ph$_2$Zn	[Cu(MeCN)$_4$]BF$_4$	3	L47a	6	Toluene, 0 °C	97	89 (R)	[128]
Et$_2$Zn	Cu(OTf)$_2$	2	L51b	2.4	MTBE, 0 °C	95	85 (R)	[132]
Me$_2$Zn	Cu(OTf)$_2$	0.1	L44a	0.25	CH$_2$Cl$_2$, −20 °C	88	98 (S)	[125]
Me$_2$Zn	Cu(OTf)$_2$	0.1	L44b	0.25	CH$_2$Cl$_2$, −20 °C	78	98 (S)	[125]
Et$_2$Zn	Cu(OTf)$_2$	0.1	L44a	0.25	CH$_2$Cl$_2$, −40 °C	99	98 (S)	[125]
Et$_2$Zn	Cu(OTf)$_2$	0.1	L44b	0.25	CH$_2$Cl$_2$, −40 °C	99	96 (S)	[125]

Cyclohept-2-enone (→ 3-R-cycloheptanone)

R$_2$Zn	Cu source	mol%	Ligand	t (h)	Solvent, T	Yield[a]	ee (config.)	Ref.
Et$_2$Zn	[Cu(MeCN)$_4$]BF$_4$	3	L47a	6	Toluene, 20 °C	93	97 (R)	[128]
Et$_2$Zn	Cu(OTf)$_2$	2	L51b	2.4	MTBE, 0 °C	96	81 (R)	[132]
Me$_2$Zn	Cu(OTf)$_2$	0.1	L44a	0.25	CH$_2$Cl$_2$, −20 °C	44	88 (S)	[125]
Me$_2$Zn	Cu(OTf)$_2$	0.1	L44b	0.25	CH$_2$Cl$_2$, −20 °C	32	83 (S)	[125]
Et$_2$Zn	Cu(OTf)$_2$	0.1	L44a	0.25	CH$_2$Cl$_2$, −40 °C	61	79 (S)	[125]
Et$_2$Zn	Cu(OTf)$_2$	0.1	L44b	0.25	CH$_2$Cl$_2$, −40 °C	47	68 (S)	[125]

[a] Yield or conversion.

Table 2.19 Enantioselective conjugated addition of R_2Zn to β-silyl-α,β-unsaturated ketones in the presence of **L55a**.

Enone	R_2Zn	Temperature (°C)	Product	Yield (%)[a]	ee (%)
(cyclohexenone with COOt-Bu, Me)	Et_2Zn	0	(product with COOt-Bu, R, Me)	89	90
	$(n-Bu)_2Zn$	−15		92	95
	$(i-Pr)_2Zn$	0		98	90
	$[AcO(CH_2)_4]_2Zn$	−15		74	95
(cyclopentenone with COOMe, Me)	Et_2Zn	01	(product with COOMe, R, Me)	80	77
	$(n-Bu)_2Zn$	−15		85	86
	$(i-Pr)_2Zn$	−15		82	66
(cyclopentenone with COOMe, n-Bu)	Me_2Zn	0	(product with COOMe, Me, n-Bu)	70	82

[a] Isolated yields.

2.4.7
Selected Experimental Procedures

The following examples of experimental procedures for the enantioselective conjugated addition of organozincs or organoaluminums to enones were taken directly from the original literature, and used as first presented.

Kinetic resolution of racemic 5-methylcyclohex-2-enone by Cu-catalyzed enantioselective conjugated addition of Et_2Zn in the Presence of L2a (100 mmol scale) [136]

In flame-dried glassware under an argon atmosphere, $[Cu(OTf)_2]$ (18 mg, 0.05 mmol) and **L2a** (54 mg, 0.10 mmol) were dissolved in dry toluene (100 ml). After stirring at room temperature (r.t.) for 1 h, the colorless solution was cooled to −30 °C and racemic 5-methylcyclohex-2-enone (11.0 g, 100 mmol) and n-dodecane (4.0 ml) were added. After stirring for 10 min, Et_2Zn (50 ml of 1.1 M solution in toluene, 55 mmol) was added dropwise via a syringe over a 5 min period. A sample (0.1 ml) was taken after reaction overnight and analyzed by chiral gas chromatography (GC); this showed the enone with 93% ee at 51% conversion. Extra Et$_2$Zn (3.6 ml, 1.1 M solution in toluene) was added and after 3 h another sample was taken. GC analysis showed >99% ee and 55% conversion. The reaction mixture was quenched with 1 M HCl aq. (150 ml), the aqueous layer extracted with Et$_2$O (3 × 100 ml), after which the combined organic layers were washed with brine and dried over Na$_2$SO$_4$. Filtration and evaporation of the solvents yielded a mixture of 5-methylcyclohex-2-enone, addition product, and n-dodecane which were separated by column chromatography (SiO$_2$, hexanes/diethyl ether 4:1) to give (R)-(−)-5-methylcyclohex-2-enone (3.6 g, 33 mmol, 33%). $[\alpha]^{20}_D$ −87.38 (c. 0.81, CHCl$_3$). ^{1}H NMR (200 MHz, CDCl$_3$): δ 1.06 (d, J = 6.1 Hz, 3H), 1.9–2.5 (m, 5H), 6.0 (m, 1H), 6.9 (m, 1H); ^{13}C NMR (300 MHz, CDCl$_3$): δ 21.06 (q), 30.20 (d), 33.87 (t), 46.12 (t), 129.44 (d), 149.80 (d), 199.93 (s). GC analysis (Chiraldex G-TA) showed >99% ee (no (S)-5-methylcyclohex-2-enone was detected).

Cu-catalyzed enantioselective addition of Et$_2$Zn to N-[4-chlorophenyl (toluene-4-sulfonyl)methyl]formamide in the presence of L2c [78]

Cu(OTf)$_2$ (3.6 mg, 0.01 mmol) and ligand **L2c** (10.8 mg, 0.02 mmol) were dissolved in anhydrous THF (10 ml) and the solution was stirred for 30 min at rt. The mixture was cooled to −50 °C and N-[4-chlorophenyl(toluene-4-sulfonyl)methyl]formamide (162 mg, 0.50 mmol) added. A solution of a Et$_2$Zn (1.25 ml of 1 M in hexane, 1.25 mmol) was added dropwise and the reaction mixture stirred for 16 h at −50 °C, quenched with saturated aqueous NH$_4$Cl (10 ml), and then extracted with EtOAc (3 × 5 ml). The combined organic extracts were washed with brine, dried (Na$_2$SO$_4$), filtered, and concentrated. The crude product was purified by column chromatography (SiO$_2$; EtOAc/pentane 1:1) and afforded compound N-[1-(4-chlorophenyl)propyl]formamide in 94% isolated yield (R$_f$ = 0.28) as a colorless oil which slowly solidified, mp 94.0–94.8 °C. Chiral GC-CP Chiralsil Dex CB, 25 m × 0.25 mm × 0.25 μm; He-flow: 1 ml min^{-1}; oven: 60 °C for 10 min, then 1 °C min^{-1} until 180 °C; t$_R$ (S) 117.57 min (minor), t$_R$ (R) 118.06 min (major); 97% ee. $[\alpha]_D$ + 149.5 (c 1.06, CHCl$_3$). The ^{1}H and ^{13}C NMR spectra (CDCl$_3$) showed a 4:1 mixture of two rotamers (rotation of the N-formyl group). Major rotamer: 1H NMR (400 MHz, CDCl$_3$) δ 8.14 (s, 1H), 7.31–7.25 (m, 2H), 7.19–7.15 (m, 2H), 6.54 (br d, J = 7.2 Hz, 1H), 4.86 (q, J = 7.6 Hz, 1H), 1.84–1.71 (m, 2H), 0.92–0.84 (m, 3H) ppm. ^{13}C NMR (50 MHz, CDCl3) δ 160.8, 140.0, 133.1, 128.7, 127.9, 53.3, 28.9, 10.5 ppm. Minor rotamer: ^{1}H NMR (400 MHz, CDCl$_3$) δ 8.04 (d, J = 11.9 Hz, 1H), 7.31–7.25 (m, 2H), 7.19–7.15 (m, 2H), 7.06 (br t, J = 10.0 Hz, 1H), 4.31 (q, J = 7.6 Hz, 1H), 1.84–1.71 (m, 2H), 0.92–0.84 (m, 3H) ppm. ^{13}C NMR (50 MHz, CDCl$_3$) δ 164.5, 140.2, 133.5, 129.0, 127.6, 57.7, 30.1, 10.5 ppm. HRMS calculated for C$_{11}$H$_{12}$C$_l$NO 197.0607, found 197.0604. Analysis calculated for C$_{11}$H$_{12}$C$_l$NO: C 60.76, H 6.12, N 7.09. Found: C 60.60, H 6.13, N 6.97.

Cu-catalyzed asymmetric conjugated addition of Me₃Al to 5-methylpent-3-en-2-one in the presence of L2d [74]

To a solution of copper thiophenecarboxylate (CuTC) (1.5 mg, 0.008 mmol) in Et$_2$O (1 ml) at r.t. under nitrogen, was added **L2d** (17 mg, 0.032 mmol) and 1 ml of Et$_2$O. The solution was stirred at 25 °C for 30 min and then cooled to −30 °C. Me$_3$Al (5.5 ml of 2 M in hexanes, 1.12 mmol) was added dropwise so that the temperature did not rise above −30 °C. The solution was stirred for 5 min, after which 5-methylpent-3-en-2-one (90 mg, 0.8 mmol) was added dropwise, either neat or in solution in 0.5 ml toluene. The reaction mixture was stirred at −30 °C for 12 h and then quenched by 2 M, HCl/Et$_2$O. Flash chromatography [pentane:Et$_2$O, 4:1] yielded 90 mg (88%) of (R)-3,5-dimethylpentan-2-one. ^{1}H NMR (400 MHz, CDCl$_3$) 0.83 (d, 3 H, J = 6.8, CHMe$_{2\alpha}$), 0.84 (d, 3 H, J = 6.8, CHMe$_{2\beta}$), 0.85 (d, 3 H, J = 6.8, CHMe), 1.50 (dbrseptet, 1 H, J = 6.8, 2.1, CHMe2), 1.93 (m, 1 H, CHMe), 2.14 (s, 3 H, COMe), 2.20 (dd, 1 H, J = 15.7, 9.2, CH$_{2\alpha}$COMe), 2.41 (dd, 1 H, J = 15.7, 4.7, CH$_{2\beta}$COMe). The enantiomeric excess was determined by chiral GC (6-Me-2,3-pe-γ-CD, isothermal 70 °C). Retention times: S 4.7 min; R 4.9 min (96% ee).

Cu-catalyzed conjugated addition of Et₂Zn to (E)-1-[(3-Phenyl)acryloyl]pyrrolidin-2-one in the presence of L2d [62]

A flame-dried Schlenk flask was charged with Cu(OTf)$_2$ (2.71 mg, 0.0075 mmol) and **L2d** (8.1 mg, 0.015 mmol) in anhydrous toluene (1.0 ml), and the mixture stirred at r.t. for 40 min. Subsequently, a solution of the imide (E)-1-[(3-phenyl)acryloyl]pyrrolidin-2-one (107.5 mg, 0.4 mmol) in toluene (0.5 ml) was added under an Ar atmosphere. The resulting solution was cooled to −78 °C, followed by the addition of Et$_2$Zn (1.5 mmol, 1.36 ml of a 1.1 M solution in toluene). The reaction was quenched after 3 h at 0 °C; the usual work-up afforded a crude reaction mixture which was subjected to flash chromatography using hexanes containing 30% AcOEt as the eluant (R$_f$ = 0.28) to give pure (3S)-1-(3-phenylpentanoyl)pyrrolidin-2-one. ^{1}H NMR (200 MHz, CDCl$_3$) δ: 7.10–7.36 (m, 5H), 3.59–3.82 (m, 2H), 3.30–3.47 (m, 1H), 3.03–3.27 (m, 2H), 2.47–2.60 (m, 2H), 1.84–2.04 (m, 2H), 1.58–1.83 (m, 2H), 0.82 (t, 3H, J = 7.3 Hz). ^{13}C NMR (50 MHz, CDCl3) δ: 175.9, 173.6, 144.9, 128.8, 128.3, 126.7, 46.0, 43.8, 43.7, 34.3, 29.8, 17.7, 12.7. The enantiomeric excess of compound (3S)-1-(3-phenylpentanoyl)pyrrolidin-2-one (99.9%) was determined by chiral HPLC (ADH) (hexanes/IPA = 95:5): t$_R$ 13.6 (minor), t$_R$ 11.1 (major).

2.4 Ligands for Cu-Catalyzed Enantioselective Conjugated Additions

Cu-catalyzed conjugated addition of (E)-(2-phenyl-1-propen-1-yl)dimethylaluminum to cyclohexenone in the presence of L2d [74]

Phenylacetylene (100 µl, 1 mmol) in dichloromethane (4 ml) was treated sequentially with Cp_2ZrCl_2 (73 mg, 25 mol%) and $AlMe_3$ (1 ml of 2.0 M hexane solution, 2 mmol), and the orange mixture stirred at 28–30 °C for 32 h. The solvent was removed by vacuum and the residue taken up in Et_2O (2 ml); the resultant solution added drop-wise to a Schlenk tube containing CuTC (3.8 mg, 2 mol%), ligand **L2d** (4 mol%) and 2-cyclohexenone (122 µl, 1.1 mmol) at −30 °C. The reaction was monitored by GC, and after 15 h was quenched with 2 M HCl aq. Extraction with diethylether (3 × 10 ml), drying ($MgSO_4$) and evaporation gave an oil. Purification by column chromatography (Petrol 40–60 °C : CH_2Cl_2, 1 : 1) afforded (+)-3-(2-phenyl-1-propenyl)cyclohexanone as a colorless oil. The yield was 54% (77% ee). $[\alpha]_D$ + 9.4(c = 0.95, $CHCl_3$); IR(film) ν_{max}: 2929m, 2853w, 1705s, 1600w, 1446w, 907m cm^{-1}; ^1H NMR (500.1 MHz, $CDCl_3$) 1.64–1.68 (m, 1 H, aliphatic), 1.77–1.86 (m, 1 H, aliphatic), 1.99–2.02 (m, 1 H, aliphatic), 2.13 (s, 3 H, Me), 2.19–2.21 (m, 1 H, aliphatic), 2.28–2.34 (m, 1 H, aliphatic), 2.40–2.45 (m, 1 H, aliphatic), 2.48–2.55 (m, 2 H, aliphatic), 2.95–2.97 (m, 1 H, CHCH = CMePh), 5.71 (brdd, 1 H, J = 8.6, 1.7, CH = CMePh), 7.32–7.34 (m, 1 H, Ar), 7.40 (app t, 2 H, J = 7.4, 2.0, Ar), 7.44–7.46 (m, 2 H, Ar); ^{13}C NMR (125 MHz, $CDCl_3$) 16.2, 25.6, 31.8, 39.0, 41.5, 47.9, 125.0, 125.9, 127.1, 128.4 (2 C), 131.0, 135.1, 143.4, 211.3; HRMS (EI) Found: 214.1358, $C_{15}H_{18}O$ requires 214.1358. The ee-value was determined by Diacel HPLC (OD-H column) (hexane : iPrOH 95 : 5, 0.8 ml min^{-1}) Retention times: (−) 11.7 min; (+) 15.5 min. The (+) antipode was assigned to the (R) stereoisomer.

Cu-catalyzed asymmetric conjugated addition of Et_2Zn to cycloheptenone in the presence of L8 [70]

A solution of CuOAc (3.6 mg, 0.01 mmol) and ligand **L8** (38 mg, 0.06 mmol) in toluene was stirred at r.t. for 1 h under N_2. After cooling to −20 °C, cyclohepten-2-enone (36 mg, 0.33 mmol) and $ZnEt_2$ (0.5 ml of 1 M solution in hexanes, 0.5 mmol) were added sequentially. The reaction was run for 3 h at this temperature, and then quenched with saturated aqueous NH_4Cl solution. The mixture was extracted twice with ether, and the combined organic phases dried with anhydrous Na_2SO_4 and concentrated. Chromatography on silica gel (ether : pentane, 1 : 6) gave 43 mg of (S)-3-ethylcycloheptanone as a colorless oil.

Cu-catalyzed conjugated addition of i-Pr₂Zn to 1-acetylcyclopentene in the presence of L12 [80]

A flame-dried 13 × 100 mm test tube was charged with Cu(OTf)₂·PhH (13.65 mg, 0.0028 mmol) and **L12** (29.5 mg, 0.0066 mmol). An aliquot (2.5 ml) of a toluene solution containing decane (internal standard) (10.0 µl, 0.0513 mmol) and 1-acetylcyclopentene (24 mg, 0.217 mmol) was then added at 0 °C, followed by the dropwise addition of i-Pr₂Zn (0.651 mmol). The resulting bright yellow solution was stirred at 0 °C for 12 h, and the reaction then quenched by the addition of 1.0 ml of saturated solution of NH₄Cl. The product was extracted from the resultant aqueous layer with 3 × 2 ml of 4 : 1 pentane : Et₂O, followed by one portion of Et₂O. The combined organic layers were concentrated to a yellow solution of product in toluene, which was purified by silica gel chromatography (100% pentane to 30 : 1 pentane/Et₂O) to afford 106 mg (69%) (−)-1-(2-isopropyl-cyclopentyl)-ethanone (2 : 1 *anti*/*syn* diastereoisomeric mixture). IR (neat): 2955 (s), 2867 (s), 1709 (s), 1470 (m), 1369 (m), 1363 (s), 1162 (m), 966 (w); ^{1}H NMR (400 MHz): δ 3.10–3.06 (1H, m, CHCO$_{(syn\ diastereoisomer)}$), 2.63 (1H, ddd, $J = 8.0, 7.6, 6.0$ Hz, CHCO$_{(anti\ diastereoisomer)}$), 2.17 (3H, s, COCH$_{3(anti\ diastereoisomer)}$), 2.16 (3H, s, COCH$_{3(syn\ diastereoisomer)}$), 2.07 (1H, dddd, $J = 8.0, 8.0, 7.6, 7.6$ Hz, aliphatic CH$_{(anti\ diastereoisomer)}$), 1.92–1.72 (3H, m. aliphatic CH), 1.71–1.55 (5H, m. aliphatic CH), 1.50 (1H, dddd, $J = 13.2, 6.8, 6.4, 6.4$ Hz, aliphatic CH), 1.35–1.26 (5H, m. aliphatic CH), 0.89 and 0.87 (3H each, d, $J = 4.8$ Hz, CH(CH$_3$)$_{2(syn\ diastereoisomer)}$), 0.87 and 0.84 (3H each, d, $J = 6.8$ Hz, CH(CH$_3$)$_{2(anti\ diastereoisomer)}$), ^{13}C NMR (100 MHz) (*anti* diastereoisomer): δ 212.0, 56.3, 48.8, 32.3, 31.0, 30.3, 25.6, 21.6, 20.2; ^{13}C NMR (100 MHz) (*syn* diastereoisomer): δ 212.4, 55.3, 53.2, 31.2, 29.7, 29.3, 29.2, 29.0, 22.2; HRMS calculated for C$_{10}$H$_{18}$O: 154.1358, Found 154.1360; Analysis calculated for C$_{10}$H$_{18}$O: C. 77.87; 11.76; Found: C, 77.86, H, 11.84; [α]$_D^{20}$ −33.86° (c = 1.027, CHCl₃) for a 97% ee and 96% de sample.

Cu-catalyzed conjugated addition of Et₂Zn to [3-[6-(*tert*-butyldimethylsilanyloxy)hex-2-enoyl]oxazolidin-2-one in the presence of L14 [83]

A 13 × 100 mm test tube was charged with (CuOTf)₂·PhH (1.3 mg, 2.5 µmol) and **L14** (3.7 mg, 6.0 µmol) under an atmosphere of nitrogen in a glove box. The tube was sealed with a septum and removed from the glove box. [3-[6-(*tert*-Butyldimethylsilanyloxy)hex-2-enoyl]oxazolidin-2-one (78 mg in 1.0 ml of toluene, 0.25 mmol) was added via a cannula, rinsing with an additional 0.7 ml of toluene. After cooling the tube to 0 °C, Et$_2$Zn (**caution, pyrophoric**; 77 μl, 0.75 mmol) was added with stirring. The mixture was stirred for 2.5 h at 0 °C before dilution with Et$_2$O (2 ml), followed by aqueous 1 M HCl (2 ml) to quench. After removal of the organic layer, the aqueous layer was washed with EtOAc (3 × 2 ml). The combined organic extracts were concentrated under reduced pressure to yield a pale yellow oil that was purified by silica gel chromatography (3 : 1 hexanes/EtOAc) to yield (+)-3-[6-(*tert*-butyldimethylsilanyloxy)-3-ethylhexanoyl]oxazolidin-2-one as a colorless oil (81 mg, 24 mmol, 95%). IR (neat): 2955 (s), 2928 (s), 2857 (s), 1785 (s), 1699 (s), 1472 (m), 1458 (m), 1385 (s); ^{1}H NMR (400 MHz): δ 4.42 − 4.37 (2H, m, OCH$_2$CH$_2$N), 4.01 (2H, dd, J = 8.4, 8.0, NCH$_2$CH$_2$), 3.58 (2H, t, J − 6.4 Hz, SiOCH$_2$), 2.87 (2H, ABqd, J_{AB} = 16.8, J_d = 6.8 Hz, CH$_2$CO), 1.95 (1H, m, CHEt), 1.60−1.25 (6H, m, aliphatic CH), 0.88 (9H, s, SiC(CH$_3$)$_3$), 0.87 (3H, t, J = 7.3 Hz, CH$_2$CH$_3$), 0.04 (6H, s, Si(CH$_3$)$_2$); ^{13}C NMR (100 MHz): δ 173.4, 153.6, 63.5, 62.0, 42.8, 39.3, 35.5, 30.0, 29.4, 26.3, 26.1, 18.5 (3C), 10.9, −5.1 (2C); Analysis calculated for C$_{17}$H$_{33}$NO$_4$Si: C, 59.44; H, 9.68; Found C, 59.70; H, 9.65; [α]$_D^{20}$ +3.0° (c = 1.4, CHCl$_3$) for a >98% ee sample.

Cu-catalyzed conjugated addition of Et$_2$Zn to 4,4-(dimethyl)cyclohexenone in the presence of L25 [92]

A 25 ml round-bottomed flask was charged with Cu(OTf)$_2$ (7.2 mg, 0.02 mmol), ligand L25 (20.8 mg, 0.050 mmol) and 5 ml CH$_2$Cl$_2$. The resulting suspension was stirred for 20 min, after which time neat R$_2$Zn (3.0 mmol) and enone (124 mg, 1.0 mmol) were added, 10 min apart. When the reaction was complete (as monitored by TLC), 1 M HCl (5 ml) was added (*dropwise at first!*) and the mixture extracted with 3 × 10 ml of ether. The combined organics were dried over MgSO$_4$, filtered, and concentrated. The residue was purified by flash chromatography (19 : 1 ether : pentane) to give 139 mg (90%) of 3-(*R*)-ethyl-4,4-dimethylcyclopentanone. The ee-value (96%) was determined by GCAstec Chiraldex G-TA (30 m × 0.25 mm × 0.125 μm) with an Astec Retention Gap Guard Column (5 m × 0.25 mm), 1.0 ml min^{-1} He flow, 100 °C.

Cu-catalyzed conjugated addition of diethylzinc to 3,3-dimethoxy-1-nitro-1-propene in the presence of L5a [67]

In a Schlenk tube with a magnetic stirring bar under an argon atmosphere, a mixture of 0.005 mmol of Cu(OTf)$_2$ (1.8 mg, 0.005 mmol) and phosphoramidite ligand **L5a** (6.2 mg, 0.01 mmol) was dissolved in 5 ml toluene, and the solution stirred at r.t. for 30 min. To this solution was added 3,3-dimethoxy-1-nitro-1-propene (78 mg, 0.5 mmol). The mixture was cooled to −65 °C, and Et$_2$Zn (0.75 ml of 1.0 M solution in toluene, 0.75 mmol) added slowly via a syringe; the mixture was maintained at this temperature for 6 h, with stirring. The reaction mixture was then poured into saturated aqueous NH$_4$Cl and extracted three times with ether; the combined organic layers were dried over MgSO$_4$. The crude product (1,1-dimethoxy-2-(nitromethyl)butane was concentrated under reduced pressure and purified by column chromatography on silica gel using hexanes/ethyl acetate as eluant. Supelco ß-Dex-120 column (30 m × 0.25 mm), 110 °C isothermic, t_R = 33.5 min (+)/34.3 min (−). ^{1}H NMR (300 MHz, CDCl$_3$) δ 4.56 (m, 1H), 4.31(m, 2H), 3.38 (s, 3H), 3.37 (s, 3H), 2.50 (m, 1H), 1.64 (m, 1H), 1.43 (m, 1H), 0.99 (t, J = 7.5 Hz, 3H). ^{13}C NMR (75 MHz, CDCl$_3$) δ 105.3, 75.2, 55.9, 54.7, 42.0, 20.4, 10.9.

Cu-catalyzed conjugated addition of Ph$_2$Zn to 3-methylcyclohexenone in the presence L38b [116]

An oven-dried 13 × 100 mm test tube was charged with chiral Ag complex **L38b** (5.6 mg, 0.0045 mmol), (CuOTf)$_2$C$_6$H$_6$ (2.7 mg, 0.0045 mmol), 3-methyl-2-cyclohexenone (20 mg, 0.18 mmol), and Ph$_2$Zn (120 mg, 0.55 mmol), which were weighed out under a N$_2$ atmosphere in a glove box (in the precise order mentioned above). The test tube was sealed with a septum and wrapped with Parafilm. The reaction vessel was removed from the glove box, after which diethyl ether (1.0 ml) was slowly added to the mixture in a dropwise manner (syringe) at −78 °C. After addition of the Et$_2$O, the solution was allowed to warm to −30 °C. After 48 h at −30 °C, the reaction was quenched by the addition of a saturated aqueous solution of ammonium chloride (1.0 ml), followed immediately by H$_2$O (1.0 ml). The mixture was washed with Et$_2$O (3 × 1 ml) and the combined organic layers were passed through a short plug of silica by elution with Et$_2$O. The volatiles were removed *in vacuo* and the resulting mixture was purified by silica gel column chromatography (hexanes/Et$_2$O, 10 : 1) to afford 32.5 mg (0.173 mmol, 95.0% yield) of (S)-3-methyl-3-phenylcyclohexanone as a colorless oil. **Important note**: To ensure high efficiency and enantioselectivity, the reactions must be set up in exactly the order described above. IR (neat): 3087 (w), 3058 (w), 3024 (w), 3022 (w), 2962 (s), 2937 (s), 2870 (m), 2357 (w), 2332 (w), 1715 (s), 1602 (w) cm^{-1}; ^{1}H NMR (400 MHz, CDCl$_3$): δ 7.35 − 7.30 (4H, m), 7.26−7.18 (1H, m), 2.89 (1H, d, J = 14.4 Hz), 2.44 (1H, d, J = 14.4 Hz), 2.31 (2H, t, J = 6.8 Hz), 2.22−2.16 (1H, m), 1.96−1.83 (2H, m), 1.72−1.61 (1H, m), 1.33 (3H, s); ^{13}C

NMR (100 MHz, CDCl$_3$): δ 211.6, 147.5, 128.6, 126.3, 125.7, 53.2, 42.9, 40.9, 38.0, 29.9, 22.1; Analysis calculated for C$_{13}$H$_{16}$O: C, 82.94; H, 8.57; Found C, 82.67; H, 8.28; HRMS calculated for C$_{13}$H$_{16}$O (EI+): 188.1201, Found: 188.1199. Optical rotation: [α]$_D^{20}$ +70.18 ($c = 1.000$, CHCl$_3$) for a 97% ee sample. Optical purity was determined by chiral GLC analysis in comparison with authentic racemic material [97% ee sample below; conditions: CDGTA column, 95 °C (70 min), 0.5 °C min^{-1} to 110 °C (30 min), 20 °C min^{-1} to 140 °C (20 min), 15 psi].

Cu-catalyzed conjugated addition of [AcO(CH$_2$)$_4$]$_2$Zn to *tert*-butyl 2-methyl-6-oxocyclohex-1-enecarboxylate in the presence L55a [136]

CuCN (0.3 mg, 3 μmol) and ligand **L55a** (1.4 mg, 3.0 μmol) were weighed into a 13 × 100 mm test tube. Toluene (1 ml) was added and the tube sealed with a rubber septum and purged with nitrogen. The suspension was stirred at 23 °C for 3 h and then cooled to 0 °C. [AcO(CH$_2$)$_4$]$_2$Zn (0.23 mmol) was added and the pale yellow solution stirred at 0 °C for 5 min. A solution of *tert*-butyl 2-methyl-6-oxocyclohex-1-enecarboxylate (31 mg, 0.148 mmol) in toluene (0.5 ml) was added via a cannula. The septum was wrapped with Teflon tape and stirred for a further 41 h at −15 °C. The reaction solution was cooled to −78 °C, diluted with EtOAc (1 ml), and then quenched with aqueous 1 M HCl (1 ml). The upper organic layer was separated and the aqueous layer washed with EtOAc (3 × 1 ml). The combined organic layers were concentrated *in vacuo*, and the residue was purified by silica gel chromatography (6 : 1 hexanes : Et$_2$O) to yield (+)-*tert*-butyl 2-(4-acetoxybutyl)-2-methyl-6-oxocyclohexanecarboxylate as a pale yellow oil (36.0 mg, 0.11 mmol, 74%). IR (neat): 2947 (m), 2874 (w), 1741 (s), 1708 (s), 1460 (w), 1367 (m), 1242 (s), 1151 (m); ^{1}H NMR (400 MHz, CDCl$_3$), 1 : 1 : 0.4 mixture of two keto diastereomers and enol tautomer; NMR data reported only for keto diastereomers: δ 4.07 − 4.02 (2H, m, CH$_2$OAcdiast 1,2 and enol), 3.11 (1H, s, C(= O)CHCO$_2$*t*-Bu), 3.08 (1H, s, C(= O)CHCO$_2$*t*-Bu), 2.79 (1H, ddd, *J* = 14.3, 10.6, 7.1 Hz, aliphatic CH), 2.66 (1H, ddd, *J* = 14.3, 7.8, 6.2 Hz, aliphatic CH), 2.34−2.20 (2H, m, aliphatic CH), 2.09 (1H, ddd, *J* = 14.0, 11.0, 4.5 Hz, aliphatic CH), 2.04 and 2.04 (3H, s, C(= O)CH$_3$), 2.04 (3H, s, C(= O)CH$_3$), 1.96−1.74 (4.4H, m, aliphatic CH), 1.66−1.52 (6.5H, m, aliphatic CH), 1.45 (9H, s, OC(CH$_3$)$_3$), 1.44 (9H, s, OC(CH3)3), 1.42−1.20 (8.6H, m, aliphatic CH), 1.04 (3H, s, CH$_2$CCH$_3$), 0.92 (3H, s, CH$_2$CCH$_3$); ^{13}C NMR (100 MHz, CDCl$_3$) δ 207.7, 207.6, 171.2 (2), 168.1, 168.0, 81.8, 81.7, 68.2, 67.2, 64.5, 64.4, 41.5, 41.3, 39.8, 39.6, 39.1, 38.7, 33.3, 32.6, 29.4, 29.3, 28.2 (3), 28.2 (3), 23.6, 22.7, 21.7, 21.5, 21.1 (2), 19.9, 19.8; Analysis calculated for C$_{18}$H$_{30}$O$_5$: C, 66.23; H, 9.26; Found C, 66.28; H, 9.47; HRMS calculated for C$_{18}$H$_{30}$O$_5$ (M): 327.2171, Found: 327.2169; [α]$_D$20 +4.6 ° ($c = 1.0$, CHCl$_3$) for a sample of 72% ee.

2.5
Ligands for Ni-Catalyzed Enantioselective Conjugated Additions

Compounds used as ligands in Ni-catalyzed conjugated addition can be classified according to their structural features, and comprise the following classes: amino alcohols; aminothiolates; pyridino-alcohols; diamines; aminoamides; and sulfoximines. In all of these cases, the conjugated addition of (alkyl)$_2$Zn (usually Et$_2$Zn) to chalcone and or its substituted derivatives was studied. The use of substituted organozincs or other substrates (cycloenones, etc.) has not been reported. In almost all cases, the reactions were run in polar MeCN. A probable reason for this might be that Ni-catalysts are insoluble in non-polar, non-coordinating solvents such as toluene. In general, the Ni-catalyzed reaction has not reached such high levels of enantioselectivity as have been observed for copper-catalyzed reactions.

2.5.1
Amino Alcohols and Aminothiolates

Amino alcohol functionality attached to various scaffolds, along with its *N*- or *O*-substituted derivatives, constitutes a large group of compounds that have been used as chiral ligands for enantioselective conjugated additions (Figure 2.18; Table 2.20).

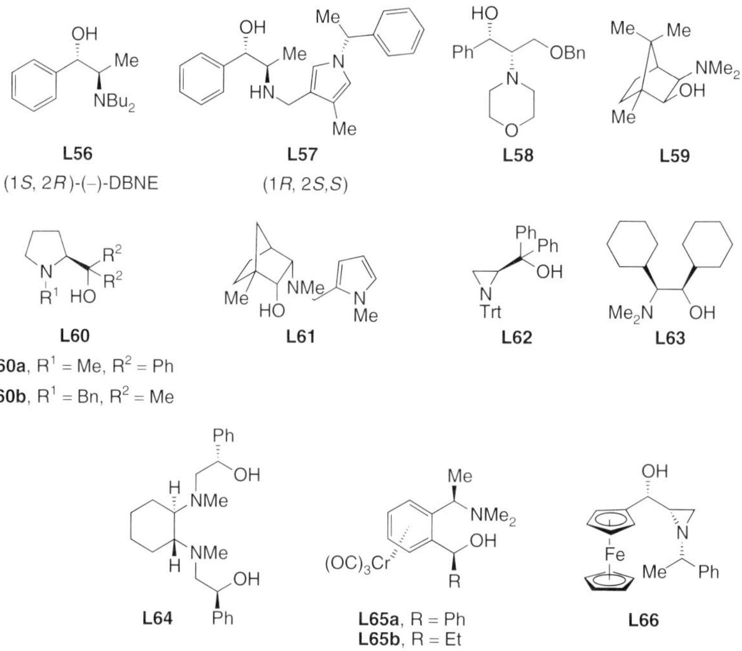

Figure 2.18 Amino alcohols used as ligands in Ni-catalyzed conjugated addition.

2.5 Ligands for Ni-Catalyzed Enantioselective Conjugated Additions

Table 2.20 Conjugated addition in the presence of various chiral amino alcohol ligands L56–L66.

$$Ph\diagup\!\!\!\diagdown\!\!\!\diagdown C(O)Ph + Et_2Zn \xrightarrow[-30-23\,°C,\,MeCN]{Ni\text{-salt, ligand }\mathbf{L56\text{-}L66}} Ph\diagup\!\!\!\diagdown CH(Et)\!\!-\!\!CH_2\!\!-\!\!C(O)Ph$$

Ni-salt	mol %	Ligand	mol %	Additive	Conditions	Yield (%)[a]	ee (%)	Reference
NiBr$_2$	25	L56	30		Toluene, −30 °C	32	55 (S)	[137a]
Ni(acac)$_2$	7	L56	17	bipy	MeCN, −30 °C	47	90 (R)	[137b]
Ni(acac)$_2$	1	L57	20		MeCN, −30 °C	70	53 (R)	[138]
Ni(acac)$_2$	1	L58	20		MeCN, −40 °C	64	88 (R)	[139]
Ni(acac)$_2$	7	L59	16		MeCN, −30 °C	94	59 (R)	[140]
Ni(acac)$_2$	7	L60a	16		MeCN, −30 °C	94	25 (R)	[140]
Ni(acac)$_2$	10	L60b	20	bipy	MeCN, −30 °C	87	59 (S)	[141]
Ni(acac)$_2$	7	L61	16		MeCN, −25 °C	83	83 (S)	[142]
Ni(acac)$_2$	9	L62	10		MeCN, −30 °C	75	78 (R)	[143]
Ni(acac)$_2$	7	L63b	17	bipy	MeCN, −30 °C	89	82 (S)	[144]
Ni(acac)$_2$	16	L64	16		MeCN, −30 °C	72	37 (R)	[145]
Ni(acac)$_2$	5	L65a	50		MeCN, −30 °C	90	62 (R)	[146]
Ni(acac)$_2$	5	L65b	50		MeCN, −30 °C	91	43 (R)	[146]
Ni(acac)$_2$	1	L66	35		MeCN, −35 °C	78	82 (R)	[147]

[a] Isolated yields.

Ephedrine derivatives such as N,N-dibutylnorephedrine **L56** were used in combination with NiBr$_2$ [137a] or Ni(acac)$_2$ [137b] for the catalytic addition of Et$_2$Zn and Bu$_2$Zn to unsaturated ketones. In the former case, the asymmetric induction was rather mediocre, with ee-values in the range of 14 to 55%. However, the combination of Ni(acac)$_2$/bipyridine/**L56** led to an improved enantioselectivity that yielded products with ee-values up to 90%. A later attempt to modify the norephedrine framework by attaching the pyrrole ring bearing an additional center of chirality produced ligand **L57**; however, this led only to a mediocre enantioselectivity of 53% ee [138]. On the other hand, a high enantioselectivity in the addition of Et$_2$Zn was achieved with the ligand **L58** derived from readily available (1S,2S)-2-amino-1-phenyl-1,3-propanediol [139]. The addition product to chalcone was obtained in 88% ee.

Enantioselective Ni-catalyzed conjugated addition was intensively studied by Ferringa et al. who screened a number of various coordinating compounds, including amino alcohols, as potential chiral ligands [140]. Among the tested compounds the best results in the addition of Et$_2$Zn to chalcone were obtained with the amino alcohol ligand having cis-exo-borneol **L59** (59% ee) and prolinol **L60a** (25% ee) frameworks. Interestingly, when the reaction was carried out in PrCN instead of MeCN and at −50 °C, the enantioselectivity rose to 84%. Variously substituted prolinols **L60** (R^1 = H, Me, Bn, R^2 = Me, Et, Ph) were also prepared by Wang *et al.*, and tested as ligands for the conjugated addition of Et$_2$Zn to chalcone [141]. The highest enantioselectivity (59% ee) was achieved with **L60b** in the presence of bipyridine. Ferringa also used another natural product, (+)-camphor, as the scaffold for synthesis of tridentate and terdentate ligands [142]. In this case, the highest ee-value (83%) in the addition of Et$_2$Zn to chalcone was obtained with *cis-endo*-3-aminoborneol derivative **L61** bearing *N*-methylpyrrole moiety in the side chain. The change to the pyridine derivative or the use of tetradentate ligands did not lead to any improvement in asymmetric induction.

Nayak *et al.* introduced the aziridine methanol **L62** as a chiral ligand, which yielded the addition product to chalcone in 78% ee [143]. Chan *et al.* prepared a chiral ligand with 1,2-aminoalcohol moiety **L63** bearing two cyclohexane rings [144], and this showed a considerably higher enantioselectivity in the addition of Et$_2$Zn to chalcone (82% ee) in the presence of 2,2'-bipyridine as an additive. It should be added that its congener, bearing two phenyl rings instead of the cyclohexane rings, exhibited a much lower enantioselectivity under the same reaction conditions (33% ee). A series of ligands containing two amino alcohol moieties based on chiral 1,2-diaminocyclohexane framework were synthesized by Marson *et al.* [145]. Unfortunately, their use resulted only in low asymmetric induction, with only 37% ee being achieved with **L64**.

Chiral amino alcohol ligands bearing an organometallic framework were also used. The chiral (1,2-disubstituted arene)chromium complexes **L65a** and **L65b** were used as ligands for the Ni-catalyzed conjugated addition [146]. The asymmetric induction was seen to depend heavily on the amount of ligand, as well as the nickel:ligand ratio. A series of chiral diastereoisomeric aziridinyl ferrocenemethanols was prepared by Dogan *et al.* and tested as ligands for conjugated addition [147]. In this case, the best asymmetric induction (82% ee) was obtained with the (*R,S,S*)-diastereoisomer **L66**.

2.5.2
Aminothiolates and Thioethers

A similar level of enantioselectivity in the Ni-catalyzed addition of Et$_2$Zn to chalcone was also achieved with aminothiolates **L67a** or disulfides **L67b** prepared by Gibson [148] (Figure 2.19). Aminothiol **L68**, prepared by Kang *et al.* [149], was tested in a similar fashion. Although, the enantioselectivity of the conjugated addition with ligands **L67** could not match those obtained with amino alcohols, the use of ligand **L68** gave comparable results (Table 2.21).

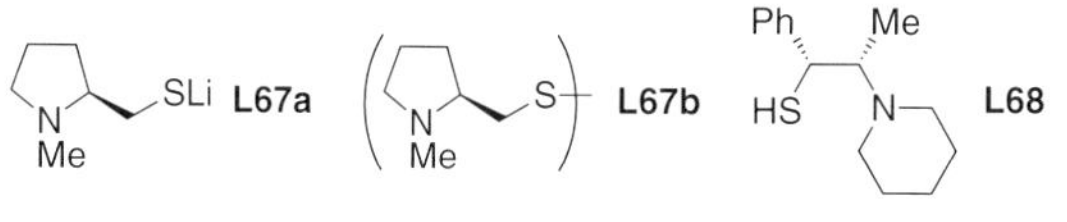

Figure 2.19 Aminothiolates and disulfides used in the Ni-catalyzed conjugated addition.

2.5 Ligands for Ni-Catalyzed Enantioselective Conjugated Additions

Table 2.21 Enantioselective conjugated addition in the presence of ligands **L67–L73**.

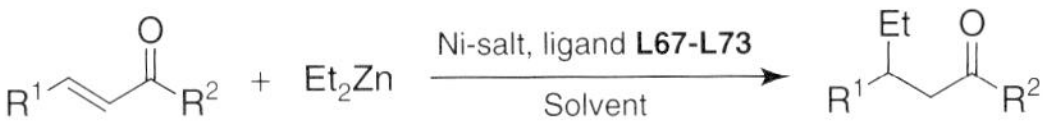

$$R^1\text{—CH=CH—C(O)—}R^2 + Et_2Zn \xrightarrow[\text{Solvent}]{\text{Ni-salt, ligand L67-L73}} R^1\text{—CH(Et)—CH}_2\text{—C(O)—}R^2$$

Ni-salt	mol (%)	Ligand	mol (%)	Conditions	R¹	R²	Yield^a (%)^a	ee (%)	Reference(s)
Ni(acac)₂	7	L67a	17	MeCN, −30 °C	Ph	Ph	87	37 (S)	[148]
Ni(acac)₂	7	L67b	17	MeCN, −30 °C	Ph	Ph	85	50 (S)	[148]
NiCl₂	25	L68	25	MeCN, −30 °C	Ph	Ph	50	74 (S)	[149]
Ni(acac)₂	1	L69	20	MeCN, −30 °C	Ph	Ph	75	72 (R)	(150, 151)
Ni(acac)₂	1	L69	10	MeCN, −30 °C	4-MeOPh	Ph	68	74 (-)	[150]
Ni(acac)₂	1	L70a	20	MeCN, −30 °C	Ph	Ph	62	86 (S)	[151]
NiCl₂	1.5	L71	30	MeCN, −30 °C	Ph	Ph	75	82 (S)	[152]
Ni(acac)₂	5	L72a	10	MeCN, −20 °C	Ph	Ph	97	78 (S)	[154]
Ni(acac)₂	5	L72a	10	MeCN, −20 °C	4-MeOPh	Ph	99	84 (S)	[154]
Ni(acac)₂	5	L72b	10	Hexane/THF, −10 °C	Ph	Ph	80	91 (R)	[155]
Ni(acac)₂	5	L72b	10	Hexane/THF, −10 °C	Ph	Me	74	95 (R)	[155]
Ni(acac)₂	1	L73	20	PrCN, −30 °C	Ph	Ph	71	70 (R)	[155]

^{a} Isolated yields.

2.5.3
Pyridino-Alcohols

Another group consisting of bidentate 2,2'-bipyridine with bulky hydroxymethyl groups in the 3,3'-positions, **L69**, was prepared by Bolm [150, 151]. The use of these compounds in the Ni-catalyzed enantioselective addition of Et₂Zn gave rise to products with ee-values up to 74%. Studies using similar, but monodentate, chiral 3-hydroxymethyl-6-arylpyridines **L70a** and **L70b** led to an improvement of the enantioselectivity, up to 86% ee [151, 152] (Figure 2.20; Table 2.21).

2.5.4
Diamines

The obvious choice for the next ligand class were diamines derived from natural amino acids, by Asami *et al.* [153] (Figure 2.21). Here, the highest enantioselectivity (82% ee) was achieved with the *n*-pentyl-substituted

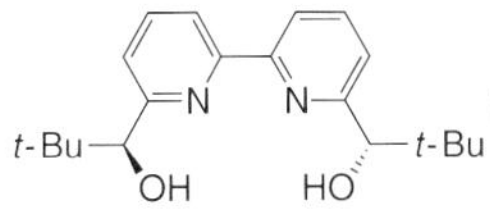

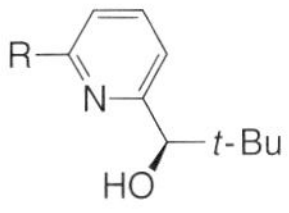

Figure 2.20 Pyridine derivatives used in the Ni-catalyzed conjugated addition.

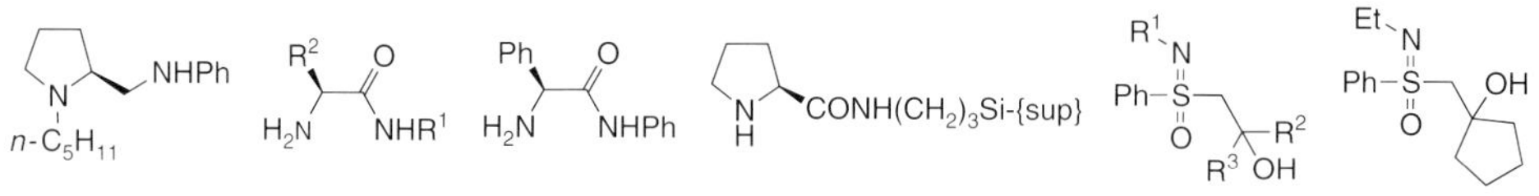

Figure 2.21 Diamines, aminoamides, and sulfoximines used in Ni-catalyzed conjugated addition.

ligand **L71**. Interestingly, the use of $NiCl_2$, rather than the more commonly used $Ni(acac)_2$ as the starting salt for the complex preparation, gave better results with respect to enantioselectivity and yields (Table 2.21).

2.5.5
Aminoamides

Chiral α-aminoamides **L72**, derived from the corresponding natural and man-made amino acids, were also tested as ligands for the Ni-catalyzed asymmetric conjugated addition of Et_2Zn [154]. However, despite the fact that a vast number of ligands with side chain modifications was synthesized, the asymmetric induction did not exceed 78% ee in the case of addition to chalcone in the presence of **L72a** (Table 2.21). The highest ee-value (84%) was achieved in the addition to p-methoxy-substituted chalcone. Ni-complexes with proline amides supported on silica gel or zeolites were also tested [155], with a high ee-value (91%) being achieved in the conjugated addition to chalcone with the **Ni/L72b** complex supported on zeolite.

2.5.6
Sulfoximines

A number of sulfoximines **L73** bearing various functional groups were tested in conjugated addition to chalcone by Bolm [156]. The highest asymmetric induction (70% ee) was obtained with the cyclopentane derivative **L73a**.

2.5.7
Cyanobisoxazolines

Chiral cyanobisoxazolines **L74** were reported to be good ligands for the asymmetric conjugated addition of alkynylaluminums to cycloenones by Corey et al. [157]. Using a cyanobisoxazoline nickel complex, the addition of (trimethylsilylethynyl)alane to cyclohexenone proceeded with good enantioselectivity of 88% ee (Scheme 2.26). Although this was the sole example, it constituted a promising strategy for the further development of asymmetric alkynylation.

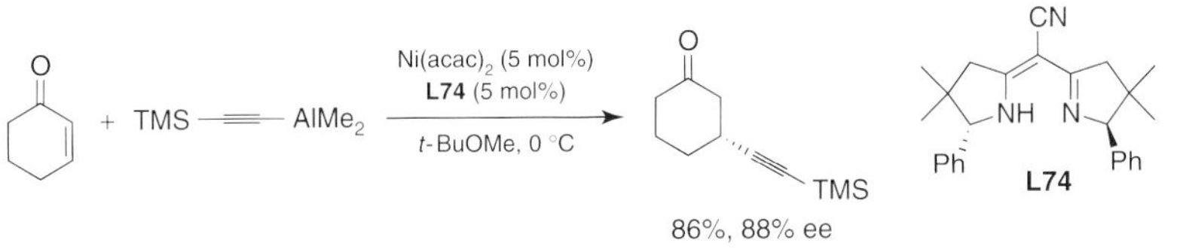

Scheme 2.26 Enantioselective alkynylation of cyclohexenone.

2.6
Application of Conjugated Additions in the Synthesis of Natural Compounds

The synthesis of natural or biologically active compounds often constitutes the culmination and final verification of a methodology development. The successful application of a synthetic method confirms it as a convenient and highly effective tool with, at least potentially, a broad application in the area of organic synthesis. In this regard, the conjugated addition constitutes in general a powerful method for constructing C–C bonds and creating centers of chirality.

The catalytic conjugated addition of organozinc and organoaluminum compounds conveyed by Cu-or Ni-compounds represents only a small part of all possible reaction variants, but has huge synthetic potential. It has been used in a number of syntheses of natural or biologically active compounds, either in nonchiral or enantioselective reactions. Of these modes, the latter is especially attractive as it enables the construction of chiral tertiary or quaternary carbon centers with high asymmetric induction.

One of the main challenges in organic synthesis is the implementation of multicomponent sequential reactions so as to achieve multiple bond formation in one operation since, by using this strategy, it is possible to avoid the use of protecting groups and the isolation of intermediates. One advantage of a conjugated addition is that it gives rise to an enolate that itself could be used in a reaction with a nucleophile (e.g., aldehydes [54], alkyl halide [158], and imine [159]). In such a manner, the one-pot conjugated addition/alkylation reaction sequence gives rise to a product with two new functionalities. Moreover, the reaction sequence usually proceeds with a high diastereoselectivity. Such an approach has been applied successfully to the total syntheses of natural products, typical examples of which include the prostaglandins, pumiliotoxin, and phthiocerol (see below).

2.6.1
Application of Non-Asymmetric Conjugated Additions

2.6.1.1 Synthesis of (±)-β-Cuparenone

Cuparenone is sesquiterpene natural product which is present, for example, in *Thuya orientallis* and *Mannia fragrans*. A rare application of the Ni-catalyzed conjugated addition of organoaluminums was used in the synthesis of (±)-β-cuparenone [48], where the final step of the cuparenone synthesis was based on the conjugated addition of Me_3Al to sterically hindered 3-(p-tolyl)-4,4-dimethylcyclopentenone in

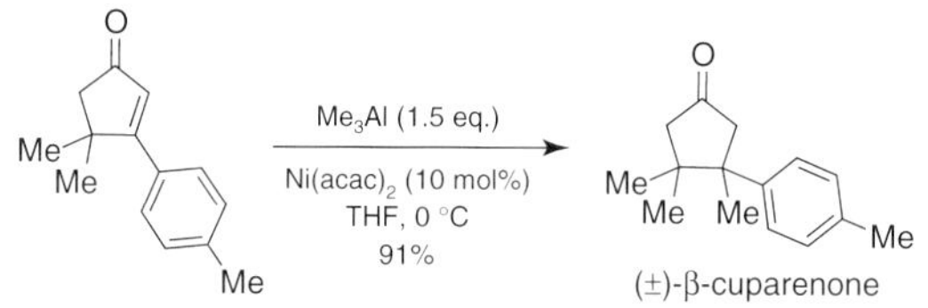

Scheme 2.27 Flemming's synthesis of (±)-β-cuparenone.

the presence of Ni(acac)$_2$, resulting in the construction of one of two quaternary carbon centers. The reaction furnished the target compound in high yield (91%) (Scheme 2.27).

Two syntheses of (±)-β-cuparenone based on the Ni-catalyzed conjugated addition of organozincs were reported by Luche et al. The first of these was based on the addition of (p-tolyl)$_2$Zn to 3,4,4-trimethylcyclopentenone [160], and the second on the addition of Me$_2$Zn to 3-(p-tolyl)-4,4-dimethylcyclopentenone [161]. Recently, the Ni-catalyzed conjugated addition of Me$_2$Zn has been also applied in a new synthesis of (±)-β-cuparenone [162]. A nice example of the Ni-catalyzed conjugated addition of a diarylzinc to a sterically hindered enone was demonstrated by Sakai et al. [163]. For this, the addition of (p-tolyl)$_2$Zn catalyzed by Ni(acac)$_2$ to a chiral trisubstituted cyclopentenone proceeded stereoselectively to give the chiral cis-substituted (hydroxy and p-tolyl groups) cyclopentanone. It is assumed that the cis-stereoselectivity of the addition was controlled by the chelation between the organozinc compound and the hydroxy group on C4. Further functional group manipulations furnished chiral cyclopentenol that served as an important intermediate for the syntheses of (−)-α-cuparenone, (+)-β-cuparenone, cuparene, and laurane (Scheme 2.28).

2.6.1.2 Synthesis of a Guanacastepene Intermediate

Guanacastepene is the fungal-derived diterpene antibiotic that was isolated from an extract of a fungus from the branch of the *Daphnopsis americana* tree. Conjugated

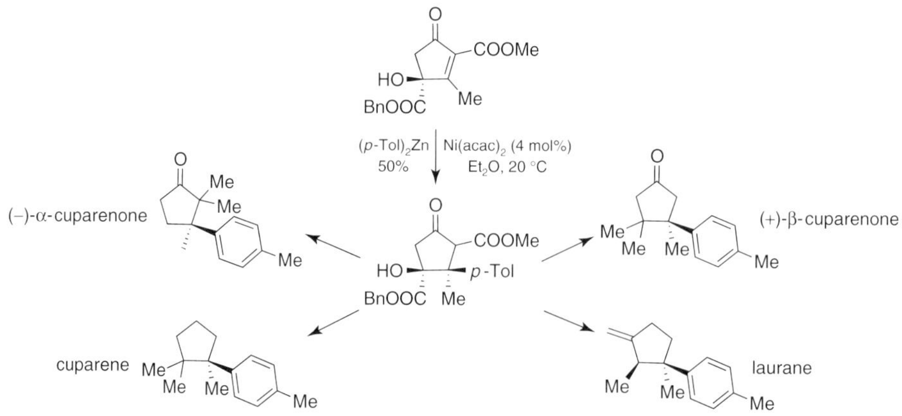

Scheme 2.28 Sakai's synthesis of (-)-α-cuparenone, (+)-β-cuparenone, cuparene, and laurane.

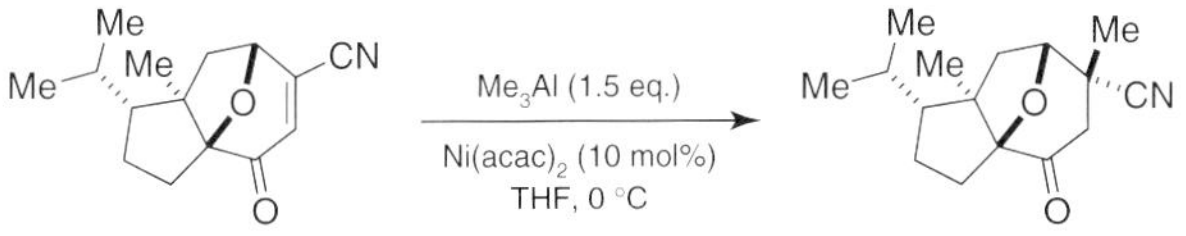

Scheme 2.29 Magnus' synthesis of a guanacastepene intermediate.

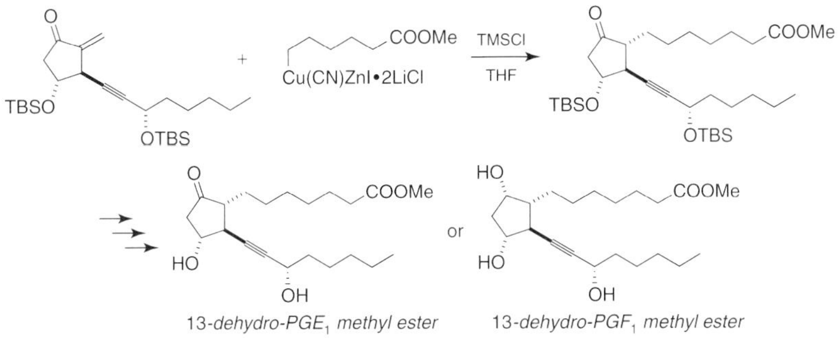

Scheme 2.30 Sato's synthesis of prostaglandins (PGs).

addition, which was one of the important steps in the synthesis of its functionalized hydroazulene (bicyclo[5.3.0]decane) portion, consisted of the addition of methyl group to the α,β-unsaturated moiety of the cycloheptenone ring [164]. Among several possible synthetic methods that brought about the conjugated addition, the best selectivity for 1,4-addition and yields was obtained by the addition of Me$_3$Al catalyzed by Ni(acac)$_2$ (Scheme 2.29).

2.6.1.3 Synthesis of Prostaglandins

The attachment of a side chain onto the cyclopentene framework of a prostaglandin (PG) is often conveniently achieved by carrying out a conjugated addition to a suitable unsaturated ketone. The conjugated addition of a functionalized organometallic reagent to methylenecyclopentanone was the key step in the synthesis of 13-dehydro-PGE$_1$ and PGF$_1$ [165]. The organozinc compound was prepared from the corresponding diorganozinc compound and CuCN·2LiCl (Scheme 2.30). Although this reaction should be rather considered as an addition of the zinc-based cuprate rather than the Cu-catalyzed process, it deserves mention at this point because it is a representative example of the synthetic application of organozinc compounds.

2.6.1.4 Synthesis of (±)-Scopadulcic Acid

A conjugated addition constitutes an important step in the synthesis of (±)-scopadulcic acid, notably because it was responsible for the creation of a quaternary carbon center. The conjugated addition of Me$_2$Zn catalyzed by Ni(acac)$_2$ to sterically hindered ketone proceeded stereoselectively from the less-hindered β-face of the enone, giving rise to the intermediate with a correct relative spatial

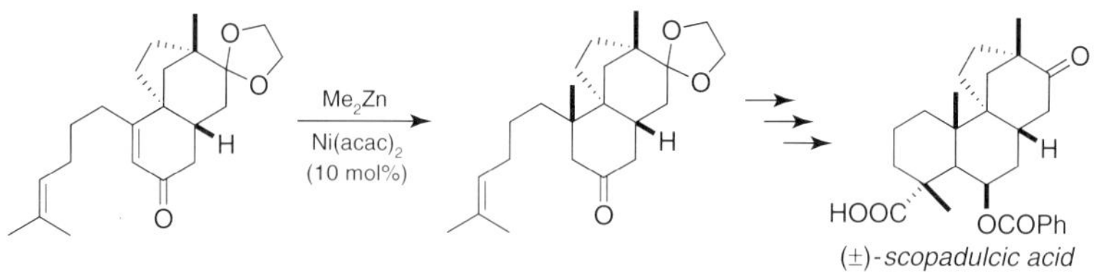

Scheme 2.31 Ziegler's synthesis of (±)-scopadulcic acid.

arrangement [166] (Scheme 2.31). Notably, another Cu-catalyzed diastereoselective conjugated addition was also used during the course of the synthesis.

2.6.2
Asymmetric Conjugated Additions

2.6.2.1 Synthesis of Prostaglandins

The core framework of PGE_1 was constructed by a three-component coupling of an enone, a functionalized organozinc compound, and an aldehyde. This was based on the relay 1,4-addition–enolate trapping reaction [167], in which the enantioselective conjugated addition of the ester functionalized organozinc reagent to cyclopentenone was catalyzed by a combination of $Cu(OTf)_2$ with the phosphoramidite ligand **L2b**, after which the formed enolate was trapped by reaction with the aldehyde (Scheme 2.32). This reaction afforded the required substituted cyclopentanone in 60% isolated yield as a mixture of *trans-threo*/*trans-erythro* diastereoisomers in 83/17 ratio. After reduction of the ketone group to alcohol, a crucial intermediate with the structural and stereochemical features of PGE1 was obtained in 94% ee.

2.6.2.2 Formal Synthesis of Clavukerins

Another sequential process based on conjugated addition–cyclopropanation reactions was used for the enantioselective preparation of a key intermediate

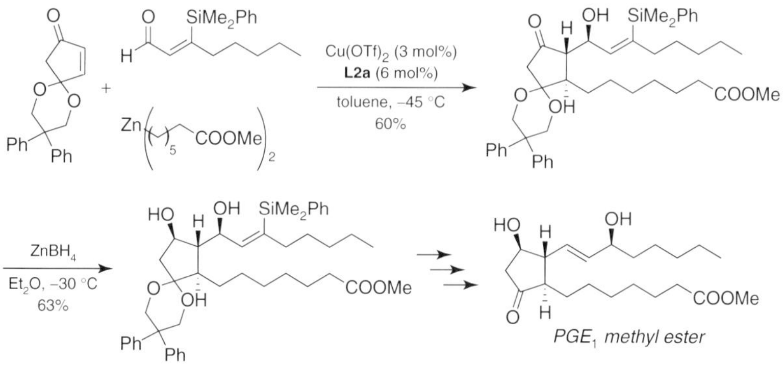

Scheme 2.32 Feringa's synthesis of prostaglandins (PGs).

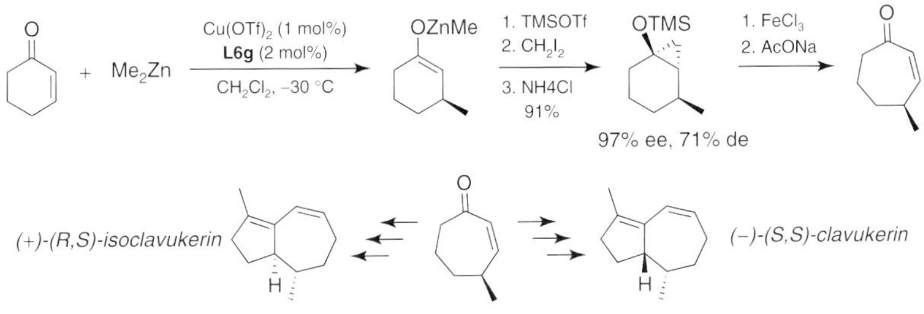

2.6 Application of Conjugated Additions in the Synthesis of Natural Compounds

Scheme 2.33 Alexakis' synthesis of clavukerins.

suitable for the synthesis of the sesquiterpenoids, (−)-(S,S)-clavukerin and (+)-(R,S)-isoclavukerin [168]. The synthesis started from cyclohexenone and the conjugated addition of Me$_2$Zn that was catalyzed by Cu(OTf)$_2$ in the presence of the chiral phosphoramidate ligand **L6g**. Silylation of the obtained zinc enolate was followed by cyclopropanation with diiodomethane, giving rise to a bicyclo[4.1.0]bicycloheptane in high yield (91%) and enantioselectivity (97% ee). Although the diastereoselectivity of the cyclopropanation was rather low (71% de), the disappearance of the stereocenters in the next step – that is, a ring expansion reaction to cycloheptenone – made this value unimportant (Scheme 2.33).

2.6.2.3 Synthesis of Muscone

A catalytic conjugated addition was used twice in a straightforward enantioselective synthesis of (R)-muscone, the principal component of musk. The first approach relied on the conjugated addition of Me$_2$Zn to (E)-cyclopentadecenone, catalyzed by Cu(OTf)$_2$ in the presence of phosphite **L30a**; unfortunately, this proceeded with a rather moderate enantioselectivity of 78% ee [98]. Although a similar result was obtained with the related phosphite **L30b** (63% ee) [99], the second approach used a combination of Me$_3$Al and Cu(MeCN)$_4$PF$_6$ in the presence of phosphine **L-26** [93]. In this instance, the desired product was also formed with only a modest enantioselectivity (77% ee) (Scheme 2.34).

2.6.2.4 Synthesis of (+)-Ibuprofen

(+)-Ibuprofen is a well-known, non-steroid anti-inflammatory drug, the feature synthetic step of which involves construction of the benzylic chiral center. In this case, such construction was implemented by the conjugated addition of Me$_3$Al to a suitably substituted nitrostyrene catalyzed by CuTC in the presence of the

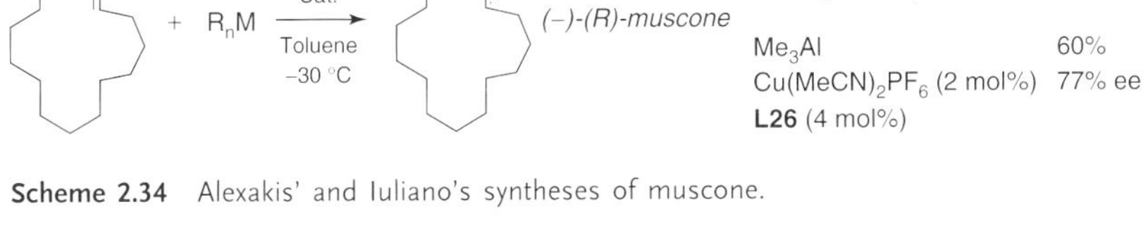

Scheme 2.34 Alexakis' and Iuliano's syntheses of muscone.

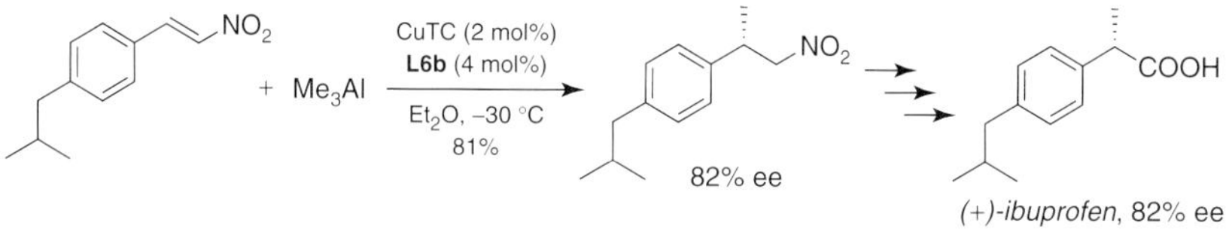

Scheme 2.35 Alexakis' synthesis of (+)-ibuprofen.

chiral phosphoramidite ligand **L6b** [169]. The reaction proceeded in good yield (81%) with acceptable enantioselectivity (82% ee). Further conversion of the nitro moiety resulted in the target compound formation without any loss of optical purity (82% ee) (Scheme 2.35).

2.6.2.5 Synthesis of Erogorgiaene

Erogorgiaene is antimycobacterial agent (an inhibitor of *Mycobacterium tuberculosis*), during the synthesis of which two enantioselective conjugated addition reactions are exploited. The first reaction was used to introduce a center of chirality into a benzylic position (this later became part of the tetrahydronaphthalene framework), while the second reaction was used to establish the center of chirality in the side chain [82]. Both conjugated additions were carried out with Me$_2$Zn in the presence of a catalytic amount of Cu(OTf)$_2$·C$_6$H$_6$ and the chiral phosphinopeptide **L13b**, with an excellent degree of enantioselectivity (Scheme 2.36).

2.6.2.6 Synthesis of (−)-Pumiliotoxin C

(−)-Pumiliotoxin C, a potent neurotoxin which acts as a noncompetitive blocker of acetylcholine receptor channels, was first isolated from *Dendrobates pumilio*, a poison dart frog. One of the key steps of its synthesis was the conjugated addition of Me$_2$Zn to cyclohexenone, catalyzed by Cu(OTf)$_2$ in the presence of chiral phosphoramidate ligand **L2c** [170]. The formed zinc enolate was then trapped with allyl acetate in the presence of [Pd(PPh$_3$)$_4$] to give the chiral 2,3-disubstituted

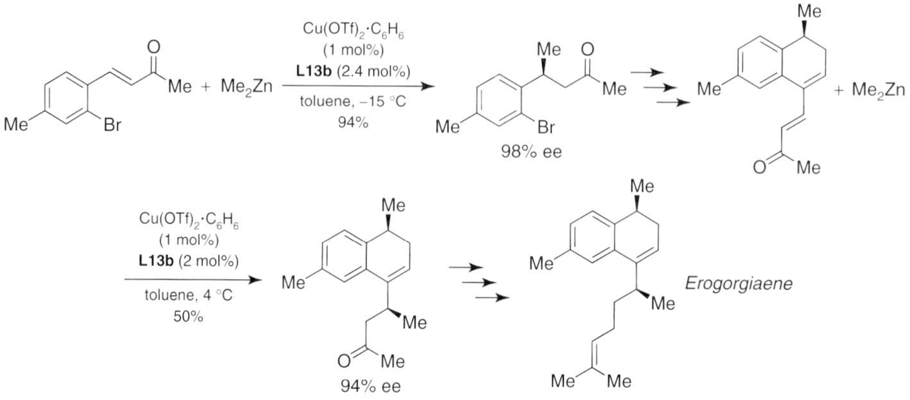

Scheme 2.36 Hoveyda's synthesis of erogorgiaene.

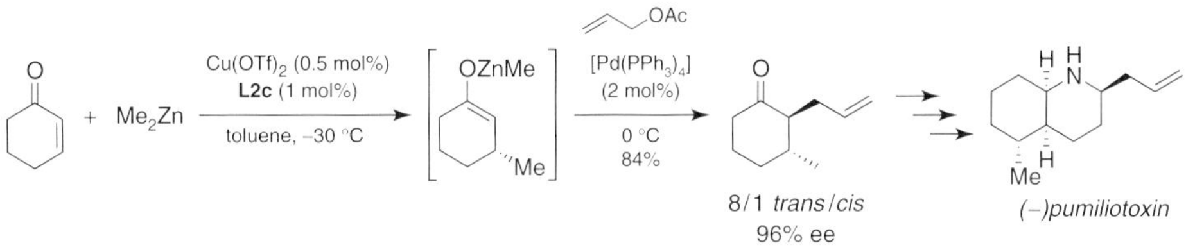

Scheme 2.37 Feringa's synthesis of pumiliotoxin C.

cyclohexanone as an 8 : 1 *trans* : *cis* mixture in 84% yield and with an excellent stereoselectivity (96% ee) (Scheme 2.37). Further C–C bond-formation reactions and functional group transformations yielded the target compound.

2.6.2.7 Synthesis of Phthiocerol

Phthiocerol dimycocerosate A (PDIM A), a lipid material from which the cell envelope of *Mycobacterium tuberculosis* is constructed, consists of two tetramethyl-substituted saturated acids (mycocerasic acid) and phthiocerol. Its synthesis was achieved by the means of a relay-conjugated addition/trapping reaction [171]. The conjugated addition of Me_2Zn to cycloheptenone, catalyzed by $Cu(OTf)_2$ in the presence of the chiral phosphoramidate ligand **L2a**, was followed by ethylation by trapping it with EtI. The reaction afforded 2,3-disubstituted cycloheptanone with excellent *trans* selectivity (20 : 1, *trans* : *cis*) and ee-value (95%) in 83% yield (Scheme 2.38).

2.6.2.8 Synthesis of Leaf Miner Pheromones

The apple leaf miner (*Lyonetia prunifiliella*) is a pest that is endemic to the eastern regions of North America. A twofold asymmetric conjugated addition was used for the desymmetrization of cyclo-octadienone, giving rise to the substituted cyclo-octanone [172]. Both conjugated additions of Me_2Zn in the presence of a catalytic amount of $Cu(OTf)_2$ were carried out in the presence of the same ligand **2d**. The first reaction yielded a cyclooctenone of high optical purity (>99% ee). After the second conjugated addition, and under the same reaction conditions, the reaction mixture was quenched with TMSOTf to yield a chiral enolether with excellent stereochemistry (>99% ee, >98% de) (Scheme 2.39). Ozonolysis, followed by reduction and esterification, gave rise to a

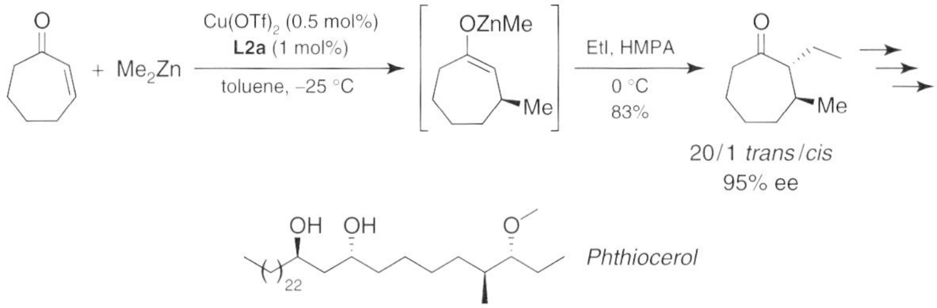

Scheme 2.38 Feringa's synthesis of phthiocerol.

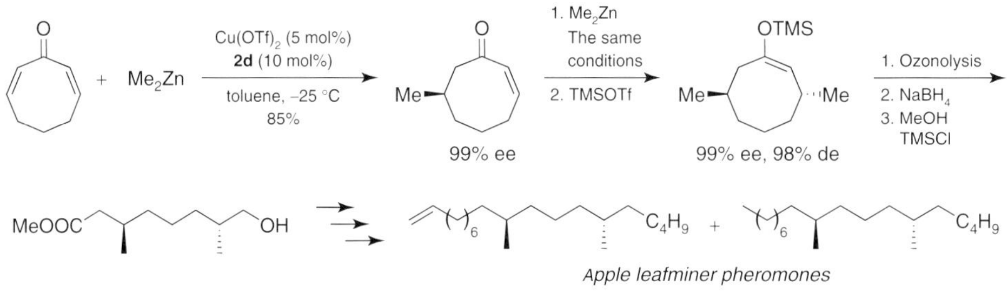

Scheme 2.39 Feringa's synthesis of the leaf miner pheromone.

α,ω-hydroxycarboxylic acid ester that was further converted to the apple leaf miner pheromones.

2.6.2.9 Synthesis of β-Amino Acids

The conjugated addition to properly substituted nitroalkenes represents a key step in the synthesis of enantiomeric β-amino acids. To date, two approaches to this reaction have been reported, simultaneously. The first report, made by Sewald et al., was based on the conjugated addition of Et_2Zn to 3-nitropropenoate [173]. In this case, the enantioselectivity depended heavily on the relative configuration of the phosphoramidite ligands, the solvent, and the reaction temperature, with the best result being obtained with ligand **L2e** at −78 °C in Et_2O. The corresponding substituted 3-nitropropanoate was formed in good enantiomeric excess (92%). The corresponding amino acid was obtained Following reduction, protection of the amino group, and hydrolysis (Scheme 2.40).

The second report, from Feringa et al., relied on the conjugated addition to nitroacetals [174], although a greater generality of this approach was demonstrated by the addition of several different dialkylzincs to furnish the corresponding nitroacetals in high yields and enantioselectivity (88–98% ee) (Scheme 2.41). The use of nitroacetals also proved to be beneficial from a synthetic point of view, since a further elaboration of the aminoacetals, after reduction of the nitro group, gave rise to various chiral amino alcohols, amino aldehydes, or amino acids.

2.6.2.10 Synthesis of Clavularin B

Clavularin B is present in the Okinawa soft coral *Clavularin koellikeri*, and demonstrates significant cytotoxic and anticarcinogenic properties. The synthetic strategy was based on the conjugated addition of Me_2Zn to cycloheptenone, catalyzed by $Cu(OTf)_2·C_6H_6$ in the presence of the peptidic ligand **L11** [79]. The subsequent alkylation of the formed enolate with 4-idodo-1-butene yielded 2,3-disubstituted

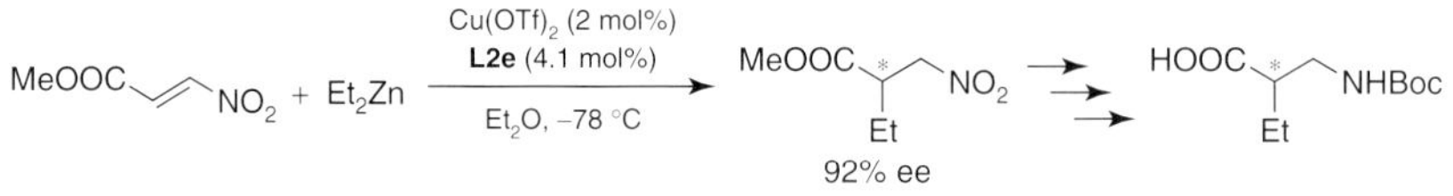

Scheme 2.40 Sewald's synthesis of β-amino acids.

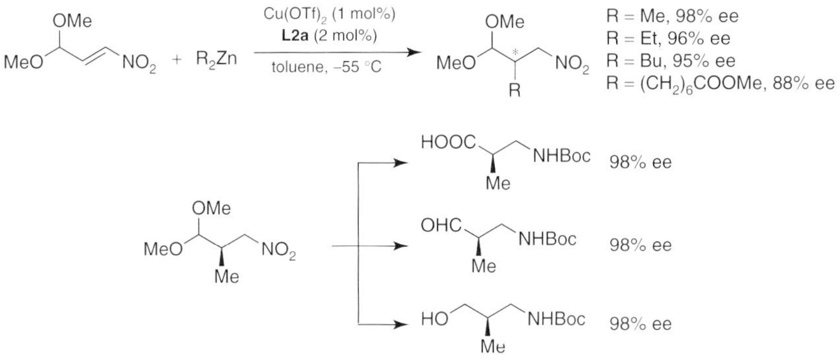

Scheme 2.41 Feringa's synthesis of β-amino acids.

cycloheptanone with excellent enantioselectivity (97%) and high diastereoselectivity (>15:1) (Scheme 2.42). Subsequent functional group transformations furnished the target compound, clavularin B.

2.6.2.11 Formal Synthesis of Axanes

Axanamides are natural compounds with a [4.3.0] bicyclononane skeleton, and have been isolated from the marine sponge, *Axinella cannabia*. It was shown that a copper-catalyzed enantioselective conjugated addition of organoaluminums to trisubstituted enones functions well with six-membered rings [75b], and this enabled the addition of Et$_3$Al to cyclohexanone bearing a side chain with the protected aldehyde moiety in the presence of **L11** to produce cyclohexanone with a chiral quaternary center in 95% yield, with 95% ee (Scheme 2.43). The subsequent acidic hydrolysis was followed by an intramolecular aldol condensation to yield bicyclo [4.3.0] nonenone in 68% yield, that constituted the crucial intermediate for the synthesis of axanes.

2.7
Conclusions

The copper- and Ni-catalyzed conjugated addition of organozinc and organoaluminum compounds represents a powerful tool for the construction of new C–C bonds. In particular, the enantioselective variant has become a reliable and

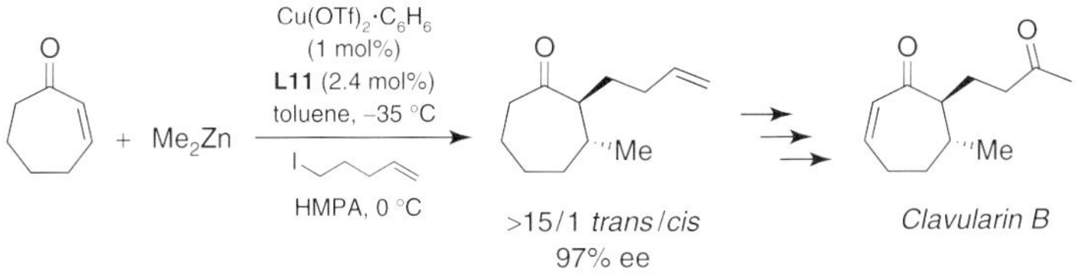

Scheme 2.42 Hoveyda's synthesis of clavularin B.

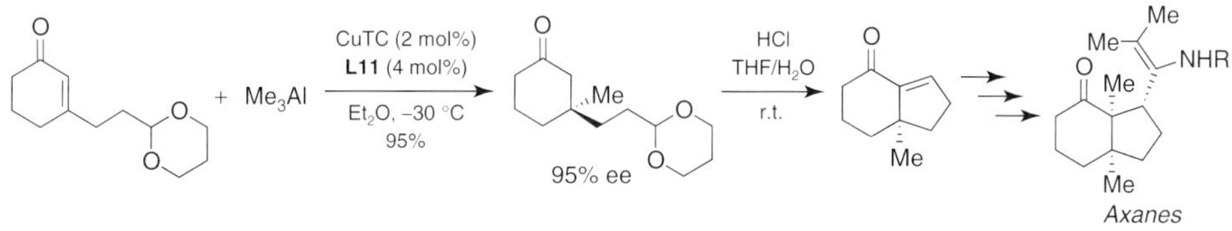

Scheme 2.43 Alexakis' synthesis of axanes.

convenient method for synthesizing chiral stereocenters, with a high asymmetric induction. Given the currently available plethora of simple and selective methods for the synthesis of alkyl, alkenyl, and aryl organozincs bearing various functional groups within its molecular framework, conjugated addition should enable the preparation of highly functionalized molecules. In addition, intermediate enolates can be utilized in sequential processes, allowing further functionalization by reaction with various electrophiles.

Although, today, it might seem that catalytic asymmetric conjugated additions represent a "mature" field of synthetic organic chemistry, this perception is far from the truth. The current state of methodology allows the addition of a range of organozincs and organoaluminums to activated double bonds (α,β-unsaturated carbonyl compounds, nitroalkenes, etc.) with enantioselectivities reaching 99% ee, although it should be noted that the majority of these cases involve alkylzincs or alkylaluminums. With regards to the conjugated addition of alkenyl-, aryl-, and alkynyl metals, very few examples have been reported. Consequently, this area is open to extensive development, and this will surely provide interesting information that will benefit organic synthesis as a whole. Yet, similar issues also concern the further design and development of new chiral ligands. Despite the fact that many chiral ligands have been created during the past two decades, there remains a vast scope for their further development with respect to substrate generality. There is no doubt that advances in these areas will lead to effective methodologies that can be applied to the various fields of organic chemistry.

References

1. Komnenos, T. (1883) *Liebigs Ann.*, **218**, 145–169.
2. Michael, A. (1887) *J. Prakt. Chem.*, **35**, 349–356.
3. Claisen, L. (1887) *J. Prakt. Chem.*, **35**, 413–415.
4. For reviews, see: Perlmutter, P. (1992) *Conjugate Addition Reactions in Organic Synthesis*, Pergamon Press.
5. For reviews, see: Trost, B.M. and Flemming, I. (eds) (1991) *Comprehensive Organic Synthesis*, vol. 2, Chapter 1.1-1.6, Pergamon Press, Oxford, pp. 1–268.
6. Knochel, P. and Winter, R.D. (1993) *Chem. Rev.*, **93**, 2117–2188.
7. Togni, A. and Venanzi, L.M. (1994) *Angew. Chem., Int. Ed.*, **33**, 497–526.
8. Fache, F., Schulz, E., Tommasino, M.L., and Lemaire, M. (2000) *Chem. Rev.*, **100**, 2159–2231.
9. Kwong, H.-L., Yeung, H.-L., Yeung, C.-T., Lee, W.-S., Lee, C.-S., and

Wong, W.-L. (2007) *Coord. Chem. Rev.*, **251**, 2188–2222.
10. Fujimori, S., Knöpfel, T.F., Zarotti, P., Ichikawa, T., Boyall, D., and Carreira, E.M. (2007) *Bull. Chem. Soc. Jpn*, **80**, 1635–1657.
11. Trost, B.M. and Weiss, A.H. (2009) *Adv. Synth. Catal.*, **351**, 963–983.
12. Von Zezschwitz, P. (2008) *Synthesis*, 1809–1831.
13. Alexakis, A., Bäckvall, J.E., Krause, N., Pàmies, O., and Diéguez, M. (2008) *Chem. Rev.*, **108**, 2796–2823.
14. Jerphagnon, T., Pizzuti, M.G., Minnaard, A.J., and Feringa, B.L. (2009) *Chem. Soc. Rev.*, **38**, 1039.
15. Thaler, T. and Knochel, P. (2009) *Angew. Chem., Int. Ed.*, **48**, 645–649.
16. (a) Rossiter, B.E. and Single, N.M. (1992) *Chem. Rev.*, **92**, 771–806; (b) Flemming, F.F. and Wang, Q. (1993) *Chem. Rev.*, **93**, 2035–2077; (c) Christoffers, J., Koripelly, G., Rosiak, A., and Rössle, M. (2007) *Synthesis*, 1279–1300.
17. Frankland, E. (1849) *Liebigs Ann. Chem.*, **71**, 171–213.
18. For a detailed historical account see: Seifert, D. (2001) *Organometallics*, **20**, 2940–2955.
19. (a) Wilkinson, G., Stone, F.G.A., and Abel, E.W. (eds) (1982) *Comprehensive Organometallic Chemistry*, vol. 2, Chapter 16, Pergamon Press, Oxford; (b) Abel, E.W., Wilkinson, G., and Stone, F.G.A. (eds) (1995) *Comprehensive Organometallic Chemistry II*, vol. 3. Chapter 10, Elsevier.
20. (a) Wilkinson, G., Stone, F.G.A., and Abel, E.W. (eds) (1982) *Comprehensive Organometallic Chemistry*, vol. 1, Chapter 6, Pergamon Press, Oxford; (b) Abel, E.W., Wilkinson., G., and Stone F.G.A. (eds) (1995) *Comprehensive Organometallic Chemistry II*, vol. 2, Chapter 4, Elsevier.
21. Zhang, H. and Gschwind, R.M. (2007) *Chem. Eur. J.*, **13**, 6691–6700.
22. Vuagnoux-d'Augustin, M. and Alexakis, A. (2007) *Eur. J. Org. Chem.*, 5832–5860.
23. Li, K. and Alexakis, A. (2006) *Angew. Chem., Int. Ed.*, **45**, 7600–7603.
24. (a) Jeffery, E.A., Meisters, A., and Mole, T. (1974) *J. Organomet. Chem.*, **74**, 365–371; (b) Bagnell, L., Jeffery, E.A., Meisters, A., and Mole, T. (1975) *Austr. J. Chem.*, **28**, 801–815; (c) Bagnell, L., Meisters, A., and Mole, T. (1975) *Austr. J. Chem.*, **28**, 817–820.
25. Knochel, P. (ed.) (2005) *Handbook of Functionalized Organometallics*, vol. 1, Chapter 7, Wiley-VCH Verlag GmbH, Weinheim, pp. 251–346.
26. (a) Zhu, L. and Rieke, R.D. (1991) *Tetrahedron Lett.*, **32**, 2865–2866; (b) Rieke, R.D., Uhm, S.J., and Hudnall, P.M. (1973) *J. Chem. Soc., Chem. Commun.*, 269–270; (c) Rieke, R.D., Li, P.T.-J., Burns, T.P., and Uhm, S.T. (1981) *J. Org. Chem.*, **46**, 4323–4324; (d) Rieke, R.D. and Uhm, S.J. (1975) *Synthesis*, 452–453; (e) Rieke, R.D. (1989) *Science*, **246**, 1260–1264.
27. For a review, see: Rieke, R.D. (2000) *Aldrichim. Acta*, **33**, 52–60.
28. Petrise, C., Luche, J.-L., and Dupuy, C. (1984) *Tetrahedron Lett.*, **25**, 3463–3466.
29. Furukawa, J., Kawabata, N., and Nishimura, J. (1966) *Tetrahedron Lett.*, **7**, 3353–3354.
30. (a) Hayashi, T., Konishi, M., and Kumada, M. (1979) *Tetrahedron Lett.*, **21**, 1871–1874; (b) Hayashi, T., Konishi, M., Kobori, Y., Kumada, M., Higuchi, T., and Hirotau, K. (1984) *J. Am. Chem. Soc.*, **106**, 158–163.
31. Thiele, K.-H. and Zdunneck, P. (1966) *J. Organomet. Chem.*, **4**, 10–17.
32. (a) Zakharin, L.I. and Okhlobystin, O.Y. (1960) *Z. Obsc. Chim.*, **30**, 2134–2138; (b) Thiele, K.-H., Engelhardt, G., Kohler, J., and Amatedt, M. (1967) *J. Organomet. Chem.*, **9**, 385–393.
33. (a) Srebnik, M. (1991) *Tetrahedron Lett.*, **32**, 2449–2452; (b) Oppolzer, W. and Radinov, R.N. (1992) *Helv. Chim. Acta*, **75**, 170–173.
34. (a) Rozema, M.J., Rajagopal, D., Tucker, C.E., and Knochel, P. (1992) *J. Organomet. Chem.*, **438**, 11–27; (b) Tucker, C.E. and Knochel, P. (1991) *J. Am. Chem. Soc.*, **113**, 9888–9890; (c) Tucker, C.E., Davidson, J., and

Knochel, P. (1992) *J. Org. Chem.*, **57**, 3482–3485.
35. Al-Batta, A. and Bergdahl, M. (2007) *Tetrahedron Lett.*, **48**, 1761–1765.
36. For a recent review, see: Uhl, W. (2008) *Coord. Chem. Rev.*, **252**, 1540–1563, and references therein.
37. Lipshutz, B.H. and Sengupta, S. (1992) *Org. React.*, **41**, 135–631, and references therein.
38. Knochel, P., Yeh, M.C.P., Berk, S.C., and Albert, J. (1988) *J. Org. Chem.*, **53**, 2390–2392.
39. (a) Tamaru, Y., Tanigawa, H., Yamamoto, T., and Yoshida, Z. (1989) *Angew. Chem.*, **101**, 358–360; (b) Tamaru, Y., Tanigawa, H., Yamamoto, T., and Yoshida, Z. (1989) *Angew. Chem., Int. Ed. Engl.*, **28**, 351–353.
40. (a) Yeh, M.C.P., Knochel, P., Butler, W.M., and Berk, S.C. (1988) *Tetrahedron Lett.*, **29**, 6693–6696; (b) Knochel, P., Chou, T.-S., Chen, H.G., Yeh, M.C.P., and Rozema, M.J. (1989) *J. Org. Chem.*, **54**, 5202–5204; (c) Zhu, L., Wehmeyer, R.M., and Rieke, R.D. (1991) *J. Org. Chem.*, **56**, 1445–1453.
41. (a) Bergdahl, M., Lindstedt, E.-L., Nilsson, M., and Olsson, T. (1988) *Tetrahedron*, **44**, 2055–2062; (b) Bergdahl, M., Lindstedt, E.-L., Nilsson, M., and Olsson, T. (1989) *Tetrahedron*, **45**, 535–543; (c) Bergdahl, M., Lindstedt, E.-L., and Olsson, T. (1989) *J. Organomet. Chem.*, **365**, C11; (d) Bergdahl, M., Nilsson, M., and Olsson, T. (1990) *J. Organomet. Chem.*, **391**, C19.
42. Kitamura, M., Miki, T., Nakano, K., and Noyori, R. (1996) *Tetrahedron Lett.*, **37**, 5141–5144.
43. Fraser, P.K. and Woodward, S. (2001) *Tetrahedron Lett.*, **42**, 2747–2749.
44. Kabbara, J., Flemming, S., Nickisch, K., Neh, H., and Westermann, J. (1994) *Chem. Ber.*, **127**, 1489–1493.
45. Kabbara, J., Flemming, S., Nickisch, K., Neh, H., and Westermann, J. (1994) *Synlett*, **127**, 679–680.
46. Bagnell, L., Meisters, A., and Mole, T. (1975) *Austr. J. Chem.*, **28**, 821–824.
47. Ashby, E.C. and Heinsohn, G. (1974) *J. Org. Chem.*, **39**, 3297–3299.
48. Flemming, S., Kabbara, J., Nicklish, K., Neh, H., and Westermann, J. (1995) *Synthesis*, 317–320.
49. Westermann, J., Neh, H., and Nickisch, K. (1996) *Chem. Ber.*, **129**, 963–966.
50. Westermann, J., Imbery, U., Nguyen, A.T., and Nicklish, K.K. (1998) *Eur. J. Org. Chem.*, 295–298.
51. Pérez, I., Sestelo, J.P., Maestro, M.A., Mouriño, A., and Sarandeses, L.A. (1998) *J. Org. Chem.*, **63**, 10074–10076.
52. Miura, T. and Murakami, M. (2005) *Chem. Commun.*, 5676–5677.
53. de Vries, A.H.M., Meetsma, A., and Feringa, B.L. (1996) *Angew. Chem., Int. Ed.*, **35**, 2374–2376.
54. Feringa, B.L., Pinesci, M., Arnold, L.A., Imbos, R., and de Vries, A.H.M. (1997) *Angew. Chem., Int. Ed.*, **36**, 2620–2623.
55. Mandolin, A., Calamante, M., Feringa, B.L., and Salvatori, P. (2003) *Tetrahedron: Asymmetry*, **14**, 3647–3650.
56. Zhou, D.-Y., Duan, Z.-C., Hu, X.-P., and Zheng, Z. (2009) *Tetrahedron: Asymmetry*, **20**, 235–239.
57. Naasz, R., Arnold, L.A., Minnaard, A.J., and Feringa, B.L. (2001) *Angew. Chem., Int. Ed.*, **40**, 927–930.
58. Urbaneja, L.M., Alexakis, A., and Krause, N. (2002) *Tetrahedron Lett.*, **43**, 7887–7890.
59. Imbos, R., Minnaard, A.J., and Feringa, B.L. (2001) *Tetrahedron*, **57**, 2485–2489.
60. (a) Fillion, E. and Wilsily, A. (2006) *J. Am. Chem. Soc.*, **126**, 2774–2775; (b) Wilsily, A., Lou, T., and Fillion, E. (2009) *Synthesis*, 2066–2072.
61. Watanabe, T., Knöpfel, T.F., and Carreira, E.M. (2003) *Org. Lett.*, **5**, 4557–4558.
62. Pinesci, M., Del Moro, F., Di Bussolo, V., and Macchia, F. (2006) *Adv. Synth. Catal.*, **348**, 301–304.
63. García, J.M., González, A., Kardak, B.G., Odriozola, J.M., Oiarbide, M., Razkin, J., and Palomo, C. (2008) *Chem. Eur. J.*, **14**, 8767–8771.
64. Palacios, F. and Vicario, J. (2006) *Org. Lett.*, **8**, 5405–5408.
65. Esquivias, J., Arrays, R.G., and Carretero, J.C. (2005) *J. Org. Chem.*, **70**, 7451–7454.

66. Duursma, A., Minnaard, A.J., and Feringa, B.L. (2002) *Tetrahedron*, **58**, 5773–5778.
67. Choi, H., Hua, Z., and Ojima, I. (2004) *Org. Lett.*, **6**, 2689–2691.
68. (a) Alexakis, A., Polet, D., Rosset, S., and March, S. (2004) *J. Org. Chem.*, **69**, 5660–5667; (b) Alexakis, A., Polet, D., Benhaim, C., and Rosset, S. (2004) *Tetrahedron: Asymmetry*, **15**, 2199–2203.
69. Zhou, H., Wang, W.-H., Fu, Y., Xie, J.-H., Shi, W.-J., Wang, L.-X., and Zhou, Q.-L. (2003) *J. Org. Chem.*, **68**, 1582–1584.
70. Zhang, W., Wang, C.-J., Gao, W., and Zhang, X. (2005) *Tetrahedron Lett.*, **46**, 6087–6090.
71. (a) Alexakis, A., Vastra, J., Burton, J., Benhaim, C., and Mangeney, P. (1998) *Tetrahedron Lett.*, **39**, 7869–7872; (b) Alexakis, A., Burton, J., Vastra, J., Benhaim, C., Fournioux, X., van den Heuvel, A., Leveque, J.M., Maze, F., and Rosset, S. (2000) *Eur. J. Org. Chem.*, 4011–4027.
72. Arena, C.G., Casilli, V., and Faraone, F. (2003) *Tetrahedron: Asymmetry*, **14**, 2127–2131.
73. Pineschi, M., Del Moro, F., Gini, F., Minnaard, A.J., and Feringa, B.L. (2004) *Chem. Commun.*, 1244–1245.
74. Alexakis, A., Albrow, V., Biswas, K., d'Augustin, M., Prieto, O., and Woodward, S. (2005) *Chem. Commun.*, 2843–2844.
75. (a) d'Augustin, M., Calais, L., and Alexakis, A. (2005) *Angew. Chem., Int. Ed.*, **44**, 1376–1378; (b) Vuagnoux-d'Augustin, M. and Alexakis, A. (2007) *Chem. Eur. J.*, **13**, 9647–9662; (c) Vuagnoux-d'Augustin, M., Odhrli, S., and Alexakis, A. (2007) *Synlett*, 2057–2060.
76. Palais, L., Mikhel, I.S., Bournaud, C., Micouin, L., Falciola, C.A., Vuagnoux-d'Augustin, M., Rosset, S., Bernardinelli, G., and Alexakis, A. (2007) *Angew. Chem., Int. Ed.*, **46**, 7462–7465.
77. Polet, D. and Alexakis, A. (2005) *Tetrahedron Lett.*, **46**, 1529–1532.
78. Pizzuti, M.G., Minnaard, A.J., and Feringa, B.L. (2008) *J. Org. Chem.*, **73**, 940–947.
79. Degrado, S.J., Mizutani, H., and Hoveyda, A.H. (2001) *J. Am. Chem. Soc.*, **123**, 755–756.
80. Degrado, S.J., Mizutani, H., and Hoveyda, A.H. (2002) *J. Am. Chem. Soc.*, **124**, 13362–13363.
81. Mizutani, H., Degrado, S.J., and Hoveyda, A.H. (2002) *J. Am. Chem. Soc.*, **124**, 779–781.
82. Cesati, J.A., De Armas, J., and Hoveyda, A.H.III (2004) *J. Am. Chem. Soc.*, **126**, 96–101.
83. Hird, A.W. and Hoveyda, A.H. (2003) *Ang. Chem., Int. Ed.*, **42**, 1276–1279.
84. (a) Mampreian, D.M. and Hoveyda, A.H. (2004) *Org. Lett.*, **6**, 2829–2832; (b) Luchaco-Culis, C.A. and Hoveyda, A.H. (2002) *J. Am. Chem. Soc.*, **124**, 8192–8193.
85. Wu, J., Mampreian, D.M., and Hoveyda, A.H. (2005) *J. Am. Chem. Soc.*, **127**, 4584–4585.
86. Kacprzynski, M.A., Kazane, S.A., May, T.L., and Hoveyda, A.H. (2007) *Org. Lett.*, **9**, 3187–3190.
87. (a) Soeta, T., Selim, K., Kuriyama, M., and Tomioka, K. (2007) *Adv. Synth. Catal.*, **349**, 629–635; (b) Soeta, T., Selim, K., Kuriyama, M., and Tomioka, K. (2007) *Tetrahedron*, **63**, 6573–6576.
88. Valleix, F., Nagai, K., Soeta, T., Selim, K., Kuriyama, M., Yamada, K., and Tomioka, K. (2005) *Tetrahedron*, **61**, 7420–7424.
89. Soeta, T., Kuriyama, K.M., and Tomioka, K. (2005) *J. Org. Chem.*, **70**, 297–300.
90. Malkov, A.V., Hand, J.B., and Kočovský, P. (2003) *Chem. Commun.*, 1948–1949.
91. Liu, L.-T., Wang, M.-C., Zhao, W.-X., Zhou, Y.-L., and Wang, X.-D. (2006) *Tetrahedron: Asymmetry*, **17**, 136–141.
92. Krauss, I.J. and Leighton, J.L. (2003) *Org. Lett.*, **5**, 3201–3203.
93. Fuchs, N., D'Augustin, M., Humam, M., Alexakis, A., Taras, R., and Gladiali, S. (2005) *Tetrahedron: Asymmetry*, **16**, 3143–3146.
94. Hajra, A., Yoshikai, N., and Nakamura, E. (2006) *Org. Lett.*, **8**, 4153–4155.

95. Shintani, R. and Fu, G.C. (2002) *Org. Lett.*, **4**, 3699–3702.
96. Smith, A.J., Abbott, L.K., and Martin, S.F. (2009) *Org. Lett.*, **11**, 4200–4203.
97. Alexakis, A., Vastra, J., Buton, J., and Mangeney, P. (1997) *Tetrahedron: Asymmetry*, **8**, 3193–3196.
98. Scafato, P., Pavano, S., Cunsolo, G., and Rosini, C. (2003) *Tetrahedron: Asymmetry*, **14**, 3873–3877.
99. Iuliano, A., Scafato, P., and Torchia, R. (2004) *Tetrahedron: Asymmetry*, **15**, 2533–2538.
100. Iuliano, A. and Scafato, P. (2003) *Tetrahedron: Asymmetry*, **14**, 611–618.
101. Facchetti, S., Losi, D., and Juliáno, A. (2006) *Tetrahedron: Asymmetry*, **17**, 2993–3003.
102. Yan, M., Yang, L.-W., Wong, K.-Y., and Chan, A.S.C. (1999) *Chem. Commun.*, 11–12.
103. Yan, M., Zhou, Z.-Y., and Chan, A.C.S. (2000) *Chem. Commun.*, 115–116.
104. Liang, L., Yan, M., Li, Y.-M., and Chan, A.C.S. (2004) *Tetrahedron: Asymmetry*, **15**, 2575–2578.
105. Wang, L., Li, Y.-M., Yip, C.-W., Qiu, L., Zhou, Z., and Chan, A.C.S. (2004) *Adv. Synth. Catal.*, **346**, 947–953.
106. Mata, Y., Diéguez, M., Pámies, O., Biswas, K., and Woodward, S. (2007) *Tetrahedron: Asymmetry*, **18**, 1613–1617.
107. Zhao, Q.-L., Wang, L.-L., Kong, F.Y., and Chan, A.S.C. (2007) *Tetrahedron: Asymmetry*, **18**, 1899–1905.
108. Liang, L., Au-Yeung, T.L.-Y., and Chan, A.C.S. (2002) *Org. Lett.*, **4**, 3799–3801.
109. (a) Liang, L. and Chan, A.C.S. (2002) *Tetrahedron: Asymmetry*, **13**, 1393–1396; (b) Su, L., Li, X., Chan, W.L., Jia, X., and Chan, A.C.S. (2003) *Tetrahedron: Asymmetry*, **14**, 1865–1869.
110. Guillen, F., Winn, C.L., and Alexakis, A. (2001) *Tetrahedron: Asymmetry*, **12**, 2083–2086.
111. Pytkowitz, J., Roland, S., and Mangeney, P. (2001) *Tetrahedron: Asymmetry*, **12**, 2087–2089.
112. Alexakis, A., Winn, C.L., Guillen, F., Pytkowitz, J., Roland, S., and Mangeney, P. (2003) *Adv. Synth. Catal.*, **345**, 345–348.
113. Arnold, P.L., Rodden, M., Davis, K.M., Scarisbrick, A.C., Blake, A.J., and Wilson, C. (2004) *Chem. Commun.*, 1612–1613.
114. Clavier, H., Coutable, L., Guillemin, J.-C., and Mauduit, M. (2005) *Tetrahedron: Asymmetry*, **16**, 921–924.
115. Rix, D., Labat, S., Tupet, L., Crévisy, C., and Mauduit, M. (2009) *Eur. J. Org. Chem.*, 1989–1999.
116. Lee, K., Brown, K., Hird, A.W., and Hoveyda, A.H. (2006) *J. Am. Chem. Soc.*, **128**, 7182–7184.
117. Alexakis, A., Frutos, J., and Mangeney, P. (1993) *Tetrahedron: Asymmetry*, **4**, 2427–2430.
118. Hu, Y., Liang, X., Wang, J., Zheng, Z., and Hu, X. (2003) *J. Org. Chem.*, **68**, 4542–4545.
119. Luo, X., Hu, Y., and Hu, X. (2005) *Tetrahedron: Asymmetry*, **16**, 1227–1231.
120. Wan, H., Hu, Y., Liang, Y., Gao, S., Wang, J., Zheng, Z., and Hu, X. (2003) *J. Org. Chem.*, **68**, 8277–8280.
121. Cramer, N., Laschat, S., and Baro, A. (2006) *Organometallics*, **25**, 2284–2291.
122. Xie, Y., Huang, H., Mo, W., Fan, X., Shen, Z., Shen, Z., Sun, N., Hu, B., and Hu, X. (2009) *Tetrahedron: Asymmetry*, **20**, 1425–1432.
123. Hu, X., Chen, H., and Zhang, X. (1999) *Angew. Chem., Int. Ed.*, **38**, 3518–3521.
124. Liang, X., Gao, S., Wan, H., Hu, Y., Chen, H., Zheng, X., and Hu, X. (2003) *Tetrahedron: Asymmetry*, **14**, 3211–3217.
125. Kawamura, K., Fukuzawa, H., and Hayashi, M. (2008) *Org. Lett.*, **10**, 3509–3512.
126. Biradar, D.B. and Gau, H.-M. (2008) *Tetrahedron: Asymmetry*, **19**, 733–738.
127. Albrow, V.A., Blake, A.J., Fryatt, R., Wilson, C., and Woodward, S. (2006) *Eur. J. Org. Chem.*, 2549–2557.
128. Shi, M., Wang, C.-J., and Zhang, W. (2004) *Chem. Eur. J.*, **10**, 5507–5516.
129. Shi, M., and Zhang, W. (2004) *Tetrahedron: Asymmetry*, **15**, 167–176.
130. Shi, M. and Zhang, W. (2005) *Adv. Synth. Catal.*, **347**, 535–540.

131. Guimet, E., Diéguez, M., Ruiz, A., and Claver, C. (2005) *Tetrahedron: Asymmetry*, **16**, 2161–2165.
132. Kang, J., Lee, J.H., and Lim, D.S. (2003) *Tetrahedron: Asymmetry*, **14**, 305–315.
133. (a) Bennett, S.M.W., Brown, S.M., Conole, G., Denis, M.R., Fraser, P.K., Radojevic, S., McPartlin, M., Topping, C.M., and Woodward, S. (1999) *J. Chem. Soc., Perkin Trans 1*, 3127–3132; (b) Bennett, S.M.W., Brown, S.M., Cunningham, A., Denis, M.R., Muxworthy, J.P., Oakley, M.A., and Woodward, S. (2000) *Tetrahedron*, **56**, 2847–2855; (c) Fraser, P.K. and Woodward, S. (2003) *Chem. Eur. J.*, **9**, 776–783.
134. Bonini, B.F., Capitó, E., Comes-Franchini, M., Ricci, A., Bottoni, A., Berbardi, F., Miscione, G.P., Giordano, L., and Cowley, A.R. (2004) *Eur. J. Org. Chem.*, 4442–4451.
135. Arink, A.M., Braam, T.W., Keeris, R., Jastrzebski, J.T.B.H., Benhaim, C., Rosset, S., Alexakis, A., and van Koten, G. (2004) *Org. Lett.*, **6**, 1959–1962.
136. Hird, A.W. and Hoveyda, A.H. (2005) *J. Am. Chem. Soc.*, **127**, 14988–14989.
137. (a) Soai, K., Hayasaka, T., Ogajin, S., and Yokoyama, S. (1988) *Chem. Lett.*, 1571–1572; (b) Soai, K., Hayasaka, T., Ogajin, S., and Yokoyama, S. (1989) *J. Chem. Soc., Chem. Commun.*, 516–517; (c) Soai, K., Yokoyama, S., Hayasaka, T., and Ebihara, K. (1988) *J. Org. Chem.*, **53**, 4148–4149.
138. Unaleroglu, C., Aydin, A.E., and Demir, A.S. (2006) *Tetrahedron: Asymmetry*, **17**, 742–749.
139. Fujisawa, T., Itoh, S., and Shimizu, M. (1994) *Chem. Lett.*, 1777–1780.
140. de Vries, A.H.M., Jansen, J.F.G.A., and Feringa, B.L. (1994) *Tetrahedron*, **50**, 4479–4491.
141. Wang, H.-S., Da, C.-S., Xin, Z.-Q., Su, W., Xiao, Y.-N., Liu, D.-X., and Wang, R. (2003) *Chin. J. Chem.*, **21**, 1105–1107.
142. de Vries, A.H.M., Imbos, R., and Feringa, B.L. (1997) *Tetrahedron: Asymmetry*, **8**, 1467–1473.
143. Shadakshari, A. and Nayak, S.K. (2001) *Tetrahedron*, **57**, 8185–8188.
144. Tong, P.-E., Li, P., and Chan, A.S.C. (2001) *Tetrahedron: Asymmetry*, **12**, 2301–2304.
145. Cobb, A.J.A. and Marson, C.M. (2005) *Tetrahedron*, **61**, 1269–1279.
146. Uemura, M., Miyake, R., Nakayama, K., and Hayashi, Y. (1992) *Tetrahedron: Asymmetry*, **3**, 713–714.
147. Isleyen, A. and Dogan, Ö. (2007) *Tetrahedron: Asymmetry*, **18**, 679–684.
148. Gibson, C.L. (1996) *Tetrahedron: Asymmetry*, **7**, 3357–3356.
149. Kang, J., Kim, J.I., Lee, J.H., Kim, H.J., and Byun, Y.H. (1998) *Bull. Korean Chem. Soc.*, **19**, 601–603.
150. Bolm, C. and Ewald, M. (1990) *Tetrahedron Lett.*, **31**, 5011–5012.
151. Bolm, C., Ewald, M., and Felder, M. (1992) *Chem. Ber.*, **125**, 1205–1215.
152. Bolm, C. (1991) *Tetrahedron: Asymmetry*, **2**, 701–704.
153. Asami, M., Usui, K., Higuchi, S., and Inoue, S. (1994) *Chem. Lett.*, 297–298.
154. Escorihuela, J., Burguete, M.I., and Luis, S.V. (2008) *Tetrahedron Lett.*, **49**, 6885–6888.
155. Corma, A., Inglesias, M., Martín, M.V., Rubio, J., and Sánchez, F. (1992) *Tetrahedron: Asymmetry*, **3**, 845–848.
156. Bolm, C., Felder, M., and Müller, J. (1992) *Synlett*, 439–441.
157. Kwak, Y.-S. and Corey, E.J. (2004) *Org. Lett.*, **6**, 3385–3388.
158. Naasz, R., Arnold, L.A., Minnaard, A.J., and Feringa, B.L. (2001) *Chem. Commun.*, 735–736.
159. González-Gómez, J.C., Foubelo, F., and Yus, M. (2009) *J. Org. Chem.*, **74**, 2547–2553.
160. Greene, A.E., Lansard, J.-P., Luche, J.-L., and Petrier, C. (1984) *J. Org. Chem.*, **49**, 931–932.
161. Petrier, C., de Souza Barbosa, J.C., Dupuy, C., and Luche, J.-L. (1985) *J. Org. Chem.*, **50**, 5761–5765.
162. Matsuda, T., Tsuboi, T., and Murakami, M. (2007) *J. Am. Chem. Soc.*, **129**, 12596–12597.
163. Okano, K., Suemune, H., and Sakai, K. (1988) *Chem. Pharm. Bull.*, **36**, 1379–1385.
164. Magnus, P., Waring, M.J., Ollivier, C., and Lynch, V. (2001) *Tetrahedron Lett.*, **42**, 4947–4950.

165. (a) Tsujiyama, H., Ono, N., Yoshino, T., Okamoto, S., and Sato, F. (1990) *Tetrahedron Lett.*, **31**, 4481–4484; (b) Yoshino, T., Okamoto, S., and Sato, F. (1991) *J. Org. Chem.*, **56**, 3205–3207.
166. Ziegler, F.E. and Owen, O.B. (1995) *J. Org. Chem.*, **60**, 3626–3636.
167. Arnold, L.A., Naasz, R., Minnaard, A.J., and Feringa, B.L. (2002) *J. Org. Chem.*, **67**, 7244.
168. Alexakis, A. and March, S. (2002) *J. Org. Chem.*, **67**, 8753–8757.
169. Polet, D. and Alexakis, A. (2005) *Tetrahedron Lett.*, **46**, 1529–1532.
170. Dijk, E.W., Panella, L., Pinho, P., Naasz, R., Meetsma, A., Minnaard, A.J., and Feringa, B.L. (2004) *Tetrahedron*, **60**, 9687–9693.
171. Casa-Arce, A., ter Horst, B., Feringa, B.L., and Minaard, A.J. (2008) *Chem. Eur. J.*, **14**, 4157–4159.
172. van Summeren, R.P., Reijmer, S.J.W., Feringa, B.L., and Minnaard, A.J. (2005) *Chem. Commun.*, 1387–1389.
173. Rimkus, A. and Sewald, N. (2003) *Org. Lett.*, **5**, 79–80.
174. Duursma, A., Minnaard, A.J., and Feringa, B.L. (2003) *J. Am. Chem. Soc.*, **125**, 3700–3711.

3
ECAs of Organolithium Reagents, Grignard Reagents, and Examples of Cu-Catalyzed ECAs

Gui-Ling Zhao and Armando Córdova

3.1
Introduction

Enantioselective conjugate addition (ECA) [1] is one of the most important C–C bond-forming reactions, and has been employed frequently as the key step in the synthesis of numerous complex biologically active molecules [2]. Conjugate additions involve the addition of a nucleophile to activated alkenes or alkynes, followed by the protonation of the anionic intermediate. The nucleophiles can be carbon (such as malonate, nitroalkanes) [3] or organometallic reagents (e.g., organolithium [4], organomagnesium (Grignard reagents) [5], organocuprate [6], organoaluminum [7], organoboron [8], and organozinc [9] reagents). The increasing demand for enantiopure compounds in the fine chemical and pharmaceutical industries has led to considerable efforts toward the development of highly ECA with these reagents. In this chapter, attention is focused on stoichiometric and catalytic ECAs using organolithium reagents and Grignard reagents. Examples of ECAs catalyzed by Cu–bisoxazoline complexes are also described.

3.2
Enantioselective Conjugate Addition of Lithium Reagents

Organolithium reagents are of major importance in organic synthesis. They are very powerful nucleophiles, and widely used in conjugate additions to α,β-unsaturated compounds. In this context, impressive results have been accomplished in the field of ECAs using chiral ligands. In 1989, the ECA of lithium reagents was pioneered by Tomioka and coworkers [10], who designed and applied a family of C_2-symmetric chiral ligands (see Figure 3.1) for the addition of organolithium compounds to α,β-unsaturated aldimines (Scheme 3.1). The ECAs with chiral ligands **1** and **2** gave the highest enantioselectivity, forming chiral complexes such as **6** (Scheme 3.1). It was noted by the authors that the chiral ligand not only controls the absolute stereochemistry but also promotes the reaction. Moreover, ligands **3–5** exhibited a poor enantioselectivity, which suggests that the methyl group on the ether oxygen

Catalytic Asymmetric Conjugate Reactions. Edited by Armando Córdova
Copyright © 2010 WILEY-VCH Verlag GmbH & Co. KGaA, Weinheim
ISBN: 978-3-527-32411-8

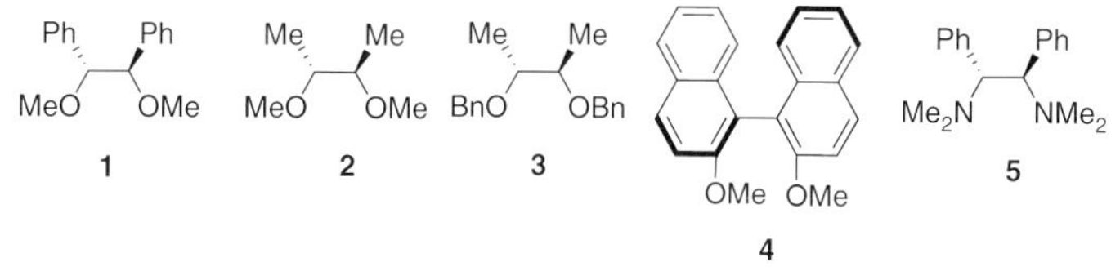

Figure 3.1 Selected ligands developed for the ECA of lithium reagents.

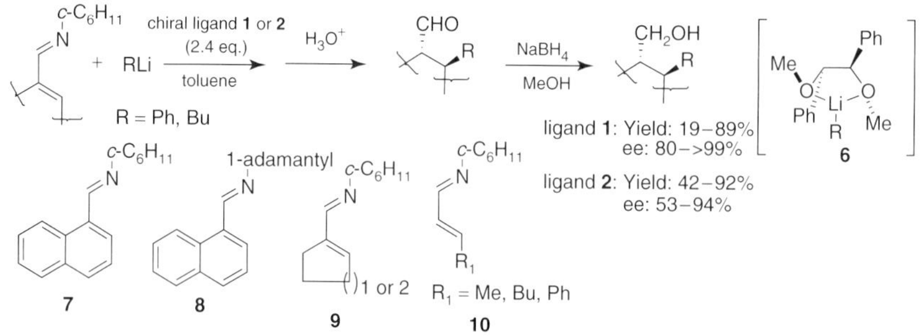

Scheme 3.1 Enantioselective conjugate additions of organolithiums to α,β-unsaturated aldimines.

is critical for achieving a high stereoselectivity during the addition. Thus, using **1** or **2** as ligands for the conjugate addition of PhLi or BuLi to imines **7–10** proceeds smoothly and gives the corresponding products in moderate to high yields and excellent enantiomeric excess (ee) values (Scheme 3.1).

Tomioka and coworkers [11] later employed 2,6-di-*tert*-butyl-4-methoxylphenyl (BHA) esters as acceptors in ECAs of aryllithium reagents using chiral ligand **1**. Increasing the steric bulk of the ester substituent increases the chemoselectivity (1,4- versus 1,2-addition) of the transformation. When a stoichiometric amount of ligand **1** was used, the reaction gave the corresponding products in high yields and ee-values. Moreover, the catalytic ECA version was also successful (Scheme 3.2).

In addition, the ECAs of organolithium reagents was expanded to other acyclic or cyclic α,β-unsaturated BHA esters (Scheme 3.3) [12]. In this case, (−)-Sparteine **11** was also investigated as a chiral ligand. It is noteworthy that the transformations

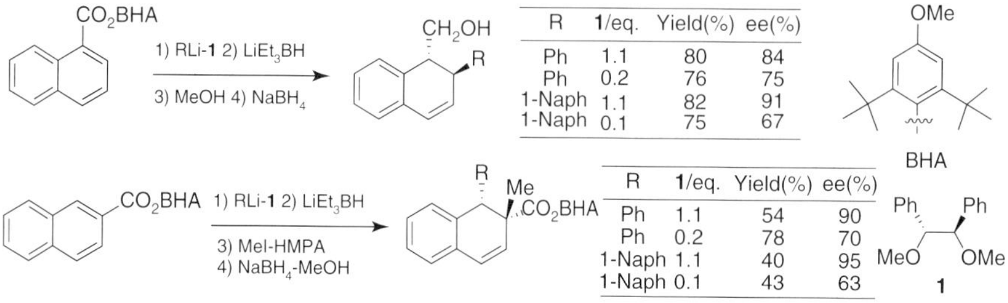

R	1/eq.	Yield(%)	ee(%)
Ph	1.1	80	84
Ph	0.2	76	75
1-Naph	1.1	82	91
1-Naph	0.1	75	67

R	1/eq.	Yield(%)	ee(%)
Ph	1.1	54	90
Ph	0.2	78	70
1-Naph	1.1	40	95
1-Naph	0.1	43	63

Scheme 3.2 Enantioselective conjugate additions of aryllithium to naphthalene BHA-esters.

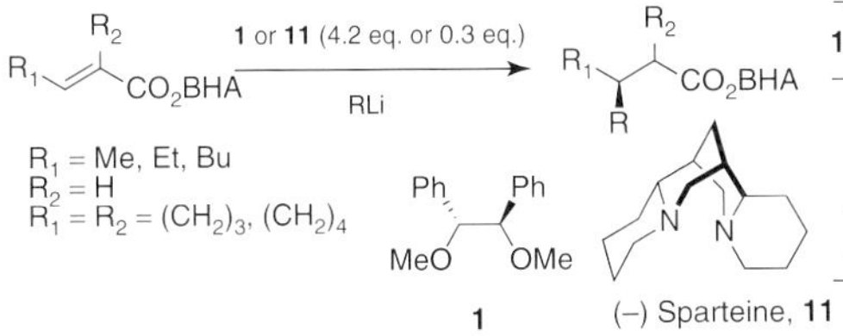

3.2 Enantioselective Conjugate Addition of Lithium Reagents | 147

1/11	R	Yield (%)	ee (%)
1	Ph	62–97	84–93[a] (36–49)[b]
1	vinyl	90	64[a]
1	Bu, Et	78–89	44–85[a]
11	Ph	89–98	42–43[a]
11	Bu, Et	73–87	91–99[a] (53–85)[b]

[a] 4.2 eq. **1** or **11** was used
[b] 0.3 eq. **1** or **11** was used

Scheme 3.3 Enantioselective conjugate additions of organolithium to BHA-esters.

with ligand **1** gives the highest stereoselectivity in ECAs using sp^2 carbanions (e.g., vinyl- and phenyllithium) nucleophiles, whereas the chiral diamine **11**, which has a higher coordination ability to lithium, exhibits the highest enantioselectivity when sp^3 carbanions (e.g., ethyl- and butyllithium) are employed as nucleophiles (Scheme 3.3).

Xu and coworkers [13] employed **1** or **11** as ligands for the ECAs of aryllithium to α,β-unsaturated *tert*-butyl esters, and the corresponding products were obtained in high yields and moderate to high ee-values (Scheme 3.4). In most cases, the transformations with chiral ligand **1** gave the corresponding products with higher ee-values as compared to (−)-Sparteine **11**.

Tomioka and coworkers [14] performed the asymmetric addition of thiophenol to methyl esters of α,β-unsaturated acids using a combination of lithium thiophenolate and an external chiral ligand. This approach gave the corresponding products in good yields and up to 99% ee (Scheme 3.5). The chiral amino ether **12** proved

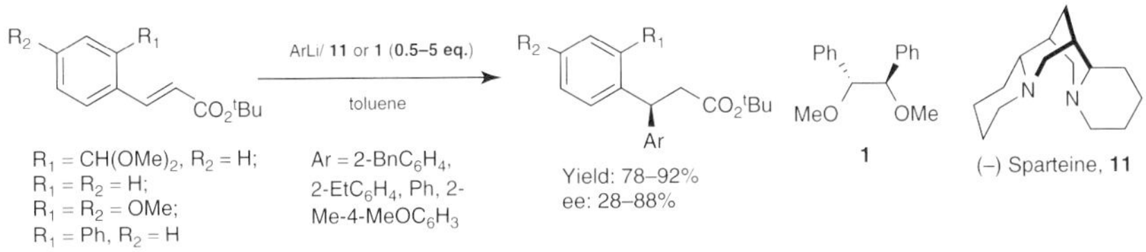

Scheme 3.4 Enantioselective conjugate additions of aryl-lithium to α,β-unsaturated *tert*-butyl esters.

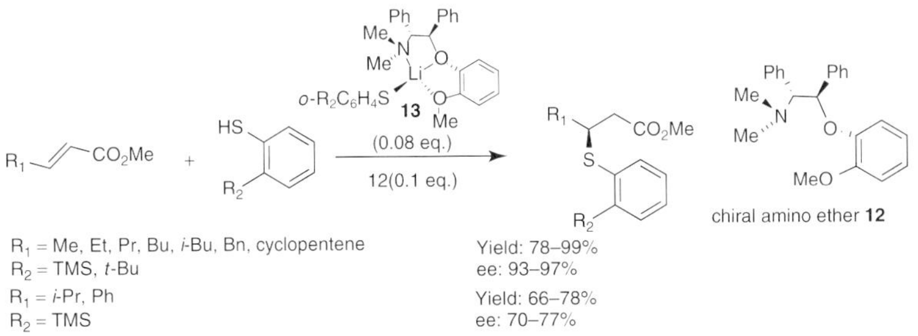

Scheme 3.5 ECA of thiophenol to enoates in the presence of chiral ether **12** and lithium thiophenolate.

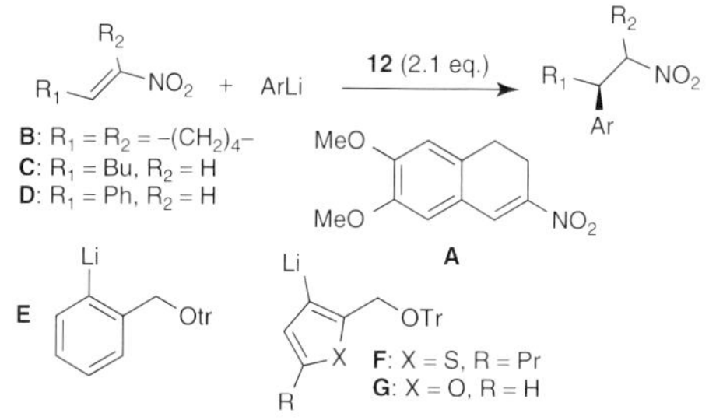

Scheme 3.6 ECA of aryllithium to nitroalkenes in the presence of chiral ether **12**.

Nitroalkene	ArLi	Yield (%)	trans:cis	ee(%)
A	E	98	69:33	95
B	E	99	12:88	89/90
B	F	94	6:94	95/94
B	G	99	11:89	91/91
C	E	93		85
D	F	66		91
A	E	92	37:63	97

to be the best ligand with respect to stereoselectivity, and increasing the size of the 2-substituent on the lithium thiophenolate improved the reactivity and enantioselectivity of the transformation. The authors explained that the inhibition of the thiolate anion and the formation of the sterically defined monomeric chelated complex **13** could be the reason for the higher enantioselectivity of these bulky nucleophiles.

Later, Tomioka and coworkers [15] reported that 2-trityloxymethylaryllithium can add to cyclic and acyclic nitroalkenes in the presence of chiral amino ether **12** with high enantioselectivity, to give the corresponding conjugate addition products in high yields and up to 97% ee (Scheme 3.6). They also elegantly applied this methodology to the synthesis of a dopamine D1 agonist, A-86929.

Beak and coworkers reported an elegant ECA based on the strategy of generating configurationally stable organolithium species **14** and **15** by the enantioselective deprotonation of N-Boc-N-arylbenzylamine and N-Boc-N-arylcinnamylamine, respectively, with a complex of n-BuLi/**11** [16]. The coordination of lithium to the bulky ligand **11** prevents inversion of the organolithium species. Next, the organolithium species **14** and **15** are employed as nucleophiles for the highly ECA to activated olefins (Scheme 3.7).

Thus, several activated olefins [e.g., 2-cyclohexenone, 2-cyclopentenone, 5,6-dihydro-2H-pyran-2-one, furan-2(5H)-one, 2-butenone, acrolein, and tert-butyl 4-oxopyridine-1(4H)-carboxylate] (Table 3.1) and trans-p-bromocinnamate [16d] (Scheme 3.8) can be employed as acceptors [16].

In addition, doubly activated acyclic olefins [17a, b] and nitro olefins [17c, d, e] were investigated as acceptors for ECAs involving species **14** or **15** (Table 3.2).

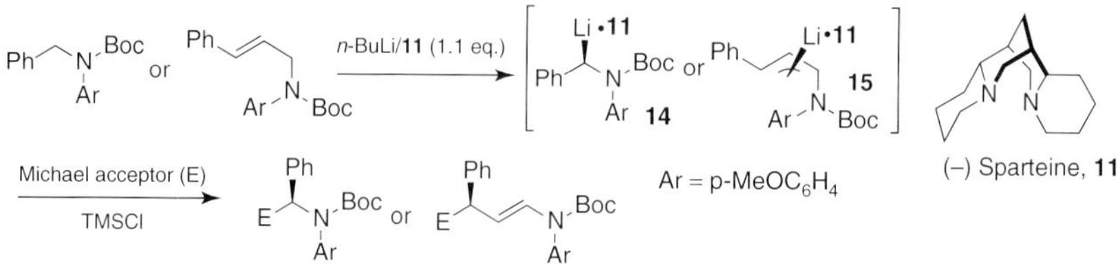

Scheme 3.7 ECA of configurationally stable organolithium species.

3.2 Enantioselective Conjugate Addition of Lithium Reagents

Table 3.1 Conjugate addition of lithium reagents **14** or **15** to different acceptors.

Entry	Acceptor (E)		Lithium reagent	dr	Yield (%)	ee (%)
1	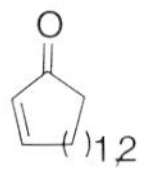 cyclopentenone/cyclohexenone	$n=1$	14	>99:1	86	92
		$n=2$	14	>99:1	82	92
		$n=2$	15	80:20	62	96
2	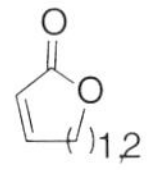 butenolide/pentenolide	$n=1$	15	96:4	66	94
		$n=2$	15	89:11	80	92
3	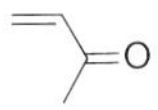 methyl vinyl ketone		14	–	63	94
4	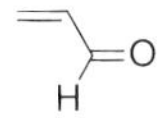 methacrolein		14	–	72	94
5	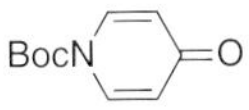 BocN-dihydropyridinone		15	93:7	80	80

dr = diastereomeric ratio.

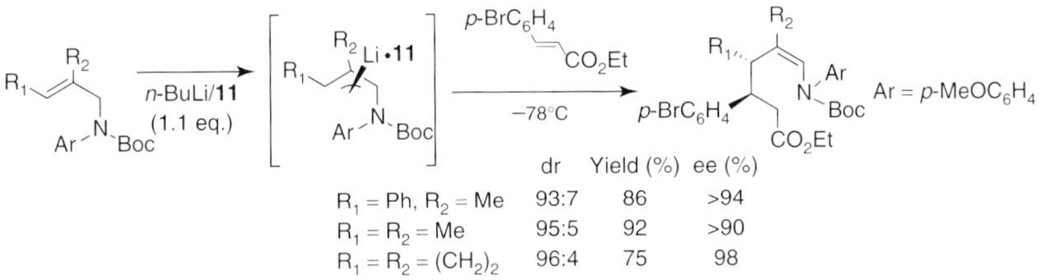

Scheme 3.8 Enantioselective lithiation–conjugate addition sequences.

	dr	Yield (%)	ee (%)
$R_1 = Ph, R_2 = Me$	93:7	86	>94
$R_1 = R_2 = Me$	95:5	92	>90
$R_1 = R_2 = (CH_2)_2$	96:4	75	98

Although high enantioselectivity is obtained for most transformations, the diastereoselectivity varies from low to high, and appears to be electrophile-dependent. The absolute configuration also varies between different electrophiles. The reactions with nonlithium-coordinating electrophiles give an inversion of the configuration, and the reactions with lithium complexation proceed with retention of the configuration. The use of a stannylation/lithiation sequence allows for the preparation of both (*R*) and (*S*) epimers of **14** and **15**, respectively. This strategy allows for the synthesis of both enantiomers of the conjugate addition product. Moreover, a series of compounds such as substituted piperidines, pyrrolidines, pyrimidinones, cyclopentanoids, and azepanes can be synthesized using this reaction as the key step. The organolithium species **15** was isolated and analyzed using X-ray analysis, which provides an explanation for the asymmetric induction of the conjugate addition [18]. Furthermore, when racemic α,β-unsaturated lactones were used, a kinetic resolution was observed and the corresponding

Table 3.2 Conjugate addition of lithium reagents **14** or **15** to activated olefins.

Entry	Acceptor (E)	14			15		
		Dr	Yield (%)	ee (%)	Dr	Yield (%)	ee (%)
1	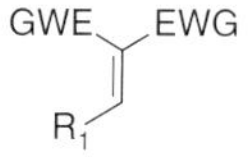 GWE⟵EWG, R_1	75:15–99:1	72–92	92–99	66:34–94:6	80–93	88–96
2	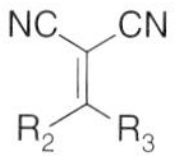 NC⟵CN, R_2, R_3	78:22–99:1	60–94	90–98	55:45–97:3	70–96	66–94
3	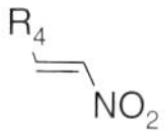 R_4⟵NO_2	–	–	–	93:7–98:2	73–90	80–94
4	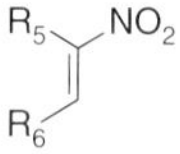 R_5⟵NO_2, R_6; R_5 = Me, R_6 = Ph, R_5,R_6 = $(CH_2)_4$	60:40, 68:32	72, 77	66, 96	53:47, 84:16	68, 65	>99, >98

EWG = CO_2Et, CN; R_1 = Me, Cy, Ph; R_2 = Ph, Me, n-Pr, CH_2OTIPS; R_4 = Ph, Cy, i-Bu, o-MeOPh, 2-furyl.

products with three contiguous stereogenic centers were formed and isolated in high yields, good diastereomeric ratios, and excellent ee-values (Scheme 3.9) [19].

Tomioka and coworkers [20] discovered that the aza–Michael addition of lithium amides **15–18** to tert-butyl cinnamate produced the corresponding β-amino esters in high yields and up to 99% ee in the presence of chiral diether **1** (Scheme 3.10). For the addition of lithium amide **15** [20a], 5 equiv. of TMSCl was needed in order to achieve a high enantioselectivity. The allyl group in amide **16** [20b] and the anthracen-9-ylmethyl group in amide **17** [20c] were designed because they can easily be removed after the reaction by rhodium chloride or hydrogenation, respectively. When the bulky amine **18** [20d] was introduced to this reaction, the corresponding product was isolated in 90% yields and 99% ee. The excellent enantioselectivity was explained by the fact that the bulky group of amine **18** prevents the formation of mixed aggregates. The ECA has

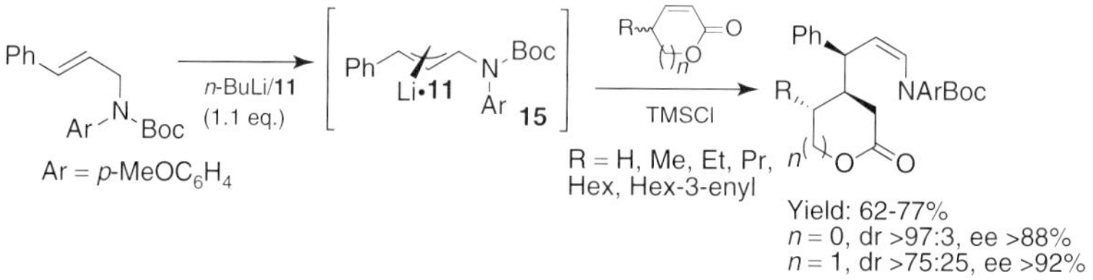

Scheme 3.9 Kinetic resolution of racemic lactones by conjugate additions of allylic organolithium species.

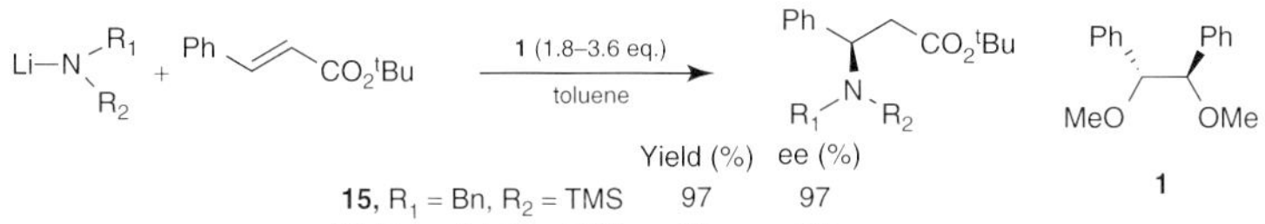

	Yield (%)	ee (%)
15, R_1 = Bn, R_2 = TMS	97	97
16, R_1 = allyl, R_2 = TBS	95	89
17, R_1 = anthracen-9-ylmethyl, R_2 = TMS	90	91
18, R_1 = Mes-CH_2, R_2 = TMS	90	99

Scheme 3.10 Asymmetric conjugate amination of lithium amides with *tert*-butyl cinnamate.

a broad substrate scope, and several enoates – including both acyclic (aromatic and aliphatic) and cyclic α,β-unsaturated *tert*-butyl esters – can be used as acceptors, providing the corresponding products in high yields and good to high ee-values.

The ECAs of simple unstabilized organolithium enolates to activated olefins are relatively challenging due to unwanted side reactions. In this context, Tomioka and coworkers [21] reported the addition of lithium enolate **19a** [21a] to acyclic and cyclic enones in the presence of chiral ligand **1**. The corresponding Michael products were obtained in fairly good yields and up to 74% ee (Scheme 3.11). The addition of lithium enolate **19b** [21b] to α,β-unsaturated *tert*-butyl esters gave the corresponding product in high ee (up to 97%) (Scheme 3.11). For this, 3 equiv. of lithium diisopropylamide (LDA) were needed to increase the enantioselectivity, which suggests that a ternary complex of enolate (**19**)-**1**-lithium amide might be formed during the reaction.

The joint research collaboration of Maddaluno, Tomioka, and coworkers [22a] led to the development of ECAs of lithium ester enolate **19b** to α,β-unsaturated *tert*-butyl esters. The reaction is catalyzed by chiral lithium amide **20** instead of chiral diether **1**. The conjugate addition products were isolated in good yields and up to 76% ee (Scheme 3.12). A mixed *endo*- or *exo*-aggregate between the lithium ester enolate and lithium amides **20** was characterized by nuclear magnetic resonance (NMR) data and density functional theory (DFT) calculations [22b].

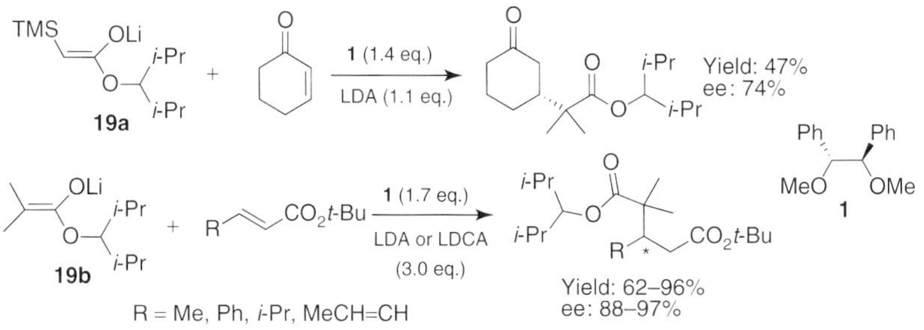

Scheme 3.11 Asymmetric conjugate addition of lithium enolate **19** to activated olefins.

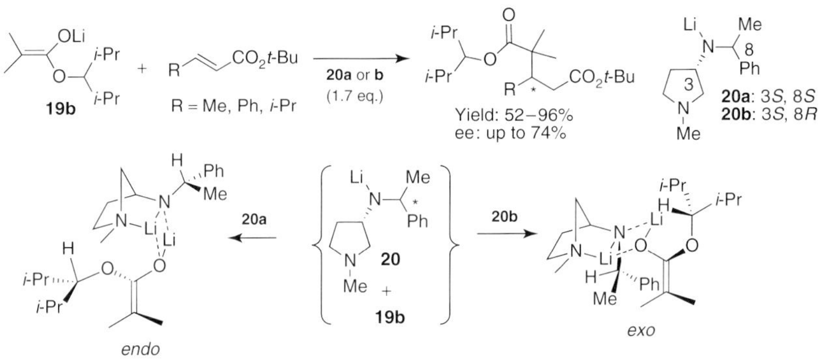

Scheme 3.12 Asymmetric conjugate additions of lithium enolate **19b** to α,β-unsaturated *tert*-butyl esters.

3.3
Catalytic Enantioselective Conjugate Addition of Grignard Reagents

Organomagnesium halides were discovered by the French chemist Victor Grignard, and named "Grignard reagents" in his honor. Compared to other organometallic reagents, the readily availability, high reactivity of these reagents, as well as their ability to transfer all of their alkyl groups, led to their wide use in organic synthesis. The ECAs of Grignard reagents to α,β-unsaturated compounds is one of the most powerful carbon–carbon bond-forming reactions. Remarkable efforts have been made to develop highly efficient catalytic systems by exploring several ligand types and reaction conditions.

The first catalytic ECA of a Grignard reagent to 2-cyclohexen-1-one was reported by Lippard and coworkers in 1988 [23], who reported the ECA of *n*-BuMgBr to cyclohexenone using catalytic amounts of Cu-amide complexes formed from chiral ligands **21** (Figure 3.2), CuBr·SMe, and PhLi or *n*-BuLi. The transformation gave

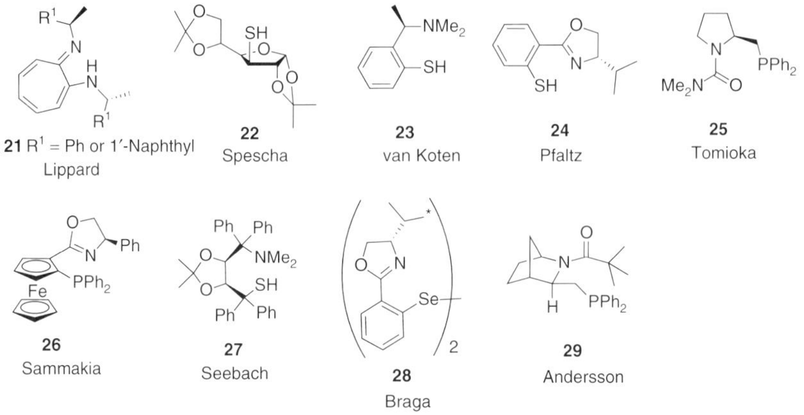

Figure 3.2 Selected ligands developed for the ECA of Grignard reagents to enones.

3.3 Catalytic Enantioselective Conjugate Addition of Grignard Reagents

Table 3.3 Cu-catalyzed ACA of Grignard reagents using different chiral ligands.

Entry	Substrate	Ligand (mol%)	Cu (mol%)	Grignard reagent	Solvent	Temp. (°C)	Additive	ee (%)
1	30a	21 (2–4)	CuBr-SMe$_2$ (2–4)	n-BuMgBr	THF	−78	HMPA	74
2	30a	22 (4)	CuI-SBu$_2$ (4)	n-BuMgCl	Et$_2$O	−78	–	60
3	30e	23 (9)	CuCl (9)	MeMgI	Et$_2$O	0	–	76
4	30a	24 (5–10)	CuI (5–10)	n-BuMgCl	THF	−78	HMPA	60
5	30b	24 (5–10)	CuI (5–10)	n-BuMgCl	THF	−78	HMPA	83
6	30c	24 (5–10)	CuI (5–10)	n-BuMgCl	THF	−78	HMPA	16
7	30a	25 (32)	CuI (8)	n-BuMgCl	Et$_2$O	−78	–	90
8	30b	25 (32)	CuI (8)	n-BuMgCl	Et$_2$O	−78	–	81
9	30c	25 (32)	CuI (8)	n-BuMgCl	Et$_2$O	−78	–	42
10	30d	25 (32)	CuI (8)	n-BuMgCl	Et$_2$O	−78	–	91
11	30a	25 (32)	CuI (8)	n-HexylMgCl	Et$_2$O	−78	–	92
12	30a	26 (12)	CuI (10)	n-BuMgCl	Et$_2$O	−78	–	83
13	30b	26 (12)	CuI (10)	n-BuMgCl	Et$_2$O	−78	–	92
14	30c	26 (12)	CuI (10)	n-BuMgCl	Et$_2$O	−78	–	65
15	30e	26 (12)	CuI (10)	n-BuMgCl	Et$_2$O	−78	–	81
16	30a	28 (10)	CuI (10)	n-BuMgCl	THF	−78	HMPA	60
17	30a	29 (4.7)	CuI (10)	n-BuMgCl	Et$_2$O	V78	–	71

good chemoselectivity (1,4- versus 1,2-addition) and modest enantioselectivity (up to 14%). The addition of silyl reagents and hexamethylphosphoric triamide (HMPA) to the reaction mixture increased the enantioselectivity (up to 74%; Table 3.3, entry 1). Later, the chiral thiosugar derivative **22** (Figure 3.2) was introduced by Spescha and coworkers [24] to this reaction, and the corresponding product was obtained with up to 60% ee (Table 3.3, entry 2). Another type of catalyst, the chiral bidentate aryl-thiolate-ligated Cu-complex **23** (Figure 3.2) was developed by van Koten and coworkers [25]. This complex catalyzed the reaction between methylmagnesium iodide and benzylideneacetone with high enantioselectivity, to give the corresponding product with 76% ee (Table 3.3, entry 3). ECA using the oxazoline aminothiol **24** developed by Pfaltz and coworkers as the ligand gave a high enantioselectivity (up to 83% ee) in the addition of RMgCl to cyclic enones (Table 3.3, entries 4–6). [26] Tomioka and coworkers [27] investigated the conjugate addition of Grignard reagents to cyclic enones in the presence of copper iodide and a chiral bidentate amidophosphine **25** (Figure 3.2) (Table 3.3, entries 7–11). The addition of n-hexylMgCl to cyclohexenone gave the corresponding Michael product with up to 92% ee (Table 3.3, entry 11). Sammakia and coworkers [28] combined planar chirality and a stereogenic center in ligand **26** (Figure 3.2). By changing the size of the alkyl group in the oxazoline ring, good chemoselectivity and enantioselectivity are accomplished in the addition of n-BuMgCl to cyclic enones (up to 92% ee; Table 3.3, entries 12–14). Moreover, the ECAs to acyclic enones give similar results (81% ee; Table 3.3, entry 15). The chiral compounds **27** [29a], **28** [29b], and **29** [29c] (Figure 3.2) are also good ligands for the conjugate addition of Grignard reagents, with good chemoselectivity and moderate enantioselectivity being obtained (Table 3.3, entries 16–17).

A breakthrough was made when Feringa and coworkers [30a] introduced ferrocenyl-based diphosphine ligands to the ECA of Grignard reagents. The addition of Grignard reagents with linear alkyl chains to cyclohexenone using 5 mol% of CuCl, 6 mol% of Taniaphos **31**, and 1.15 equiv. of RMgBr in Et$_2$O at 0 °C gave a high chemoselectivity (81 : 19–95 : 5; 1,4- to 1,2-addition) and enantioselectivity (90–96% ee) (Table 3.4, entries 1–4). In the cases of ECAs with α- and β-branched Grignard reagents, the reaction resulted in a poor stereoselectivity (Table 3.4, entries 5–6). However, the γ-branched reagent gave the corresponding 1,4-addition product with 95% ee (Table 3.4, entry 7). A good to high enantioselectivity was achieved for α- and β-branched Grignard reagents by using Josiphos ligand **32a** (Table 3.4, entries 8–9). Moreover, other cyclic enones and lactones were also used

Table 3.4 Cu-catalyzed ACA of Grignard reagents to cyclic enones using chiral ligands **31** and **32a**.

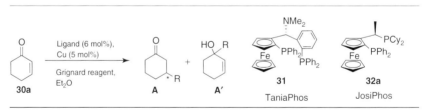

Entry	Ligand	Cu	Grignard reagent	Temp. (°C)	A : A′	ee (%) of A
1	31	CuCl	EtMgBr	0	95 : 5	96
2	31	CuCl	MeMgBr	0	83 : 17	90
3	31	CuCl	n-PrMgBr	0	81 : 19	94
4	31	CuCl	n-BuMgBr	0	88 : 12	96
5	31	CuCl	i-PrMgBr	0	78 : 22	1
6	31	CuCl	i-BuMgBr	0	62 : 38	33
7	31	CuCl	(CH$_3$)$_2$CH(CH$_2$)MgBr	0	76 : 24	95
8	32	CuBr•SMe$_2$	i-PrMgBr	−60	99 : 1	54
9	32	CuBr•SMe$_2$	i-BuMgBr	−60	99 : 1	92
10	32	CuBr•SMe$_2$	EtMgBr	−60	99 : 1	56

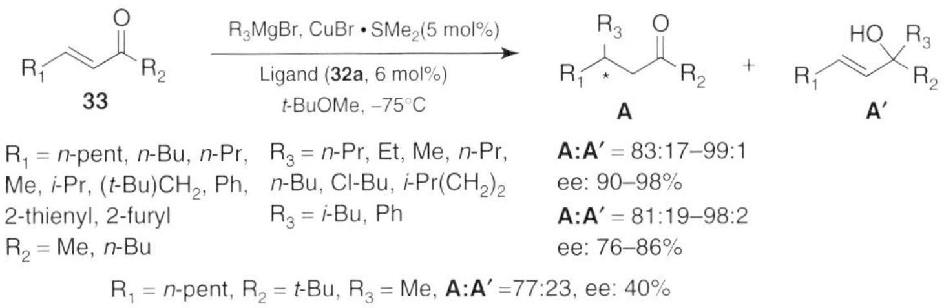

3.3 Catalytic Enantioselective Conjugate Addition of Grignard Reagents | 155

Scheme 3.13 Cu-catalyzed ACA of Grignard reagents to acyclic enones using chiral ligand **32a**.

R_1 = n-pent, n-Bu, n-Pr, Me, i-Pr, (t-Bu)CH$_2$, Ph, 2-thienyl, 2-furyl
R_2 = Me, n-Bu
R_3 = n-Pr, Et, Me, n-Pr, n-Bu, Cl-Bu, i-Pr(CH$_2$)$_2$
R_3 = i-Bu, Ph

A:A' = 83:17–99:1, ee: 90–98%
A:A' = 81:19–98:2, ee: 76–86%

R_1 = n-pent, R_2 = t-Bu, R_3 = Me, **A:A'** =77:23, ee: 40%

as Michael acceptors in this reaction, giving high levels of chemoselectivity and enantioselectivity.

Feringa and coworkers [30b] also investigated the ECA of Grignard reagents to acyclic enones catalyzed by copper with ferrocenyl-based diphosphine ligands. The optimized condition was Josiphos **32a** as ligand, CuBr•SMe$_2$ as copper source, and t-BuOMe as solvent. Grignard reagents with different linear alkyl chains and a variety of aliphatic linear enones can be used as substrates for this ECA, providing the corresponding products with moderate to high ee-values (Scheme 3.13). Both, β-substituted enones and aromatic enones can be employed as acceptors, while the sterically hindered t-Bu ketone gave a drastic decrease in enantioselectivity. Moreover, β-branched and aryl Grignard reagents provided only moderate enantioselectivities.

The Feringa group [31] also used ligand **32a** in combination with Cu to catalyze the ECA of Grignard reagents to α,β-unsaturated esters. The catalyst loading could be lowered to as little as 1 mol%, and an air-stable dinuclear Cu-complex **34a** which served as an active catalyst was isolated (Scheme 3.14). Several different Grignard reagents were tested, and high yields and good enantioselectivities were achieved except for MeMgBr, which gave a low conversion. For the ECAs to esters of cinnamic acid, the use of ligand **34a** gave the highest enantioselectivity (Scheme 3.15).

The proposed catalytic pathways of the ECAs are shown in Scheme 3.15b. The transformation involves the formation of an active complex **BF** from the dimeric precatalysts **34** via transmetallation with the Grignard reagents. Next, the complex **BF** forms a π-complex with the Michael acceptor and subsequent formation of a Mg enolate is achieved via a Cu(III) complex. This latter step is the rate-limiting reductive elimination step, regenerating the catalytic complex BF as well as

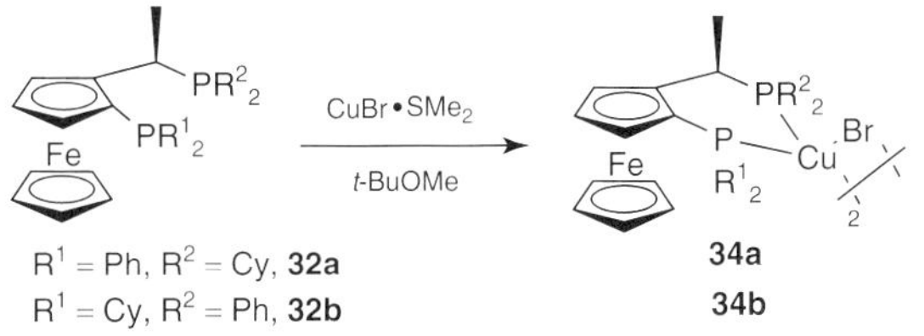

R^1 = Ph, R^2 = Cy, **32a**
R^1 = Cy, R^2 = Ph, **32b**

34a
34b

Scheme 3.14 The dinuclear Cu-complex isolated from the reaction.

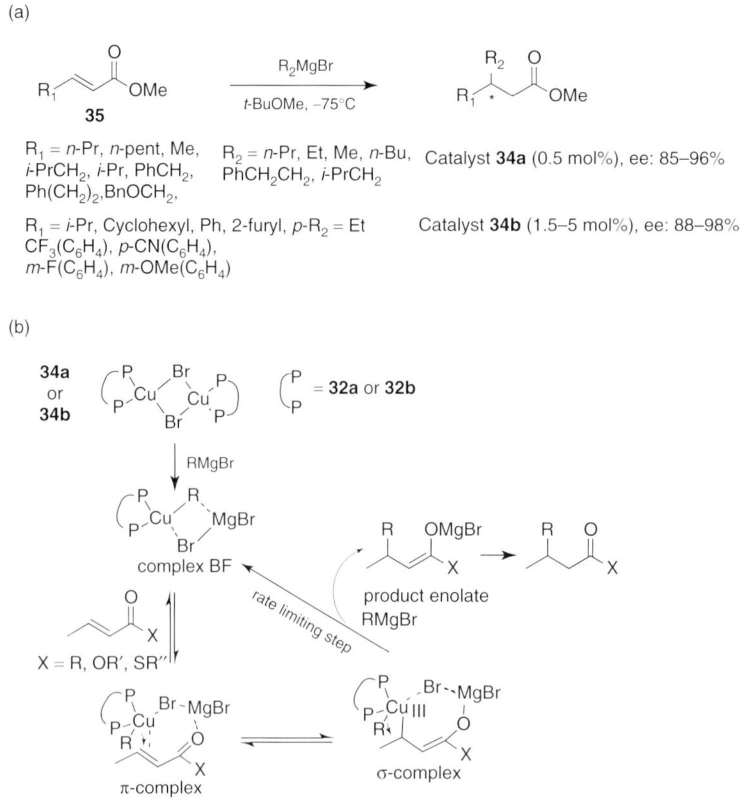
(a)
OMe
35
R2MgBr
t-BuOMe, −75 °C
R1 = n-Pr, n-pent, Me, i-PrCH2, i-Pr, PhCH2, Ph(CH2)2,BnOCH2,
R2 = n-Pr, Et, Me, n-Bu, PhCH2CH2, i-PrCH2
Catalyst 34a (0.5 mol%), ee: 85–96%
R1 = i-Pr, Cyclohexyl, Ph, 2-furyl, p-CF3(C6H4), p-CN(C6H4), m-F(C6H4), m-OMe(C6H4)
R2 = Et
Catalyst 34b (1.5–5 mol%), ee: 88–98%
(b)
34a or 34b
= 32a or 32b
RMgBr
complex BF
rate limiting step
product enolate
RMgBr
X = R, OR′, SR″
π-complex
σ-complex

Scheme 3.15 (a) Cu-catalyzed ACA of Grignard reagents to α,β-unsaturated esters; (b) Mechanism for the reactions catalyzed by complexes **34a** and **34b**.

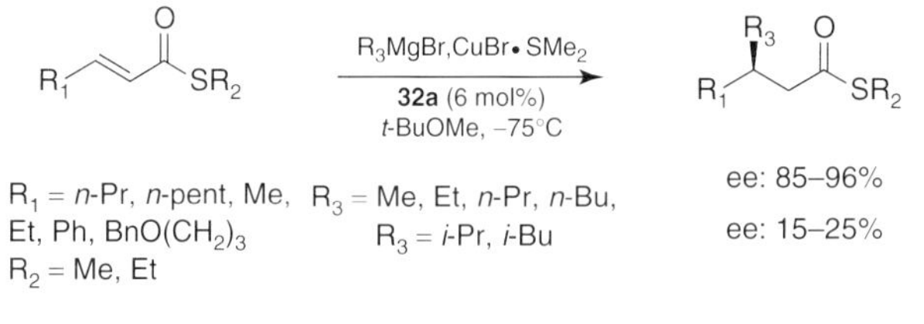
R3MgBr,CuBr• SMe2
32a (6 mol%)
t-BuOMe, −75 °C
R1 = n-Pr, n-pent, Me, Et, Ph, BnO(CH2)3
R2 = Me, Et
R3 = Me, Et, n-Pr, n-Bu,
R3 = i-Pr, i-Bu
ee: 85–96%
ee: 15–25%

Scheme 3.16 Cu-catalyzed ACA of Grignard reagents to α,β-unsaturated thioesters.

rendering the C–C bond-formation. In order to improve the reactivity of the ECA with MeMgBr, Feringa and coworkers employed more reactive α,β-unsaturated thioesters as acceptors [32a]. The *in situ*-prepared complex of $CuBr{\bullet}SMe_2$ and Josiphos ligand **32a** catalyzed the 1,4-addition of linear Grignard reagents (e.g., MeMgBr) to α,β-unsaturated thioesters, and the corresponding products were isolated in 88–94% yield and 85–96% ee (Scheme 3.16). The method was successfully applied to the synthesis of both *syn-* and *anti-*1,3-dimethyl

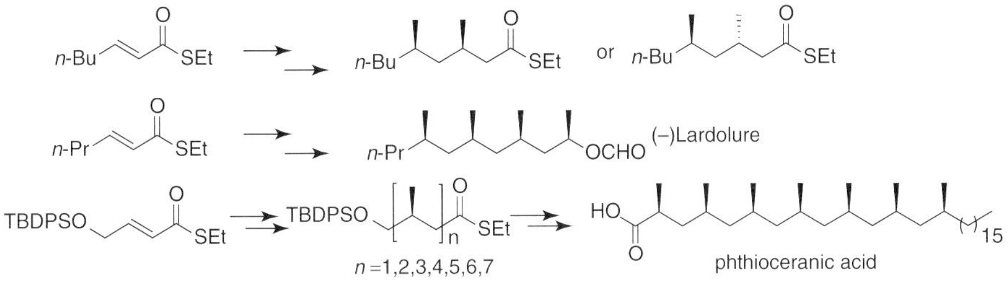

Scheme 3.17 Application of the ACA of Grignard reagents to α,β-unsaturated thioesters.

compounds and deoxypropionate chains (Scheme 3.17). The 12-step, 26% overall yield synthesis of (−)-lardolure (Scheme 3.17) and the 24-step, 4% overall yield synthesis of phthioceranic acid [32b] (Scheme 3.17) demonstrated the versatility of this method.

Loh and Ji [33] investigated the ECA of Grignard reagents to α,β-unsaturated esters by using commercially available chiral Tol-BINAP **36**. The catalytic system is prepared from CuI/Tol-BINAP (1/1.5) in t-BuOMe. Notably, a 0.5 equiv. excess of Tol-BINAP to CuI is essential to obtain a high enantioselectivity for the transformation, and the corresponding ECA products were isolated with 86–98% ee (Scheme 3.18). The presence of a double bond in the Grignard reagent did not affect the enantioselectivity as compared to the system using CuBr·SMe$_2$ and Josiphos ligand **32**. The opposite enantiomer of the product can be obtained by changing the stereochemistry of the ligand, or by changing the geometry of the staring α,β-unsaturated esters. The catalytic system was further extended to the ECA of MeMgBr to α,β-unsaturated esters, providing an entry to both *syn*- and *anti*-deoxypropionates and natural products such as Siphonarienal and Siphonarienone (Scheme 3.19).

Subsequently, Feringa and coworkers [34] investigated the conjugate addition of Grignard reagents to aromatic or aliphatic α,β-unsaturated thioesters by using

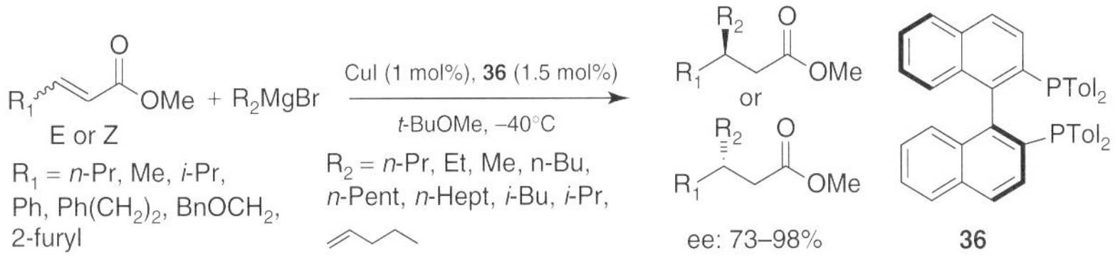

Scheme 3.18 Cu-Tol-BINAP-catalyzed ACA of Grignard reagents to α,β-unsaturated esters.

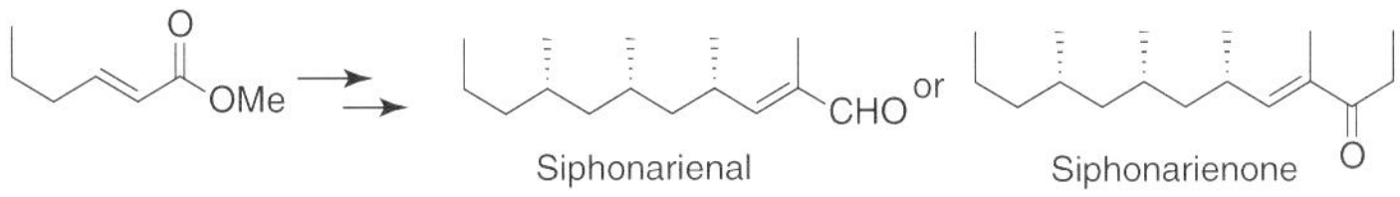

Scheme 3.19 Application of Cu-Tol-BINAP-catalyzed ACA of Grignard reagents to α,β-unsaturated esters.

Table 3.5 Cu-catalyzed ACA of Grignard reagents to α,β-unsaturated thioesters.

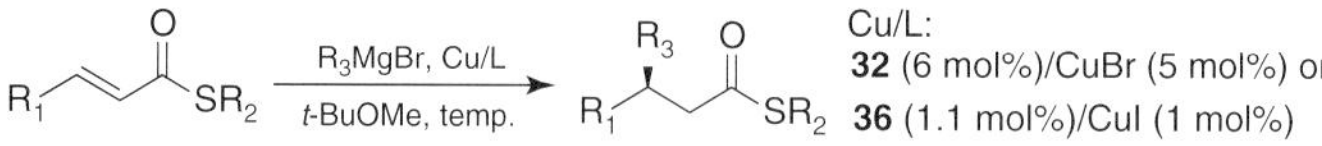

Entry	R_1	R_2	R_3	32/CuBr		36/CuI	
				Yield (%)	ee (%)	Yield (%)	ee (%)
1	Ph	Me	Me	65	95	88	94
2	p-Me-C$_6$H$_4$	Et	Me	33	>99	34	99
3	Me	Et	n-Bu	90	90	94	74
4	n-Pent	Et	Me	90	96	90	93
5	n-Pent	Et	i-Pr	93	25	89	65
6	n-Pent	Et	i-Bu	80	15	95	94

two different systems: CuBr/Josiphos or CuI/Tol-BINAP. These catalytic systems were found to be complementary: the use of Josiphos ligand **32** gave a slightly higher enantioselectivity in ECA with linear Grignard reagents, while Tol-BINAP **36** provided more stereoselective transformations with secondary or bulky Grignard reagents or crowded β-substituted aliphatic thioesters (Table 3.5).

Alexakis and coworkers [35] reported the ECA of Grignard reagents to trisubstituted enones catalyzed by copper and N-heterocyclic carbenes (NHCs) (Scheme 3.20). Two different families of chiral imidazolidinium salts (ImH$^+$) (**37** and **38**), together with copper (II) triflate in Et$_2$O, gave the highest enantioselectivities. Both, primary and secondary Grignard reagents can be used as nucleophiles to give the corresponding products in good yields and moderate to high ee, but the sterically hindered t-BuMgBr did not react. Moreover, various trisubstituted cyclohexenones – even poorly reactive enones (e.g., phenyl cyclohexenone) – can be used as acceptors, and the corresponding products are obtained in excellent yields and good ee-values. However, the ECAs of EtMgBr to cyclopentenones and cycloheptenones gives the corresponding Michael products with moderate ee-values.

Recently, Feringa and coworkers [36] reported the conjugate addition of Grignard reagents to α,β-unsaturated sulfones by using Cu/Tol-BINAP **36** as catalyst.

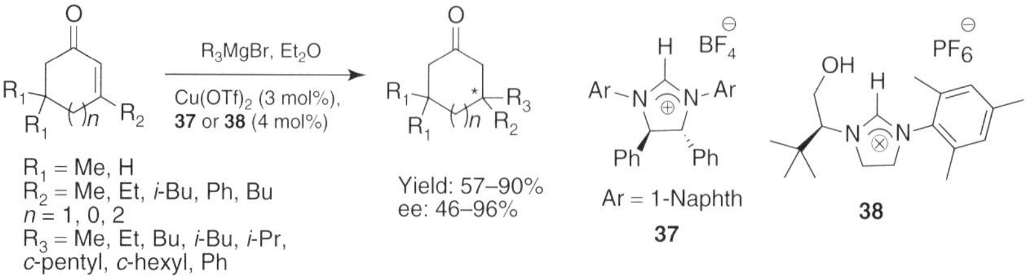

Scheme 3.20 Cu-catalyzed ACA of Grignard reagents to trisubstituted enones.

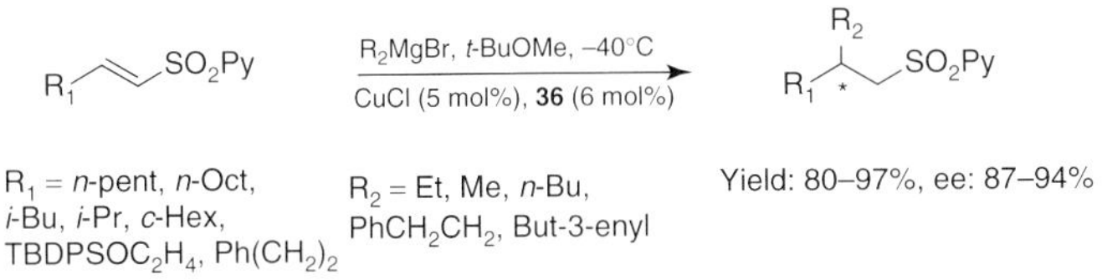

R$_2$MgBr, t-BuOMe, −40 °C
CuCl (5 mol%), **36** (6 mol%)

R$_1$ = n-pent, n-Oct, i-Bu, i-Pr, c-Hex, TBDPSOC$_2$H$_4$, Ph(CH$_2$)$_2$

R$_2$ = Et, Me, n-Bu, PhCH$_2$CH$_2$, But-3-enyl

Yield: 80–97%, ee: 87–94%

Scheme 3.21 Cu-catalyzed ACA of Grignard reagents to α,β-unsaturated sulfones.

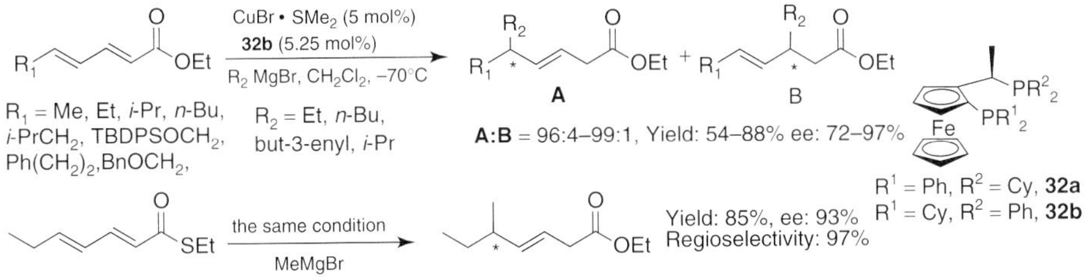

CuBr·SMe$_2$ (5 mol%)
32b (5.25 mol%)
R$_2$MgBr, CH$_2$Cl$_2$, −70 °C

R$_1$ = Me, Et, i-Pr, n-Bu, i-PrCH$_2$, TBDPSOCH$_2$, Ph(CH$_2$)$_2$, BnOCH$_2$,

R$_2$ = Et, n-Bu, but-3-enyl, i-Pr

A:B = 96:4–99:1, Yield: 54–88% ee: 72–97%

R^1 = Ph, R^2 = Cy, **32a**
R^1 = Cy, R^2 = Ph, **32b**

the same condition
MeMgBr

Yield: 85%, ee: 93%
Regioselectivity: 97%

Scheme 3.22 Cu-catalyzed 1,6-CA of Grignard reagents to linear dienoates.

Grignard reagents with different linear alkyl chains or having a double bond can be used as nucleophiles for the ECA, with high enantioselectivities, but transformations with PhMgBr do not exhibit stereoselectivity (Scheme 3.21). The ECA also tolerates linear aliphatic α,β-unsaturated sulfones as acceptors to give the corresponding products in high yields and ee-values. Although the ECAs with aromatic α,β-unsaturated sulfones as acceptors exhibited only a moderate enantioselectivity, the transformations with acceptors having a 2-pyridyl group exhibited good enantioselectivity and reactivity.

The first catalytic enantioselective 1,6-addition of Grignard reagents to *bis*-unsaturated esters was also reported by Feringa and coworkers [37], and the reversed Josiphos ligand **32b** was shown to be the best choice for this 1,6-addition. The reaction was slower as compared to the 1,4-addition (1,4-addition at −75 °C was 2 h, compared to 1,6-addition at −70 °C being 16 h). Grignard reagents possessing a longer alkyl chain or a homoallylic moiety can be used as nucleophiles for this 1,6-addition, which exhibit excellent regioselectivity and enantioselectivity. The use of an *i*-Pr Grignard reagent resulted in a slight decrease in enantioselectivity (72% ee). Moreover, several types of dienoates (e.g., linear alkyl dienoates) can be used as acceptors for this transformation (Scheme 3.22). Furthermore, the ECAs with less-reactive MeMgBr can also be used successfully for 1,6-addition to the corresponding thioester (Scheme 3.22).

3.4
Cu-Complexes as Catalysts for Enantioselective Conjugate Additions

During the past 20 years, major breakthroughs have been made in Cu-catalyzed ECA such that, today, both organometallic reagents and organic nucleophiles

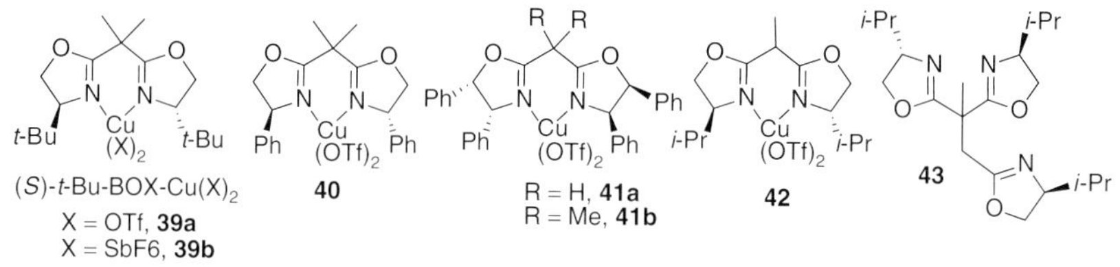
(S)-t-Bu-BOX-Cu(X)2
X = OTf, 39a
X = SbF6, 39b
40
R = H, 41a
R = Me, 41b
42
43

Figure 3.3 Selected copper catalysts used for enantioselective conjugate additions.

can be added to Michael acceptors in this type of transformation. Whilst the important use of diorganozinc reagents has been summarized in Chapter 2 of this book, attention in this chapter is focused on the Cu-catalyzed conjugate additions of organic nucleophiles. In this area, the C_2-symmetric bisoxazoline copper (II) complexes introduced by Evans and coworkers for Diels–Alder reactions are the most commonly used catalysts (Figure 3.3) [38]. In this context, Katsuki and coworkers [39] reported that the chiral catalyst, $Cu(OTf)_2$-bis(oxazoline) **39a** can promote the ECA of 2-(trimethylsilyloxy)furans to oxazolidinone enoates with good chemical yields and high enantioselectivities (Scheme 3.23). The ECA with a crotonoyl derivative as the acceptor exhibited excellent enantioselectivity, whilst the addition of 1 equiv. of hexafluoro-2-propanol (HFIP) was essential to obtain high chemical yields.

The combination of copper (II) and bis(oxazoline) ligands was investigated extensively by Evans and coworkers [40], with the addition of silylketene acetal to the β-substituted alkylidene malonates affording the corresponding products with high yields and excellent ee-values in the presence of catalyst **39b**. Both, aromatic and sterically demanding alkyl substituents on the malonate were well tolerated (Scheme 3.24). The structure of the catalyst–alkylidene malonate complex was confirmed by single-crystal X-ray diffraction (XRD) studies. It is also important to note that a significant distortion of the bis(oxazoline) ligand out of the plane, which

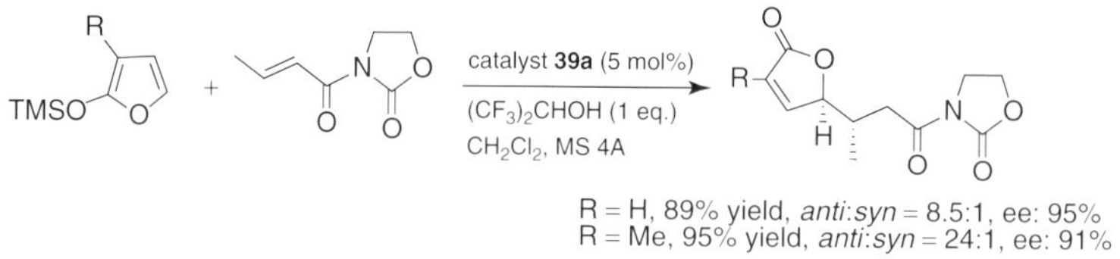
catalyst 39a (5 mol%)
(CF3)2CHOH (1 eq.)
CH2Cl2, MS 4A
R = H, 89% yield, anti:syn = 8.5:1, ee: 95%
R = Me, 95% yield, anti:syn = 24:1, ee: 91%

Scheme 3.23 ECA of 2-(trimethylsilyloxy)furans to oxazolidinone enoates in the presence of catalyst **39a**.

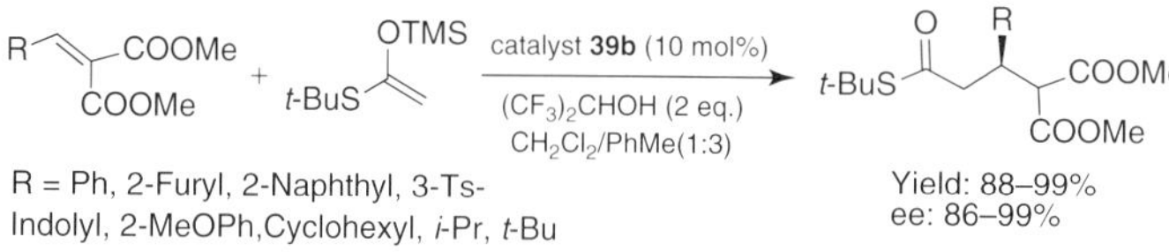
catalyst 39b (10 mol%)
(CF3)2CHOH (2 eq.)
CH2Cl2/PhMe(1:3)
R = Ph, 2-Furyl, 2-Naphthyl, 3-Ts-Indolyl, 2-MeOPh,Cyclohexyl, i-Pr, t-Bu
Yield: 88–99%
ee: 86–99%

Scheme 3.24 ECA of enolsilanes to alkylidene malonates catalyzed by **39b**.

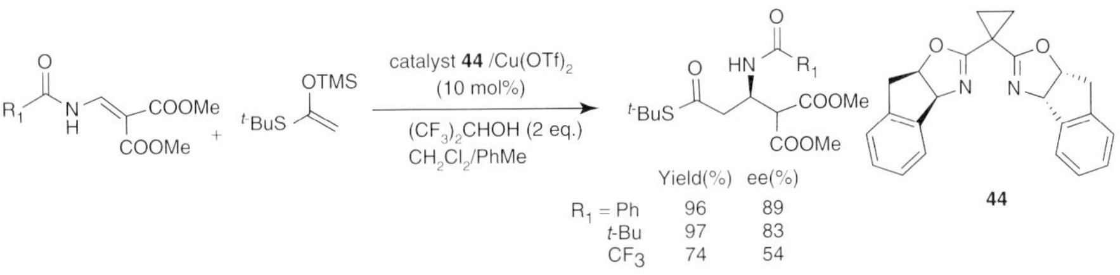

Scheme 3.25 ECA of enolsilanes to β-enamidomalonates catalyzed by **44**.

was observed in the X-ray crystal structure, may be essential for the asymmetric induction.

Sibi and coworkers [41] reported the ECA of silylketene acetals to β-enamidomalonates in the presence of catalyst **44** (Scheme 3.25). The effects of the *N*-acyl substituent and nucleophiles were investigated, with the best results being obtained when R_1 = Ph and the O, S-ketene silyl acetal was used.

When Evans and coworkers [42] used catalyst **39b** in the conjugate addition of enolsilanes to fumaroyl oxazolidinone, the Michael addition products were obtained in high yields and ee-values (Scheme 3.26). The diastereoselectivities were directly related to the enolsilane geometry, and impacted upon by the size of the enolsilane substituent. A hetero-Diels–Alder reaction product dihydropyran **A** (Scheme 3.26) was observed as an intermediate.

Jørgensen and coworkers [43] extended the use of copper catalyst **39a** to the catalytic enantioselective alkylation of heteroaromatic compounds (Scheme 3.27). In the presence of catalyst **39a**, the reactions between heteroaromatic compounds

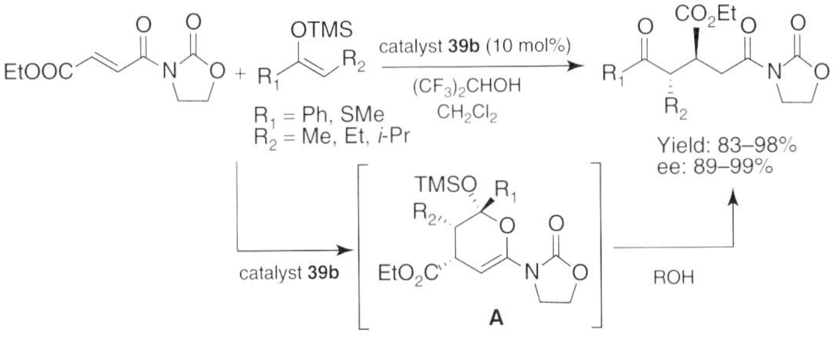

Scheme 3.26 ECA of enolsilanes to fumaroyl oxazolidinone catalyzed by **39b**.

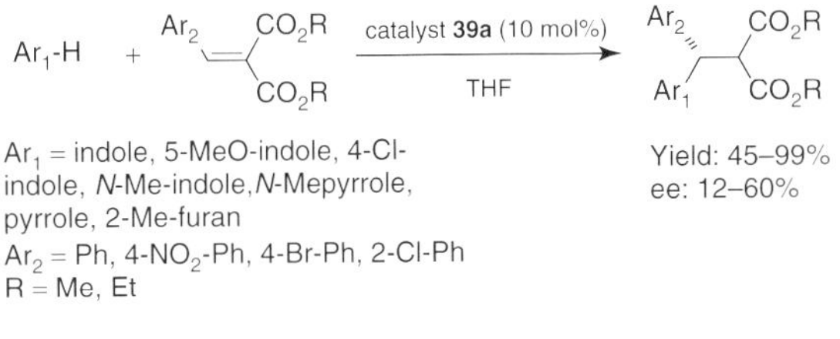

Scheme 3.27 Enantioselective Friedel–Crafts alkylation of heteroaromatic compounds catalyzed by **39a**.

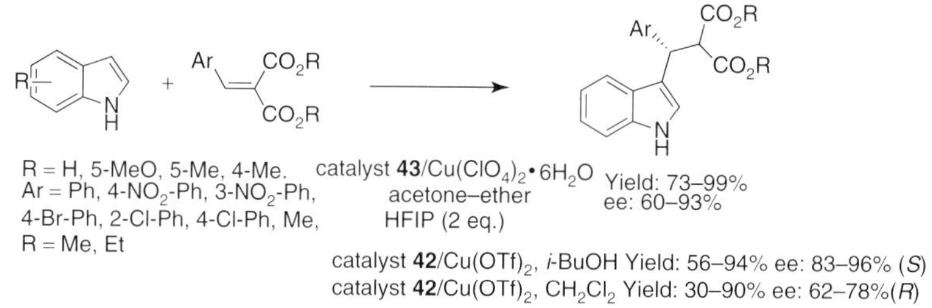

Scheme 3.28 Enantioselective conjugate addition of indoles to alkylidene malonate.

R = H, 5-MeO, 5-Me, 4-Me.
Ar = Ph, 4-NO$_2$-Ph, 3-NO$_2$-Ph, 4-Br-Ph, 2-Cl-Ph, 4-Cl-Ph, Me,
R = Me, Et

catalyst **43**/Cu(ClO$_4$)$_2$•6H$_2$O acetone–ether HFIP (2 eq.) Yield: 73–99% ee: 60–93%

catalyst **42**/Cu(OTf)$_2$, *i*-BuOH Yield: 56–94% ee: 83–96% (S)
catalyst **42**/Cu(OTf)$_2$, CH$_2$Cl$_2$ Yield: 30–90% ee: 62–78% (R)

and various β-substituted alkylidene malonates proceeded with high yields and moderate ee-values. Unfortunately, the addition of HFIP to the reaction did not provide any better results.

A significant improvement of the enantiomeric excess was achieved when Tang and coworkers [44] introduced pseudo-C_3-symmetric trioxazoline **43** to these asymmetric conjugate additions (ACAs) of indoles to alkylidene malonates (Scheme 3.28). The authors pointed out that the trioxazoline **43** could induce a more stable pseudo-C_3-symmetric chiral space by coordination of three nitrogen atoms to the copper center. In the presence of the catalytic complex **43**/Cu(ClO$_4$)$_2$•6H$_2$O, however, the Michael addition of indoles to alkilidene malonates afforded the adducts in high yields and with excellent ee-values. The absolute stereochemistry of this reaction was seen to depend on the solvent, while the addition of HFIP improved the reactivity, without any loss in ee-value. Moreover, the even more simple bisoxazoline **42** was also shown to be a powerful ligand for this reaction, with both enantiomers of the alkylation adducts capable of being prepared by the subtle choice of solvent (Scheme 3.28).

Later, Jørgensen and coworkers [45] investigated the asymmetric Friedel–Crafts alkylation of β,γ-unsaturated α-ketoesters catalyzed by chiral Lewis acid **39a**. The reactions between various indole derivatives and different β,γ-unsaturated α-ketoesters proceeded very well, and the corresponding products were isolated in excellent yields and up to >99% ee (Scheme 3.29). When 2-methylfuran and 1,3-dimethoxybenzene were tested in this reaction, the products were obtained in good yields and 60–89% ee.

In 2003, Jørgensen and coworkers [46] reported the catalytic ECA of cyclic 1,3-dicarbonyl compounds to unsaturated 2-ketoesters in the presence of a Lewis

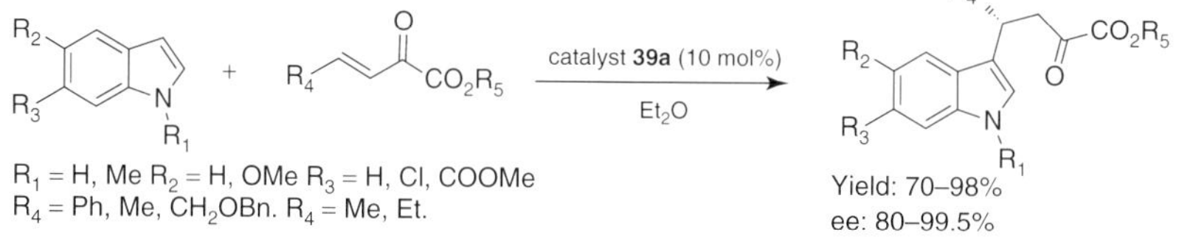

R_1 = H, Me R_2 = H, OMe R_3 = H, Cl, COOMe
R_4 = Ph, Me, CH$_2$OBn. R_4 = Me, Et.

catalyst **39a** (10 mol%) Et$_2$O
Yield: 70–98%
ee: 80–99.5%

Scheme 3.29 Enantioselective addition of aromatic C–H bonds to alkenes catalyzed by **39a**.

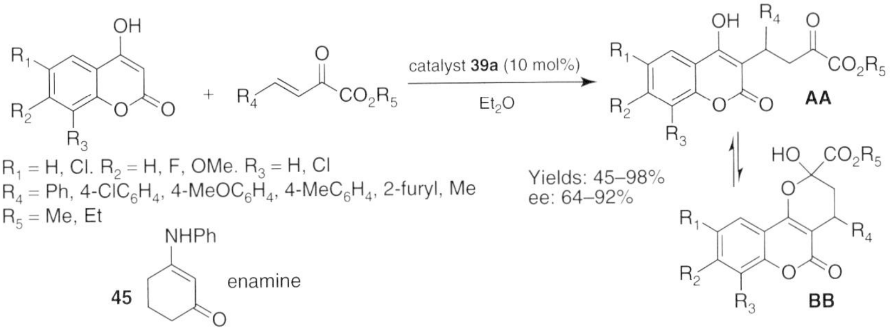

Scheme 3.30 Enantioselective Michael addition of cyclic 1,3-dicarbonyl compounds catalyzed by **39a**.

acid catalyst **39a**. The corresponding products were formed in high yields and with moderate to high ee-values (Scheme 3.30). Furthermore, the products **AA** were found to be in a rapid equilibrium with the cyclic hemiketal **BB**. Notably, enamines such as **45** could be added to the α,β-unsaturated 2-ketoesters to give the products in practically quantitative yields and with excellent ee-values.

Palomo and coworkers found that α′-hydroxy enones can coordinate to Evans bisoxazoline–copper complexes through a 1,4-metal-binding pattern [47]. In this case, they successfully applied α′-hydroxy enones to the conjugate addition of carbamates (Scheme 3.31) and Friedel–Crafts alkylation of pyrroles and indoles (Scheme 3.32) in the presence of a Cu(II)-bisoxazoline catalyst **39a**. Both, linear and branched aliphatic chains were tolerated in both reactions. In the conjugate addition of carbamates, those substrates with a bulky *tert*-butyl group gave the product in 65% yield and 94% ee, but enones bearing β-aryl substituents were unreactive. For the Friedel–Crafts alkylation, the indole derivatives reacted equally

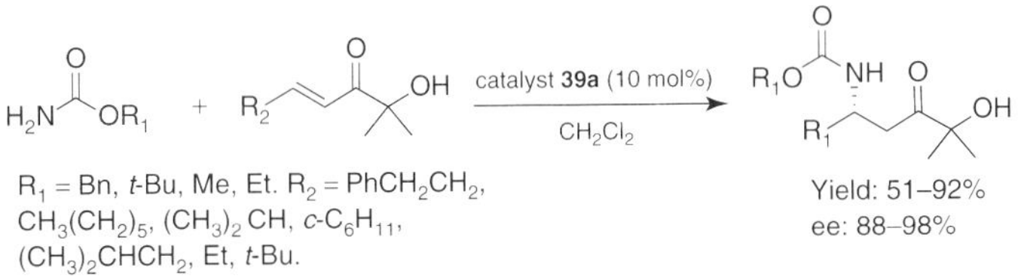

Scheme 3.31 Enantioselective conjugate addition of carbamates catalyzed by **39a**.

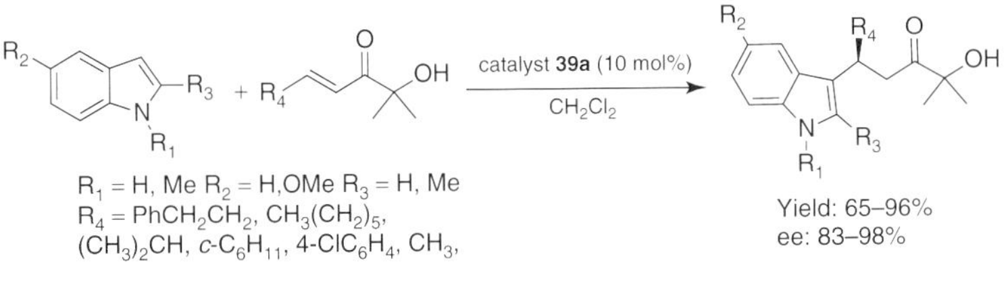

Scheme 3.32 Enantioselective Friedel–Crafts alkylation of indoles catalyzed by **39a**.

well as the pyrroles, and provided the corresponding products in excellent yields and ee-values. Notably, the catalyst loading could be lowered to 5 and 2 mol%, without any significant loss of enantioselectivity and/or yield.

3.5
Conclusions

During the past 30 years, an enormous growth has occurred in the ECA of organolithium and organomagnesium reagents to different activated α,β-unsaturated compounds and this, in turn, has opened up the use of organometallic reagents in ACAs. However, several challenges remain that include the use of catalytic amounts of chiral catalysts for ECA with organolithium reagents, and the use of sterically demanding Grignard reagents and functionalized Grignard reagents. The use of mechanistic studies and theoretical chemistry studies (e.g., DFT) will surely also contribute to the discovery of appropriate chiral ligands and/or catalytic systems in order to tune the high reactivity of organolithium reagents and Grignard reagents. The development of copper-catalyzed reactions is an important step towards this goal, and in this context the pioneering investigations of Evans and colleagues into Cu-catalyzed C–C bond-forming reactions has led to the recent development of highly enantioselective conjugation additions. Clearly, further developments will be made as a result of the stimulating research currently being conducted in these areas.

References

1. For reviews on enantioselective conjugate additions: (a) Sibi, M.P. and Manyem, S. (2000) *Tetrahedron*, **56**, 8033–8061; (b) Krause, N. and Hoffmann-Röder, A. (2001) *Synthesis*, 171; (c) Alexakis, A. and Benhaim, C. (2002) *Eur. J. Org. Chem.*, 3221; (d) Christoffers, J., Koripelly, G., Rosiak, A., and Rössle, M. (2007) *Synthesis*, **9**, 1279.
2. (a) Jacobsen, E.N., Pfaltz, A., and Yamamoto, H. (1999) *Comprehensive Asymmetric Catalysis*, Springer-Verlag, Berlin; (b) Noyori, R. (1994) *Asymmetric Catalysis in Organic Synthesis*, John Wiley & Sons, Inc., New York; (c) Blaser, H.-U. and Schmidt, E. (2004) *Asymmetric Catalysis on Industrial Scale: Challenges, Approaches and Solutions*, Wiley-VCH Verlag GmbH, Weinheim.
3. (a) Berner, O.M., Tedeschi, L., and Enders, D. (2002) *Eur. J. Org. Chem.*, 1877; (b) Fleming, F.F. and Wang, Q. (2003) *Chem. Rev.*, **103**, 2035; (c) Ballini, R., Bosica, G., Fiorini, D., Palmieri, A., and Petrini, M. (2005) *Chem. Rev.*, **105**, 933; (d) Sulzer-Mosse, S. and Alexakis, A. (2007) *Chem. Commun.*, 3123; (e) Tsogoeva, S.B. (2007) *Eur. J. Org. Chem.*, 1701.
4. For reviews on asymmetric reactions of organolithium: Tomioka, K. (2004) *Yakugaku Zasshi*, **124**, 43–53.
5. For reviews on asymmetric conjugate additions of Grignard reagents: (a) López, F., Minnaard, A.J., and Feringa, B.L. (2007) *Acc. Chem. Res.*, **40**, 179–188; (b) Harutyunyan, S.R., den Hartog, T., Geurts, K., Minnaard, A.J., and Feringa, B.L. (2008) *Chem. Rev.*, **108**, 2824–2852.
6. Alexakis, A., Bäckvall, J.E., Krause, N., Pàmies, O., and Diéguez, M. (2008) *Chem. Rev.*, **108**, 2796–2823.
7. Fraser, P.K. and Woodward, S. (2003) *Chem. Eur. J.*, **9**, 776.

8. Takaya, Y., Ogasawara, M., Hayashi, T., Sakai, M., and Miyaura, N. (1998) *J. Am. Chem. Soc.*, **120**, 5579–5580.
9. Alexakis, A., Frutos, J.C., and Mangeney, P. (1993) *Tetrahedron: Asymmetry*, **4**, 2427.
10. (a) Tomioka, K., M., Shindo; M., and Koga, K. (1989) *J. Am. Chem. Soc.*, **111**, 8266–8268; (b) Tomioka, K., Okamoto, T., Kanai, M., and Yamataka, H. (1994) *Tetrahedron Lett.*, **35**, 1891–1892; (c) Shindo, M., Koga, K., and Tomioka, K. (1998) *J. Org. Chem.*, **63**, 9351–9357.
11. (a) Tomioka, K., Shindo, M., and Koga, K. (1993) *Tetrahedron Lett.*, **34**, 681–682; (b) Shindo, M., Yamamoto, Y., Yamada, K., and Tomioka, K. (2009) *Chem. Pharm. Bull.*, **57**, 752–754.
12. (a) Asano, Y., Iida, A., and Tomioka, K. (1997) *Tetrahedron Lett.*, **38**, 8973–8976; (b) Asano, Y., Iida, A., and Tomioka, K. (1998) *Chem. Pharm. Bull.*, **46**, 184–186.
13. Xu, F., Tillyer, R.D., Tschaen, D.M., Grabowski, E.J.J., and Reider, P.J. (1998) *Tetrahedron: Asymmetry*, **9**, 1651–1655.
14. (a) Nishimura, K., Ono, M., Nagaoka, Y., and Tomioka, K. (1997) *J. Am. Chem. Soc.*, **119**, 12974–12975; (b) Nishimura, K. and Tomioka, K. (2002) *J. Org. Chem.*, **67**, 431–434.
15. Yamashita, M., Yamada, K., and Tomioka, K. (2004) *J. Am. Chem. Soc.*, **126**, 1954–1955.
16. (a) Park, Y.S., Weisenburger, G.A., and Beak, P. (1997) *J. Am. Chem. Soc.*, **119**, 10537–10538; (b) Park, Y.S. and Beak, P. (1997) *J. Org. Chem.*, **62**, 1574–1575; (c) Lim, S.H., Curtis, M.D., and Beak, P. (2001) *Org. Lett.*, **3**, 711–714; (d) Lee, S.J. and Beak, P. (2006) *J. Am. Chem. Soc.*, **128**, 2178–2179.
17. (a) Curtis, M.D. and Beak, P. (1999) *J. Org. Chem.*, **64**, 2996–2997; (b) Jang, D.O., Kim, D.D., Pyun, D.K., and Beak, P. (2003) *Org. Lett.*, **5**, 4155–4157; (c) Johnson, T.A., Curtis, M.D., and Beak, P. (2001) *J. Am. Chem. Soc.*, **123**, 1004–1005; (d) Johnson, T.A., Curtis, M.D., and Beak, P. (2002) *Org. Lett.*, **4**, 2747–2749; (e) Johnson, T.A., Jang, D.O., Slafer, B.W., Curtis, M.D., and Beak, P. (2002) *J. Am. Chem. Soc.*, **124**, 11689–11698.
18. Pippel, D.J., Weisenburger, G.A., Wilson, S.R., and Beak, P. (1998) *Angew. Chem., Int. Ed.*, **37**, 2522–2524.
19. Lim, S.H. and Beak, P. (2002) *Org. Lett.*, **4**, 2657–2660.
20. (a) Doi, H., Sakai, T., Iguchi, M., Yamada, K., and Tomioka, K. (2003) *J. Am. Chem. Soc.*, **125**, 2886–2887; (b) Doi, H., Sakai, T., Yamada, K., and Tomioka, K. (2004) *Chem. Commun.*, 1850–1851; (c) Sakai, T., Doi, H., Kawamoto, Y., Yamada, K., and Tomioka, K. (2004) *Tetrahedron Lett.*, **45**, 9261–9263; (d) Sakai, T., Doi, H., and Tomioka, K. (2006) *Tetrahedron*, **62**, 8351–8359; (e) Suzuki, M., Kawamoto, Y., Sakai, T., Yamamoto, Y., and Tomioka, K. (2009) *Org. Lett.*, **11**, 653–655; (f) Sakai, T., Kawamoto, Y., and Tomioka, K. (2006) *J. Org. Chem.*, **71**, 4706–4709.
21. (a) Iguchi, M., Doi, H., Hata, S., and Tomioka, K. (2004) *Chem. Pharm. Bull.*, **52**, 125–129; (b) Yamamoto, Y., Suzuki, H., Yasuda, Y., Iida, A., and Tomioka, K. (2008) *Tetrahedron Lett.*, **49**, 4582–4584; (c) Yamada, K., Yamashita, M., Sumiyoshi, T., Nishimura, K., and Tomioka, K. (2009) *Org. Lett.*, **11**, 1631–1633; (d) Yamamoto, Y., Yasuda, Y., Nasu, H., and Tomioka, K. (2009) *Org. Lett.*, **11**, 2007–2009.
22. (a) Duguet, N., Harrison-Marchand, A., Maddaluno, J., and Tomioka, K. (2006) *Org. Lett.*, **8**, 5745–5748; (b) Lecachey, B., Duguet, N., Oulyadi, H., Fressigné, C., Harrison-Marchand, A., Yamamoto, Y., Tomioka, K., and Maddaluno, J. (2009) *Org. Lett.*, **11**, 1907–1910.
23. (a) Villacorta, G.M., Rao, Ch.P., and Lippard, S.J. (1988) *J. Am. Chem. Soc.*, **110**, 3175–3182; (b) Ahn, K.-H., Klassen, R.B., and Lippard, S.J. (1990) *Organometallics*, **9**, 3178–3181.
24. Spescha, M. and Rihs, G. (1993) *Helv. Chim. Acta*, **76**, 1219–1230.
25. (a) Lambert, F., Knotter, D.M., Janssen, M.D., van Klaveren, M., Boersma, J., and van Koten, G. (1991) *Tetrahedron: Asymmetry*,

2, 1097–1100; (b) Knotter, D.M., Grove, D., Smeets, W.J.J., Spek, A.L., and van Koten, G. (1992) *J. Am. Chem. Soc.*, **114**, 3400–3410; (c) van Koten, G. (1994) *Pure Appl. Chem.*, **66**, 1455–1462; (d) van Klaveren, M., Lambert, F., Eijkelkamp, D.J.F.M., Grove, D.M., and van Koten, G. (1994) *Tetrahedron Lett.*, **35**, 6135–6138.

26. (a) Zhou, Q.-L. and Pfaltz, A. (1993) *Tetrahedron Lett.*, **34**, 7725–7728; (b) Zhou, Q.-L. and Pfaltz, A. (1994) *Tetrahedron Lett.*, **50**, 4467–4478.

27. (a) Kanai, M. and Tomioka, K. (1995) *Tetrahedron Lett.*, **36**, 4273–4274; (b) Nakagawa, Y., Kanai, M., Nagaoka, Y., and Tomioka, K. (1998) *Tetrahedron*, **54**, 10295–10307; (c) Kanai, M., Nakagawa, Y., and Tomioka, K. (1999) *Tetrahedron*, **55**, 3843–3854.

28. Stangeland, E.L. and Sammakia, T. (1997) *Tetrahedron*, **53**, 16503–16510.

29. (a) Seebach, D., Jaeschke, G., Pichota, A., and Audergon, L. (1997) *Helv. Chim. Acta*, **80**, 2515–2519; (b) Braga, A.L., Silva, S.J., Lüdtke, D.S., Drekener, R.L., Silveira, C.C., Rocha, J.B.T., and Wessjohann, L.A. (2002) *Tetrahedron Lett.*, **43**, 7329–7331; (c) Modin, S.A., Pinho, P., and Andersson, P.G. (2004) *Adv. Synth. Catal.*, **346**, 549–553.

30. (a) Feringa, B.L., Badorrey, R., Peña, D., Harutyunyan, S.R., and Minnaard, A.J. (2004) *Proc. Natl Acad. Sci. USA*, **101**, 5834–5838; (b) López, F., Harutyunyan, S.R., Minnaard, A.J., and Feringa, B.L. (2004) *J. Am. Chem. Soc.*, **126**, 12784–12785.

31. (a) López, F., Harutyunyan, S.R., Lopez, F., Browne, W.R., Correa, A., Peña, D., Badorrey, R., Meetsma, A., Minnaard, A.J., and Feringa, B.L. (2005) *Angew. Chem., Int. Ed.*, **44**, 2752–2756; (b) Harutyunyan, S.R., Meetsma, A., Minnaard, A.J., and Feringa, B.L. (2006) *J. Am. Chem. Soc.*, **128**, 9103–9118.

32. (a) Mazery, R.D., Pullez, M., López, F., Harutyunyan, S.R., Minnaard, A.J., and Feringa, B.L. (2005) *J. Am. Chem. Soc.*, **127**, 9966–9967; (b) ter Horst, B., Feringa, B.L., and Minnaard, A.J. (2007) *Org. Lett.*, **9**, 3013–3015;

(c) Harutyunyan, S.R., Zhao, Z., den Hartog, T., Bouwmeester, K., Minnaard, A.J., Feringa, B.L., and Govers, F. (2008) *Proc. Natl Acad. Sci. USA*, **105**, 8507–8512.

33. (a) Wang, S.-Y., Ji, S.-J., and Loh, T.-P. (2007) *J. Am. Chem. Soc.*, **129**, 276–277; (b) Lum, T.-K., Wang, S.-Y., and Loh, T.-P. (2008) *Org. Lett.*, **10**, 761–764.

34. Ruiz, B.M., Geurts, K., Fernández-Ibáñez, M.Á., ter Horst, B., Minnaard, A.J., and Feringa, B.L. (2007) *Org. Lett.*, **9**, 5123–5126.

35. Martin, D., Kehrli, S., d'Augustin, M., Clavier, H., Mauduit, M., and Alexakis, A. (2006) *J. Am. Chem. Soc.*, **128**, 8416–8417.

36. Bos, P.H., Minnaard, A.J., and Feringa, B.L. (2008) *Org. Lett.*, **10**, 4219–4222.

37. den Hartog, T., Harutyunyan, S.R., Font, D., Minnaard, A.J., and Feringa, B.L. (2008) *Angew. Chem., Int. Ed.*, **47**, 398–401.

38. Johnson, J.S. and Evans, D.A. (2000) *Acc. Chem. Res.*, **33**, 325–335.

39. (a) Kitajima, H., Ito, K., and Katsuki, T. (1997) *Tetrahedron*, **53**, 17015–17028; (b) Nishikori, H., Ito, K., and Katsuki, T. (1998) *Tetrahedron: Asymmetry*, **9**, 1165–1170.

40. (a) Evans, D.A., Rovis, T., Kozlowski, M.C., and Tedrow, J.S. (1999) *J. Am. Chem. Soc.*, **121**, 1994–1995; (b) Evans, D.A., Rovis, T., Kozlowski, M.C., Downey, C.W., and Tedrow, J.S. (2000) *J. Am. Chem. Soc.*, **122**, 9134–9142.

41. Sibi, M.P. and Chen, J. (2002) *Org. Lett.*, **4**, 2933–2936.

42. (a) Evans, D.A., Willis, M.C., and Johnston, J.N. (1999) *Org. Lett.*, **1**, 865–868; (b) Evans, D.A., Rovis, T., and Johnson, J.S. (1999) *Pure Appl. Chem.*, **71**, 1407–1415; (c) Evans, D.A., Scheidt, K.A., Johnston, J.N., and Willis, M.C. (2001) *J. Am. Chem. Soc.*, **123**, 4480–4491.

43. (a) Zhuang, W., Hansen, T., and Jørgensen, K.A. (2001) *Chem. Commun.*, 347–348; (b) Jørgensen, K.A. (2003) *Synthesis*, 1117–1125.

44. (a) Zhou, J. and Tang, Y. (2002) *J. Am. Chem. Soc.*, **124**, 9030–9031; (b) Zhou, J., Tang, Y. (2004) *Chem. Commun.*

432–433; (c) Zhou, J., Ye, M.-C., Huang, Z.-Z., and Tang, Y. (2004) *J. Org. Chem.*, **69**, 1309–1320.

45. Jensen, K.B., Thorhauge, J., Hazell, R.G., and Jørgensen, K.A. (2001) *Angew. Chem., Int. Ed.*, **40**, 160–163 (see also Ref. [44b]).

46. Halland, N., Velgaard, T., and Jørgensen, K.A. (2003) *J. Org. Chem.*, **68**, 5067–5074.

47. (a) Palomo, C., Oiarbide, M., Halder, R., Kelso, M., Gómez-Bengoa, E., and García, J.M. (2004) *J. Am. Chem. Soc.*, **126**, 9188–9189; (b) Palomo, C., Oiarbide, M., Kardak, B.G., García, J.M., and Linden, A. (2005) *J. Am. Chem. Soc.*, **127**, 4151–4155.

4
Asymmetric Bifunctional Catalysis Using Heterobimetallic and Multimetallic Systems in Enantioselective Conjugate Additions
Armando Córdova

4.1
Introduction

The conjugate addition of a nucleophile to an α,β-unsaturated carbonyl compound (commonly referred to as *Michael addition*) is a classic carbon–carbon bond-forming reaction in organic synthesis [1]. It is also an important transformation for the formation of carbon-heteroatom bonds. As these reactions often result in the creation of a stereogenic center, the importance of controlling the stereoselectivity of such conjugate addition reactions has sparked intense and impressive developments by chemists in both academia and industry. The use of stoichiometric amounts of chiral reagents, ligands, and auxiliaries has led to the development of several asymmetric Michael reactions [2]. However, the need to develop more effective and atom-economic methods has led to the utilization of asymmetric catalysis in enantioselective conjugate additions (ECAs) [3]. In Nature, natural catalysts – enzymes – mainly use general base and acid catalysis to achieve carbon–carbon bond-forming reactions, with almost exclusive stereocontrol [4]. This is accomplished in the active site of the enzyme by the perfect arrangement and dual activation of the electrophilic acceptor and nucleophilic donor substrate. Sometimes, a metallic cofactor in cooperation with the neighboring amino acids of the active site is necessary to achieve C–C bond-formation. In this context, type II aldolase enzymes employ a Zn ion (which is a Lewis acid) to activate dihydroxyacetone phosphate, in cooperation with an amino acid residue (which acts as a Brønsted base) to generate, regioselectively, an enolate (Figure 4.1a) [4]. At the same time, the electrophilic aldehyde acceptor is a Brønsted acid activated by an amino acid residue, such that stereoselective C–C bond-formation will occur with excellent enantioselectivity. In Nature, millions of years of evolution were required order to achieve this perfectly orchestrated transformation in an aqueous medium. Consequently, in taking their inspiration from Nature, research chemists have developed asymmetric catalysts that may reach impressive levels of enantioselectivity in carbon–carbon bond-forming reactions [5]. One way to achieve such enantioselectivity is to employ a strategy similar to that utilized by type II aldolase enzymes, which includes asymmetric bifunctional catalysis and metal activation (Figure 4.1b)

Catalytic Asymmetric Conjugate Reactions. Edited by Armando Córdova
Copyright © 2010 WILEY-VCH Verlag GmbH & Co. KGaA, Weinheim
ISBN: 978-3-527-32411-8

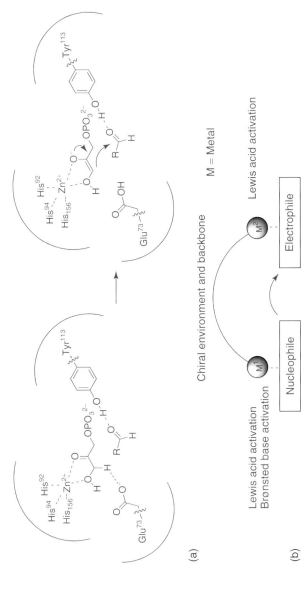

Figure 4.1 (a) Active site of a Type II aldolase enzyme; (b) Bifunctional chiral bimetallic catalyst.

[6]. In this chapter, attention will be focused on the relatively recent progress that has been made in ECA, using di-metallic and multi-metallic bifunctional catalysts.

4.2
Dinuclear Zn-Complexes in Catalytic ECAs

The pioneering investigations of Shibasaki and coworkers on the use of self-assembled multimetallic chiral catalysts in asymmetric catalysis, has led to the development of chiral dinuclear Zn complexes as catalysts for direct aldol, Mannich, and Michael reactions [6a–c]. Inspired by the mechanism of type II aldolase enzymes, the Et$_2$Zn/linked-BINOL complex **1** derived from BINOL ligand **2** and Et$_2$Zn was developed as a catalyst for catalytic asymmetric C–C bond-forming reactions (Figure 4.2) [7, 8].

The Et$_2$Zn/linked-complex **1** was investigated as the catalyst due to its high selectivity in direct asymmetric *syn*-selective aldol reactions with aryl hydroxyketones as the donors [9]. With respect to the direct catalytic conjugate addition of carbonyl compounds, complex **1** was recently investigated for the reaction between aryl hydroxyketone **3** and vinyl ketones [Eq. 4.1] [10].

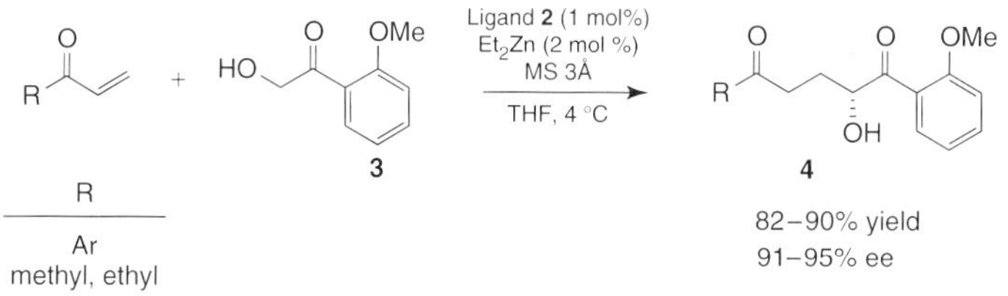

R
Ar
methyl, ethyl

82–90% yield
91–95% ee

(4.1)

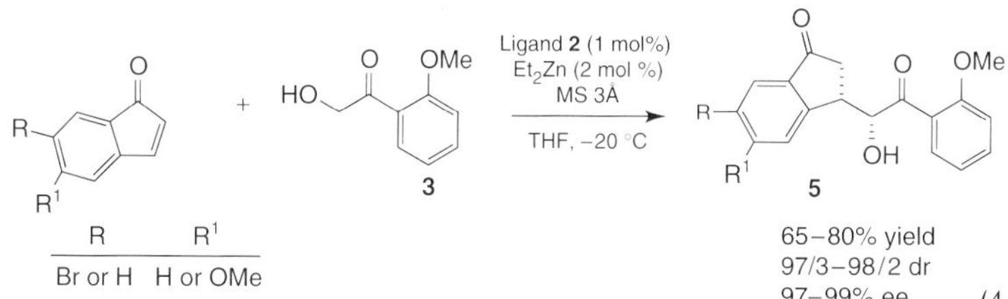

R	R^1
Br or H	H or OMe

65–80% yield
97/3–98/2 dr
97–99% ee

(4.2)

The catalytic dinuclear Zn complex **1** exhibited excellent chemoselectivity towards the donor hydroxy ketone **3**, and gave the corresponding Michael addition products **4** in high yields and 91–95% enantiomeric excess (ee). The catalyst loading could

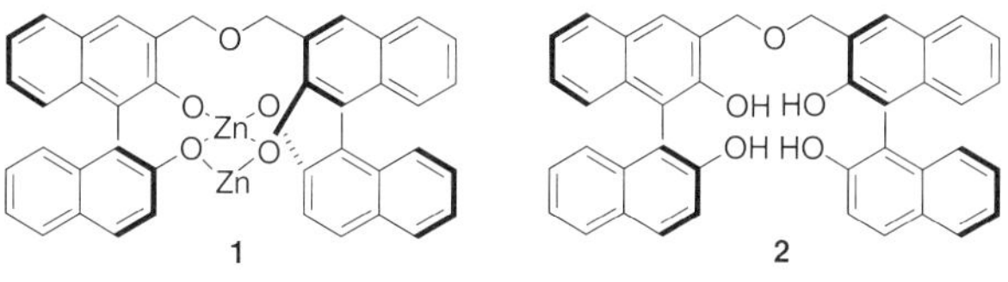

Figure 4.2 (S,S)-Zn-Zn-linked BINOL complex **1** and (S,S)-linked-BINOL ligand **2**.

also be reduced to as little as 0.01 mol%, without significantly affecting the efficiency and enantioselectivity of the reaction. The scope of reaction was further expanded to the direct catalytic asymmetric Michael addition of ketone **3** to indenones, giving the corresponding Michael adducts **5** in excellent enantioselectivity [Eq. 4.2] [10b]. An explanation for the stereoselectivity of these conjugation reactions was that the *Re* face of the dinuclear Zinc enolate in transition state **6** was approached by the enone acceptor.

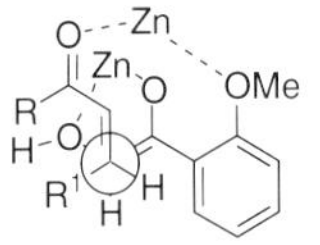

6

The reaction also exhibits good stereoselectivity for reactions with β-substituted enones, providing the corresponding products **7** with good *syn*-selectivity and 74–97% ee [Eq. 4.3]. By employing hydroxy arylketone **3a**, the reaction can be used to create tertiary alcohol stereocenters with high enantioselectivity [Eq. 4.4]. The catalytic complex **1** was also highly stereoselective for the addition of ketone **3** to *N*-benzylmaleimide [Eq. 4.5] [10b]. One important expansion of the direct catalytic asymmetric Michael reaction with complex **1** is the possibility of employing α,β-unsaturated acylpyrroles as substrates [Eq. 4.6] [11], in which case the acylpyrrole acts as a monodentate ester surrogate in transition state **10**, giving rise to the corresponding Michael products **11** with good *syn*-diastereoselectivity and enantioselectivity [Eq. 4.6].

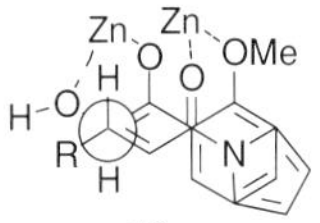

10

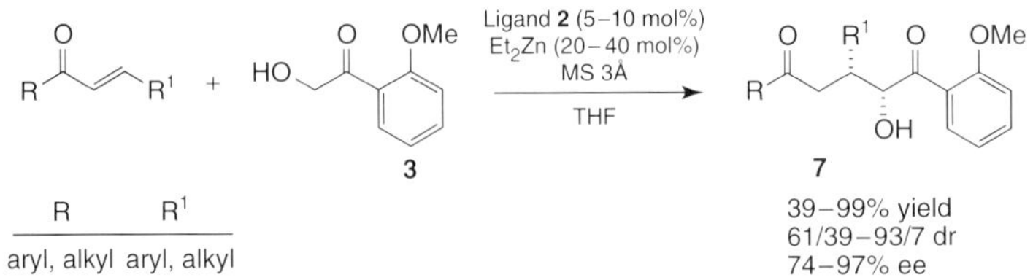

R	R¹
aryl, alkyl	aryl, alkyl

39–99% yield
61/39–93/7 dr
74–97% ee   (4.3)

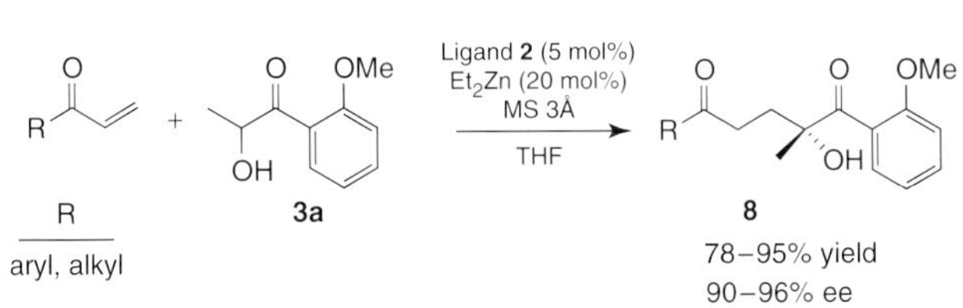

R
aryl, alkyl

78–95% yield
90–96% ee   (4.4)

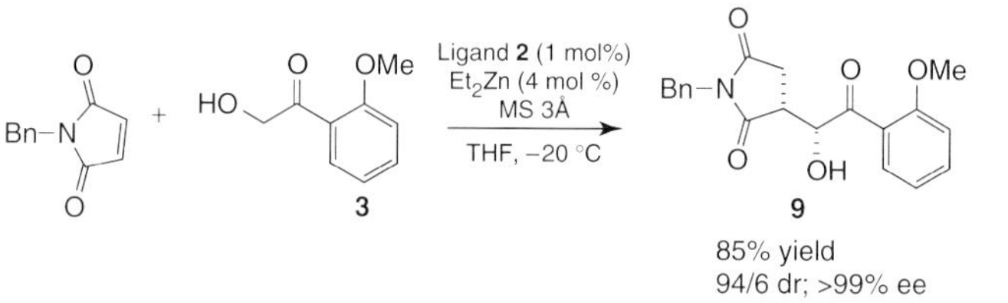

4.2 Dinuclear Zn-Complexes in Catalytic ECAs

(4.5)

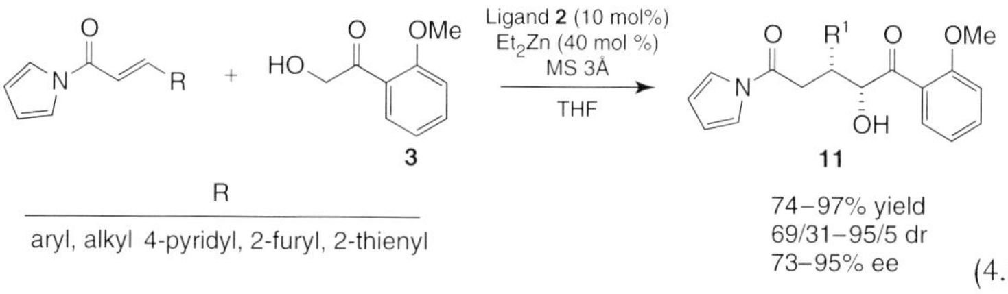

(4.6)

The *N*-acylpyrrole moiety of the triethylsilane (TES)-protected Michael product **11** is readily converted to the corresponding ethyl ester group (Scheme 4.1).

Trost *et al.* have discovered a novel design of dinuclear zinc catalysts that can catalyze diastereoselective and enantioselective direct aldol and Mannich reactions [12]. The dinuclear zinc catalyst **12a** is generated *in situ* by exposing the appropriate ligand **12** with 2 equiv. of diethylzinc in tetrahydrofuran (THF) (Scheme 4.2).

Recently, Trost and coworkers reported that the dinuclear zinc–complex **12a** may catalyze direct Michael reactions between acetophenone **3a** and nitrostyrene,

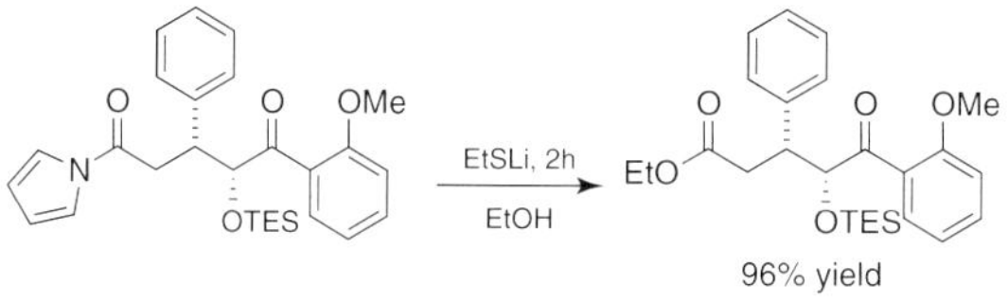

Scheme 4.1 Conversion of the *N*-acylpyrrole group to an ethyl ester group.

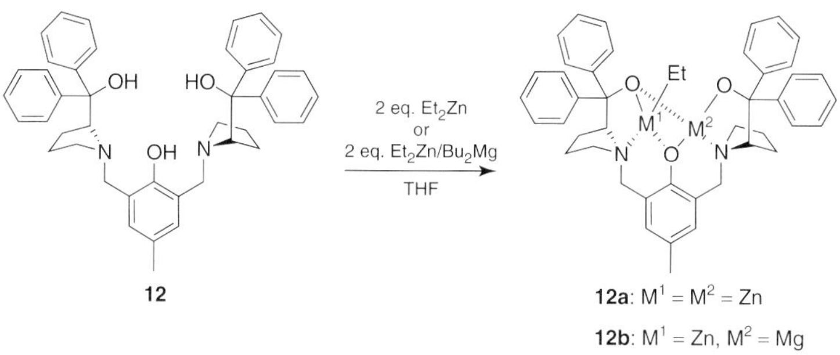

Scheme 4.2 Generation of di-nuclear Zn catalysts **12a**.

with excellent enantioselectivity but moderate diastereoselectivity, to form the corresponding Michael product **13a** in 89% yield and 99% ee [Eq. 4.7] [13]. By employing a mixed Zn/Mg metal system in the generation of chiral complex **12b**, the *anti*-selectivity could be improved and the scope of reaction expanded to include both aliphatic and aromatic nitroolefin acceptors [Eq. 4.8]. The corresponding Michael products **13** were isolated in moderate to high yields, poor to excellent diastereomeric ratio (dr), and 76–99% ee.

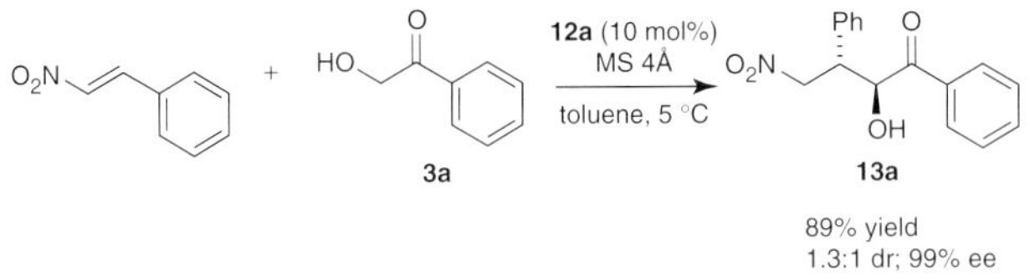

$$\text{(4.7)}$$

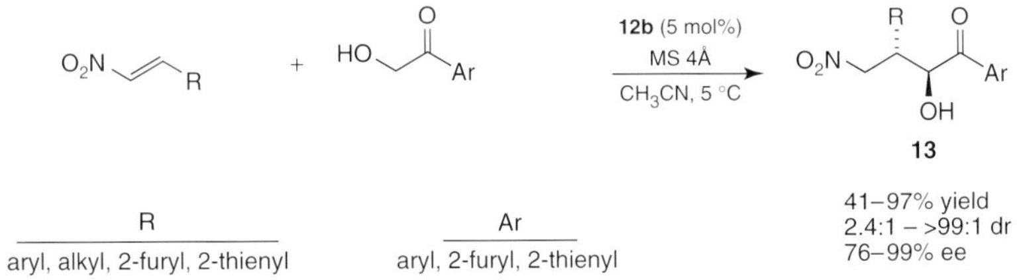

89% yield
1.3:1 dr; 99% ee

$$\text{(4.8)}$$

41–97% yield
2.4:1 – >99:1 dr
76–99% ee

R	Ar
aryl, alkyl, 2-furyl, 2-thienyl	aryl, 2-furyl, 2-thienyl

Trost and coworkers have recently expanded the use of dinuclear Zn complex **12a** to other catalytic ECAs to nitroolefins. For example, pyrroles were successfully employed as substrates in asymmetric Friedel–Crafts alkylation reactions with nitroolefins [Eq. 4.9] [14].

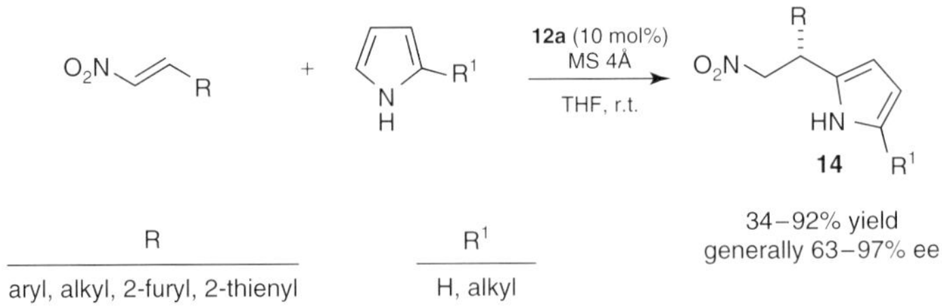

$$\text{(4.9)}$$

34–92% yield
generally 63–97% ee

R	R^1
aryl, alkyl, 2-furyl, 2-thienyl	H, alkyl

The dinuclear Zn complex **12a**-catalyzed transformations proceed generally with good to high enantioselectivity, providing the corresponding optically active alkylated pyrroles **14** in moderate to high yields. The reaction tolerates aromatic, aliphatic, and ester functionalities on the nitroolefin. An explanation for the stereochemistry of the reaction was that the *Re*-face of the nitroolefin (R = Ar) approached the pyrrole substrate in structure **15**.

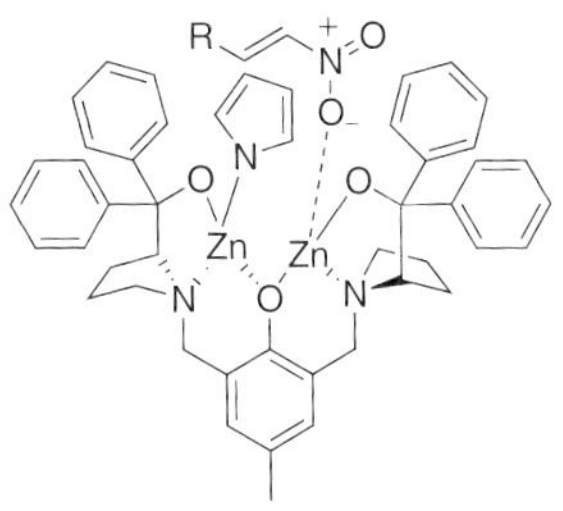

15

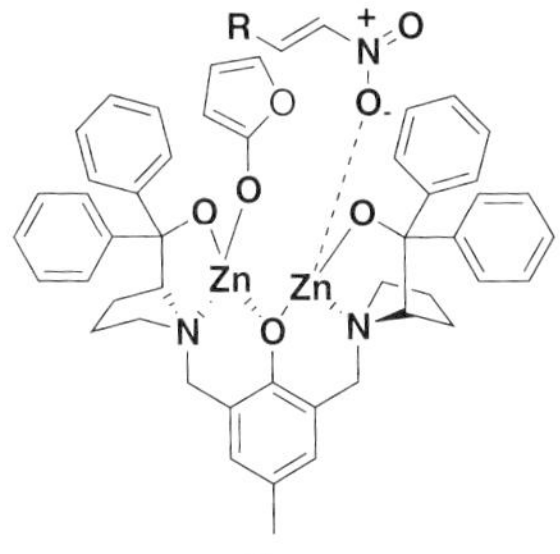

15aa

The dinuclear zinc catalyst **12a** is also a highly enantioselective catalyst for the vinologous conjugate addition of 2(5H)-furanone **16** to nitroolefins [Eq. 4.10] [15]. In fact, according to Mukaiyama, this donor typically requires activation as a siloxydiene. However, with the help of dinuclear Zn catalysis, unmodified lactone **16** could be employed directly as the nucleophile.

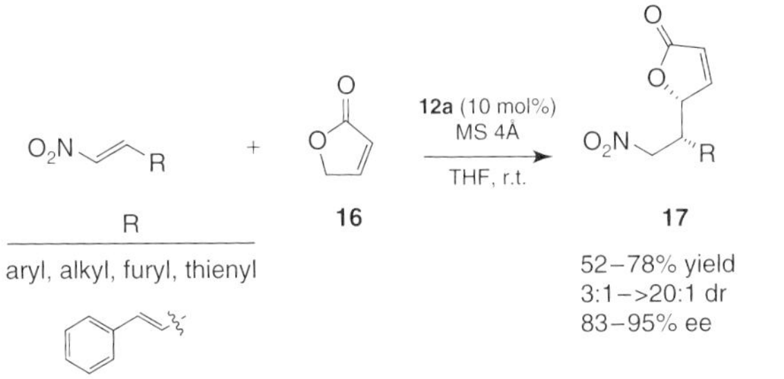

R
aryl, alkyl, furyl, thienyl

52–78% yield
3:1–>20:1 dr
83–95% ee

(4.10)

The reaction has a broad acceptor range, and both aromatic and aliphatic nitroolefins are tolerated without significantly affecting the enantioselectivity of the reaction. The corresponding vinologous products **17** were isolated in good to high yields with good to excellent dr and 83–95% ee. The stereoselectivity of the reaction can be explained by fact that, in structure **15aa** the Re-face of the nitroolefin (R = Ar) approached the activated 2(5H)-furanone **16**.

4.3
Heterobimetallic Rare-Earth–Alkali Metal-Binol Complexes in ECAs

Shibasaki and coworkers reported the first example of the use of bimetallic rare-earth lithium BINOL complexes (LnLB: Ln = rare earth, L = lithium, and B = BINOL, respectively) as catalysts for catalytic asymmetric nitroaldol reactions [16]. This breakthrough has led to development of heterobimetallic complexes for several important C−C bond-forming reactions. Heterobimetallic catalysts contain two metal centers: the first center is Lewis acidic (lanthanides or Group 13 elements) that activate the acceptor, while the second metal center (alkali metals bound to a Brønsted base) coordinates to the donor. Heterobimetallic-complexes were initially investigated as catalysts with respect to asymmetric Michael reactions by Shibasaki and coworkers, and employed for the catalytic ECAs of malonates to enones (Scheme 4.3) [17].

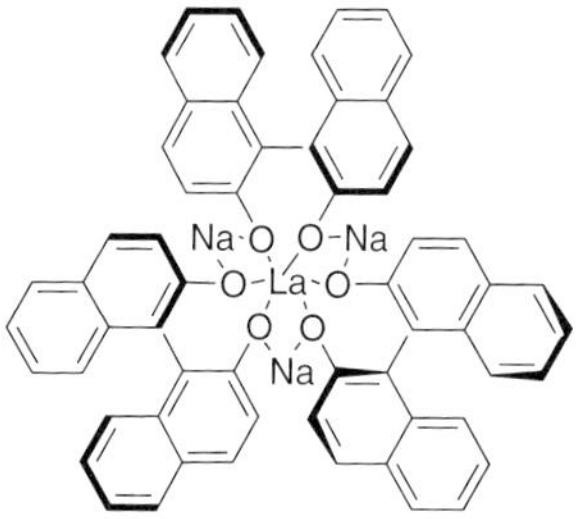

In this case, the lanthanum–sodium–BINOL complex (LSB) **18** was found to exhibit the highest stereoselectivity for the reaction. Complex **18** catalyzed the reaction between malonates and cyclic enones with high enantioselectivity, providing the corresponding Michael products **20** in high yield and 88–92% ee. It should be noted that, by employing lithium instead of sodium as the alkali metal, near-racemic Michael products **20** are produced [17]. The LSB complex **18** is also an effective catalyst for the ECAs to chalcones and enones (Scheme 4.3) [17, 18].

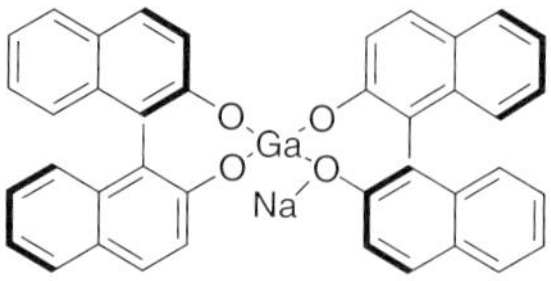

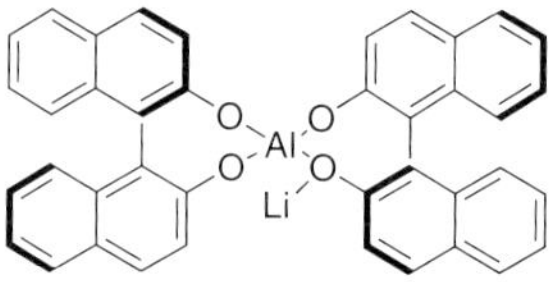

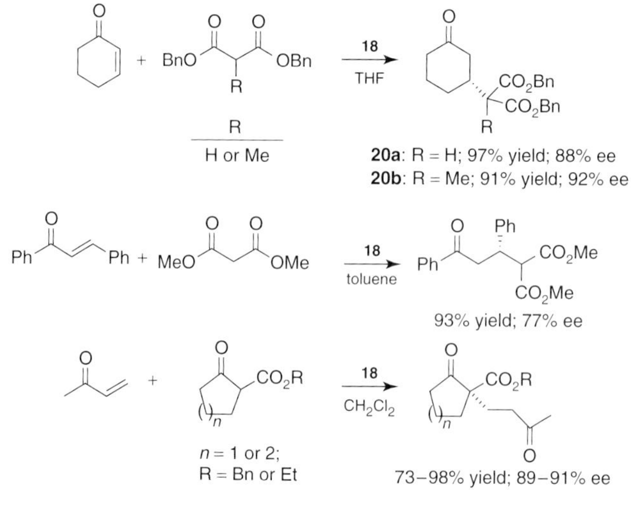

Scheme 4.3 LSB complex **18**-catalyzed ECAs.

4.3 Heterobimetallic Rare-Earth–Alkali Metal-Binol Complexes in ECAs

The enantioselectivity for the ECAs to cyclic enones could be further improved by using the gallium–sodium complex (GASB) **18b**, while the best catalyst was seen to be provided by the aluminum–lithium complex (ALB) **18c**, which yielded the corresponding products **20**, respectively, and with excellent ee-values [Eqs 4.11 and 4.12] [19].

20a: $n = 2$; 87% yield; 98% ee
20c: $n = 1$; 96% yield; 98% ee
20d: $n = 3$; 79% yield; >99% ee   (4.11)

20a: $n = 2$; 99% yield; 98% ee
20d: $n = 3$; 83% yield; 96% ee   (4.12)

18d: M = K
18e: M = H

Shibasaki's group also investigated complexes **18d** and **18e** as catalysts for the ECAs displayed in Eqs 4.11 and 4.12 [20]. In the case of complex **18e**-catalyzed transformations, the corresponding products **20** were obtained almost enantiomerically pure (>99% ee).

The power of such methodology was further demonstrated in the complex *ent*-**18c**-catalyzed domino Michael-aldol reaction between enone **21**, a malonate ester donor, and an aliphatic acceptor aldehyde (Scheme 4.4) [21]. In addition to the domino reaction, complex *ent*-**18c**-catalyzes a simultaneous kinetic resolution of enone **21** to form the corresponding three-component coupling product **22** with 12:1 dr and 97% ee. Ketone **22** is an important starting material for the synthesis of biologically significant compounds such as prostaglandin (PG) $F_{1\alpha}$ **23**.

Bimetallic complex *ent*-**18c** may also serve as a catalyst for the chemoselective addition of Horner–Wadsworth–Emmons reagent **25** to cyclopentenone [Eq. 4.13] [22]. The catalytic asymmetric reaction gave the corresponding ketone **26** in 94% ee, this being a key building block in the synthesis of coronafic acid **27**. The

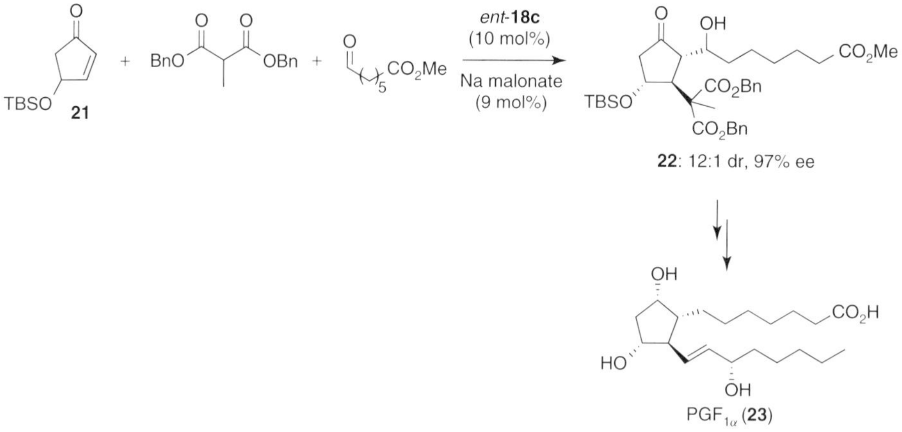

Scheme 4.4 The first heterobimetallic complex-catalyzed domino reaction and kinetic resolution.

complex was also employed for the catalytic asymmetric synthesis of the alkaloid, tubifolidine [23].

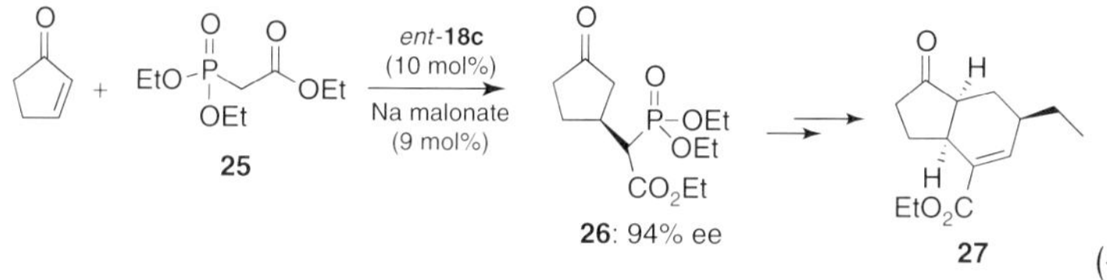

(4.13)

Heterobimetallic rare-earth complexes such as **24**, which are derived from readily available amino alcohols, may catalyze the catalytic ECAs of malonates to enones with high enantioselectivity, as reported by Sundarajan and Manickam [Eq. 4.14] [24].

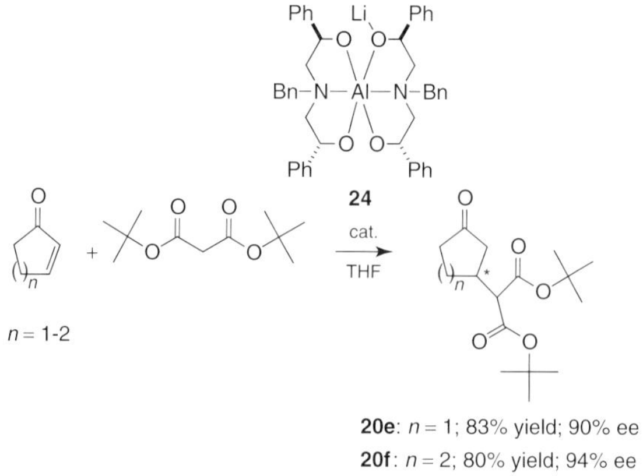

20e: $n = 1$; 83% yield; 90% ee
20f: $n = 2$; 80% yield; 94% ee

(4.14)

Furthermore, Sasai and coworkers have shown that heterobimetallic BINOL-derived dendrimer complexes exhibit excellent enantioselectivity for the catalytic ECAs shown in Eqs 4.11 and 4.12 [25]. These dendrimer complexes may also be employed as efficient supports, with the catalyst capable of being recycled at least three times.

Heterobimetallics can also catalyze the conjugate additions of nitro compounds. In this context, Feringa and coworkers reported the asymmetric addition of α-nitro esters to enones to give the corresponding products **28** with up to 80% ee, using complex **18c** as the catalyst [Eq. 4.15] [26]. Shibasaki et al. were able to employ a potassium rather than a sodium complex of **18** as a catalyst for the highly ECA of nitromethane to chalcones; the corresponding product **29** was isolated with 97% ee [Eq. 4.16] [27].

$$(4.15)$$

$$(4.16)$$

4.4
Heterobimetallic Rare-earth–Alkali Metal-Binol Complexes in ECAs of Heteroatom Nucleophiles

The 1,4-addition of thiols to Michael acceptors is an important transformation of organic synthesis. In this context, Shibasaki and coworkers have employed the LSB complex **18** for the conjugate addition of thiols to cyclic enones to provide the corresponding Michael products in good yields, and with moderate to high enantioselectivities [28]. For example, benzylthiol was added to cyclopentenone, cyclohexenone, and cycloheptenone, and the corresponding products were isolated with 56%, 90%, and 83% ee, respectively [Eq. 4.17].

$$(4.17)$$

LSB complex **18** and samarium–sodium–Binol complex (SSB) **31** also catalyze the addition of 4-*tert*-butyl(thiophenol) to α,β-unsaturated thioesters to give the corresponding Michael products **32**, in high yields and enantioselectivities [Eq. 4.18].

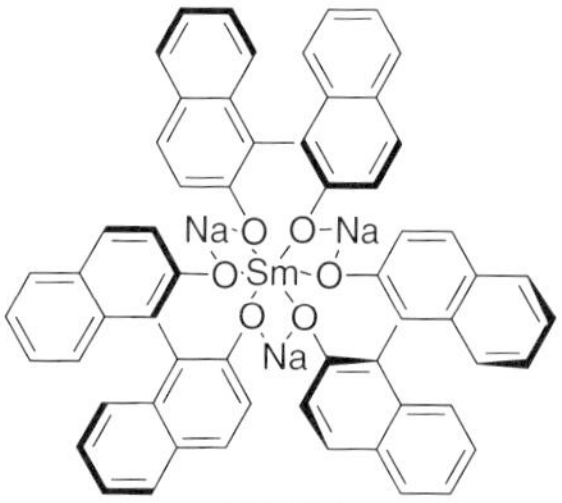

78–98% yield
84–93% ee   (4.18)

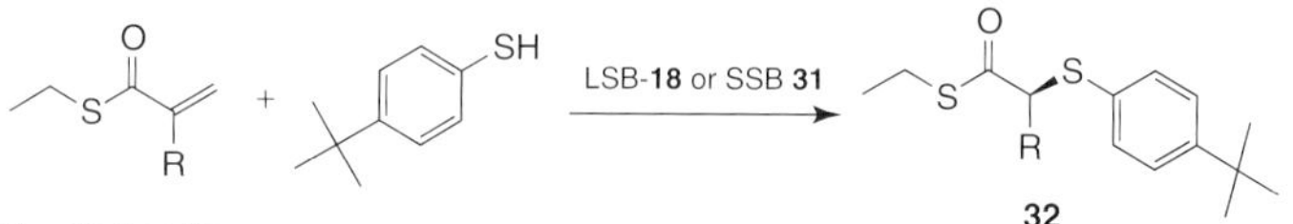

These authors proposed that the thia-Michael reaction would proceeds via an asymmetric protonation step of enolate complex **33** generated by the conjugate addition in complex **33a**, thus releasing product **32** and regenerating the catalyst SSB complex **31** (Scheme 4.5).

The heterobimetallic rare-earth–alkali complex-catalyzed thia-Michael addition can also be employed in the kinetic resolution of racemic 5-methylbicyclo[3.3.0]oct-1-ene-3,6-dione **34** [Eq. 4.19] [29]. In this case, the reaction with ALB-complex **18c** gave the best result, and the enantiomer (R)-**34** was furnished with 79% ee, together with the Michael product **35**. Oxidation and a basic work-up produced (S)-**34** in 78% ee.

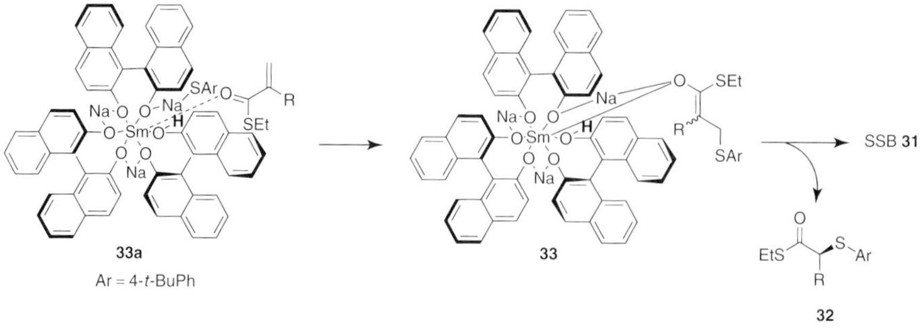

Scheme 4.5 The conjugate addition step and asymmetric protonation in the SSB **31** catalyzed thia-Michael reaction.

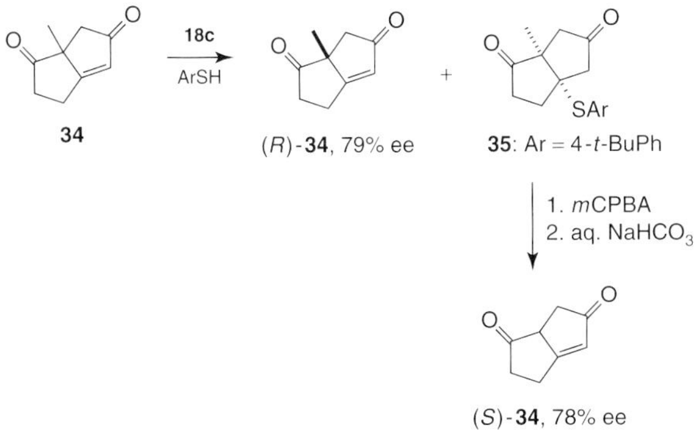

$$\text{34} \xrightarrow[\text{ArSH}]{\text{18c}} (R)\text{-34, 79\% ee} + \text{35: Ar} = 4\text{-}t\text{-BuPh} \xrightarrow[\text{2. aq. NaHCO}_3]{\text{1. }m\text{CPBA}} (S)\text{-34, 78\% ee} \quad (4.19)$$

Heterobimetallic cooperative catalysis can be applied for the aza–Michael reaction. In this case, a Lewis acid/Lewis acid catalysis is combined to achieve activation of the acceptor and also of the donor (Figure 4.3) [30]. By employing this strategy Shibasaki and coworkers found that the yttrium-based YLi$_3$tris(binaphthoxide) complex (YLB **35**) would catalyze the addition of alkoxyamines to enones and α,β-unsaturated acylpyrroles [Eqs 4.20–4.22], respectively, to form the corresponding products **36** and **37**, respectively, in good to high yields and enantioselectivities [30]. In order to reduce the catalyst loading, CaSO$_4$ was used as the desiccant. Moreover, the dysprosium-based complex DyLB **35a** catalyzed the ECA to α,β-unsaturated acylpyrroles with high enantioselectivity [Eq. 4.22].

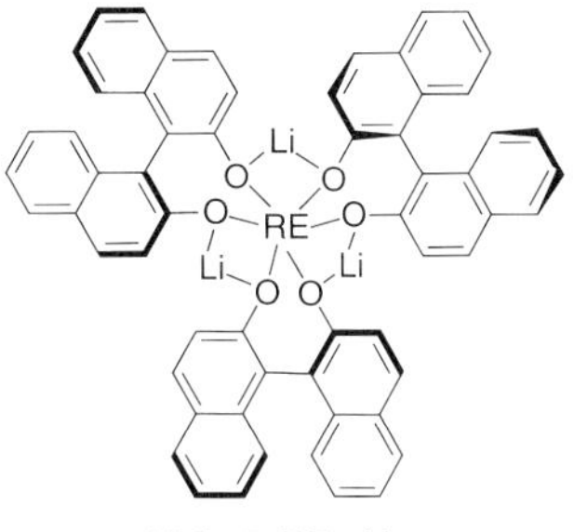

YLB **35**: RE = Y
DyLB **35a**: RE = Dy

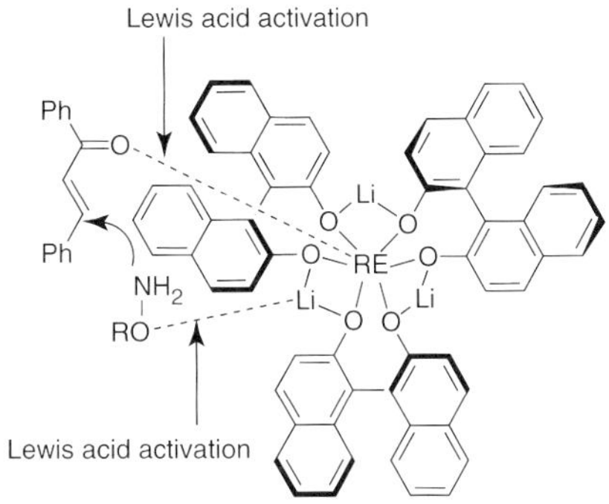

Figure 4.3 Lewis acid–Lewis acid cooperative catalysis in enantioselective aza–Michael reaction of alkoxyamines. RE = rare earth.

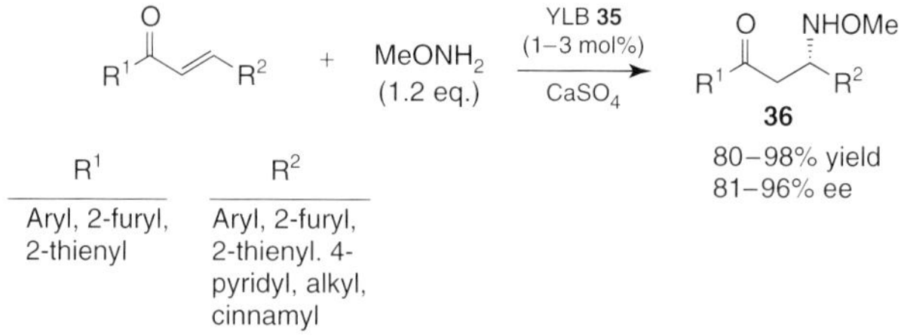
YLB 35
(1–3 mol%)
CaSO4
MeONH2
(1.2 eq.)
O
NHOMe
R1
R2
36
80–98% yield
81–96% ee
R1
Aryl, 2-furyl,
2-thienyl
R2
Aryl, 2-furyl,
2-thienyl. 4-
pyridyl, alkyl,
cinnamyl

(4.20)

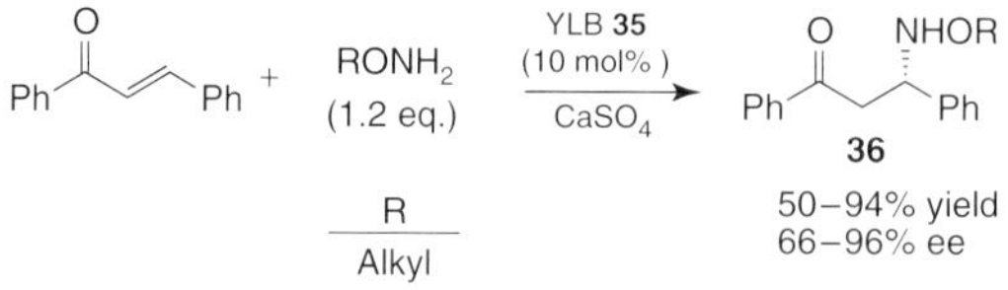
Ph
RONH2
(1.2 eq.)
YLB 35
(10 mol%)
CaSO4
NHOR
36
50–94% yield
66–96% ee
R
Alkyl

(4.21)

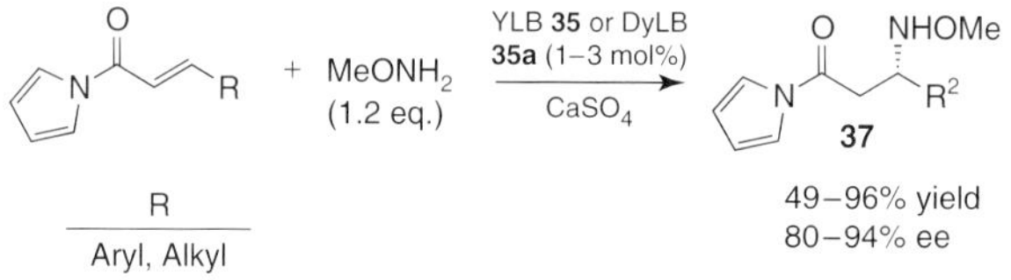
MeONH2
(1.2 eq.)
YLB 35 or DyLB
35a (1–3 mol%)
CaSO4
NHOMe
R2
37
49–96% yield
80–94% ee
R
Aryl, Alkyl

(4.22)

The reaction exhibited significant positive nonlinear effects, and the reaction rate was decreased significantly when non-optically pure YLB **35** was used as the catalyst. An explanation for this was that a thermodynamically more stable (*S,S,R*)-complex, as compared to (*S,S,S*)-YLB-**35**, was formed. This complex was not active for the conjugate addition, and only a small amount of (*S,S,S*)-YLB-**35** would be needed to catalyze the asymmetric transformation. The β-amino carbonyl compounds **36** and **37** could be readily converted to the corresponding aziridines **38** and **39**, respectively, without any loss of enantioselectivity by treatment with $TiCl_4$ and Et_3N [Eq. 4.23].

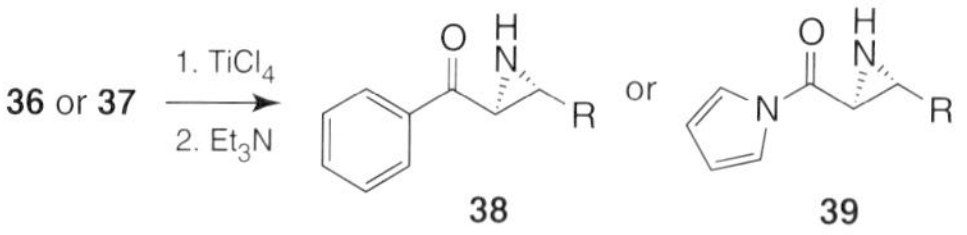
36 or 37
1. TiCl4
2. Et3N
38
or
39

(4.23)

With regards to the use of oxygen nucleophiles in ECAs, Shibasaki and coworkers have developed a beautiful asymmetric epoxidation of α,β-unsaturated acylpyrroles [Eq. 4.24] [11]. In this case, the enantioselective transformation proceeds via a domino oxa-Michael/intramolecular substitution mechanism (Scheme 4.6). The same group has previously reported the catalytic asymmetric epoxidation of enones, which also follows the mechanism depicted in Scheme 4.6.

The reaction was performed in one-pot fashion by first forming the α,β-unsaturated acylpyrroles *in situ*; this produced the corresponding epoxides **40**

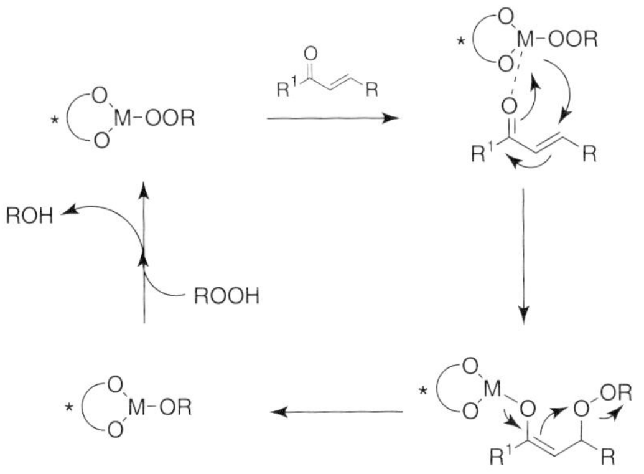

Scheme 4.6 The proposed reaction mechanism for Shibasaki's epoxidation.

with high yields and enantioselectivities [Eq. 4.24].

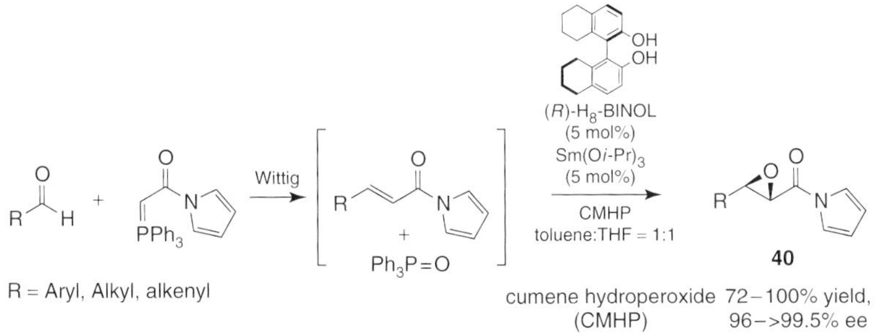

$$(4.24)$$

The reaction has been used successfully as part of the enantioselective synthesis of aeruginosin 298-A and its analogs [31]. Yet, the methodology can also be applied to the enantioselective epoxidation of α,β-unsaturated esters, employing rare-earth metals and **41** or binol **42** as chiral ligands [32].

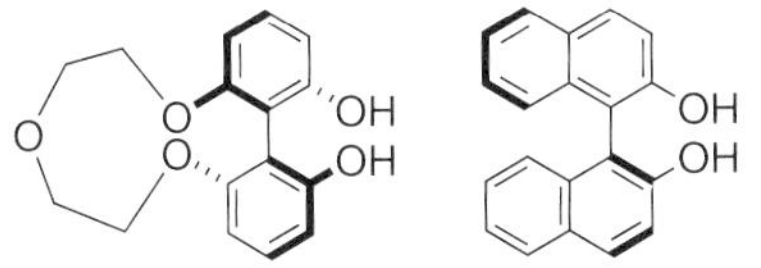

For example, the combination of ligand **41**, Y(OiPr)$_3$ and Ph$_3$As=O using *tert*-butylhydroperoxide (TBHP) as the oxidant gave the corresponding epoxides **43** in high yields and enantioselectivities [32]. The use of ligand **42** improved the enantioselectivity when aliphatic α,β-unsaturated esters were used as substrates [Eq. 4.25].

R = Aryl, Alkyl, 2-furyl, 2-thienyl. 3-pyridyl

62–97% yield, 71–99% ee

(4.25)

4.5
Miscellaneous

Based on the importance of the cyclopropane unit, and the pioneering investigations of Aggarwal and Gaunt, Matsunaga, Shibasaki and coworkers began to investigate the possible use of rare-earth metals for the asymmetric cyclopropanation of enones and α,β-unsaturated acylpyrroles [33]. In addition, their previous development of catalytic enantioselective epoxidation reactions of the same substrates supported the suggestion that a domino Michael/intramolecular substitution mechanism would be applicable to asymmetric reactions with stabilized ylides. Thus, they embarked on the development of a catalytic Corey–Chaykovsky cyclopropanation transformation. By screening complexes of La-Li$_3$-(ligand **41** or **42**)$_3$ for the reaction between dimethyloxosulfonium methylide and chalcone, the complex derived from ligand **41**, using NaI as an additive, was found to give the corresponding product **44a** with the highest enantioselectivity (69% yield, 82% ee) [Eq. 4.26]. Exchanging the oxygen of **41** to a methylene group (as in ligand **45**), and adding the enone slowly to the reaction mixture, led to significant improvements in both the yield and ee of **44a** (97% yield, 97% ee) [Eq. 4.27]. Based on these results, the catalytic complex (S)-La-Li$_3$-(**45**)$_3$ was chosen to test the utility of the catalytic reaction. The transformation provided excellent results, and the corresponding cyclopropanes could be isolated in high yields and enantioselectivities [Eq. 4.28]. The addition of NaI was seen to be important for improving the enantioselectivity, an explanation being that a partial alkali metal exchange would occur *in situ* to afford complex **46**, which exhibited a higher enantioselectivity and reactivity when compared to (S)-La-Li$_3$-(**45**)$_3$.

69% yield, 82% ee

(4.26)

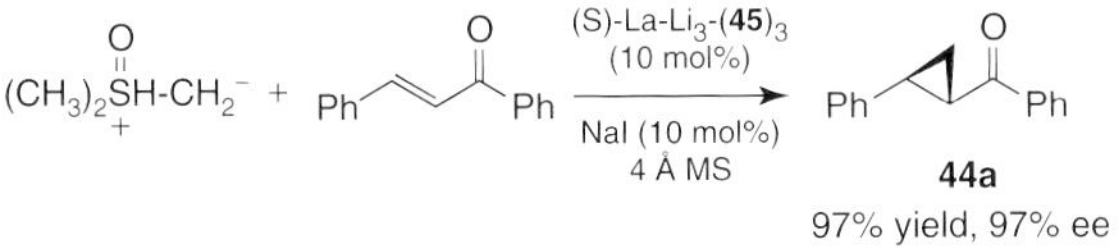

$$(CH_3)_2\overset{+}{S}H\text{-}CH_2^- + Ph\overset{O}{\diagup\diagdown}Ph \xrightarrow[\text{NaI (10 mol\%)}]{\substack{(S)\text{-La-Li}_3\text{-}(\mathbf{45})_3 \\ (10\text{ mol\%})}}_{4\text{ Å MS}} Ph\overset{O}{\triangle}Ph$$

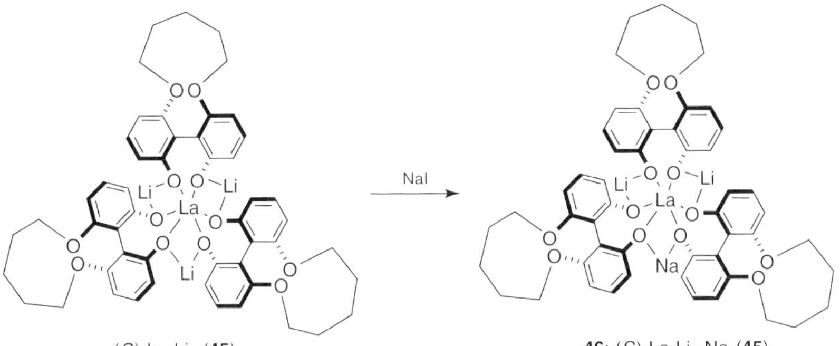

44a
97% yield, 97% ee (4.27)

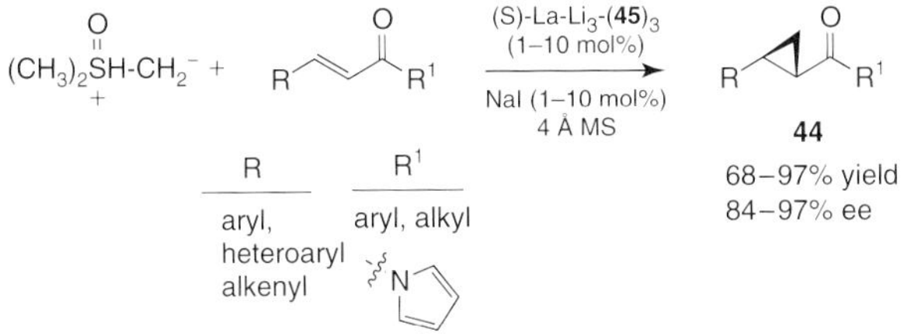

$$(CH_3)_2\overset{+}{S}H\text{-}CH_2^- + R\diagup\diagdown\overset{O}{}R^1 \xrightarrow[\text{NaI (1–10 mol\%)}]{\substack{(S)\text{-La-Li}_3\text{-}(\mathbf{45})_3 \\ (1\text{–}10\text{ mol\%})}}_{4\text{ Å MS}} R\overset{O}{\triangle}R^1$$

R	R¹
aryl, heteroaryl alkenyl	aryl, alkyl, N-pyrrolyl

44
68–97% yield
84–97% ee (4.28)

(S)-La-Li₃-(**45**)₃ →[NaI] **46**: (S)-La-Li₂-Na-(**45**)₃

The group of Shibasaki have also developed a set of ligands such as **47a** and **47b** derived from D-glucose. In combination with rare-earth metals such as gadolinium, these form multi-metallic complexes that are excellent catalysts for the enantioselective conjugate addition of cyanide to α,β-unsaturated acylpyrroles [Eq. 4.29] [34]. The corresponding products **48** are isolated in high yields and with high ee-values. Further elaboration of compounds **48** provided useful γ-amino acid derivatives. In addition to the ligands **47**, second-generation ligands **49** without the ring oxygen were also developed.

The active catalysts for the transformation are polymetallic complexes **50** of higher-order structures created by modular self-assembly. The complexes are multifunctional gadolinium cyanide (or isocyanide), formed by transmetallation from trimethylsilyl cyanide (TMSCN), which acts as a nucleophile, and an acidic proton from the protic additives, which acts as a turnover accelerator.

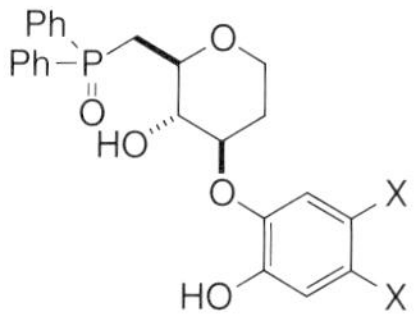

47a: X = H
47b: X = F

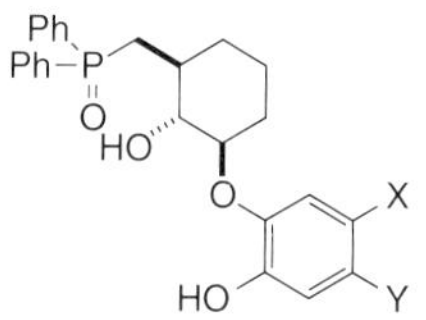

49a: X = Y = F
49b: X = CN, Y = F

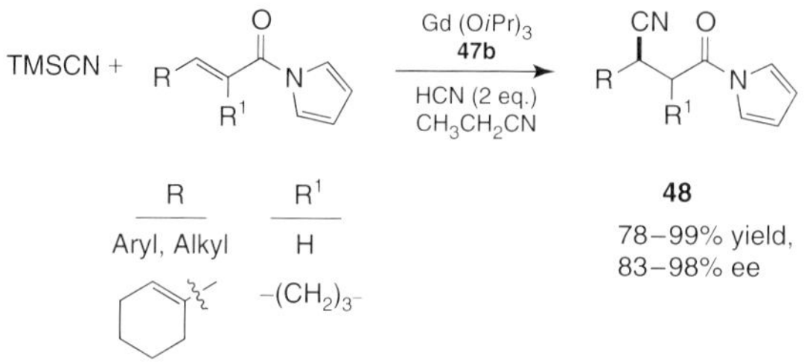

R	R¹
Aryl, Alkyl	H
cyclohexenyl	–(CH$_2$)$_3$–

48
78–99% yield,
83–98% ee

(4.29)

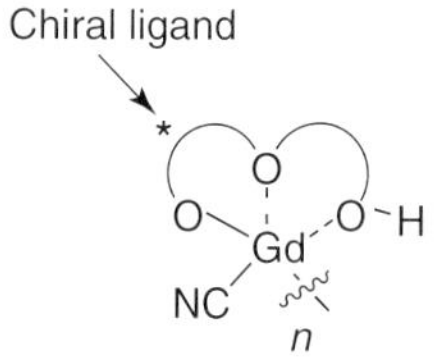

Chiral ligand

50

Based on these results, Shibasaki and coworkers began to investigate an even more challenging transformation, namely the catalytic enantioselective addition of cyanide to enones [35]. This reaction has two possible pathways – 1,2-addition and 1,4-addition – which must be differentiated one from another in order to create enantioselection. To achieve this goal catalytically, even for one of the factors, was not an easy task. However, extensive screening to identify the reaction conditions and ligands **47** and **49** led to the creation of a catalytic system based on ligands **49a** or **49b**, gadolinium, tributylsilyl cyanide (TBSCN), and 2,6-dimethylphenol (DMP) in THF, that provided excellent 1,4-selectivity and high enantioselectivity. Moreover, the scope of the reaction was broad, and both aliphatic linear and cyclic enones could be used as acceptors [Eqs 4.30–4.32].

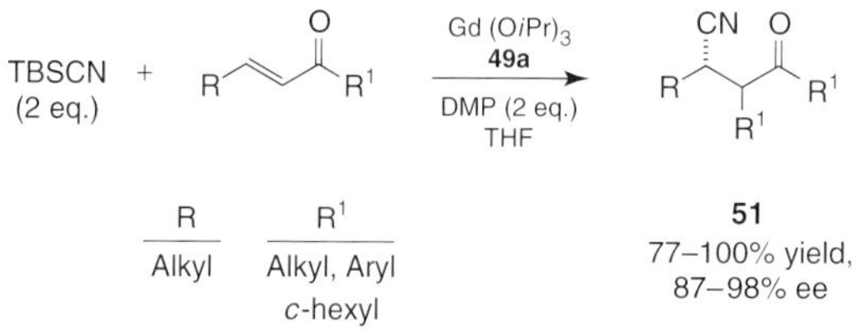

R	R¹
Alkyl	Alkyl, Aryl
	c-hexyl

51
77–100% yield,
87–98% ee

(4.30)

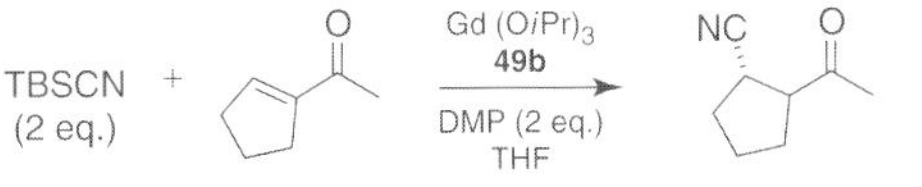

$$\text{TBSCN (2 eq.)} + \text{cyclopentenyl methyl ketone} \xrightarrow[\text{THF}]{\substack{\text{Gd}(OiPr)_3 \\ \text{49b} \\ \text{DMP (2 eq.)}}} \text{51a}$$

51a: 67% yield, 4:1 dr, 95% ee (4.31)

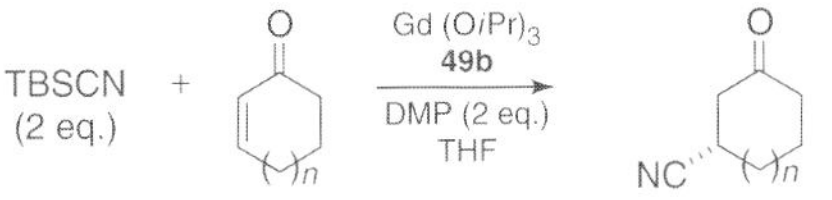

$$\text{TBSCN (2 eq.)} + \text{cyclohexenone (}n=1\text{-}3\text{)} \xrightarrow[\text{THF}]{\substack{\text{Gd}(OiPr)_3 \\ \text{49b} \\ \text{DMP (2 eq.)}}} \text{51b-d}$$

51b: $n = 1$, 90% yield, 81% ee
51c: $n = 2$, 73% yield, 52% ee
51d: $n = 3$, 74% yield, 81% ee (4.32)

In the case of linear enones, catalyst **49a** exhibited the highest enantioselectivity for the transformation, and the corresponding products **51** were isolated in high yields and 87–98% ee [Eq. 4.30]. With respect to cyclic enones, catalyst **49b** gave the highest enantioselectivity for the ECAs [Eqs 4.31 and 4.32].

Multi-metallic chiral Pd aqua complexes [36] can also be used for the direct catalytic ECAs by a Brønsted acid–base cooperative system (Scheme 4.7) [37].

In this context, Sodeoka and coworkers have developed ECAs of β-ketoesters to methylvinyl ketone and acrolein [Eqs 4.33 and 4.34], using the chiral Pd aqua complex **52** as the catalyst [37a]. The reaction generates an all-carbon quaternary stereocenter, and the corresponding products are formed with high enantioselectivity. The reaction was also expanded to the ECAs to other α,β-unsaturated ketones and crotonaldehyde, and gave the corresponding products in moderate dr and high enantioselectivity [Eqs 4.35 and 4.36].

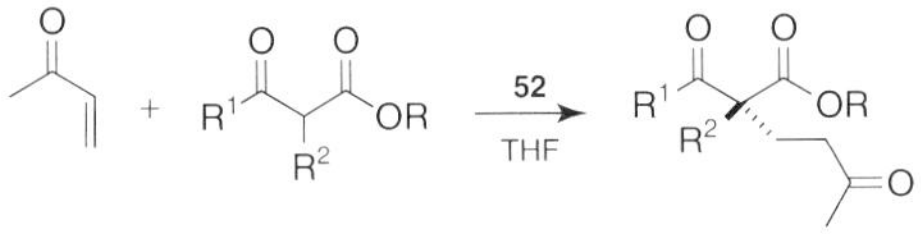

82–93% yield, 90–94% ee (4.33)

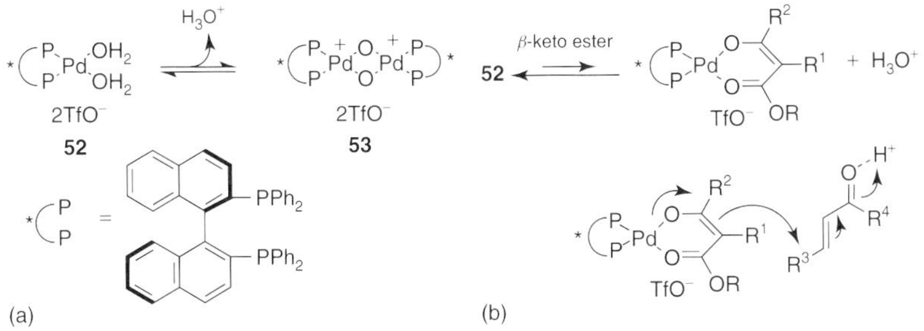

Scheme 4.7 (a) Chiral Pd aqua complexes **52** and **53**; (b) Cooperative activation of the Pd-complexes.

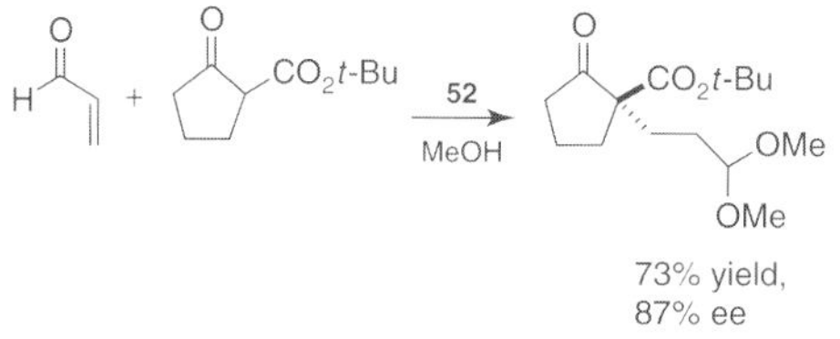

188 | 4 Asymmetric Bifunctional Catalysis

(4.34)

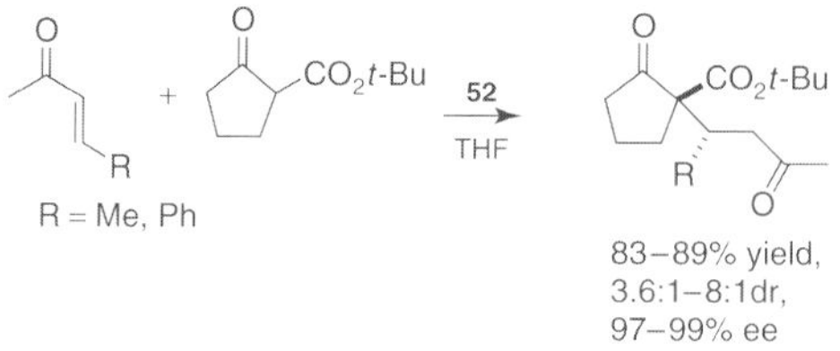

73% yield, 87% ee

(4.35)

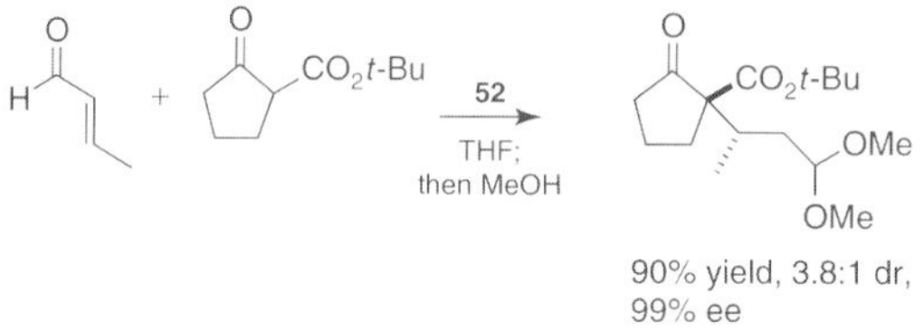

R = Me, Ph

83–89% yield, 3.6:1–8:1 dr, 97–99% ee

(4.36)

90% yield, 3.8:1 dr, 99% ee

Notably, the chiral Pd aqua complex **53** can catalyze the aza–Michael addition of anilines and benzylamine to give the corresponding β-amino acid derivatives in high yields, and up to 98% ee [Eqs 4.37 and 4.38] [38].

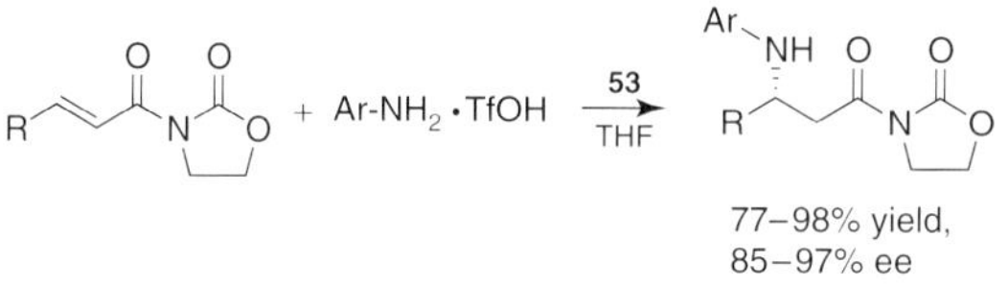

(4.37)

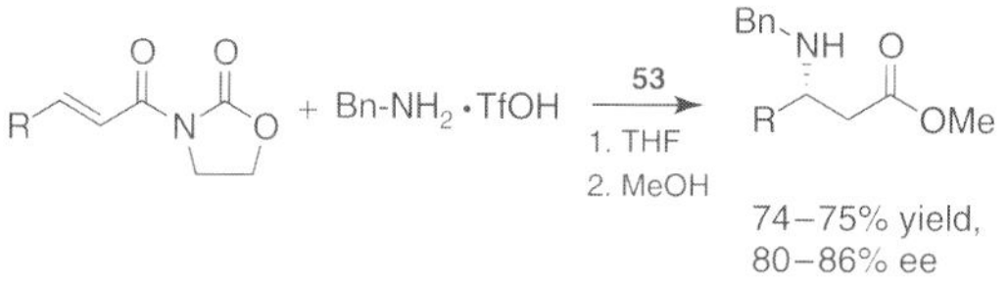

77–98% yield, 85–97% ee

(4.38)

74–75% yield, 80–86% ee

4.6
Conclusion

Dimetallic and multimetallic bifunctional complexes are import catalysts for the catalytic enantioselective conjugate reaction. As reported in this chapter, there are several multi-metallic bifunctional complexes that catalyze different asymmetric Michael-type reactions with high stereoselectivity. The modularity and versatility of the catalyst described, and the complementarity of the described catalytic methods, causes the asymmetric bifunctional catalyzed ECAs of to become of paramount importance in the construction of stereodefined, complex small molecules. Today, this area of research field has continued to expand rapidly from the pioneering investigations of the 1990s and, indeed, many more equally important advances are to be expected during the next decade (2010–2020) within the field and development of ECAs, using multi-metallic bifunctional complexes as catalysts.

References

1. (a) Perlmutter, P. (1992) *Conjugate Addition Reactions in Organic Synthesis*, Pergamon, Oxford; (b) Regmann, E.D., Ginsburg, D., and Rappo, R. (1959) *Org. React.*, **10**, 179; (c) Krause, N. and Gerold, A. (1997) *Angew. Chem., Int. Ed.*, **36**, 186.
2. (a) Oppolzer, W. (1987) *Tetrahedron*, **43**, 1969; (b) Oare, D.A. and Heathcock, C.H. (1991) *Top. Stereochem.*, **20**, 87; (c) Rossiter, B.E. and Swingle, N.M. (1992) *Chem. Rev.*, **92**, 771; (d) Juaristi, E., Beck, A., Hansen, J., Matt, T., Mukhopadhyay, T., Simson, M., and Seebach, D. (1993) *Synthesis*, 1271; (e) Alexakis, A. (1994) in *Organocopper Reagents, A Practical Approach* (ed. R.J.K. Taylor), Oxford University Press, Oxford, p. 159.
3. (a) Feringa, B.L. and De Vries, A.H.M. (1995) in *Advances in Catalytic Processes*, vol. 1 (ed. M.P. Doyle), JAI, Connecticut, p. 151; (b) Sibi, M.P. and Manyem, S. (2000) *Tetrahedron*, **56**, 8033; (c) Krause, N. and Hoffmann-Röder, A. (2001) *Synthesis*, 171; (d) Leonard, J., Diez-Barra, E., and Merino, S. (1998) *Eur. J. Org. Chem.*, 2051.
4. (a) Fessner, W.-D. (2000) in *Stereoselective Biocatalysis* (ed. R.N.Patel), Marcel Dekker, New York, p. 239; (b) Machajewski, T.D. and Wong, C.-H. (2000) *Angew. Chem., Int. Ed.*, **39**, 1352.
5. (a) Mahrwald, R. (2004) in *Modern Aldol Reactions*, vols 1 and 2, Wiley-VCH Verlag GmbH, Weinheim; (b) Kagan, H.B. (1999) in *Comprehensive Asymmetric Catalysis* (eds E.N. Jacobsen, A. Pfaltz, and H. Yamamoto), Springer, Heidelberg, p. 9; (c) Carreira, E.M. (1999) in *Comprehensive Asymmetric Catalysis* (eds E.N.Jacobsen, A. Pfaltz, and H. Yamamoto), Springer, Heidelberg, p. 997; (d) Johnson, J.S. and Evans, D.A. (2000) *Acc. Chem. Res.*, **33**, 325; (e) Palomo, C., Oiarbide, M., and García J.M. (2004) *Chem. Soc. Rev.*, **33**, 65.
6. (a) Matsunaga, S. and Shibasaki, M. (2008) *Bull. Chem. Soc. Jpn*, **81**, 60; (b) Shibasaki, M., Kanai, M., Matsunaga, S., and Kumagai, N. (2009) *Acc. Chem. Res.*, **42**, 1117; (c) Shibasaki, M. and Matsunaga, S. (2006) *Chem. Soc. Rev.*, **35**, 269; (d) Sodeoka, M. and Hamashita, Y. (2009) *Chem. Commun.*, 5787.
7. Matsunaga, S., Kumagai, N., Harada, N., Harada, S., and Shibasaki, M. (2003) *J. Am. Chem. Soc.*, **125**, 4712.
8. Matsunaga, S., Yoshida, T., Morimoto, H., Kumagai, N., and Shibasaki, M. (2004) *J. Am. Chem. Soc.*, **126**, 8777.
9. (a) Matsunaga, S., Das, J., Roels, J., Vogel, E.M., Yamomoto, N., Iida, T., Yamaguchi, K., and Shibasaki, M. (2000) *J. Am. Chem. Soc.*, **122**, 2252; (b) Kumagai, N., Matsunaga, S., Kinoshita, T., Harada, S., Okada, S., Sakamoto, S., Yamaguchi, K., and Shibasaki, M. (2003) *J. Am. Chem. Soc.*, **125**, 2169, and references therein; (c) Harada, S., Kumagai, N., Kinoshita, T., Matsunaga, S., and Shibasaki, M. (2003) *J. Am. Chem. Soc.*, **125**, 2582, and references therein.
10. (a) Kumagai, N., Matsunaga, S., and Shibasaki, M. (2001) *Org. Lett.*, **3**, 4251; (b) Harada, S., Kumagai, N., Kinoshita, T., Matsunaga, S., and Shibasaki, M. (2003) *J. Am. Chem. Soc.*, **125**, 2582.
11. Matsunaga, S., Kinoshita, T., Okada, S., Harada, S., and Shibasaki, M. (2004) *J. Am. Chem. Soc.*, **126**, 7559.

12. (a) Trost, B.M. and Ito, H. (2000) *J. Am. Chem. Soc.*, **122**, 12003; (b) Trost, B.M., Ito, H., and Silcoff, E. (2001) *J. Am. Chem. Soc.*, **123**, 3367; (c) Trost, B.M., Silcoff, E., and Ito, H. (2001) *Org. Lett.*, **3**, 2497.
13. Tros, B.M. and Hisaindee, S. (2009) *Org. Lett.*, **8**, 6003.
14. Tros, B.M. and Müller, C. (2008) *J. Am. Chem. Soc.*, **130**, 2438.
15. Trost, B.M. and Hitce, J. (2009) *J. Am. Chem. Soc.*, **131**, 4572.
16. (a) Sasai, H., Suzuki, T., Arai, S., Arai, T., and Shibasaki, M. (1992) *J. Am. Chem. Soc.*, **114**, 4418; (b) Sasai, H., Suzuki, T., Itoh, N., and Shibasaki, M. (1993) *Tetrahedron Lett.*, **34**, 855.
17. (a) Sasai, H., Arai, T., and Shibasaki, M. (1994) *J. Am. Chem. Soc.*, **116**, 1571; (b) Sasai, H., Arai, T., Satow, Y., Houk, K.N., and Shibasaki, M. (1995) *J. Am. Chem. Soc.*, **117**, 6194.
18. (a) Sasai, H., Emori, E., Arai, T., and Shibasaki, M. (1996) *Tetrahedron Lett.*, **37**, 5561.
19. (a) Arai, T., Sasai, H., Aoe, K., Okamura, K., Date, T., and Shibasaki, M. (1996) *Angew. Chem., Int. Ed.*, **35**, 104; (b) Arai, T., Yamada, Y.M.A., Yamamoto, N., Sasai, H., and Shibasaki, M. (1996) *Chem Eur. J.*, **2**, 1368.
20. (a) Kim, Y.S., Matsunaga, S., Das, J., Sekinem, A., Oshima, T., and Shibasaki, M. (2000) *J. Am. Chem. Soc.*, **122**, 6506; (b) Matsunaga, S., Oshima, T., and Shibasaki, M. (2000) *Tetrahedron Lett.*, **41**, 8473.
21. Yamada, K., Arai, T., Sasai, H., and Shibasaki, M. (1998) *J. Org. Chem.*, **63**, 3666.
22. Arai, T., Sasai, H., Yamaguchi, K., and Shibasaki, M. (1998) *J. Am. Chem. Soc.*, **63**, 7547.
23. Shimizu, S., Oroi, K., Arai, T., Sasai, H., and Shibasaki, M. (1998) *J. Org. Chem.*, **63**, 3666.
24. (a) Manickan, G. and Sundarajan, G. (1999) *Tetrahedron*, **55**, 2721; (b) Manickam, G. and Sundarajan, G. (1997) *Tetrahedron: Asymmetry*, **8**, 2271.
25. Arai, T., Sekiguti, T., Iizuka, Y., Takizawa, S., Sakamoto, S., Yamaguchi, K., and Sasai, H. (2002) *Tetrahedron: Asymmetry*, **12**, 2083.
26. Keller, E., Veldman, N., Spek, A.L., and Feringa, B.L. (2002) *Tetrahedron: Asymmetry*, **12**, 2083; **8**, 3403.
27. Funabashi, K., Saida, Y., Kanai, M., Arai, T., Sasai, H., and Shibasaki, M. (1998) *Tetrahedron Lett.*, **39**, 7557.
28. Emori, E., Arai, T., Sasai, H., and Shibasaki, M. (1998) *J. Am. Chem. Soc.*, **120**, 4043.
29. Emori, E., Iida, T., and Shibasaki, M. (1999) *J. Org. Chem.*, **64**, 5318.
30. Yamagiwa, N., Qin, H., Matsunaga, S., and Shibasaki, M. (2005) *J. Am. Chem.*, **127**, 13419.
31. Fukuta, Y., Ohshima, T., Gnanadeskian, V., Shibuguchi, T., Nemoto, T., Kisugi, T., Okino, T., and Shibasaki, M. (2004) *Proc. Natl Acad. Sci. USA*, **101**, 5433.
32. Kakei, H., Tsuji, R., Ohshima, T., Morimoto, H., Matsunaga, S., and Shibasaki, M. (2007) *Chem. Asian J.*, **2**, 257.
33. Kakei, H., Sone, T., Sohtome, Y., Matsunaga, S., and Shibasaki, M. (2007) *J. Am. Chem. Soc.*, **129**, 13410.
34. Mita, T., Sasaki, K., and Shibasaki, M. (2005) *J. Am. Chem. Soc.*, **127**, 514.
35. Tanaka, Y., Kanai, M., and Shibasaki, M. (2008) *J. Am. Chem. Soc.*, **130**, 6072.
36. For a review see: Reference [6d]. For examples see: (a) Sodeoka, M., Ohrai, M., and Shibasaki, M. (1995) *J. Org. Chem.*, **60**, 2648; (b) Sodeoka, M. and Shibasaki, M. (1998) *Pure Appl. Chem.*, **70**, 411; (c) Hagiwara, E., Fujii, A., and Sodeoka, M. (1998) *J. Am. Chem. Soc.*, **120**, 2474; (d) Fujii, A., Hagiwara, E., and Sodeoka, M. (1999) *J. Am. Chem. Soc.*, **121**, 5450.
37. (a) Hamashima, Y., Hotta, D., and Sodeoka, M. (1999) *J. Am. Chem. Soc.*, **121**, 5450; (b) Hamashima, Y., Takano, H., Hotta, D., and Sodeoka, M. (2003) *Org. Lett.*, **5**, 3225; (c) Hamashima, Y., Hotta, D., Umebayashi, N., Tsuchiya, Y., Suzuki, T., and Sodeoka, M. (2005) *Adv. Synth. Catal.*, **347**, 1576.
38. Hamashima, Y., Somei, H., Shimura, Y., Tamura, T., and Sodeoka, M. (2004) *Org. Lett.*, **6**, 1861.

5
Enamines in Catalytic Enantioselective Conjugate Additions
Ramon Rios and Albert Moyano

5.1
Introduction and Background

The use of chiral secondary amines as asymmetric catalysts (asymmetric aminocatalysis; AA) was pioneered during the early 1970s both by Hajos and Parrish [1] and, independently, by Eder *et al.* [2] in their reports on the first examples of highly enantioselective, proline-catalyzed intramolecular aldol cyclizations. In spite of some subsequent significant applications of this transformation – most notably, in Woodward's asymmetric total synthesis of erythromycin [3] – the potential of AA was largely overlooked during the following three decades. Following the rediscovery of proline in 2000 as a catalyst in the intermolecular aldol reaction by List *et al.* [4] and, soon after, the introduction of iminium catalysis by D. W. C. MacMillan [5] (Scheme 5.1), the development of AA – and of the broader field of asymmetric organocatalysis (AO) – has been impressive. Today, AO has become a powerful and well-established synthetic tool for the enantioselective construction of chiral organic compounds [6], that complements – and, in some instances, also improves – the more traditional approaches based on metal-catalyzed reactions [7] and on biocatalysis [8].

Carbon nucleophiles that contain relatively acidic methylene groups have been widely applied in direct Michael additions, whereas simple enolizable carbonyl compounds need generally to be converted into more reactive species such as enol ethers or enamines prior to their use. Since chemical transformations that avoid additional reagents, the generation of waste, and extended working times are highly desirable, a more promising, and atom-economic strategy would involve the direct addition of unmodified carbonyl compounds to Michael acceptors. Covalently bonded aminocatalysts operate through two different mechanisms, by transforming the carbonyl substrates either into activated nucleophiles (enamine intermediates), or into electrophiles (iminium intermediates). Very recently, MacMillan and coworkers have disclosed a third way of activation that takes place via cation-radical chemistry singly occupied molecular orbital (SOMO) catalysis (Scheme 5.2) [9].

Catalytic Asymmetric Conjugate Reactions. Edited by Armando Córdova
Copyright © 2010 WILEY-VCH Verlag GmbH & Co. KGaA, Weinheim
ISBN: 978-3-527-32411-8

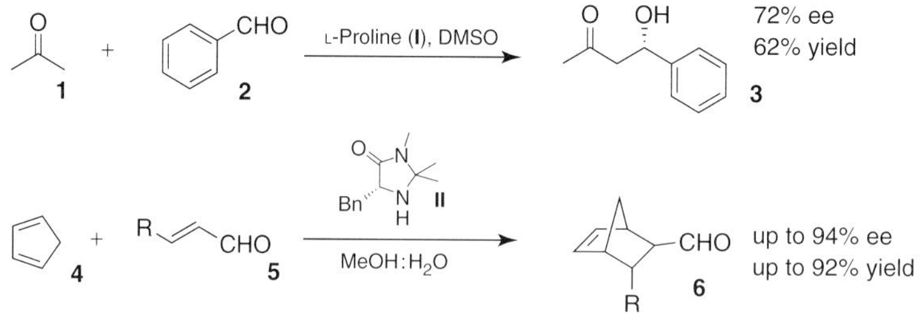

5 Enamines in Catalytic Enantioselective Conjugate Additions

Scheme 5.1 Asymmetric organocatalytic reactions reported in 2000.

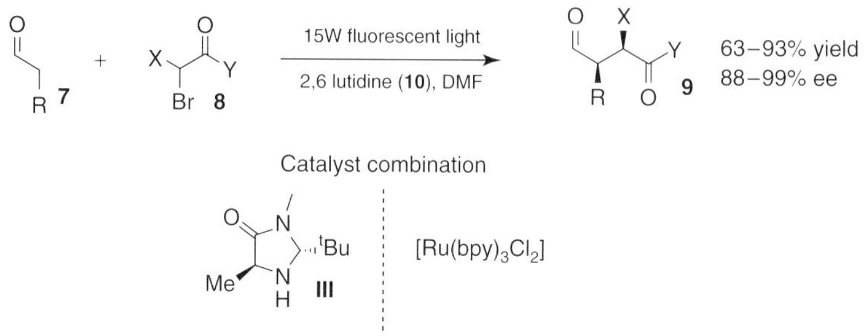

Scheme 5.2 SOMO reaction reported by MacMillan.

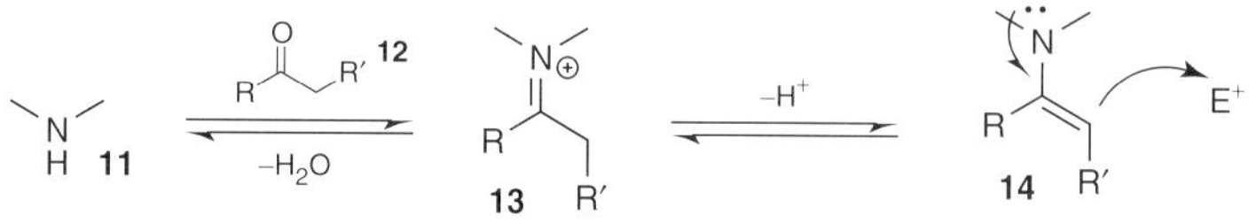

Scheme 5.3 General enamine catalysis.

In the field of organocatalysis, enantioselective conjugate additions have lately received widespread attention. In enamine catalysis, the addition of the organocatalyst generates an iminium ion as the initial species, the subsequent deprotonation of which provides the enamine nucleophilic intermediate. This species can attack a wide range of electrophiles, as shown in Scheme 5.3.

The efficiency and scope of organocatalysis, and particularly of aminocatalysis, have been broadly established by the development of highly reliable synthetic methods. The remarkable selectivity of asymmetric aminocatalysis has made this approach a common objective for many research groups, enhancing and enriching the know-how of the field.

In particular, conjugate or Michael addition represents one of the most important methodologies in organic chemistry. Since the pioneering investigations of Stork et al. [10], chemists have devoted much effort to the discovery of new enantioselective methodologies in order to achieve the "perfect" 1,4 or conjugate addition.

The aim of this chapter is to describe the state of the art of asymmetric conjugate additions via enamine activation, with the intention of detailing the latest trends in this burgeoning field of research.

5.2 Mechanistic Considerations

Small chiral primary or secondary amines can easily form enamines with carbonyl compounds such as ketones or aldehydes, and the enamines can then react with a broad diversity of electrophiles. Michael addition is among the main reactions that can be triggered by enamine catalysis; the general mechanistic features of the reaction are outlined in Scheme 5.4. In the first step, an iminium ion (**13a**) is generated by the reversible reaction between a chiral amine catalyst (**IV**) and a carbonyl compound (**12**); **13a** can be easily deprotonated to form the enamine nucleophilic intermediate (**14a**), due the increase of C–H acidity. This enolate equivalent **14a** can react with an electron-deficient olefin **15** in order to create a new C–C bond. Subsequent hydrolysis of the α-modified iminium ion affords the Michael adduct **17** and restores the aminocatalyst **IV**, which is ready to participate in a new catalytic cycle (as shown in Scheme 5.4).

One danger of enamine chemistry is the possibility that the amine catalyst (more often a secondary amine) can be trapped by the electrophilic substrate, thus diminishing the availability of the catalyst (Scheme 5.5). For this reason, the reversibility of this trapping plays a crucial role in the efficiency of the Michael addition.

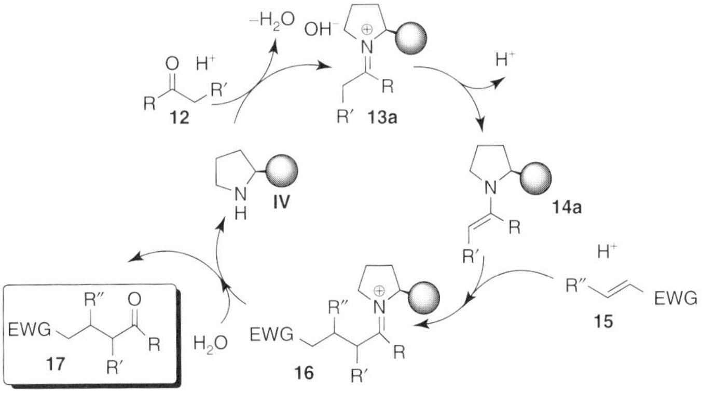

Scheme 5.4 General mechanism of enantioselective conjugate addition (ECA).

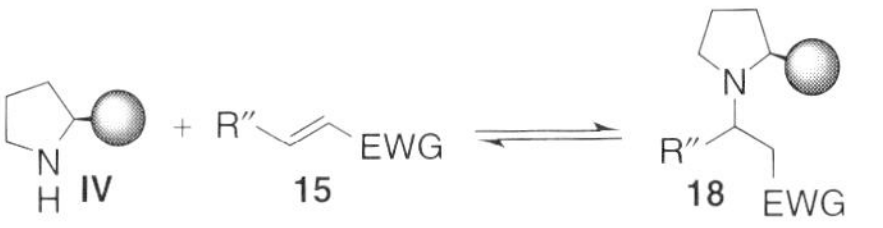

Scheme 5.5 Catalyst trapping.

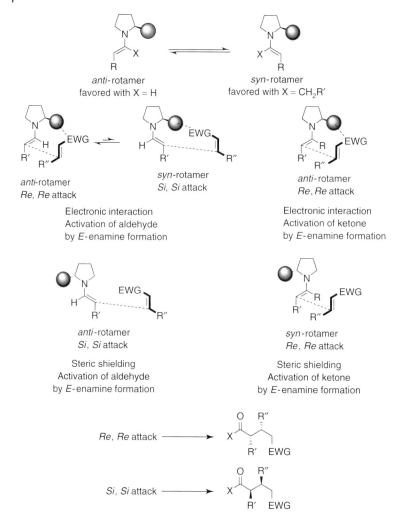

Figure 5.1 Transition states.

The stereoselectivity of the intermolecular conjugate Michael addition could be explained by the limiting "electronic" or "steric" transition states (Figure 5.1). One of the most important mechanistic features is the geometry of the enamine. This geometry, whether *E* or *Z*, is essentially determined by the catalyst structure. Due to steric hindrance, the *E*-enamine (which is thermodynamically favored and derives from either an aldehyde or a ketone) would be predominant unless other interactions were to favor the *Z*-enamine. The α-substituent of the catalyst governs the position of the equilibrium between rotamers, which therefore influences the facial selectivity of the electrophilic attack to the enamine. The relative size of the two sides of the enamine depends on the carbonyl substrate. The smallest group – hydrogen for aldehydes – leads to the formation of the relatively more stable *anti*-rotamer whereas, in the case of ketones, the less-hindered moiety is the double bond which gives preferentially the *syn*-rotamer, as long as other interactions are not involved.

Finally, the nucleophilic *E*-enamine may attack the Michael acceptor via an acyclic synclinal transition state [11] through two different pathways. Hence, the face selectivity is determined by electronic or steric interactions, as shown respectively in Figure 5.1. Electronic interactions would be enforced by hydrogen bonding in L-proline, tetrazole, or thiourea catalysts. Consequently, the conjugate addition to the Michael acceptor would take place from the face of the enamine occupied by the α-substituent. Actually, stabilizing H-bonding in the transition state could counterbalance any repulsive steric interactions and force the *syn*-rotamer in the case of aldehydes and the *anti*-rotamer in the case of ketones.

In contrast, steric shielding could be involved in the determination of the facial selectivity. Indeed, the bulky group on the catalyst framework could prevent the H-bonding interaction, forcing the attack from the face opposite to that occupied by the substituent, as depicted in Figure 5.1. In this scenario, the less-hindered *si, si* transition state via an *anti*-enamine is usually well favored for aldehydes, compared to the *re, re* approach for the ketones via a *syn*-enamine.

In conclusion, the preferred diastereoselectivity and enantioselectivity of the addition relies on electronic or steric interactions and, obviously, also on the absolute configuration of the chiral aminocatalyst.

5.3
Ketone Conjugate Additions

5.3.1
Ketone Conjugate Additions to Nitroolefins

In 2001, List and coworkers [12] and Barbas and coworkers [13] reported, almost at the same time, the first asymmetric intermolecular Michael-type addition of ketones to nitroolefins (Scheme 5.6). Both research groups used proline to catalyze the reaction via enamine formation; the reaction proceeded with high chemical yields and diastereoselectivity, albeit affording nearly racemic products in dimethylsulfoxide (DMSO). Shortly afterwards, Enders demonstrated that the use of methanol as solvent was beneficial in order to increase the enantioselectivity of the reaction [14].

In spite of the low degree of asymmetric induction achieved in these early examples, the utility and relevance of the reaction impelled several groups to seek new catalysts that would afford higher enantioselectivities. The most common approach has been via to chemically modify the structure of the proline catalyst, which would lead to a more efficient catalyst that showed better selectivity and an improved synthetic scope. In this context, much effort has been devoted to the synthesis of pyrrolidine derivatives with an *N*-containing side chain or heterocyclic

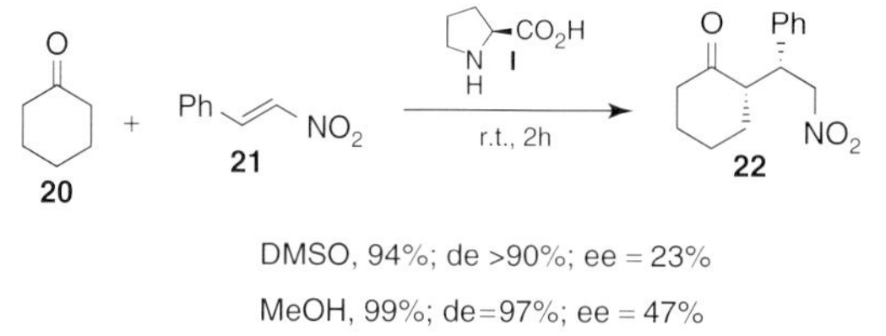

Scheme 5.6 Proline-catalyzed reaction of cyclohexanone with nitrostyrene.

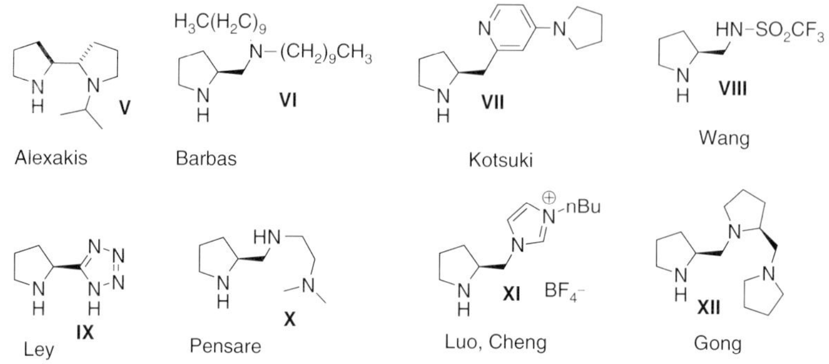

DMSO, 94%; de >90%; ee = 23%
MeOH, 99%; de = 97%; ee = 47%

Alexakis **V** Barbas **VI** Kotsuki **VII** Wang **VIII**

Ley **IX** Pensare **X** Luo, Cheng **XI** BF$_4^-$ Gong **XII**

Scheme 5.7 Different organocatalysts.

moiety (some examples are illustrated in Scheme 5.7). The free amines or the corresponding salts promote the highly *syn*-selective addition of α-alkyl-substituted cyclic or acyclic ketones to nitrostyrenes. The acid co-catalyst is thought to facilitate the formation of the enamine intermediate, and this leads to an overall rate acceleration and increased conversion. On the other hand, although the presence of bulky substituents in the pyrrolidine ring leads to a considerable improvement in the selectivity of the process, it also decreases the reactivity of the catalyst, making formation of the iminium intermediate more difficult.

An important issue of this reaction is that, while the selectivity profile of the catalysts has been considerably improved during the past few years, the selectivity of the addition has remained highly substrate-dependent. Very often, the best enantioselectivities were obtained with cyclohexanone (ee-values up to 99%), while considerably lower enantiomeric purities were achieved with cyclopentanone, or with acyclic ketones.

Another issue that lies ahead is the use of a large excesses of the ketone reactant, in order to achieve convenient kinetics and conversions. Moreover, it should be noted that the solvent is crucially important not only for the reaction rate but also for the diastereoselectivity and enantioselectivity of asymmetric organocatalyzed conjugate additions. The reactions are usually run in CHCl$_3$ or tetrahydrofuran (THF), and the presence of different additives such as water is sometimes beneficial [15].

5.3.1.1 Secondary Amines

Barbas *et al.* showed that (*S*)-1-(2-pyrrolidinylmethyl)pyrrolidine (**XIII**) was a privileged catalyst for the addition of ketones to nitroolefins in comparison with

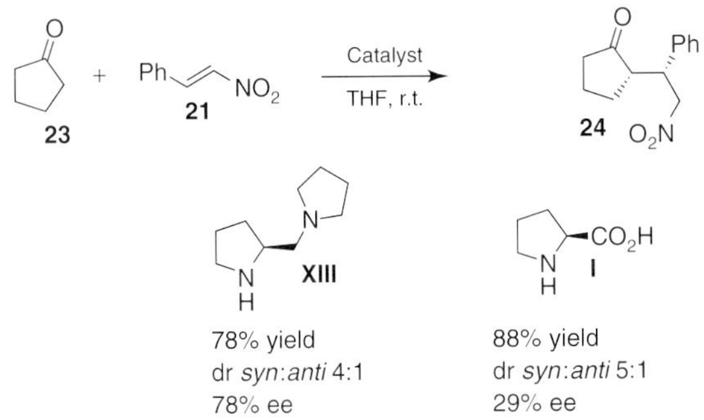

Scheme 5.8 Cyclopentanone addition to nitrostyrene, as reported by Barbas.

Catalyst **XIII**
78% yield
dr syn:anti 4:1
78% ee

Catalyst **I**
88% yield
dr syn:anti 5:1
29% ee

L-proline (**I**) (Scheme 5.8) [13b]. This suggests that L-proline (**I**) requires a lone pair on the atom adjacent to the electrophilic center in order to induce high stereocontrol through H-bonding with its carboxylic acid moiety, as occurs in aldol, Mannich, and related reactions. The conclusion is that, in general, chiral diamines derived from pyrrolidine-bearing secondary and tertiary amine groups would be more efficient than proline in Michael additions to nitroalkenes, by favoring the "steric" transition states previously described (see Figure 5.1).

Since then, several research groups have devoted their efforts to the synthesis of diamine catalysts. For example, Alexakis and coworkers developed N-alkyl-2S-2S′-bypirrolidine derivatives that showed a high efficiency and stereoselectivity in this reaction [16].

According to Alexakis, the N-iPr-2,2′-bipyrrolidine (**V**) appeared to be the most effective catalyst. As mentioned above, the acid co-catalyst accelerated the C–C bond-forming process by increasing the rate of enamine formation [17]. The syn selectivity obtained was in accordance with Seebach's model based on steric shielding (see Figure 5.1), in which there are additional favorable electrostatic interactions between the nitrogen of the enamine and the nitro group (Scheme 5.9). The isopropyl substituent would promote the selective formation of the E-enamine and induce a marked bias toward the re, re approach.

The alkylation of asymmetric acyclic ketones (a challenging problem in organic synthesis) takes place regioselectively at the most-substituted carbon, thus affording the syn isomers as major products. α-Hydroxyketones showed anti selective additions, and a stereochemical outcome similar to that observed in related aldol and Mannich-type additions. Such selectivity is due to the preferred formation of the Z-enamine intermediate, stabilized by intramolecular hydrogen bonding between the hydroxy group and the tertiary amine of the catalyst (Scheme 5.10) [18].

All of these precedents provided the basis for the development of various chiral pyrrolidine organocatalysts, as illustrated above (see Scheme 5.7).

Barbas et al. reported the promotion of the conjugate addition of cyclohexanone to nitrostyrene by using diamine **VI**, bearing hydrophobic alkyl groups [19]. Later, the same group demonstrated the excellent reactivity and selectivity of diamine **VI** in combination with trifluoroacetic acid (TFA) and in brine as the solvent [20].

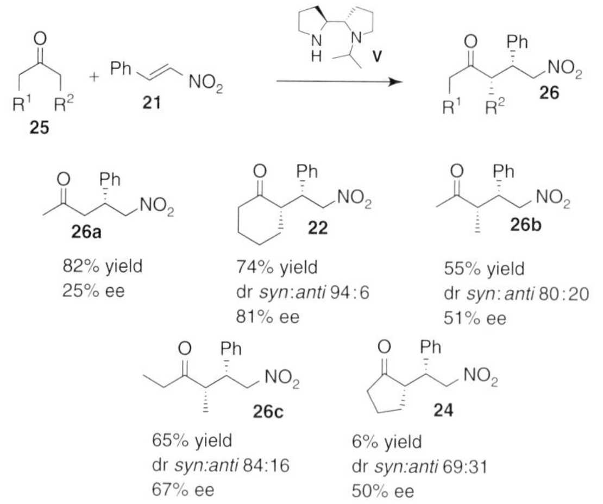

Scheme 5.9 Conjugate addition of ketones to nitroolefins, as reported by Alexakis.

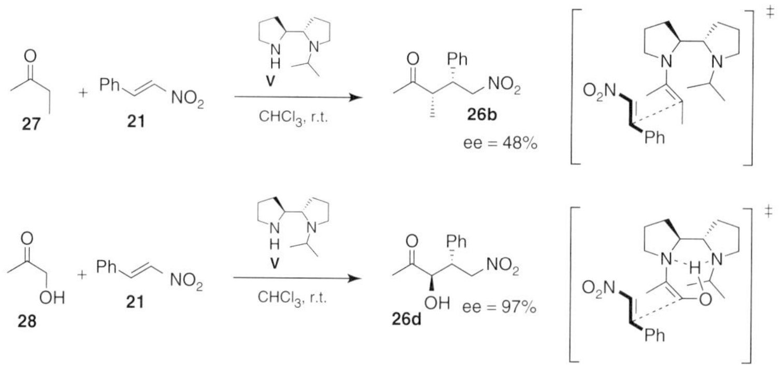

Scheme 5.10 Variation of syn:anti diastereoselectivity of the conjugate addition of ketones to nitrostyrenes in function of the substrate.

Kotsuki et al. designed a new pyrrolidine–pyridine-based catalyst such as **VII**, which can easily be prepared from L-prolinol [21]. Given the importance of the proximity of the pyrrolidine–pyridine functionality, it was postulated that this basic functionality could facilitate the enamine formation, while the resulting pyridinium ring could effectively shield one face of the enamine double bond. Hence, the reaction would be highly efficient in terms of yield and stereocontrol (up to 99% ee).

Shortly afterwards, Ley et al. replaced the carboxylic acid group of proline with tetrazole, a bioisostere of this functionality [22], and successfully applied the resultant catalyst **IX** to the conjugate addition of ketones to nitroolefins [23]. This organocatalyst far outperformed L-proline (**I**) in every aspect and, interestingly, the reaction performed well when using only a relatively small amount of ketone (1.5 equiv.). The improved activity could be associated to the difference in the hydrogen-bonding strengths between the tetrazole and the carboxylic acid functionality, to the increased size of the tetrazole moiety, and/or to an enhancement of the solubility of the tetrazole analog. Further optimization from the same group led to the development of a homoproline tetrazole catalyst, which afforded high enantioselectivities for a wide range of ketones and nitroolefins [23b].

More recently, protonated proline-derived triamine catalysts **X** were developed by Pansare and Pandya for the highly enantioselective conjugate addition of cyclic six-membered ketones to nitroalkenes, and showed high levels of enantioselectivity [24].

Wang *et al.* reported an interesting new pyrrolidine sulfonamide **VIII** which efficiently mediated the asymmetric conjugate addition of ketones to nitroolefins [25]. The enhanced catalytic activity and selectivity of catalyst **VIII** relative to L-proline (**I**) were a consequence of the acidic and steric bulk properties of the NHTf group, that enables the formation of intramolecular and intermolecular hydrogen bonds. The selectivity appeared to originate from the steric hindrance between the bulky sulfonamide group and the alkyl substituent of the enamine intermediate. The same group also developed a recyclable and reusable fluorous pyrrolidine sulfonamide for promoting highly enantioselective and diastereoselective additions of ketones and aldehydes to nitroolefins in water [26].

One of the most intriguing and exciting approximations for the synthesis of new catalysts for the conjugate addition of ketones to nitrostyrenes consisted of the *in situ* formation of supramolecular assemblies to improve a pre-existing catalyst. For example, in 2008, Zhao and coworkers introduced a modularly designed proline–thiourea assembly as an organocatalyst for the direct nitro-Michael addition [27]. In this approach, it was proposed that ionic interactions could be applied for the self-assembly of different units. In order to prove this concept, Zhao used simple proline as the primary organocatalyst, which interacts with a second unit of catalyst (in particular, a tertiary amine) carrying on a thiourea moiety. It was proposed that these two units in the reaction mixture would undergo an acid–base reaction between the carboxylic acid and the tertiary amine, leading to an ammonium salt. Ionic interactions between the ammonium and the carboxylate should cause these two modules to self-assemble [28], thus affording a new organocatalyst and incorporating the proline reactive center and a stereocontrolling moiety based on a chiral thiourea unit (Scheme 5.11).

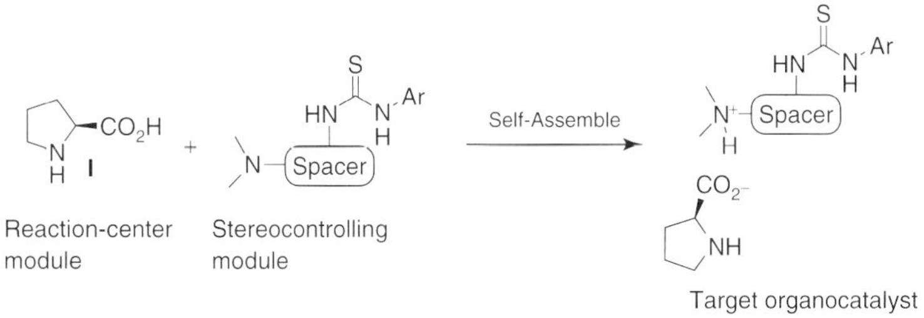

Scheme 5.11 Zhao's self-assembled organocatalyst.

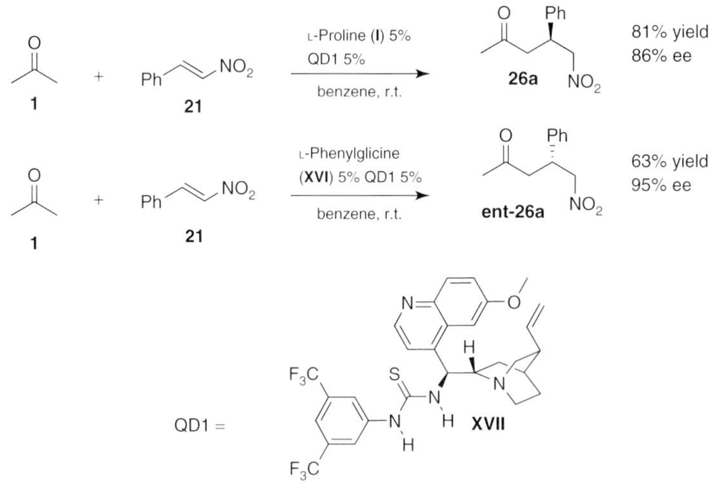

Scheme 5.12 Michael addition catalyzed by assembled organocatalysts, as reported by Zhao.

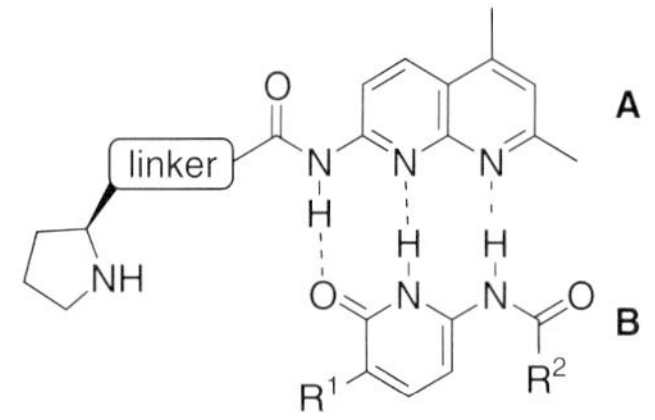

Scheme 5.13 Supramolecular assembled organocatalyst, as proposed by Clarke.

In order to demonstrate the power of this approximation, several self-assembled organocatalysts were tested in the direct nitro-Michael addition of ketones to nitrostyrenes. The authors reported that those improved organocatalysts rendered the addition product in high yields and excellent diastereoselectivity and enantioselectivity, as shown in Scheme 5.12. Remarkably, the stereoinduction of the reaction was determined by the nature of the amino acid in the assembled organocatalysts; consequently, L-proline and L-phenylglycine gave the desired adducts, but with opposite enantioselectivities.

The first example of an improved organocatalyst by supramolecular interaction with achiral additives was reported by M. L. Clarke in 2007 [29]. In this case, Clarke disclosed that achiral additives bearing hydrogen bond donors such as pyridinones (**B**) could strongly associate with amidonaphthyridines (**A**) in apolar solvents, as shown in Scheme 5.13. This association was previously reported by Kelly *et al.* to template a S_N2 reaction [30].

Clarke and coworkers chose the enantioselective Michael addition of ketones to nitrostyrenes as a test reaction, and showed that the assembly of pyridinones with amidonaphthyridines caused a dramatic increase in the enantioselectivity of this reaction. Moreover, it was shown that the complex could easily be prepared and was adjustable to the requirements of the reactions under study.

Table 5.1 Reactions as reported by Clarke.

Entry	B	Yield (%)	dr	ee (%)
1	None	87	15 : 1	15
2	27	82	31 : 1	47
3	28	74	33 : 1	34
4	29	59	41 : 1	35
5	30	63	58 : 1	79

As shown in Table 5.1, when (S)-ProNap (**XVIII**) is used on its own, the reaction was efficiently catalyzed, but the products were obtained in almost racemic form. Gratifyingly, the addition of different pyridinones increased the enantioselectivity up to 79% (entry 5; Table 5.1).

When Cheng et al. and Luo identified chiral ionic liquids such as **XI** as suitable catalysts for the ketone addition to nitrostyrenes [31], they reported that the catalyst would maintain the unique properties of ionic liquids and present a high efficiency as a catalyst for the Michael reaction (up to 99% ee). The use of an acidic co-catalyst is essential to accelerate the reaction rate and to increase the turnover of the catalyst.

A triamine bearing three pyrrolidine cores (**XII**), has been used by Gong et al. in this reaction [32]. As usual, the efficiency and the selectivity was enhanced dramatically when (+)-camphorsulfonic acid was used as co-catalyst. Remarkably, the chirality of the stereogenic center of the acid had little effect on the enantioselectivity of the reaction but, once again, steric shielding by the pyrrolidine substituent was invoked to rationalize the observed stereochemical outcome.

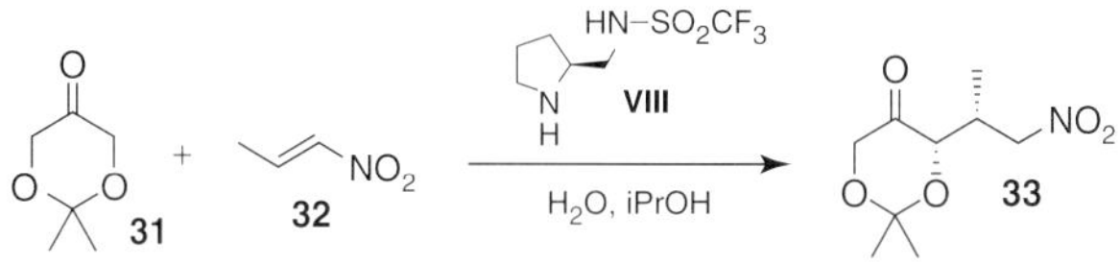

Scheme 5.14 Conjugate addition of dihydroxyacetone.

Another interesting approximation was made by Benaglia et al. [33], which involved the use of a poly(ethyleneglycol)-supported proline as a recyclable aminocatalyst for the asymmetric synthesis of γ-nitroketones. However, these could be obtained in lower yields and similar enantioselectivities in comparison with L-proline (**I**).

While cyclohexanone has been extensively used in the addition to nitroolefins, Enders made an interesting move by selecting a protected dihydroxyacetone [34] (DHA) derivative **32** instead of cyclohexanone. As shown in Scheme 5.14, the Michael reaction of DHA with various nitroalkenes works extremely well when pyrrolidine sulfonamide **VIII** is used as the organocatalyst, providing 1,3-adducts in high diastereoselectivity and enantioselectivity (up to 86% ee). However, the addition of water caused an acceleration in the reaction rate and increased the yield, with shorter reaction times. In this case, the stereochemical outcome of the reaction could be accounted for by electronic interactions via hydrogen bonding.

5.3.1.2 Primary Amines

Primary amines represent an interesting and almost unexplored option as catalysts for the activation of ketones as enamines. In this area, only the use of peptides, as reported for the first time by List and Martin [35], has relied on the power of primary amines. Córdova and coworkers screened several simple dipeptides to catalyze the Michael reaction between ketones and nitroalkenes [36]. The alanine–alanine dipeptide **XX**, bearing a primary amine, promoted the addition of a wide range of cyclic ketones to nitroolefins with excellent enantioselectivities (up to 98% ee). A small excess of water was added to increase the efficiency of the transformation, and N-methyl-2-pyrrolidone (NMP) was used as a co-solvent. In this reaction, hydrogen bonding, due to the presence of acid and amide groups and water, was plausibly assumed to be involved in the transition state, based on Seebach's acyclic synclinal model. Soon thereafter, the same group also used amino acid-derived amides such as **XXI** as an efficient organocatalyst for the conjugate addition [37]. (Scheme 5.15).

5.3.2
Amine–Thiourea Catalysts

The groups of Tsogoeva, Tang, and Jacobsen pioneered, independently, the development of new bifunctional catalysts that combined a nucleophilic amine and a thiourea-based Brønsted acid. For example, Jacobsen and Huang reported

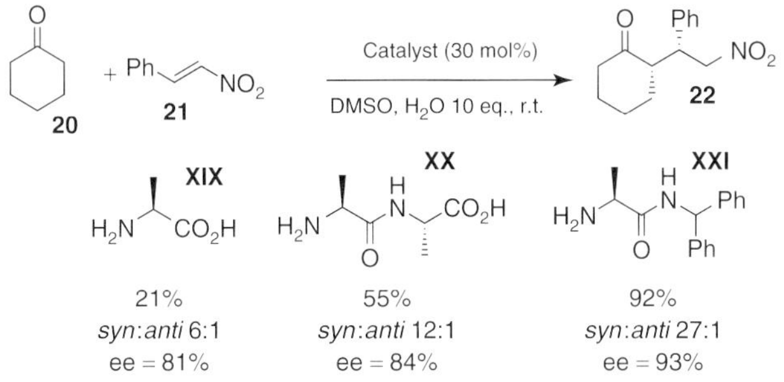

Scheme 5.15 Enantioselective conjugate addition catalyzed by primary amines.

a new bifunctional catalyst **XXII** to induce high selectivity for a broad substrate scope with respect to nucleophilic and electrophilic reacting partners. An acid additive such as benzoic acid, n-butyric acid, or acetic acid was used as co-catalyst, which allowed the double alkylation to be avoided, and also speeded up the reaction. In the case of the Jacobsen catalyst **XXII** [38], the selectivity was a consequence of the preferred Z-enamine formation in the transition state; the catalyst also activated the acceptor and orientated it in space by binding the nitro group.

Tsogoeva and coworkers developed a similar catalyst (**XXIII**) [39], best results for which were obtained with chiral naphthylamine-based thiourea catalysts having free NH_2 functions in the side chains. This led to *syn* adducts with cyclic ketones, whereas acyclic ketones afforded the *anti* adducts (Scheme 5.16).

The Tang group combined the chiral pyrrolidine core with a thiourea function [40]. The optimal reaction conditions were obtained with **XXIV** under solvent-free conditions and in the presence of n-butyric acid as additive at 0 °C. The high selectivity of the addition was attributed to the formation of a rigid three-dimensional (3-D) H-bonding network which was consistent with proline and tetrazole hydrogen bond interactions (Scheme 5.17).

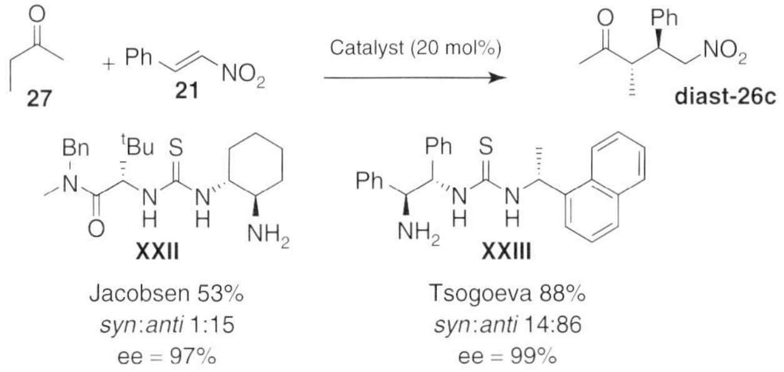

Scheme 5.16 Enantioselective conjugate addition catalyzed by primary amine–thiourea bifunctional catalysts.

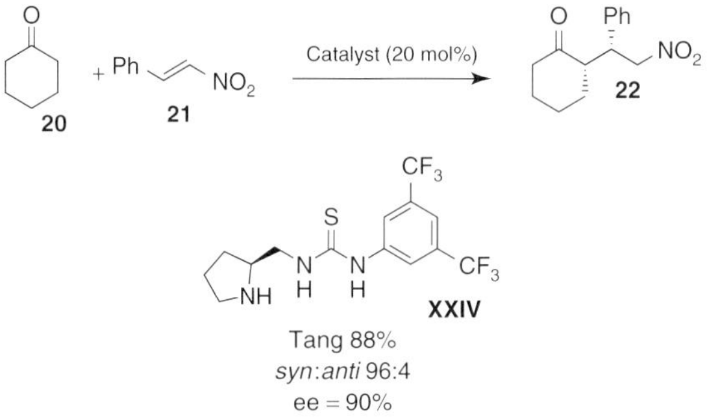

Scheme 5.17 Enantioselective conjugate addition, as reported by Tang.

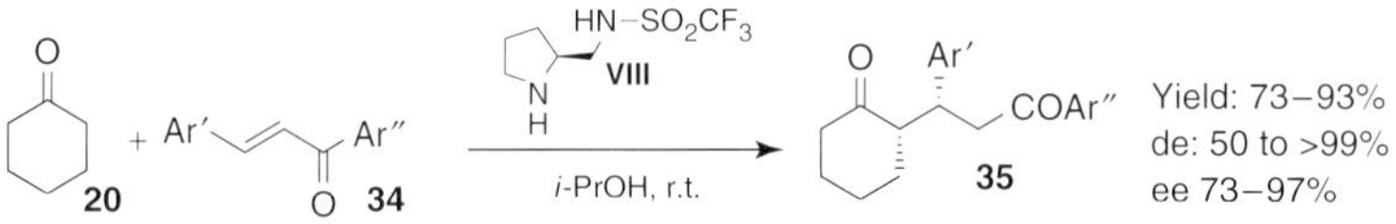

Scheme 5.18 Ketone addition to unsaturated ketones.

5.3.3
Ketone Conjugate Additions to α,β-Unsaturated Carbonyl Compounds

Currently, the organocatalytic Michael reaction of ketones and enones remains little explored, due both to the unreactive nature of the enone acceptors and to the difficulty of the formation of the enamine intermediate from ketones.

Only Wang and coworkers have reported that **VIII** catalyzed the Michael reaction between cyclic ketones and 1,4-diaryl acyclic enones with remarkably good results (Scheme 5.18). These authors described a mechanism which involved exclusively an enamine-type activation of the nucleophile, with no contribution of any intermediate iminium species that could eventually activate the electrophile [41].

Another reaction that currently remains almost unexplored is the enantioselective addition of ketones to alkylidene malonates, although pyrrolidine–amine-based catalysts have been shown to mediate this process. In 2007, Tang and coworkers [42] used catalyst **VIII** in order to catalyze the addition of ketones to alkylidene malonates. As usual, the addition of n-butyric acid proved beneficial in terms of yield and selectivity, and the reaction afforded the addition products in good yields and with excellent enantioselectivities. The stereochemical outcome of the reaction was explained by hydrogen bond interactions in the transition state (Scheme 5.19).

5.3.4
Other Reactions

In 2008, Barros et al. reported the use of vinylphosphonates as acceptors in the Michael addition of ketones, catalyzed by a simple diamine **XIII** [43]. The reaction works well for cyclic ketones, affording the desired Michael adducts in good yields

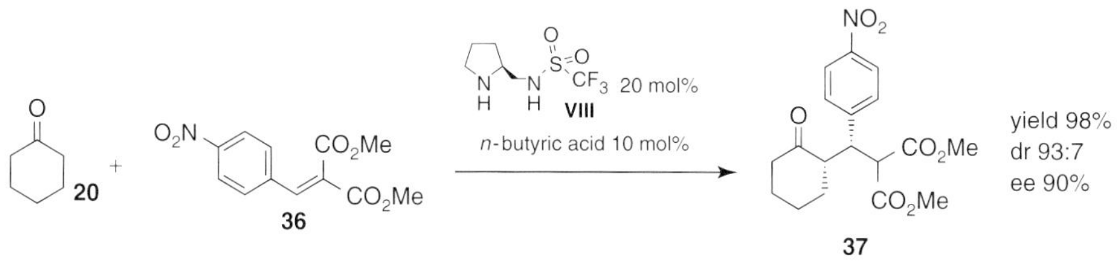

Scheme 5.19 Ketone addition to alkylidene malonates.

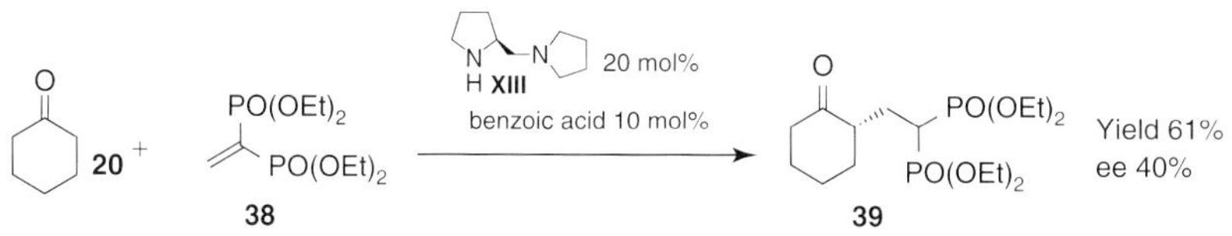

Scheme 5.20 Ketone addition to vinylphosphonates, as reported by Barros et al.

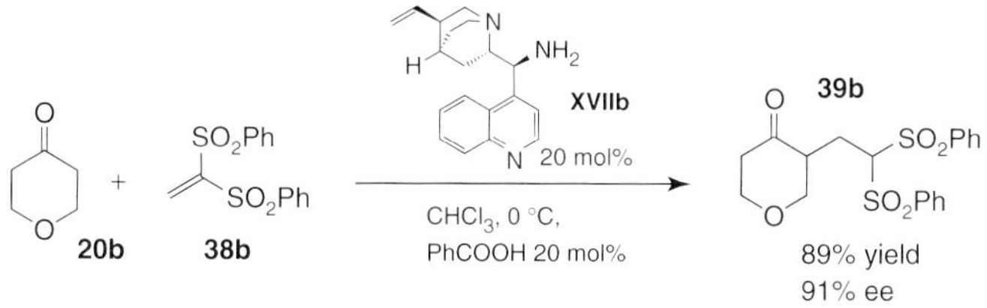

Scheme 5.21 Conjugate addition of ketones to vinyl sulfones.

but with moderate enantioselectivities, as shown in Scheme 5.20. However, when acyclic ketones were used, the enantioselectivity fell dramatically.

Very recently, Lu and coworkers reported the first enantioselective addition of ketones to vinylsulfones via an enamine intermediate [44]. The reaction was seen to be efficiently catalyzed by primary amines derived from cinchona alkaloids, and afforded the Michael adducts in excellent yields and enantioselectivities (Scheme 5.21).

5.4
Aldehyde Conjugate Additions

Aldehyde-derived enamines are more reactive than ketone enamines, and may add to nitroolefins, alkylidene malonates, vinylsulfones, or vinylphosphonates at lower temperatures than those at which ketones would react. Most of the catalysts discussed above are well-suited to the conjugate addition of aldehydes. Furthermore, the high reactivity shown by aldehydes allows the use of more bulky catalysts that are not active for ketone activation, such as diphenylprolinol derivatives. Similar to studies with ketones, acid co-catalysts suppressed any side reactions such as base-catalyzed homoaldol coupling, and fostered the Michael reaction. Typically, both aldehydes and ketones showed *syn*-selectivity but gave the opposite enantiomers, in accordance with the transition states described above (see Figure 5.1). This fact could be rationalized by differences in the steric demands of the enamine forms.

5.4.1
Aldehyde Conjugate Additions to Nitroolefins

The first example of an aldehyde Michael addition to nitroolefins was reported by Barbas and coworkers, where the reaction was efficiently catalyzed by secondary amines [45]. A large number of pyrrolidine derivatives were evaluated, and under substoichiometric conditions (20 mol%) up to 86% ee was obtained with (S)-2-morpholinomethyl)pyrrolidine **XXV**. In contrast, L-proline (**I**) proved to be a poor catalyst for this class of Michael additions (Scheme 5.22).

According to the observed stereochemical outcome, the approach takes place from the less-hindered *si*-face of the enamine. Soon afterwards, the same research group reported the formation of quaternary centers by using catalyst **XIII** with TFA as cocatalyst (Scheme 5.23) [46]. Subsequently, the reaction was run successfully using brine as the solvent [15].

When Alexakis *et al.* used their 2,2′-bypyrrolidine catalyst **V** in the same reaction [47], the catalyst showed very good yields and enantioselectivities with linear

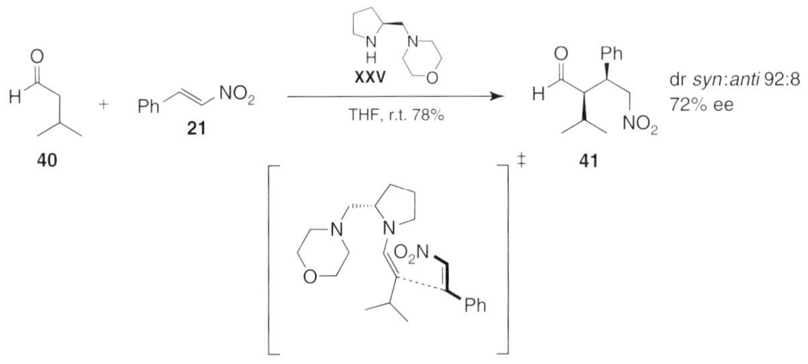

Scheme 5.22 Aldehyde addition to nitrostyrenes, as reported by Barbas.

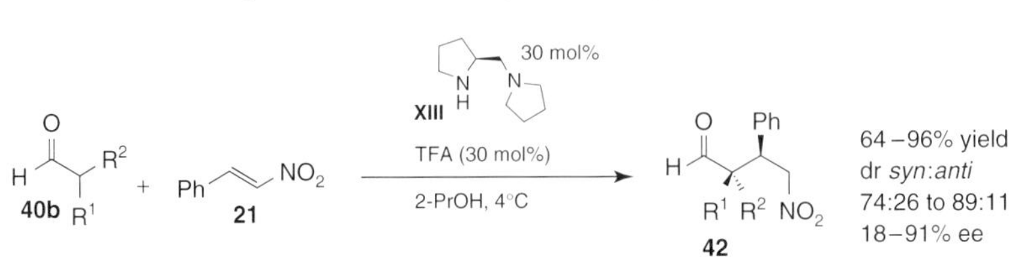

Scheme 5.23 Diamine **XIII**-catalyzed conjugate addition of α,α-disubstituted acetaldehydes to nitrostyrenes.

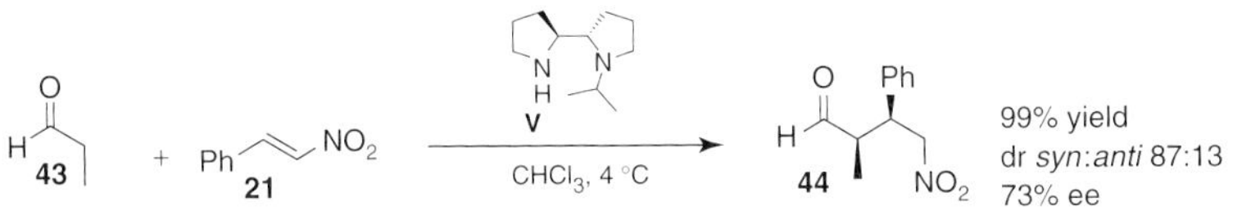

Scheme 5.24 Enantioselective conjugate addition, as reported by Alexakis.

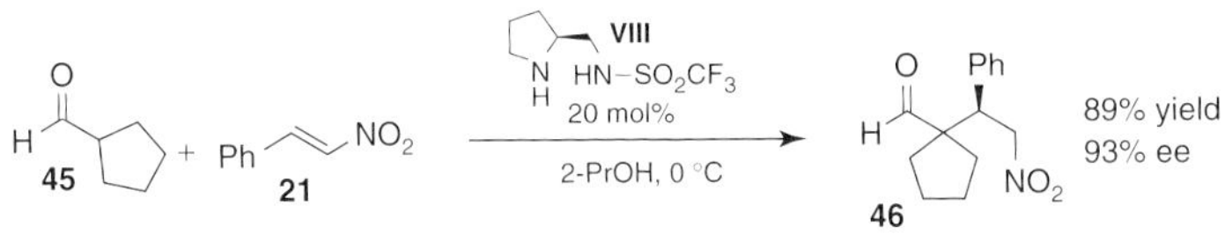

Scheme 5.25 Enantioselective conjugate addition, as described by Wang.

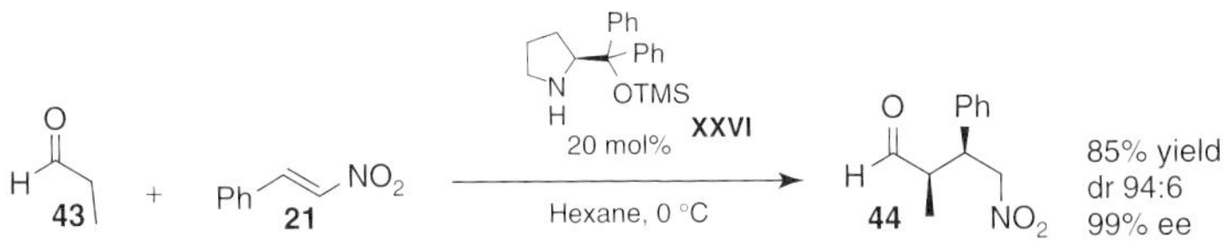

Scheme 5.26 Diphenylprolinol silyl ether-catalyzed conjugate addition.

aliphatic aldehydes. However, more hindered, branched aldehydes, such as isovaleraldehyde, provided worse enantioselectivities (Scheme 5.24).

As in the examples above, the stereochemical outcome was seen to be dictated by steric hindrance, favoring the *si, si* approach.

As highlighted in the case of ketones, the chiral pyrrolidine sulfonamide catalyst **VIII** developed by Wang [48] produced the desired Michael adducts between aldehydes and nitroolefins, with high levels of enantioselectivity and diastereoselectivity (Scheme 5.25).

As noted previously, diphenylprolinol silyl ether **XXVI** was found to be an exceptional catalyst for the addition of aldehydes to nitroalkenes, according to a report by Hayashi and coworkers (Scheme 5.26) [49].

Shortly afterwards, Wang and coworkers developed a recyclable diphenylprolinol silyl ether bearing an n-C_8F_{17} fluorous tag, and achieved excellent results [50].

Later, Palomo and coworkers developed a new type of catalyst that included both a bulky α-group and a hydrogen bond-donor, directing γ-group (Scheme 5.27) [51]. This catalyst was efficiently used in the enantioselective synthesis of γ-butyrolactones, a common scaffold in natural products synthesis.

Recently, Barros *et al.* have reported the use of chiral 2,5-disubstituted piperazines **XXVIII** as organocatalysts to synthesize γ-nitroaldehydes (Scheme 5.28) [52]. The best results were obtained when butyraldehyde was the donor substrate, as this achieved the highest enantioselectivities.

Jacobsen and coworkers also used their thiourea–amine bifunctional catalysts to add α-branched aldehydes to nitroalkenes, with excellent results [53]. The remarkable stereoinduction was achieved by the simultaneous activation of both the

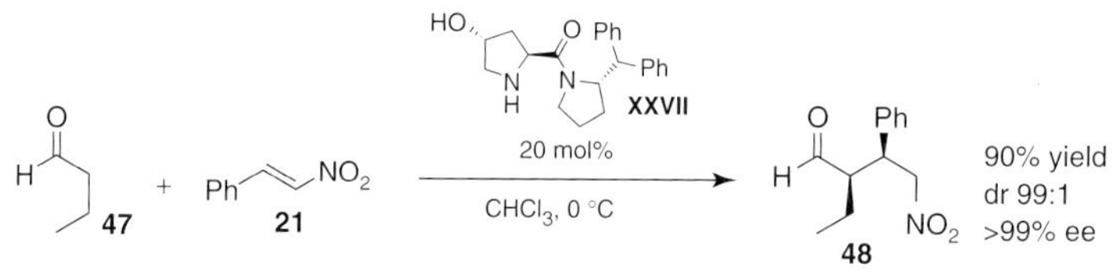

Scheme 5.27 Reaction as described by Palomo.

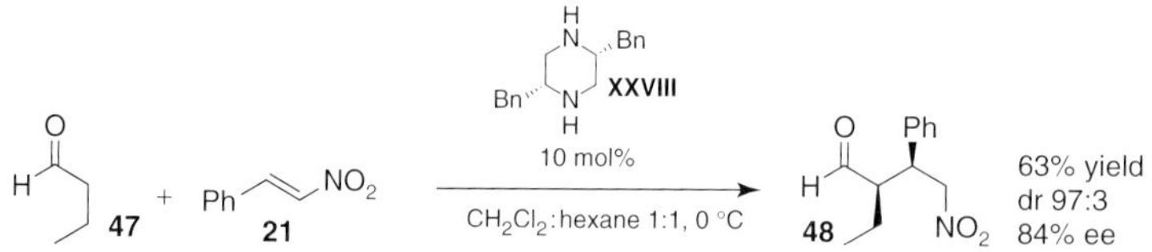

Scheme 5.28 Enantioselective conjugate addition, as reported by Barros.

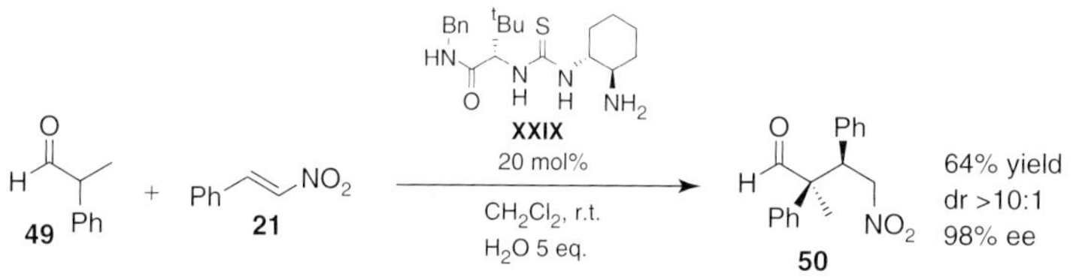

Scheme 5.29 Aldehyde addition to nitrostyrene catalyzed by chiral thioureas.

nucleophile and electrophile through covalent *E*-enamine catalysis and hydrogen bonding, respectively (Scheme 5.29).

A similar approach was developed in 2008 by Yan and coworkers, who described the use of chiral bifunctional sulfonamides (**XXX**) as suitable organocatalysts for the addition of aldehydes to nitroolefins [54]. The most important advantage of this methodology was that the catalysts bore a primary amine that allowed the use of α-branched aldehydes, and afforded the final adducts with excellent yields and enantioselectivities (Scheme 5.30). It is worthy of note here, that the addition of bases as additives enhanced both the rate and enantioselectivity of the reaction.

5.4.2
Aldehyde Conjugate Additions to Vinyl Sulfones

Very recently, several research groups have turned their attention to the enantioselective addition of aldehydes to vinyl sulfones, the latter being excellent Michael acceptors (see Section 5.3.3).

In 2005, Alexakis reported the first enantioselective addition of aldehydes to vinyl disulfones (1,1-disulfonylethenes), catalyzed by simple amines [55]. The reaction was efficiently catalyzed by amine **V**, and the 1,4 adducts were obtained with good yields and enantioselectivities. Remarkably, the more hindered the aldehyde, the

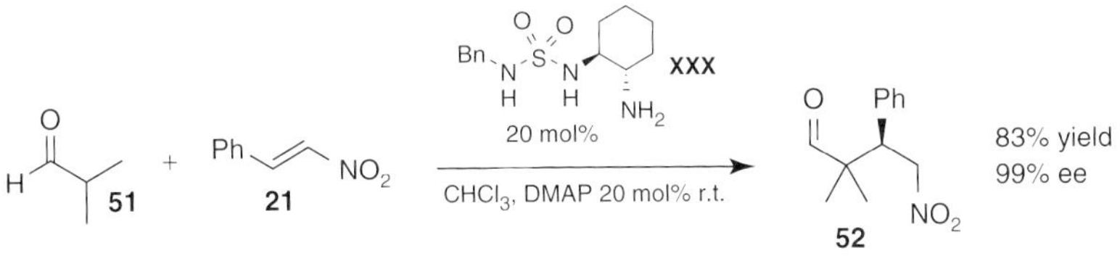

Scheme 5.30 α-Branched aldehyde additions to nitrostyrene.

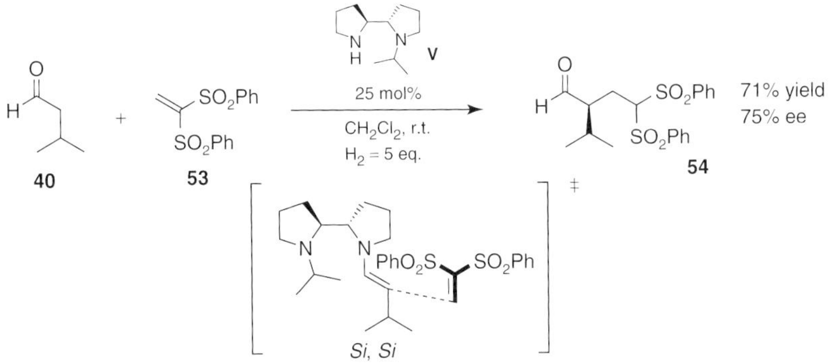

Scheme 5.31 Aldehyde addition to vinyl disulfones, as reported by Alexakis.

better the stereoinduction (Scheme 5.31). The absolute configuration of the final product could be accounted for by a *si, si* transition-state model, as shown previously for nitroolefins.

In 2008, Lu and coworkers reported an improved procedure using diphenylprolinol derivatives as catalysts [56]; in this case, the addition products were obtained in good yields and, essentially, in enantiopure form.

Soon afterwards, Palomo and coworkers described the addition of aldehydes to vinyl disulfones, E-α-ethoxycarbonyl vinyl sulfones (**57**), and E-α-cyano vinyl sulfones (**59**), promoted by diphenylprolinol derivative **XXVI** [57]. The reaction afforded the 1,4-adducts with good yields and enantioselectivities, and with moderate diastereoselectivities (Scheme 5.32).

5.4.3
Other Reactions

Jorgensen and coworkers described an aldehyde addition to enones, catalyzed by different pyrrolidine-derived compounds [58]. When proline was used as catalyst, only moderate selectivity was obtained, whereas unhindered prolinol derivatives such as **XXXI** led to good conversions and enantioselectivities in the addition of linear aldehydes to methyl vinyl ketone. However, cyclic enones afforded very poor results. Shortly after this, Gellman showed that a lower catalyst loading would result in a higher ee-value of the product, most likely by minimizing the racemization of the final Michael adduct (Scheme 5.33).

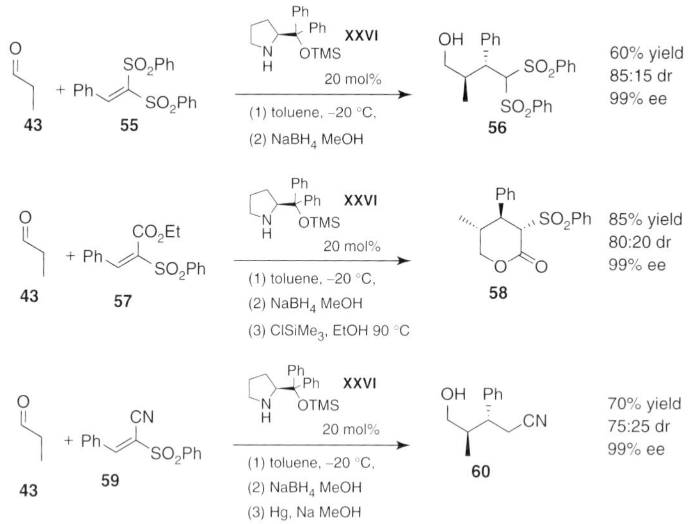

Scheme 5.32 Reactions as reported by Palomo.

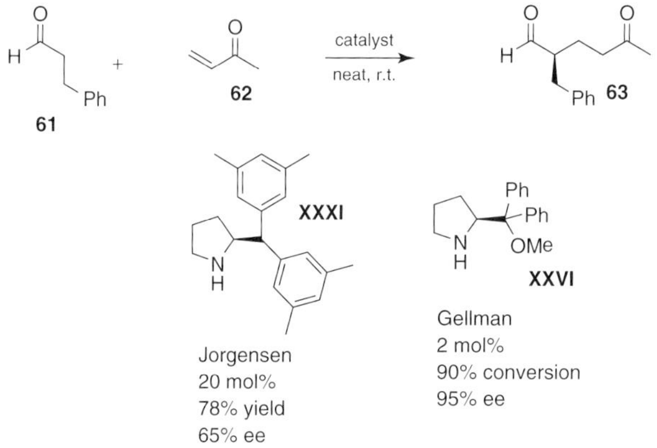

Scheme 5.33 Aldehyde addition to enones.

In 2009, Gellman developed a powerful aldehyde addition to enones catalyzed by imidazolidinones [59]. In this report, Gellman produced evidence for an enamine-based mechanism, and showed that the use of additives such as hydrogen bond donors dramatically enhanced the efficiency of the reaction.

In 2005, List and coworkers developed a clean and straightforward intramolecular addition of aldehydes to enones, catalyzed by MacMillan's organocatalysts **II** [60]. This reaction was highly *trans* selective in the formation of 2-substituted cyclopentanecarbaldehydes (**65**), and also tolerated aryl and alkyl substituents on the enone, as well as the presence of heteroatoms (Scheme 5.34).

Soon afterwards, the same group disclosed the combination of the domino 1,4-hydride reduction addition, followed by intramolecular Michael addition [61].

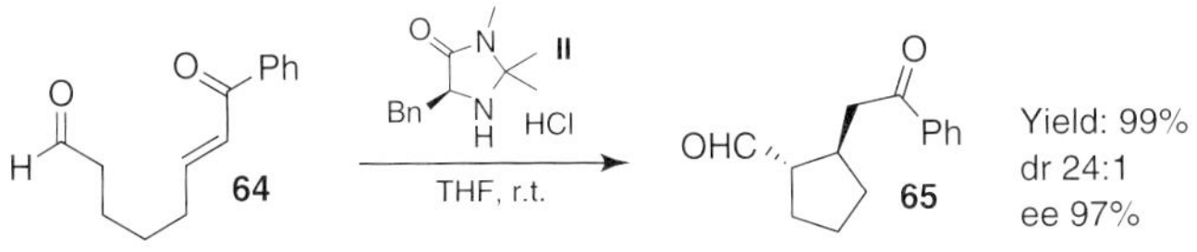

Scheme 5.34 Intramolecular reaction, as reported by List.

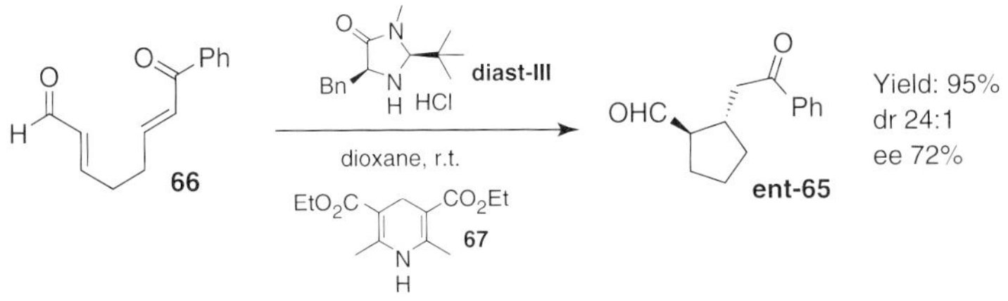

Scheme 5.35 Tandem reaction, as reported by List.

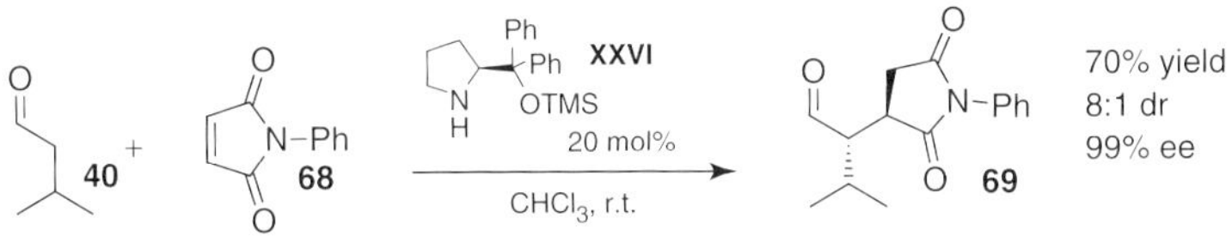

Scheme 5.36 Aldehyde addition to maleimides.

The reaction was shown to be totally chemoselective and to afford the final products in excellent yields and enantioselectivities (Scheme 5.35). Hayashi, soon after, reported a similar approach in which a cysteine-derived prolinamide analog was used as the catalyst [62].

Córdova and coworkers, in 2007, studied the addition of different aldehydes to maleimides, catalyzed by different amines as organocatalysts [63]. The best results were obtained with the diphenylprolinol derivative **XXVI**, which afforded the corresponding adducts in excellent yields and enantioselectivities (Scheme 5.36).

Palomo and coworkers reported the aldehyde addition to enals catalyzed by the same diphenylprolinol derivative **XXVI** [64]. This reaction bore an additional mechanistic interest, as it most likely involved a double imine-enamine activation process. In all cases, a very high diastereoselectivity in favor of the *anti* adduct and very high enantioselectivities were obtained with catalyst **XXVI**.

In 2008, Córdova and coworkers developed an elegant aldehyde addition to alkylidene malonates, catalyzed by the diphenylprolinol derivative **XXVI** [65]. The reaction furnished the corresponding 1,4 addition products in outstanding yields and enantioselectivities, and with good diastereoselectivities (Scheme 5.37). Remarkably, a wide range of aldehydes and alkylidene malonates could be used, without any perceptible loss of enantioselectivity.

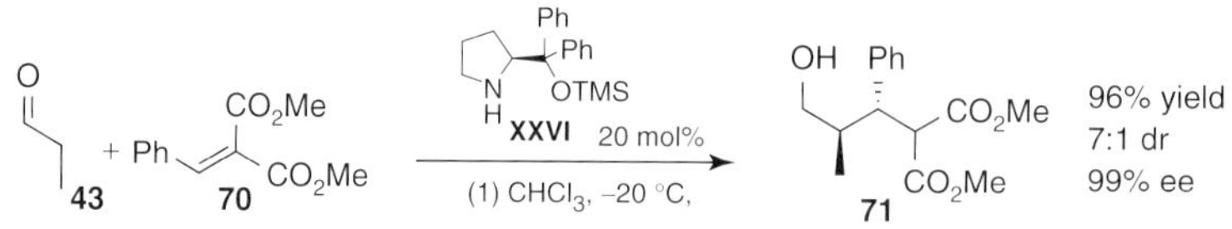

Scheme 5.37 Aldehyde addition to alkylidene malonates, as reported by Córdova.

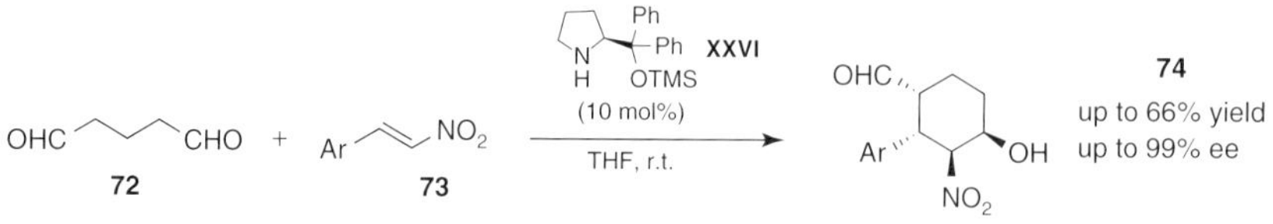

Scheme 5.38 Synthesis of cyclohexanes (**74**), as reported by Hayashi.

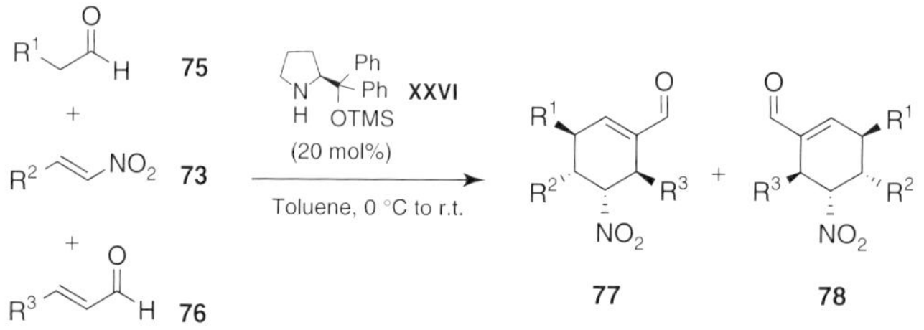

Scheme 5.39 Asymmetric organocatalytic triple cascade.

5.5
Tandem or Cascade Reactions

In 2007, Hayashi and coworkers disclosed an impressive tandem Michael/Henry reaction that builds chiral cyclohexanes (**74**) with total control of four stereocenters [66]. The reaction between 2,6-dihydroxy-tetrahydropyran (a hydrated form of pentanedial **72**) and a set of nitrostyrenes (**73**) was efficiently catalyzed by the trimethylsilyl (TMS)-protected diphenylprolinol derivative **XXVI**, rendering the chiral cyclohexanes (**74**) in high yields and enantioselectivities (Scheme 5.38).

Enders and coworkers developed an asymmetric organocatalytic triple cascade reaction for the construction of tetrasubstituted cyclohexene-1-carbaldehydes (**77**, **78**) (Scheme 5.39) [67]. These investigations paved the way for the sequential creation of three carbon–carbon bonds by a highly enantioselective combination of enamine–iminium–enamine catalysis in a triple cascade reaction.

This catalytic cascade is a three-component reaction comprising a linear aldehyde (**75**), a nitroalkene (**73**), an α,β-unsaturated aldehyde (**76**) and a simple chiral secondary amine (**XXVI**), which is capable of catalyzing each step of the triple

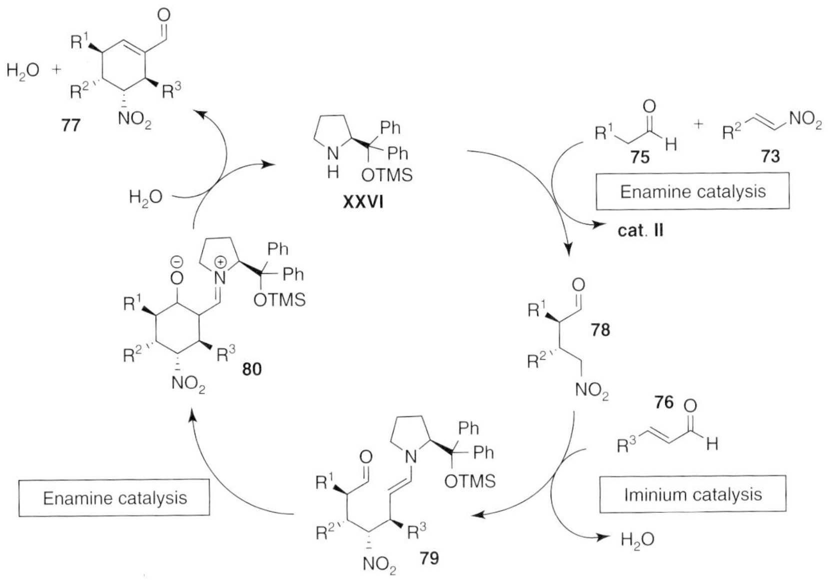

Scheme 5.40 Proposed catalytic cycle.

cascade. This multicomponent reaction proceeds through a Michael/Michael/aldol condensation sequence, affording four stereogenic centers generated in three consecutive carbon–carbon bond formations with high diastereoselectivities and complete enantioselectivities. Thus, of the final compound, only two epimers at the α-position of the nitro group are formed in a ratio ranking from 2 : 1 to 99 : 1, the minor isomer being easily separated using chromatography. Besides varying the starting materials, diverse polyfunctional cyclohexene derivatives can be obtained by employing a 1 : 1 : 1 ratio of the three substrates.

In the first step, the catalyst activates the linear aldehyde (**75**) by enamine formation to achieve a Michael-type addition to the nitroolefin (**73**). The catalyst is then liberated by hydrolysis, and is capable of forming the iminium ion with the enal (**76**) to catalyze the second conjugate addition of the nitroalkane. During this addition, a new enamine intermediate is formed which cyclizes through an intramolecular aldol condensation to afford cyclohexenes (**77, 78**) with moderate to good yields (30–58%) and complete enantioselectivity (≥99% ee; Scheme 5.40).

This multicomponent domino reaction was recently extended by means of a highly stereoselective intramolecular Diels–Alder reaction. The domino reaction is followed by a Lewis acid-mediated intramolecular [4 + 2] cycloaddition, which leads to complex tricyclic frameworks (**79, 80**) [68]. In this domino reaction, five C–C bonds are formed, with the diastereoselective and enantioselective construction of up to eight new stereogenic centers in a one-pot operation (Scheme 5.41).

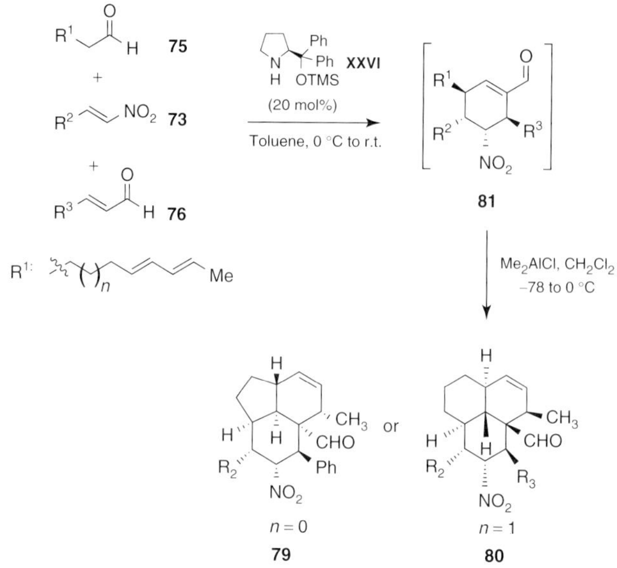

Scheme 5.41 The Michael/Michael/aldol condensation/Diels–Alder domino sequence.

5.6
Conclusions

Today, the enantioselective conjugate addition of enolizable carbonyl compounds to electron-deficient alkenes provides one of the most popular methods for obtaining α-alkyl carbonyls in nonracemic form. As seen above, several organocatalytic methods are available, using different Michael acceptors and aldehydes or ketones, that afford these privileged structures in exceptional yields and with high enantiomeric purities. This chapter has provided a summary of the most useful and selective procedures that allow this important transformation to be carried out. The versatility of the methods described, their efficiency, and the exceptional diversity of catalysts, lead to the conjugate addition via enamine catalysis being one of the best choices for the creation of C–C bonds with the concomitant formation of stereogenic centers. With research in this area continuing apace, each year leads to huge improvements being reported in the literature. From the seminal reports by List, in 2001, to the latest developments made by Palomo, in 2009, organic chemists have focused much effort on improving this reaction, their aim being to achieve exceptional levels of stereoselectivity and atom economy. As a result, the asymmetric organocatalytic Michael addition of carbonyl compounds today ranks among the most powerful tools in organic synthesis. Moreover, the recent disclosures of new tandem or cascade reactions, and the development of one-pot procedures for the formation of highly complex scaffolds with exquisite degrees of regiochemical and stereochemical control, have uncovered new perspectives that will continue to enlarge the future of this field.

5.7
Experimental

Ketone Addition to Nitrostyrenes (as reported by List [12])

A mixture of D- or L-proline (17 mg, 15 mol%), the nitroolefin (1 mmol) in 8 ml of DMSO and the ketone (10 mmol in the case of tetrahydro-thiopyran-4-one) was stirred at room temperature for 2–24 h. Ethyl acetate (10 ml) and saturated NH_4Cl solution (10 ml) were added sequentially and the aqueous layer was extracted with ethyl acetate (3 × 10 ml). The combined organic layers were dried ($MgSO_4$), filtered, and concentrated. Flash chromatography (hexanes/ethyl acetate mixtures) furnished the desired γ-nitroketones.

Asymmetric Michael Reaction of Aldehydes and Nitroalkenes (as reported by Hayashi [49])

Propanal (0.75 ml, 10 mmol) was added to a solution of nitrostyrene (154 mg, 1.0 mmol) and **XXVI** (34 mg, 0.1 mmol) in hexane (1.0 ml) at 0 °C. The mixture was stirred for 5 h at that temperature, and the reaction then quenched by the addition of aqueous 1 M HCl. The organic materials were extracted three times with AcOEt, and the combined organic phases dried (Na_2SO_4), concentrated, and purified by preparatory thin-layer chromatography (chloroform) to afford the Michael adduct (183 mg, 85%) as a clear oil : syn/anti 94 : 6 (by ^{1}H NMR spectroscopy of the crude mixture), 99% ee [by HPLC on a chiral phase: chiralcel OD-H column, $l = 254$ nm, 2-PrOH/hexane, 1 : 10, flow rate 1.0 ml min^{-1}; $R_t = 14.5$ min (minor), 19.7 min (major)].

Catalytic Conjugate Addition of Aldehydes to Vinyl Sulfones (as reported by Palomo [57])

To a mixture of the catalyst **XXVI** (10–20 mol%) and vinyl sulfone (1 mmol) in toluene (1 ml) or CH_2Cl_2 (1 ml) at −40 °C was added the aldehyde (1.5–3.0 equiv.), and the mixture stirred overnight (16 h) at the same temperature. The resulting solution was diluted with EtOH (1 ml) and a suspension of $NaBH_4$ (2 equiv.) in EtOH (2 ml) was then added drop-wise. The mixture was stirred at −40 °C for 30 min, and then quenched with H_2O (10 ml). The aqueous phase was extracted with CH_2Cl_2 (3 × 10 ml), and the combined organic layers were dried over $MgSO_4$, filtered, and concentrated under reduced pressure. The crude product was purified using flash column chromatography (hexane/EtOAc, 60 : 40).

General Procedure for the Synthesis of Cyclohexene Derivatives (as reported by Enders [67])

To a solution of catalyst (S)-**XXVI** (65 mg, 0.20 mmol) and nitroalkene B (1.00 mmol, 1.00 equiv.) in toluene (0.8 ml) was added subsequently under stirring aldehyde A

(1.20 mmol, 1.20 equiv.) and α,β-unsaturated aldehyde C (1.05 mmol, 1.05 equiv.) at 0 °C. After 1 h, the solution was allowed to reach room temperature, and stirred until complete conversion of the starting materials had occurred (16–24 h; monitored using gas chromatography). The reaction mixture was directly purified using flash column chromatography (SiO$_2$, ethyl acetate/pentane, 1 : 8 to 1 : 6) to afford the product.

References

1. (a) Hajos, Z.G. and Parrish D.R. (1971) DE 2102623; (b) Hajos, Z.J. and Parrish, D.R. (1974) *J. Org. Chem.*, **39**, 1615.
2. (a) Eder, U., Sauer, G., and Wiechert J. (1971) DE 2014757; (b) Eder, U., Sauer, G., and Wiechert, J. (1971) *Angew. Chem., Int. Ed. Engl.*, **10**, 496.
3. (a) Woodward, R.B., Logusch, E., Nambiar, K.P., Sakan, K., Ward, D.E., Au-Yeung, B.W., Balaram, P., Browne, L.J., Card, P.J., and Chen, C.H. (1981) *J. Am. Chem. Soc.*, **103**, 3210; (b) Paulmier, C., Outurquin, F., and Plaquevent, J.-C. (1988) *Tetrahedron Lett.*, **29**, 5889; (c) Kawara, A. and Taguxhi, T. (1994) *Tetrahedron Lett.*, **35**, 8805.
4. List, B., Lerner, R.A., and Barbas, C.F.III (2000) *J. Am. Chem. Soc.*, **122**, 2395.
5. Ahrendt, K.A., Borths, C.J., and MacMillan, D.W.C. (2000) *J. Am. Chem. Soc.*, **122**, 4243.
6. For authoritative reviews on the topics of AA and of AO, see: (a) Dalko, P.I. and Moisan, L. (2001) *Angew. Chem., Int. Ed.*, **40**, 3726; (b) List, B. (2001) *Synlett*, 1675; (c) List, B. (2002) *Tetrahedron*, **58**, 2481; (d) Dalko, P.I. and Moisan, L. (2004) *Angew. Chem., Int. Ed.*, **43**, 5138; (e) Berkessel, A. and Gröger H. (eds) (2005) *Asymmetric Organocatalysis: From Biomimetic Concepts to Applications in Asymmetric Synthesis*, Wiley-VCH Verlag GmbH, Weinheim; (f) Seayad, J. and List, B. (2005) *Org. Biomol. Chem.*, **3**, 719; (g) Lelais, G. and MacMillan, D.W.C. (2006) *Aldrichim. Acta*, **39**, 79; (h) Bressy, C. and Dalko, P.I. (2007) *Enantioselective Organocatalysis*, Wiley-VCH Verlag GmbH, Weinheim; (i) Mosse, S. and Alexakis, A. (2007) *Chem. Commun.*, 3123; (j) Gruttadauria, M., Giacalone, F., and Noto, R. (2008) *Chem. Rev.*, **37**, 1666; (k) Bertelsen, S. and Jorgensen, K.A. (2009) *Chem. Soc. Rev.*, **38**, 2178; For an authoritative review on asymmetric organocatalytic conjugate additions, see: (l) Almasi, D., Alonso, D.A., and Nájera, C. (2007) *Tetrahedron: Asymmetry*, **18**, 299.
7. (a) Noyori, R. (1994) *Asymmetric Catalysis in Organic Synthesis*, John Wiley & Sons, Inc., New York; (b) Jacobsen, E.N., Pfaltz, A., and Yamamoto, H. (eds) (1999) *Comprehensive Asymmetric Catalysis*, Springer, Berlin; (c) Ojima, I. (ed.) (2000) *Catalytic Asymmetric Synthesis*, 2nd edn, John Wiley & Sons, Inc., New York; (d) Beller, M. and Bolm, C. (eds) (2004) *Transition Metals for Organic Synthesis*, 2nd edn, Wiley-VCH Verlag GmbH, Weinheim.
8. (a) Roberts, S.M. (ed.) (1999) *Biocatalysts for Fine Chemicals Synthesis*, John Wiley & Sons, Inc., New York; (b) Drauz, K. and Waldmann, H. (eds) (2002) *Enzyme Catalysis in Organic Synthesis*, 2nd edn, Wiley-VCH Verlag GmbH, Weinheim; (c) Bommarius, A.S. and Riebel, B.R. (2004) *Biocatalysis*, Wiley-VCH Verlag GmbH, Weinheim.
9. For an important highlight of SOMO catalysis: Melchiorre, P. (2009) *Angew. Chem., Int. Ed.*, **48**, 1360, and references therein.
10. Stork, G., Brizzolara, A., Landesman, H., Szmuszkovicz, J., and Terrell, R. (1963) *J. Am. Chem. Soc.*, **85**, 207.
11. Seebach, D. and Golinski, J. (1981) *Helv. Chim. Acta*, **64**, 1413.
12. List, B., Pojarliev, P., and Martin, H.J. (2001) *Org. Lett.*, **3**, 2423.
13. (a) Bui, T. and Barbas, C.F. III (2000) *Tetrahedron Lett.*, **41**, 6951; (b) Betancort, J.M., Sakthivel, K.,

Thayumanavan, R., and Barbas, C.F. III (2001) *Tetrahedron Lett.*, **42**, 4441.

14. Enders, D. and Seki, A. (2002) *Synlett*, 26.
15. Mase, N., Watanabe, K., Yoda, K., Tanaka, F., and Barbas, C.F.III (2006) *J. Am. Chem. Soc.*, **128**, 4966.
16. (a) Alexakis, A. and Andrey, O. (2002) *Org. Lett.*, **4**, 3611; (b) Andrey, O., Alexakis, A., Tomasini, A., and Bernardinelli, G. (2004) *Adv. Synth. Catal.*, **346**, 1147.
17. (a) Bolm, C., Rantanen, T., Schiffers, I., and Zani, L. (2005) *Angew. Chem., Int. Ed.*, **44**, 1758; (b) Pikho, P.M., Laurikanen, K.M., Usano, A., Nyberg, A.I., and Kaavi, J.A. (2006) *Tetrahedron*, **62**, 317; (c) Pikho, P.M. (2005) *Lett. Org. Chem.*, **2**, 398.
18. (a) Andrey, O., Alexakis, A., and Bernardelli, G. (2003) *Org. Lett.*, **5**, 2559; (b) Andrey, O., Vidonne, A., and Alexakis, A. (2003) *Tetrahedron Lett.*, **44**, 7901.
19. Betancort, J.M., Sakthivel, K., Thayumanavan, R., Tanaka, F., and Barbas, C.F.III (2004) *Synthesis*, 1509.
20. Mase, N., Watanabe, K., Yoda, H., Tanabe, K., Tanaka, F., and Barbas, C.F.III (2006) *J. Am. Chem. Soc.*, **128**, 4966.
21. Ishiii, T., Fujioka, S., Sekiguchi, Y., and Kotsuki, H. (2004) *J. Am. Chem. Soc.*, **126**, 4966.
22. (a) Cobb, A.J.A., Shaw, D.M., and Ley, S.V. (2004) *Synlett*, 558; (b) Torii, H., Nakadai, M., Ishihara, K., Saito, S., and Yamamoto, H. (2004) *Angew. Chem., Int. Ed.*, **43**, 1983; (c) Hartikka, A. and Arvidsson, P.I. (2004) *Tetrahedron: Asymmetry*, **15**, 1831.
23. (a) Cobb, A.J.A., Longbottom, D.A., Shaw, D.M., and Ley, S.V. (2004) *Chem. Commun.*, 1808; (b) Mitchell, C.T., Cobb, A.J.A., and Ley, S.V. (2005) *Synlett*, 611.
24. Pansare, S.V. and Pandya, K. (2006) *J. Am. Chem. Soc.*, **128**, 9624.
25. Wang, J., Li, H., Lou, B., Zu, L., Guo, H., and Wang, W. (2006) *Chem. Eur. J.*, **12**, 4321.
26. Zu, L., Wang, J., Li, H., and Wang, W. (2006) *Org. Lett.*, **8**, 3077.
27. Mandal, T. and Zhao, C.-G. (2008) *Angew. Chem., Int. Ed.*, **47**, 7714.
28. For reviews see: (a) Gennari, C. and Piarulli, U. (2003) *Chem. Rev.*, **103**, 3071; (b) Reetz, M.T. (2001) *Angew. Chem., Int. Ed.*, **40**, 284; (c) Breit, B. (2005) *Angew. Chem., Int. Ed.*, **44**, 6816.
29. Clarke, M.L. and Fuentes, J.A. (2007) *Angew. Chem., Int. Ed.*, **46**, 930.
30. Kelly, T.R., Bridger, G.J., and Zhao, C. (1990) *J. Am. Chem. Soc.*, **112**, 8024.
31. (a) Luo, S., Mi, X., Liu, S., Xu, H., and Cheng, J.P. (2006) *Angew. Chem., Int. Ed.*, **45**, 3093; (b) Luo, S., Xu, H., Mi, X., Li, J., Zheng, X., and Cheng, P.-P. (2006) *J. Org. Chem.*, **71**, 9244.
32. Zhu, M.-K., Cun, L.-F., Mi, A.-Q., Jiang, Y.-Z., and Gong, L.-Z. (2006) *Tetrahedron: Asymmetry*, **17**, 491.
33. Benaglia, M., Cinquini, M., Cozzi, F., Puglisi, A., and Calentano, G. (2003) *J. Mol. Catal. A: Chem.*, **204-205**, 157.
34. (a) Enders, D. and Chow, S. (2006) *Eur. J. Org. Chem.*, 4578; see also: (b) Ibrahem, I., Zao, W., Xu, Y., and Cordova, A. (2006) *Adv. Synth. Catal.*, **348**, 211.
35. Martin, H.J. and List, B. (2003) *Synlett*, 1901.
36. Xu, Y., Zou, W., Sunden, H., Ibrahem, I., and Cordova, A. (2006) *Adv. Synth. Catal.*, **348**, 418.
37. Xu, Y. and Cordova, A. (2006) *Chem. Commun.*, 460.
38. Huang, H. and Jacobsen, E.N. (2006) *J. Am. Chem. Soc.*, **128**, 7170.
39. Tsogoeva, S.B. and Wei, S. (2006) *Chem. Commun.*, 1451.
40. Cao, C.-L., Ye, M.-C., Sun, X.-L., and Tang, Y. (2006) *Org. Lett.*, **8**, 2559.
41. Wang, W., Li, H., Zu, L., and Wang, W. (2006) *Adv. Synth. Catal.*, **348**, 425.
42. Cao, C.-L., Sun, X.-L., Zhou, J.-L., and Tang, Y. (2007) *J. Org. Chem.*, **72**, 4073.
43. Barros, M.T. and Faisca Phillips, A.M. (2008) *Eur. J. Org. Chem.*, 2525.
44. Zhu, Q., Cheng, L., and Lu, Y. (2008) *Chem. Commun.*, 6315.
45. Betancort, J.M. and Barbas, C.F.III (2001) *Org. Lett.*, **3**, 3737.
46. Masse, N., Thayumanavan, R., Tanaka, F., and Barbas, C.F.III (2004) *Org. Lett.*, **6**, 2527.

47. Andrey, O., Alexakis, A., Tomassini, A., and Bernardinelli, G. (2004) *Adv. Synth. Catal.*, **346**, 1147.
48. (a) Wang, J., Li, H., Lou, B., Zu, L., Guo, H., and Wang, W. (2006) *Chem. Eur. J.*, **12**, 4321; (b) Zu, L., Wang, J., Li, H., and Wang, W. (2006) *Org. Lett.*, **8**, 3077.
49. Hayashi, Y., Gotoh, H., Hayashi, T., and Shoji, M., (2006) *Angew. Chem., Int. Ed.*, **44**, 4212.
50. Zu, L., Li, H., Wang, J., Xu, X., and Wang, W. (2006) *Tetrahedron Lett.*, **47**, 5131.
51. Palomo, C., Vera, S., Mielgo, A., and Gomez-Bengoa, E. (2005) *Angew. Chem., Int. Ed.*, **25**, 5984.
52. Barros, M.T. and Phillips, A.M.F. (2007) *Eur. J. Org. Chem.*, 178.
53. Lalonde, M.P., Chen, Y., and Jacobsen, E.N. (2005) *Angew. Chem., Int. Ed.*, **45**, 6366.
54. Zhang, X.-J., Liu, S.-P., Li, X.-M., Yan, M., and Chan, A.S.C. (2009) *Chem. Commun.*, 833.
55. (a) Mosse, S. and Alexakis, A. (2005) *Org. Lett.*, **7**, 4361; (b) Mosse, S., Andrey, O., and Alexakis, A. (2006) *Chimia*, **60**, 216.
56. Zhu, Q. and Lu, Y. (2008) *Org. Lett.*, **10**, 4083.
57. Landa, A., Maestro, M., Masdeu, C., Puente, A., Vera, S., Oiarbide, M., and Palomo, C. (2009) *Chem. Eur. J.*, **7**, 1562.
58. Melchiorre, P. and Jorgensen, K.A. (2003) *J. Org. Chem.*, **68**, 4151.
59. Chi, Y. and Gellman, S.H. (2005) *Org. Lett.*, **7**, 4253.
60. Hechavarria Fonseca, M.T. and List, B. (2004) *Angew. Chem., Int. Ed.*, **43**, 3958.
61. Yang, J.W., Hechavarria Fonseca, M.T., and List, B. (2005) *J. Am. Chem. Soc.*, **127**, 15036.
62. Hayashi, Y., Gotoh, H., Tamura, T., Yamaguchi, H., Masui, R., and Shoji, M. (2005) *J. Am. Chem. Soc.*, **127**, 16028.
63. Zhao, G.-L., Xu, Y., Sundén, H., Eriksson, L., Sayah, M., and Córdova, A. (2007) *Chem. Commun.*, 734.
64. Palomo, C., Landa, A., Mielgo, A., Oiarbide, M., Puente, A., and Vera, S. (2007) *Angew. Chem., Int. Ed.*, **46**, 8431.
65. Zhao, G.-L., Vesely, J., Sun, J., Christensen, K.E., Bonneau, C., and Cordova, A. (2008) *Adv. Synth. Catal.*, **350**, 657.
66. Hayashi, Y., Okano, T., Aratake, S., and Hazelard, D. (2007) *Angew. Chem., Int. Ed.*, **46**, 4922.
67. Enders, D., Huttl, M.R.M., Grondal, C., and Raabe, G. (2006) *Nature*, **441**, 861.
68. Enders, D., Hüttl, M.R.M., Runsink, J., Raabe, G., and Wendt, B. (2007) *Angew. Chem., Int. Ed.*, **46**, 467.

6
Iminium Activation in Catalytic Enantioselective Conjugate Additions
Jose L. Vicario, Efraim Reyes, Dolores Badía, and Luisa Carrillo

6.1
Introduction

Since the seminal studies of List and Barbas in 2000, which described the proline-catalyzed enantioselective aldol reaction between acetone and aromatic aldehydes [1], the use of chiral amines as catalysts in enantioselective transformations has experienced an impressive growth, with a plethora of new reactions and transformations developed by many different research groups [2]. This concept has since become a very attractive tool by which synthetic chemists can produce highly enantioenriched molecules with a remarkable high level of complexity, in a very simple and efficient manner. In fact, one of the main advantages of this methodology is related to the operational simplicity of the laboratory operations that must be carried out when performing an amine-catalyzed reaction. This is due mainly to the fact that chiral secondary amines are fairly robust compounds and, even more important, that the intermediates participating in the catalytic cycle of an amine-catalyzed reaction are stable towards oxygen and moisture, which in turn allows the reaction to be carried out without a need for dry solvents or an inert atmosphere. In addition, most of the chiral amines employed as catalysts are commercially available compounds, or can be easily synthesized from low-cost starting materials. For all these reasons, several of the chiral amine-catalyzed methodologies developed to date can eventually compete with, or in many cases may even supersede, the corresponding metal-mediated reactions that are typically employed when carrying out the same or closely related transformations.

In this context, the conjugate addition of nucleophiles to unsaturated carbonyl compounds is an exceptional playground for the development of novel amine-catalyzed enantioselective transformations [3]. Chiral secondary amines are able to activate an α,β-unsaturated aldehyde or ketone towards a conjugate addition process via the reversible formation of an intermediate iminium ion upon condensation (*vide infra*), with the chiral information present at the amine catalyst being responsible for exerting the required stereocontrol. As a result of this, a wide number of different methodologies have been reported during the past years, describing this interesting transformation, which is one of the most important reactions available in the organic chemistry toolbox for the asymmetric formation of C–C and C–heteroatom bonds. Moreover, the inherent mechanistic profile of the conjugate addition opens the way for the development of new cascade processes initiated by this reaction, increasing even more the synthetic potential of this transformation, with attention focused on acquiring compounds of high molecular complexity, starting from simple starting materials [4]. On the other hand, the need for a condensation reaction between the catalyst and the substrate for the activation of the latter restricts the application of this methodology to conjugate acceptors capable of forming iminium ions. Unfortunately, this is a clear limitation of the approach when compared to the related metal-mediated reactions.

In this chapter, the state of the art of this methodological approach will be presented, with attention focused on the conjugate addition of stabilized carbon nucleophiles (the Michael reaction). First, the concept of activation via iminium ion formation will be introduced, together with the regiochemical and stereochemical issues that must be addressed when applying this methodology. A summary of the many developments in the area will then be presented, along with discussions of the mechanisms and selectivity models, including some illustrative applications in total synthesis. The final section is devoted to the cascade processes initiated by amine-catalyzed Michael reactions, in which this activation concept has been applied.

6.2
Mechanistic Aspects of the Iminium Activation Concept, and Factors Influencing Stereocontrol in Michael Additions

As noted above, the mechanism operating in the activation of an α,β-unsaturated aldehyde or ketone by a chiral secondary amine toward a conjugate addition process relies mainly on the reversible formation of an iminium ion by condensation. This results in a significant lowering of the energy of the lowest unoccupied molecular orbital (LUMO) orbital of the conjugate acceptor, which opens the way for the nucleophilic addition reaction, delivering an enamine intermediate which, upon hydrolysis, releases the product and the chiral amine catalyst that is ready to participate in a subsequent catalytic cycle (Scheme 6.1).

Some initial investigations in this field were made as early as 1937, when Langenbeck and coworkers reported the conjugate addition of water to crotonaldehyde, catalyzed by piperidine or sarcosine [5]. In 1981, Woodward reported a D-proline-catalyzed deracemization in the total synthesis of erythromycin that proceeded via an iminium-mediated cascade process involving a retro-sulfa–Michael/sulfa–Michael/intramolecular aldol reaction sequence (Scheme 6.2) [6]. Between 1991 and 1993, Yamaguchi reported the use of alkaline prolinate salts such as 2 in the asymmetric Michael reaction between malonates and enones and enals [7], and the first example involving a purely organic aminocatalyst was introduced in 1994 by Kawara and Taguchi, who

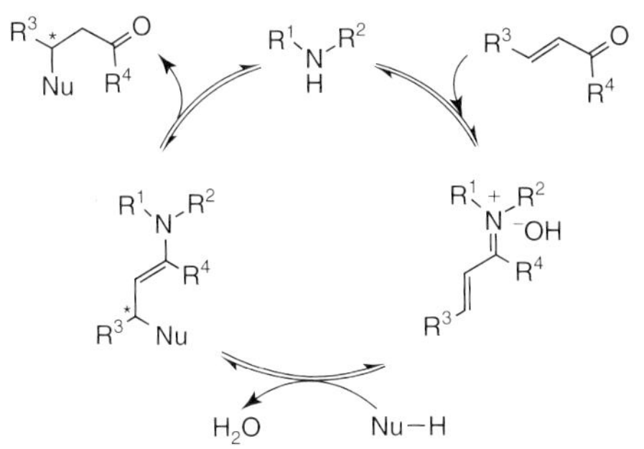

Scheme 6.1 The catalytic cycle proposed for a Michael reaction catalyzed by a secondary amine via iminium ion formation.

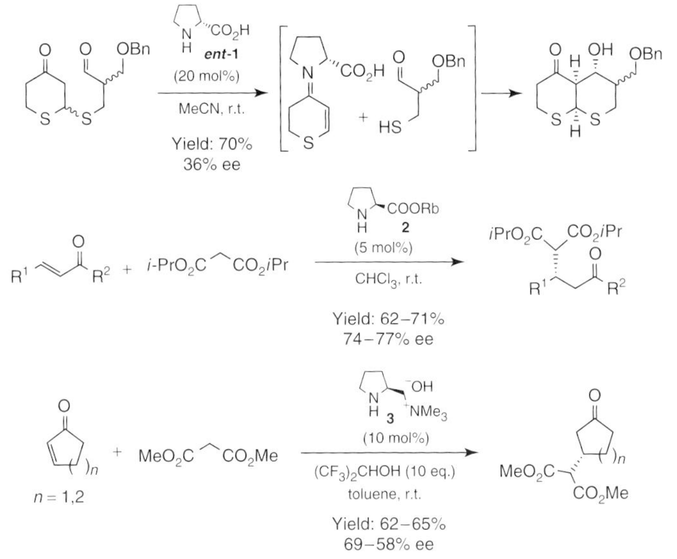

Scheme 6.2 The first iminium-mediated conjugate additions reported.

described the enantioselective Michael addition of malonates to enones catalyzed by (2-pyrrolidinylmethyl)trimethylammonium hydroxide **3** [8].

Nevertheless, the rational development and application of this concept should be attributed to MacMillan, starting from his seminal report on amine-catalyzed enantioselective Diels–Alder cycloadditions reported in 2000 (Scheme 6.3) [9]. In this report, MacMillan outlined a potential analogy between amine-catalyzed and the parent Lewis acid-mediated reactions under the hypothesis that, in the same way a Lewis acid activates an electron-deficient alkene by lowering the energy of its LUMO after complexation, the formation of an iminium ion should result in the same type of electronic activation towards nucleophilic addition. On the other hand, MacMillan also indicated that the dynamic processes involving complexation/decomplexation which allowed catalyst turnover in the Lewis acid-mediated transformations would also find a counterpart in the reversibility of the reaction operating in the formation of iminium ions from amines and

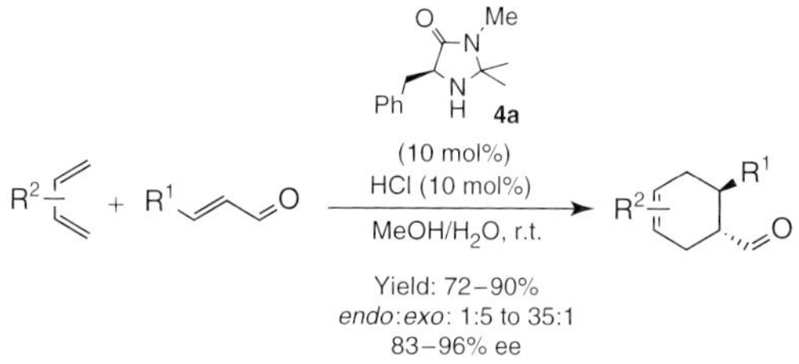

Scheme 6.3 The enantioselective Diels–Alder reaction catalyzed by chiral imidazolidinone **4a**, as developed by MacMillan.

aldehydes or ketones. As pointed out by MacMillan himself [10], the basis for this LUMO-lowering effect was taken from several experimental evidences, such as the well-known reductive amination reaction, in which the hydride-mediated reduction of an intermediate iminium ion present in equilibrium with the mixture of the aldehyde and amine starting materials takes place much faster than the reduction of the aldehyde. Other evidences supporting this effect were found on the higher reactivity displayed by α,β-unsaturated iminium ions in Diels–Alder cycloadditions compared to other unsaturated carbonyl compounds [11], and also on the fast inverse electron-demand hetero-Diels–Alder reaction observed when iminium ions were employed as dienophiles [12]. The combination of this background knowledge provided by these transformations in which amines were employed in stoichiometric amounts, together with the incorporation of an intermediate step allowing the turnover of the amine reagent (the reversibility of the condensation reaction of aldehydes with amines), led to the first example of this iminium activation concept in the shape of a very efficient enantioselective Diels–Alder reaction between enals and dienes catalyzed by chiral imidazolidinone **4a**. After this pioneering example, the concept was extended to other transformations involving α,β-unsaturated aldehydes and/or ketones, including the Michael reaction.

The covalent nature of the interaction between the catalyst and the substrate and the need for reversible reactions for the attachment and removal of the catalyst, raise the issue of the catalytic activity and low turnover frequency (TOF) numbers usually observed. The rate of the dynamic processes involved (the formation of the iminium ions and hydrolysis of the enamine intermediate formed after the conjugate addition step) very often determine the viability of the catalytic process. This is the main reason why most of the amine catalysts employed are cyclic secondary amines; they are generally based on a pyrrolidine structure, and most are derived from proline, due to its easy availability. Nevertheless, the rather long reaction times usually required to reach full conversion is a common feature shared by most of the amine-catalyzed reactions reported to date. In the past, one way to circumvent these problems has been to incorporate additives that increase the rates of these processes. Indeed, it is very often found that acid co-catalysts are included in the reaction design in order to help in the formation of the iminium ion intermediate. In other cases, water may be incorporated in variable amounts so as to assist in the final hydrolysis step.

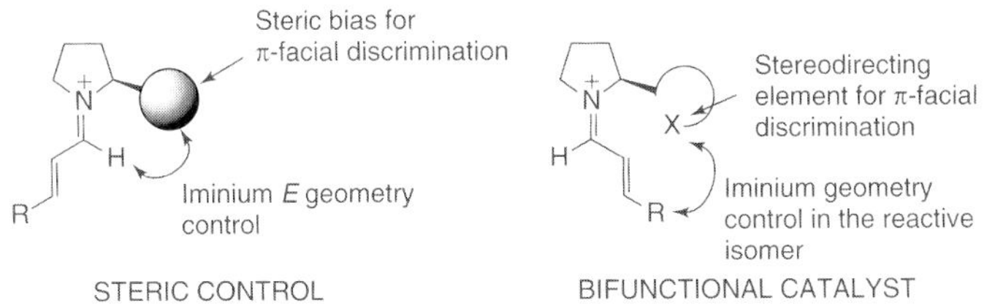

Figure 6.1 Stereochemical issues to be considered in the design of an amine catalyst.

With regards to stereochemical control, two main issues must be taken into account when designing a chiral amine catalyst useful in enantioselective reactions proceeding via iminium activation. First, the catalyst must be able to clearly differentiate between both stereotopic faces of the Michael acceptor. This stereodiscrimination can be carried out by steric bias or, alternatively, by introducing a convenient stereodirecting element (a hydrogen bond donor, a basic site, or a positively or negatively charged substituent) which interacts with the nucleophile and therefore directs its attack from a preferential position (a bifunctional catalyst). The second aspect to be considered is that the catalyst must exert an efficient geometry control on the iminium ion, because mixtures of Z/E isomers would turn into low enantioselectivity on the reaction. This is not a problem when bifunctional catalysts are employed because the reactive conformation of the iminium ion during the C–C bond-formation step is clearly determined by the stereodirecting element incorporated at the catalyst. However, it becomes crucial in the design of a catalyst exerting stereodifferentiation by steric bias, because the volume of the substituent at the pyrrolidine ring must be large enough not only to exert the required high level of π-facial discrimination but also to determine the preferred E geometry of the iminium ion (Figure 6.1).

It must be pointed out that control of the geometry of the iminium ion is rather easy in the case where α,β-unsaturated aldehydes are used as Michael acceptors. However, when enones must be employed the similarity between both substituents of the carbonyl group makes this element much more difficult to control (Figure 6.2). Indeed, this problem adds to the low catalytic activity demonstrated by secondary amines when employing ketones as Michael donors, and is mainly derived from the inherent difficulties associated with the formation

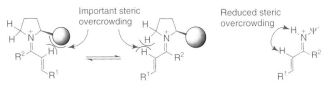

SECONDARY AMINE CATALYST
- Difficult iminium geometry control
- Unfavorable formation of iminium ion

PRIMARY AMINE CATALYST
- Easier iminium geometry control
- Favorable formation of iminium ion

Figure 6.2 Primary versus secondary amine catalysts in the iminium activation of enones.

of such sterically congested iminium ion intermediates. A solution to this problem has been provided recently with the discovery that primary amines can operate as highly active and efficient aminocatalysts in several Michael transformations [13]. The condensation reaction between an enone and a primary amine is sterically more feasible, delivering the corresponding imine which, due to its high basicity, will exist in a protonated form under acidic conditions. In addition, the geometry control on this intermediate is easier because of the higher differences in steric bulk between the coplanar substituents across the C=N bond.

Finally, there is an additional possibility of exerting stereocontrol in conjugate additions that proceed via iminium activation. This has been developed recently, and is based on the tight interaction that exists between the iminium cation and the corresponding counteranion. As a result, it has been shown that reactions can be carried out with excellent levels of stereoselection, simply by using an achiral amine catalyst (e.g., pyrrolidine or morpholine) and incorporating a chiral acid as an additive in substoichiometric amounts (a chiral phosphoric acid or an α-amino acid).

6.3
Michael Reactions

A variety of stabilized carbon nucleophiles has been studied in the amine-catalyzed conjugate addition to α,β-unsaturated aldehydes and ketones. As a general rule, the inherent characteristics of the reaction exclude the use of previously generated metal enolates, due to incompatibilities with other reactive species participating in the reaction, such as the amine catalyst or water. For this reason, the applicability of the iminum activation methodology to Michael reactions has been generally limited to the use of highly acidic compounds such as 1,3-dicarbonyl compounds and nitroalkanes, which can undergo deprotonation under the very mild reaction conditions employed. In many cases, it is the amine catalyst itself which is involved in the deprotonation of the nucleophile source.

6.3.1
1,3-Dicarbonyl Compounds as Nucleophiles

As noted above, one of the first applications of the iminium activation concept in conjugate additions was the Michael reaction of malonates to enones catalyzed by prolinate salts, as described by Yamaguchi (see Scheme 6.2). After a preliminary report in which lithium prolinate was shown to be a very effective catalyst for the conjugate addition of dimethylmalonate to several α,β-unsaturated aldehydes (no enantioselectivities reported), rubidium prolinate **2** was subsequently identified as an efficient promoter of the Michael reaction between diisopropyl malonate and α, β-unsaturated ketones. In this case, both the cyclic and acyclic enones furnished good yields and moderate enantioselectivities. The stereoselection was found to depend heavily on the structure of the Michael acceptor, with the aliphatic enone

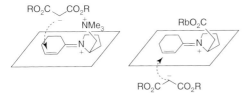

Figure 6.3 Origins of the opposite facial selectivity in the Michael reaction of malonates and cyclohexenone catalyzed by proline-derived tetra-alkylammonium salt **3** and rubidium prolinate (**2**).

substrates furnishing the best results. The Michael addition of dibenzyl malonate to cyclic enones and to benzylideneacetone using proline-based ammonium salt **3**, as reported by Kawara and Taguchi in 1994 (see Scheme 6.2), provided similar results in terms of yields and enantioselectivities (up to 71% ee for the addition to cyclohexenone, 58% ee in the addition to cyclopentenone, and 68% ee when acyclic benzylideneacetone was employed as Michael acceptor). The absolute configuration of the Michael adducts was found to be the opposite of that found for the same rubidium prolinate-catalyzed reaction. Moreover, taking into account the fact that both catalysts had similar structures and configurations, the authors provided a tentative model for the stereochemical outcome of the reaction that involved an ability of the ammonium substituent to direct the attack of the malonate enolate via electrostatic interactions (Figure 6.3). Consequently, in the reaction mediated by **2**, the role played by the carboxylate substituent should be limited to a merely steric effect.

Following these early examples [14], in 2003 Jørgensen developed a new type of chiral secondary amine with an imidazolidine structure (**5a**) for the conjugate addition of malonates to acyclic enones (Scheme 6.4) [15]. Although the reaction proceeded well in terms of yields and enantioselectivities for a wide range of differently substituted arylideneacetophenones and for cyclohexenone, the yields tended to decrease when more bulky substituents were placed around the carbonyl moiety. In addition, the enantioselectivity was seen to depend heavily on the nature of the malonate reagent, with dibenzyl malonate and diethylmalonate giving the best results. The most important drawbacks associated with this protocol were the need for a large excess of the malonate nucleophile (it was used as the reaction solvent), and the rather long reaction times required to reach to good conversions (typically 7–10 days). Alternatively, the group of Ley developed a newly designed catalyst for this transformation, with their efforts being directed towards a scalable procedure with the target being the preferable use of the more accessible and low-cost diethyl and dimethyl malonates as Michael donors, and also avoiding the use of such large amounts of nucleophile. In this context, the proline tetrazole catalyst **6** was shown to be an excellent catalyst for this transformation, furnishing excellent yields and enantioselectivities while using only a slight excess of diethylmalonate reagent, and in a three-day reaction time [16]. In this case, the most important drawbacks associated with this methodology were related to the more limited reaction scope

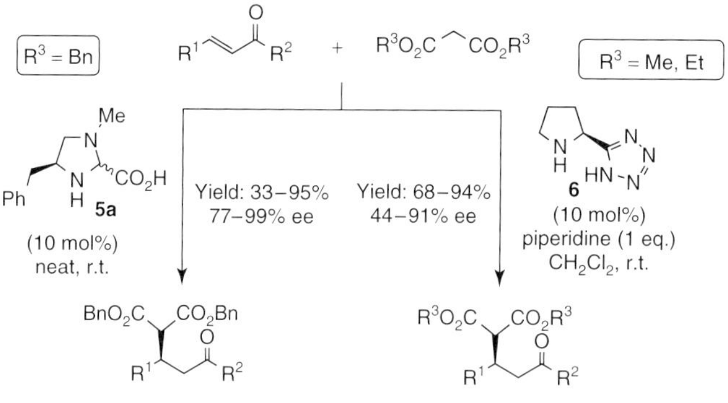

Scheme 6.4 The enantioselective Michael addition of malonates to enones, as reported by Jørgensen and Ley.

(while excellent yields and enantioselectivities were obtained for cyclohexenone and arylideneacetones, poorer results were achieved with aliphatic enones), and to the need for a large amount of piperidine as additive in order to activate the nucleophile. The dependence of yield on the alkoxide substituents of the malonate reagent, as observed by both Jørgensen and Ley, might be attributed to the different acidities of the nucleophile required for the iminium-mediated Michael reaction to occur; this indicates, in turn, that a fine-tuning of this parameter is most likely needed for a successful reaction. In relation to this topic, Barbas reported in a related study [17] that there must exist a pK_a barrier for nucleophile activation in conjugate additions of carbon nucleophiles that proceed via iminium activation, and that this should lie between pK_a values of 16 and 17.

The conjugate addition of cyclic 1,3-dicarbonyl compounds to enones has received particular attention, due mainly to the fact that warfarin – one of the most widely employed anticoagulants – can be prepared directly by using this reaction (see Scheme 6.5). Jørgensen was the first to incorporate the enantioselective amine-catalyzed Michael reaction to the synthesis of warfarin, using a method which was also applicable to a wide range of different cyclic 1,3-dicarbonyl compounds. The method also involved the use of a modified imidazolidine catalyst (**5b**), which was structurally related to **5a** but had an improved performance in this case [18]. Although the ee-values remained at about 80%, the optical purity of the Michael adducts prepared could be further increased to >99% by a single crystallization.

The nature of active catalytic species involved in this reaction is not completely clear. Jørgensen has proposed the formation of a cyclic hemiaminal as the intermediate undergoing the conjugate addition reaction (Figure 6.4). The reasons for such a proposal relied on a PM3 calculation on the possible iminium species, which suggested a poor shielding of the alkene and led the authors to propose the alternative formation of this hemiaminal species that seemed more consistent with the excellent enantioselectivities attained. In a later report by Chin and coworkers, when the reaction was studied by employing different diarylethylenediamines as catalysts, it was observed that **7** provided similar results in the reaction as were

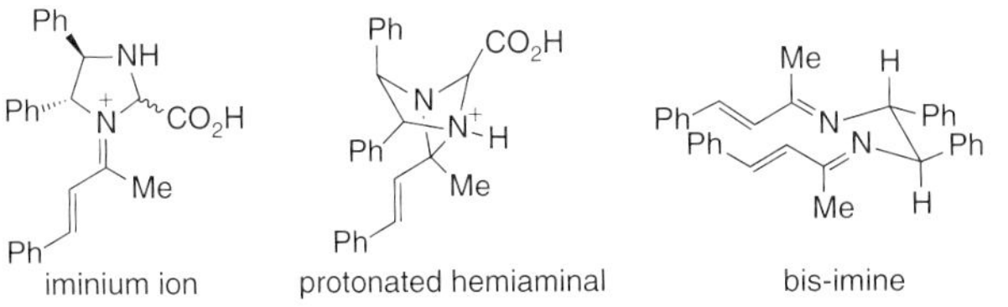

Figure 6.4 Proposed models for the **5b**-catalyzed (two on the left) and the **7**-catalyzed (right) Michael reaction of cyclic 1,3-dicarbonyl compounds to benzylideneacetone.

reported by Jørgensen when the diphenylethylenediamine-derived imidazolidine **5b** was used (Scheme 6.6) [19]. Consequently, an alternative intermediate was proposed that involved the direct formation of a bis-iminium ion intermediate by condensation of 2 equiv. of the enone with diphenylethylenediamine, which should be present in the reaction medium due to the inherent instability of catalyst **5b** towards the presence of water (Figure 6.4). It was also proposed that hydrolysis of the catalyst would generate glyoxylic acid, which might in turn function as a Brønsted acid co-catalyst in the formation of the aforementioned bis-iminium ion. This was in accordance with the observed increase in enantioselectivity for the **7**-catalyzed reaction in the presence of AcOH as additive.

Other different chiral primary amines have been tested in this reaction with different results (Figure 6.5). In general, these are bifunctional catalysts containing the primary amine functionality connected to a basic site by a chiral scaffold, such as **8a** [20] or **9** [21]. This additional basic site is required to assist in the deprotonation of the malonate reagent, in order to increase the concentration of the nucleophile species. In other cases, such as **10a** [22], a fragment containing hydrogen donors

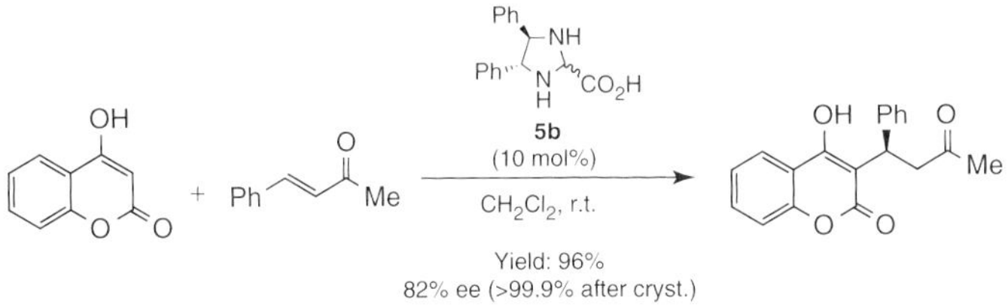

Scheme 6.5 The enantioselective synthesis of (R)-warfarin, as designed by Jørgensen.

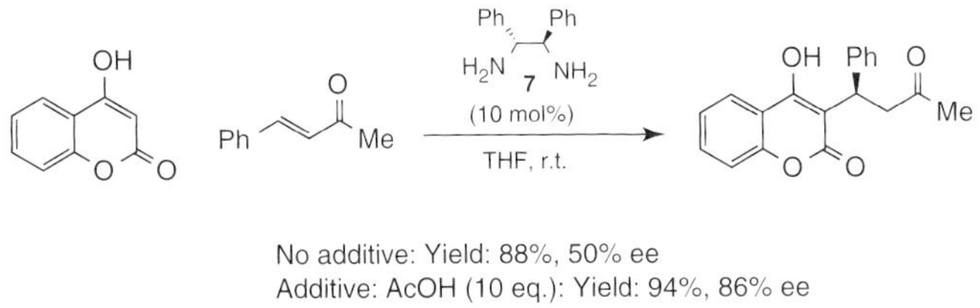

Scheme 6.6 The (R,R)-diphenylethylenediamine-catalyzed enantioselective synthesis of (R)-warfarin.

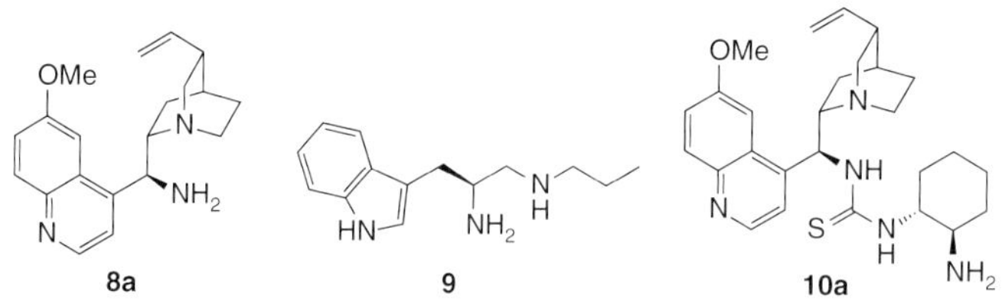

Figure 6.5 Some primary amine catalysts for the Michael reaction of 1,3-dicarbonyl compounds and enones.

has been introduced to help in the preorganization of the reactants before the reaction takes place, by establishing H-bonding interactions.

Extension of the methodology to the use of α,β-unsaturated aldehydes as Michael acceptors has been straightforward, since formation of the intermediate iminium species is much easier in this case compared to the enones. In this context, diarylprolinol trimethylsilyl ethers **11a** and **11b** (Figure 6.6) constitute, without any doubt, a privileged family of chiral secondary amine catalysts that have demonstrated an outstanding performance in this reaction, and also in many others that proceed via iminium activation [23] (as will be shown in many representative examples in this chapter). These catalysts were developed independently by the groups of Jørgensen [24] and Hayashi [25], their design being based essentially on the excellent geometry control achieved during formation of the intermediate iminium ion, and the excellent sterical bias that is exerted on the stereotopic unsaturated moiety by the bulky diaryltrimethylsilyloxy substituent. Although, today, these catalysts are commercially available, they can easily be prepared in multigram quantities, starting from proline, in only four steps.

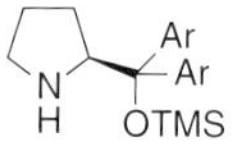

Figure 6.6 Diarylprolinol trimethylsilyl ethers.

11a: Ar = Ph
11b: Ar = 3, 5-$(CF_3)_2C_6H_3$

In case of the Michael addition of malonates to enals, both catalysts have been successfully applied to this reaction, although use of the diphenylprolinol derivative **11a** required the incorporation of either a base co-catalyst or, alternatively, an aqueous acidic medium in order for the reaction to proceed with good yields [26]. Catalyst **11b**, as employed by Jørgensen, furnished excellent results in terms of both yields and enantioselectivities, and its performance has been successfully demonstrated in the synthesis of two important chiral drugs such as (−)-paroxetine and (+)-femoxetine (Scheme 6.7) [27]. As expected, compared to the parent reaction with enones, the Michael reaction of dibenzyl malonates to α,β-unsaturated aldehydes catalyzed by **11b** was shown to proceed much faster. On the other hand, the substrate scope is limited to aromatic enals, with no data referring to alkyl-substituted Michael acceptors.

The behavior of the reaction when cyclic 1,3-dicarbonyl compounds are employed deserves special attention. In this case, following the conjugate addition reaction,

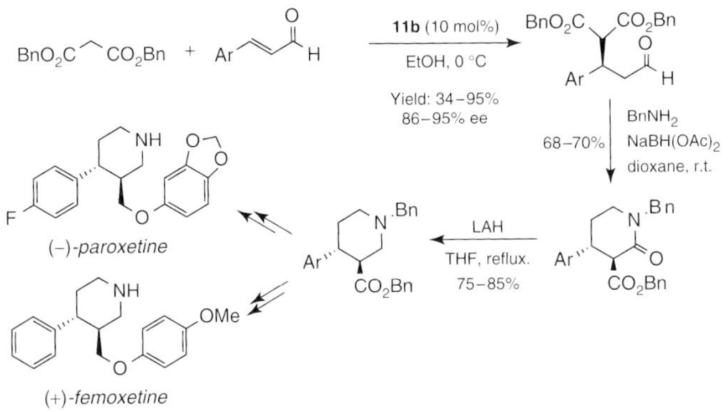

Scheme 6.7 The **11b**-catalyzed Michael reaction of malonates and enals and its application to the synthesis of (−)-paroxetine and (+)-femoxetine.

the Michael adduct can undergo intramolecular hemiacetalization because of the stabilization of the enol tautomer of the 1,3-dicarbonyl moiety. This transformation, which gives rise to compounds with a dihydropyran structure in a single step, has been studied by the groups of Rueping and Jørgensen, with catalyst **11b** emerging as the most effective promoter of the reaction [28]. The first examples were reported using 2-hydroxy-1,4-naphthoquinones (Scheme 6.8) and also cyclic 1,3-diketones; however, further studies using other different cyclic 1,3-dicarbonyl compounds permitted the preparation of many different oxygen heterocycles such as benzopyranes, chromenes, quinolones, or pyranones. The higher acidity of the cyclic 1,3-diketones compared to the acidity of malonates, is most likely the main reason why the reaction proceeds much faster and with better conversions than the corresponding Michael addition of malonates to the same substrates. In addition, the stabilization of the final Michael addition as the cyclic hemiacetal form contributes to an acceleration of the reaction by reducing the concentration of the Michael adduct in solution, thus pushing the equilibria that participate in the process towards formation of the final products.

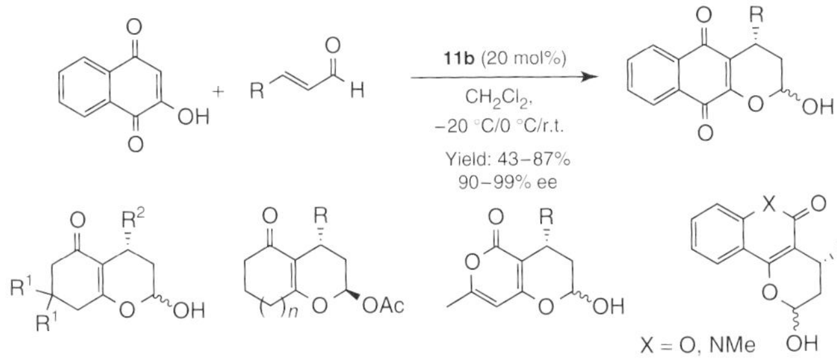

Scheme 6.8 The **11b**-catalyzed Michael reaction of 2-hydroxy-1,4-naphthoquinones with enals and different heterocycles prepared following this strategy.

6.3.2
Nitroalkanes as Nucleophiles

The ability of the nitro group to stabilize an adjacent negative charge makes nitroalkanes a very C–H acidic compounds for furnishing the corresponding nitronate under mild conditions, without a need for strong bases to exert the deprotonation process. The nitronate nucleophile formed can subsequently undergo conjugate addition to the corresponding Michael acceptor, which has been previously activated by the catalyst via the formation of a chiral iminium ion. Moreover, the rich reactivity profile displayed by the nitro group makes this transformation a very powerful method in organic synthesis for the stereoselective preparation of many different chiral building blocks.

The majority of reports concerning conjugate additions of nitroalkanes have focused on the use of α,β-unsaturated ketones as Michael acceptors and, in particular, have restricted the scope of the reaction to cyclic enones. In this context, rubidium prolinate **2** was studied by Yamaguchi in 1994 as a catalyst involved in the Michael addition of nitroalkanes to cyclic and acyclic enones [29], and led to good yields but only moderate enantioselectivities. The proline-catalyzed Michael reaction of nitroalkanes to cycloalkenones was later studied by Hanessian and Pham [30a], and shown to afford the conjugate addition products in good yields and enantioselectivities (Scheme 6.9). A basic additive such as *trans*-2,5-dimethylpiperazine was included in the reaction as co-catalyst, and was involved in deprotonation of the nitroalkane donor in order to increase the concentration of the reactive nitronate nucleophile. An improved version of the reaction was later reported by the same group [30b], using *trans*-4,5-methanoproline **12** as catalyst. Other catalysts employed in this transformation are also shown in Scheme 6.9 [31], and in all cases required the inclusion of *trans*-2,5-dimethylpiperazine as the basic additive. In all these examples, the stereochemistry of the Michael product was the same, regardless of the catalyst employed. This indicated that the role played by the substituent at the pyrrolidine ring of the secondary amine was limited exclusively to exerting a steric shield, without stereodirecting contributions by the secondary functionalities introduced at this substituent as, for example, when tripeptide **13** was employed. It should also be stressed that, in all cases, when nitroalkanes with different

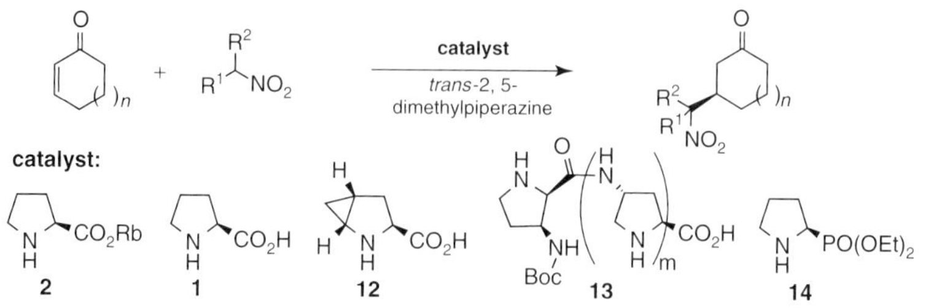

Scheme 6.9 The Michael reaction of nitroalkanes to cyclic enones.

substituents at the α-carbon were employed, the final products were typically obtained as mixtures of diastereoisomers, in ratios ranging from 1:1 to 1:3; however, the latter point remains an unresolved issue associated with this reaction.

The Michael reaction of nitroalkanes and acyclic enones has been the center of attention in studies conducted by the groups of both Ley and Jørgensen [32]. These reports were developed in parallel to the same reaction in which malonates were used as nucleophiles, and involved using the same types of catalyst as were employed previously, namely imidazolidines **5a** and **5b** by Jørgensen's group and the proline tetrazole derivative **6** by Ley's group. The first approach related to the use of imidazolines **5a** and **5b**, and showed the former to be much more effective than the latter in terms of both yield and enantioselectivity, although the ee-values of the obtained adducts did not exceed 85% in the best case [32a]. Although the use of differently substituted nitroalkenes was also evaluated, the diastereoselectivities still remained rather low, with diastereomeric ratios of between 2:1 and 1:1. Remarkably, the authors noted that the reaction was extremely clean, with an absence of byproducts, and the sole reason for the moderate yields obtained in some cases being a matter of conversion. This allowed the reaction to be carried out on the kilogram scale, with recovery and reuse of the catalyst but no reduction in either catalytic activity or enantioselectivity. Catalyst **6** afforded similar results to those furnished by **5a** and **5b**; that is, good to moderate yields (most likely due to a lack of conversion) and enantioselectivities up to 80% ee, even though the catalysts proved to be excellent for the Michael addition of nitroalkanes to cyclohexenone [32b]. Likewise, the use of substituted nitroalkanes led to mixtures of diastereoisomers in an approximately 1:1 ratio. Later, Jørgensen developed a new imidazole catalyst which incorporated the tetrazole substituent (**5c**) and showed a better performance in the reaction, although the enantioselectivities remained between 80% and 87% ee (Scheme 6.10) [32c]. In this case, the reaction was carried out using a 2:1 nitroalkane/enone ratio, with an excess of nitroalkane acting also as the solvent for the reaction.

With α,β-unsaturated aldehydes as Michael acceptors, the state of the art appears in close analogy with the parent Michael reaction using 1,3-dicarbonyl compounds, with *O*-TMS diphenylprolinol catalyst **11a** demonstrating an outstanding performance in the reaction [33]. Several groups have reported, almost simultaneously, the use of this catalyst for the reaction between nitromethane and different alkyl- and aryl-substituted α,β-unsaturated aldehydes, and have obtained the expected Michael adducts with excellent yields and enantioselectivities.

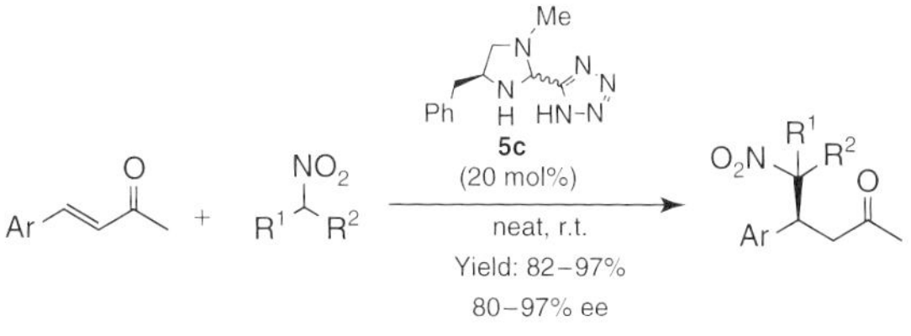

Scheme 6.10 The **5c**-catalyzed Michael reaction of nitroalkanes and acyclic enones.

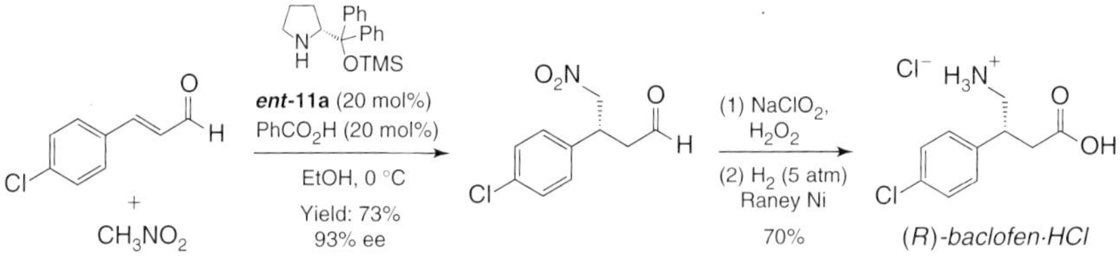

Scheme 6.11 The enantioselective synthesis of (R)-baclofen.

As occurred with enones, when nitroalkenes that generate an additional stereogenic center in the reaction (such as nitroethane) were employed, 1:1 mixtures of diastereoisomers were usually obtained, while ee-values greater than 95% were maintained for both diastereoisomers. It should be pointed out that, in all cases, it was necessary to introduce an acid or base co-catalyst into the reaction scheme. In the first of two cases, benzoic acid was incorporated in order to assist the formation of the iminium intermediate, as it is very often found in many reactions proceeding via iminium activation. In the second case, LiOAc was required to help in the formation of the more reactive nitronate nucleophile. An example of this reaction is addressed by the enantioselective synthesis of optically active baclofen (Scheme 6.11) [33c], a γ-aminobutyric acid (GABA) receptor agonist used in the treatment of spinal cord injury-induced spasm. Despite being administered as a racemic mixture, the (R) isomer of baclofen is essentially the active form, the (S) enantiomer being inactive. In order to synthesize (R)-baclofen, the enantiomer of the original catalyst, **ent-11a**, must be used.

Prior to these reports, an extensive study was conducted on the Michael reaction of nitroethane and nitropropane with aromatic enals [34]. Surprisingly, catalyst **11a** was tested in this reaction, but furnished very poor results under the conditions employed for the reaction (neat conditions at room temperature, and without the incorporation of any additive). As a consequence, a new MacMillan-type imidazolidinone catalyst **4b** (Figure 6.7) was prepared which was tested in this reaction, and provided moderate to good yields and enantioselectivities of 70–90% ee for a wide variety of aromatic α,β-unsaturated aldehydes. Disappointingly, an approximate 1:1 mixture of diastereoisomers was obtained in all cases which, once again, illustrated the difficulties associated with stereochemical control of the reaction when nitroalkanes other than nitromethane are employed. One possibility of overcoming such limitations involves the use of bifunctional catalysts that are capable of interacting with both the nucleophile and the electrophile,

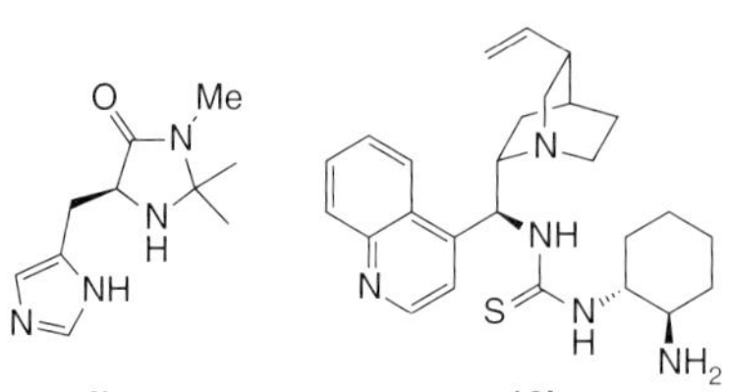

Figure 6.7 Two different catalysts employed in the Michael addition of nitroethane to enals.

leading to a well-ordered transition state in the C—C bond-formation step. In this context, a thiourea–amine catalyst **10b** (Figure 6.7) was tested in the reaction of nitroethane with cyclohexenones, and also with acyclic benzylideneacetone [35]. The incorporation of the thiourea moiety was intended to allow a preorganization of the reagents following formation of the iminium ion by hydrogen-bonding interactions with the nitro group of the Michael donor. However, the diastereoselectivity ratio remained between 3 : 1 and 3 : 2, even though the catalyst was seen to be an excellent promoter of the same reaction with simple nitromethane.

This diastereoselectivity problem when using substituted nitroalkanes was also found in the reaction shown in Scheme 6.12 [36]. This consisted of a crossed Michael reaction between a nitroalkene and an α,β-unsaturated aldehyde, in which the cooperative use of two different catalytic species capable of activating (independently) the nucleophile and the electrophile (dual activation strategy) was applied. In this case, diphenylprolinol **15** was employed as catalyst, and 1 equiv. of trimethylphosphite and 1 equiv. of AcOH were included as promoters of the reaction. The reaction proceeded through a first step in which conjugate addition of the phosphite to the nitroalkene occurred, thus generating an intermediate nitronate nucleophile which next underwent conjugate addition to the α,β-unsaturated aldehyde (activated by catalyst **15** via iminium ion formation). A final β-elimination and hydrolysis reaction released both $(MeO)_3P$ and the catalyst **15** respectively, ready to participate in a subsequent catalytic cycle. When using these conditions, a wide range of different β-methyl nitrostyrenes and α,β-unsaturated aldehydes could react with each other to furnish the corresponding Michael adducts in excellent yield and enantioselectivities, although as mixtures of diastereoisomers in 1 : 1 to 1 : 2 ratio, the *anti* isomer was the major diastereoisomer obtained.

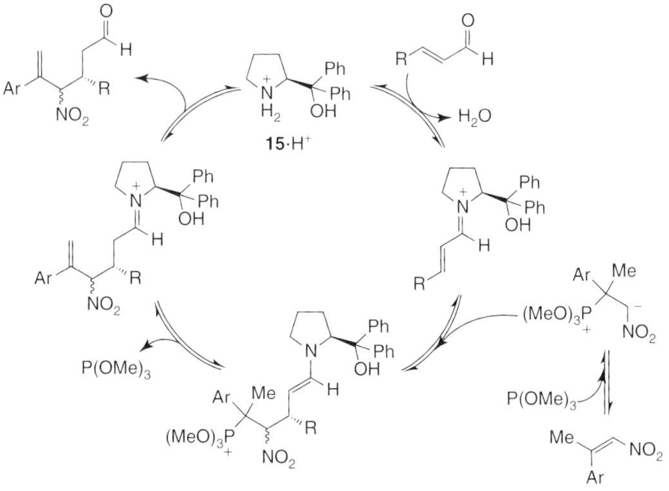

Scheme 6.12 Double activation strategy in the crossed Michael reaction between nitroalkenes and α,β-unsaturated aldehydes.

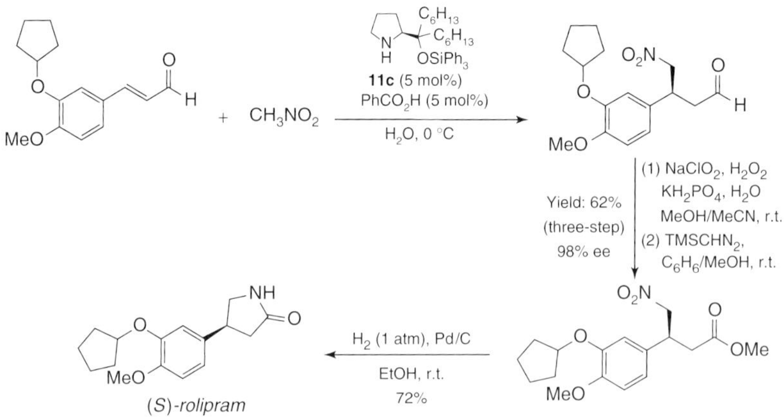

Scheme 6.13 Organocatalytic enantioselective Michael reaction of nitromethane to α,β-unsaturated aldehydes in water and its application to the synthesis of (S)-rolipram.

Another issue that has led to intensive research during recent years has been the development of more environmentally benign methodologies, notably with regards to the use of water or an aqueous medium as a solvent for the organocatalytic reaction. Indeed, this is a fairly realistic possibility for the iminium activation strategy, due to the compatibility of the intermediate species participating in the catalytic cycle towards the presence of water. In this context, an important report describing the preparation of a new O-TMS dialkylprolinol catalyst designed specifically for carrying out the reaction with water as solvent, was made by Palomo and coworkers (Scheme 6.13) [37]. The principles of the catalyst design were based on previous examples, in which the introduction of hydrophobic long alkyl side chains in the structure of several proline-based compounds resulted in an increased catalytic activity in many enamine-mediated reactions in water [38]. These side chains presumably contributed to the formation of a hydrophobic pocket, thereby assisting the incorporation of organic reagents into the catalytic site. Alternatively, the formation of micellar structures in an aqueous medium has been proposed as a reason for the excellent performance of these types of catalyst. In this case, linear alkyl side chains of different lengths were introduced to replace the diaryl groups present in the basic structure of diarylprolinols **11a** and **11b**. After some control experiments, dihexyl derivative **11c** was identified as the most efficient catalyst, with the related dimethyl, dipropyl, dinonyl, and didodecyl derivatives being less active in the model reaction. The use of this catalyst in water as the only solvent resulted in a high-yielding and very enantioselective procedure for the Michael addition of nitromethane to α,β-unsaturated aldehydes, and also for the same reaction using malonates as nucleophiles. In addition, the applicability of this methodology to the asymmetric synthesis of valuable chiral compounds was highlighted with the synthesis of (S)-rolipram, a type IV phosphodiesterase inhibitor originally commercialized as an antidepressant and also used in the treatment of asthma and chronic obstructive pulmonary disease (COPD).

6.3.3
Other Acidic Carbonyl Compounds as Nucleophiles

As noted previously [17], Barbas had suggested that there must exist a pK_a barrier for conjugate addition reactions of carbon nucleophiles which proceeded via iminium activation, and which should lie between 16 and 17. In the case of Michael donors containing fewer acidic hydrogen atoms, the formation of an active nucleophile species in the reaction medium requires an additional reagent to assist the deprotonation process. As a consequence of this, the range of nucleophiles amenable for use in these transformations will be limited to rather acidic compounds, as has been shown extensively above for the case of 1,3-dicarbonyl compounds and nitroalkanes. As an alternative, Barbas devised the use of trifluoroethyl thioesters as nucleophiles in iminium-mediated Michael reactions [17], because of the known higher acidity of thioesters compared to that of the corresponding esters, and also based on previous studies which showed that trifluoroethyl thioesters had α-proton exchange rates approximately 10-fold faster than those of phenyl thioesters [39]. These hypotheses were confirmed with the finding that catalyst **11b** was able to promote the Michael reaction of different arylacetic acid-derived trifluoroethyl thioesters and α,β-unsaturated aldehydes in a rather efficient manner (Scheme 6.14). The reaction was limited to α-aryl-substituted thioesters which appear over the pK_a barrier required for their activation and furnished ee-values between 67% and 96%, although typically mixtures of *syn/anti* diastereoisomers were obtained with a slight preference for formation of the *anti* isomer. The scope of the Michael acceptor was also restricted to aromatic enals, as the reaction using crotonaldehyde as electrophile furnished the final compound in 54% ee.

Another class of potential Michael donor containing acidic protons can be found in the dicyanoacrylate derivatives. This is a case of a vinylogous Michael addition, in which α,α-dicyanoolefins reacted stereoselectively with α,β-unsaturated aldehydes or ketones under iminium activation by a chiral amine catalyst, furnishing the corresponding γ-addition product (Scheme 6.15) [40]. For enals, diphenylprolinol **15** was seen to be an excellent promoter of the reaction, furnishing excellent yields and stereoselectivities over a wide range of differently substituted α,β-unsaturated aldehydes. The same reaction, using enones as Michael acceptors, was subsequently developed by the same authors, and involved the use of a chiral primary amine catalyst such as **8a**, which was required for a more effective formation of the

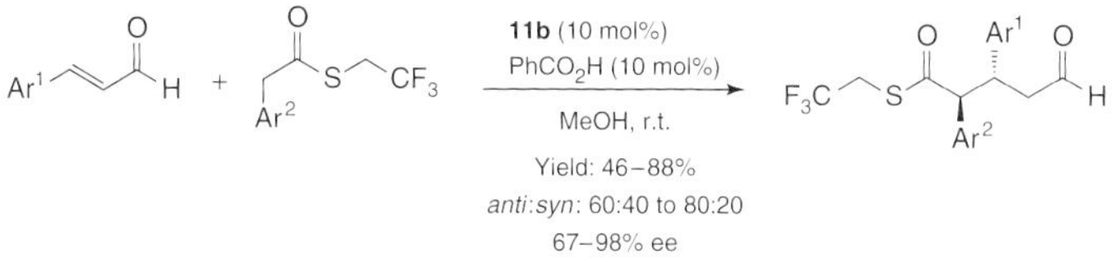

Scheme 6.14 Organocatalytic enantioselective Michael reaction of trifluoroethyl thioesters to aromatic α,β-unsaturated aldehydes.

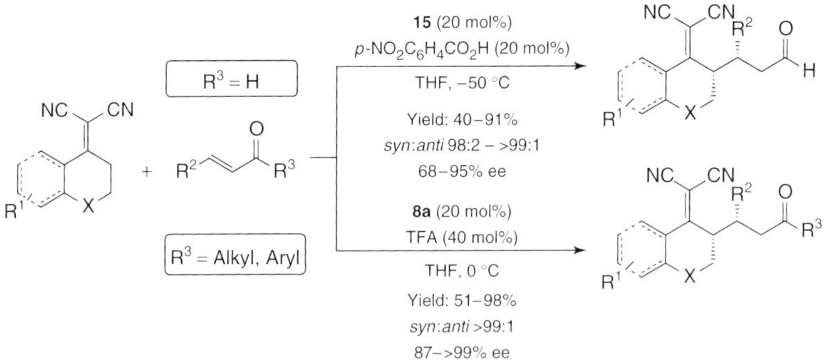

Scheme 6.15 Vinylogous Michael reaction of dicyanoacrylates with enones and enals.

intermediate iminium ion. In both cases, the inclusion of a Brønsted acid co-catalyst was reported to facilitate the condensation step between the catalyst and the Michael acceptor. Further research by other groups has led to the use of modified prolinols containing hydrophobic alkyl side chains (closely related to catalyst **11c** developed by Palomo; see Scheme 6.13), for carrying out exactly the same reaction between dicyanoacrylates and α,β-unsaturated aldehydes in water [41].

Azalactones or oxazolones have also been employed as nucleophiles in the Michael reaction with α,β-unsaturated aldehydes, because of the enhanced acidity of these compounds compared to the corresponding open-chain N-alkoxycarbonyl α-amino acid derivatives. In this context, Jørgensen has reported **11b** to be a very efficient catalyst for this transformation, and obtained the expected adducts in excellent yields and with an outstanding level of stereocontrol (Scheme 6.16) [42]. However, perhaps the most interesting feature of these investigations was the fact that, even though racemic C4-substituted oxazolones were used as Michael donors, the final adducts were not only obtained as highly enantiomerically enriched

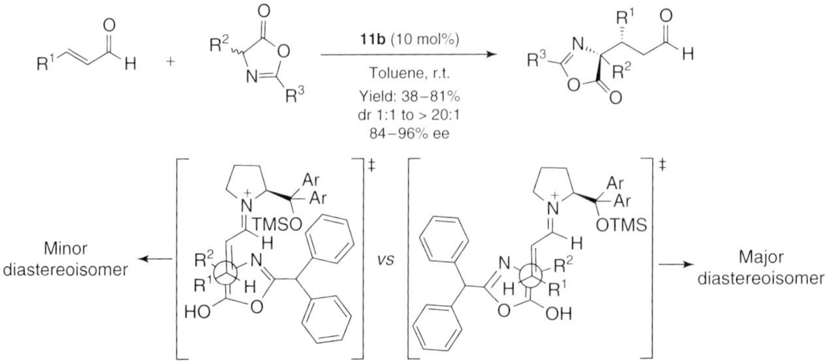

Scheme 6.16 Michael reaction of oxazolones with α,β-unsaturated aldehydes catalyzed by **11b** and the models proposed to explain the high diastereo- and enantioselectivity obtained.

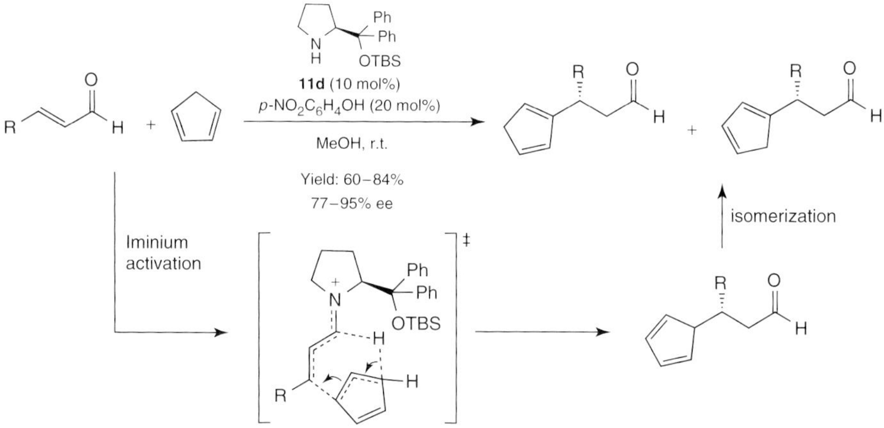

Scheme 6.17 Amine-catalyzed enantioselective formal ene reaction between cyclopentadiene and enals.

compounds, but also achieved a remarkable degree of stereocontrol at the newly generated stereogenic center formed at the center currently present at the oxazolone moiety, and which was created over the pre-existing center. The reaction appeared to be fairly tolerant to substitution at the Michael donor, with alkyl-substituted enals furnishing excellent yields of the Michael adducts, whereas only moderate yields were obtained when cinnamaldehyde was used. Nevertheless, the ee-values remained higher than 85% in all cases. With regards to the influence of the substitution pattern at the nucleophile, changing the substituent at C4 appeared not to have any negative influence in the process; however, the substituent at C2 exerted a major influence in both the yield and the diastereoselectivity of the reaction. On this basis, it was found that the installation of a benzhydryl group at this position resulted in a dramatic increase in the diastereoselectivity of the reaction, and led to the almost exclusive formation of a single diastereoisomer in very high enantiomeric excess. Density functional theory (DFT) calculations indicated that the presence of a bulky group at C2 of the oxazolone moiety determined the orientation of the nucleophile, and favored the approach whereby steric interactions between this group and the two aryl substituents of the catalyst were avoided (see Scheme 6.17). An extremely effective facial shielding of one of the stereotopic faces of the Michael acceptor was achieved by the presence of the catalyst, which provided the required geometry control in the iminium ion formation and also directed the income of the nucleophile by steric bias.

The synthetic use of the obtained adducts was studied intensely, with the result that several protocols were designed for the selective elaboration of the different functionalities present in these molecules. In particular, the obtained Michael adducts were converted into α,α-disubstituted α-amino esters, polysubstituted prolines, δ-lactams, β-amino alcohols, and tetrahydropyrans by using very simple procedures that proceeded in high yields and maintained the stereochemical integrity of the starting materials.

Finally, cyclopentadiene has also been used as an acidic C-nucleophile in conjugate additions, proceeding via iminium ion formation in what can also be considered as an ene reaction (Scheme 6.17) [43]. In this case, the bulkier O-TBS-containing diphenyl prolinol catalyst **11d** was found to be the most effective for this reaction, which proceeded with the help of p-nitrophenol as a Brønsted acid co-catalyst. The Michael adducts were obtained as mixtures of regioisomers arising after an isomerization process that furnished the more stable 1- and 2-substituted cyclopentadienes. None the less, the combined yield of the reaction was high, and excellent enantioselectivities were reached for all cases studied, although the substrate scope was restricted to the use of aromatic α,β-unsaturated aldehydes as Michael acceptors.

6.3.4
Silyl Enol Ethers and Enamides as Nucleophiles

The iminium-activation strategy has also been applied to the use of silyl enol ethers as nucleophiles in an organocatalytic version of the Mukaiyama–Michael reaction. In this context, imidazolidinone **16a**, which was expressly designed by MacMillan for the conjugate Friedel–Crafts alkylation of indoles with enals (see Section 6.4), was identified as an excellent chiral catalyst for the enantioselective conjugate addition of silyloxyfurans to α,β-unsaturated aldehydes, thus providing a direct and efficient route to the γ-butenolide architecture (Scheme 6.18) [44]. This is a clear example of the complementarity between organocatalysis and transition metal catalysis, with the latter furnishing usually the 1,2-addition product (Mukaiyama aldol) while the former proceeded via 1,4-addition. The reaction needed for the incorporation of a Brønsted acid co-catalyst such as 2,4-dinitrobenzoic acid (DNBA), assisting the formation of the intermediate iminium ion, while 2 equiv. of water had to be included as an additive for the reaction to proceed to completion. In the absence of water, low yields of the Michael addition products were obtained, and this was interpreted in terms of water loss from the catalytic cycle by the presence of the TMS cation generated during the reaction (presumably via the formation of TMS_2O), which prevents catalyst turnover. Under optimized conditions, the reaction proceeded with good yields and excellent levels of stereocontrol for a

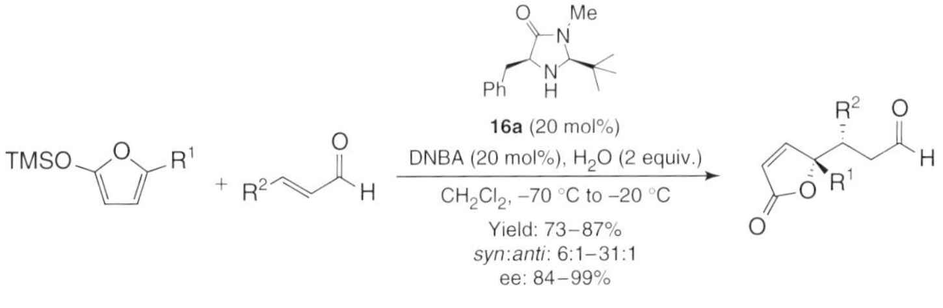

Scheme 6.18 Mukaiyama–Michael reaction of silyloxyfuranes and α,β-unsaturated aldehydes.

series of α,β-unsaturated aldehydes and different silyloxyfuranes. The outstanding performance of this imidazolidinone catalyst in this case is also related to its ability to control the iminium geometry, together with a very effective steric shielding achieved for one of the enantiotopic faces of the Michael acceptor. On the other hand, the applicability of this reaction in synthesis was addressed by MacMillan himself with the very efficient preparation of spiculisporic acid [44], a *Penicillium spiculisporum* fermentation product that has found industrial application as a biosurfactant, and also with an elegant formal total synthesis of compactin [45], an inhibitor of the enzyme hydroxymethylglutaryl coenzyme A reductase, which has a key role in endogenous cholesterol biosynthesis.

Some years later, Wang reported that the same catalyst **16a** was also capable of catalyzing the Mukaiyama–Michael reaction of acyclic silyl enol ethers with α,β-unsaturated aldehydes, thus obtaining the corresponding Michael adducts in excellent yields and enantioselectivities [46]. As occurred with the MacMillan case, DNBA had to be included as additive to assist formation of the iminium intermediate. Scavenging of the TMS cation formed as the reaction proceeded forward was achieved by using a 5 : 1 mixture of *t*-BuOH/*i*-PrOH as solvent for the reaction.

Enamides are also suitable nucleophiles in Michael additions to α,β-unsaturated aldehydes. This reaction was first reported by Hayashi, and soon also by Wang, using O-silylated diarylprolinols **11d** and **11a**, respectively, as catalysts (Scheme 6.19) [47]. The reaction proceeded in two steps, starting with a conjugate addition of the enamine to the activated form of the Michael acceptor (this can also be considered an aza–ene reaction), followed by a release of the catalyst that caused the Michael adduct to undergo intramolecular hemiaminal formation and to deliver a piperidine-2-ol which resulted in what was defined as a formal aza [3 + 3] cycloaddition. In Hayashi's report, the catalyst **11d** incorporating a bulkier O-TBS group was identified as the most efficient among other diarylprolinols tested, while the reaction conditions involved the use of N-Boc enamines as nucleophiles and the reaction being carried out at rather high temperatures (70 °C). The final piperidine adducts were obtained as mixtures of α/β isomers in excellent yields and enantioselectivities for several N-Boc-1-arylethenamines, and also for a wide variety of α,β-unsaturated aldehydes, despite being restricted to aromatic substituents at the β-position. Under the conditions reported by Wang, N-acetylenamines were employed as nucleophiles and catalyst **11a** was found to be the most effective, incorporating PhCO$_2$H as a Brønsted acid co-catalyst and carrying out the reaction in the same solvent, albeit at room temperature. In this case, the substrate scope also appeared to be restricted exclusively to aryl enamides and aromatic enals.

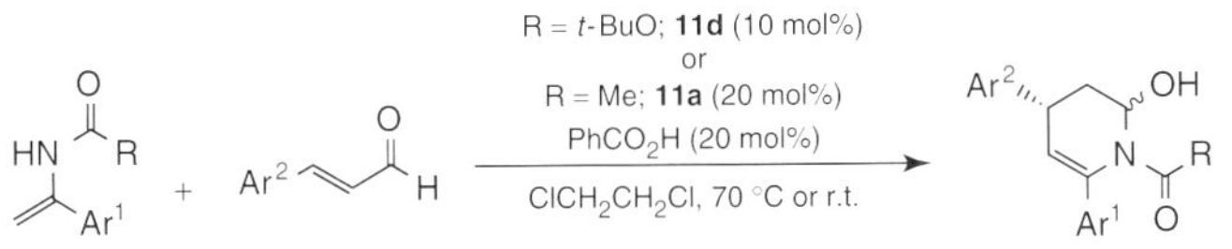

Scheme 6.19 Enantioselective Michael addition of enamides to α,β-unsaturated aldehydes.

6.3.5
Aldehydes as Nucleophiles

The use of aldehydes as Michael donors requires previous activation of the nucleophile which, in this case, it is achieved by the chiral secondary amine via formation of an enamine intermediate; the latter then is able to undergo the conjugate addition step with a variety of electrophiles. However, in the particular case in which α,β-unsaturated aldehydes and ketones are employed as electrophiles, the participation of an activated form of the Michael acceptor as an iminium ion is a matter of controversy, because the formation of a chiral enamine intermediate by itself is sufficient to explain the observed stereoselectivities, and might also be sufficient for the reaction to proceed without further activation of the Michael acceptor counterpart. In this context, the Michael reaction between aldehydes and alkyl vinyl ketones has been carried out using **17** as catalyst, affording the Michael adducts with good yields and moderate enantioselectivities (Scheme 6.20) [48a]. The authors reported herein a slight nonlinear effect in the reaction, which pointed towards the hypothesis that two molecules of the catalysts were involved in the reaction mechanism, and therefore a possible iminium-type activation of the electrophile might have occurred in addition to an activation of the aldehyde by enamine formation. Further research led to an improved protocol that involved O-silyldiarylprolinol **11b** as catalyst [48b]. On the other hand, Gellman and coworkers showed that O-methyldiphenylprolinol **11e** and imidazolidinone **4c** were also able to promote the same reaction (Scheme 6.20); however, they proposed that in this case the reaction proceeded exclusively via an enamine activation of the aldehyde [49]. The reaction also required the presence of a catechol derivative as an additive, which explained the role of this protic co-catalyst in terms of activation of the enone acceptor by hydrogen-bond donation, and also supported the absence of an iminium-type activation of the enone. The exclusive formation of an enamine intermediate in this case was also supported by carrying out control experiments using a previously generated enamine as nucleophile, under the same reaction conditions.

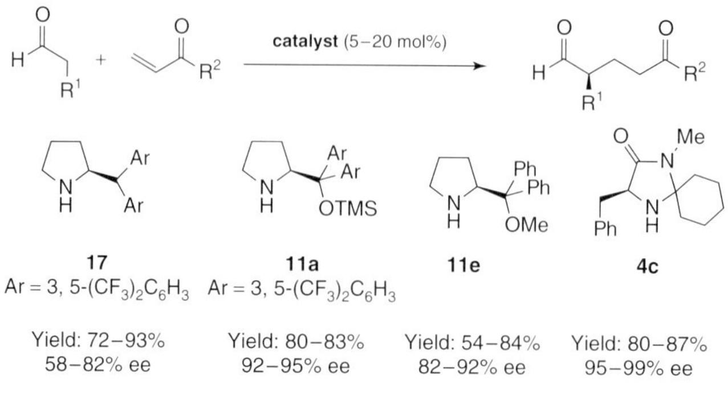

Scheme 6.20 Enantioselective intermolecular Michael reaction between aldehydes and alkyl vinyl ketones.

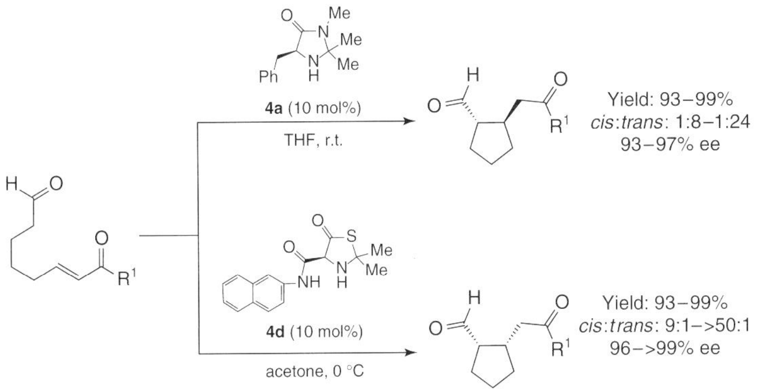

Scheme 6.21 Enantioselective intramolecular Michael reaction.

In a similar context, List and coworkers reported a very efficient intramolecular version of this reaction which used imidazolidinone **4a** as the catalyst (Scheme 6.21) [50]. In this case, the authors were unable to clearly rule out the possible contribution of iminium-type intermediates, although the fact that the lower enantioselectivity afforded by a substrate incorporating an α,β-unsaturated aldehyde as Michael acceptor on its structure also suggested some type of dual enamine/iminium activation. This methodology shows up as a very simple and efficient protocol for the preparation of functionalized chiral nonracemic cyclopentanes or related heterocyclic compounds in a stereocontrolled fashion. Interestingly, modified thiazolidinone catalyst **4d**, derived from cysteine, was also employed in the same type of transformation, and led to the corresponding cyclic compounds in excellent yield and enantioselectivity, but with an opposite relative configuration (*cis*) of the products obtained in the parent reaction reported by List (Scheme 6.21) [51].

6.4
Conjugate Friedel-Crafts Alkylations

The conjugate addition of electron-rich aromatic substrates to electron-deficient olefins, also known as *conjugate Friedel–Crafts alkylation*, is a powerful strategy for building up complex heterocyclic structures. Developments in asymmetric versions of this reaction have been dominated by transition metal catalysis, using cleverly designed templates to act as Michael acceptors capable of interacting with Lewis acids in a well-organized transition state (e.g., *N*-acyloxazolidinones or β-hydroxyenones, among others). However, the use of simple enones – and especially α,β-unsaturated aldehydes – as Michael acceptors has not found applicability in this reaction because of their tendency to undergo 1,2-addition rather than conjugate addition in the presence of Lewis acids. However, this limitation has been overcome with the appearance of an aminocatalytic approach, in which the LUMO-lowering effect associated with the formation of iminium intermediates implies a clear preference for these substrates towards the 1,4-addition pathway.

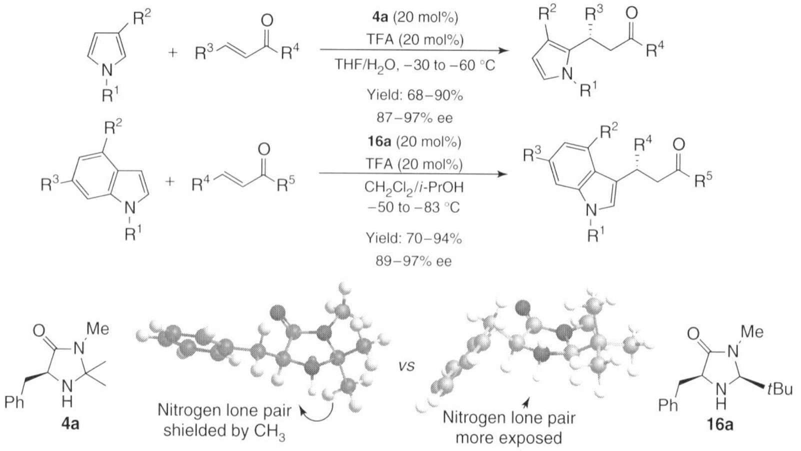

Scheme 6.22 Enantioselective conjugate Friedel–Crafts alkylation catalyzed by imidazolidinones **4a** and **16a**, and the principles applied to the evolution of the catalyst design.

This additionally opens up the possibility for the asymmetric versions, provided that chiral amines are employed as catalysts.

The first developments in this field were reported by MacMillan, who used chiral imidazolidinones as organic catalysts [52]. The first study was carried out using pyrroles as nucleophiles, with catalyst **4a** furnishing outstanding results (Scheme 6.22). However, the same reaction with heteroaromatics of significantly lower reactivity such as indoles proceeded with disappointingly low yield and enantioselectivity, and this led the authors to carry out a crucial catalyst modification based on a clever design [52b]. MacMillan hypothesized that the presence of two methyl groups at the catalyst structure would lead to a lower availability of the nitrogen lone pair, which would hamper formation of the iminium intermediate. Substituting one of the methyl groups by a smaller hydrogen would lead to a more exposed nitrogen lone pair, and hence to a higher reactivity of the catalyst. In addition, the introduction of a *tert*-butyl group in place of the other methyl group (the one *cis* to the benzyl moiety) was suggested to increase the capacity of the catalyst and to exert steric shielding of one of the enantiotopic faces of the α,β-unsaturated aldehyde. This new design proved to be extremely effective not only in this transformation but also in the addition of electron-rich benzenes [52c], providing an extraordinarily useful new approach to build up complex aromatic and heteroaromatic structures.

The exact explanation for the different behavior regarding stereocontrol provided by these two catalysts has been provided by Houk, using computational methods [53]. Calculations at a high theory level indicated that the minimum energy geometry of the intermediate iminium ion in the transition state for catalyst **4a** placed the phenyl ring of the benzyl moiety away from the reactive site, as a consequence of a stabilizing C–H···Me π-interaction between one of the methyl groups at C2 of

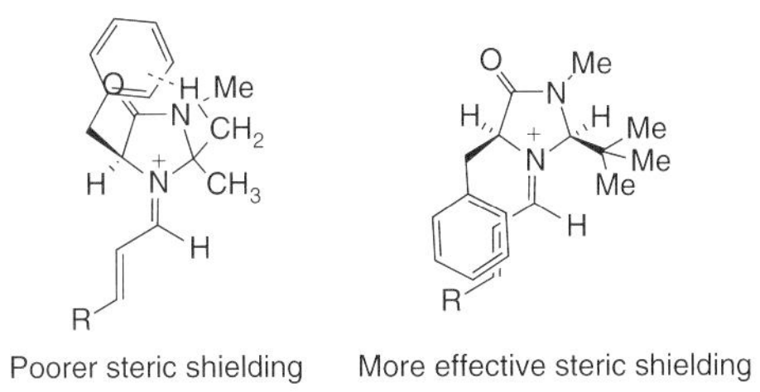

Poorer steric shielding   **More effective steric shielding**

Figure 6.8 Proposed models to account for the different enantioselectivity provided by catalysts **4a** and **16a** in conjugate Friedel–Crafts alkylations.

the imidazolidinone ring and this phenyl ring (Figure 6.8). This conformation is essentially avoided in the case of catalyst **16a** due to the additional steric hindrance provided by the *t*-Bu group. This causes the system to adopt a conformation in which the phenyl ring of the Bn moiety and the reactive α,β-unsaturated moiety are placed much closer to each other than with **4a**. This more effective shielding has also been proposed to occur by stabilizing the π-stacking interactions between the electron-rich benzyl group and the electron-poor α,β-unsaturated iminium cation [9]. Nevertheless, the former reactive conformation found for **4a** should have sufficient stereodiscriminating ability to account for the high enantioselectivities provided by this catalyst in the conjugate addition of pyrroles to enals [52a], and also in Diels–Alder reactions [10].

Intramolecular versions of this reaction have also been explored with excellent results. For example, the intramolecular Friedel–Crafts alkylation of indoles has been reported to proceed with excellent yields and enantioselectivities using MacMillan imidazolidinone **16a** as catalyst [54a]. In a different report, *O*-TMS diarylprolinols **11a** and **11b** have also demonstrated their ability in the intramolecular conjugate addition of electron-rich benzenes to α,β-unsaturated aldehydes [54b]. In addition, several modified versions of these imidazolidinone catalysts have been used in this reaction (Figure 6.9). For example, imidazolidinone **16b** has been used in the reaction of 1-formylcyclopentene with indoles, reaching moderate enantioselectivities of the corresponding Michael adducts [55], while a solid supported version of imidazolidinone **4a** has also been prepared [56] in which the catalyst was immobilized on a siliceous mesocellular foam, as well as on a polymer-coated mesocellular foam. Although both forms provided similar yields in the conjugate Friedel–Crafts alkylation of indoles to enals as were reported with the unsupported catalyst, the stereoselectivities were seen to vary according to the nature of the solid phase employed. Alternatively, catalyst **11a** has also been used successfully in the enantioselective Friedel–Crafts alkylation of indoles with α,β-unsaturated aldehydes, in this case requiring very singular conditions such as the inclusion of Et_3N as additive (this is explained in terms of activation of the indole by N–H deprotonation) and the use of methyl *tert*-butyl ether as solvent, but providing the expected alkylation products in good yields and enantioselectivities for a wide range of enals tested [57].

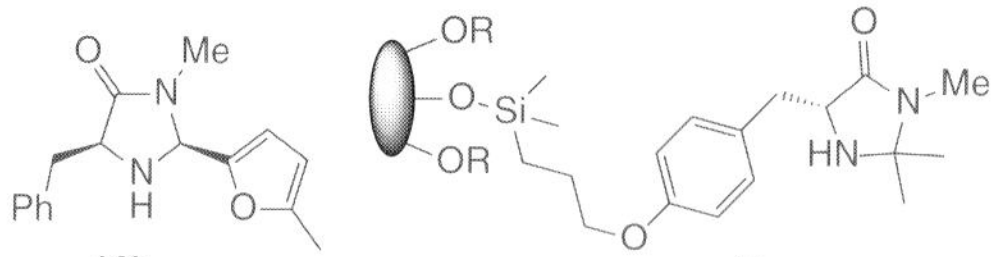

Figure 6.9 Two modified imidazolidinone catalysts employed in conjugate Friedel–Crafts alkylations of α,β-unsaturated aldehydes.

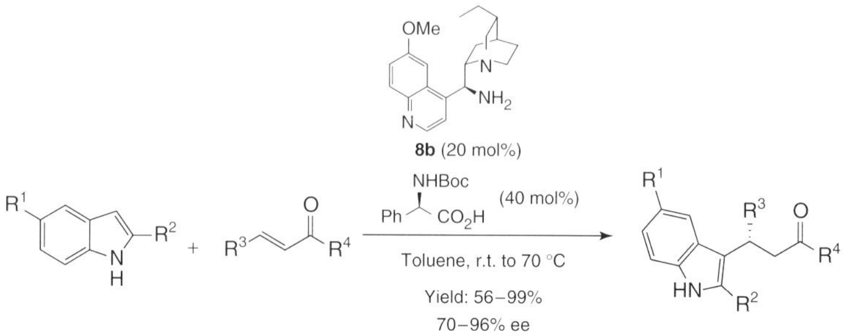

Scheme 6.23 Asymmetric counterion directed catalysis in the conjugate Friedel–Crafts alkylation of indoles.

Notably, the conjugate Friedel–Crafts alkylation of enones required a long period of development. An initial attempt using the MacMillan catalyst **16b** in the reaction of indole with 5-methyl-3-buten-2-one led to a moderate yield of the addition product which was obtained with low enantioselectivity (28% ee) [58], most likely due to difficulties associated with formation of the activated iminium intermediate. However, changing to a primary amine catalyst for which activation of the substrate, as shown previously, was much easier led to a breakthrough in this field. Initially, a primary amine derivative of cinchonine was surveyed as promoter of the reaction between indoles and a wide range of different acyclic enones. In particular, the reaction was seen to have good to moderate yields, to require between two and six days for completion, and to achieve ee-values between 50% and 85% under the best experimental conditions. In this case, the incorporation of a strong Brønsted acid co-catalyst (such as TfOH) and also a rather high catalyst loading (30 mol%) were required in order to obtain the highest yields and stereoselectivities [59]. A similar catalyst system formed by **8b** in combination with a chiral amino acid as Brønsted acid co-catalyst, was employed by Melchiorre and coworkers in the same reaction (Scheme 6.23), and reached improved yields of addition products with a higher level of enantiocontrol, and with a slightly reduced catalyst loading [60]. This strategy, in which a combination of the amine catalyst with a chiral acid co-catalyst led to a significant improvement in the performance of the original amine catalyst (asymmetric counterion-directed catalysis), was developed by List in the context of the conjugate hydrogen transfer reactions, and will be described in more detail in the following section.

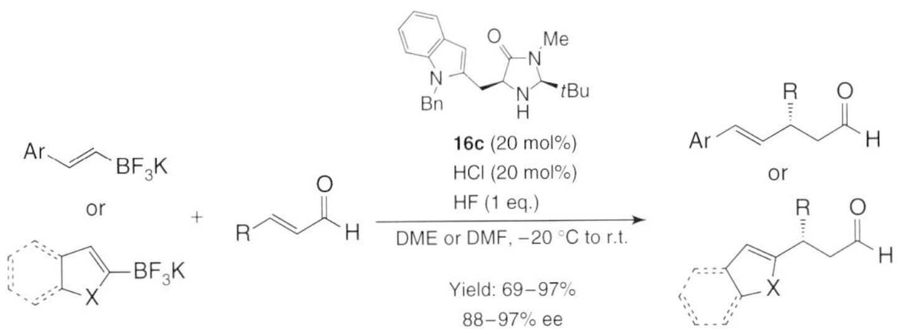

Scheme 6.24 Asymmetric conjugate addition of organo trifluoroborate salts to enals.

Unfortunately, the Friedel–Crafts approach to the alkylation of aromatic rings is not only restricted to the use of electron-rich substrates; rather, there is also a clear limitation which arises from a lack of flexibility in the regioselectivity of the reaction, which is exclusively determined by the substitution pattern or the nature of the aromatic ring (for example, indoles can only be alkylated at C3, or pyrroles at C2). In an effort to overcome this problem, MacMillan applied a clever new reaction design in which, in analogy to the Petasis reaction, boronic acids were employed as effective π-activation groups that would enable unreactive arenes to undergo 1,2-addition to iminium ions. Hence, the possibility of using aryl- and vinyl trifluoroborates as activated π-nucleophiles in the conjugate addition to α,β-unsaturated aldehydes under iminium activation was proposed [61]. This new approach allowed a complete control of the regioselectivity of the reaction by introducing the borate group at the desired position of the arene or heteroarene nucleophile. Moreover, it also allowed the use of vinylborates as Michael donors, thus greatly expanding the versatility and synthetic applicability of this methodology. A modified tryptophan-derived imidazolidinone catalyst **16c** was identified as the most efficient promoter of the reaction, producing excellent yields and enantioselectivities for several styriletrifluoroborates and also for differently substituted electron-deficient aryltrifluoroborates (Scheme 6.24); however, simple phenyltrifluoroborates were found to be unreactive under the reaction conditions employed. Nevertheless, the reaction showed a broad substrate scope with regards to the α,β-unsaturated aldehyde reagent used. Notably, the addition of 1 equiv. of hydrofluoric acid was found to be crucial for the reaction to proceed to completion, the HF being needed to sequester the boron trifluoride formed during the reaction.

6.5
Conjugate Hydrogen-Transfer Reactions

Enantioselective conjugate hydrogen-transfer reactions have also been carried out under iminium activation conditions [62]. The key to success for this reaction is the use of an organic compound, such as a Hantzsch ester, that is able to transfer, simultaneously, a hydride and a proton to the Michael acceptor. The general concept lying behind this reaction design imitates the way in which reductions are accomplished in biological processes, with the Hantzsch ester mimicking the chemical behavior of the hydride-reduction cofactors such as NADH or $FADH_2$ that are used by enzymes in these types of reaction. A nonstereoselective version was carried out by List in 2004 by reducing α,β-unsaturated aldehydes with a commercially available Hantzsch ester in the presence of dibenzylamine as catalyst [63], in a typical case of a conjugate addition proceeding via iminium activation. Shortly afterwards, the same group reported an asymmetric version of this reaction when they used MacMillan imidazolidinone **16a** as catalyst and a modified version of the Hantzsch ester (this was somewhat bulkier than the original ester used in the nonstereoselective version) as the hydride-donor reagent (Scheme 6.25) [64]. The reaction proceeded with very high efficiency, and provided excellent yields and enantioselectivities for several β-disubstituted α,β-unsaturated aldehydes, but in all cases incorporating an aromatic substituent at the β-position. Interestingly, the double-bond geometry of the Michael acceptor did not appear to have any influence on the stereochemical outcome of the reaction, which indicated that enals of different configuration would furnish the same enantiomer after reduction. This allowed the use of mixtures of Z and E enals as substrates in this reaction, thus avoiding the need for a highly diastereomerically pure Michael acceptor in order to reach to good enantioselectivity in the reduction. Simultaneously, the group of MacMillan reported independently a similar reduction protocol [65a], showing that imidazolidinone **16d** containing a single stereogenic center (the benzyl group was removed from the original catalyst **16a**) would serve as an outstanding catalyst for this transformation,

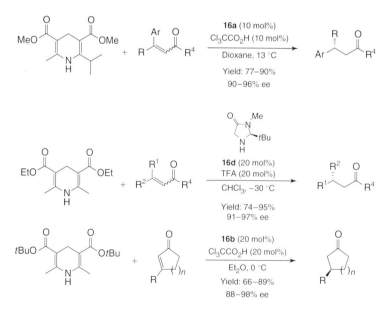

Scheme 6.25 Enantioselective transfer hydrogenation of enals and enones.

but using a different Hantzsch ester as the hydride donor (Scheme 6.25). This reaction was applicable to enals containing both aliphatic and aromatic substituents at the β-position, and even substrates containing bulky substituents (e.g., a *tert*-butyl group) were reduced with high efficiency. The same enantioconvergency was also found in this reaction, which allowed the use of Z/E mixtures of the starting Michael acceptors to furnish equally high enantioselectivities. This behavior was explained in terms of a rapid Z/E isomerization of the starting materials, catalyzed by the imidazolidinone during formation of the intermediate iminium ion. Cyclic enones could also be reduced in a very efficient manner by using, in this case, imidazolidinone **16b** as catalyst (Scheme 6.25) [65b]. However, it was necessary first to optimize the structure of the Hantzsch ester, at which point it was observed that the nature of the alkoxy substituents exerted an important influence on the enantioselectivity of the reaction. Following a structure–selectivity relationship study, a di-*tert*-butyl derivative of the original reagent Hantzsch ester was found to be the most effective hydride-donor reagent for the reaction.

Having employed this reaction mainly as a testing ground, List went on to develop the asymmetric counteranion-directed catalysis concept, which relies on the hypothesis that the counterions can exert an important influence in polar reactions proceeding via charged intermediates, when they are conducted in organic solvents in which ion pairs are closely associated with each other. This suggested the possibility of carrying out enantioselective conjugate addition reactions via iminium activation by using an achiral ammonium salt catalyst, incorporating a chiral counteranion capable of translating its stereochemical information to the iminium intermediate on the basis of their proximity to each other. This hypothesis was confirmed when it was found that the chiral phosphoric acid **18** (TRIP), together with morpholine, were able to catalyze the conjugate reduction of α,β-unsaturated aldehydes in a very efficient manner, using a Hantzsch dihydropyridine as reductant (Scheme 6.26) [66]. The so-formed catalytic salt demonstrated its ability with a wide

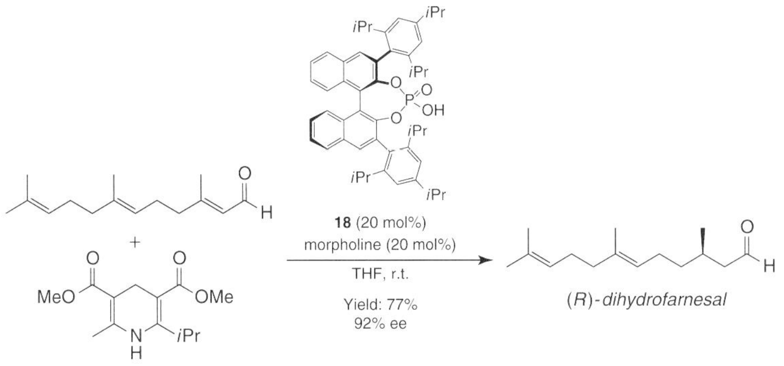

Scheme 6.26 Enantioselective synthesis of (R)-*dihydrofarnesal* by applying the asymmetric counterion directed catalysis to the transfer hydrogenation of enals.

variety of β-disubstituted enals containing both aromatic and aliphatic substituents that furnished excellent yields and enantioselectivities, although the system failed with very hindered substrates. This catalytic system also demonstrated an improved performance with respect to imidazolidinones **16a** and **16d** in the enantioselective reduction of (*E*)-citral. The reaction proceeded with a low enantioselectivity when the latter catalysts were employed, but furnished the reduction product in good yield (71%) and enantioselectivity (90% ee) when morpholinium-TRIP ammonium phosphonate was employed as the catalyst. This same methodology was applied to the synthesis of (*R*)-dihydrofarnesal (Scheme 6.26), which has been identified as a marking pheromone of several bumblebee species, and also as a constituent of the scent of several flowers such as orchids or lemon blossom.

The application of this concept to the enantioselective transfer hydrogenation of enones deserves special attention. In this case, a primary amine was employed as the iminium-forming reagent because of the above-mentioned superior ability of primary amines to form of iminium ions with enones compared to their secondary amine counterparts. In particular, chiral aminoesters have been employed as amine catalysts in this reaction design, together with the chiral phosphoric acid, TRIP (**18**) [67]. Following extensive studies, the authors found the *tert*-butyl ester of valine to be the most appropriate amine catalyst for this transformation, in terms of both yield and enantioselectivity. However, as both constituents of the catalytic salts were chiral, an interesting double stereodifferentiation phenomenon was observed, which led to the identification of the corresponding *matched* and *mismatched* combination of reagents. As can be seen in Scheme 6.27, the model reaction with 3-methyl cyclohexenone proceeded in excellent yield and enantioselectivity with **18** and (*S*)-valine, whereas changing the configuration of the chiral phosphoric acid led to an almost racemic product. The stereodirecting ability of the amine or the phosphoric acid themselves was not sufficient to achieve a good enantioselectivity, as their independent use led to poor enantioselection in the process. When these optimized conditions were extended to a wide range of enones they provided excellent results, especially for cyclic enones, with only slightly lower enantioselectivities being obtained than for acyclic enones. In addition, a remarkably low catalyst loading (5 mol%) was required to achieve such excellent results

Finally, a recyclable chiral catalyst has been developed by using a resin-supported polypeptide which incorporates a polyleucine chain as linker, and contains a terminal proline unit (Figure 6.10), which has been successfully employed in the enantioselective transfer hydrogenation of enals, using Hantzsch esters as hydrogen donors, and in combination with a Brønsted acid co-catalyst such as TFA [68]. The terminal *N*-proline motif is believed to engage in iminium activation of the substrate, while the polyleucine tether shows up as an excellent linker, contributing with the formation of a hydrophobic pocket close to the active catalytic site where organic reagents can concentrate. In this case, the catalyst was shown to be particularly active in aqueous media, while the fact that it was supported on a polystyrene resin allowed a very easy recycling. Unfortunately, the substrate scope

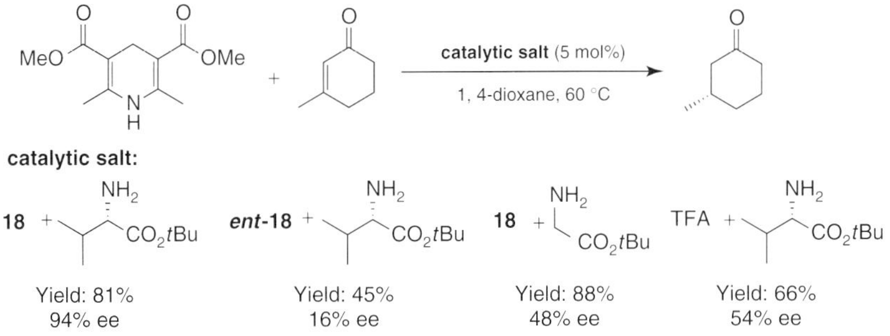

Scheme 6.27 The asymmetric counterion directed catalysis concept in the transfer hydrogenation of enones.

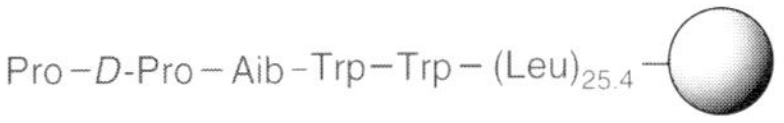

Catalytic salt	Yield	ee
18 + Val-CO$_2$tBu	81%	94% ee
ent-18 + Val-CO$_2$tBu	45%	16% ee
18 + Leu-CO$_2$tBu	88%	48% ee
TFA + Val-CO$_2$tBu	66%	54% ee

Pro–D-Pro–Aib–Trp–Trp–(Leu)$_{25.4}$–●

Figure 6.10 A resin-supported peptide for the enantioselective transfer hydrogenation of enals (Aib = 2-aminoisobutyric acid).

indicated that the yield of the reaction was greatly affected by subtle changes in the structure of the Michael acceptor, which in turn limited its applicability to a broad range of substrates. Nevertheless, the enantioselectivity appeared to be excellent in all of the cases tested.

6.6
Conjugate Additions of Heteronucleophiles

The conjugate addition of heteroatom-centered nucleophiles to α,β-unsaturated aldehydes and enones, using chiral amines as catalysts, has also been the focus of intensive research in the field of asymmetric organocatalysis. This has been especially apparent during the past two years, with the number of reports being made having increased notably. The different types of nucleophile employed, such as sulfur-, oxygen-, nitrogen-, and phosphorus-based Michael donors, will be described at this point. In general, the main problem associated with these reactions relates to the reversibility of the conjugate addition process, which leads to low conversions or may translate into a low configurational stability of the final compounds, resulting in low enantioselectivities. For this reason, most of the methods reported in this context have incorporated an additional electrophile into the reaction design in order to quench the hetero-Michael addition product in a typical cascade process, as this will override the reversibility of the reaction. For this reason, most of the methodologies described for the conjugate addition of heteronucleophiles proceeding via iminium activation are detailed in Section 6.7, which is devoted to cascade processes.

6.6.1
N-Nucleophiles

The aza–Michael reaction stands as a very important transformation in organic synthesis [69], as it provides a direct access to β-amino carbonyl compounds, which are the key constituents of many biologically active compounds and also serve as useful building blocks in total synthesis. Although research directed towards the development of enantioselective organocatalytic methods has also been intense, especially during the past few years, the enantioselective amine-catalyzed aza–Michael reaction with α,β-unsaturated aldehydes or ketones has remained undeveloped until very recently. The main reason for such a delay is associated with the additional chemoselectivity issues that must be addressed in this particular case, because both the catalyst and the nucleophile are amine species. Hence, the chiral amine chosen as a catalyst must not undergo conjugate addition reactions, as this would eventually lead to catalyst consumption. Likewise, the amine reagent selected as the nucleophile must not participate in iminium ion formation, as this would lead to the generation of an achiral intermediate and, in turn, to the formation of a racemic product.

MacMillan first provided the means to overcome these difficulties by selecting N-tert-butyldimethylsilyloxycarbamates as very convenient nitrogen nucleophiles [70]. These reagents show an enhanced nucleophilicity due to the presence of the trialkylsilyloxy moiety (the so-called α-effect), while the introduction of a carbamate functionality results in the formation of a nonbasic N-protected β-amino carbonyl compound as reaction product, thus avoiding the reversibility of the reaction which would eventually lead to racemization. As can be seen in Scheme 6.28, the chiral imidazolidinone **ent-16a** was shown to be an excellent catalyst for the aza–Michael addition of these particular nucleophiles to α,β-unsaturated aldehydes, producing excellent results for a wide variety of differently substituted enals. However, the scope of the reaction was limited to only alkyl-substituted enals, and no examples illustrating the performance of this catalytic system with aryl-substituted substrates – which demonstrate a lower reactivity towards the addition process and also an increased tendency for the retro-aza–Michael process – have been given. Shortly afterwards, Córdova showed that O-trimethylsilyldiphenylprolinol **11a** was also a very effective catalyst for the same reaction, in this case employing

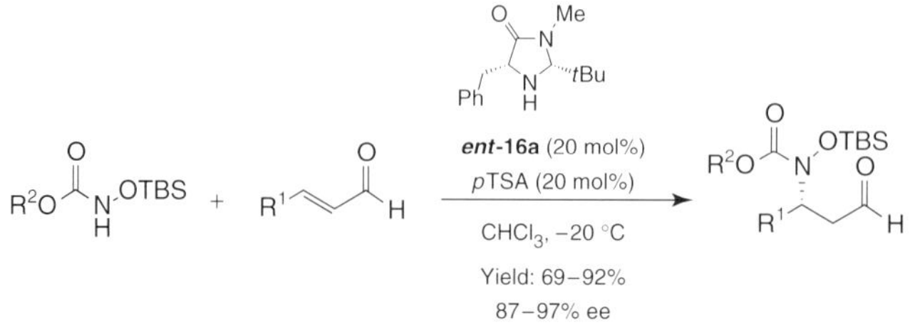

Scheme 6.28 Enantioselective aza–Michael reaction of N-silyloxycarbamates and enals catalyzed by **ent-16a**.

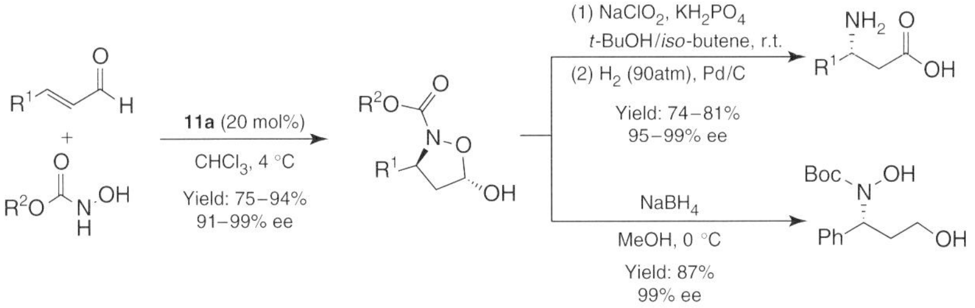

Scheme 6.29 Enantioselective aza–Michael reaction of N-hydroxycarbamates and enals.

N-methoxycarbamates as nucleophiles [71]. However, the yields were somewhat lower than those achieved when using imidazolidinone **ent-16a**, and examples were provided in which aromatic enals were used as electrophiles, obtaining significantly lower yields and enantioselectivities in those particular cases. This illustrated clearly the difficulties associated with the use of these particular Michael acceptors.

Structurally related nucleophiles such as simple N-hydroxy carbamates have found applicability as nucleophiles in this context (Scheme 6.29). For instance, the conjugate addition of these reagents to α,β-unsaturated aldehydes can easily be carried out with high yields and enantioselectivities, using **11a** as catalyst in CHCl$_3$ as solvent and at 4 °C [72]. In this case, the Michael adduct underwent subsequent intramolecular 1,2-addition following the conjugate addition step, furnishing 5-hydroxyoxazolines as single diastereoisomers. This process also assisted the reaction by avoiding the reversibility of the aza–Michael step and pushing the equilibrium towards formation of the final products. The reactivity of these 5-hydroxyoxazolines was surveyed by carrying out several transformations, such as hydrogenolytic ring-opening reactions or a reduction of the hemiacetal moiety using one adduct as model compound. This led directly to a β-amino acid and a γ-hydroxylamino alcohol respectively, and illustrated not only the chemical versatility but also the potential of these compounds as chiral building blocks in total synthesis.

Nitrogen heterocycles represent another group of Michael donors employed in enantioselective amine-catalyzed aza–Michael reactions. These compounds incorporate an acidic N–H group, which makes them suitable nucleophiles for conjugate additions under the iminium activation manifold. In this context, Jørgensen has reported a thoughtful study of the conjugate addition of 1,2,4-triazole, 1,2,3-triazole, benzotriazole, and 5-phenyltetrazole to α,β-unsaturated aldehydes, using **11b** as catalyst [73]. The best results were obtained in the reaction with 1,2,4-triazole (Scheme 6.30), which proceeded with complete regioselectivity and high enantioselectivity at room temperature, whereas in the reactions with other heterocycles the temperature of the reaction was a key parameter to be controlled due to the tendency shown by the adducts to racemize at room temperature, presumably via a retro-Michael/Michael process. This also led to a need for the *in situ* reduction/esterification of the obtained adducts for their isolation and purification. In addition, whilst the reaction with 5-phenyltetrazole occurred with complete

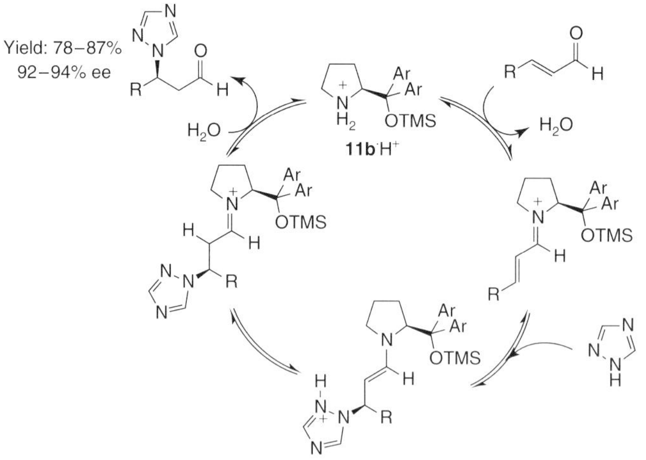

Scheme 6.30 The proposed catalytic cycle for the enantioselective aza–Michael reaction of 1,2,4-triazole with enals catalyzed by **11b**.

regioselectivity, the use of 1,2,3-triazoles led to the formation of mixtures of regioisomers, which could be separated using chromatography and isolated as highly enantioenriched compounds. In the same report, a detailed computational study was carried out for the reaction of 1,2,4-triazole with (*E*)-2-pentenal in order to gain further insight into the mechanistic pathway of the transformation. In this case, the calculated free energies of the transition state for the conjugate addition of the triazole to the iminium ion and for the subsequent protonation of the enamine intermediate were found to be very similar, with the former having a slightly lower energy barrier. In fact, a difference of only 1.4 kcal mol^{-1} was calculated between the free energy of both transition states, which was attributed to be the rate-determining step of the reaction. An energetically favored pathway for this protonation step was found which consisted of a water-mediated intramolecular proton delivery from the protonated heterocyclic moiety to the enamine unit. However, these results suggested that subtle changes in the reaction conditions, or in the structure of the reagents, might lead to changes in the energetic balance of the intermediates participating in the catalytic cycle, and this would eventually translate into different experimental results. The calculations also provided a good explanation for the high enantioselectivity of the reaction, and led to a very good agreement between the calculated difference in free Gibbs energy between the two stereoisomeric transition states in the rate-determining step and the experimental results. As had occurred for other cases in which catalyst **11b** was employed for the activation of enals via iminium formation, the bulky diaryltrimethylsilyloxy group incorporated at the pyrrolidine catalyst allowed a very effective geometry control of the iminium ion (formation of the *E*-isomer was favored over the *Z*-isomer), and provided the required steric shielding for one of the stereotopic faces of the Michael acceptor. Alternatively, when Jørgensen's group evaluated the use of succinimide

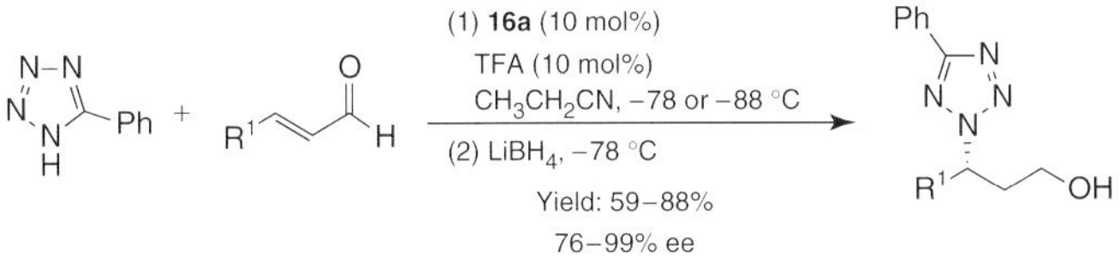

Scheme 6.31 Enantioselective aza–Michael reaction of phenyltetrazole to enals catalyzed by **16a**.

as nucleophile, it was found to provide slightly lower enantioselectivities in the reaction catalyzed by **11b** [74]. In this case, a base co-catalyst (NaOAc) had to be included in order to increase the nucleophilicity of succinimide, by promoting its deprotonation.

In an independent and almost simultaneously conducted study, imidazolidinone **16a** was also identified as a very efficient catalyst for the enantioselective aza–Michael reaction of 5-phenyltetrazole with α,β-unsaturated aldehydes (Scheme 6.31) [75]. The final compounds were also shown to be very configurationally unstable, and therefore had to be reduced *in situ* for isolation and purification to maintain their stereochemical integrity. The reaction proceeded with complete regioselectivity, affording exclusively N-2-alkyl-substituted tetrazole derivatives, with good yields and enantioselectivities being obtained for a wide range of different enals tested, although the reaction with aromatic substrates did not proceed satisfactorily. In this case, the reversibility of the reaction was avoided by carrying it out at very low temperatures (−78 or −88 °C), which resulted in very long reaction times to reach synthetically useful yields. Interestingly, it should be noted that the absolute configuration of the final adducts obtained in this case was the same as that observed for the aza–Michael reaction using N-silyloxycarbamates as nucleophiles (as reported by MacMillan), even though the same catalyst with an opposite configuration was employed in that case. Although no explanation for this behavior was provided in the original report, it is possible that protonation of the enamine intermediate formed after the conjugate addition reaction might be the rate-determining step of the process, rather than the conjugate addition.

Interesting intramolecular versions have also been studied by the group of Fustero, showing the high potential of the methodology in total synthesis with the preparation of several natural products (Scheme 6.32). The reaction design included a functionalized α,β-unsaturated aldehyde containing a protected amino group located at a convenient position, thus allowing the formation of a heterocyclic structure after the intramolecular aza–Michael reaction had taken place [76]. As in other cases, the reversibility of the reaction made the obtained adducts configurationally unstable, such that the reduction of the formyl group had to be carried out *in situ* for isolation of the final compounds. Several tethers of different lengths and the incorporation of intercalating heteroatoms were evaluated as connectors between the amino functionality and the enal moiety. Subsequently, variable yields and excellent enantioselectivities of the target compounds were achieved, with catalyst **11b** appearing as the most effective among the different chiral amines tested. A similar intramolecular reaction in which an aromatic ring

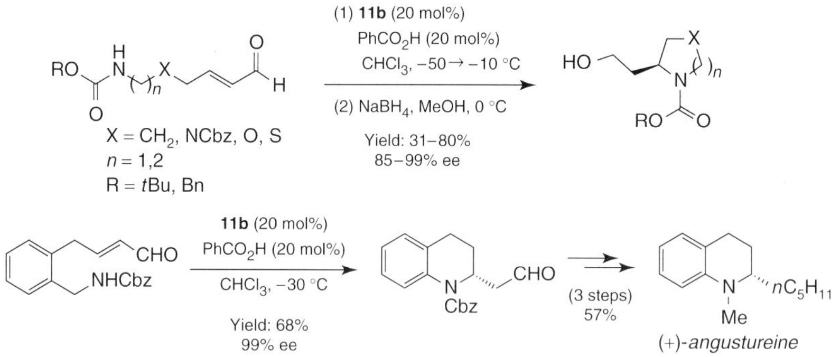

Scheme 6.32 Enantioselective intramolecular aza–Michael reactions catalyzed by **11b**.

was intercalated into the tether was also used in the asymmetric preparation of tetrahydroquinolines, tetrahydroisoquinolines, and indolines, including the biologically active alkaloid (+)-angustureine (Scheme 6.32) [76b].

The aza–Michael reaction to enones under iminium activation has been faced with the help of a bifunctional catalyst containing a primary amine moiety capable of condensing readily with the Michael acceptor. The catalyst also incorporated an additional tertiary amine basic site capable of interacting with the N-nucleophile and also assisting its deprotonation, thus activating the Michael donor in cooperative fashion in the reaction. In this way, a very well-organized transition state would be formed that should result in a very high level of stereoinduction. The single case reported in this context employed 9-amino cinchona alkaloids and N-benzyloxy carbamates as nucleophiles in the aza–Michael reaction to a wide range of acyclic enones [77]. The reaction proceeded with excellent yields and enantioselectivities in most cases, although different catalysts had to be used depending on the substitution pattern at the Michael acceptor. Catalyst **8c** furnished excellent results for β-alkyl substituted enones, but the use of enones with aromatic substituents at the β-position required use of the more active derivative **8a** as catalyst (Scheme 6.33).

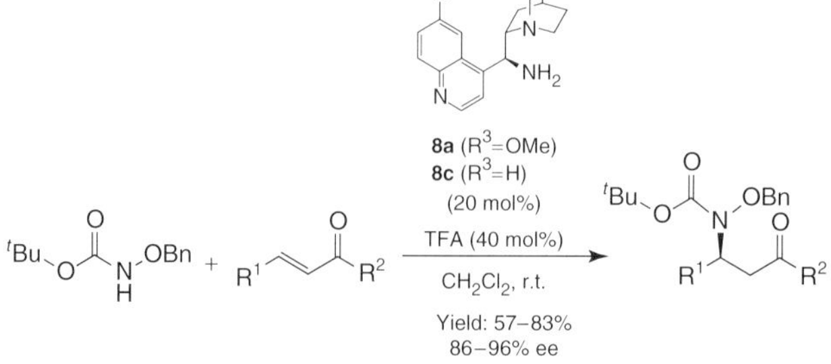

Scheme 6.33 Asymmetric aza–Michael reaction of benzyloxycarbamates and enones catalyzed by bifunctional catalysts.

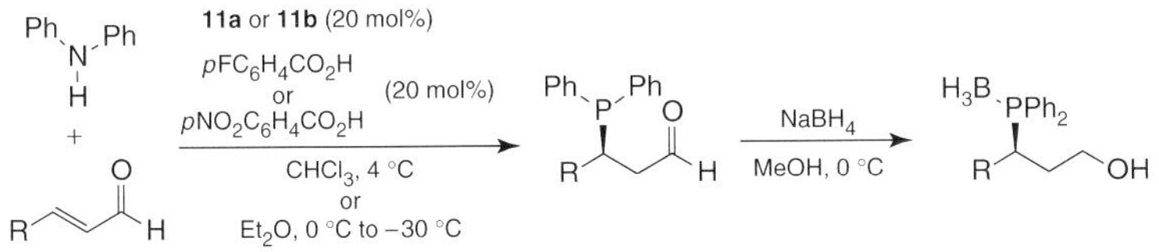

Scheme 6.34 Asymmetric phospha–Michael reaction of diphenylphosphine and enals.

6.6.2
P-Nucleophiles

Currently, the phospha–Michael reaction has undergone much less development than its aza–Michael counterpart, with only two very recent approaches having been reported. The first of these examples was developed independently by Córdova and Melchiorre, and involved the use of diphenylphosphine as nucleophile and different carboxylate ammonium salts derived from O-trimethylsilyl diarylprolinols **11a** or **11b** as catalysts (Scheme 6.34) [78]. Due to their low configurational stability, and to the tendency of alkyl phosphines to undergo oxidation in the presence of air, the phospha–Michael adducts had to be reduced *in situ* and the phosphine moiety transformed to the corresponding phosphine–borane complex for a better isolation and purification of the final products. In this reaction, both reports indicated that the nature of the Brønsted acid co-catalyst was decisive to achieve a good stereocontrol, with the slightly more acidic *p*-nitrobenzoic acid [78a] and *p*-fluorobenzoic acid [78b] as additives furnishing the best enantioselectivities. The reaction appeared to proceed much more easily with aromatic enals, whereas the alkyl-substituted α,β-unsaturated aldehydes were more problematic substrates, especially when the β-alkyl substituent was a small group such as methyl or ethyl. In such cases, experimental conditions such as dilution appeared to be crucial to achieve good results. The authors also surveyed the reactivity of the obtained β-phosphino aldehyde adducts by carrying out several modifications, such as an *in situ* reductive amination to furnish γ-phosphino amines [78a]. Alternatively, the "one-pot" phospha–Michael reaction/oxidation sequence has been studied using different oxidants. As an example, treatment of the Michael adducts with $NaClO_2$ allowed the preparation of β-phosphine oxide carboxylic acids, whilst an iodine-mediated oxidation allowed a selective oxidation of the phosphine moiety without affecting the formyl group, producing cleanly the corresponding β-phosphine oxide aldehydes [78b,c]. In all cases the stereochemical integrity of the stereogenic center previously generated in the enantioselective phospha–Michael reaction was maintained.

Other phosphorus nucleophiles have been tested in this reaction with different degrees of success. For example, Melchiorre reported that di-*tert*-butylphosphine was inert under the optimized reaction conditions, the diphenylphosphine–borane complex furnished a racemic product, and diphenylphosphine delivered only the 1,2-addition side product [78a]. Córdova also tested a wider variety of other compounds [78c], and concluded that only trivalent phosphorus reagents were applicable in this reaction, which was attributed to the higher nucleophilicity of

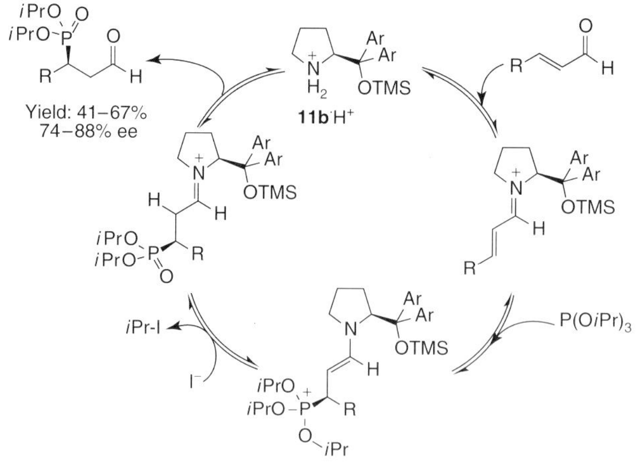

Scheme 6.35 Proposed catalytic cycle for the **11b**-catalyzed phospha–Michael reaction of P(O*i*Pr)$_3$ and enals.

these compounds. On the other hand, several trialkylphosphites such as P(OMe)$_3$ and P(OEt)$_3$ were found to provide the phospha–Michael adducts, but furnished low conversions and enantioselectivities. Interestingly, in a different report [79], Jørgensen identified improved conditions for carrying out the conjugate addition of P(O*i*Pr)$_3$ to α,β-unsaturated aldehydes using **11b** as catalyst (Scheme 6.35). In this case, the use of trialkylphosphites as nucleophiles implied a different mechanistic pathway because, after the conjugate addition step, an intermolecular nucleophilic substitution must occur in order to deliver the final phosphonate moiety. In the Jørgensen report, it was proposed that this second S_N reaction was the rate-determining step of the process and therefore, the inclusion of an external nucleophilic as an additional additive (e.g., sodium iodide) would lead to improvements in the conversion and enantioselectivity of the reaction. Under these improved conditions, a wide variety of alkyl- and aryl-substituted α,β-unsaturated aldehydes underwent conjugate addition reactions with P(O*i*Pr)$_3$, delivering the corresponding β-phosphonate aldehydes in moderate yields, and enantioselectivities in the range of 75–88% ee.

6.6.3
O-Nucleophiles

The oxa–Michael reaction to α,β-unsaturated aldehydes or ketones under the iminium activation strategy is an even more challenging reaction than the other hetero-Michael reactions presented above. The low nucleophilicity exhibited by the most common oxygen nucleophiles such as water, alcohols or carboxylates, and their tendency to undergo 1,2-addition rather than the desired conjugate addition process, makes this transformation especially difficult. In addition, when stereochemical control must be achieved, special care must be paid towards the

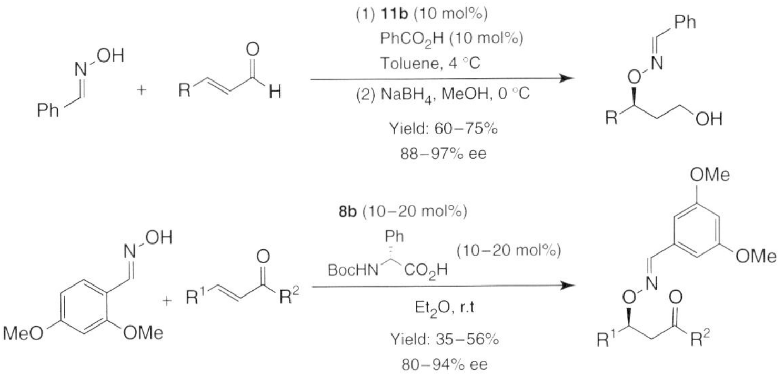

Scheme 6.36 Enantioselective oxa–Michael reaction of aromatic oximes with enals and enones.

high tendency of the oxa–Michael adducts to engage in the retro-addition reaction, which eventually would lead to a racemization of the final compounds. In the past, one way to overcome this problem has been to carry out a cascade process; this makes the reaction proceed forward, pushing the equilibria that are operating in the catalytic cycle towards formation of the final product. These particular cases are described in Section 6.7.

For these reasons, the initial attempts in this field refer to nonstereoselective versions, including the above-mentioned very early example by Langenbeck in 1937 [5], in which the conjugate addition of water to crotonaldehyde was carried out using morpholine as catalyst. Attempts by Maruoka to carry out the conjugate addition of aliphatic alcohols to enals in a stereoselective manner, using axially chiral amines derived from BINOL as catalysts, also proceeded with low levels of enantioselection, but with satisfactory yields in several examples [80]. Nevertheless, the first general and effective aminocatalytic procedure for carrying out enantioselective intermolecular oxa–Michael reactions was presented by Jørgensen, who solved the problem of the low reactivity of the oxygen nucleophile and reversibility of the reaction by choosing aromatic oximes as Michael donors (Scheme 6.36) [81]. Under the optimized conditions, **11b** catalyzed the conjugate addition of benzaldehyde oxime to α,β-unsaturated aldehydes in good yields and excellent enantioselectivities. The reversibility of the reaction was found to be problematic with regards to the configurational stability of the adducts; consequently, they were subjected to *in situ* reduction yielding the corresponding alcohols, which could be more easily manipulated for their purification and characterization. It should also be pointed out here that a large excess of nucleophile (up to 3 equiv.) was required in order to reach full conversion as rapidly as possible, and to avoid any prolonged reaction times that might allow racemization of the adducts to occur. The reaction was shown to have a rather broad substrate scope for alkyl-substituted enals, but failed in the case of β-aryl-substituted substrates. Moreover, the structure of the oxime moiety also had an important influence on the reaction, with the bulkier mesitylaldehyde oxime requiring very long reaction times to reach full conversion.

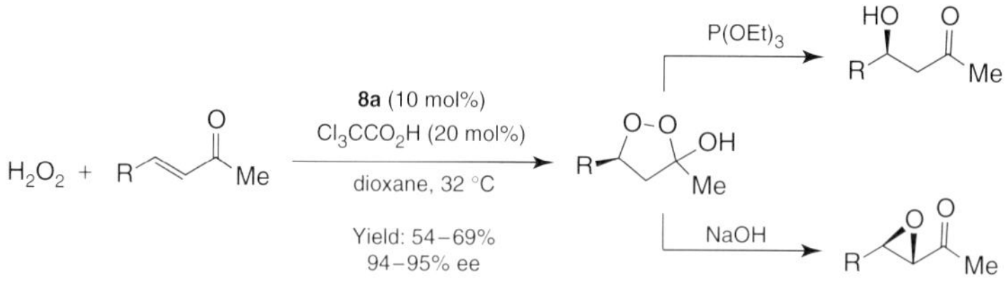

Scheme 6.37 Enantioselective hydroperoxidation of enones.

One important advantage of the methodology proved to be the easy cleavage of the O–N bond, which allowed a very simple procedure for the asymmetric synthesis of 1,3-diols, thus increasing the synthetic applicability of the methodology. The strategy of using oximes as nucleophiles was later applied by Melchiorre and coworkers to the oxa–Michael addition to enones, again using the asymmetric counterion-directed catalysis approach (Scheme 6.36). In this case, the combination of 9-aminocinchona alkaloid **8b** with (R)-N-Boc-phenylglycine was shown to be an appropriate catalyst for the reaction, reaching moderate yields of conjugate addition products, albeit with ee-values between 80% and 94% ee [82]. As occurred in the reaction with enals, the scope of the reaction was limited to aliphatic enones.

Another approach to the β-hydroxylation of α,β-unsaturated enones has been developed in which hydrogen peroxide is employed as the oxygen nucleophile (Scheme 6.37) [83]. In this case, the reaction delivered a β-hydroperoxide which underwent spontaneous intramolecular addition, furnishing a 1,2-dioxolan-3-ol derivative as the final reaction product, mixed with minor amounts of an α,β-epoxy ketone side product. The formation of this cyclic derivative operates as the driving force for the reaction, avoiding the retro-addition process, and therefore accounting for the high conversion observed and lack of racemization of the final product. In this context, the reaction of several enones with H_2O_2 in the presence of primary amine catalyst **8a** proceeded with moderate yields but with excellent enantioselectivities, although the substrate scope was limited exclusively to methyl alkenyl ketones incorporating alkyl substituents at the β-position. Substrates with aromatic substituents or trisubstituted enones were shown to be unreactive under the reaction conditions. One interesting feature of this methodology was the possibility of converting the 1,2-dioxolane adducts into β-hydroxy ketones by reduction with $P(OEt)_3$, and also the easy access to epoxides by carrying out a base-promoted sequential hemiacetal cleavage/intramolecular substitution process.

6.6.4
S-Nucleophiles

The iminium-activation strategy has been successfully applied in a couple of examples of conjugate additions of thiols to conjugated aldehydes and enones. The Michael-type addition of thiols to α,β-unsaturated aldehydes to enals has been reported to proceed very efficiently using O-TMS-protected diarylprolinol **11b** as catalyst (Scheme 6.38) [84]. As occurred with other hetero-Michael reactions,

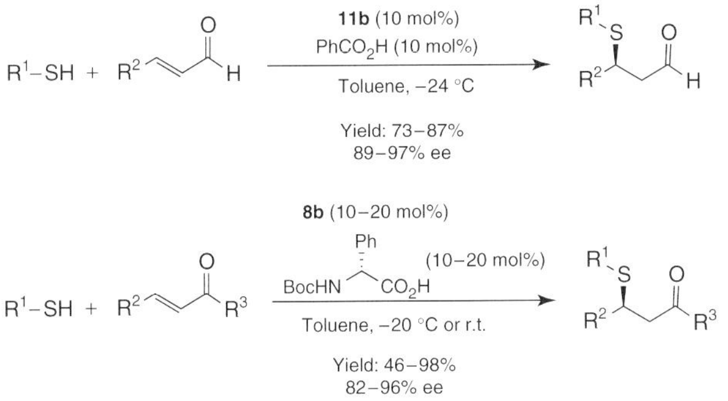

Scheme 6.38 Enantioselective sulfa-Michael reaction of thiols with enals and enones.

these authors observed that the conjugate addition products were not configurationally stable at room temperature, due to the competitive retro-addition reaction. However, this could be avoided by carrying out the reaction at −24 °C and by reducing the adducts *in situ*, thus furnishing the corresponding γ-thio alcohols in excellent yields and enantioselectivities. Those substrates amenable to undergo this reaction included both alkyl- and aryl-substituted enals of different structures. In addition, different thiols such as *tert*-butyl mercaptan, benzyl mercaptan, or even a functionalized thiol, were evaluated and shown, in all cases, to be equally effective nucleophiles for use in this reaction. These conditions were later applied to a domino sequence (see below). The same reaction involving α,β-unsaturated ketones as Michael acceptors has been reported by Melchiorre and coworkers, using the same primary ammonium-*N*-Boc-phenylglycinate salt catalyst as was previously used in the parent oxa–Michael reaction (Scheme 6.38) [85]. In this case, the addition of both, *tert*-butyl- and benzyl mercaptan to enones was found to proceed with moderate to good yields and with ee-values between 84% and 95% for a wide range of different acyclic enones, although the reaction seemed to be more problematic when aromatic substituents were introduced at the enone moiety. In contrast, when a cyclic enone such as cyclohexenone was evaluated as a Michael acceptor, excellent results were obtained.

6.7
Cascade Reactions

Domino or cascade reactions represent a highly efficient approach for the straightforward construction of complex molecules in the shortest and most efficient way, because they allow the build-up of complex molecules in a very efficient manner, while minimizing the number of laboratory operations required and the generation of waste materials [86]. In addition, when stereochemical control is a fundamental parameter to be controlled, domino processes emerge as a very effective approach for building up the target molecule with good stereoselectivity. In this context, the use of chiral amines as catalysts in domino processes can be envisaged as a very useful and competitive method for creating molecular complexity from readily available, low-cost starting materials, while displaying an exceptional performance with regards to stereochemical control [87]. A particularly interesting situation arises when the iminium activation strategy is applied to cascade processes initiated by Michael-type reactions, mainly because the intermediate enamine generated after conjugate addition can participate in a subsequent reaction, provided that a conveniently functionalized substrate or an additional electrophile is included in the reaction design. In some other cases, the cascade process may occur when the catalyst is released; the functionalities present in the Michael adduct may then react intramolecularly with each other, such that the stereochemistry of the newly generated stereogenic centers will be controlled by the chirality present at the substrate.

6.7.1
Michael/Aldol Cascade Reactions

A number of early reports have described cascade processes consisting of a Michael reaction to α,β-unsaturated aldehydes or ketones (activated as the corresponding iminium ion), followed by an intramolecular aldol reaction step. Most of these reactions have been focused on the self-dimerization of enals or enones catalyzed by proline (or analogs derived therefrom), and generally proceed with low enantioselectivities [88]. Although no clear and definitive mechanistic pathways have been confirmed for these reactions, the most widely accepted suggestion for the dimerization of enals (Scheme 6.39) involves a sequential activation by the catalyst of one molecule of substrate as a dienamine (Michael donor), and another molecule as an iminium ion (Michael acceptor). Following the Michael reaction step, an intermediate containing an iminium and an enamine moiety

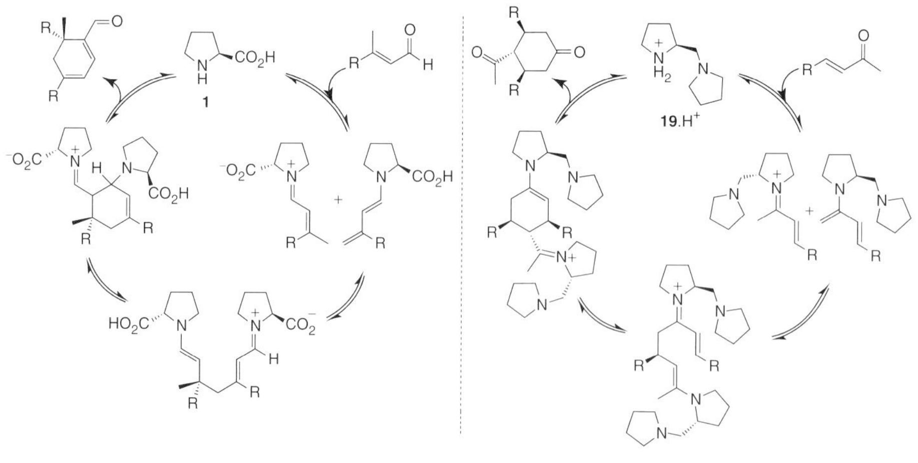

Scheme 6.39 Possible catalytic cycles for the self-dimerization of enals and enones catalyzed by chiral secondary amines.

is generated; the direct reaction between these two moieties (Mannich reaction), followed by an elimination step (which is required to release the catalyst) has been proposed to account for the generation of the final compound. Alternatively, an intramolecular Morita–Baylis–Hillman process has been claimed to be the final step to furnish the cyclic products [88c]. An example has also been reported of a crossed reaction between two different aldehydes, allowing the introduction of different substituents at the final cyclohexadiene product, although in those cases where high enantioselectivities were observed the yields were only moderate at best [88d]. Although, for the dimerization of enones, a Diels–Alder reaction has been proposed to explain the results [88e], a Michael/Michael cascade process might also account for formation of the final products. In this case, the reaction would start with the formation of a more stable 2-aminodiene nucleophile intermediate; this would undergo a first Michael addition step to the iminium ion derived from a second molecule of the substrate, delivering in this case an intermediate incorporating an enamine and an α,β-unsaturated iminium moiety in the correct situation so as to engage in an intramolecular Michael reaction; a final hydrolysis step would then allow release of the catalyst (Scheme 6.39). When this reaction was carried out using pyrrolidinomethylpyrrolidine **19** as the chiral catalyst, it was seen to proceed with moderate to high diastereoselectivity, but to favor the *meso* isomer as the major compound. The minor optically active diastereoisomer was isolated with a low ee-value in a single case.

The *Robinson annulations* is considered to be the classical Michael/aldol cascade reaction, and consequently it has been the subject of many attempts to exploit the iminium activation approach, in particular the L-proline-catalyzed condensation of 1,3-diketones and methyl vinyl ketone. In this case, activation of the nucleophile is not necessary because of the highly acidic nature of the Michael donor, which delivers spontaneously the required enolate nucleophile in the reaction medium. Hence, the role played by the amine catalyst is thought to be limited exclusively to activation of the electrophile via iminium ion formation. The first example of this reaction, which was reported by Barbas in 2000 [14], showed that a Robinson annulation reaction between methyl vinyl ketone and a 2-methylcyclohex-1,3-dione could be carried out in an enantioselective manner, using proline as the catalyst (49% yield, 76% ee) (Scheme 6.40). A more elaborate version of the reaction has since been reported that consisted of a Knoevenagel/hydrogenation/Robinson annulation cascade process for the asymmetric synthesis of Wieland–Miescher and Hajos–Parrish ketone analogs with different substitution patterns [89].

Following a similar reaction design, a cascade Michael/aldol reaction has been investigated using other different 1,3-dicarbonyl compounds to build up

Scheme 6.40 L-Proline-catalyzed enantioselective Robinson annulation.

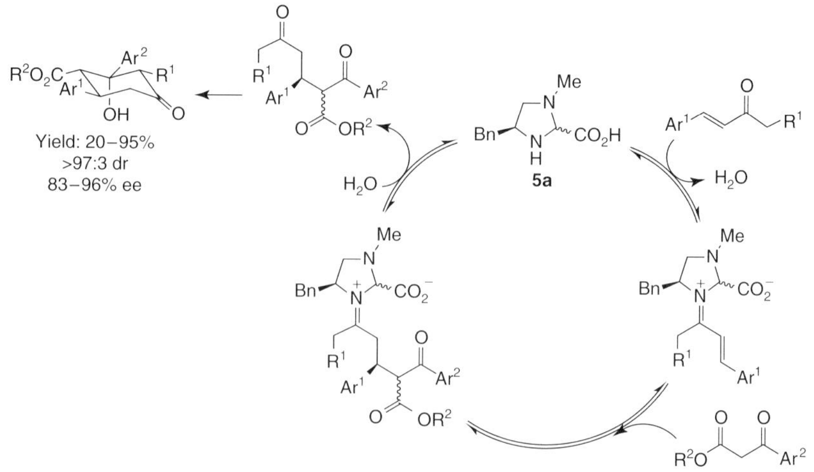

Scheme 6.41 Cascade Michael/aldol reaction of β-ketoesters and α,β-unsaturated ketones.

highly functionalized cyclohexanes containing multiple stereogenic centers. In this context, Jørgensen applied the imidazoline catalyst **5a** (which had previously demonstrated an ability to activate enones in simple Michael reactions) to the reaction between β-ketoesters and enones (Scheme 6.41) [90a]. The proposed mechanism involved a first step in which the Michael reaction between both reagents under iminium activation established the first stereogenic center; the catalyst was then released and a mixture of diastereomeric Michael adducts obtained. An intramolecular aldol reaction would then take place, in which participation of the imidazoline catalyst as a base would assist deprotonation of the α-hydrogen of the ketone; however, the possible participation of an intermediate enamine species was not completely ruled out. Interestingly, the final compounds were obtained as single diastereoisomers, even though a mixture of diastereoisomers was suggested to have been formed in the initial Michael reaction. This situation was interpreted in terms of the formation of the most stable diastereoisomer, which contained all bulky groups in equatorial bonds in the final cyclohexane derivatives. The final compounds were obtained in variable yields, and showed a dependence on the nature of the ester moiety (better yields were obtained with benzyl esters than with ethyl or methyl esters), although the range of substrates tested had been limited to β-aryl-substituted enones as Michael acceptors and to 3-aryl-substituted β-ketoesters as Michael donors. This reaction was further extended to other acidic carbon nucleophiles such as β-ketosulfones and 1,3-diketones [90b], the latter reaction also being possible using L-proline as catalyst, albeit with a lower enantioselectivity [90c].

When this strategy was applied to α,β-unsaturated aldehydes as Michael acceptors, the reaction proceeded in a similar way. In this context, Jørgensen set up an interesting variant which involved the use of *tert*-butyl β-ketoesters as nucleophiles in the Michael reaction that would initiate the cascade process (Scheme 6.42) [91]. Under the optimal reaction conditions, the Michael/aldol cascade process took

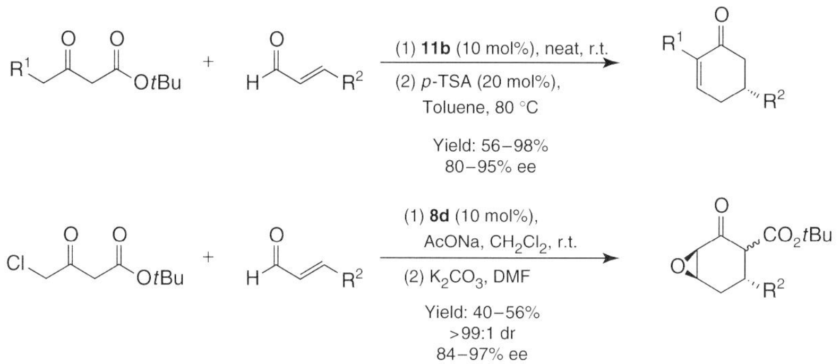

Scheme 6.42 Cascade Michael/aldol reactions of *tert*-butyl β-ketoesters and α,β-unsaturated aldehydes.

place as expected, with *O*-trimethylsilyl diarylprolinol **11b** being the most efficient catalyst for the transformation. When the substrate consumption was complete, however, it became necessary to add *p*-TSA to the crude reaction mixture and to apply heat for several hours. This induced a dehydration reaction and hydrolysis of the *tert*-butoxycarbonyl moiety, followed by a decarboxylation process to deliver, in a one-pot operation, 5-alkyl or 2,5-dialkyl substituted 2-cyclohexenones containing a single stereogenic center, in excellent yields and with ee-values ranging from 80% to 96%. Alternatively, when the reaction between γ-chloro-β-ketoesters and enals using the same catalyst was carried out, a one-pot Michael/Darzens reaction was observed which delivered α,β-epoxycyclohexanones in excellent yields and both diastereoselectivity and enantioselectivity (Scheme 6.42) [92]. In this case, the treatment of both reagents in the presence of a catalyst at room temperature initiated the Michael/intramolecular aldol cascade reaction. Subsequently, it became necessary to add K_2CO_3 and dimethylformamide (DMF) to the reaction mixture, followed by stirring for several hours to accomplish the final intramolecular nucleophilic substitution required to form the epoxide moiety. The high diastereoselectivity of the reaction could be explained in terms of the reversibility of the intramolecular aldol reaction step. Hence, the stereochemical constraints required for the intramolecular S_N2 reaction to occur (OH group and chlorine atom in *anti*-periplanar disposition), together with formation of the most stable diastereoisomer after epoxide formation (bulkier substituents in equatorial positions), would account for the high diastereoselectivity observed. As the stereogenic center between both carbonyl groups was not configurationally unstable, the products could be isolated as an equilibrating mixture of the two possible epimers. It should be noted that, in all of these cases, it was suggested that the chiral amine catalyst had not participated in the intramolecular aldol reaction.

The high performance of the amine-catalyzed cascade processes in the formation of highly functionalized chiral compounds in an easy and stereocontrolled manner has been clearly demonstrated with a cleverly designed cascade

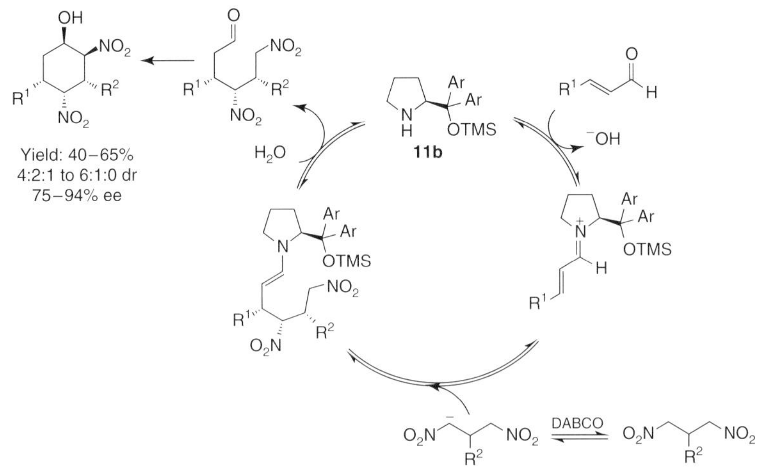

Scheme 6.43 Cascade nitro-Michael/Henry reaction for the formation of cyclohexanes with five stereogenic centers.

nitro-Michel/nitroaldol cascade, as shown in Scheme 6.43 [93]. In this reaction, the use of a 2-substituted 1,3-dinitroalkane reagent in combination with an α,β-unsaturated aldehyde in the presence of O-TMS diarylprolinol **11b** led to the direct formation of 2,4-dinitrocyclohexanols containing five stereogenic centers, and which were generated with a very high degree of stereocontrol. The reaction proceeded first, by a nitro-Michael reaction of the dinitro compound to form the activated iminium ion, generated by condensation between the enal and the catalyst. Based on the configuration of the final adducts, it was assumed that the cyclization step would proceed without catalyst control; consequently, it was proposed that hydrolysis of the intermediate enamine would occur after the nitro-Michael reaction, delivering a dinitroaldehyde intermediate which finally would engage in an intramolecular Henry reaction. Remarkably, one major diastereoisomer was obtained from 32 possibilities, with other two minor diastereoisomers being formed in variable amounts. These minor diastereoisomers were isolated for a single case and identified as the C-3 epimer and a third diastereoisomer with opposite configurations at C1, C2, and C3, that was formed in almost negligible amounts. These results indicated that the catalysts could exert an excellent stereodifferentiation in the first Michael addition reaction with respect to the stereocenters at C4 and C5, whereas the stereocontrol C-3, which would be created due to desymmetrization of the starting dinitroalkane, would be controlled by the nature of the substituents incorporated at both the Michael donor and acceptor (R^1 and R^2). The formation of the third, very minor, diastereoisomer was due to a different orientation of the substituents in the intramolecular Henry reaction. In general, the reaction showed a remarkably high substrate scope with regards to the enal reagent, and allowed the incorporation not only of alkyl-, aryl-, and functionalized alkyl substituents, but also of the 2-aryl substituted dinitroalkane reagent. As a consequence, the final compounds demonstrated enantioselectivities in the range of 75 to 94% ee.

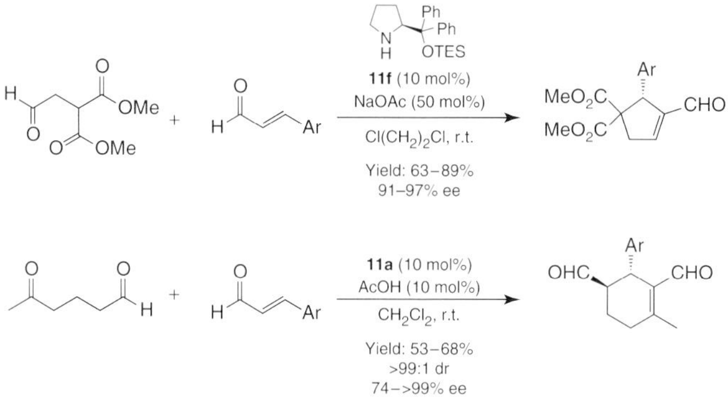

Scheme 6.44 Cascade Michael/aldol reactions for the asymmetric synthesis of polysubstituted cyclopentenes and cyclohexenes.

In contrast, an interesting cascade Michael/aldol process has been developed for the preparation of cyclopentenes which makes use of the iminium/enamine manifold. In this reaction, a malonate reagent containing a functionalized side chain, which also incorporates a formyl group at the appropriate position, has been used as the nucleophile that initiates the cascade process (Scheme 6.44) [94]. The reaction was started with a Michael addition of the malonate to the enal under iminium activation, after which the intermediate enamine underwent an intramolecular aldol reaction with the formyl moiety present at the functionalized malonate reagent employed. The catalyst was then released by hydrolysis, after which a final dehydration reaction occurred to deliver a conjugate cyclopentene adduct as the reaction product. In contrast to previous cascades reported by Jørgensen (Schemes 6.41–6.43), in this case the amine catalyst also participated in the cyclization step by activating the nucleophile via enamine formation. The optimal conditions found for the reaction involved O-TES diphenyl-prolinol **11s** as catalyst, and the inclusion of a basic additive such as NaOAc, which was claimed to activate the malonate nucleophile by assisting its deprotonation. Under these conditions, the reaction took place smoothly for a wide range of α,β-unsaturated aldehydes containing aromatic substituents, and provided excellent yields and enantioselectivities of the final compounds. However, the reaction failed completely when aliphatic enals were tested as substrates. A similar reaction has been developed using 5-oxohexanal as the functionalized reagent undergoing Michael addition/intramolecular aldol reaction with aromatic enals (Scheme 6.44). This led to the formation of functionalized cyclohexenes in moderate yields, but with excellent enantioselectivities and as single diastereoisomers [95]. Although no mechanistic proposal was provided in this case, a dual activation of the 5-oxohexanal (via enamine formation) and the α,β-unsaturated aldehyde (via iminium ion formation) might be operating in this reaction.

6.7.2
Michael/Knoevenagel Cascade Reactions

Other 1,3,5-tricarbonyl compounds which differ from the 1,3-dinitroalkane used by Jørgensen in the nitro-Michael–Henry cascade reaction shown in Scheme 6.43 have been used as potential 1,3-dinucleophiles capable of undergoing a first Michael addition step with enals, which subsequently react with the remaining formyl group to form a cyclohexane ring. This has led to the development of a couple of cascade processes that consists of a Michael reaction that proceeds in an asymmetric fashion by the activation of the enal via chiral iminium ion formation; this is then followed by a release of the catalyst and intramolecular Knoevenagel condensation. By following this reaction design, Hayashi and Jørgensen employed dimethyl 3-oxo-pentanodioate and ethyl 5-diethoxyphosphoryl-3-oxobutanoate, respectively, as dinucleophiles, using O-silyl diarylprolinols as the best catalyst for the reaction in both cases (Scheme 6.45) [96]. The reaction reported by Hayashi provided the final compounds with excellent yields and enantioselectivities following a diastereoselective reduction of the cascade product which, presumably, was obtained as an equilibrating mixture of diastereoisomers due to the inherent configurational instability of the stereogenic center at the β-ketoester moiety. Unfortunately, the reaction scope was limited to the use of aromatic enals, and reported poorer results with the β-alkyl-substituted substrates. In the reaction described by Jørgensen, the final compounds were directly isolated as single diastereoisomers with excellent yields and enantioselectivities, and for a wide range of different alkyl- and

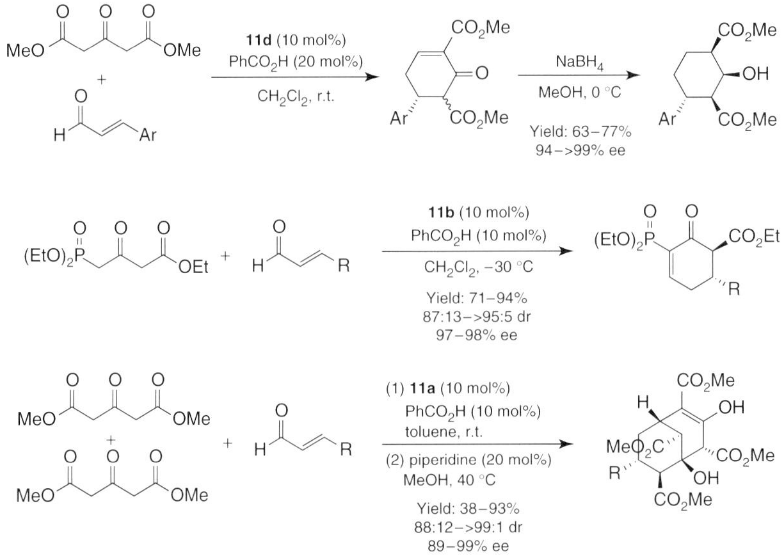

Scheme 6.45 Michael/Knoevenagel cascade reactions for the asymmetric synthesis of functionalized cyclohexanols, cyclohexenones, and bicyclo[3.3.1]non-2-enes.

aryl-substituted α,β-unsaturated aldehydes. However, it was necessary to identify modified conditions when using aliphatic enals, which included the incorporation of hydroquinine as co-catalyst (5 mol%) and also the use of a smaller amount of Brønsted acid additive (2.5 mol%). In a different study, Jørgensen reported that to perform the reaction between 2 equiv. of dimethyl 3-oxo-pentanodioate and an α,β-unsaturated aldehyde, using **11a** as catalyst, resulted in the fully diastereoselective and enantioselective formation of a compound with a bicyclo[3.3.1]non-2-ene structure containing six stereocenters (Scheme 6.45) [96c]. This reaction consisted of two cascade reactions that proceeded in a one-pot manner. Starting with a Michael/Knoevenagel cascade process, piperidine was then added to the reaction mixture; this promoted a second Michael/aldol cascade process that delivered the final tricyclic compounds.

6.7.3
Cascade Michael/N-Acyliminium Cyclization Reaction

A very interesting approach to densely functionalized indolo[2,3a]quinolizidine systems has been developed that involved the reaction of an α,β-unsaturated aldehyde and a modified tryptamine-derived malonate-type reagent, catalyzed by O-TMS diphenylprolinol **11a** [97]. The reaction consisted of an initial cascade process, starting with a Michael addition of the malonate reagent to the enal under iminium activation. This was followed by a release of the catalyst by hydrolysis, and delivery of the addition product in the form of a more stable cyclic hemiaminal, which was generated after intramolecular addition of the amide nitrogen to the formyl group. Next, in a typical one-pot operation, a strong Brønsted acid such as TFA was added; this promoted the formation of an N-acyliminium ion by dehydration of the hemiacetal intermediate, which finally underwent intramolecular reaction with the electron-rich indole moiety. The reaction proceeded with the generation of a tetracyclic system following the formation of two C–C and one C–N bond, and also involved the creation of three new stereogenic centers (Scheme 6.46). Moderate yields of the final compounds were obtained for a variety of different aromatic α,β-unsaturated aldehydes. It was also shown that the introduction of methoxy substituents at the indole ring could be tolerated, although no examples were provided to illustrate the use of aliphatic enals in this transformation. The products were also obtained as highly enantioenriched

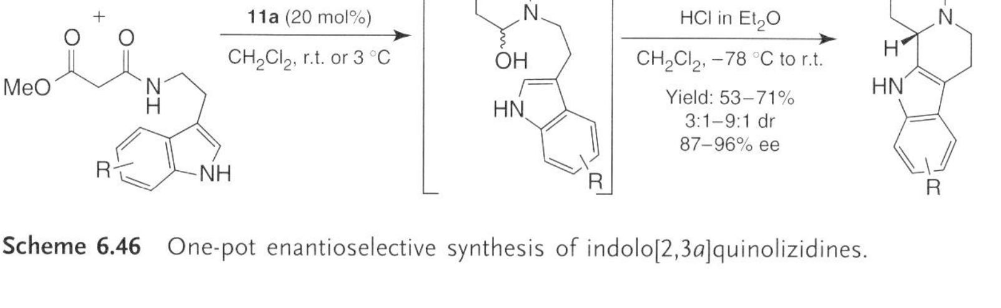

Scheme 6.46 One-pot enantioselective synthesis of indolo[2,3a]quinolizidines.

compounds, and as mixtures of diastereoisomers ranging from 3 : 1 to 9 : 1 that had resulted from a less-effective stereochemical control in the cyclization reaction via the *N*-acyliminium ion intermediate (Scheme 6.46).

6.7.4
Michael/Michael Cascade Reactions

Functionalized reagents containing an acidic moiety capable of participating as a carbon nucleophile in a conjugate addition under iminium activation, and which also incorporated an activated double bond in their structure that was capable of reacting intramolecularly with the enamine intermediate generated after the conjugate addition step, have been employed as bifunctional compounds in cascade reactions. A good example of this was provided by the elegant asymmetric synthesis of highly functionalized cyclopentenes (see Scheme 6.47) [98]. In this report, *O*-TMS diphenylprolinol **11a** was identified as an excellent catalyst for the Michael/Michael cascade reaction between α,β-unsaturated aldehydes and a functionalized malonate reagent containing an α,β-unsaturated ester moiety located at a convenient position. The reaction appeared to be fairly tolerant with regards to substitution at the enal reagent, and also with respect to the nature of the alkoxide substituents of the malonate reagent. In all cases, the reaction proceeded with high diastereoselectivity and enantioselectivity, providing the final cyclopentenes with excellent yields. Furthermore, a modified version of this reaction using a δ-nitro-α,β-unsaturated ester as bifunctional reagent has also been explored [99]. In this case, the reaction consisted of a nitro-Michael/Michael cascade reaction, delivering cyclopentanes with four contiguous stereogenic centers, which were isolated in good yields and with high diastereoselectivity and enantioselectivity.

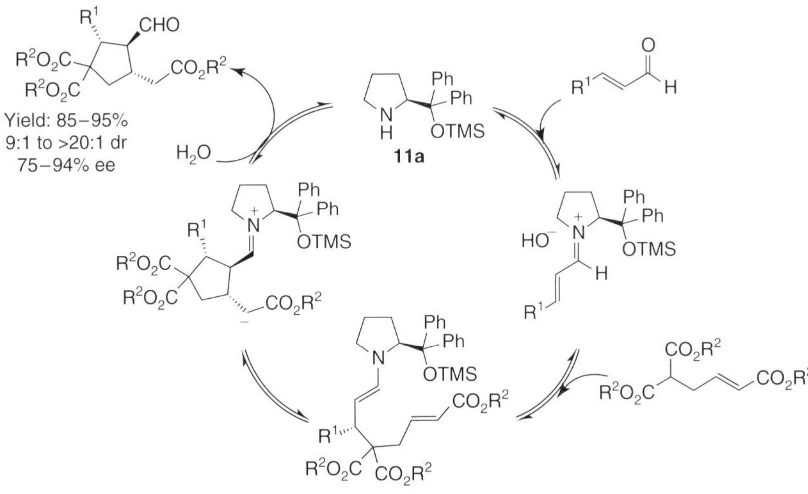

Scheme 6.47 Asymmetric synthesis of cyclopentanes by Michael/Michael cascade reaction.

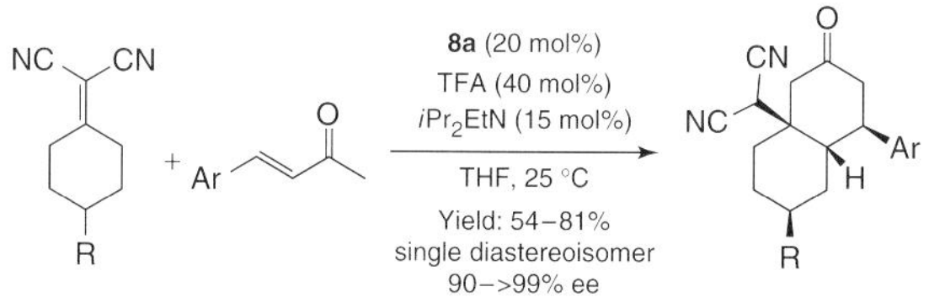

Scheme 6.48 Cascade vinylogous Michael/Michael reaction.

A very interesting vinylogous Michael/Michael cascade sequence has been reported between substituted malononitriles and enones (Scheme 6.48) [40b], [100]. In this case, a chiral primary amine derived from quinine such as **8a** was employed for the activation of the enone reagent, as is usually found for the case where α,β-unsaturated ketones must be employed as Michael acceptors. The reaction consisted of first, a Michael addition of the malononitrile reagent to the iminium ion derived from the enone, which proceeds through a deprotonation of the more acidic γ-position of the α,α-dicyanoalkene reagent. An equilibration of the two possible regioisomers of the enamine intermediate formed after the Michael addition step must then occur before the second intramolecular Michael reaction takes place, delivering the final cyclic derivatives following release of the catalyst in a final hydrolysis step. However, the possibility of the catalyst being released before the cyclization step – and, consequently, the possibility also that the intramolecular Michael reaction will take place without participation of the catalyst – should also be considered as a plausible mechanism operating in this reaction. Interestingly, this behavior was only observed when cyclohexylidenemalononitriles were employed, whereas the simple alkyl-substituted malononitriles underwent only the first vinylogous Michael step. In this context, when *para*-substituted cyclohexylidenemalononitriles were employed, the process occurred together with a desymmetrization of the starting material, thus generating a new stereogenic center with complete stereocontrol. Under optimal conditions, a wide range of different reagents could be employed to obtain the final adducts with good yields and excellent diastereo- and enantioselectivities. The applicability of this methodology was, however, limited to the use of β-aryl substituted enones.

6.7.5
Michael/Morita–Baylis–Hilman Cascade Reactions

When employing Nazarov reagents in an attempt to carry out a Michael/Michael cascade reaction with α,β-unsaturated aldehydes, Jørgensen and coworkers isolated an unexpected product, the formation of which was interpreted by assuming a Michael/Morita–Baylis–Hillman cascade process. Further experimentation led to the development of an outstanding protocol for the preparation of polysubstituted cyclohexenols containing multiple functionalities and two stereogenic centers, with good chemical efficiency and an outstanding stereochemical control [101]. As shown in Scheme 6.49, the reaction starts with the Michael reaction between the malonate moiety of the Nazarov reagent and the enal under iminium activation by

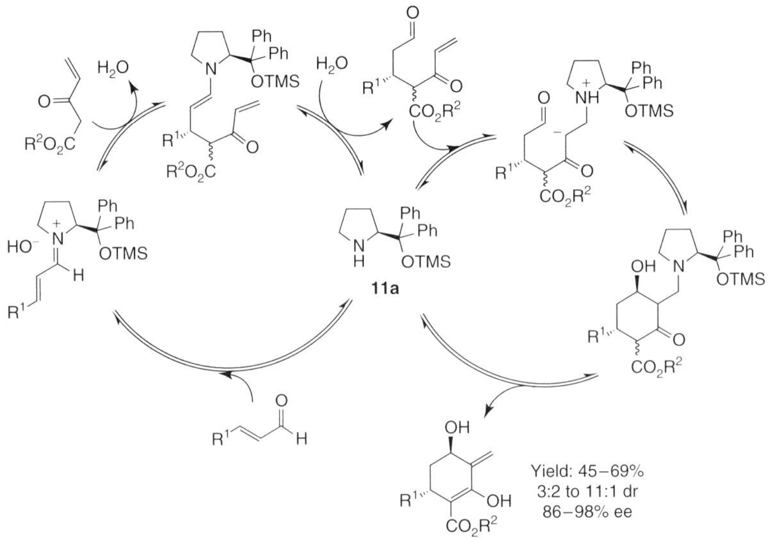

Scheme 6.49 Enantioselective Michael/Morita–Baylis–Hillman cascade reaction.

catalyst **11a**. This must be followed by an hydrolysis which affords the corresponding Michael adduct that, subsequently, should engage in a second catalytic cycle in which the amine catalyst must also participate by promoting an intramolecular Morita–Baylis–Hillman reaction. In this second cycle, the catalyst undergoes aza–Michael addition to the enone moiety, while the intermediate enolate must next react intramolecularly with the formyl group to deliver the final compound after a final retro-aza–Michael reaction that releases the amine catalyst and accounts for formation of the final compound (which is isolated as the more stable enol tautomer). The high stereochemical control attained in this reaction is due to the highly enantioselective Michael reaction occurring in the first stage of the cascade process; the formation of the stereocenter containing the OH moiety is a substrate-controlled cyclization process. Control experiments conducted to confirm the proposed mechanism led to the isolation of an intermediate Michael adduct, and also verified the existence of a fully diastereoselective intramolecular Morita–Baylis–Hillman reaction not only in the presence of **11a**, but also when other catalysts typical for this type of reaction (PPh$_3$ or DABCO) were employed. Otherwise, no cyclization occurred in the presence of acid additives. Many interesting transformations were also carried out on the adducts obtained, notably to verify their potential as chiral building blocks in the synthesis of complex cyclohexane architectures.

6.7.6
Michael/Michael/Aldol Triple Cascade Reactions

One of the most impressive demonstrations of the ability and power of enantioselective organocatalytic cascade reactions for the generation of molecular complexity

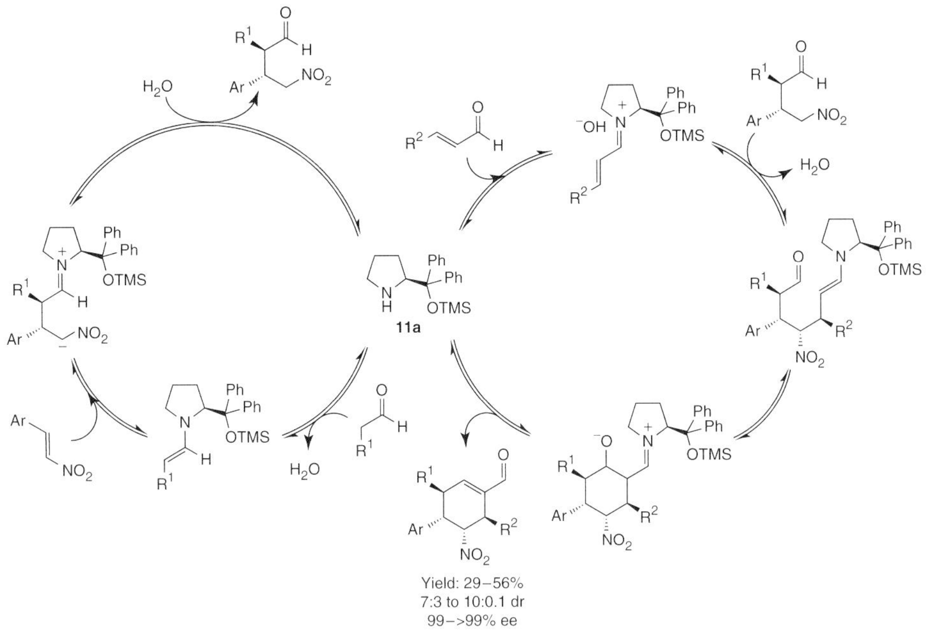

Scheme 6.50 Enantioselective triple Michael/Michael/aldol cascade reaction developed by Enders.

from very simple and low-cost starting materials can be found on the triple Michael/Michael/Aldol sequence, as developed by Enders (Scheme 6.50) [102]. In this reaction, a nitroalkene, an enolizable aldehyde and an α,β-unsaturated aldehyde condensed in the presence of chiral amine **11a** to furnish a cyclohexenecarbaldehyde product that contained four contiguous stereogenic centers. One major diastereoisomer was obtained from the eight possible diastereoisomers, in almost enantiomerically pure form, together with minor amounts of the C-5 epimer. This cascade process involved a first step in which the enolizable aldehyde underwent a Michael addition to the nitroalkene reagent that required activation of the nucleophile by the chiral amine catalyst via enamine formation. In this step, the chiral catalyst was capable of exerting a very efficient degree of stereocontrol in the two contiguous stereogenic centers generated in this step. Moreover, the reaction was shown to occur with complete chemoselectivity, and to favor reaction of the enamine Michael donor to the nitroalkene in the presence of the α,β-unsaturated aldehyde electrophile, due to the higher electrophilicity of the former. The next stage was to release the catalyst from the adduct, thereby delivering a functionalized and a highly acidic nitroalkane that could operate as a C-nucleophile in a subsequent Michael addition step to the enal; the latter compound also required activation by the catalyst via iminium ion formation. In this case, two additional stereogenic centers were also formed, at the point where formation of the minor diastereoisomer occurred, due to the inherent difficulties of controlling diastereoselectivity in the amine-catalyzed Michael addition of nitroalkanes to enals (as shown previously; see Section 6.3.2). Nevertheless, one major diastereoisomer was formed in good diastereomeric excess and, more importantly, with an excellent facial stereodiscrimination of the enantiotopic faces of the enal reagent selection being exerted by the catalyst. Finally, the enamine intermediate generated after the second Michael reaction must undergo an intramolecular aldol reaction, followed by dehydration. The final highly functionalized cyclohexenecarbaldehyde product was then delivered following a final hydrolysis step that re-released the catalyst in readiness for its participation in a subsequent catalytic cycle. This transformation proved to be rather general with respect to the enolizable aldehyde and the enal, with higher yields being obtained when aliphatic α,β-unsaturated aldehydes were employed. Unfortunately, all of the nitroalkenes tested were nitrostyrene derivatives, and no data were provided for using the more problematic aliphatic nitroalkenes. Several selective transformations were also carried out on the compounds obtained by this methodology [103], thus proving the possibility of chemical manipulation of the different functionalities present in their structure. A version involving the use of recyclable polystyrene-supported prolinol-based catalysts has also been reported, which furnished similar yields to those obtained by Enders in solution for a comparative example [104].

In an extension of this work, a tandem procedure has been optimized to convert the cyclohexenecarbaldehyde adducts obtained in this cascade reaction into complex tricyclic compounds that contain not only multiple stereogenic centers but also several functionalities amenable for modification into a wide variety of potentially useful chiral building blocks. The approach involved the use of an enolizable aldehyde reagent incorporating a conjugate diene moiety at an appropriate position. In this case, following the **11a**-catalyzed condensation with the corresponding aromatic nitroalkene and enal, an excess of a Lewis acid (e.g., Et_2AlCl) was added to the crude reaction mixture, thus inducing a highly diastereoselective intramolecular Diels–Alder reaction (Scheme 6.51) [105]. By this process, a tricyclic framework containing eight stereogenic centers was formed in moderate yield and high stereoselectivity, in a single operation. Depending on the length of the tether alkyl chain used between the formyl and the diene moieties in the initial Michael donor reagent, different tricyclic frameworks were obtained. Strangely, although the intramolecular Diels–Alder reaction conducive to the formation of an all six-membered ring proceeded with excellent diastereoselectivity, if the length of the tether led to the formation of a five-membered ring, then

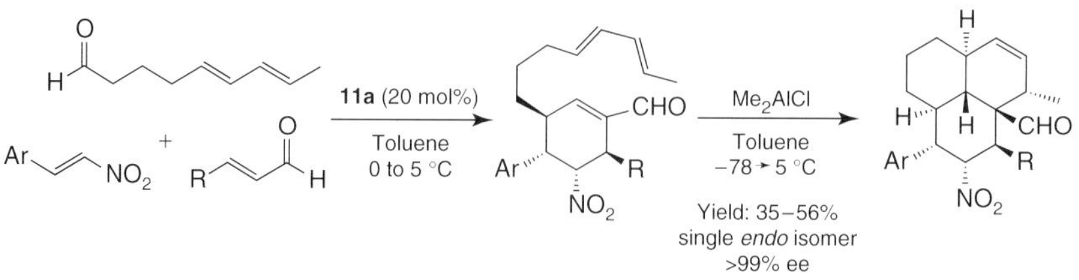

Scheme 6.51 Triple Michael/Michael/aldol cascade followed by one-pot intramolecular Diels–Alder reaction.

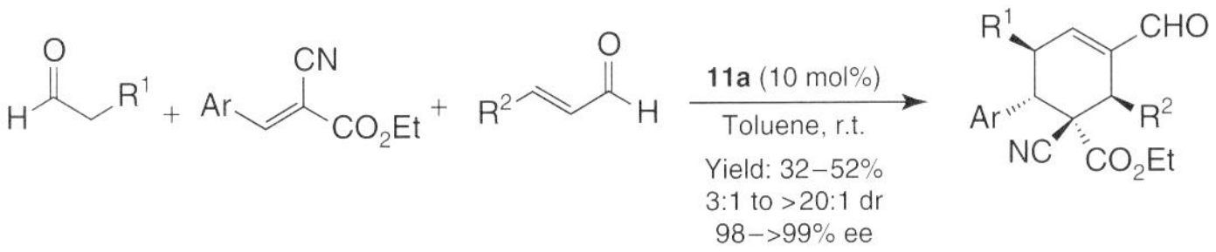

Scheme 6.52 Triple Michael/Michael/aldol cascade using 2-cyanoacrilates as highly active electrophiles.

mixtures of diastereoisomers were obtained. Unfortunately, this reaction sequence was only tested using cinnamaldehyde and acrolein as the α,β-unsaturated aldehyde reagents.

This triple Michael/Michael/aldol cascade reaction, proceeding via a sequential enamine/iminium/enamine activation mechanism, has been applied to other combinations of reagents. One such example, which involved a substitution of the nitroalkene reagent by another highly electrophilic α,β-unsaturated carbonyl compound (such as a 2-cyanoacrilate reagent) resulted in a highly efficient protocol for the preparation of cyclohexenacarbaldehydes that were similar to those obtained by Enders but contained a quaternary stereocenter (Scheme 6.52) [106]. The chemoselectivity of the process was, in this case, guaranteed because of the higher reactivity of the doubly activated 2-cyanoacrilate Michael acceptor with respect to the enal reagent. Also in this case, O-TMS diphenylprolinol **11a** was identified as the most efficient catalyst for this transformation, since it allowed the preparation of the final cyclic compounds in yields and enantioselectivities similar to those obtained by Enders in the same reaction, using nitroalkenes as the Michael acceptor undergoing the first conjugate addition step. Likewise, with respect to the diastereoselectivity of the reaction, in this case one major diastereoisomer was obtained among the eight possible candidates, together with small amounts of the C5 epimer. Much better results were achieved when crotonaldehyde was used as the enal reagent compared to the same reaction using cinnamaldehyde; in this case, the formation was observed of a single diastereoisomer for several enolizable aldehydes and 3-aryl substituted 2-cyanoacrylates employed.

In contrast, the group of Jørgensen has also reported an alternative procedure in which two molecules of an α,β-unsaturated aldehyde and an active methylene compound reacted together, yielding substituted 1-formylcyclohexenones containing up to three stereogenic centers (Scheme 6.53) [107]. In this reaction, the cascade process started with the Michael addition of the active methylene compound to the enal under iminium activation. The enamine intermediate subsequently underwent a second Michael reaction with another molecule of α,β-unsaturated aldehyde, which was also activated by the catalyst as the corresponding iminium ion. As in both of the preceding examples, the enamine formed following this second Michael reaction underwent an intramolecular aldol condensation, followed by hydrolysis in order to close the catalytic cycle. The remarkable features that illustrated the high efficiency of this process were the fact that α,β-unsaturated aldehydes with different substituents and hence, of different reactivity towards the two Michael reactions taking place in the cascade process could be used in the

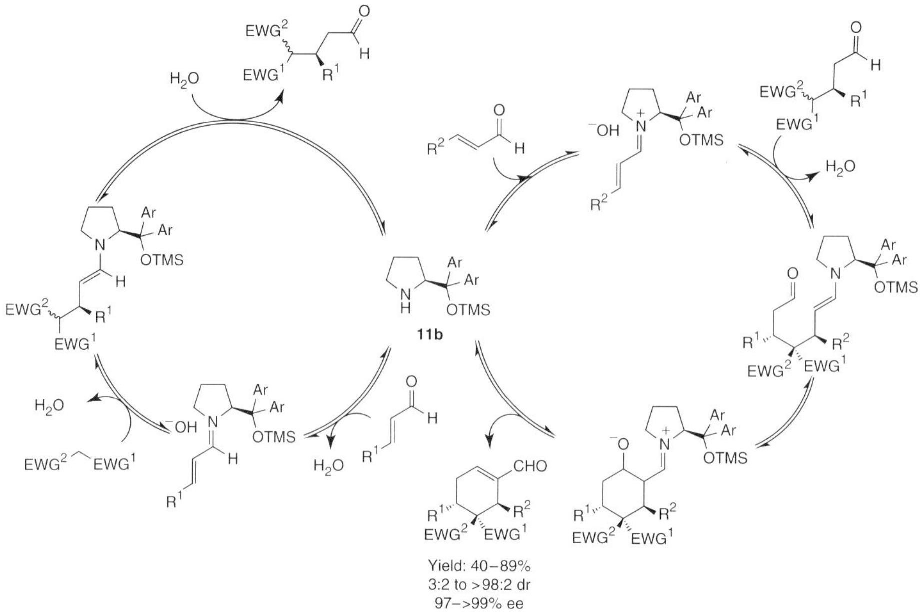

Scheme 6.53 Enantioselective triple Michael/Michael/aldol cascade reaction developed by Jørgensen.

reaction. This allowed the sequential incorporation of two different alkyl chains at the 3- and 5-positions of the cyclohexene ring, although it was necessary to modify the experimental procedure and the second enal reagent was added only when total consumption of the first α,β-unsaturated aldehyde had been observed. Remarkably, the yield and diastereoselectivity of this reaction were significantly higher than in both previously presented examples, this fact being attributed to the higher efficiency when controlling the diastereoselectivity of the second Michael reaction of the cascade. In the present case, the catalyst was involved in activation of both the donor and acceptor reagents, unlike the other cases. An additional interesting feature of this reaction was the fact that active methylene compounds with two different electron-withdrawing groups could be employed; this led to the formation of a quaternary stereocenter, although in this case the major diastereoisomer was formed together with significant amounts of the corresponding C5 epimer.

6.7.7
Michael/α-Alkylation Cascade Reactions

The use of a functionalized Michael donor containing a good leaving group on its structure, and capable of promoting an intramolecular alkylation reaction following the conjugate addition, has also been surveyed by several groups. Nevertheless, it should be noted that, when an iminium activation approach is required for this reaction, the question of chemoselectivity must first be resolved. The problem arises because the amine catalyst might react with the electrophilic moiety of the functionalized acceptor, delivering the corresponding N-alkylation product, and this would lead, eventually, to catalyst consumption. Despite these difficulties, in recent years several important and efficient methodologies that involve the iminium activation concept, all with vast potential, have been developed in the area of synthetic organic chemistry.

Within this context, the most extensively studied of these reactions has been the enantioselective cyclopropanation of α,β-unsaturated aldehydes which is, in turn, based on the pioneering investigations of Ley and Gaunt on the same reaction, but using cinchona alkaloids as organocatalysts [108]. The reaction developed by Gaunt and Ley involved the use of bromomalonates or bromonitromethane as functionalized C–H acidic Michael donors capable of undergoing a subsequent intramolecular alkylation after the conjugate addition step; this would deliver a chiral cyclopropane derivative, while exerting the required stereocontrol due to the cinchona alkaloid being present as a catalyst. Ley subsequently applied this design to the enantioselective nitrocyclopropanation of cyclohexenone catalyzed by pyrrolidine tetrazole **6** which, under the best reaction conditions, furnished the final compound in 80% yield and 77% ee, although crystallization allowed the optical purity to be increased to >98% ee [109a]. Remarkably, although single diastereoisomer was obtained under all conditions tested, when the reaction was extended to an acyclic enone mixtures of diastereoisomers were obtained. Slightly modified conditions allowed the enantioselectivity of the reaction to be increased to 90% ee for the reaction with cyclohexenone (Scheme 6.54) [109b]; however, poorer results were obtained when other cyclic enones were used, especially in the case of cyclopentenones. An improved version of this reaction was later reported where changing to a primary amine catalyst **8a** led to a highly efficient reaction, furnishing the nitrocyclopropanes in excellent yields and diastereo- and enantioselectivities when reacting cyclic enones of different ring sizes [110]. The use of acyclic enones required the conditions to be modified. One such modification was to perform the reaction in two steps; this proved to be equally efficient, but the substrate scope was restricted to 1,3-diaryl-substituted enones (Scheme 6.53).

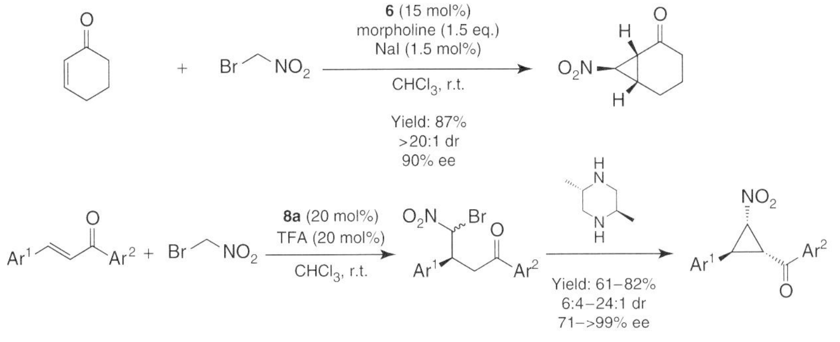

Scheme 6.54 Amine-catalyzed enantioselective nitrocyclopropanation of enones.

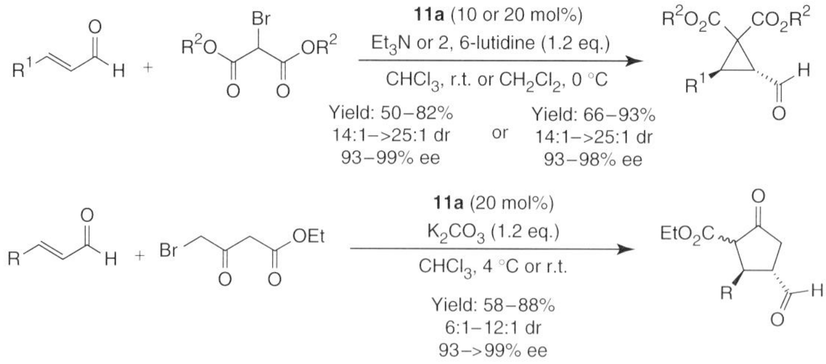

Scheme 6.55 Synthesis of cyclopropanecarbaldehdyes and cyclopentanecarbaldehydes via Michael/α-alkylation cascade reaction.

Following this reaction design, the nitrocyclopropanation of α,β-unsaturated aldehydes has also been reported, in this case using O-TMS diphenylprolinol **11a** as the best catalyst. However, whilst excellent enantioselectivities were generally obtained, the yields were moderate and mixtures of diastereoisomers in almost 1 : 1 ratio were obtained in most cases [111a]. In contrast, the same catalyst allowed the cyclopropanation reaction to be carried out using dialkyl bromomalonates as functionalized Michael donors (Scheme 6.55) [111b–d]. This reaction was developed simultaneously by the groups of Córdova and Wang, and proceeded with good yields, high diastereoselectivities, and excellent enantioselectivities for a wide range of α,β-unsaturated aldehydes – especially when aryl-substituted enals were used as substrates. The corresponding aliphatic α,β-unsaturated aldehydes furnished slightly lower yields, and required the use of a twofold excess of the enal, although the reaction proceeded with similarly high diastereoselectivity and enantioselectivity. The nature of the alkyl substituents of the malonate reagent was also shown to influence the reaction; notably, an increase in size led to a need for longer reaction times, and even resulted in a significant decrease in the reaction yield when diisopropyl bromomalonate was employed. Alternatively, this concept has been applied to the synthesis of cyclopentane structures, by using γ-bromo-β-ketoesters as functionalized Michael donors, and with excellent results (Scheme 6.55) [111c].

It should be noted that each of these cyclopropanation reactions required the incorporation of a base additive (such as morpholine, *trans*-2,5-dimethylmorpholine, Et_3N, K_2CO_3, or 2,6-lutidine) in order for the reaction to proceed, or to reach full conversion within a reasonable time. The additive was considered to function as a scavenger of the HBr generated as the reaction proceeded. Unfortunately, however, the presence of a basic additive was shown to promote a competitive retro-Michael process on the cyclopropane products obtained in the reaction of bromomalonates with enals. This led to the formation of α,β-disubstituted α,β-unsaturated aldehydes, and might explain for the lower yields observed in some of the above-described reactions [111d].

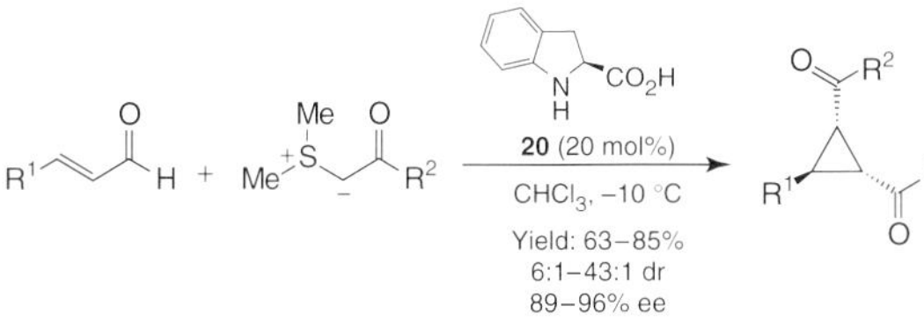

Scheme 6.56 Enantioselective cyclopropanation of α,β-unsaturated aldehydes using sulfonium ylides.

A different approach to the cyclopropanation of α,β-unsaturated aldehydes has been reported by MacMillan, which involved the use of sulfonium ylides as functionalized Michael donors and of indolinecarboxylic acid **20** as a chiral secondary amine catalyst (Scheme 6.56) [112]. The reaction was proposed to occur via conjugate addition of the ylide reagent, followed by an intramolecular substitution reaction. In addition, the authors proposed an interesting novel activation mechanism whereby the formation of a carboxylate iminium ion, together with the use of a zwitterionic reagent (such as the ammonium ylide), would permit the catalyst to exert a stereodirecting ability via electrostatic interactions between the carboxylate group and the ylide. The first experiments, which were carried out using proline as catalyst, achieved moderate enantioselectivities. Consequently, a new catalyst design was proposed based on the introduction of a fused aromatic ring at the catalyst structure, to which the authors attributed the provision of additional steric requirements that should allow a better control of the iminium geometry – and hence a much higher enantioselectivity compared to proline. Ultimately, their hypothesis proved to be correct when, by using indolinecarboxylic acid **20** as catalyst, cyclopropanation products were prepared in excellent yields and diastereo- and enantioselectivities. Moreover, the reaction was shown to be highly tolerant towards substitution at the enal reagent, and provided excellent results independent of the nature of the β-substituent. Arsonium ylides have also been used in this reaction, and have achieved good results, in this case by using *O*-TMS diphenylprolinol **11a** as catalyst [113].

6.7.8
Cascade Processes Initiated by Conjugate Friedel–Crafts Reaction

An extension of the studies of MacMillan and colleagues in the enantioselective conjugate Friedel–Crafts reactions catalyzed by chiral imidazolidinones to cascade processes led to the development of a very interesting and highly efficient protocol for carrying out a sequential conjugate addition/chlorination sequence (Scheme 6.57) [114]. In this reaction design, two consecutive reactions take place involving the iminium/enamine activation of an α,β-unsaturated aldehyde substrate, leading to β-aryl-α-chloro-substituted aldehydes in excellent yields and stereoselectivities. Based on previous reports, a pentachloroquinone reagent was employed as the source of the electrophilic chlorine atoms required for halogenation of the enamine intermediate. Several electron-rich aromatic compounds, as

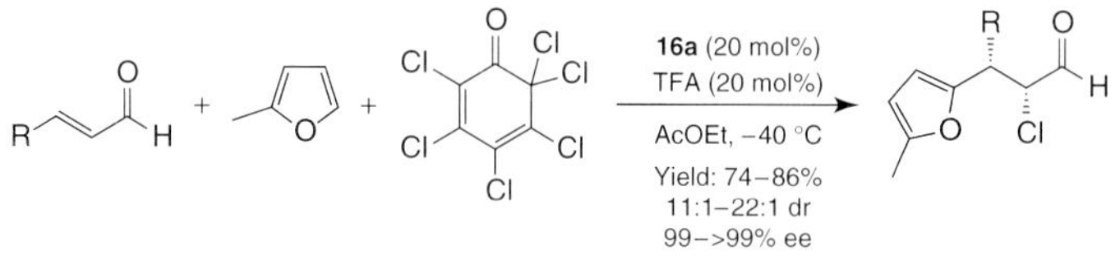

Scheme 6.57 An example of the Friedel–Crafts/chlorination cascade reaction developed by MacMillan.

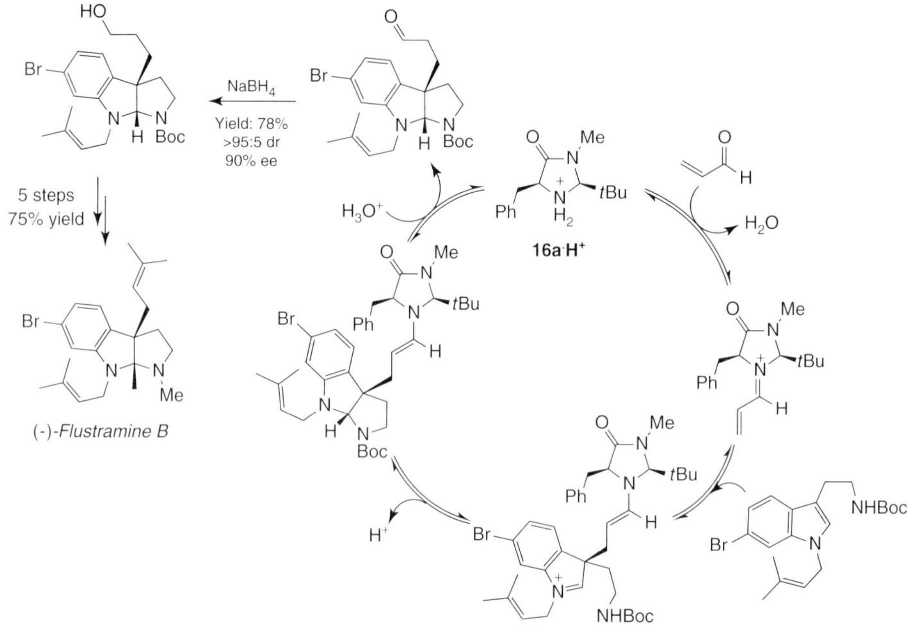

Scheme 6.58 Enantioselective total synthesis of (−)-flustramine-B.

well as differently substituted enals, were tested in the reaction, with satisfactory results. It should be noted that this is one of the few reported examples in which the amine-catalyzed cascade process takes place in an intermolecular manner, rather than the more favored (and thus easier) approach which involves intramolecularly occurring cascade reactions, as shown in all of the above-described examples.

In a different approach, MacMillan also developed a cascade conjugate Friedel–Crafts/intramolecular hemiaminal reaction for the enantioselective total synthesis of (−)-flustramine B (Scheme 6.58) [115]. The reaction design relied on the use of a 3-substituted indole reagent derived from tryptamine as an electron-rich heterocycle undergoing enantioselective conjugate Friedel–Crafts reaction with acrolein, under iminium activation. The intermediate formed after conjugate addition could not aromatize in this case because of the substitution pattern, and therefore underwent an intramolecular nucleophilic addition of the

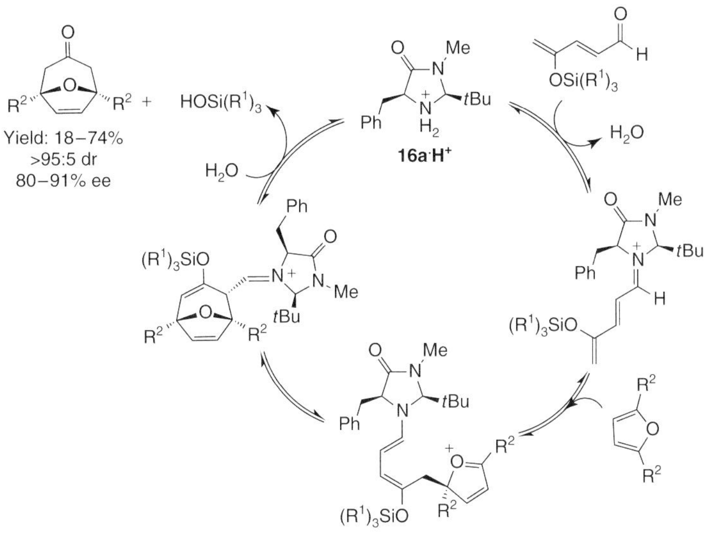

Scheme 6.59 Formal [4 + 3] cycloaddition reaction between silyloxydienals and furans catalyzed by **16a**.

amine group incorporated at the functionalized side chain. This led to the creation of the tricyclic basic skeleton of the target compound in excellent yield, as a single diastereoisomer and with excellent enantiomeric excess. In this case, the final compound was reduced *in situ*, furnishing directly the corresponding alcohol that was then converted in only five steps into the target compound, using very high-yielding procedures. More recently, this protocol has been extended to other similar heterocyclic derivatives of tryptamine with different substituents at the aromatic core, and with a variety of other enals different from acrolein. In all cases, this cascade process was observed to proceed with excellent results.

A very interesting cascade process between a conjugated 4-silyloxy-2,4-dienal and 2,5-dialkyl substituted furans which resulted in a formal [4 + 3] cycloaddition, was also developed in an asymmetric version, using imidazolidinone **16a** as catalyst (Scheme 6.59) [116]. This reaction consisted first of a conjugate Friedel–Crafts alkylation that proceeded under iminium activation by catalyst **16a**, followed by intramolecular aldol reaction where the oxonium ion remained at the furan moiety. The latter could not recover its aromaticity due to the presence of two alkyl substituents at C2 and C5; this which prevented elimination and therefore stabilized the oxonium ion for sufficient time to allow the intramolecular aldol reaction. When different trialkylsilyl groups were tested at the silyloxydienal reagent, and several alkyl substituents were introduced at the furan reagent, the final cycloaddition products could be obtained in low to good yields, with excellent endo-selectivity and enantioselectivities of 80–90% ee. Aromatic substituents at the furan led to the formation of mixtures of diastereoisomers, and to a significant decrease in enantioselectivity.

6.7.9
Cascade Processes Initiated by Conjugate Hydrogen-Transfer Reaction

The use of Hantzsch esters as hydride donors for carrying out conjugate hydrogen transfer reactions under iminium activation has also been extended to initiate cascade processes. An interesting application of this concept to the synthesis of substituted cyclohexanes and cyclopentanes was developed by List in a cascade reaction which consisted of a conjugate reduction, followed by an intramolecular Michael reaction, using substrates which contained both an enone and an enal that were communicated by an all-carbon tether (Scheme 6.60) [117]. The chemoselectivity of the reaction relied on the greater reactivity of the enal moiety towards formation of the iminium ion, and this in turn resulted in a preferential activation for undergoing a conjugate reduction with the Hantzsch ester. The next stage was for an intramolecular Michael reaction of the intermediate enamine with the enone moiety to take place, delivering the final cyclic compounds. Substrates incorporating an arene moiety as the linker between the enone and the enal reacted very efficiently, and conducive to the formation of a cyclopentane ring with excellent yields and diastereo- and enantioselectivities. Notably, when the arene tether was changed by a saturated alkyl chain, the reaction proceeded satisfactorily when a six-membered ring was formed; however, the reaction that was conducive to the formation of a cyclopentane proved to be significantly less enantioselective. Alternatively, a related intermolecular process that involved the conjugate hydrogen-transfer reduction of α,β-unsaturated aldehydes, followed by a Mannich reaction, has been developed by Córdova for the synthesis of highly functionalized acyclic α-amino acid derivatives that contain up to three stereocenters (Scheme 6.60) [118]. In this case, the imine electrophile was added following consumption of the starting enal, which indicated that the reaction was sequential tandem in nature, rather than a cascade process.

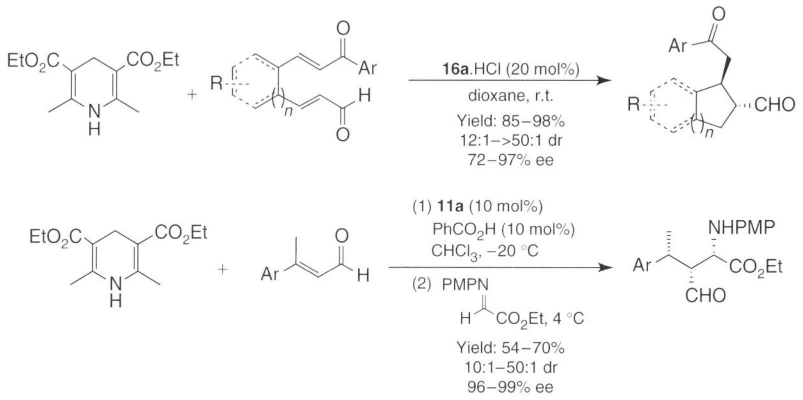

Scheme 6.60 Examples of two cascade processes initiated by conjugate hydrogen-transfer reaction.

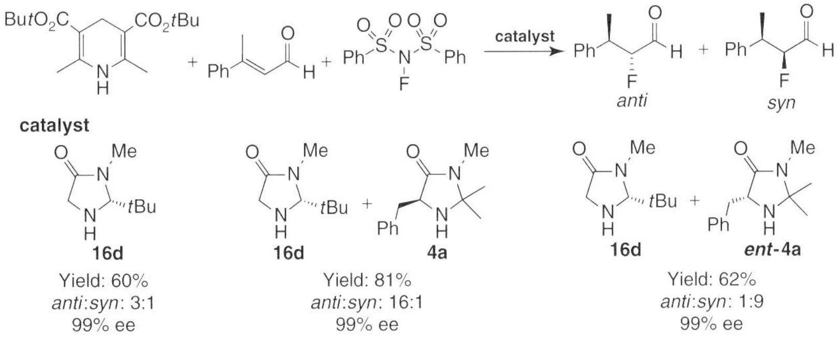

Scheme 6.61 Tandem conjugate hydrogen-transfer/electrophilic fluorination using cycle-specific imidazolidinone catalysts.

This reaction manifold was brilliantly exploited by MacMillan in an intermolecular conjugate reduction/halogenation reaction, that resulted in a formal addition of HCl or HF to α,β-unsaturated aldehydes [114]. In this case, the imidazolidinone catalyst **16d** provided an efficient promotion of the reaction for a model substrate. However, perhaps the most interesting concept developed when studying these reactions was the possibility to employ two different imidazolidinone catalysts in this transformation, each of which could promote its own catalytic reaction via different routes of selective activation. For example, the reaction between 3-phenyl-2-butenal and a Hantzsch ester, incorporating N-fluorobenzenesulfonimide as an electrophilic fluorine source in the presence of different amounts of imidazolidinone catalysts **4a** and **16d**, furnished the expected addition product in much higher yield and diastereoselectivity than was provided for the same reaction in which only catalyst **16d** was employed (Scheme 6.61). This indicated that two different catalytic cycles were taking place in the reaction, with each involving its own catalyst; it also meant that the imidazolidinones were able to engage selectively in a specific type of activation. In this case, it was believed that the less sterically demanding imidazolidinone **16d** had promoted the conjugate reduction of the substrate via iminium activation, while imidazolidinone **4a** had catalyzed the α-fluorination reaction by activating the saturated aldehyde generated following conjugate reduction via enamine formation. Also in this case, the addition of reagents was carried out separately, incorporating both the second imidazolidinone catalyst **4a** and the electrophilic fluorine source when substrate consumption had been observed, and hence classifying the transformation as a tandem process rather than as a cascade reaction. Confirmation of the specificity of the catalysts towards different pathways of activation was obtained by carrying out the tandem sequence separately with the two enantiomers of imidazolidinone catalyst **4a** and ***ent*-4a**. This led to a separate isolation of the two different diastereoisomers, and indicated that the second electrophilic fluorination step had taken place by an exclusive catalyst control exerted by **4a**. This strategy was later extended to cascade processes involving conjugate reduction followed by amination, hydroxylation, and Mannich [119].

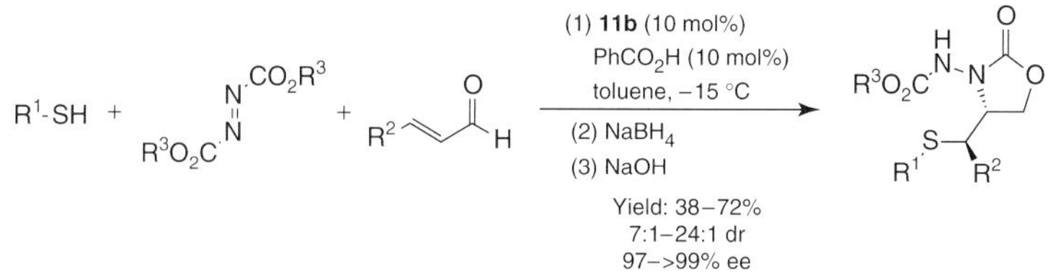

Scheme 6.62 Intermolecular sulfa–Michael/electrophilic amination reaction catalyzed by **11b**.

6.7.10
Cascade Processes Initiated by Hetero-Michael Reactions

As noted above, carrying out a cascade process avoids the inherent reversibility of the hetero-Michael reactions, and pushes forward all of the equilibria participating in the catalytic cycle; this will result in better conversions and solve the problem of the configurational instability of the hetero-Michael adducts. For this reason, intensive research has been carried out in developing such cascade processes, typically employing heteronucleophiles that contain an electrophilic functionality capable of reacting with the enamine intermediate that is generated after the conjugate addition to the iminium ion.

One of the first attempts in this field refers to a multicomponent intermolecular sulfa-Michael/amination cascade process. In this case, an α,β-unsaturated aldehyde and a thiol were reacted in the presence of a dialkyl azodicarboxylate reagent, giving access to β-amino-γ-thioalcohol derivatives with excellent yields and enantioselectivities [84]. Isolation of the final products was accomplished by *in situ* reduction, followed by base-promoted cyclization (Scheme 6.62). O-trimethylsilyl diarylprolinol **11b** was identified as the best catalyst for this transformation. In fact, the authors started to optimize the reaction with the initial sulfa–Michael step, but found that hydrolysis of the enamine intermediate generated following the iminium-mediated sulfa–Michael reaction proceeded very slowly, leading to low concentrations of free catalyst and therefore slowing the process. It was found subsequently that an addition of the azodicarboxylate electrophile caused a remarkable acceleration of the sulfa–Michael process; this was attributed to an easier release of the catalyst, which was favored by the formation of a highly sterically encumbered iminium intermediate after the electrophilic amination step. With all such evidence in hand, a cascade sequence was developed that led, in turn, to a highly efficient process for the generation of chemical complexities from very simple and low-cost reagents.

A different concept has been devised very recently, however, in which – in contrast to the above-described and other related reactions – the intermolecular cascade process does not involve the use of two different reagents (a nucleophile for the hetero-Michael reaction and an electrophile for the subsequent reaction with the enamine intermediate). Rather, a single reagent is involved which, when the Michael process has started, participates as the electrophile reagent

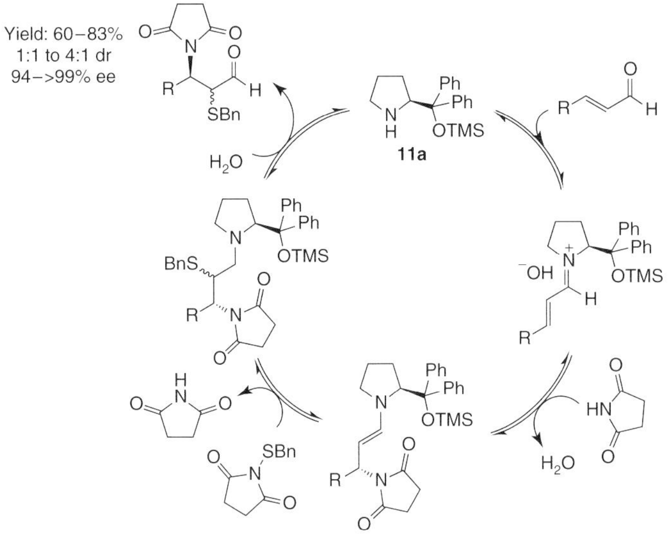

Scheme 6.63 Enantioselective aminosulfenylation of α,β-unsaturated aldehydes.

but incorporates the nucleophile counterpart as the leaving group. Following the electrophilic addition step, the nucleophile reagent is regenerated *in situ* in the reaction medium, and incorporated into the subsequent catalytic cycle, without need for any further addition (1 equiv.) of the reagent. This approach also leads to a reduction in the amount of waste chemicals generated in the process. The design has been applied to the enantioselective aminosulfenylation of α,β-unsaturated aldehydes catalyzed by *O*-TMS diphenylprolinol **11a**, using an *N*-alkylthiosuccinimide as electrophilic reagent to regenerate the nucleophile (succinimide) after the electrophilic addition step (Scheme 6.63) [120]. Although the reaction required a catalytic quantity of free succinimide in order to be kick-started, the nucleophile was delivered continuously to the medium when reaction of the intermediate enamine with the *N*-alkylthiosuccinimide electrophile had taken place. The final highly functionalized acyclic products were obtained with good yields, and very high enantioselectivities; mixtures of *anti/syn* diastereoisomers were obtained in ranges which varied from 1 : 1 to 4 : 1.

Despite these two impressive examples of intermolecular cascade processes initiated by hetero-Michael reactions, the majority of the developed methodologies in this field have involved intramolecular versions leading to the formation of complex cyclic structures. A very illustrative example of this is shown in Scheme 6.64, which consists of an enantioselective sulfa–Michael/aldol reaction using α,β-unsaturated aldehydes and 2-mercapto-1-phenylethanone as the functionalized heteronucleophile, leading to the formation of enantioenriched tetrahydrothiophenes in a single step (Scheme 6.64) [121]. Interestingly, it was found that two different regioisomers could be selectively obtained by introducing slight modifications in the reaction conditions. Carrying out the reaction in the presence of an acid co-catalyst

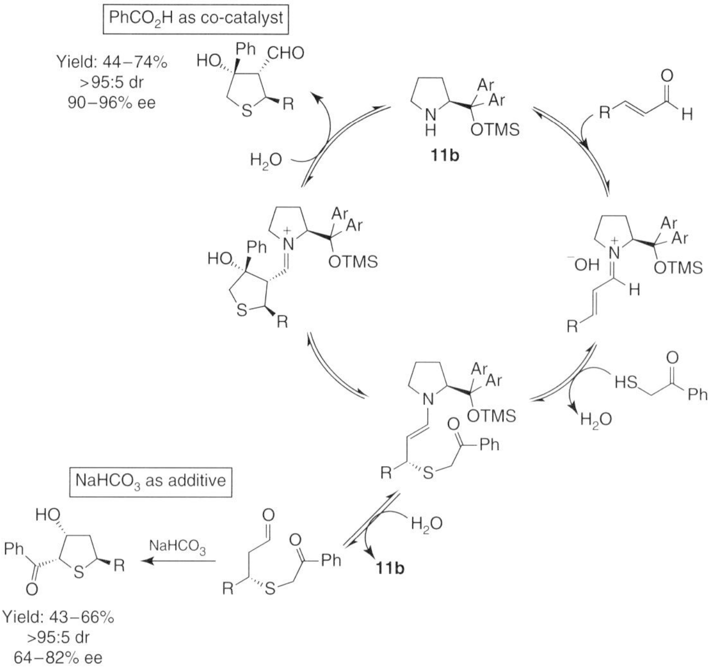

Scheme 6.64 Sulfa–Michael/aldol cascade for the synthesis of tetrahydrothiophenes.

led to the formation of 2-benzoyl-5-alkyltetrahydrotiophene-4-ols, yet under basic conditions a completely different isomer was obtained. These results were interpreted in terms of an iminium-enamine manifold operating in the first case, with the acid co-catalyst facilitating formation of the iminium intermediate. However, under basic conditions a hydrolysis of the enamine intermediate might occur, such that a final intramolecular base-catalyzed aldol reaction with the more reactive aldehyde moiety as electrophile, without participation of the catalyst, would lead to the final product. In the latter case, the basic conditions employed could also promote the uncatalyzed sulfa–Michael reaction, and this was interpreted as the reason for the slightly lower enantioselectivity observed.

Hetero-Michael/aldol cascade reactions have been studied by several research groups using *ortho*-substituted benzaldehydes as functionalized reagents. In this context, different benzo-fused heterocyclic architectures have been prepared by the reaction of phenols, thiophenols, and anilines incorporating a formyl or even an acetyl group at the *ortho* position, in combination with α,β-unsaturated aldehydes and using a chiral amine catalyst, usually a diarylprolinol silyl ether such as **11a** or **11b** (Scheme 6.65). Due to the particular structure which is built up, these reactions normally finish with a dehydration reaction after the intramolecular aldol step. Using this approach, benzopyrans have been prepared using a cascade oxa–Michael/aldol/dehydration sequence [122], benzothiopyrans by means of a sulfa–Michael/aldol/dehydration cascade [123], while access to

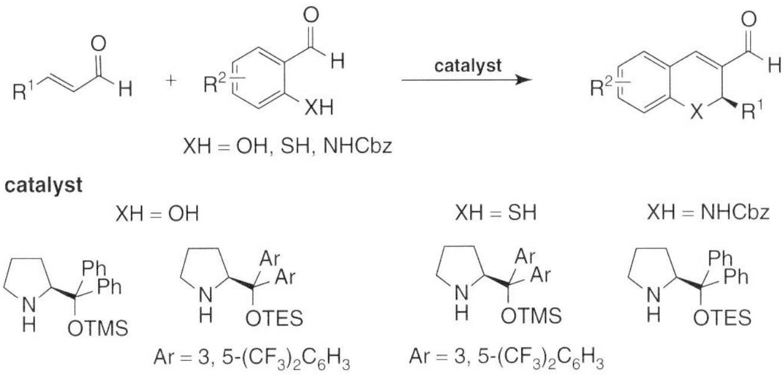

Scheme 6.65 Hetero-Michael/aldol cascade for the asymmetric synthesis of benzo-fused heterocycles.

tetrahydroquinolines has been achieved using an aza–Michael/aldol/dehydration approach [124]. In general, all of these reactions have proceeded with excellent yields and enantioselectivities, and with good substrate scope with regards to the enal reagent employed and tolerating the incorporation of aromatic, aliphatic, and other functionalized substituents at the β-position. There is a single example, in which a ketone moiety has been employed as an internal electrophilic moiety. In this case, as the final dehydration reaction could not take place, the final products were formed with the concomitant generation of three contiguous stereogenic centers, with remarkably high diastereoselectivities in all of the cases tested [125]. In contrast, the reactions involving enones as the initial electrophile have been limited to the use of cyclohexenone. It has also been shown that tetrahydroxanthenones and tetrahydrothioxanthenones may be obtained using this reaction design, whereas other less bulky secondary amines such as pyrrolidinomethylpyrrolidine **20** or the simple prolinol showed improved catalytic abilities for activating enones [126]. Surprisingly, no reports have been made evaluating the use of primary amines as catalysts in these transformations.

Cascade sulfa–Michael/Michael and aza–Michael/Michael reactions have also been developed for the synthesis of densely functionalized chiral tetrahydrothiophenes and pyrrolidines (Scheme 6.66) [127]. This approach involved the use of 4-mercapto or 4-amino-2-butenoate as a functionalized nucleophile which, after the initial hetero-Michael reaction under iminium activation, underwent a subsequent intramolecular Michael reaction whereby the intermediate enamine moiety reacted with the enoate moiety incorporated at the functionalized reagent; this resulted in the formation of a five-membered ring heterocycle. O-TMS diphenylprolinol **11a** was identified as the best catalyst for these reactions, which proceeded with good yields and excellent diastereoselectivities and enantioselectivities for a wide variety of different α,β-unsaturated aldehydes as substrates, although they appeared to be limited exclusively to aromatic substituents at the β-position. Importantly, the aza–Michael-initiated cascade reaction was shown to depend heavily on the protecting group incorporated at the amine nucleophile, with a consequent limitation to the use of N-tosyl-protected substrates. Moreover, the reaction also required the

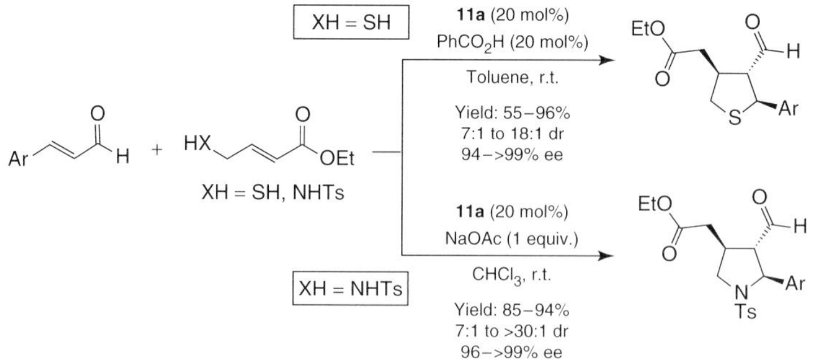

Scheme 6.66 Hetero-Michael/Michael cascade reaction for the asymmetric synthesis of tetrahydrothiophenes and pyrrolidines.

addition of 1 equiv. of a basic additive (e.g., NaOAc) to assist deprotonation of the nitrogen nucleophile.

Finally, it should be noted that the enantioselective epoxidation of α,β-unsaturated compounds has also been considered by some groups to occur via a domino oxa–Michael/intramolecular nucleophilic substitution pathway; hence, the iminium activation concept has appeared to represent a highly successful approach to this particularly important transformation. In this context, following the initial report by Jørgensen [128a], who used hydrogen peroxide as oxidant and O-TMS diarylprolinol **11b** as catalyst (Scheme 6.67), several other secondary amine catalysts and alternative oxidants have been used with different degrees of success in this reaction [128b–e]. The epoxidation of enones has also been described recently using a primary amine catalyst [129], or by applying the asymmetric counterion-directed catalysis concept with a combination of an α-amino acid and chiral phosphoric acid **18** as catalytically active species [130]. In a similar way, the

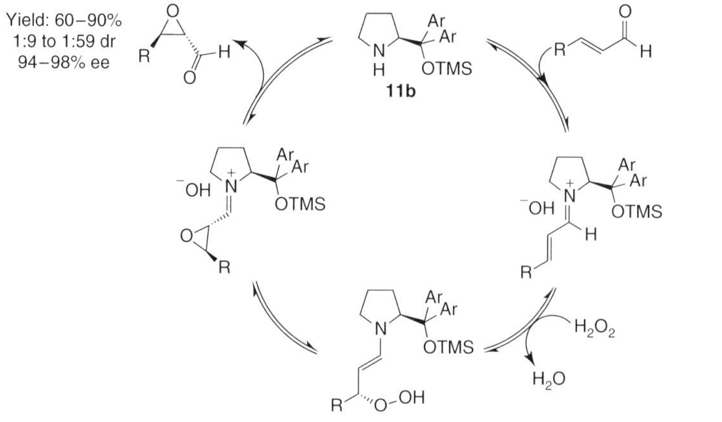

Scheme 6.67 Enantioselective epoxidation of enals with H_2O_2 catalyzed by **11b**.

enantioselective aziridination of α,β-unsaturated aldehydes has been achieved via a cascade aza–Michael intramolecular nucleophilic displacement; in this case, N-acetoxy carbamates were used as functionalized Michael donors, with O-TMS diphenylprolinol **11a** being shown as the most efficient catalyst [131].

6.8
Concluding Remarks and Outlook

The iminium activation concept has been identified as a very reliable alternative for carrying out enantioselective conjugate additions of many different nucleophiles to α,β-unsaturated aldehydes and ketones. In general, excellent yields and stereoselectivities can be obtained with operationally simple procedures, reflecting the true advantage of this methodology compared to other metal-catalyzed systems. The potential of the concept with regards to applications in synthesis has been well addressed in many of the examples presented in this chapter. In particular, the impressively high degree of chemoselectivity and stereoselectivity control attained in cascade reactions, enabling the construction of highly complex structures starting from low-cost and readily available starting materials, has been recognized. Whilst the ability to use enals as Michael acceptors represents another interesting feature of these reactions, which proceed via iminium intermediates, this may become a limitation when certain possible Michael acceptors such as α,β-unsaturated esters, amides or even nitroalkenes may be unsuited for use as substrates. Another current problem relates to the very high catalyst loadings required for most of the reactions developed to date, and this is especially important when scaling-up for industrial purposes. In this context, the development of recyclable catalysts will surely attract increasing attention in the years to come. Finally, certain conjugate addition reactions remain elusive to this methodological approach, including some hetero-Michael reactions where the inherent reversibility of the reaction implies that special nucleophiles designed expressly for each transformation must be employed; alternatively, shortcuts must be taken to isolate the conjugate addition compounds in good yields and stereoselectivities as chemically equivalent entities. As a consequence, additional synthetic steps must be included to convert the initial products into target hetero-Michael adducts, thus reducing the efficiency of the methodology.

References

1. List, B., Lerner, R.A., and Barbas, C.F. III (2000) *J. Am. Chem. Soc.*, **122**, 2395.
2. For some general reviews covering amine-catalyzed enantioselective reactions see: (a) Mukherjee, S., Yang, J.W., Hoffmann, S., and List, B. (2007) *Chem. Rev.*, **107**, 5471; (b) Erkkilä, A., Majander, I., and Pihko, P.M. (2007) *Chem. Rev.*, **107**, 5416; (c) Bertelsen, S., Nielsen, M., and Jørgensen, K.A. (2007) *Angew. Chem., Int. Ed.*, **46**, 7356; (d) List, B. (2004) *Acc. Chem. Res.*, **37**, 548; (e) Notz, W., Tanaka, F., and Barbas, C.F.III (2004) *Acc. Chem. Res.*, **37**, 580; (f) List, B. (2001) *Synlett*, 1675; (g) List, B. (2006) *Chem. Commun.*, 819; (h) Lelais, G. and MacMillan, D.W.C. (2006) *Aldrichim.*

Acta, **39**, 79; (i) Mukherjee, S. and List, B. (2007) *Nature*, **129**, 7498; For some more general reviews on asymmetric organocatalysis see: (j) Melchiorre, P., Marigo, M., Carlone, A., and Bartoli, G. (2008) *Angew. Chem., Int. Ed.*, **47**, 6138; (k) Dondoni, A. and Massi, A. (2008) *Angew. Chem., Int. Ed.*, **47**, 4638; (l) Special Issue on Organocatalysis 2007, *Chem. Rev.*, **107**, 12; (m) List, B. and Yang, J.-W. (2006) *Science*, **313**, 1584; (n) Dalko, P.I. (2007) *Enantioselective Organocatalysis*, Wiley-VCH Verlag GmbH, Weinheim; (o) Seayad, J. and List, B. (2005) *Org. Biomol. Chem.*, **3**, 719; (p) Berkessel, A. and Gröger, H. (2004) *Asymmetric Organocatalysis*, Wiley-VCH Verlag GmbH, Weinheim; (q) Gaunt, M.J., Johansson, C.C.C., McNally, A., and Vo, N.C. (2007) *Drug Discovery Today*, **2**, 8; (r) Pellissier, H. (2007) *Tetrahedron*, **63**, 9267; (s) MacMillan, D.W.C. (2008) *Nature*, **455**, 304.

3. For some reviews on organocatalytic conjugate additions see: (a) Vicario, J.L., Badia, D., and Carrillo, L. (2007) *Synthesis*, 2065; (b) Almaçi, D., Alonso, D.A., and Najera, C. (2007) *Tetrahedron: Asymmetry*, **18**, 299; (c) Tsogoeva, S.B. (2007) *Eur. J. Org. Chem.*, 1701; (d) Sulzer-Mosse, S. and Alexakis, A. (2007) *Chem. Commun.*, 3123. For a specific review covering organocatalytic reactions via iminium activation, see also Ref. [2b].

4. (a) Yu, X. and Wang, W. (2008) *Org. Biomol. Chem.*, **6**, 2037; (b) Enders, D., Grondal, C., and Huettl, M.R.M. (2007) *Angew. Chem., Int. Ed.* **46**, 1570; (c) Guillena, G., Ramon, D.J., and Yus, M. (2007) *Tetrahedron: Asymmetry*, **18**, 693.

5. Langebeck, W. and Sauerbier, R. (1937) *Chem. Ber.*, **70**, 1540.

6. Woodward, R.B., Logusch, E., Nambiar, K.P., Sakan, K., Ward, D.E., Au-Yeung, B.-W., Balaram, P., Browne, L.J., Card, P.J., Chen, C.H., Chenevert, R.B., Fliri, A., Frobel, K., Gais, H.-J., Garratt, D.G., Hayakawa, K., Heggie, W., Hesson, D.P., Hoppe, D., Hoppe, I., Hyatt, J.A., Ikeda, D., Jacobi, P.A., Kim, K.S., Kobuke, Y., Kojima, K., Krowicki, K., Lee, V.J., Lautert, T., Malchenko, S., Martens, J., Matthews, R.S., Ong, B.S., Press, J.B., Rajan Babu, T.V., Rousseau, G., Sauter, H.M., Suzuki, M., Tatsuta, K., Tolbert, L.M., Truesdale, E.A., Uchida, I., Ueda, Y., Uyehara, T., Vasella, A.T., Vladuchick, W.C., Wade, P.A., Williams, R.M., and Wong, H.N.-C. (1981) *J. Am. Chem. Soc.*, **103**, 3210.

7. (a) Yamaguchi, M., Yokota, N., and Minami, T. (1991) *J. Chem. Soc. Chem. Commun.*, 1088; see also (b) Yamaguchi, M., Shiraishi, T., and Hirama, M. (1993) *Angew. Chem., Int. Ed. Engl.*, **32**, 1176; (c) Yamaguchi, M., Shiraishi, T., and Hirama, M. (1996) *J. Org. Chem.*, **61**, 3520.

8. Kawara, A. and Taguchi, T. (1994) *Tetrahedron Lett.*, **35**, 8805.

9. Ahrendt, K.A., Borths, C.J., and MacMillan, D.W.C. (2000) *J. Am. Chem. Soc.*, **122**, 4243.

10. Lelais, G. and MacMillan, D.W.C. (2007) in *Enantioselective Organocatalysis: Reactions and Experimental Procedures* Chapter 3, (ed. P.I.Dalko), Wiley-VCH Verlag GmbH, Weinheim.

11. (a) Baum, J.S. and Viehe, H.G. (1976) *J. Org. Chem.*, **41**, 183; (b) Jung, M.E., Baccaro, W.D., and Buszek, K.R. (1989) *Tetrahedron Lett.*, **30**, 1893.

12. (a) Grieco, P.A. and Larsen, S.D. (1986) *J. Org. Chem.*, **51**, 3553; (b) Waldman, H. (1988) *Angew. Chem., Int. Ed. Engl.*, **27**, 274.

13. (a) Xu, L.-W. and Lu, Y. (2008) *Org. Biomol. Chem.*, **6**, 2047; (b) Bartoli, G. and Melchiorre, P. (2008) *Synlett*, 1759; (c) Chen, Y.-C. (2008) *Synlett*, 1919.

14. In 1993, Belokon reported the L-prolinol-catalyzed addition of methyl nitroacetate to crotonaldehyde proceeding with low enantioselectivity: (a) Belokon, Y.N., Kochetkov, K.A., Churkina, T.D., Ikonnikov, N.S., Orlova, S.A., Kuzmina, N.A., and Bodrov, D.E. (1993) *Russ. Chem. Bull.*, **9**, 1591; In 2000, Barbas carried out the Robinson annulation between 2-methyl-1,3-cyclohexanedione and methyl vinyl ketone catalyzed by several chiral amines, and reported that the process stopped in many cases after

the conjugate addition step. However, neither yields nor enantioselectivities were given for this case. (b) Bui, T. and Barbas, C.F.III (2000) *Tetrahedron Lett.*, **41**, 6951.
15. Halland, N., Aburel, P.S., and Jørgensen, K.A. (2003) *Angew. Chem., Int. Ed.*, **42**, 661.
16. (a) Knudsen, K.R., Mitchell, C.E.T., and Ley, S.V. (2006) *Chem. Commun.*, 66; (b) Wacholowski, V., Knudsen, K.R., Mitchell, C.E.T., and Ley, S.V. (2008) *Chem. Eur. J.*, **14**, 6155.
17. Alonso, D.A., Kitagaki, S., Utsumi, N., and Barbas, C.F.III (2008) *Angew. Chem., Int. Ed.*, **47**, 4588.
18. Halland, N., Hansen, T., and Jørgensen, K.A. (2003) *Angew. Chem. Int. Ed.*, **42**, 4955.
19. Kim, H., Yen, C., Preston, P., and Chin, J. (2006) *Org. Lett.*, **8**, 5239.
20. Xie, J.-W., Yue, L., Chen, W., Du, W., Zhu, J., Deng, J.-G., and Chen, Y.-C. (2007) *Org. Lett.*, **9**, 413.
21. Yang, Y.-Q. and Zhao, G. (2008) *Chem. Eur. J.*, **14**, 10888.
22. Li, P., Wen, S., Yu, F., Liu, Q., Li, W., Wang, Y., Liang, X., and Ye, J. (2009) *Org. Lett.*, **11**, 753.
23. For some reviews see (a) Lattanzi, A. (2009) *Chem. Commun.*, 1452; (b) Mielgo, A. and Palomo, C. (2008) *Chem. Asian J.*, **3**, 922; (c) Palomo, C. and Mielgo, A. (2006) *Angew. Chem., Int. Ed.*, **45**, 7876.
24. Marigo, M., Wabnitz, T.C., Fielenbach, D., and Jørgensen, K.A. (2005) *Angew. Chem., Int. Ed.*, **44**, 794.
25. Hayashi, Y., Gotoh, H., Hayashi, T., and Shoji, M. (2005) *Angew. Chem., Int. Ed.*, **44**, 4212.
26. (a) Ma, A., Zhu, S., and Ma, D. (2008) *Tetrahedron Lett.*, **49**, 3075; (b) Wang, Y., Li, P., Liang, X., and Ye, J. (2008) *Adv. Synth. Catal.*, **350**, 1383.
27. (a) Brandau, S., Landa, A., Franzen, J., Marigo, M., and Jørgensen, K.A. (2006) *Angew. Chem., Int. Ed.*, **45**, 4305; For an extension of this methodology to the one-pot asymmetric synthesis of substituted 1,4-dihydropyridines see (b) Franke, P.T., Johansen, R.L., Bertelsen, S., and Jørgensen, K.A. (2008) *Chem. Asian J.*, **3**, 216; (c) See also Jiang, J., Yu, J., Sun, X.-X., Rao, Q.-Q., and Gong, L.-Z. (2008) *Angew. Chem., Int. Ed.*, **47**, 2458.
28. (a) Rueping, M., Sugiono, E., and Merino, E. (2008) *Angew. Chem., Int. Ed.*, **47**, 3046; (b) Rueping, M., Sugiono, E., and Merino, E. (2008) *Chem. Eur. J.*, **14**, 6329; (c) Franke, P.T., Richter, B., and Jorgensen, K.A. (2008) *Chem. Eur. J.*, **14**, 6317; (d) Rueping, M., Merino, E., and Sugiono, E. (2008) *Adv. Synth. Catal.*, **350**, 2127.
29. Yamaguchi, M., Shiraishi, T., Igarashi, Y., and Hirama, M. (1994) *Tetrahedron Lett.*, **35**, 8233.
30. Hanessian, S. and Pham, V. (2000) *Org. Lett.*, **2**, 2975; (b) Hanessian, S., Shao, Z., and Warrier, J.S. (2006) *Org. Lett.*, **8**, 4787.
31. (a) Tsogoeva, S.B., Jagtap, S.B., Ardemasova, Z.A., and Kalikhevich, V.N. (2004) *Eur. J. Org. Chem.*, 4014; (b) Malmgren, M., Granander, J., and Amedjkouh, M. (2008) *Tetrahedron: Asymmetry*, **19**, 1934.
32. Halland, N., Hazell, R.G., and Jorgensen, K.A. (2002) *J. Org. Chem.*, **67**, 8331; (b) Mitchell, C.E., Brenner, S.E., and Ley, S.V. (2005) *Chem. Commun.*, 5346; (c) Prieto, A., Halland, N., and Jorgensen, K.A. (2005) *Org. Lett.*, **7**, 3897.
33. Zu, L., Xie, H., Li, H., Wang, J., and Wang, W. (2007) *Adv. Synth. Catal.*, **349**, 2660; (b) Gotoh, H., Ishikawa, H., and Hayashi, Y. (2007) *Org. Lett.*, **9**, 5307; (c) Wang, Y., Li, P., Liang, X., Zhang, T.Y., and Ye, J. (2008) *Chem. Commun.*, 1232.
34. Hojabri, L., Hartikka, A., Moghaddam, F.M., and Arvidsson, P.I. (2007) *Adv. Synth. Catal.*, **349**, 740.
35. Li, P., Wang, Y., Liang, X., and Ye, J. (2008) *Chem. Commun.*, 3302.
36. Zhong, C., Chen, Y., Petersen, J.L., Akhmedov, N.G., and Shi, X. (2009) *Angew. Chem., Int. Ed.*, **48**, 1279.
37. Palomo, C., Landa, A., Mielgo, A., Oiarbide, M., Puente, A., and Vera, S. (2007) *Angew. Chem., Int. Ed.*, **46**, 8431.
38. (a) Mase, N., Nakai, Y., Ohara, N., Yoda, H., Takabe, K., Tanaka, F., and Barbas, C.F.III (2006) *J. Am. Chem.*

Soc., **128**, 734; For a review see (b) Paradowska, J., Stodulski, M., and Mlynarski, J. (2009) *Angew. Chem., Int. Ed.*, **48**, 4288.

39. Um, P.-J. and Drueckhammer, D.G. (1998) *J. Am. Chem. Soc.*, **120**, 5605.
40. (a) Xie, J.-W., Yue, L., Xue, D., Ma, X.-L., Chen, Y.-C., Wu, Y., Zhu, J., and Deng, J.-G. (2006) *Chem. Commun.*, 1563; (b) Xie, J.-W., Chen, W., Li, R., Zeng, M., Du, W., Yue, L., Chen, Y.-C., Wu, Y., Zhu, J., and Deng, J.-G. (2007) *Angew. Chem., Int. Ed.*, **46**, 389.
41. Lu, J., Liu, F., and Loh, T.-P. (2008) *Adv. Synth. Catal.*, **350**, 1781.
42. Cabrera, S., Reyes, E., Alemán, J., Milelli, A., Kobbelgaard, S., and Jorgensen, K.A. (2008) *J. Am. Chem. Soc.*, **130**, 12031; For a later report see (b) Hayashi, Y., Obi, K., Ohta, Y., Okamura, D., and Ishikawa, H. (2009) *Chem. Asian J.*, **4**, 246.
43. Gotoh, H., Matsui, R., Ogino, H., Shoji, M., and Hayashi, Y. (2006) *Angew. Chem., Int. Ed.*, **45**, 6853.
44. Brown, S.P., Goodwin, N.C., and MacMillan, D.W.C. (2003) *J. Am. Chem. Soc.*, **125**, 1192.
45. Robichaud, J. and Tremblay, F. (2006) *Org. Lett.*, **8**, 597.
46. Wang, W. and Wang, J. (2005) *Org. Lett.*, **7**, 1637.
47. (a) Hayashi, Y., Gotoh, H., Matsui, R., and Ishikawa, H. (2008) *Angew. Chem., Int. Ed.*, **47**, 4012; (b) Zu, L., Xie, H., Li, H., Wang, J., Yu, X., and Wang, W. (2008) *Chem. Eur. J.*, **14**, 6333.
48. Melchiorre, P. and Jorgensen, K.A. (2003) *J. Org. Chem.*, **68**, 4151; (b) Franzén, J., Marigo, M., Fielenbach, D., Wabnitz, T.C., Kjaersgaard, A., and Jorgensen, K.A. (2005) *J. Am. Chem. Soc.*, **127**, 18296.
49. (a) Chui, Y. and Gellman, S.H. (2005) *Org. Lett.*, **7**, 4253; (b) Peelen, T.J., Chi, Y., and Gellman, S.H. (2005) *J. Am. Chem. Soc.*, **127**, 11598.
50. Hechavarria-Fonseca, M.T. and List, B. (2004) *Angew. Chem., Int. Ed.*, **43**, 3958.
51. Hayashi, Y., Gotoh, H., Tamura, T., Yamaguchi, H., Masui, R., and Shoji, M. (2005) *J. Am. Chem. Soc.*, **127**, 16028.
52. (a) Paras, N.A. and MacMillan, D.W.C. (2001) *J. Am. Chem. Soc.*, **123**, 4370; (b) Austin, J.F. and MacMillan, D.W.C. (2002) *J. Am. Chem. Soc.*, **124**, 1172; (c) Paras, N.A. and MacMillan, D.W.C. (2002) *J. Am. Chem. Soc.*, **124**, 7894.
53. (a) Gordillo, R., Carter, J., and Houk, K.N. (2004) *Adv. Synth. Catal.*, **346k**, 1175; See also (b) Brazier, J.B., Evans, G., Gibbs, T.J.K., Coles, S.J., Hursthouse, M.B., Platts, J.A., and Tomkinson, N.C.O. (2009) *Org. Lett.*, **11**, 133.
54. (a) Li, C.-F., Liu, H., Liao, J., Cao, Y.-J., Liu, X.-P., and Xiao, W.-J. (2007) *Org. Lett.*, **9**, 1847; (b) Liu, H.-H., Liu, H., Wu, W., Wang, X.-F., Lu, L.-Q., and Xiao, W.-J. (2009) *Chem. Eur. J.*, **15**, 2742.
55. King, H.D., Meng, Z., Denhart, D., Mattson, R., Kimura, R., Wu, D., Gao, Q., and Macor, J.E. (2005) *Org. Lett.*, **7**, 3437.
56. Zhang, Y., Zhao, L., Lee, S.S., and Ying, J.Y. (2006) *Adv. Synth. Catal.*, **348**, 2027.
57. Hong, L., Wang, L., Chen, C., Zhang, B., and Wang, R. (2009) *Adv. Synth. Catal.*, **351**, 772.
58. Li, D.-P., Guo, Y.-C., Ding, Y., and Xiao, W.-J. (2006) *Chem. Commun.*, 799.
59. Chen, W., Du, W., Yue, L., Li, R., Wu, Y., Ding, L.-S., and Chen, Y.-C. (2007) *Org. Biomol. Chem.*, **5**, 816.
60. Bartoli, G., Bosco, M., Carlone, A., Pesciaioli, F., Sambri, L., and Melchiorre, P. (2007) *Org. Lett.*, **9**, 1403.
61. Lee, S. and MacMillan, D.W.C. (2007) *J. Am. Chem. Soc.*, **129**, 15438; For a related intramolecular reaction see (b) Kim, S.-G. (2008) *Tetrahedron Lett.*, **2008**, 6148.
62. For a general review see Adolfsson, H. (2005) *Angew. Chem., Int. Ed.*, **44**, 3340.
63. Yang, J.W., Hechavarria-Fonseca, M.T., and List, B. (2004) *Angew. Chem., Int. Ed.*, **43**, 6660.
64. Yang, J.W., Hechavarria-Fonseca, M.T., Vignola, N., and List, B. (2005) *Angew. Chem., Int. Ed.*, **44**, 108.

65. (a) Ouellet, S.G., Tuttle, J.B., and MacMillan, D.W.C. (2005) *J. Am. Chem. Soc.*, **127**, 32; (b) Tuttle, J.B., Ouellet, S.G., and MacMillan, D.W.C. (2006) *J. Am. Chem. Soc.*, **128**, 12662.
66. Mayer, S. and List, B. (2006) *Angew. Chem., Int. Ed.*, **45**, 4193.
67. Martin, N.J.A. and List, B. (2006) *J. Am. Chem. Soc.*, **128**, 13368.
68. Akagawa, K., Akabane, H., Sakamoto, S., and Kudo, K. (2008) *Org. Lett.*, **10**, 2035.
69. For some reviews on the asymmetric aza–Michael reaction see: (a) Vicario, J.L., Badía, D., Carrillo, L., Etxebarria, J., Reyes, E., and Ruiz, N. (2005) *Org. Prep. Proc. Int.*, **37**, 513; (b) Xu, L.-W. and Xia, C.-G. (2005) *Eur. J. Org. Chem.*, 633.
70. Chen, Y.K., Yoshida, M., and MacMillan, D.W.C. (2006) *J. Am. Chem. Soc.*, **128**, 9328.
71. (a) Vesely, J., Ibrahem, I., Rios, R., Zhao, G.-L., Xu, Y., and Córdova, A. (2007) *Tetrahedron Lett.*, **48**, 2193; For a related example using a chiral sulfonyl hydrazine as organocatalyst see (b) Chen, L.-Y., He, H., Pei, B.-J., Chan, W.-H., and Lee, A.W.M. (2009) *Synthesis*, 1573.
72. Ibrahem, I., Rios, R., Vesely, J., Zhao, G.-L., and Córdova, A. (2007) *Chem. Commun.*, 849.
73. Dinér, P., Nielsen, M., Marigo, M., and Jorgensen, K.A. (2007) *Angew. Chem., Int. Ed.*, **46**, 1983.
74. Jiang, H., Nielsen, J.B., Nielsen, M., and Jorgensen, K.A. (2007) *Chem. Eur. J.*, **13**, 9068.
75. Uria, U., Vicario, J.L., Badia, D., and Carrillo, L. (2007) *Chem. Commun.*, 2509.
76. (a) Fustero, S., Jiménez, D., Moscardó, J., Catalán, S., and del Pozo, C. (2007) *Org. Lett.*, **9**, 5283; (b) Fustero, S., Moscardó, J., Jiménez, D., Pérez-Carrión, M.D., Sánchez-Roselló, M., and del Pozo, C. (2008) *Chem. Eur. J.*, **14**, 9868; For a later report involving similar substrates and products see (c) Carlson, E.C., Rathbone, L.K., Yang, H., Collett, N.D., and Carter, R.G. (2008) *J. Org. Chem.*, **73**, 5155.
77. Lu, X. and Deng, L. (2008) *Angew. Chem., Int. Ed.*, **47**, 7710.
78. (a) Carlone, A., Bartoli, G., Bosco, M., Sambri, L., and Melchiorre, P. (2007) *Angew. Chem., Int. Ed.*, **46**, 4504; (b) Ibrahem, I., Rios, R., Vesely, J., Hammar, P., Eriksson, L., Himo, F., and Córdova, A. (2007) *Angew. Chem., Int. Ed.*, **46**, 4507; (c) Ibrahem, I., Hammar, P., Vesely, J., Rios, R., Eriksson, L., and Córdova, A. (2008) *Adv. Synth. Catal.*, **350**, 1875.
79. Maerten, E., Cabrera, S., Kjaersgaard, A., and Jorgensen, K.A. (2007) *J. Org. Chem.*, **72**, 8893.
80. Kano, T., Tanaka, Y., and Maruoka, K. (2007) *Tetrahedron*, **63**, 6858.
81. Bertelsen, S., Dinér, P., Johansen, R.L., and Jorgensen, K.A. (2007) *J. Am. Chem. Soc.*, **129**, 1536.
82. Carlone, A., Bartoli, G., Bosco, M., Pesciaioli, F., Ricci, P., Sambri, L., and Melchiorre, P. (2007) *Eur. J. Org. Chem.*, 5492.
83. Reisinger, C.M., Wang, X., and List, B. (2008) *Angew. Chem., Int. Ed.*, **47**, 8112.
84. Marigo, M., Schulte, T., Franzén, J., and Jorgensen, K.A. (2005) *J. Am. Chem. Soc.*, **127**, 15710.
85. Ricci, P., Carlone, A., Bartoli, G., Bosco, M., Sambri, L., and Melchiorre, P. (2008) *Adv. Synth. Catal.*, **350**, 49.
86. For some selected reviews see (a) Ismabery, N. and Lavila, R. (2008) *Chem. Eur. J.*, **14**, 8444; (b) Albert, B. and Scott, K. (2007) *Tetrahedron*, **63**, 5341; (c) Chapman, C.J. and Frost, C.G. (2007) *Synthesis*, 1; (d) Miyabe, H. and Takemoto, Y. (2007) *Chem. Eur. J.*, **13**, 7280; (e) Patil, N.T. and Yamamoto, Y. (2007) *Synlett*, 994; (f) Zhu, J. and Bienayme, H. (eds) (2005) *Multicomponent Reactions*, Wiley-VCH Verlag GmbH, Weinheim; (g) Tejedor, D. and Garcia-Tellado, F. (2007) *Chem. Soc. Rev.*, **36**, 484; (h) Tietze, L.F., Brasche, G., and Gerike, K. (2006) *Domino Reactions in Organic Chemistry*, Wiley-VCH Verlag GmbH, Weinheim; (i) Pellissier, H. (2006) *Tetrahedron*, **62**, 1619; (j) Nicolaou, K.C., Edmonds, D.J., and Bulger, P.G. (2006) *Angew. Chem.*, **118**, 7292; *Angew.*

Chem., Int. Ed., (2006), **45**, 7134; (k) Pellissier, H. (2006) *Tetrahedron*, **62**, 2143; (l) Doemling, A. (2006) *Chem. Rev.*, **106**, 17; (m) Guo, H.-C. and Ma, J.-A. (2006) *Angew. Chem.*, **118**, 362; *Angew. Chem., Int. Ed.*, (2006), **45**, 354; (n) Ramon, D.J. and Yus, M. (2005) *Angew. Chem.*, **117**, 1628; *Angew. Chem. Int. Ed.* (2005), **44**, 1602; (o) Wasilke, J.-C., Obrey, S.J., Baker, R.T., and Bazan, G.C. (2005) *Chem. Rev.*, **105**, 1001.

87. (a) Enders, D., Grondal, C., and Huettl, M.R.M. (2007) *Angew. Chem., Int. Ed.*, **46**, 1570; (b) Guillena, G., Ramon, D.J., and Yus, M. (2007) *Tetrahedron: Asymmetry*, **18**, 693.

88. Enals: (a) Asato, A.E., Watanabe, C., Li, X.-Y., and Liu, R.S.H. (1992) *Tetrahedron Lett.*, **33**, 3105; (b) Bench, B.J., Liu, C., Evett, C.R., and M. H., C. (2006) *J. Org. Chem.*, **71**, 6458; (c) Hong, B.-C., Wu, M.-F., Tseng, H.-C., and Liao, J.-H. (2006) *Org. Lett.*, **8**, 2217; (d) Hong, B.-C., Wu, M.-F., Tseng, H.-C., Huang, G.-F., Su, C.-F., and Liao, J.-H. (2007) *J. Org. Chem.*, **72**, 8459; Enones: (e) Ramachary, D.B., Chowdari, N.S., and Barbas, C.F.III (2002) *Tetrahedron Lett.*, **43**, 6743.

89. (a) Ramachary, D.B. and Kishor, M. (2007) *J. Org. Chem.*, **72**, 5056; (b) Ramachary, D.B. and Kishor, M. (2008) *Org. Biomol. Chem.*, **6**, 4176.

90. (a) Halland, N., Aburel, P.S., and Jorgensen, K.A. (2004) *Angew. Chem., Int. Ed.*, **43**, 1272; (b) Pulkkinen, J., Aburel, P.S., Halland, N., and Jorgensen, K.A. (2004) *Adv. Synth. Catal.*, **346**, 1077; (c) Gryko, D. (2005) *Tetrahedron: Asymmetry*, **16**, 1377.

91. (a) Carlone, A., Marigo, M., North, C., Landa, A., and Jorgensen, K.A. (2006) *Chem. Commun.*, 4928; (b) Bolze, P., Dickmeiss, G., and Jorgensen, K.A. (2008) *Org. Lett.*, **10**, 3753.

92. Marigo, M., Bertelsen, S., Landa, A., and Jorgensen, K.A. (2006) *J. Am. Chem. Soc.*, **128**, 5475.

93. Reyes, E., Jiang, H., Milelli, A., Elsner, P., Hazell, R.G., and Jorgensen, K.A. (2007) *Angew. Chem., Int. Ed.*, **46**, 9202.

94. Wang, J., Li, H., Xie, H., Zu, L., Shen, X., and Wang, W. (2007) *Angew. Chem., Int. Ed.*, **46**, 9050.

95. Hong, B.-C., Nimje, R.Y., Sadani, A.A., and Liao, J.-S. (2008) *Org. Lett.*, **10**, 2345.

96. (a) Hayashi, Y., Toyoshima, M., Gotoh, H., and Ishikawa, H. (2009) *Org. Lett.*, **11**, 45; (b) Albrecht, L., Richter, B., Vila, C., Krawczyk, H., and Jorgensen, K.A. (2009) *Chem. Eur. J.*, **15**, 3093; (c) Bertelsen, S., Johansen, R.L., and Jorgensen, K.A. (2008) *Chem. Commun.*, 3016.

97. Franzén, J. and Fisher, A. (2009) *Angew. Chem., Int. Ed.*, **48**, 787.

98. Zu, L., Li, H., Xie, H., Wang, J., Jiang, W., Tang, Y., and Wang, W. (2007) *Angew. Chem., Int. Ed.*, **46**, 3732.

99. Zhao, G.-L., Ibrahim, I., Dziedzic, P., Sun, J., Bonneau, C., and Córdova, A. (2008) *Chem. Eur. J.*, **14**, 10007.

100. Kang, T.-R., Xie, J.-W., Du, W., Feng, X., and Chen, Y.-C. (2008) *Org. Biomol. Chem.*, **6**, 2673.

101. Cabrera, S., Alemám, J., Bolze, P., Bertelsen, S., and Jorgensen, K.A. (2008) *Angew. Chem., Int. Ed.*, **47**, 121.

102. Enders, D., Hüttl, M.R.R., Grondal, C., and Raabe, G. (2006) *Nature*, **441**, 861.

103. Enders, D., Hüttl, M.R.M., Raabe, G., and Bats, J.W. (2008) *Adv. Synth. Catal.*, **350**, 267.

104. Varela, M.C., Dixon, S.M., Lam, K.S., and Schore, N.E. (2008) *Tetrahedron*, **64**, 10087.

105. Enders, D., Hüttl, M.R.R., Runsink, J., Raabe, G., and Wendt, B. (2007) *Angew. Chem., Int. Ed.*, **46**, 467.

106. Penon, O., Carlone, A., Mazzanti, A., Locatelli, M., Sambri, L., Bartoli, G., and Melchiorre, P. (2008) *Chem. Eur. J.*, **14**, 4788.

107. Carlone, A., Cabrera, S., Marigo, M., and Jorgensen, K.A. (2007) *Angew. Chem., Int. Ed.*, **46**, 1101.

108. (a) Bremeyer, N., Smith, S.C., Ley, S.V., and Gaunt, M.J. (2004) *Angew. Chem., Int. Ed.*, **43**, 2681; (b) Papageorgiou, C.D., Ley, S.V., and Gaunt, M.J. (2003) *Angew. Chem., Int. Ed.*, **42**, 828.

109. (a) Hansen, H.M., Longbottom, D.A., and Ley, S.V. (2006) *Chem. Commun.*,

4838; (b) Wascholowski, V., Hansen, H.M., Longbottom, D.A., and Ley, S.V. (2008) *Synthesis*, 1269.

110. Lv, J., Zhang, J., Lin, Z., and Wang, Y. (2009) *Chem. Eur. J.*, **15**, 972.

111. (a) Vesely, J., Zhao, G.-L., Bartoszewicz, A., and Córdova, A. (2008) *Tetrahedron Lett.*, **49**, 4209; (b) Rios, R., Sundén, H., Vesely, J., Zhao, G.-L., Dziedzic, P., and Córdova, A. (2007) *Adv. Synth. Catal.*, **349**, 1028; (c) Ibrahem, I., Zhao, G.-L., Rios, R., Vesely, J., Sundén, H., Dziedzic, P., and Córdova, A. (2008) *Chem. Eur. J.*, **14**, 7867; (d) Xie, H., Zu, L., Li, H., Wang, J., and Wang, W. (2007) *J. Am. Chem. Soc.*, **129**, 10886.

112. Kunz, R.K. and MacMillan, D.W.C. (2005) *J. Am. Chem. Soc.*, **127**, 3240.

113. (a) Zhao, Y.-H., Zheng, C.-W., Zhao, G., and Cao, W.-G. (2008) *Tetrahedron: Asymmetry*, **19**, 701; (b) Zhao, Y.-H., Zhao, G., and Cao, W.-G. (2007) *Tetrahedron: Asymmetry*, **18**, 2462.

114. Huang, Y., Walji, A.M., Larsen, C.H., and MacMillan, D.W.C. (2005) *J. Am. Chem. Soc.*, **127**, 15051.

115. Austin, J.F., Kim, S.-G., Sinz, C.J., Xiao, W.-J., and MacMillan, D.W.C. (2004) *Proc. Natl Acad. Sci. USA*, **101**, 5482.

116. Harmata, M., Ghosh, S.K., Hong, X., Wacharasindhu, S., and Kirchhoefer, P. (2003) *J. Am. Chem: Soc.*, **125**, 2058.

117. Yang, J.W., Hechavarria-Fonseca, M.T., and List, B. (2005) *J. Am. Chem. Soc.*, **127**, 15036.

118. Zhao, G.-L. and Córdova, A. (2006) *Tetrahedron Lett.*, **47**, 7417.

119. Simmons, B., Walji, A.M., and MacMillan, D.W.C. (2009) *Angew. Chem., Int. Ed.*, **48**, 4349.

120. Zhao, G.-L., Rios, R., Vesely, J., Eriksson, L., and Córdova, A. (2008) *Angew. Chem., Int. Ed.*, **47**, 8468.

121. Brandau, S., Maerten, E., and Jorgensen, K.A. (2006) *J. Am. Chem. Soc.*, **128**, 14986.

122. (a) Govender, T., Hojabri, L., Moghaddam, F.M., and Arvidsson, P.I. (2006) *Tetrahedron: Asymmetry*, **17**, 1763; (b) Sundén, S., Ibrahem, I., Zhao, G.-L., Eriksson, L., and Córdova, A. (2007) *Chem. Eur. J.*, **13**, 574; (c) Li, H., Wang, J., E-Nunu, T., Zu, L., Jiang, W., Wei, S., and Wang, W. (2007) *Chem. Commun.*, 507.

123. (a) Rios, R., Sundén, H., Ibrahem, I., Zhao, G.-L., Eriksson, L., and Córdova, A. (2006) *Tetrahedron Lett.*, **47**, 8547; (b) Wang, W., Li, H., Wang, J., and Zu, L. (2006) *J. Am. Chem. Soc.*, **128**, 10354.

124. (a) Li, H., Wang, J., Xie, H., Zu, L., Jiang, W., Duesler, E.N., and Wang, W. (2007) *Org. Lett.*, **9**, 965; (b) Sunden, H., Rios, R., Ibrahem, I., Zhao, G.-L., Eriksson, L., and Córdova, A. (2007) *Adv. Synth. Catal.*, **349**, 827.

125. Zhao, G.-L., Vesely, J., Rios, R., Ibrahem, I., Sundén, H., and Córdova, A. (2008) *Adv. Synth. Catal.*, **350**, 237.

126. (a) Rios, R., Sundén, H., Ibrahem, I., Zhao, G.-L., and Córdova, A. (2006) *Tetrahedron Lett.*, **47**, 8679; (b) Rios, R., Sundén, H., Ibrahem, I., and Córdova, A. (2007) *Tetrahedron Lett.*, **48**, 2181.

127. (a) Li, H., Zu, L., Xie, H., Wang, J., Jiang, W., and Wang, W. (2007) *Org. Lett.*, **9**, 1833; (b) Li, H., Zu, L., Xie, H., Wang, J., and Wang, W. (2008) *Chem. Commun.*, 5636.

128. (a) Marigo, M., Franzén, J., Poulsen, T.B., Zhuang, W., and Jørgensen, K.A. (2005) *J. Am. Chem. Soc.*, **127**, 6964; (b) Lattanzi, A. (2006) *Adv. Synth. Catal.*, **348**, 339; (c) Sundén, H., Ibrahem, I., and Córdova, A. (2006) *Tetrahedron Lett.*, **47**, 99; (d) Li, Y., Liu, X., Yang, Y., and Zhao, G. (2007) *J. Org. Chem.*, **72**, 288; (e) Lee, S. and MacMillan, D.W.C. (2006) *Tetrahedron*, **62**, 11413; (f) Sparr, C., Schweizer, W.B., Senn, H.M., and Gilmour, R. (2009) *Angew. Chem., Int. Ed.*, **48**, 3065.

129. Lu, X., Liu, Y., Sun, B., Cindric, B., and Deng, L. (2008) *J. Am. Chem. Soc.*, **130**, 8134.

130. (a) Wang, X., Reisinger, C.M., and List, B. (2008) *J. Am. Chem. Soc.*, **130**, 6070; This concept was also applied to the use of enals as substrates: (b) Wang, X. and List, B. (2008) *Angew. Chem., Int. Ed.*, **47**, 1119.

131. Vesely, J., Ibrahem, I., Zhao, G.-L., Rios, R., and Córdova, A. (2007) *Angew. Chem., Int. Ed.*, **46**, 778.

7
Organocatalytic Enantioselective Conjugate Additions of Heteroatoms to α,β-Unsaturated Carbonyl Compounds

Shilei Zhang and Wei Wang

7.1
Introduction

The power of conjugate addition reactions is fueled by a wide variety of substances that can serve as Michael donors and acceptors and, consequently, a diverse array of products can be generated [1–4]. Recent intensive efforts have focused on the development of organocatalytic methods to perform direct, asymmetric Michael addition reactions, and significant progress has been made in the area of asymmetric conjugate additions of heteroatoms such as "N," "O," "S," and "P" to α,β-unsaturated carbonyl systems. In this chapter, attention is focused on key developments, classified on the basis of the different nucleophiles "N," "O," "S," and "P." For reasons of space limitation, it is impossible at this point to discuss every single case; hence, for additional detailed information that is not discussed in this chapter, the reader is referred to several excellent reviews [1–10].

7.2
"N" as Nucleophiles

Owning to the trivalent nature of nitrogen atom and the resulting structural diversity, various versions of nitrogen-centered nucleophiles have been explored for organocatalytic asymmetric aza–Michael reactions. The nucleophilic "N" species have been demonstrated in both intermolecular and intramolecular and cascade formats. These diversified Michael adducts serve as very useful building blocks in organic synthesis.

7.2.1
Intermolecular aza–Michael Reactions

At the early stage of organocatalytic reaction development, Miller and coworkers first used β-tune peptides as catalysts to promote the Michael addition of TMSN$_3$ to α,β-unsaturated imides (Scheme 7.1) [11, 12]. It was found that more rigid

Catalytic Asymmetric Conjugate Reactions. Edited by Armando Córdova
Copyright © 2010 WILEY-VCH Verlag GmbH & Co. KGaA, Weinheim
ISBN: 978-3-527-32411-8

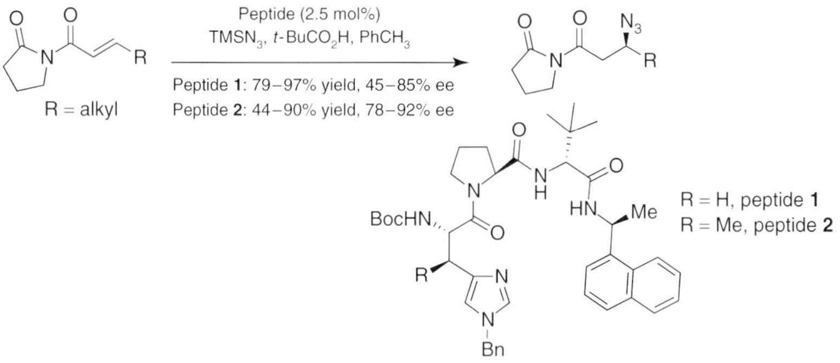

Scheme 7.1 Peptides-catalyzed Michael addition of TMSN$_3$ to α,β-unsaturated imides.

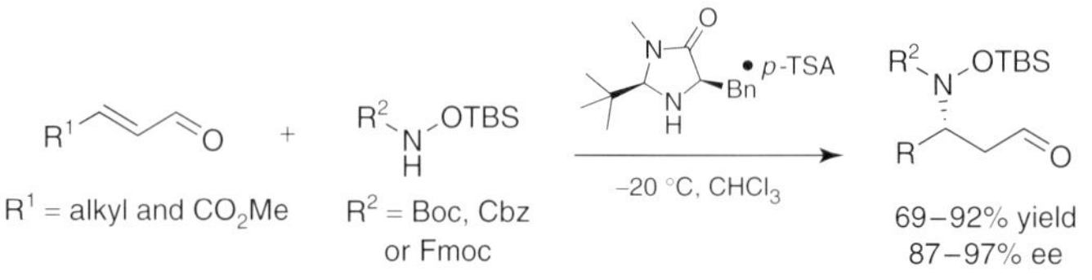

Scheme 7.2 Chiral imidazolidinone-catalyzed Michael addition of N-silyloxycarbamates to enals.

β-substituted histidine residue was critical for improving the enantioselectivities of the process [12].

MacMillan *et al.* developed the first enantioselective organocatalytic conjugate addition of amines to α,β-unsaturated aldehydes (Scheme 7.2) [13]. This reaction was catalyzed by MacMillan's imidazolidinone catalyst. Due to the weak nucleophilicity, the use of highly reactive N-silyloxycarbamates was essential for the aza–Michael reaction. Moreover, the specific nitrogen species can also prevent the formation of an iminium ion with the enals and, impressively, the reaction gave products with good to excellent enantioselectivities. It was realized, however, that the reactions were limited to the alkyl enals, while the Michael products, after simple transformations, could generate very useful β-amino acids or their derivatives. A similar study was also reported by the group of Córdova, whereas a diarylprolinol silyl ether served as an effective promoter [14].

Deng and coworkers disclosed the conjugate addition of protected hydroxylamines to enones (Scheme 7.3) [15], where the process was promoted by cinchona alkaloid-derived primary amines, which are effective activators for enones. The benzyloxy protecting group could easily be removed under mild reductive conditions (H$_2$/Raney Ni) to furnish the corresponding N-Boc-protected β-amino ketones, without any significant deterioration in optical purity.

The simple and commercially available benzylhydroxylamine was also exploited in the aza–Michael reaction with chalcones in the presence of a quinine-derived thiourea [16]. Although only poor enantioselectivities (27–60% ee) were attained, a subsequent significantly improved protocol was successfully developed by Sibi

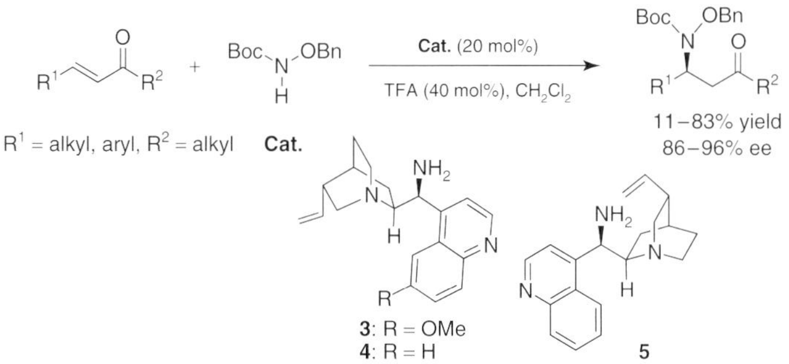

Scheme 7.3 Chiral cinchona alkaloid-derived primary amines-catalyzed Michael addition of Boc-protected N-benzyloxyamine to enones.

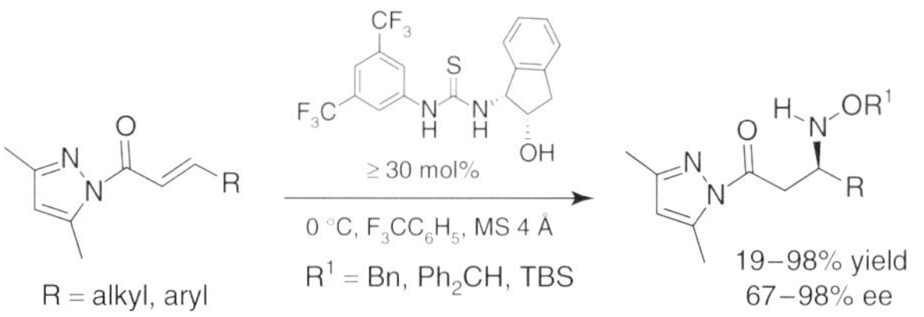

Scheme 7.4 Chiral aminoindanol thiourea-catalyzed Michael addition of benzylhydroxylamine to α,β-unsaturated pyrazoles.

and coworkers (Scheme 7.4) [17]. Importantly, the use of α,β-unsaturated pyrazole templates and the aminoindanol thiourea catalyst was crucial to achieve good yields and high enantioselectivities, despite the relatively high catalyst loading needed (≥ 30 mol%).

Jørgensen *et al.* reported nucleophilic hydrazones for a cinchona alkaloid dehydroquinine (DHQN)-catalyzed asymmetric aza–Michael reaction with cyclic enones, but low to moderate enantioselectivities were observed (Scheme 7.5) [18]. In a recent study, even simple less-reactive aniline was directly employed for the aza–Michael addition to chalcones catalyzed by cinchonine [19]. Unfortunately, the low level of enantioselectivity (11–58% ee) limited the application of this methodology.

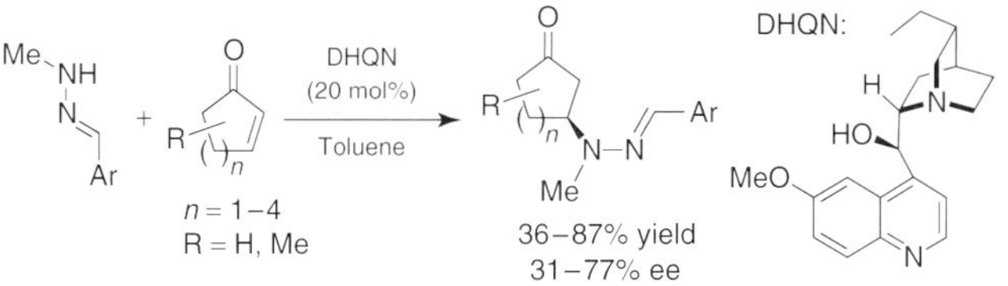

Scheme 7.5 Cinchona alkaloid DHQN-catalyzed Michael addition of hydrazones to cyclic enones.

Succinimides as nitrogen-centered nucleophiles were exploited in an diaryl-prolinol trimethylsilyl (TMS) ether promoted highly enantioselective two- and three-component aza-Michael reactions with enals [Scheme 7.6; Eqs (7.1) and (7.2)] [20].

Jørgensen and coworkers also developed organocatalytic conjugate additions of N-heterocycles, including 1,2,4-triazole, 5-phenyltetrazole, and 1,2,3-benzotriazole, to α,β-unsaturated aldehydes in moderate to high yields and with good to high enantioselectivities, using the same catalyst [Scheme 7.6; Eqs (7.3–7.5)] [21]. High regioselectivities for 1,2,4-triazole and 5-phenyltetrazole, but low for 1,2,3-benzotriazole were observed. A similar process was also catalyzed by MacMillan's catalyst [22].

Pyrazoles as nucleophiles have been employed in amine-mediated aza–Michael reactions (Scheme 7.7) [23, 24]. In first case, a chiral primary amine was used for enones [Eq. (7.6)], whereas in the latter case the enals were activated by a secondary

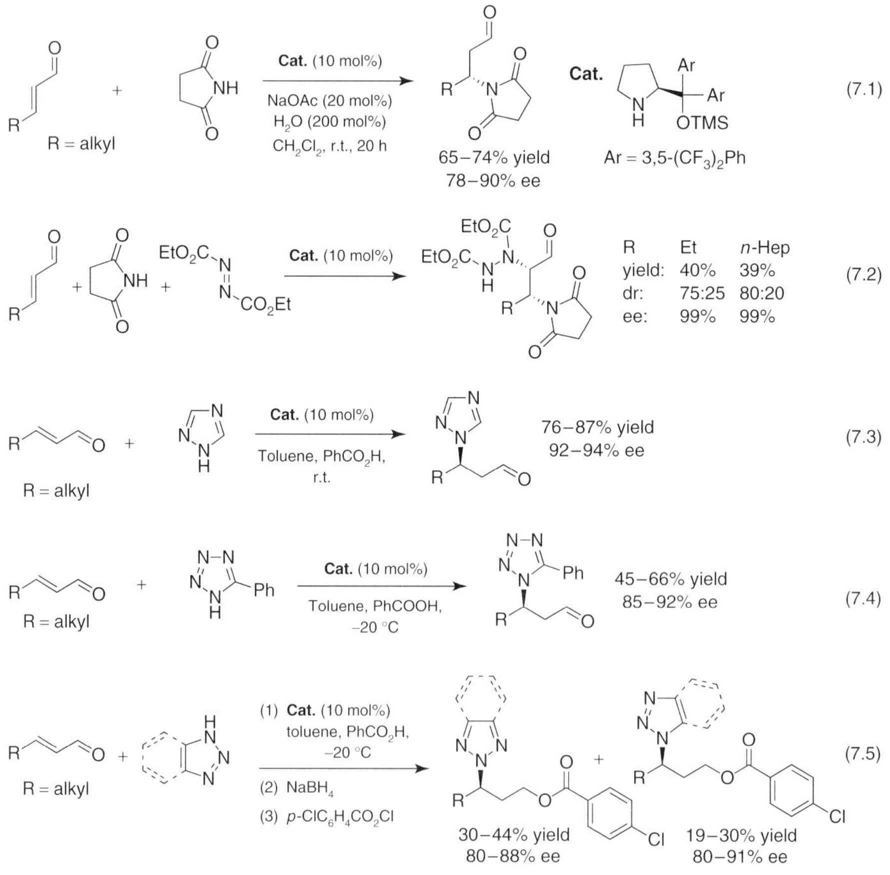

Scheme 7.6 Diarylprolinol TMS ether-catalyzed Michael addition of succinimides and N-heterocycles to enals.

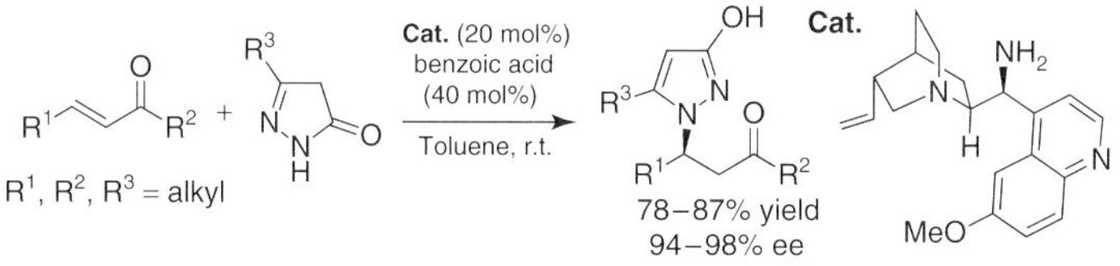

Scheme 7.7 Organocatalytic enantioselective Michael addition of pyrazoles to α,β-unsaturated carbonyls.

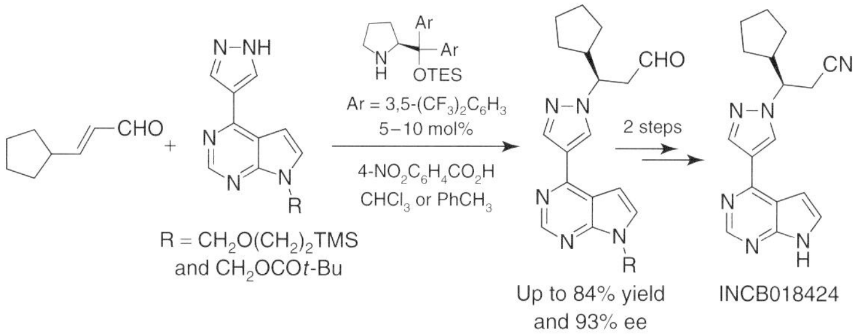

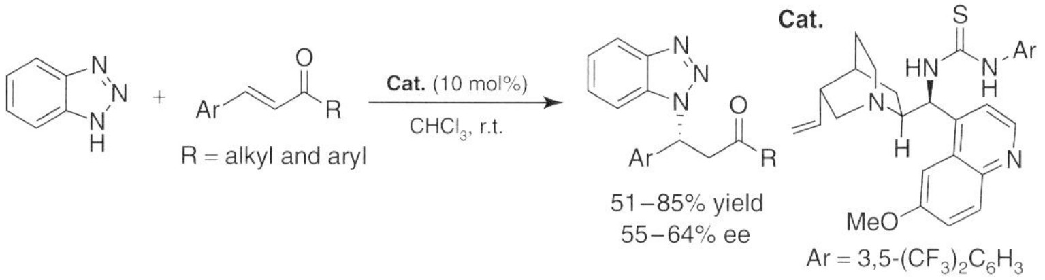

Scheme 7.8 Amine thiourea-catalyzed Michael addition of 1,2,3-benzotriazole to enone.

amine [Eq. (7.7)]. This chemistry has been applied successfully in the synthesis of the Janus-associated kinase (JAK) inhibitor INCB018424 [24].

Finally, Wang and coworkers developed a cinchona alkaloid thiourea-catalyzed conjugate addition of *N*-heterocycles to enones (Scheme 7.8) [25]. In spite of the moderate enantioselectivities, these reactions were highly regioselective.

7.2.2
Intramolecular aza–Michael Reactions

The organocatalytic intramolecular aza–Michael reactions have been less widely documented than their intermolecular counterparts. Hsung's synthesis of quinolizidines provided a series of aza-[3 + 3] cycloaddition products (Scheme 7.9) [26], where the reaction proceeded through an intramolecular cascade sequence. In this case, the key step in the process involved the first *N*-1,4-addition of vinylogous amine to α,β-unsaturated aldehydes, and low to moderate ee-values were obtained, even after an intensive optimization of the reaction conditions.

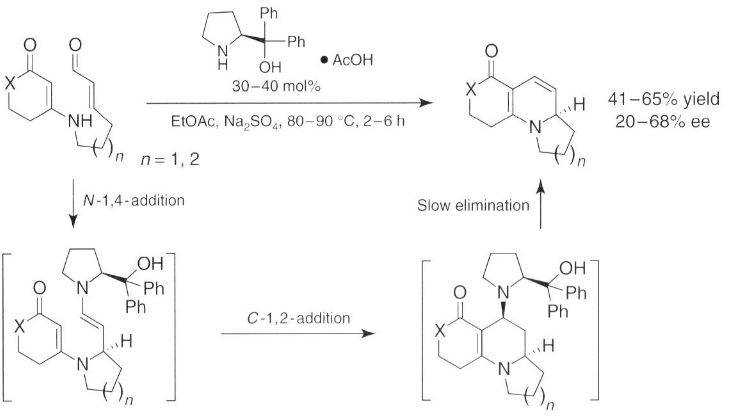

Scheme 7.9 Diphenylprolinol-catalyzed aza-[3 + 3] cycloadditions.

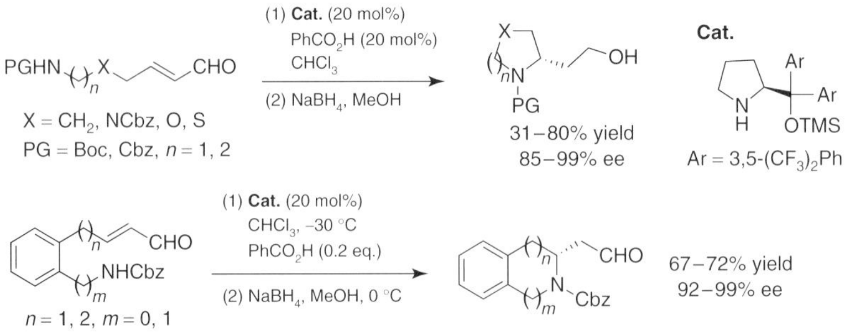

Scheme 7.10 Enantioselective organocatalytic intramolecular aza–Michael reactions.

In two related studies, Fustero *et al.* disclosed an efficient and general approach to constructing five- and six-membered nitrogen-containing rings, such as pyrrolidine, piperidine, indolines, isoindolines, tetrahydroquinolines, and tetrahydroisoquinolines, by employing diarylpyrrolidinol TMS ether-catalyzed intramolecular aza–Michael reactions (Scheme 7.10) [27, 28]. The process gave good yields and good to excellent ee-values; moreover, the reaction products could be used in the synthesis of many natural products, including (+)-sedamine, (+)-allosedamine, (+)-coniine, and (S)-(+)-angustureine. A related strategy was also developed by Carter and colleagues [29].

7.2.3
Nitrogen Initiated aza–Michael Cascade Reactions

Recently, organocatalyzed asymmetric cascade reactions have become powerful tools for the efficient construction of complex molecular architectures [30, 31]. In these cascade processes, only a single reaction solvent, work-up procedure and purification step is required to create a product that would otherwise require several steps. Therefore, cascade reactions with the significant improvement of synthetic efficiency, the avoidance of toxic agents, and the reduction of waste and hazardous byproducts, fall under the banner of "green chemistry."

Inspired by the successes of their previous investigations, Itoh described an (S)-proline-catalyzed cascade Mannich–aza–Michael process for the assembly of indole alkaloid scaffolds (Scheme 7.11) [32]. This efficient method has been demonstrated in the efficient total synthesis of the natural product *ent*-dihydrocorynantheol.

Both Wang [33] and Córdova [34] have, respectively, reported a catalytic cascade aza–Michael–aldol reaction between *o*-amino benzaldehydes and α,β-unsaturated aldehydes, generating 1,2-dihydroquinolines in high yields and with high levels of enantioselectivities (Scheme 7.12). The difference between two methodologies lay in the fact that Wang used Cbz-protected amino benzaldehydes, while Córdova employed the unprotected counterparts. These developments were of great synthetic and biological significance, since functionalized 1,2-dihydroquinolines serve as very useful building blocks and are found in a large number of biologically and pharmaceutically interesting compounds. For instance, Hamada recently succeeded with the first catalytic enantioselective synthesis of the martinelline chiral core by using this efficient strategy [Eq. (7.9)] [35].

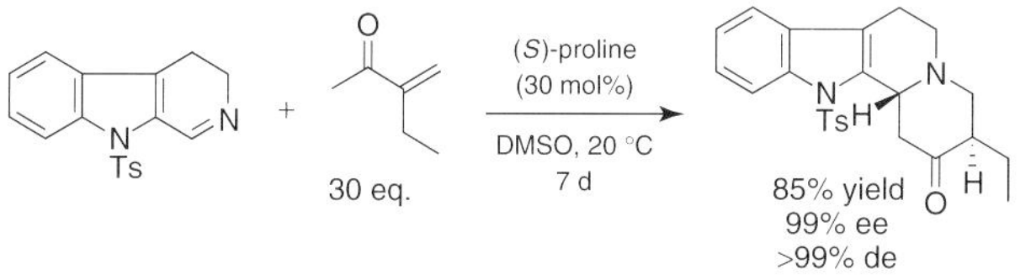

Scheme 7.11 (S)-Proline-catalyzed cascade Mannich–aza–Michael reactions.

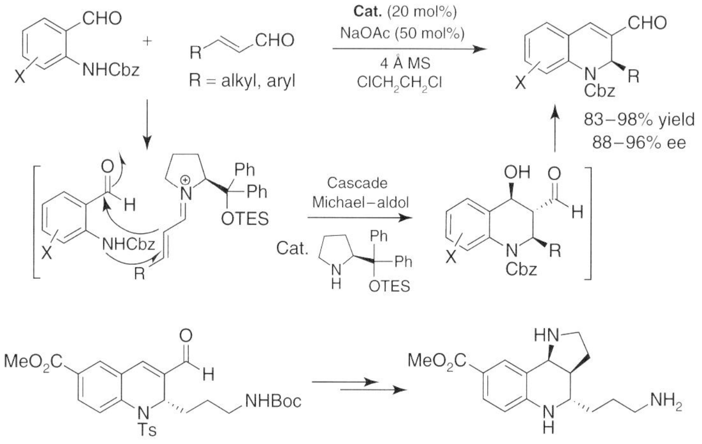

Scheme 7.12 Organocatalytic enantioselective cascade aza–Michael–aldol–dehydration reactions.

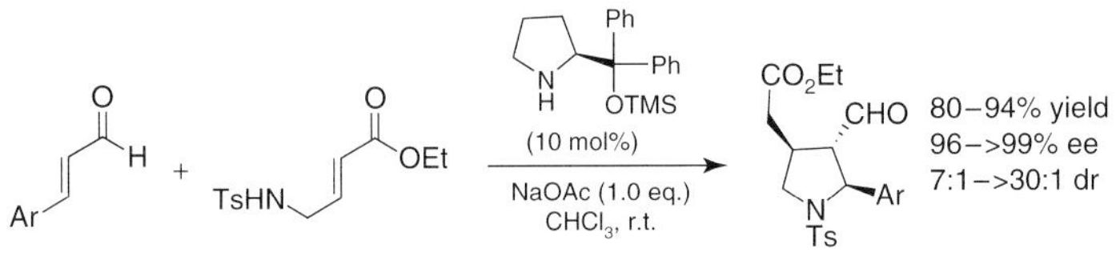

Scheme 7.13 Organocatalytic enantioselective cascade aza–Michael/Michael reactions.

Wang also developed a cascade aza–Michael–Michael reaction of α,β-unsaturated aldehydes with *trans*-γ-Ts-protected amino α,β-unsaturated ester (Scheme 7.13) [36]. The highly functionalized trisubstituted chiral pyrrolidines products were generated in high yields and with high stereo- and enantioselectivities.

Aziridines represent important structures that exist in a range of natural products and which have been used widely as building blocks in organic synthesis. The first organocatalytic aziridination reaction of α,β-unsaturated aldehydes was reported by Córdova (Scheme 7.14) [37], in a process which involved a Michael–amination sequence. The first step was similar to that of MacMillan's aza–Michael protocol, except that acylated hydroxycarbamates was used as the nucleophile instead. The acetyl group served as the leaving group for a subsequent intramolecular substitution by an attack of the aldehyde-derived enamine. Notably, the reaction yields were moderate but the enantioselectivities were good.

Córdova *et al.* also used hydroxycarbamates as nucleophiles for the cascade Michael–hemiacetal reaction (Scheme 7.15) [38]. In this case, the aza–Michael reaction products reacted intramolecularly between the aldehyde and the hydroxyl group to form cyclic hemiacetals. The 5-hydroxyisoxazilidones products could then be oxidized to 5-oxazilidinones by treatment with $NaClO_2$. In addition, β-amino acid and γ-amino alcohol derivatives could also be obtained after simple transformations.

The use of enones as Michael acceptors for the cascade reactions was reported by Melchiorre (Scheme 7.16) [39], where the key to the process was the use of a cinchona primary amine salt catalyst to activate the substrate enones. The chiral

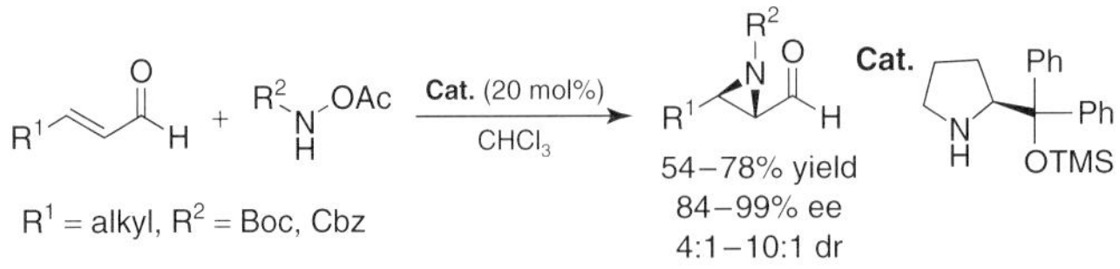

Scheme 7.14 Organocatalytic enantioselective aziridination of enals.

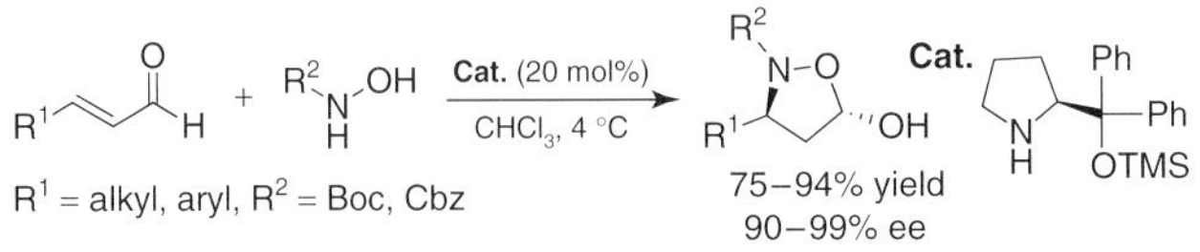

Scheme 7.15 Organocatalytic enantioselective cascade aza–Michael–hemiacetal reaction.

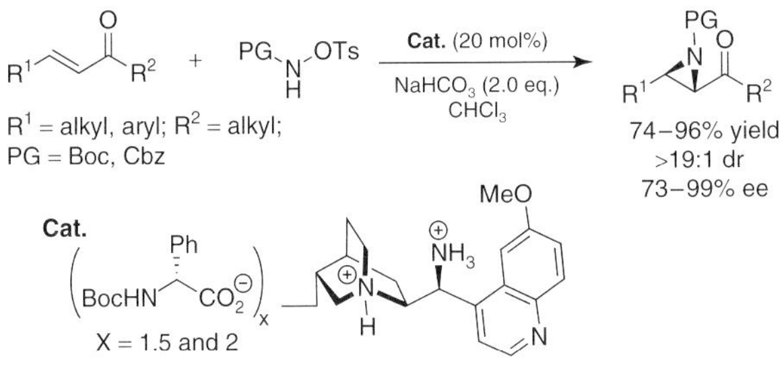

7.2 "N" as Nucleophiles

Scheme 7.16 Organocatalytic enantioselective cascade aza–Michael reactions with enones.

anions also played an important role in governing the enantioselectivity of the reaction, and both acyclic and cyclic enones afforded good results.

An asymmetric example of aziridination of chalcones was reported by Armstrong using O-(diphenylphosphinyl)hydroxylamine as nucleophile, promoted by quinine [40].

Córdova et al. also reported a new type of cascade reaction, in which both components of N-(benzylthio)succinimide were incorporated into the enals in a single operation (Scheme 7.17) [41]. This economic and efficient aminosulfenylation of α,β-unsaturated aldehydes provided useful β-amino-α-mercaptoaldehydes in high yields and with high ee-values, although the diastereoselectivities were poor.

Based on the L-proline-catalyzed α-aminoxylation reaction of aldehydes with arylnitroso [42], Zhong et al. developed a new cascade aminoxylation/aza–Michael reaction by using specific aldehyde α,β-unsaturated esters (Scheme 7.18) [43].

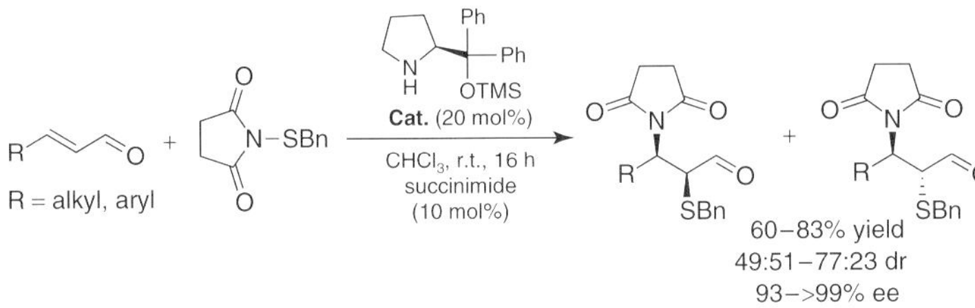

Scheme 7.17 Organocatalytic enantioselective cascade aza–Michael sulfenylation reactions of enals with N-(benzylthio)succinimide.

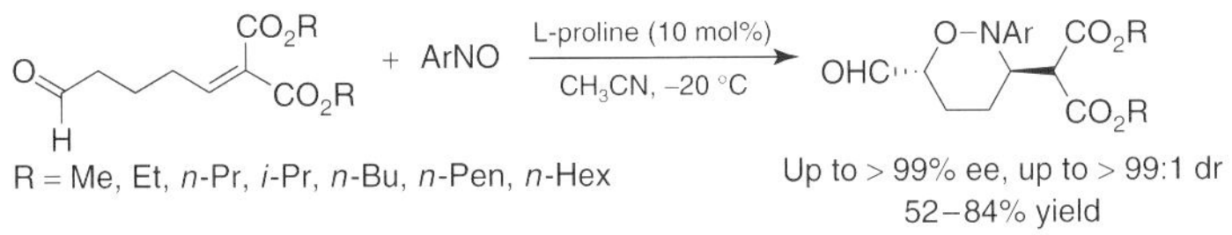

Scheme 7.18 Organocatalytic enantioselective cascade aminoxylation/aza–Michael reactions.

In this case, highly enantioselective and diastereoselective tetrahydro-1,2-oxazines were synthesized.

7.3
"O" as Nucleophiles

The asymmetric conjugate addition of oxygen-centered nucleophiles to α,β-unsaturated carbonyl compounds provides enantioenriched β-hydroxyl carbonyl compounds and their derivatives, all of which serve as very important chiral building blocks in organic synthesis. For these successful organocatalytic enantioselective oxa–Michael reactions, it was shown that due to their weak nucleophilicity, both the oximes and hydroperoxides proved to be excellent nucleophilic species in intermolecular reactions. As in the case of the intramolecular and cascade reactions, phenols held the overwhelming position as effective nucleophiles.

7.3.1
Intermolecular oxa–Michael Reactions

Simple alcohol, when employed as a nucleophile in a direct oxa–Michael reaction, presents a major challenge, the only such example having been reported by Maruoka's group in 2007 (Scheme 7.19) [44]. In this case, despite having designed and utilized a bifunctional catalyst based on a biphenyldiamine structure, low ee-values of only up to 53% were achieved, even under optimal catalytic conditions. These findings also confirmed that alcohol is not an ideal nucleophile.

However, the problem was overcome by Jørgensen, who used more-reactive oximes as the nucleophile in the oxa–Michael addition to α,β-unsaturated aldehydes (Scheme 7.20) [45]. The reaction was promoted by a diarylprolinol TMS ether, and gave moderate yields and good to excellent ee-values. The resultant aldehydes were reduced to the corresponding alcohols and, following removal of the oxime moiety by hydrogenation under $Pd(OH)_2/C$ conditions, provided very useful 1,3-diol scaffolds. Unfortunately, the one limitation of this methodology was that only aliphatic-substituted enals could be used as suitable substrates, and no aromatic system was reported.

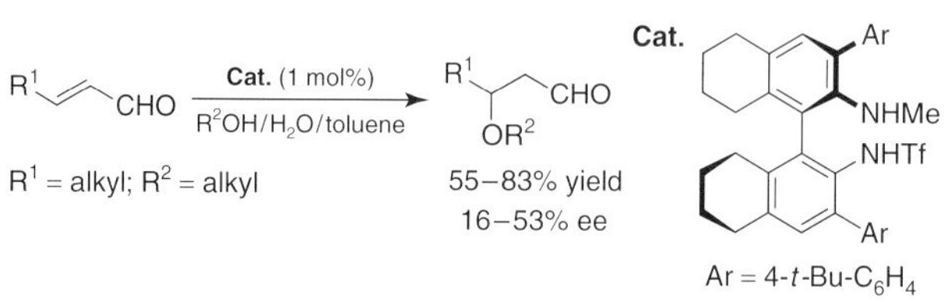

Scheme 7.19 Organocatalytic enantioselective oxa–Michael reaction of enals with alcohols.

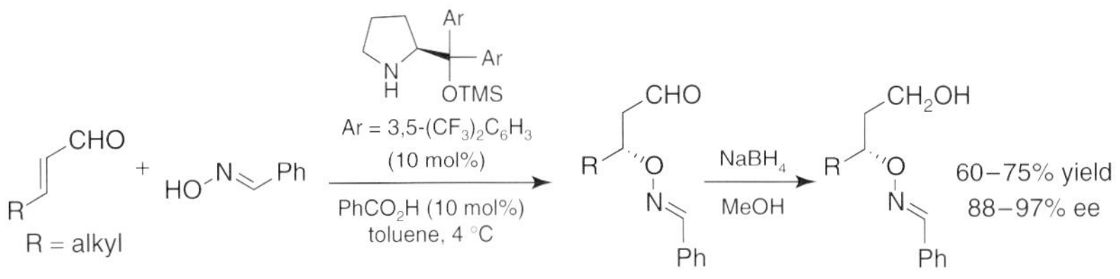

Scheme 7.20 Organocatalytic enantioselective oxa–Michael reaction of enals with oximes.

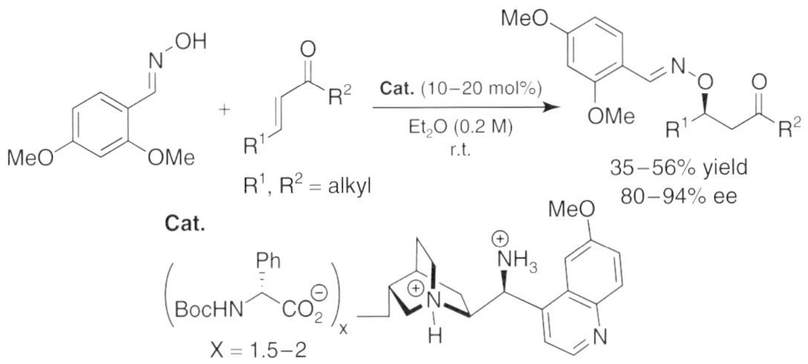

Scheme 7.21 Organocatalytic enantioselective oxa–Michael reaction of enones with oximes.

When the use of enones was reported subsequently (Scheme 7.21) [46], the cinchona alkaloid-derived primary amine salt, which was an effective catalyst in aza–Michael reactions [39], was found also to be the best promoter for the reaction of α,β-unsaturated ketones with 2,4-dimethoxybenzaldoxime. Whilst the yields were low due to the reversibility of the process (retro-Michael), the enones used were again limited to the aliphatic variety.

Peroxides represent very useful building blocks in organic synthesis and, indeed, Jørgensen and colleagues subsequently developed a highly organocatalytic enantioselective epoxidation of the enals, using H_2O_2 as the nucleophile [Scheme 7.22; Eq. (7.10)] [47]. Inspired by these findings, Deng [48] and List [49] each reported (independently) the details of an enantioselective peroxidation of aliphatic enones, using the same quinine-derived primary amine catalyst. When Deng used alkyl peroxides as a nucleophile for the conjugate reaction [Eq. (7.11)], the product formation was seen to depend heavily on the reaction temperature. Typically, a low temperature (e.g., room temperature) led to β-peroxide ketones as the major products, whereas at a higher temperature epoxides became the dominant products. In the studies conducted by List, however, hydrogen peroxide was used and the corresponding cyclic peroxyhemiketal obtained [Eq. (7.12)]. The resultant intermediates could be transformed into epoxides under basic conditions, or into β-hydroxyketones by a $P(OEt)_3$-mediated reduction. Most impressively, high to excellent ee-values were achieved in all cases.

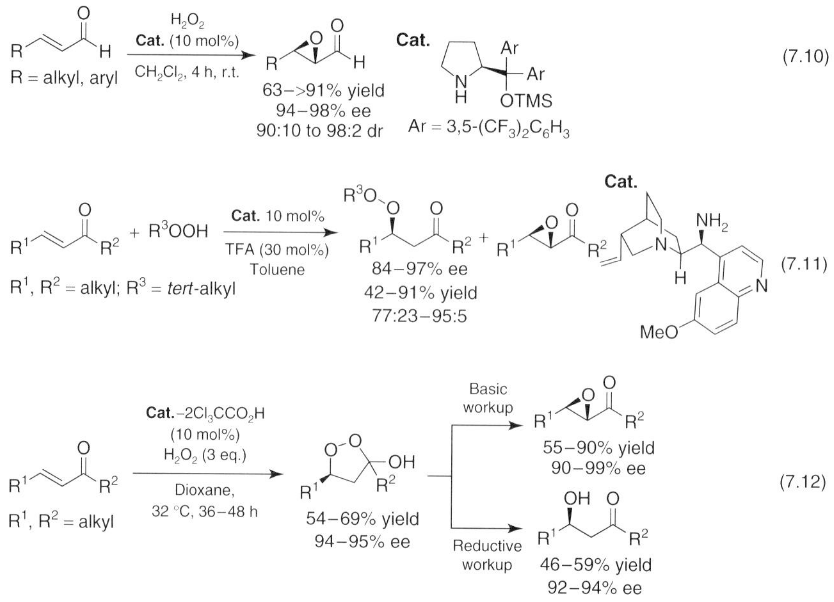

Scheme 7.22 Organocatalytic enantioselective epoxidation of enals and enones.

7.3.2
Intramolecular oxa–Michael Reactions

Flavanone and chroman(one) structures are widely distributed in a large number of natural products with a broad spectrum of biological activities. The most straightforward method to synthesize this class of compounds is via an intramolecular oxa–Michael reaction of 2′-hydroxy chalcones. For the preparation of the anti-HIV-1 the natural products (+)-inophyllum B and (+)-calanolide A bearing chromane frameworks, Ishikawa's group first developed an organocatalytic intramolecular oxa–Michael reaction promoted by (−)-quinine [Scheme 7.23; Eq. (7.13)] [50, 51], and produced chromanes with ee-values of up to 97%. Despite a similar strategy being applied by Hintermann three years later for the synthesis of flavanones [52], a more general protocol was recently developed by Scheidt [Eq. (7.14)] [53]. In this case, the introduction of a *tert*-butyl ester group at the α-position of ketones not only enhanced the reactivity of the conjugate acceptors, but also afforded a high enantioselectivity by steric effect, and an extra interaction with the cinchona alkaloid-based bifunctional thiourea catalyst. Subsequently, the *tert*-butyl ester group was readily removed using *p*-TsOH in a one-pot operation so as to produce highly enantioenriched flavanones or chromanones, in high yields.

A similar synthetic strategy for the synthesis of chromanes with a β-disubstituted α,β-unsaturated ester was reported, where a chiral guanidine was used as the catalyst [51, 54]. As a consequence, a quaternary chiral carbon center was produced in a moderate enantioselectivity.

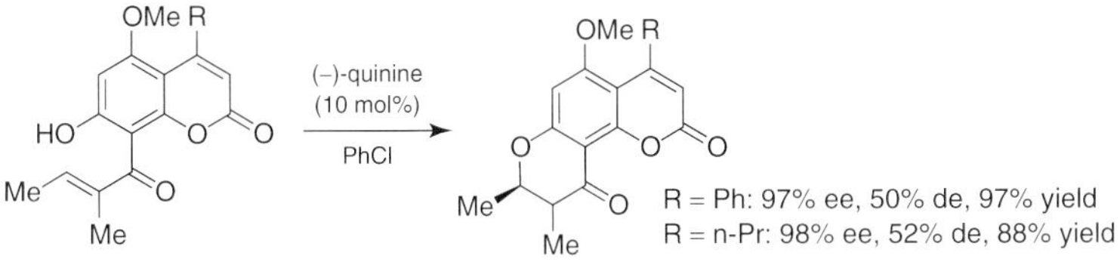

(7.13)

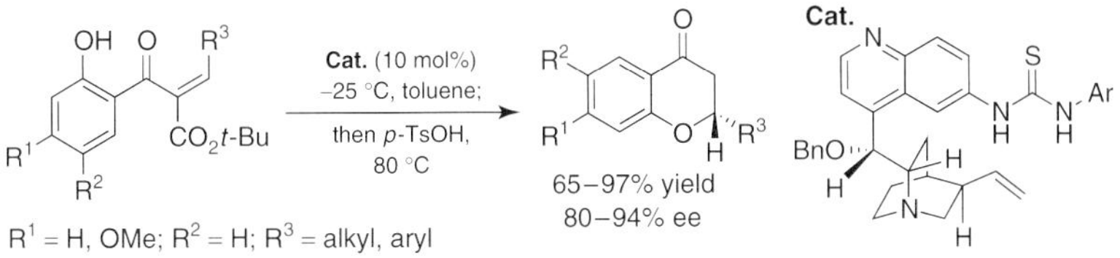

R = Ph: 97% ee, 50% de, 97% yield
R = n-Pr: 98% ee, 52% de, 88% yield

(7.14)

R^1 = H, OMe; R^2 = H; R^3 = alkyl, aryl

65–97% yield
80–94% ee

Scheme 7.23 Organocatalytic enantioselective intramolecular oxa–Michael reactions.

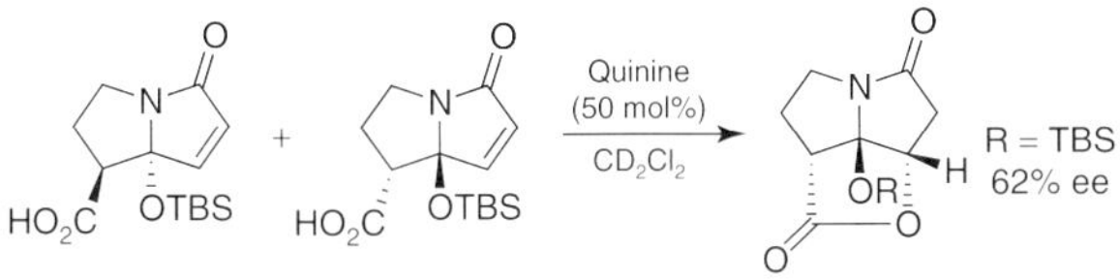

R = TBS
62% ee

Scheme 7.24 Quinine-mediated kinetic resolution via an intramolecular oxa–Michael reaction.

Examples of the use of a carboxylic acid as a nucleophile in asymmetric oxa–Michael additions are extremely rare, with only one study having been reported for the synthesis of the key intermediate of the telomerase inhibitor UCS1025A (Scheme 7.24) [55]. For this reaction, a quinine-mediated kinetic resolution of the racemic acid precursors afforded the lactone via an intramolecular Michael addition reaction in 62% ee. However, the optical purity of the product was improved to 98.2% following a simple trituration.

Boronic acid is an interesting and useful "O"-derived species, and generally considered to be the equivalent of hydroxyl groups. Falck *et al.* reported an elegant cinchona alkaloid thiourea-catalyzed intramolecular oxa–Michael addition of boronic acids to γ,δ-hydroxy-α,β-enones (Scheme 7.25) [56] in which, mechanistically, the substrates were first reacted with phenylboronic acid to form boronic acid hemiesters. Subsequently, a bifunctional cinchona alkaloid thiourea was used to activate and direct the intramolecular asymmetric oxa–Michael reaction to yield highly enantioselective dioxaborolanes that could be transformed to diols by H$_2$O$_2$ oxidation. These processes were applicable to a wide range of substrates, and led to the production of synthetically valuable 1,2-diols or 1,3-diols as building blocks.

7.3.3
Oxygen-Initiated oxa–Michael Cascade Reactions

By using the Michael–aldol–dehydration approach [33, 34], Wang [57], Córdova [58], and Arvidsson [59] have each respectively extended the oxa– and

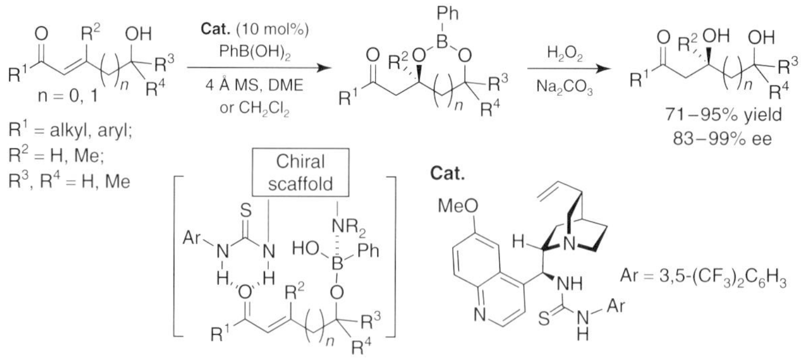

Scheme 7.25 Cinchona alkaloid-derived thiourea-catalyzed intramolecular oxa–Michael addition of boronic acids to enones.

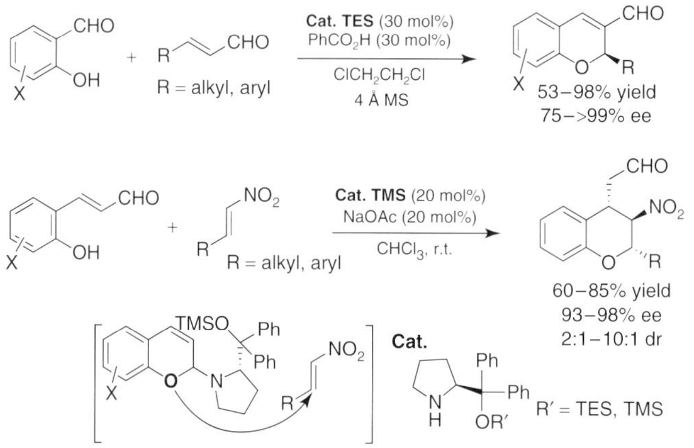

Scheme 7.26 Diphenylpyrrolidinol silyl ether-catalyzed oxa–Michael–aldol/Michael reactions.

thia–Michael–aldol–dehydration reactions (see below) of α,β-unsaturated aldehydes, catalyzed by diphenylpyrrolidinol silyl ethers [Scheme 7.26; Eq. (7.15)]. The powerful cascade provided a simple approach to the synthesis of enantioenriched 2,3-disubstituted chromenes from simple, achiral starting materials. Moreover, the strategy was subsequently and successfully extended by Córdova to include cyclohexenone systems in a one-pot synthesis of tetrahydroxanthenones [60].

While attempting recently to create synthetically useful scaffolds with stereochemical and functional diversities, Wang and coworkers discovered an unprecedented organocatalyzed asymmetric cascade oxa–Michael–Michael reaction, which would afford chiral highly functionalized chromanes with the creation of three new stereogenic centers [Scheme 7.26; Eq. (7.16)] [61]. The cascade process

7.3 "O" as Nucleophiles | 309

was efficiently catalyzed by a diphenylprolinol silyl ether from simple achiral substances, and provided a one-pot access to enantioenriched chromanes. Most significantly, a novel activation mode of the chiral amine-catalyzed cascade process involving the formation of aminal, which serves as a nucleophile, rather than a free phenol–OH group for the Michael addition, was identified for the first time.

The cascade strategy has been successfully applied in the efficient synthesis of α-tocopherol by Woggon (Scheme 7.27) [62]. The key step here involved an organocatalytic aldol–Michael–cyclization sequence, in which a dienamine served as a nucleophile for an aldol reaction, followed by an intramolecular oxy–Michael reaction and subsequent hemiacetalation to give the *tert*-carbone-containing chromanols. This product could be converted very conveniently to the natural product, α-tocopherol.

Hong reported a three-component oxa–Michael/Michael/Michael–aldol cascade reaction which had only a moderate yield but excellent ee-values (>99%) (Scheme 7.28) [63]. The reaction was initially triggered by an intermolecular oxa–Michael reaction between 2-((E)-2-nitrovinyl)phenol and α,β-unsaturated aldehyde, but then underwent an intramolecular Michael reaction and final Michael–aldol reaction with a second α,β-unsaturated aldehyde.

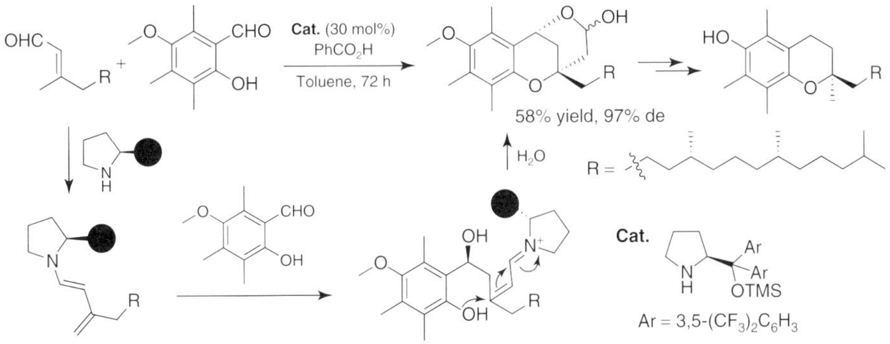

Scheme 7.27 Diarylpyrrolidinol silyl ether-catalyzed cascade aldol/oxa–Michael reaction.

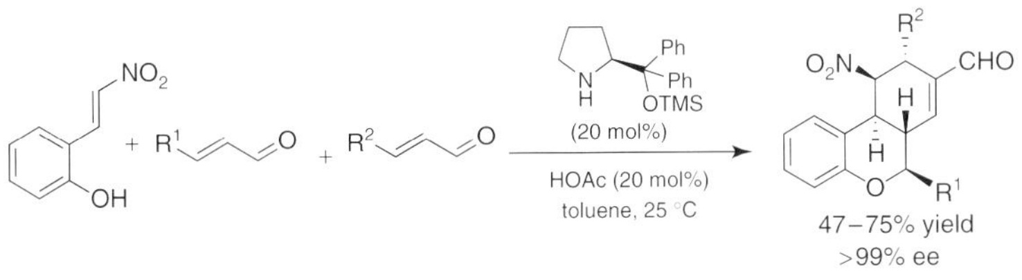

Scheme 7.28 Diphenylpyrrolidinol silyl ether-catalyzed three-component oxa–Michael/Michael/Michael–aldol cascade reaction.

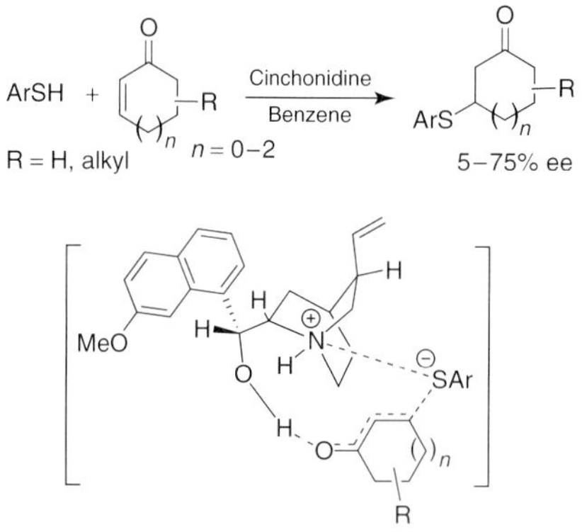

Scheme 7.29 Cinchona alkaloids-catalyzed thia–Michael reaction with cycloenones.

7.4
"S" as Nucleophiles

7.4.1
Intermolecular thia–Michael Reactions

The first example of organocatalytic intermolecular Michael reactions of thiols to α,β-unsaturated carbonyl compounds can be traced back some thirty years when, in 1981, Wynberg et al. reported the addition of thiophenols to cycloalkenones catalyzed by β-amino alcohols, such as cinchona alkaloids and their derivatives (Scheme 7.29) [64]. Following intensive investigations of the nucleophiles and reaction conditions, ee-values of up to 75% were obtained. An activation model between substrates and the bifunctional catalyst cinchona alkaloid served was proposed. In the transition state, the hydroxyl group of the catalyst activated the substrate enone by hydrogen-bond interaction, while the amino group interacted with the thiol by forming ion pairs. Although only moderate ee-values were obtained, these seminal studies promoted great inspiration in the development of noncovalent, bifunctional catalysis.

Shortly thereafter, Mukaiyama and coworkers used a 4-hydroxypyrrolidine-derived secondary amine as the catalyst for the same process (Scheme 7.30) [65, 66]. These authors also found that a catalyst without an hydroxyl group or with an opposite hydroxyl configuration produced poor enantioselectivity. such that a similar transition state model to that of Wynberg's was proposed. By using this catalytic system, however, an optical yield of up to 88% was achieved.

Some three years later, the same group reported a Michael addition of thiolphenol to diisopropyl maleate, promoted by cinchonine [67], in which the Michael product could be converted into synthetically useful 3,4-epoxy-1-butanol. Subsequently, almost no attention was paid to these potential catalytic systems for almost 20 years until 2001, when Skarżewski and colleagues reported the details of a cinchonine Michael addition of thiophenol to chalcones, applying the same strategy as used

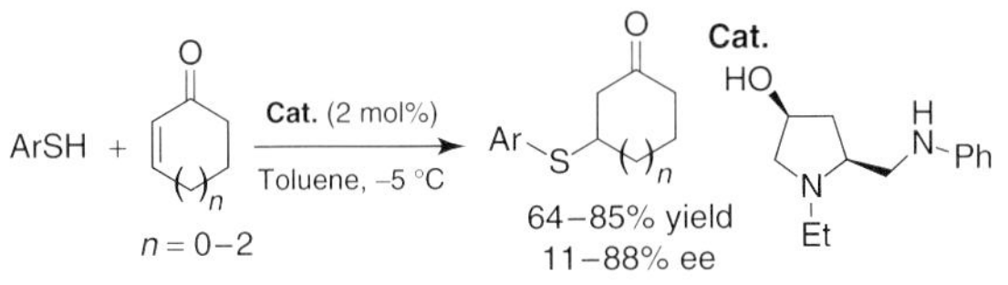

Scheme 7.30 4-Hydroxypyrrolidine derivative-catalyzed thia–Michael reaction with cycloenones.

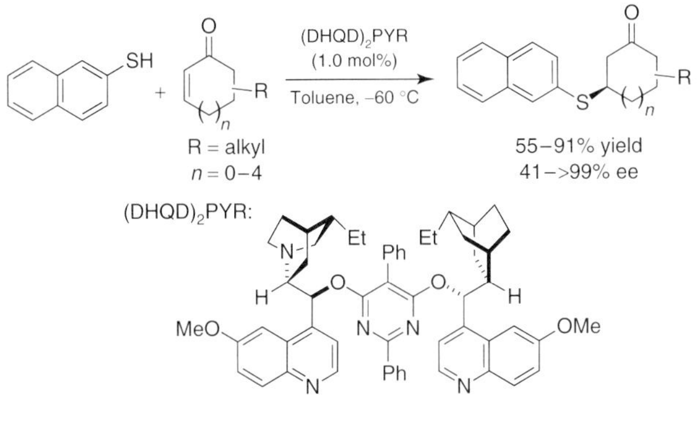

Scheme 7.31 Asymmetric 1,4-addition of 2-thionaphthol to cyclic enones, catalyzed by (DHQD)2PYR.

by Wynberg and Mukaiyama [68]. Notably, the reaction provided good yields but again, showed only moderate enantioselectivities.

A significant improvement of the conjugate addition reaction was made by Deng and coworkers one year later (Scheme 7.31) [69], who utilized the modified cinchona alkaloids but without hydroxy groups. Impressively good yields and excellent enantioselectivities were achieved with the catalyst for a wide range of cyclic enones, at which point it was realized that use of the bulky 2-thionaphthol was crucial to achieve a high enantioselectivity. The authors appreciated that their catalytic system must have different reaction mechanisms, as the lack of a hydrogen donor compared with the bifunctional cinchona alkaloids. The most solid evidence was that, by using hydroquinidine-2,5-diphenyl-4,6-pyrimidinediyl diether [(DHQD)$_2$PYR] as a promoter, the Michael addition product had been generated with an opposite stereochemical configuration to that obtained with natural cinchona alkaloids, although no transition state was proposed.

During the past few years, (thio)ureas have been demonstrated as efficient hydrogen bond donors and catalysts, as the incorporation of such a moiety can form strong hydrogen-bonding interactions with substrates containing carbonyl and nitro groups [70–73]. In 2005, the group of Chen reported a Michael addition of arylthiols to α,β-unsaturated imides and ketones catalyzed by thiourea bifunctional organocatalysts (Scheme 7.32) [74]. In this case, Takemoto's catalyst was found to be the best promoter for the addition of thiols to α,β-unsaturated imides, though

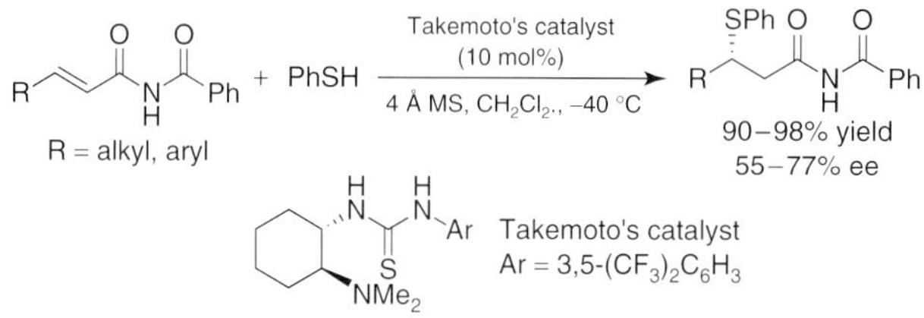

Scheme 7.32 Takemoto's catalyst-catalyzed asymmetric Michael addition of arylthiols to α,β-unsaturated carbonyls.

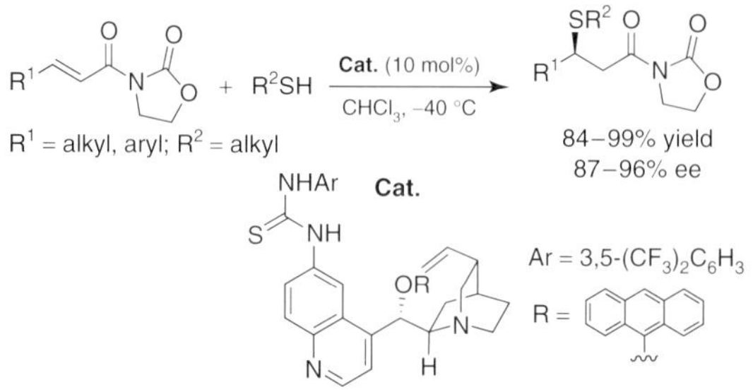

Scheme 7.33 Catalytic asymmetric Michael addition of alkylthiols to α,β-unsaturated oxazolidinones.

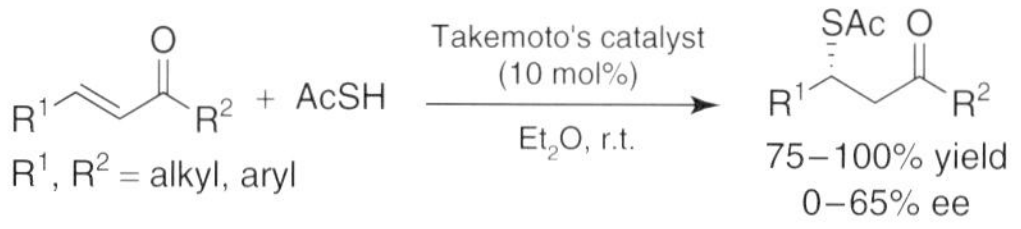

Scheme 7.34 Takemoto's catalyst-catalyzed asymmetric Michael addition of thioacetic acid to enones.

only moderate enantioselectivities were obtained. Cyclic enones were also probed as Michael receptors, and produced excellent yields with ee-values up to 85%.

Deng's group also disclosed a Michael reaction similar to that described by Chen et al., but which gave a much improved enantioselectivity (Scheme 7.33) [75]. Critically, the use of α,β-unsaturated N-acylated oxazolidinones was important for the 6'-thiourea cinchona alkaloid-catalyzed highly enantioselective thia–Michael reaction.

Sulfur-centered nucleophiles such as thioacetic acid have been rarely used in Michael additions, despite their ready transformation into versatile SH groups under various and mild reaction conditions. When Wang and coworkers reported the first organocatalytic Michael addition of thioacetic acid to enones in the presence of Takemoto's catalyst (Scheme 7.34) [76], the best result was obtained when chalcones were used as substrates, with good yields and moderate ee-values.

The conjugate addition of thiols to α,β-unsaturated aldehydes was disclosed by Jørgensen et al. (Scheme 7.35) [77], in a reaction that was activated with diarylpyrrolidinol TMS ether by forming an iminium ion intermediate. Various alkyl thiols

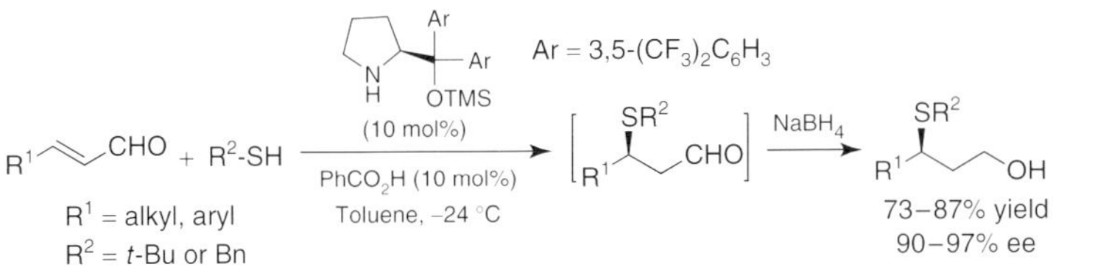

Scheme 7.35 Diarylprolinol TMS ether-catalyzed enantioselective conjugated addition of thiols to α,β-unsaturated aldehydes.

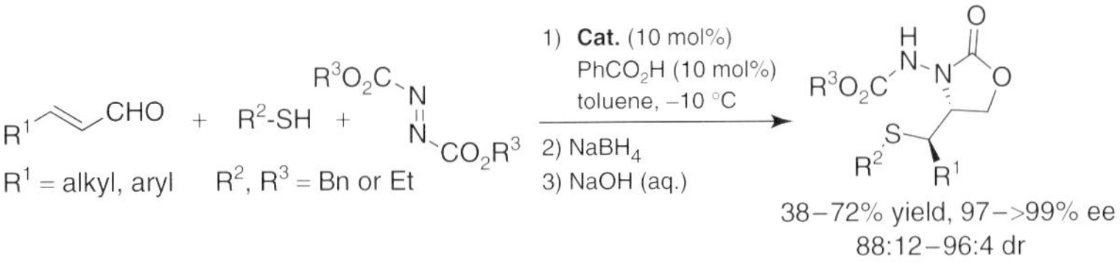

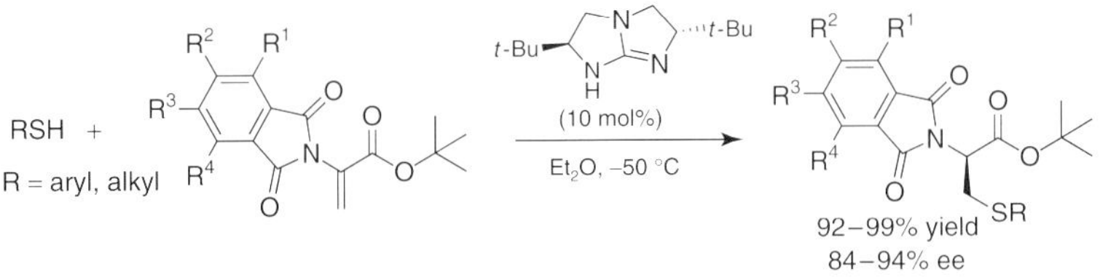

Scheme 7.36 A chiral guanidine-promoted Michael reaction of thiols to *tert*-butyl 2-phthalimidoacrylates.

may effectively participate in the process with good yields, and good to excellent ee-values [Eq. (7.17)]. It should be noted that, in the process of optimization, the authors observed the formation of stable enamine species between the catalyst and adduction product, which attributed to a slow reaction rate because of the slow turnover. By using this strategy, a three-component cascade process was successfully developed and the cascade products were conveniently transformed into valuable chiral oxazolidinones by *in situ* reduction and a spontaneous cyclization reaction [Eq. (7.18)].

A cinchona alkaloid-derived primary amine-chiral acid as cocatalyst was also used for a thia–Michael addition to enones [78].

The asymmetric protonation of enolates represents an important strategy for the preparation of optically active compounds. Recently, Tan and coworkers developed a tandem reaction that involved a conjugate addition of thiol to *tert*-butyl 2-phthalimidoacrylates, followed by enantioselective protonation or deuteration promoted by a guanidine catalyst (Scheme 7.36) [79]. This reaction afforded cysteine or deuterated cysteine derivatives and analogs in excellent yields, and with good ee-values. The approach could be applied for the addition of phosphine species and *tert*-butylthiol to *N*-substituted itaconimides.

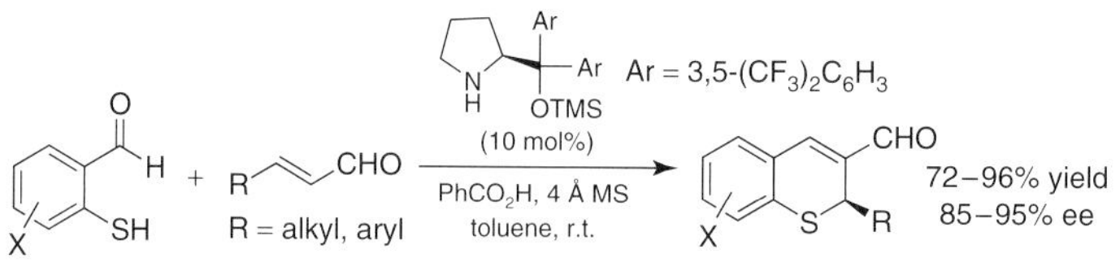

Scheme 7.37 Organocatalytic enantioselective thia–Michael–aldol–dehydration cascade.

7.4.2
Sulfur-Initiated thia–Michael Cascade Reactions

Inspired by the previous discoveries of enamine and iminium ion chemistries, several cascade processes have been developed using the thiol-triggered Michael reactions to assemble enantiomerically enriched thiochromenes, thiochromanes, or tetrahydrothiophenes, in a simple and efficient fashion.

The Michael–aldol–dehydration reactions were also successfully applied between 2-mercaptobenzaldehydes and α,β-unsaturated aldehydes by Wang [80] and Córdova [81], respectively (Scheme 7.37). Again, the reaction was efficiently promoted by diarylpyrrolidinol silyl ether catalysts and proceeded via the iminium and enamine pathway to afford 2,3-disubstituted thiochromenes in good to excellent yields and with high ee-values.

The Córdova group extended the cascade strategy for enones reacting with 2-mercaptobenzaldehyde using chiral diamine catalyst to afford thioxanthones with moderate yields and ee-values [82]. Moreover, Michael–aldol cascade reactions between 2-mercaptoacetophenones and α,β-unsaturated aldehydes were also successfully carried out in the presence of the diarylpyrrolidinol silyl ether [83]. Notably, at −25 °C, nondehydrated thiochromanes with three contiguous stereocenters and a tertiary aldol moiety were generated with up to >15 : 1 diastereomeric ratio (dr) and 96 to >99% ee.

The cascade reaction strategy can also be applied to the preparation of tetrahydrothiophenes, through the design of suitable substrates. For example, Jørgensen *et al.* reported the details of organocatalytic domino Michael–aldol reactions between enals and 2-mercapto-1-phenylethanone (Scheme 7.38) [84], whereby the additives were found to be very important in the formation of the tetrahydrothiophene products. If benzoic acid was used as an additive, the reaction gave rise to 2,3,4,4-tetrasubstituted tetrahydrothiophene carbaldehydes as a single isomer in good yields and with excellent ee-values [Eq. (7.19)]; however, the use of $NaHCO_3$ as additive led to (tetrahydrothiophen-2-yl)phenyl methanones as the main products, with moderate yields and ee-values [Eq. (7.20)].

The group of Wang also reported two methods for the assembly of multiple-functionalized tetrahydrothiophenes, using a new cascade Michael/Michael sequence between α,β-unsaturated aldehydes and *trans*-ethyl 4-mercapto-2-butenoate, and Michael–aldol reactions between α,β-unsaturated aldehydes and ethyl 3-mercapto-2-oxopropanoate (Scheme 7.39) [85, 86]. In both cases, good to excellent diastereo- and enantioselectivities were observed.

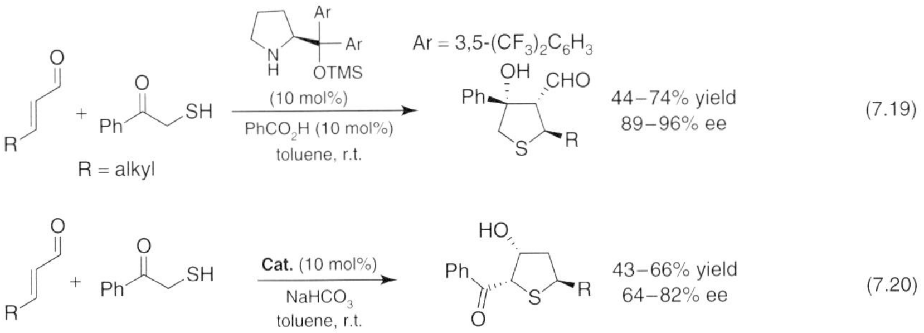

7.4 "S" as Nucleophiles

(7.19)

(7.20)

Scheme 7.38 Organocatalytic domino Michael–aldol reactions between enals and 2-mercapto-1-phenylethanone.

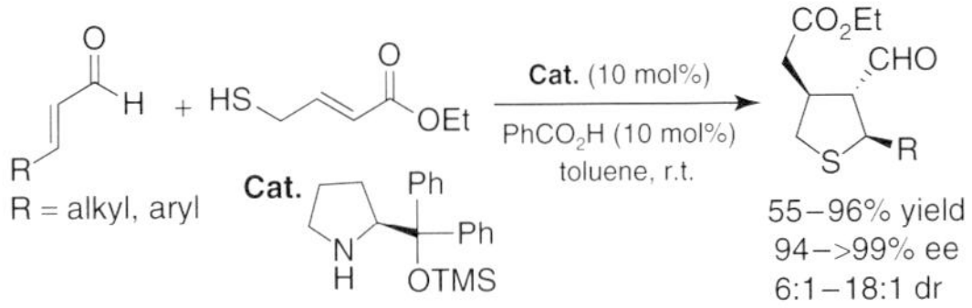

Scheme 7.39 Organocatalytic enantioselective thia–Michael/Michael reactions.

Previously, organocatalyzed enantioselective cascade processes have been dominated by enamine and iminium ion activation modes. Nonetheless, the use of a noncovalent bond activation force has proved to be a major challenge, and few examples have been reported. By employing bifunctional tertiary amine thiourea catalysts to activate α,β-unsaturated systems, Wang et al. identified the highly efficient cascade Michael–aldol reactions of 2-mercaptobenzaldehydes with oxazolidinones/maleimides (Scheme 7.40) [87, 88], both of which provided highly enantio- and stereoselective products, with the generation of three new stereogenic centers. Remarkably, a catalyst loading as low as 1 mol% was sufficient to catalyze both reactions.

When using a similar strategy, Zhao and colleagues reported a cascade Michael–Knoevenagel reaction between benzylidenemalonates and 2-mercaptobenzaldehyde [89].

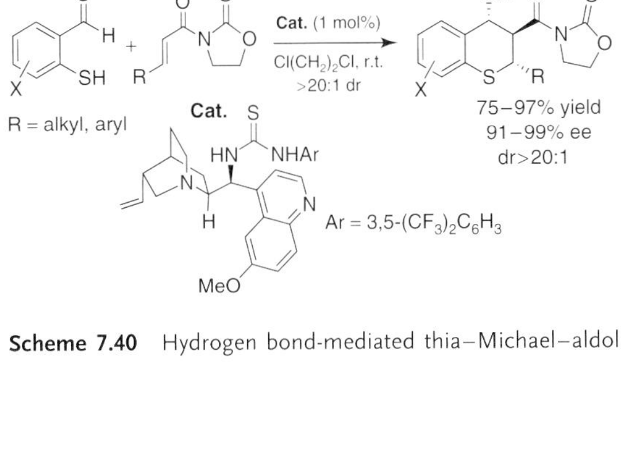

Scheme 7.40 Hydrogen bond-mediated thia–Michael–aldol cascade reaction.

7.5
"P" as Nucleophiles

Chiral phosphines are very important ligands for organometallic catalysis, and usually are prepared either by resolution or by the use of a stoichiometric amount of chiral auxiliaries. In contrast to "N," "O," and "S," the use of "P" as nucleophile in organocatalytic Michael addition reactions has been studied to a much lesser extent. Nonetheless, organocatalytic Michael reactions of P-centered nucleophiles to α,β-unsaturated compounds have recently provided a simple yet useful method for the construction of this class of compound.

Both, Melchiorre [90] and Córdova [91, 92] have reported independently details of the organocatalytic hydrophosphination of α,β-unsaturated aldehydes (Scheme 7.41). In this case, the use of diarylprolidinol silyl ethers as promoters for the conjugate addition of diphenylphosphine to α,β-unsaturated aldehydes led to highly enantioselective adducts, which could be reduced *in situ* by $NaBH_4$ to produce air-stable phosphineborane-alcohol derivatives, or oxidized to β-phosphine oxide carboxylic acid with sodium chlorite for compound characterization and data analysis.

In contrast, Jørgensen *et al.* used diphenyl phosphite as a nucleophile in the similar addition reactions, with moderate yields and moderate to good ee-values (Scheme 7.42) [93]. The products could be readily converted into useful chiral building blocks, such as glutamic acid analogs.

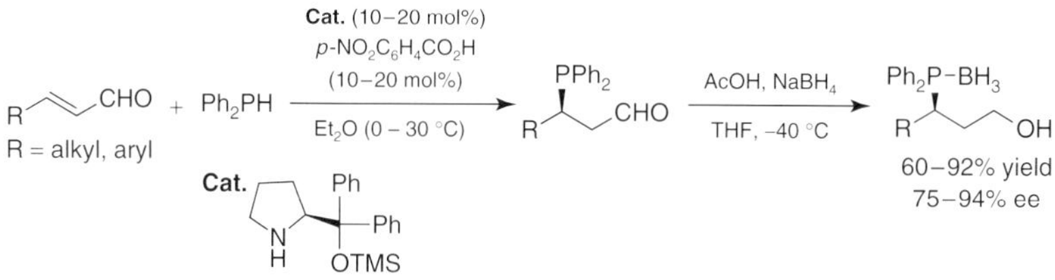

Scheme 7.41 Organocatalytic enantioselective conjugate addition of hydrophosphination to enals.

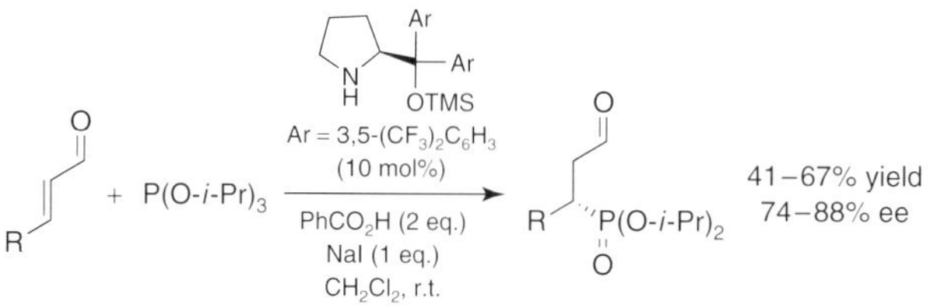

Scheme 7.42 Organocatalytic enantioselective conjugate addition of trimethyl phosphite to enals.

7.6
Concluding Remarks

Within a relatively short period of time, impressive progress has been made in the organocatalytic enantioselective Michael addition of heteroatoms, including "N," "O," "S," and "P." More recently, several useful Michael reactions have been developed and are currently serving as efficient methods in organic synthesis, notably in the production of natural products and biologically significant molecules. There remain, however some significant problems, and synthetic efficiencies in terms of reaction yields and enantio- and/or diastereoselectivities need to be improved in many cases. In comparison with strong covalent-bond catalysis, the state-of the art of noncovalent bond-mediated conjugate addition reactions is limited, with only strong sulfur-derived nucleophiles effectively participating in these processes. Consequently, whilst the development of new organocatalysts and activation modes may be considered by some as critical, there is no doubt that, in the very near future, a range of innovative systems will be developed within the field of organocatalysis.

References

1. Berkessel, A. and Groger, H. (2005) *Asymmetric Organocatalysis-from Biomimetic Concepts to Applications in Asymmetric Synthesis*, Wiley-VCH Verlag GmbH & Co. KGaA, Weinheim.
2. Enders, D., Lüttgen, K., and Narine, A.A. (2007) *Synthesis*, 959.
3. Tsogoeva, S.B. (2007) *Eur. J. Org. Chem.*, 1701.
4. Almasi, D., Alonso, D.A., and Najera, C. (2007) *Tetrahedron: Asymmetry*, **18**, 299.
5. Vicario, J.L., Badía, D., and Carrillo, L. (2007) *Synthesis*, 2065–2092.
6. Pellissier, H. (2007) *Tetrahedron*, **63**, 9267.
7. Erkkilä, A., Majander, I., and Pihko, P.M. (2007) *Chem. Rev.*, **107**, 5416.
8. Bartoli, G. and Melchiorre, P. (2008) *Synlett*, 1759.
9. Mielgo, A. and Palomo, C. (2008) *Chem. Asian. J.*, **3**, 922.
10. Nising, C.F. and Bräse, S. (2008) *Chem. Soc. Rev.*, **37**, 1218.
11. Horstmann, T.E., Guerin, D.J., and Miller, S.J. (2000) *Angew. Chem., Int. Ed.*, **39**, 3635.
12. Guerin, D.J. and Miller, S.J. (2002) *J. Am. Chem. Soc.*, **124**, 2134.
13. Chen, Y.K., Yoshida, M., and MacMillan, D.W.C. (2006) *J. Am. Chem. Soc.*, **128**, 9328.
14. Vesely, J., Ibrahem, I., Rios, R., Zhao, G.-L., and Córdova, A. (2007) *Tetrahedron Lett.*, **48**, 2193.
15. Lu, X. and Deng, L. (2008) *Angew. Chem., Int. Ed.*, **47**, 7710.
16. Pettersen, D., Piana, F., Bernardi, L., Fini, F., Fochi, M., Sgarzani, V., and Ricci, A. (2007) *Tetrahedron Lett.*, **48**, 7805.
17. Sibi, M.P. and Itoh, K. (2007) *J. Am. Chem. Soc.*, **129**, 8064.
18. Perdicchia, D. and Jørgensen, K.A. (2007) *J. Org. Chem.*, **72**, 3565.
19. Scettri, A., Massa, A., Palombi, L., Villano, R., and Acocella, M.R. (2008) *Tetrahedron: Asymmetry*, **19**, 2149.
20. Jiang, H., Nielsen, J.B., Nielsen, M., and Jørgensen, K.A. (2008) *Chem. Eur. J.*, **13**, 9068.
21. Diner, P., Nielsen, M., Marigo, M., and Jørgensen, K.A. (2007) *Angew. Chem., Int. Ed.*, **46**, 1983.
22. Uria, U., Vicario, J.L., Badía, D., and Carrillo, L. (2007) *Chem. Commun.*, 2509.

23. Gogoi, S., Zhao, C.-G., and Ding, D. (2009) *Org. Lett.*, **11**, 2249.
24. Lin, Q., Meloni, D., Pan, Y., Xia, M., Rodgers, J., Shepard, S., Li, M., Galya, L., Metcalf, B., Yue, T.-Y., Liu, P., and Zhou, J. (2009) *Org. Lett.*, **11**, 1999.
25. Wang, J., Zu, L.-S., Li, H., Xie, H.-X., and Wang, W. (2007) *Synthesis*, 2576.
26. Gerasyuto, A.I., Hsung, R.P., Sydorenko, N., and Slafer, B. (2005) *J. Org. Chem.*, **70**, 4248.
27. Fustero, S., Jimnez, D., Moscard, J., Catalan, S., and del Pozo, C. (2007) *Org. Lett.*, **9**, 5283.
28. Fustero, S., Moscardo, J., Jimenez, D., Perez-Carrion, M.D., Sanchez-Rosello, M., and del Pozo, C. (2008) *Chem. Eur. J.*, **14**, 9868.
29. Carlson, E.C., Rathbone, L.K., Yang, H., Collett, N.D., and Carter, R.G. (2008) *J. Org. Chem.*, **73**, 5155.
30. Enders, D., Grondal, C., and Huttl, M.R.M. (2007) *Angew. Chem., Int. Ed.*, **46**, 1570.
31. Yu, X.-H. and Wang, W. (2008) *Org. Biomol. Chem.*, **6**, 2037.
32. Itoh, T., Yokoya, M., Miyauchi, K., Nagata, K., and Ohsawa, A. (2006) *Org. Lett.*, **8**, 1533.
33. Li, H., Wang, J., Xie, H.-X., Zu, L.-S., Jiang, W., Duesler, E.N., and Wang, W. (2007) *Org. Lett.*, **9**, 965.
34. Sunden, H., Rios, R., Ibrahem, I., Zhao, G.-L., Eriksson, L., and Cordova, A. (2007) *Adv. Synth. Catal.*, **349**, 827.
35. Yoshitomi, Y., Arai, H., Makino, K., and Hamada, Y. (2008) *Tetrahedron*, **64**, 11568.
36. Li, H., Zu, L.-S., Xie, H.-X., Wang, J., and Wang, W. (2008) *Chem. Commun.*, 5636.
37. Vesely, J., Ibrahem, I., Zhao, G.-L., Rios, R., and Córdova, A. (2007) *Angew. Chem., Int. Ed.*, **46**, 778.
38. Ibrahem, I., Rios, R., Vesely, J., Zhao, G.-L., and Cordova, A. (2007) *Chem. Commun.*, 849.
39. Pesciaioli, F., De Vincentis, F., Galzerano, P., Bencivenni, G., Bartoli, G., Mazzanti, A., and Melchiorre, P. (2008) *Angew. Chem., Int. Ed.*, **47**, 8703.
40. Armstrong, A., Baxter, C.A., Lamont, S.G., Pape, A.R., and Wincewicz, R. (2007) *Org. Lett.*, **9**, 351.
41. Zhao, G.-L., Rios, R., Vesely, J., Eriksson, L., and Cordova, A. (2008) *Angew. Chem., Int. Ed.*, **47**, 8468.
42. Zhong, G. (2003) *Angew. Chem., Int. Ed.*, **42**, 4247.
43. Zhu, D., Lu, M., Chua, P.J., Tan, B., Wang, F., Yang, X., and Zhong, G.-F. (2008) *Org. Lett.*, **10**, 4585.
44. Kano, T., Tanaka, Y., and Maruoka, K. (2007) *Tetrahedron*, **63**, 8658–8664.
45. Bertelsen, S., Diner, P., Johansen, R.L., and Jørgensen, K.A. (2007) *J. Am. Chem. Soc.*, **129**, 1536.
46. Carlone, A., Bartoli, G., Bosco, M., Pesciaioli, F., Ricci, P., Sambri, L., and Melchiorre, P. (2007) *Eur. J. Org. Chem.*, 5492.
47. Marigo, M., Franzen, J., Poulsen, T.B., and Jorgensen, K.A. (2005) *J. Am. Chem. Soc.*, **127**, 6964.
48. Lu, X., Liu, Y., Sun, B., Cindric, B., and Deng, L. (2008) *J. Am. Chem. Soc.*, **130**, 8134.
49. Reisinger, C.M., Wang, X., and List, B. (2008) *Angew. Chem., Int. Ed.*, **47**, 8112.
50. Sekino, E., Kumamoto, T., Tanaka, T., Ikeda, T., and Ishikawa, T. (2004) *J. Org. Chem.*, **69**, 2760.
51. Merschaert, A., Delbeke, P., Dalozeb, D., and Dive, G. (2004) *Tetrahedron Lett.*, **45**, 4697.
52. Dittmer, C., Raabe, G., and Hintermann, L. (2007) *Eur. J. Org. Chem.*, 5886.
53. Biddle, M.M., Lin, M., and Scheidt, K.A. (2007) *J. Am. Chem. Soc.*, **129**, 3830.
54. Saito, N., Ryoda, A., Nakanishi, W., Kumamoto, T., and Ishikawa, T. (2008) *Eur. J. Org. Chem.*, 2759.
55. de Figueiredo, R.M., Frohlich, R., and Christmann, M. (2007) *Angew. Chem., Int. Ed.*, **46**, 2883.
56. Li, D.R., Murugan, A., and Falck, J.R. (2008) *J. Am. Chem. Soc.*, **130**, 46.
57. Li, H., Wang, J., E-Nunu, T., Zu, L.-S., Jiang, W., Wei, S.-H., and Wang, W. (2007) *Chem. Commun.*, 507.
58. Sunden, H., Ibrahem, I., Zhao, G.-L., Eriksson, L., and Cordova, A. (2007) *Chem. Eur. J.*, **13**, 574.
59. Govender, T., Hojabri, L., Moghaddamb, F.M., and Arvidsson, P.I. (2006) *Tetrahedron: Asymmetry*, **17**, 1763.

60. Rios, R., Sunden, H., Ibrahem, I., and Cordova, A. (2007) *Tetrahedron Lett.*, **48**, 2181.
61. Zu, L.-S., Zhang, S.-L., Xie, H.-X., and Wang, W. (2009) *Org. Lett.*, **11**, 1627.
62. Liu, K., Chougnet, A., and Woggon, W.-D. (2008) *Angew. Chem., Int. Ed.*, **47**, 5827.
63. Kotame, P., Hong, B.-C., and Liao, J.-H. (2009) *Tetrahedron Lett.*, **50**, 704.
64. Hiemstra, H. and Wynberg, H. (1981) *J. Am. Chem. Soc.*, **103**, 417.
65. Mukaiyama, T., Ikegawa, A., and Suzuki, K. (1981) *Chem. Lett.*, 165.
66. Suzuki, K., Ikegawa, A., and Mukaiyama, T. (1982) *Bull. Chem. Soc. Jpn*, **55**, 3277.
67. Yamashita, H. and Mukaiyama, T. (1985) *Chem. Lett.*, 363.
68. Skarzewski, J., Zielinska-Błajeta, M., and Turowska-Tyrk, I. (2001) *Tetrahedron: Asymmetry*, **12**, 1923.
69. McDaid, P., Chen, Y.-G., and Deng, L. (2002) *Angew. Chem., Int. Ed.*, **41**, 338.
70. Takemoto, Y. (2005) *Org. Biomol. Chem.*, **3**, 4299.
71. Doyle, A.G. and Jacobsen, E.N. (2007) *Chem. Rev.*, **107**, 5713.
72. Connon, S.J. (2006) *Chem. Eur. J.*, **12**, 5419.
73. Yu, X.-H. and Wang, W. (2008) *Asian J. Chem.*, **3**, 516–532.
74. Li, B.-L., Jiang, L., Liu, M., Chen, Y.-C., Ding, L.-S., and Wu, Y. (2005) *Synlett*, 603.
75. Liu, Y., Sun, B., Wang, B., Wakem, M., and Deng, L. (2009) *J. Am. Chem. Soc.*, **131**, 418.
76. Li, H., Zu, L.-S., Wang, J., and Wang, W. (2006) *Tetrahedron Lett.*, **47**, 3145.
77. Marigo, M., Schulte, T., Franzen, J., and Jorgensen, K.A. (2005) *J. Am. Chem. Soc.*, **127**, 15710.
78. Ricci, P., Carlone, A., Bartoli, G., Bosco, M., Sambri, L., and Melchiorre, P. (2008) *Adv. Synth. Catal.*, **350**, 49.
79. Leow, D., Lin, S., Chittimalla, S.K., Fu, X., and Tan, C.-H. (2008) *Angew. Chem., Int. Ed.*, **47**, 5641.
80. Wang, W., Li, H., Wang, J., and Zu, L.-S. (2006) *J. Am. Chem. Soc.*, **128**, 10354.
81. Rios, R., Sunden, H., Ibrahem, I., Zhao, G.-L., Eriksson, L., and Cordova, A. (2006) *Tetrahedron Lett.*, **47**, 8547.
82. Rios, R., Sunden, H., Ibrahem, I., Zhao, G.-L., and Cordova, A. (2006) *Tetrahedron Lett.*, **47**, 8679.
83. Zhao, G.-L., Vesely, J., Rios, R., Ibrahem, I., Sunden, H., and Cordova, A. (2008) *Adv. Synth. Catal.*, **350**, 237.
84. Brandau, S., Maerten, E., and Jorgensen, K.A. (2006) *J. Am. Chem. Soc.*, **128**, 14986.
85. Li, H., Zu, L.-S., Xie, H.-X., Wang, J., Jiang, W., and Wang, W. (2007) *Org. Lett.*, **9**, 1833.
86. Luo, G., Zhang, S.-L., Duan, W.-H., and Wang, W. (2009) *Tetrahedron Lett.*, **50**, 2946.
87. Zu, L.-S., Wang, J., Li, H., Xie, H.-X., Jiang, W., and Wang, W. (2007) *J. Am. Chem. Soc.*, **129**, 1036.
88. Zu, L.-S., Xie, H.-X., Li, H., Wang, J., Jiang, W., and Wang, W. (2007) *Adv. Synth. Catal.*, **349**, 1882.
89. Dodda, R., Mandal, T., and Zhao, C.-G. (2008) *Tetrahedron Lett.*, **49**, 1899.
90. Carlone, A., Bartoli, G., Bosco, M., Sambri, L., and Melchiorre, P. (2007) *Angew. Chem., Int. Ed.*, **46**, 4504.
91. Ibrahem, I., Rios, R., Vesely, J., Hammar, P., Eriksson, L., Himo, F., and Cordova, A. (2007) *Angew. Chem., Int. Ed.*, **46**, 4507.
92. Ibrahem, I., Hammar, P., Vesely, J., Rios, R., Eriksson, L., and Cordova, A. (2008) *Adv. Synth. Catal.*, **350**, 1875.
93. Maerten, E., Cabrera, S., Kjærsgaard, A., and Jørgensen, K.A. (2007) *J. Org. Chem.*, **72**, 8893.

8
Domino Reactions Involving Catalytic Enantioselective Conjugate Additions

Lutz F. Tietze and Alexander Düfert

8.1
Introduction

In modern synthetic chemistry, there is a great need for the development of new procedures that will allow the efficient formation of complex structures in a few steps, starting from simple substrates. Moreover, new ways must be sought to reduce the burden on the environment and to conserve natural resources. Yet, these new synthetic procedures methods must incorporate economic advantages such that they can be used by industry in areas of both development and production.

Over the past years, *domino reactions* – which are defined as "... processes of two or more bond-forming reactions under identical reaction conditions, in which the latter transformations take place at the functionalities obtained in the former bond-forming reactions ..." [1] have demonstrated their huge potential in synthetic organic chemistry. These reactions represent a novel concept in synthesis that enables the efficient preparation of complex compounds, starting from simple substrates. Moreover, such reactions are economically and ecologically beneficial, as they not only allow reductions to be made in the generation of waste materials, but also guarantee a diligent conduct with the Earth's resources.

In this chapter, a description is provided of the enantioselective domino-1,4-addition to α,β-unsaturated carbonyl compounds, followed by reaction with an electrophile or another transformation. The 1,4-addition of nucleophiles to α,β-unsaturated aldehydes, ketones, and carboxylic acid derivatives, followed by an interception of the intermediately formed enolate with electrophiles, represents a well-known and highly established synthetic procedure. Whereas, the details of many diastereoselective Michael additions of chiral substrates have been reported in the literature, enantioselective 1,4-additions under the control of a chiral catalyst – whether a transition metal or an organocatalyst – are much less common and have been established only relatively recently [2]. This "new" approach has been used not only for the creation of small molecules; rather, it has also demonstrated its versatility in the enantioselective total synthesis of several natural products.

Two different procedures have been developed, both of which allow the introduction of a first stereocenter with excellent enantiomeric excess (ee)-values;

Catalytic Asymmetric Conjugate Reactions. Edited by Armando Córdova
Copyright © 2010 WILEY-VCH Verlag GmbH & Co. KGaA, Weinheim
ISBN: 978-3-527-32411-8

in most cases, good diastereoselectivities are also observed for the introduction of any subsequent stereogenic centers. In a first approach, a transition metal-catalyzed reaction is used with cuprates, either in the presence of monodentate ligands of the Feringa type or bidentate ligands of the Noyori type. In a second, contrasting, approach, an increasing success has been observed using organocatalysts of the proline type, although more recently modified peptides have also been used. Accordingly, this chapter has been divided into four sections of: (i) metal-mediated domino Michael–aldol reactions; (ii) metal-mediated domino Michael reactions/electrophilic trapping with noncarbonyl compounds; (iii) organocatalytic domino Michael reactions/electrophilic trapping; and (iv) Michael reactions followed by a cycloaddition, hydrogenation, rearrangement, or other reactions.

Whilst this chapter incorporates published reports and information up until July 2009, details of transformations have also been included which do not follow the given definition of domino reactions in all aspects. A notable example of this is the "one-pot" reaction, where the reaction temperature may be changed or a reagent added after a certain time.

8.2
Metal-Mediated Domino Michael/Aldol Reactions

The metal-mediated domino Michael/aldol reaction of **1** to give **3** via the enolate **2** is the most widely utilized domino process (Scheme 8.1), whereby three stereogenic centers can be formed [3]. The reaction is an exceedingly reliable transformation, and allows for a great variety of reaction conditions [4].

One of the earliest examples of an enantioselective Michael/aldol reaction was reported by Shibasaki *et al*. This employed an enantiopure AlLibis[(S (ALB))-binaphthoxide] (ALB) complex, **7**, as catalyst [5]. Such heterobimetallic compounds demonstrate Lewis acidity and Brønsted basicity, and can catalyze both Michael and aldol reactions [5–7].

In a three-component domino reaction, cyclopentenone **4**, aldehyde **5**, and dibenzyl methylmalonate **6** led to β-hydroxy ketones **8** at room temperature in the presence of 5 mol% of **7** as a mixture of diastereomers in 84% yield. The transformation of **8** by a mesylation/elimination sequence afforded **9** with 92% ee, and this was recrystallized to give enantiopure **9**. This compound was used for the synthesis of the prostaglandin (PG), 11-deoxy-PGF$_{1\alpha}$ (**10**) (Scheme 8.2).

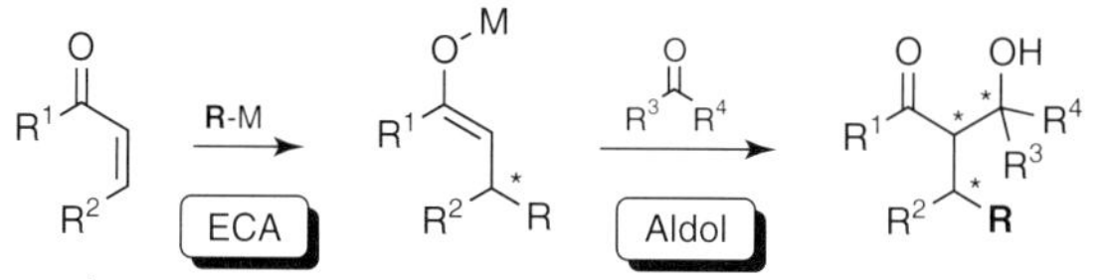

Scheme 8.1 Domino Michael–aldol reaction.

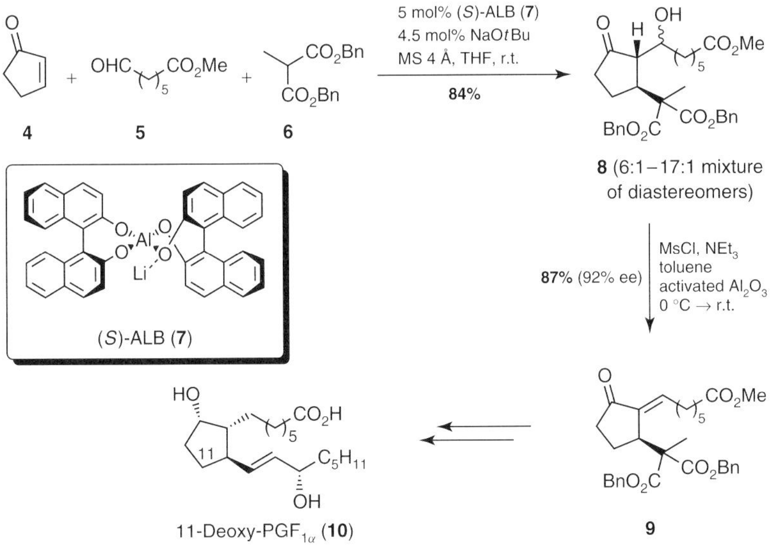

8.2 Metal-Mediated Domino Michael/Aldol Reactions

Scheme 8.2 Synthesis of 11-deoxy-PGF$_{1\alpha}$ (**10**).

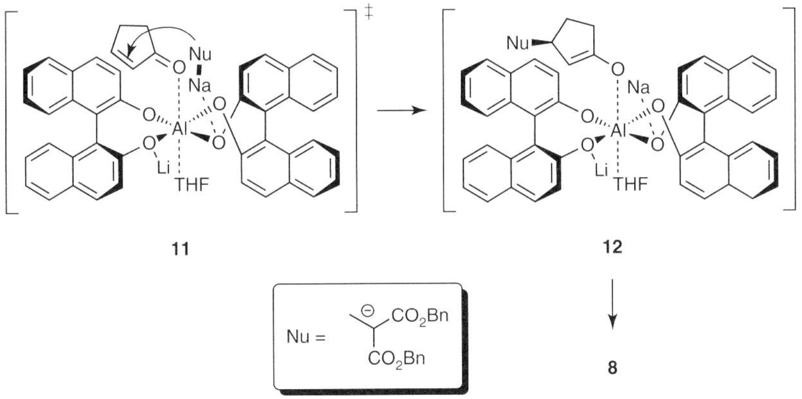

Scheme 8.3 Proposed transition states for the synthesis of **8**.

The transition state **11** and the intermediate **12** illustrate the stereochemical result (Scheme 8.3). The coordination of the enone to the aluminum atom not only results in its activation, but also fixes its position for the Michael addition, as depicted in **11**. The subsequent intermolecular aldol reaction of **12** occurs more rapidly than a protonation of the intermediately formed enolate moiety, and this results in the formation of **8**. Furthermore, the same methodology was also used for an approach towards enantiopure PGF$_{1\alpha}$ through a catalytic kinetic resolution using (S)-ALB (**7**) [5].

Due to their inherent reactivity profile and ability to undergo regioselective 1,4-additions, copper compounds have in the past been the dominating metal for both enantioselective conjugate additions (ECAs) and domino reactions involving the former [8].

The research group of Feringa and coworkers reported on the application of cyclopentene-3,5-dione monoacetal **13** as substrate in an asymmetric, three-component domino Michael/aldol reaction with dialkyl zinc reagents **14** and aromatic aldehydes **15** (Scheme 8.4) [9]. In the presence of 2 mol% of the *in situ*-generated catalyst Cu(OTf)$_2$/phosphoramidite **16**, the cyclopentanone derivative **18** was formed via **17** almost exclusively with 64–76% yield and ≥94% ee. Small amounts (<5%) of the epimer **19** were detected in some reactions [10].

The selectivity of the aldol addition following the conjugate addition can be rationalized in terms of a Zimmerman–Traxler transition-state model, with **17** having the lowest energy. Thus, it can be assumed that, due to the high reactivity of the zinc enolate, the chiral copper complex has no influence on the selectivity of this step. The mixture of the two products was oxidized to yield the corresponding diketone **20**, clearly showing that **18** and **19** are epimers. This methodology was later adapted to include nonaromatic aldehydes and successfully applied for the synthesis of **21**. This compound served as a precursor for the synthesis of (−)-prostaglandin E$_1$ methyl ester (**21**), which was accomplished in 7% overall yield and 94% ee in seven steps, based on **13** [11].

A conjugate addition/intramolecular aldol sequence of **22** and Et$_2$Zn was used by Krische *et al.* for the formation of cyclic β-hydroxy ketones **23** in almost quantitative yield and good to excellent ee-values, using Feringa's ligand **16** and copper triflate (Scheme 8.5) [12]. However, the intramolecular aldol reaction of the enolate formed in the ECA had only a low selectivity such that two diastereomers **23a** and **23b** were obtained in a 2.3 : 1 ratio.

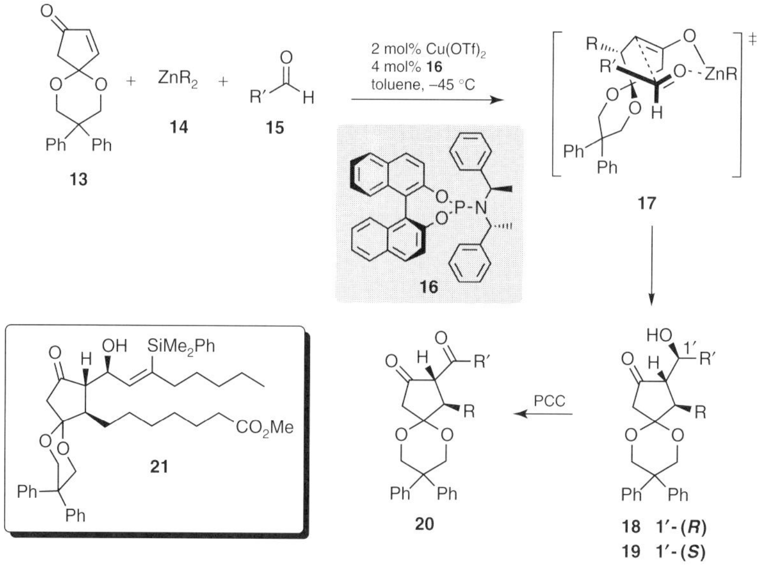

Scheme 8.4 Enantioselective synthesis of **20**.

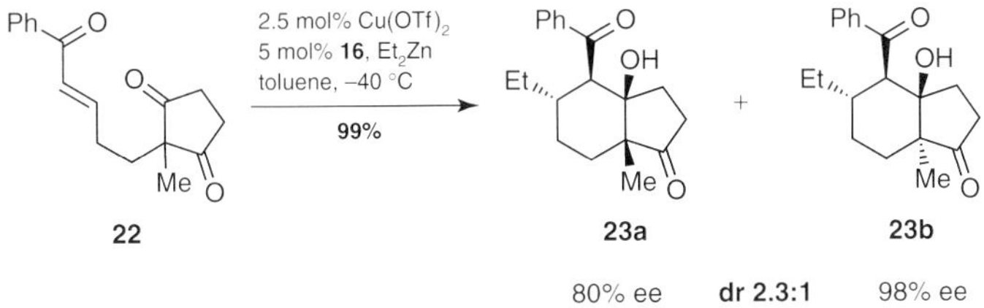

Scheme 8.5 Synthesis of **23a** and its epimer **23b**.

Besides the use of an aldol reaction as the second step in the domino process, Dieckmann condensations and Blaise-type reactions employing esters and nitriles, respectively, have also been investigated. For example, Liu and coworkers used a copper-mediated conjugate addition/nitroso aldol reaction to construct different α-keto hydroxyl amines [13]. The exposure of ketone **24** to Et$_2$Zn in the presence of catalytic amounts of Cu(OTf)$_2$ and phosphoramidite **27** gave the intermediate zinc enolate, which was captured by nitrosobenzene to yield α-hydroxyamino ketone **25** in good yield. A subsequent treatment with HCl gave the *p*-chloroaniline derivative **26** as a mixture of diastereomers in a 1 : 2.1 ratio (Scheme 8.6). Since both the diastereomeric ratio (dr) and ee-values were determined at the stage of **26** due to the instability of **25**, it was unclear whether the almost 1 : 1 ratio of the two diastereomers formed was caused by a poor stereoinduction in the aldol reaction, or by an epimerization during the ensuing reaction with HCl.

Recently, Shibasaki reported on the three-component assembly of allenic esters **29** with ketones **28** and alkyl-Zn derivatives in the presence of copper acetate and the bidentate phosphane Difluorophos (**31**), that led to the formation of highly functionalized δ-lactones **30** with good to excellent yield and ee-values (Scheme 8.7) [14].

Since, in the process, small amounts of α-hydroxy-β′, γ′-ketones (**36-α**, α-product) were formed that can also be converted into **30** (γ-product), two competing catalytic cycles were proposed (Scheme 8.8). Following formation of the catalytically active species **31**, either the formation of a mixed copper/zinc enolate (**32-α**) or a transmetallation to an organocopper species (**32-γ**) can occur. **32-α** initiates the reversible α-cycle, leading ultimately to the formation of **36-α**, while the copper organyl **32-γ** acts as an extended enolate equivalent, giving rise to γ-aldol product

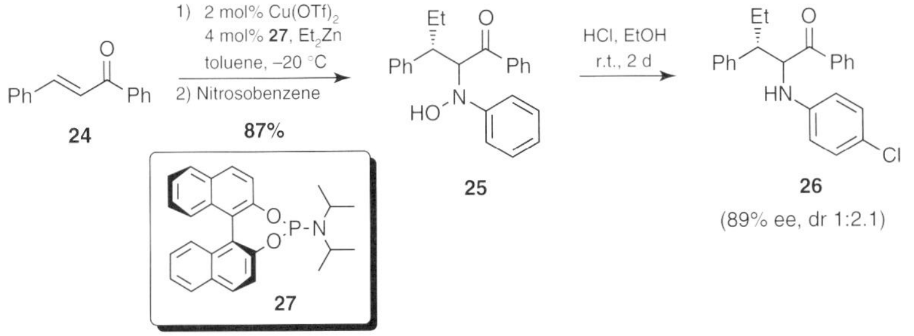

Scheme 8.6 Synthesis of amino ketone **26**.

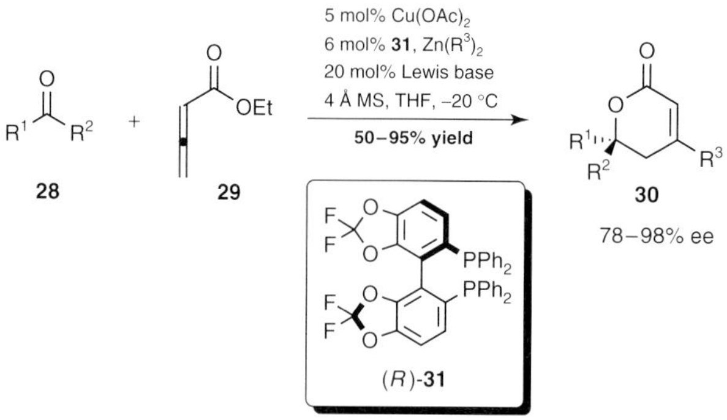

Scheme 8.7 Three-component domino reaction for the formation of lactone **30**.

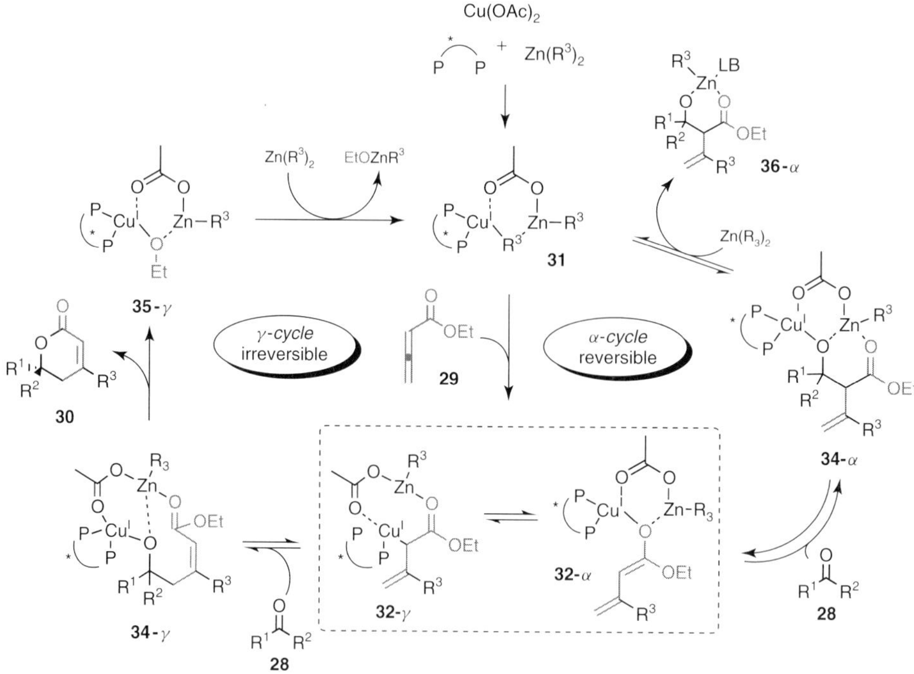

Scheme 8.8 Proposed catalytic cycles.

34-γ. The latter readily cyclizes to give the lactone **30** and the bimetallic intermediate **35-γ** which, upon reaction with dialkyzinc, is transformed again into **31**. As the α-cycle is reversible, **36-α** can also yield **30**, whereas the lactone – once formed – is stable under the reaction conditions.

Besides the use of zinc, aluminum, or magnesium organyls as stoichiometric group transfer reagents, hydrides may also be used in the ECA, which is followed by an aldol reaction [15]. Normally, organosilanes serve as the hydride source, as the direct hydride transfer from a stoichiometric copper hydride source such as

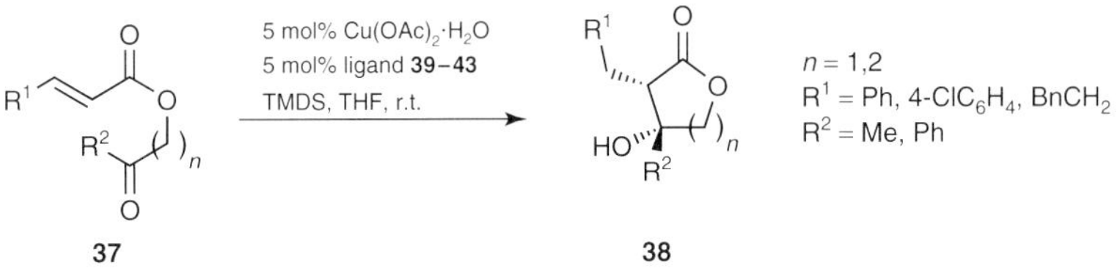

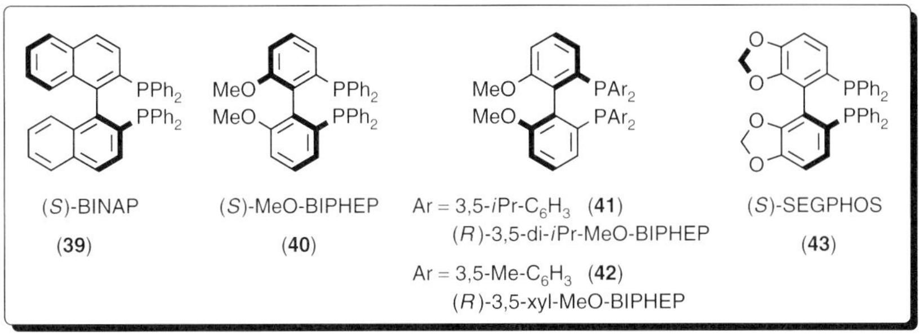

Scheme 8.9 Intramolecular reductive domino Michael–aldol reaction.

Stryker's reagent [16] [CuH(PPh$_3$)]$_6$, or the use of molecular hydrogen, have not yet proved to be successful [17].

Recently, Lam reported on the first enantioselective intramolecular reductive domino conjugate addition/aldol reaction (Scheme 8.9) [18]. Reduction of the α,β-unsaturated ester **37** was performed using air-stable copper(II) acetate and tetramethyldisiloxane (TMDS) as the hydride source. Various axially chiral biaryl phosphines such as the BINAP (**39**), BIPHEP (**39–42**), or SEGPHOS (**43**) systems were tested, whereby **43** produced the best results with 74–82% ee; no further activation of the intermediately formed silyl ether was required for the aldol cyclization to obtain **38** as a single diastereomer.

Shortly afterwards, the groups of Riant and Shibasaki produced their seminal reports on intermolecular domino reductive Michael/aldol reactions (Scheme 8.10) [19, 20]. Whereas, Riant used Taniaphos (**47**) as ligand for the reaction of **44** and **45** to give **46** with both excellent yields and ee-values, Shibasaki employed Tol-BINAP **50** for the reductive addition of **48** to **45** to afford **49** with only reasonable yields and low ee-values.

Very recently, B. H. Lipshutz and colleagues, who have been investigating copper hydride chemistry for many years, produced an intramolecular variant of a copper-mediated domino conjugate reduction/aldol reaction of **51** to give **52** in the presence of the ligand **53**, with up to 97% ee (Scheme 8.11) [21].

The other predominant metal for enantioselective Michael additions is *rhodium* which, although still lagging behind in terms of application range, has recently produced results showing much promise, opening up the possibility of its use in domino processes. As opposed to copper-based reactions in which only alkyl groups can be introduced, the Rh-based reactions provide a complementary method for the enantioselective introduction of both aryl and alkenyl groups, and also a hydrogen atom at the β-position of electron-deficient double bonds [22].

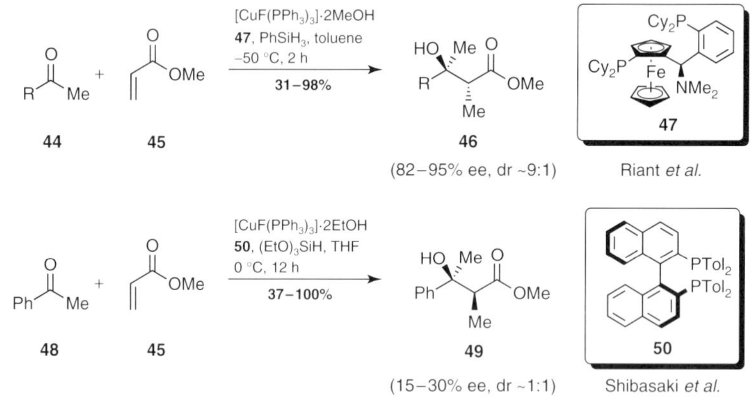

Scheme 8.10 Intermolecular reductive domino Michael–aldol reaction.

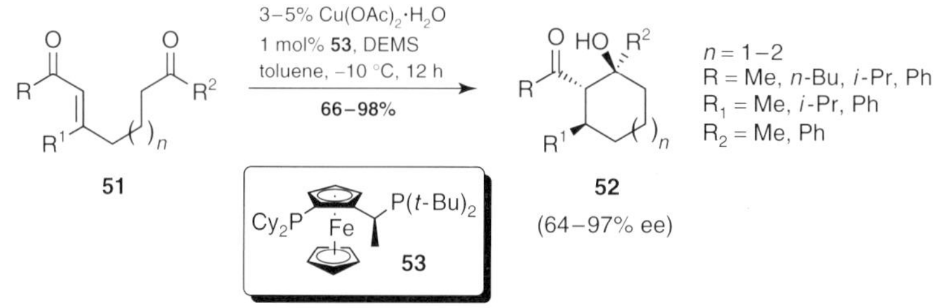

Scheme 8.11 Intramolecular domino ECA/aldol reaction.

The first asymmetric rhodium-catalyzed conjugate addition/aldol sequence was reported by Morken et al. in 2000 [23]. Based on a complex derived from (R)-BINAP (**56**) and [Rh(cod)Cl]$_2$ (**57**), a moderately diastereoselective reductive aldol sequence between acrylate esters **54** and aldehydes **15** with good enantioselectivity was achieved (Scheme 8.12).

Morken later reported the details of several improvements with better diastereoselectivity and enantioselectivity, as well as a wider range of substrates [24].

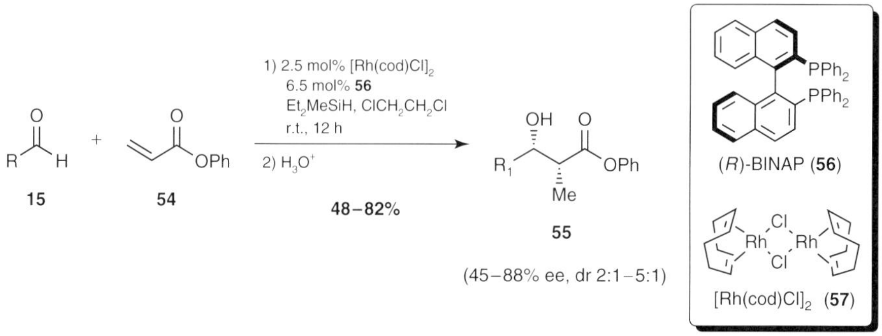

Scheme 8.12 Rhodium-catalyzed reductive aldol reaction sequence.

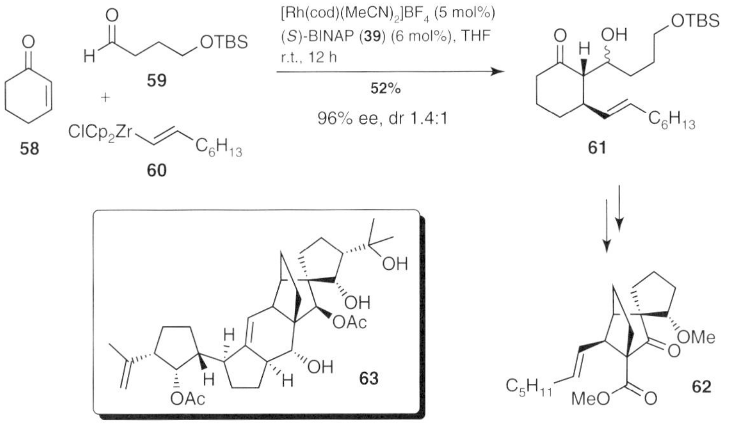

Scheme 8.13 Synthesis of spirane fragment **62** of natural product vannusal A (**63**).

However, one drawback of these new conditions was the need to use a large excess of aldehyde (5.0 equiv.), which might prevent the application of such methodology in complex target syntheses with precious starting materials.

K. C. Nicolaou employed an asymmetric rhodium-catalyzed domino Michael/aldol process in a synthesis of the spirocyclic skeleton **62** of vannusal A (**63**), a structurally complex marine natural product isolated from the ciliate species *Euplotes vannus* (Scheme 8.13) [25]. Thus, the reaction of cyclohexenone (**58**), the aldehyde **59**, and the vinyl zirconium derivative **60** in the presence of catalytic amounts of [Rh(cod)(MeCN)$_2$]BF$_4$ and (*S*)-BINAP (**39**) led to **61** in 52% yield and 96% ee as a mixture of two diastereomers (1.4 : 1).

Krische and coworkers developed a domino reduction/aldol reaction for the synthesis of substituted cyclopentanols and cyclohexanols. In this process, three contiguous stereogenic centers, including a quaternary center, were formed with excellent diastereo- and enantioselectivity (Scheme 8.14) [26]. Thus, by using an enantiopure Rh-BINAP catalyst system and phenyl boronic acid **66**, the substrates **64** and **65**, respectively, were converted into the corresponding cyclized products **67** and **68** in 69–88% yield and with 94 and 95% ee. The reaction was considered likely to proceed through the (oxa-π-allyl)rhodium intermediate **69**, as established by detailed mechanistic studies performed on related Rh-catalyzed enone conjugate additions (Scheme 8.15) [27]. As the intramolecular reaction of the intermediate

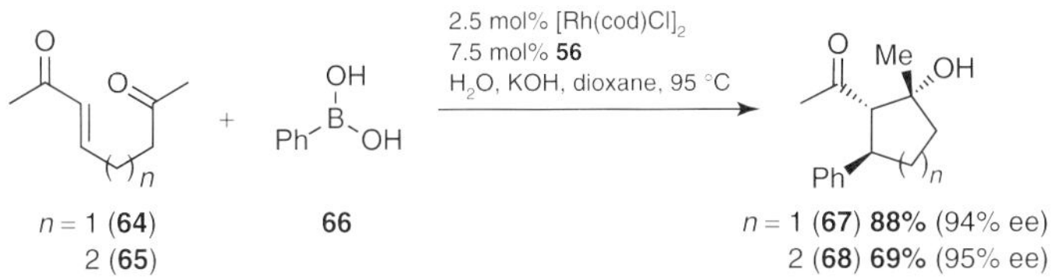

Scheme 8.14 Rh-catalyzed domino boronic acid/enone coupling/intramolecular aldol reaction.

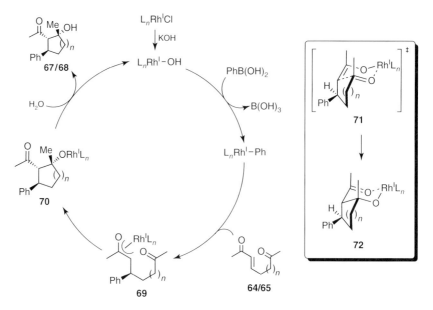

Scheme 8.15 Proposed catalytic cycle and stereochemical model.

with the ketone moiety is faster than protonolysis, the aldol product **70** was obtained in high yields, even in the presence of water. A proposed simplified mechanism for the conjugate addition/aldol cyclization which accounted for the observed relative stereochemistry called for a Zimmerman–Traxler-type transition state with a (Z)-enolate, as shown in **71**, to give **72**.

In light of these results, Krische and coworkers explored the applicability of such methodology for the desymmetrization of prochiral diketones (**73**, $R^2 = R^3$), as well as for the kinetic resolution of racemic substrates (**73**, $R^2 \neq R^3$) using arylboronic acid **74** in the presence of catalytic amounts of (S)-BINAP (**39**) and the methoxy-bridged dimer [Rh(cod)(OMe)]$_2$ as precatalytic species (Scheme 8.16) [28]. The Rh-enolate generated on enone carbometallation effectively discriminates among four diastereotopic π-faces of the dione, providing products that embody four contiguous stereocenters, including two quaternary centers, with excellent levels of diastereoselectivity (>99:1 dr) and enantioselectivity (85–92% ee).

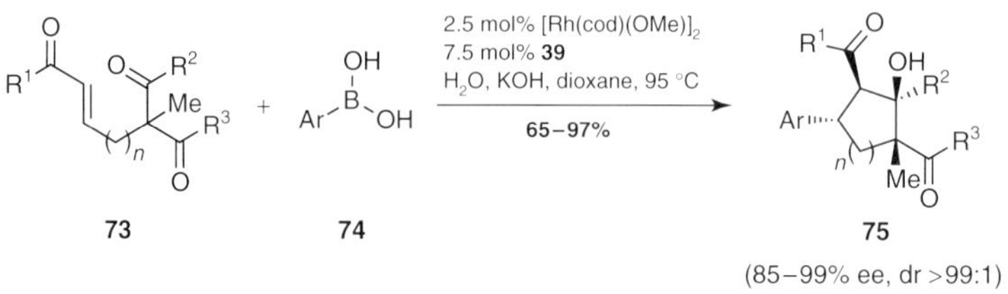

Scheme 8.16 Rh-catalyzed desymmetrization ($R^2 = R^3$) and kinetic resolution ($R^2 \neq R^3$) of diketoenones.

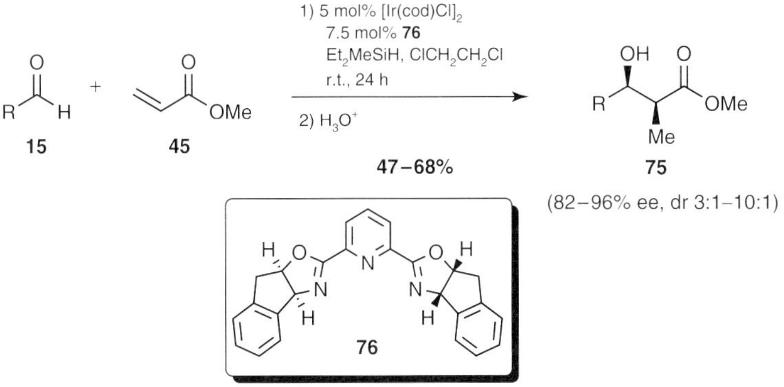

Scheme 8.17 Ir-catalyzed reductive aldol reaction sequence.

Besides the aforementioned metal catalysts, systems based on other metals have also been successfully employed; these catalytic species usually rely on iridium or magnesium. For example, Morken and coworkers extended the enantioselective reductive aldol methodology employing iridium salts and the aminoindanol-derived Indane-PyBox ligand **76** [29] (Scheme 8.17). Suitable substrates for the transformations are benzaldehyde, as well as *tert*-butyldimethylsilyl (TBS)- or benzyl-protected α- and β-hydroxy aldehydes. Propanal and cinnamaldehyde did not lead to any product, thus substantially limiting the scope of substrates.

Besides the use of carbon nucleophiles and hydrides in the conjugate addition step, other nucleophiles may also be employed as phenols. In this way, functionalized chromanes may become accessible through a methodology introduced by the group of Jørgensen, utilizing a Mg-Box-species **80** as catalyst (Scheme 8.18) [30]. This complex, which is formed *in situ* from Mg(OTf)$_2$ and **80**, is able to catalyze the domino transformation of phenols **77** containing a methoxy group in

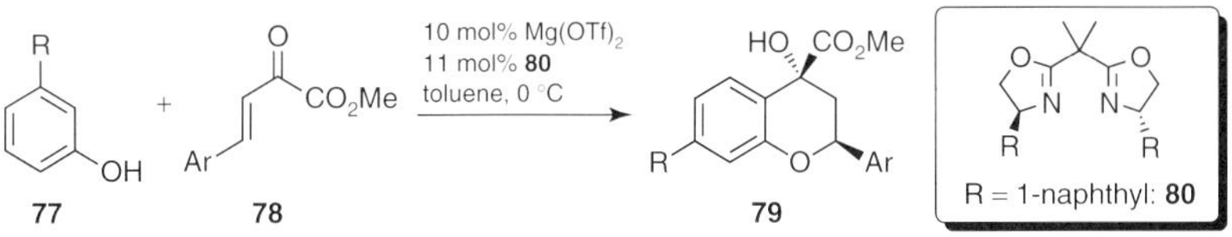

Entry	R	Ar	Yield (%)	ee (%)
1	OMe	Ph	67	73
2	OMe	Ph	77	80
3	OMe	p-FC$_6$H$_4$	43	74
4	OMe	p-BrC$_6$H$_4$	45	66
5	NMe$_2$	p-ClC$_6$H$_4$	>95	<18
6	NMe$_2$	Ph	>95	13

The reaction was performed in the presence of p-methyl-N,N,dimethylaniline

Scheme 8.18 Mg-Box-catalyzed formation of chiral chromanes.

the *meta*-position and the β,γ-unsaturated α-ketoesters **78** to give the corresponding chromanes **79**, with excellent diastereo- and enantioselectivity of up to 80%.

Interestingly, replacing the methoxy group by a NMe$_2$ group in substrate **77** led to chromanes **79** with a NMe$_2$ group (Scheme 8.18, entries 5 and 6), again with excellent diastereoselectivity and an even higher yield, but with ee-values of <18%. Most likely, due to the higher electron-donating character of the dimethylamine substituent as compared to the methoxy group, the nucleophile is sufficiently reactive to give an uncatalyzed, nonselective oxa–Michael addition, resulting in a dramatic decrease in enantioselectivity. Replacement of the aryl substituent by a simple methyl group resulted in the formation of a mixture of diastereomers (4 : 1) with low enantioselectivity. The same chromanes have previously been prepared using copper-based catalysts, but with a considerably lower yield and ee-values [31].

8.3
Metal-Mediated Domino Michael Reaction/Electrophile Trapping with Noncarbonyl Compounds

Although, until now, the ECAs described have been followed by an aldol or and aldol-type reaction, it is possible that instead of an aldol reaction, an alkylation might also take place. As the intermediately formed metal enolates have a rather low reactivity, only highly electrophilic alkylating agents such as allyl halides can be used. In addition to the allylation, there are examples where orthoesters and acetals

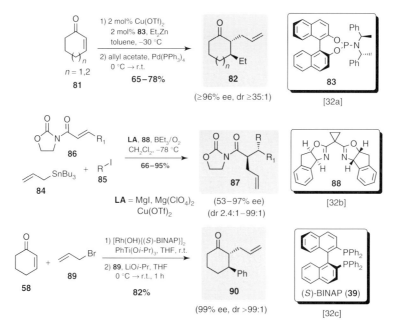

Scheme 8.19 Various protocols utilizing allyl reagents I.

8.3 Metal-Mediated Domino Michael Reaction/Electrophile Trapping with Noncarbonyl Compounds

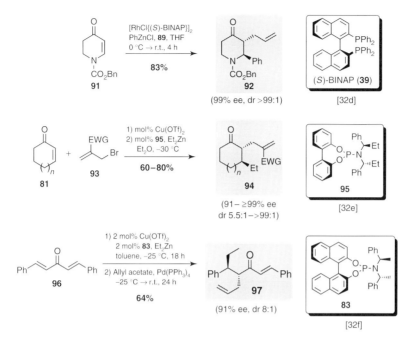

Scheme 8.20 Various protocols utilizing allyl reagents II.

in the presence of a Lewis acid have been used as electrophiles. Several examples of the conjugate addition/allylation sequence are presented in Schemes 8.19 and 8.20, starting from cyclic as well as acyclic enones and enamides (**81**, **86**, **91**, **96**) with ee- and dr-values ranging up to 99% using biphenyl-, binaphthyl-, and bisoxazoline-derived ligands (**39**, **83**, **88**, **95**) [32]. As products, the compounds **82**, **87**, **90**, **92**, **94**, and **97**, respectively, were obtained.

Feringa *et al.* have used this approach also for the enantioselective synthesis of the neurotoxin (−)-pumiliotoxin C (**99**), which functions as a noncompetitive blocker of acetylcholine receptor channels. The synthesis started from cyclohexenone **58** to give the intermediate **98** with 90% ee and 8.5 : 1 dr (Scheme 8.21) [33].

Alexakis *et al.* reported on the ECAs of Et$_2$Zn to cyclohexenone and cycloheptenone in the presence of the ligand **100**, followed by the BF$_3$·OEt$_2$-catalyzed reaction of the transient zinc enolate **101** to afford the aldol-type products **102** (Scheme 8.22) [34]. When the prochiral electrophiles (R^1 ≠ R^2) were used, a poor diastereocontrol was observed. Thus, in order to overcome this hurdle and obtain

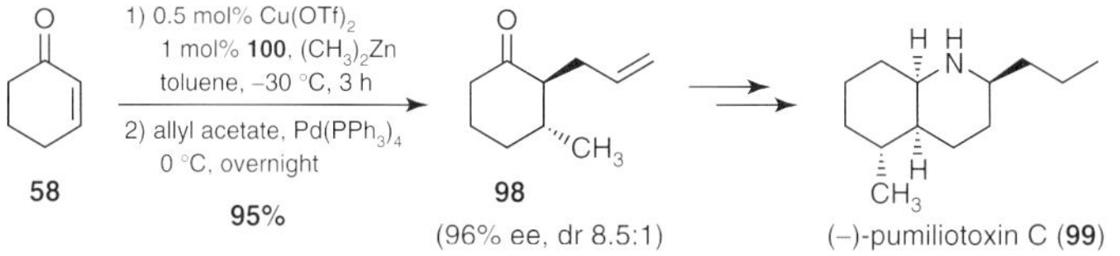

Scheme 8.21 Enantioselective synthesis of (−)-pumiliotoxin C.

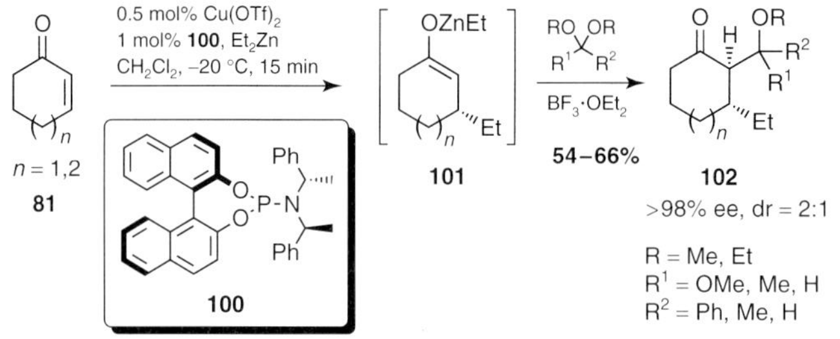

Scheme 8.22 Domino conjugate addition/trapping of carbonyl analogs.

complete stereocontrol, the intermediate **101** was treated with chiral acetals in the presence of TMSOTf. Following removal of the chiral auxiliary, the *anti,anti*-aldol product was obtained in quantitative yield.

Some examples have also been reported of ECAs followed by an aldol reaction, in which the intermediately formed metal enolate is first trapped as a silyl enol ether, and then transformed into a reactive lithium or boron enolate [35]. Thus, the reaction of **103** with AlMe$_3$ and Et$_3$SiOTf in the presence of catalytic amounts of Cu(OTf)$_2$ and the chiral silver complex **107** led to **104** with 82% yield and 84% ee [36]. The treatment of **104** with *n*BuLi and BEt$_3$ to give a boron enolate, followed by addition of the aldehyde **105**, led to **106** as a mixture of diastereomers (dr = 1.5 : 1). This compound was used as a substrate in the synthesis of clavirolide C, a marine diterpene isolated from the Pacific soft coral *Clavularia viridis* (Scheme 8.23).

In a similar fashion, following the ECA to, for example, cyclohexenone (**58**) with the formation of a silyl enol ether **108**, other reagents can be used as electrophiles; these include ozone, *meta*-chloroperbenzoic acid, MeLi followed by PhN(OTf)$_2$, and H$_2$CNMe$_2$I followed by *meta*-chloroperbenzoic acid to give **109–112**, as described by Alexis and coworkers (Scheme 8.24) [37]. For the conjugate addition, several Feringa-type ligands (**95**, **113–115**) [10] were used to yield the products with good yields and excellent ee-values (>99%). Besides cyclohexenone (**58**) several acyclic substrates were also reacted, and in each case the influence of the solvent polarity on both yield and ee-value was determined.

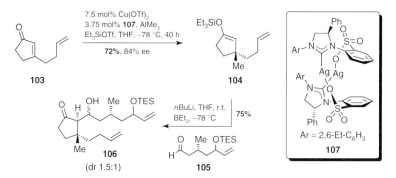

Scheme 8.23 Formation of **106** as intermediate in the total synthesis of clavirolide C.

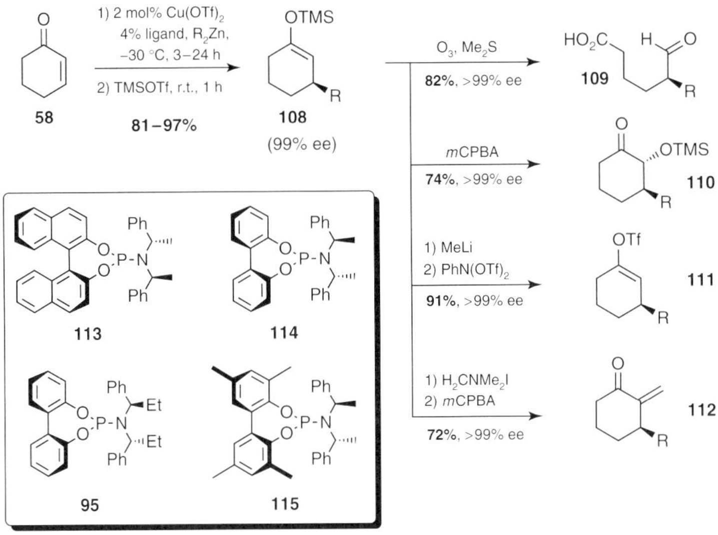

Scheme 8.24 Domino Michael addition–enolate silylation and reaction scope of **108**.

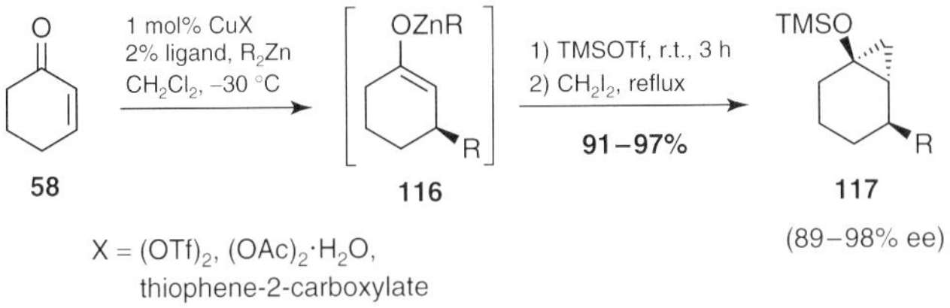

Scheme 8.25 Domino conjugate addition–silylation–cyclopropanation reaction.

This sequence was also used later by the same group for a threefold reaction which consisted of the former process followed by a subsequent cyclopropanation to achieve **117** via **116** (Scheme 8.25) [38]. In this procedure, ligands **95**, **114**, and **115** were tested in the presence of catalytic amounts of several copper salts, and all allowed a good discrimination of the enantiotopic double bond with 89–98% ee and 91–97% yield. Moreover, the regioselectivity of the ring opening of the three-membered ring could be precisely controlled, giving access to a key intermediate in the enantioselective synthesis of (−)-(S,S)-clavukerin A and (−)-(R,S)-isoclavukerin.

8.4
Organocatalytic Domino Michael Reactions/Electrophilic Trapping

There are two distinct types of organocatalytic process. In the first type, the catalytic rate enhancement occurs by activation through noncovalent interactions such as hydrogen bonding [39], while in the second type it occurs through covalent bond formation. An example of this is the enamine/iminium ion pathway of proline catalysis [40].

The group of Terashima developed an organocatalytic domino Michael/aldol process using the cinchona alkaloid (−)-cinchonidine (**124**) to prepare an intermediate for the synthesis of the natural product (−)-huperzine A (**123**) (Scheme 8.26) [41]. The reaction of β-ketoester **118** and methacrolein **119** in the presence of **124** led to the desired product **121** in 45% yield. A transition-state model based on non-covalent hydrogen-bonding and ion-pairing (**120**) has been postulated, as reported for similar asymmetric transformations [42]. The diastereomeric mixture of **121** was transformed into **122** by mesylation and subsequent elimination. Although the addition proceeded only with 64% ee, **122** could be obtained in enantiopure form by recrystallization from hexane.

In contrast, several domino reactions employing proline-based organocatalysts have been reported in recent years, and the field of organocatalysis has witnessed vast improvements not only in substrate scope and yield but also in enantioselectivity and diastereoselectivity. These advances have been reflected by the growing number of domino reactions that employ this approach.

One of the earliest examples was reported by Barbas *et al.*, who used (S)-1-(2-pyrrolidinylmethyl)pyrrolidine (**128**) to catalyze the domino Knoevenagel–Michael

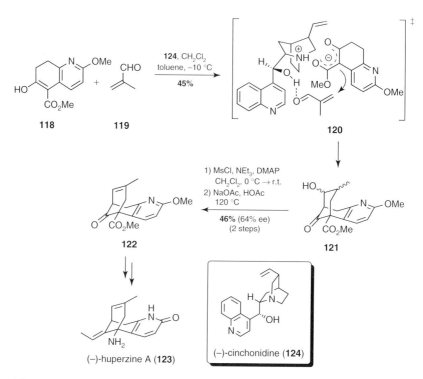

Scheme 8.26 Enantioselective synthesis of (−)-huperzine A.

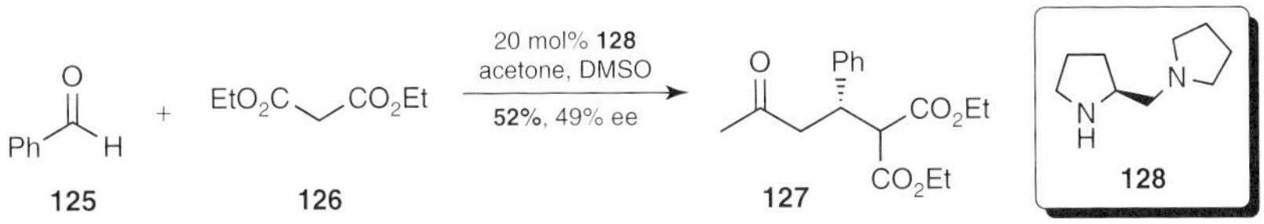

Scheme 8.27 Organocatalytic domino Knoevenagel–Michael addition.

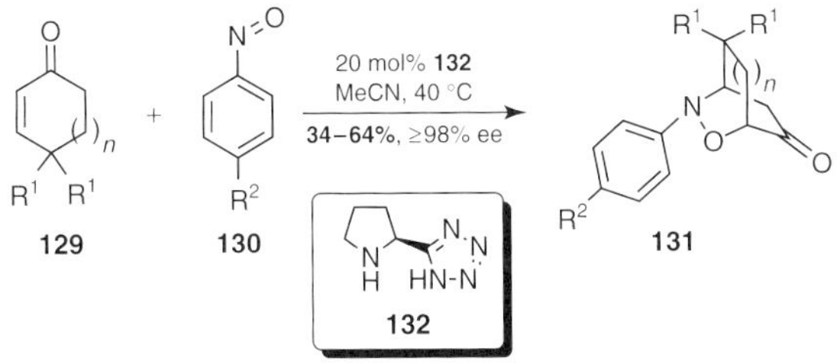

Scheme 8.28 Organocatalytic domino nitroso–aldol/Michael reaction.

addition of benzaldehyde (**125**), acetone, and diethyl malonate (**126**) to give **127** in 52% yield and 49% enantioselectivity [43] (Scheme 8.27).

Yamamoto's group reported on a highly enantioselective domino O-nitroso aldol/Michael reaction of **129** and **130** (Scheme 8.28) [44]. As products, the formal Diels–Alder adducts **131** were obtained with ≥98% ee, which arose from the selective attack of an enamine, temporarily formed from the Ley catalyst **132** and the enone **129**, onto the nitroso functionality.

A conjugate addition/aldol cyclization/dehydration reaction was also developed by Wang and coworkers for the synthesis of 1,2-dihydroquinolones [45], while Córdova et al. also described an organocatalytic entry to formal aza–Diels–Alder products via a domino Mannich–Michael addition reaction (Scheme 8.29) [46].

Following formation of the enamine of enone **133** with (S)-proline (**138**), a Mannich reaction with formaldehyde (**135**) and the amine **134** took place to produce **136**. The initially introduced stereocenter in **136** then served as a control element in the ensuing conjugate addition, thereby forming **137** in varying yield (10–90%), but with excellent enantioselectivity (96 to >99% ee).

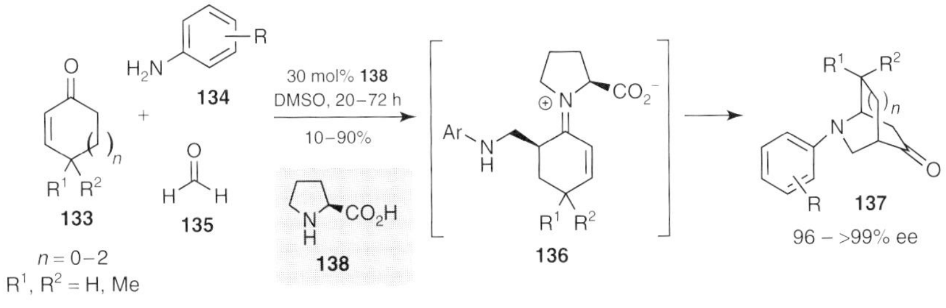

Scheme 8.29 Organocatalytic formation of Diels–Alder products via a domino Mannich–Michael addition.

Several examples of conjugate addition/cyclization sequences of small building blocks have also been reported recently. Thus, chiral cyclohexenones (**141**) [47], secondary cyclohexyl-substituted amines (**146**) [48], dihydropyrones (**150**) [49], cyclohexenes (**153**) [50], and piperidines (**156**) [51] have been efficiently prepared by organocatalytic means in domino reactions with excellent ee-values, as shown in Scheme 8.30.

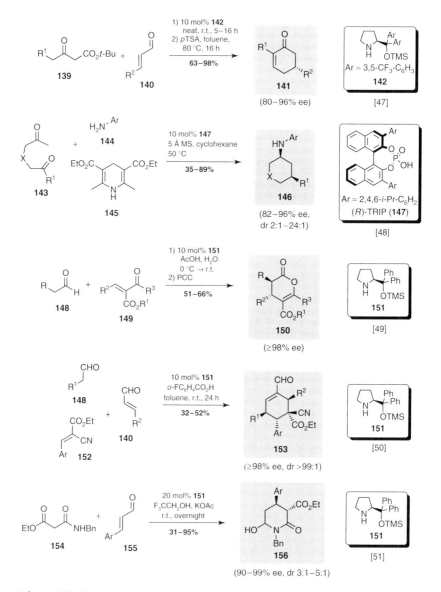

Scheme 8.30 Various organocatalytic domino reactions leading to chiral small molecule building blocks.

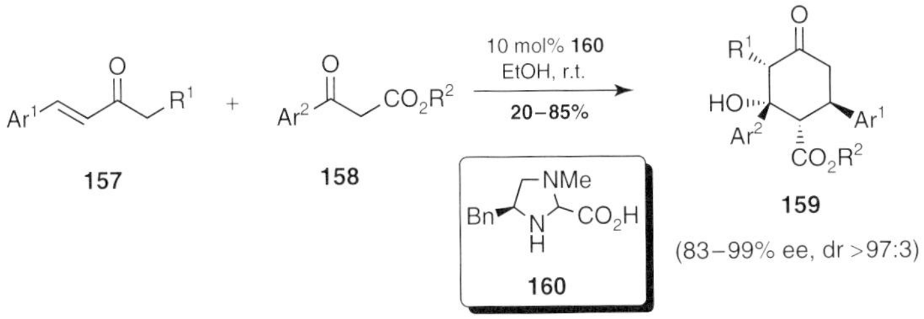

Scheme 8.31 Domino Michael–aldol reaction of α,β-unsaturated ketones with β-ketoesters.

The details of an impressive organocatalytic asymmetric two-component domino Michael/aldol reaction to give cyclohexanones with four stereogenic centers has been reported by Jørgensen and coworkers (Scheme 8.31) [52]. Inspired by the proline-catalyzed Robinson annulation pioneered by Wiechert, Hajos, Parrish, and coworkers [53], Jørgensen et al. constructed compounds of type **159** with excellent enantioselectivity and diastereoselectivity, starting from unsaturated ketones **157** and acyclic β-ketoesters **158** in the presence of 10 mol% phenylalanine-derived imidazolidine catalyst **160**. The final products could easily be converted into useful cyclohexanediols, as well as γ- and ε-lactones.

The enantioselective synthesis of addition products from substituted acrolein has been intensively investigated, with two recent reports describing the potential and simplicity of the said transformation by organocatalytic means. The research groups of both MacMillan and Jørgensen disclosed (independently) their successful discovery in quick succession in 2005, using the organocatalysts **142**, **168**, **173**, and **174** (Scheme 8.32) [54].

In this case, Jørgensen's group employed thiols as nucleophiles and protons as "electrophiles," followed by reduction of the carbonyl moiety to give the corresponding alcohols **162** from **161** and **140**. Azodicarboxylates (**163**) were also applied as co-substrate which, following capture of the nitrogen and reduction of the aldehyde to give the alcohol, led to formation of oxazolidinone **164** with yields of 38–72% and excellent diastereoselectivity (dr 7:1–24:1) and enantioselectivity (97–99%). In another approach, MacMillan and coworkers employed electron-rich heteroaromatic compounds **165** as nucleophiles and chlorinated dienone **167** as chlorine source, to produce compounds of type **166**. In addition, a Hantzsch pyridine ester **170** was applied as a hydride transfer reagent and, in combination with either dienone **167** or **172** as molecular chlorine/fluorine source, a hydrochlorination and hydrofluorination (respectively) was achieved with acceptable levels of diastereoselectivity and excellent ee-values (99%). In MacMillan's studies, either **173** was utilized as a catalyst, or a combination of **173** with either enantiomer of **174**, so as to allow the stereoselective formation of both the syn- and anti-diastereomers of **171**. This methodology was very recently expanded to include olefin hydroaminations, hydro-oxidations, and reductive Mannich reactions [55].

In a report from MacMillan's group on the enantioselective synthesis of natural products containing a pyrroloindoline skeleton, such as (−)-flustramine B

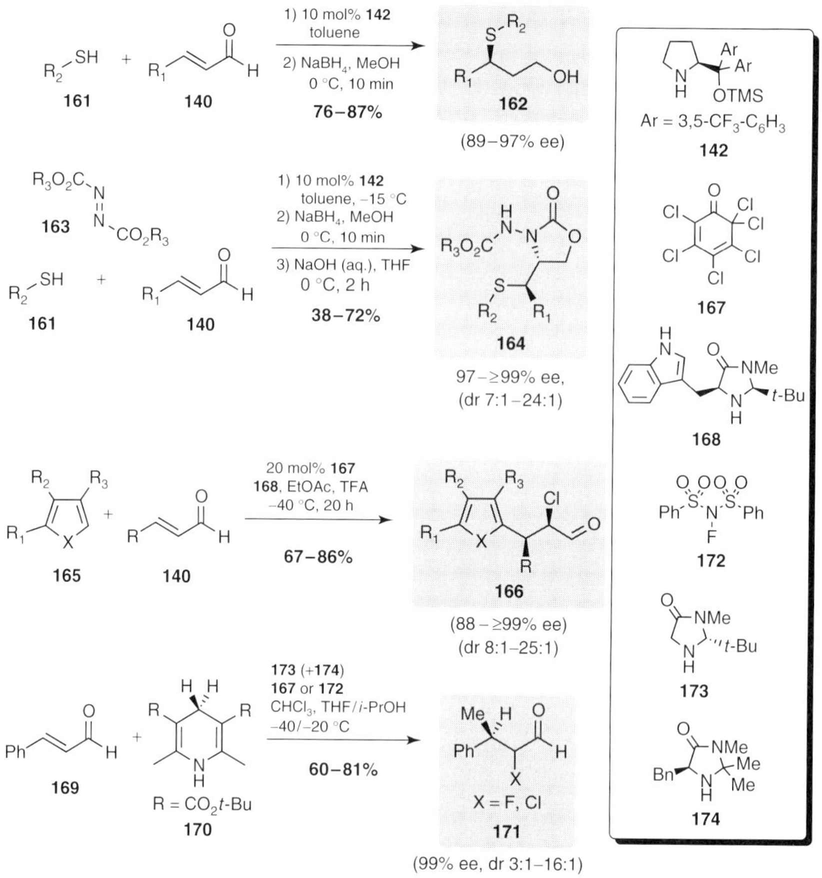

Scheme 8.32 α,β-Functionalization of acrolein derivatives by organocatalytic means.

(**181**) [56], enantiopure amines such as the imidazoline **177** were used as organocatalysts to promote a facial-selective domino Michael addition/cyclization sequence (Scheme 8.33) [57]. Thus, the reaction of **175** and acrolein in the presence of **177** led to **180**, most likely via the intermediate **178**, which cyclized with subsequent hydrolysis of the enamine moiety and reconstitution of the imidazolidinone. Following reduction of the aldehyde functionality in **179** with NaBH₄ the flustramine precursor **180** was isolated in very good enantioselectivity (90% ee) and 78% yield.

Franzén reported on the synthesis of quinolizidine derivatives (**184** and **185**) via an imposing three-step domino Michael addition/double cyclization sequence of a mono-amide of a malonate containing an electron-rich aromatic moiety **182** with cinnamic aldchydes **155** (Scheme 8.34) [58]. Although several catalytic systems such as proline (**138**) and the MacMillan amine (**177**) were screened and evaluated with respect to the enantiomeric excess and conversion, only the Jørgensen catalyst **151** gave useful yields and good to excellent ee-values.

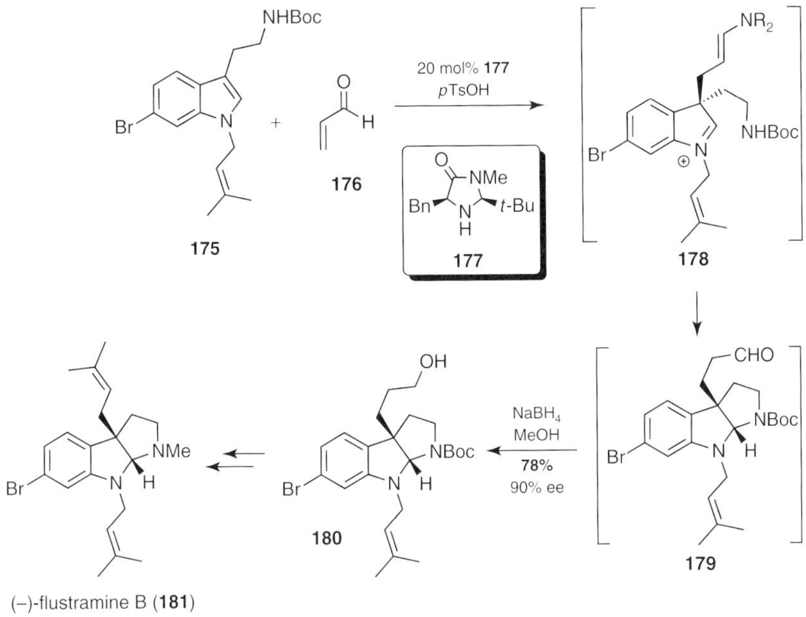

8.4 Organocatalytic Domino Michael Reactions/Electrophilic Trapping

Scheme 8.33 Synthesis of (−)-flustramine B (**181**).

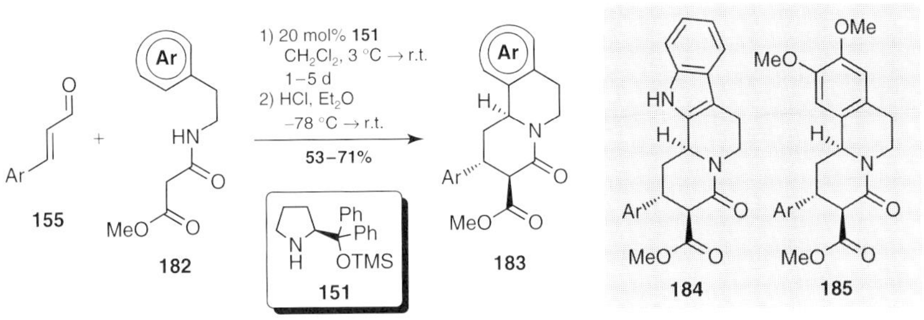

Scheme 8.34 Synthesis of quinolizidine derivatives **184** and **185**.

As already known from enantioselective metal-catalyzed 1,4-conjugate additions, oxygen nucleophiles as well as carbon nucleophiles and hydride can also be used in the corresponding organocatalytic domino reactions. For the first time, Córdova et al. and Rueping et al. described such a reaction using 2-hydroxy- or 2-amino-benzaldehydes (**186**) and 2-hydroxy-1,4-naphthoquinone **188**, respectively, with α,β-unsaturated aldehydes **140** as substrates to give 1,2-dihydroquinolines or 2H-chromenes (**187**) and 1,4-pyranonaphtho-quinones (**189**). Whereas, Córdova obtained **187** in 31–95% yield with up to 99% ee in the presence of the chiral diphenylprolinol ether **151**, Rueping et al. were able to isolate **189** in 43–87% yield and up to 99% ee using the diarylprolidinol ether **142** (Scheme 8.35) [59–61]. Subsequently, **189** could easily be transformed into biologically interesting 1,2-pyranonaphtho-quinones by reduction with sodium borohydride, followed by

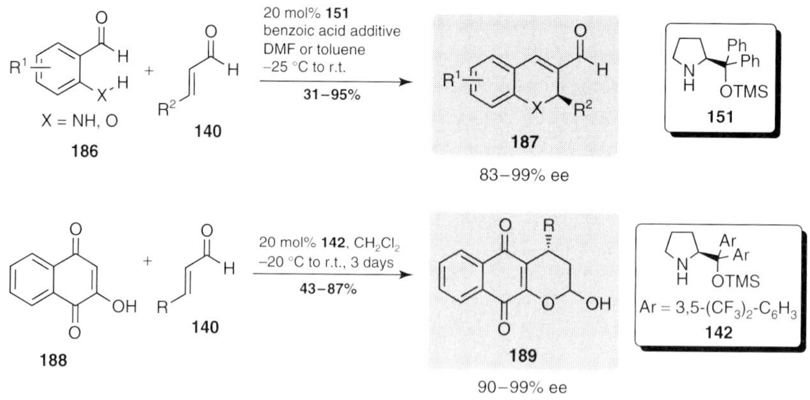

Scheme 8.35 Organocatalytic formation of 1,2-dihydroquinolines, 2H-chromenes, and 1,4-pyranonaphthoquinones **181**.

treatment with a strong acid and without any loss of stereointegrity. Furthermore, cyclohexenone was employed by Córdova to receive enantioenriched tetrahydroxanthenones [62].

One recently reported and very interesting transformation was the enantioselective organocatalytic domino reaction of cyclohexane-1,2-dione **190** with the α,β-unsaturated aldehydes **140**, for the formation of bicyclo[3.2.1]octanones **191** in the presence of a proline-derived catalyst **151**, which provided ee-values of 90–98% (Scheme 8.36) [63]. Within this process, an initial a keto–enol tautomerization of

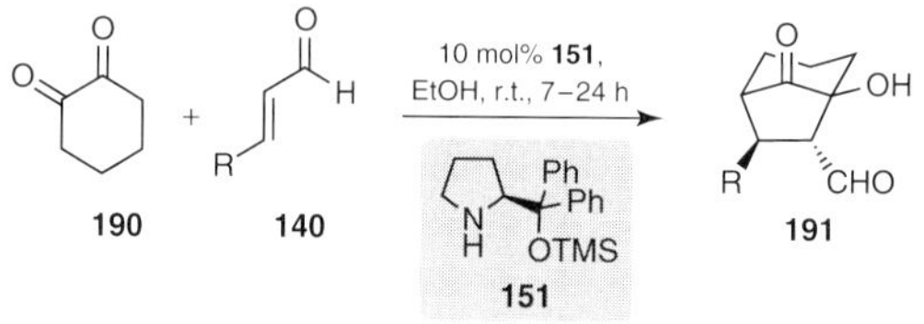

Entry	R	Yield (%)	ee (%)
1	Ph	77	96
2	4-MeOC$_6$H$_4$	79	95
3	4-BrC$_6$H$_4$	80	95
4	3-BrC$_6$H$_4$	62	96
5	2-NO$_2$C$_6$H$_4$	66	98
6	4-NMe$_2$C$_6$H$_4$	60	93
7	Benzo[1,3]dioxol-5-yl	68	96
8	2-Furanyl	81	95
9	2-Thiophenyl	77	90
10	n-Butyl	44	98

Scheme 8.36 Organocatalytic formation of chiral bicyclo[3.2.1]octanones.

one of the carbonyl groups of **190** could be assumed to produce the corresponding α-keto-enol. This could then act as a nucleophile in the Michael addition of an iminium ion formed from **140** and **151**, followed by an intramolecular aldol reaction. The products **191** were further transformed to obtain polysubstituted cycloheptanones and tetrahydrochromenones.

One of the most exciting enantioselective domino conjugate additions was developed by Enders and coworkers [64], who reported the details of a convergent triple-domino reaction of aldehydes **148**, α,β-unsaturated aldehydes **140**, and nitroalkenes **192** to form highly substituted cyclohexenylcarbaldehydes **194** in medium yield, albeit with excellent enantioselectivity (>99% ee) and diastereoselectivity (99% de) (Scheme 8.37). Mechanistically, following the first ECA of **148** and **192**, catalyzed by the proline derivative **142**, the intermediate **193** was formed. The latter could act as a nucleophile in the second Michael addition with the

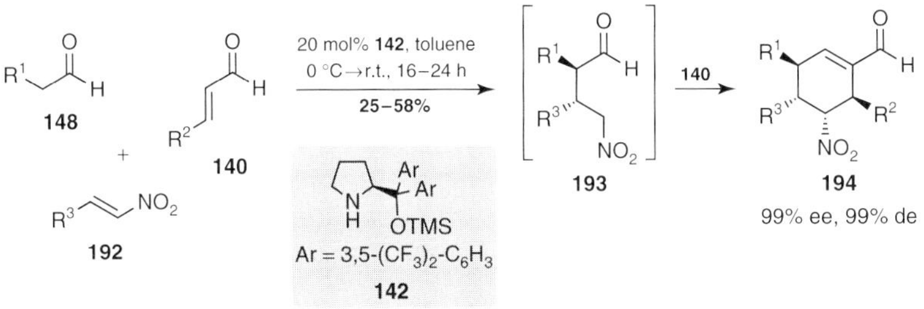

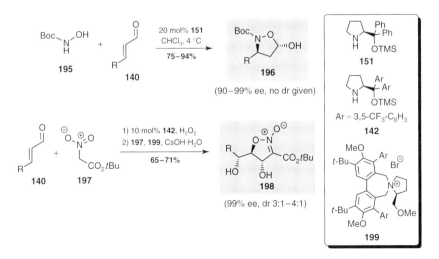

Scheme 8.37 Triple domino-conjugate addition of aldehydes, α,β-unsaturated aldehydes, and nitroalkenes.

Scheme 8.38 Organocatalytic formation of isoxazolidine and isoxazoline derivatives, according to Córdova and Jørgensen.

α,β-unsaturated aldehyde **140**, followed by the terminating ring closure through an aldol condensation.

Moreover, Córdova and Jørgensen reported on the organocatalytic synthesis of isoxazolidines and isoxazoline derivatives **196** and **198** (Scheme 8.38) [65, 66]. In this case, the formation comprised either a conjugate addition of hydroxyl amine **195** followed by intramolecular cyclization to give **196**, or a conjugate addition of HO_2^- to **140**, intramolecular epoxide formation and a terminating epoxide opening/ring closure to obtain **198**. Whereas, Córdova employed proline derivative **151** to receive the cyclic product in good yields and very good to excellent ee, Jørgensen and colleagues utilized **142** and a Lygo-type chiral ammonium salt **199** as a phase-transfer catalyst to obtain **198** in good yields and with near-perfect enantiomeric excess [67].

8.5
1,4-Conjugate Additions Followed by a Cycloaddition, Hydrogenation, Rearrangement, or Other Reactions

Although, the majority of the enantioselective domino conjugate addition/ electrophilic reactions reported to date have fallen into the three categories discussed so far, there are some reactions which follow a different scheme. One such reaction is a conjugate addition/Huisgen cycloaddition that was developed by Miller and coworkers, that allows the regioselective and enantioselective synthesis of substituted triazoles **202**, starting from **200** via **201** (Scheme 8.39) [68]. In this case, the use of a modified peptide **203** as catalyst enabled the reaction and subsequent either intramolecular or intermolecular 1,3-dipolar cycloaddition to proceed with good to excellent enantiomeric excess (82–92% ee) and good overall yield (73–85%).

Based on results of Noyori's group, and their reports concerning enantioselective hydrogenations, Morris and coworkers employed a chiral ruthenium complex **206**

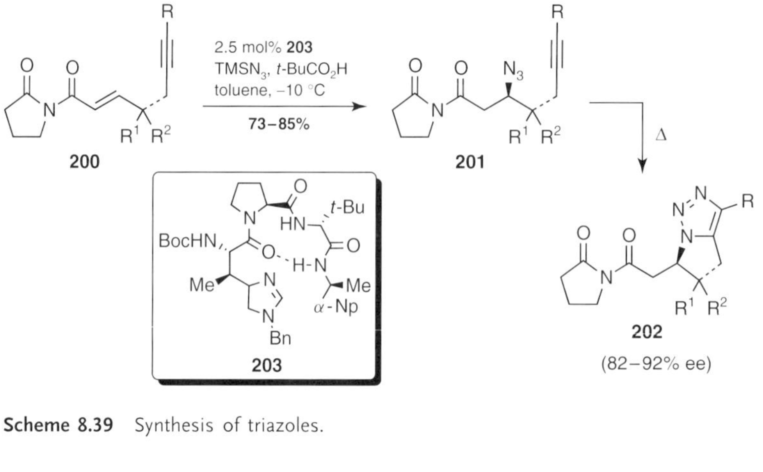

Scheme 8.39 Synthesis of triazoles.

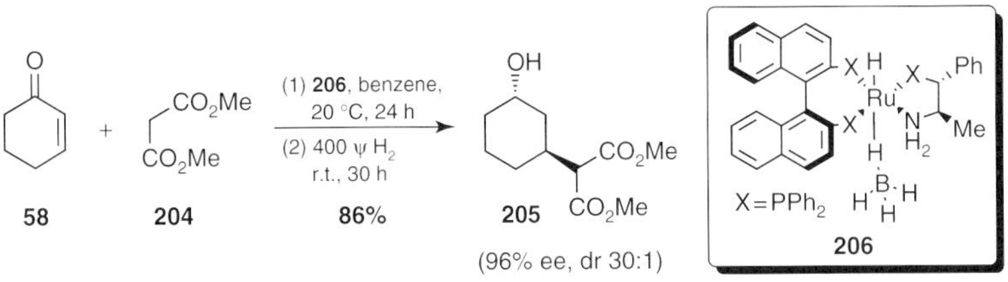

8.5 Michael additions followed by cycloadditions and other reactions

Scheme 8.40 Domino conjugate addition/reduction reaction.

for a Michael addition/ketone hydrogenation sequence (Scheme 8.40) [69]. The ECA of dimethyl malonate (**204**) to cyclohexenone (**58**) was proposed to involve a metal–NH bifunctional effect related to that proposed for the asymmetric transfer hydrogenation of ketones [70]. Following the conjugate addition, hydrogen pressure was applied to allow an asymmetric hydrogenation to produce the secondary alcohol **205** in 86% yield with 96% ee, as a single diastereomer.

Shortly thereafter, Feringa and coworkers reported on a successive conjugate addition/ketone allylation reaction which utilized both a rhodium complex with the monodentate ligand **207** and elementary indium in the second step (Scheme 8.41) [71]. In this case, reaction of the cyclic enones **81** with boronic acid **74** in the presence of catalytic amounts of **209**, followed by the addition of allyl, methallyl, crotyl, or prenyl bromide **207**, led to production of the carbocycles **208** in 46–75% yield with 85–99% ee and dr of 15:1–99:1.

Recently, Beak described a domino reaction that comprised an enantioselective lithiation catalyzed by (−)-sparteine (**214**) and a 1,4-addition to attach prochiral amines to α,β-unsaturated esters **212** to give compounds of type **213** (Scheme 8.42) [72]. Following formation of the chiral π-lithium-allyl intermediate **211** from **210** and **214**, the Michael system in **212** was attacked to give **213** with 80–90% ee and a very good diastereoselectivity. The products were used to prepare various 3,4,5,6-substituted azepanes in a further three to six steps.

A domino sequence in which 3-(2-hydroxyphenyl)cyclobutanones as **215** were reacted with a rhodium–BINAP complex to yield 3,4-dihydrocoumarins as **217** was developed by Murakami and coworkers. Deuterium-labeling experiments

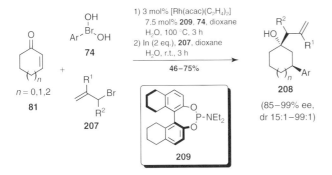

Scheme 8.41 Rhodium-catalyzed conjugate addition/allylation reaction.

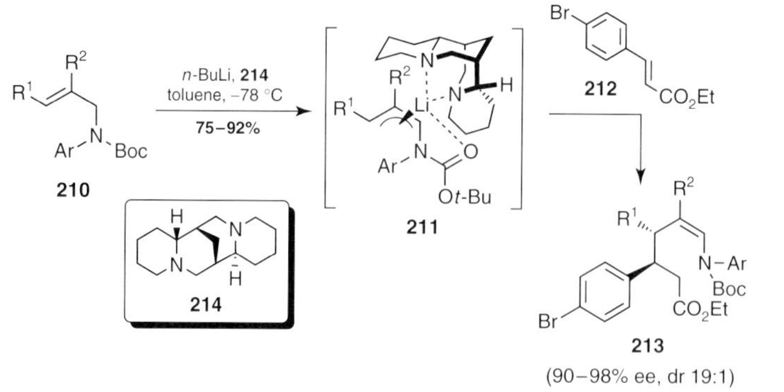

Scheme 8.42 Conjugate addition reaction of a sparteine–allylamine complex.

showed that, after formation of the organo-rhodium intermediate **220**, a rhodium-hydrogen-shift occurred. Thus, the reaction of **215** with [Rh(OH)cod]$_2$ in the presence of the chiral ligand **216** led to the α,β-unsaturated lactone **223**, presumably via the intermediates **218–222**. The lactone **223** then underwent a conjugate addition of hydrogen via the rhodium complex **224**, thereby forming

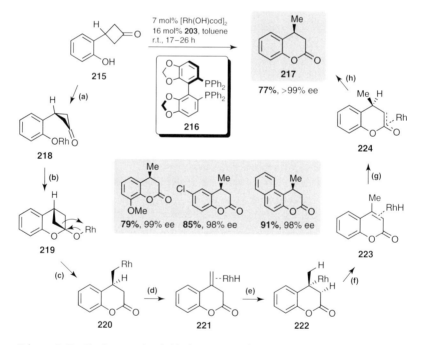

Scheme 8.43 Rhodium-catalyzed dihydrocoumarin formation and proposed mechanism. (a) Rhodium aryloxide formation; (b) Carbonyl addition; (c) β-carbon elimination; (d) β-H elimination; (e) Oxidation addition; (f) β-H elimination; (g) Conjugate addition; (h) Hydrolysis.

the almost enantiopure 4-substituted dihydrocoumarin **217** (Scheme 8.43) [73]. Besides the unsubstituted cyclobutanone **215**, several other aryl-substituted substrates were employed, giving also good yields (79–91%) and excellent ee-values (98–99%).

8.6
Conclusion

The 1,4-conjugate addition to α,β-unsaturated carbonyl derivatives, followed by reaction of the intermediately formed enolate with an electrophile in a domino fashion, allows the highly efficient synthesis of products with up to three new stereogenic centers and a greatly increased complexity as compared to the substrates used.

Whereas, in the past a facial-selective addition as the first step was only possible by using chiral substrates, today several catalytic systems have been developed that permit the introduction of a primary stereogenic center, with excellent ee-values. In many cases, the second and third stereogenic centers are also formed with good selectivities. The majority of these studies have been based on two different approaches, notably as a cuprate addition in the presence of a chiral ligand, or the use of organocatalysts. These new enantioselective domino reactions have also demonstrated their great potential for the synthesis of several natural products.

References

1. (a) Tietze, L.F. and Beifuss, U. (1993) *Angew. Chem.*, **105**, 137–170; (1993) *Angew. Chem., Int. Ed. Engl.*, **32**, 131–163; (b) Tietze, L.F. (1996) *Chem. Rev.*, **96**, 115–136; (c) Tietze, L.F. and Rackelmann, N. (2004) *Pure Appl. Chem.*, **76**, 1933–1983; (d) Tietze, L.F. and Rackelmann, N. (2005) in *Multicomponent Reactions* (ed. J. Zhu), Wiley-VCH Verlag GmbH, Weinheim; (e) Tietze, L.F., Brasche, G., and Gericke, K. (2006) *Domino Reactions in Organic Synthesis*, Wiley-VCH Verlag GmbH, Weinheim; (f) Tietze, L.F. and Levy, L. (2009) in *The Mizoroki-Heck Reaction* (ed. M. Oestreich), Wiley-VCH Verlag GmbH, Chichester, pp. 281–344; (g) Tietze, L.F., Kinzel, T., and Brazel, C. (2009) *Acc. Chem. Res.*, **42**, 367–378; (h) Tietze, L.F., Spiegl, D.A., and Brazel, C.C. (2009) in *Experiments in Green and Sustainable Chemistry* (ed. H.W. Roesky and D.K. Kennepohl), Wiley-VCH Verlag GmbH, Weinheim, pp. 158–167.
2. (a) Tomioka, K. and Nagaoka, Y. (1999) in *Comprehensive Asymmetric Catalysis I-III*, vol. 3 (eds E.N. Jacobsen, A. Pfaltz, and H. Yamamoto), Springer, Berlin, pp. 1105–1120; (b) Guo, H.-C. and Ma, J.-A. (2006) *Angew. Chem., Int. Ed.*, **45**, 354–366.
3. Kataoka, T. and Kinoshita, H. (2005) *Eur. J. Org. Chem.*, 45–58.
4. (a) Palomo, C., Oiarbide, M., and Garcia, J.M. (2004) *Chem. Soc. Rev.*, **33**, 65–75; (b) Machajewski, T.D., Wong, C.-H., and Lerner, R.A. (2000) *Angew. Chem., Int. Ed.*, **39**, 1352–1374.
5. Yamada, K., Arai, T., Sasai, H., and Shibasaki, M. (1998) *J. Org. Chem.*, **63**, 3666–3672.
6. Shibasaki, M., Sasai, H., and Arai, T. (1997) *Angew. Chem., Int. Ed. Engl.*, **36**, 1237–1256, and references cited therein.

7. Arai, T., Sasai, H., Aoe, K., Okamura, K., Date, T., and Shibasaki, M. (1996) *Angew. Chem., Int. Ed. Engl.*, **35**, 104–106.
8. (a) Thaler, T. and Knochel, P. (2009) *Angew. Chem., Int. Ed.*, **48**, 645–648; (b) Alexakis, A., Bäckvall, J.E., Krause, N., Pàmies, O., and Diéguez, M. (2008) *Chem. Rev.*, **108**, 2796–2823; (c) Breit, B. and Schmidt, Y. (2008) *Chem. Rev.*, **108**, 2928–2951.
9. Arnold, L.A., Naasz, R., Minnaard, A.J., and Feringa, B.L. (2002) *J. Org. Chem.*, **67**, 7244–7254.
10. Feringa, B.L. (2000) *Acc. Chem. Res.*, **33**, 346–353.
11. Arnold, L.A., Naasz, R., Minnaard, A.J., and Feringa, B.L. (2001) *J. Am. Chem. Soc.*, **123**, 5841–5842.
12. Agapiou, K., Cauble, D.F., and Krische, M.J. (2004) *J. Am. Chem. Soc.*, **126**, 4528–4529.
13. Xu, Y.-J., Liu, Q.-Z., and Dong, L. (2007) *Synlett*, 273–277.
14. Oisaki, K., Zhao, D., Kanai, M., and Shibasaki, M. (2007) *J. Am. Chem. Soc.*, **129**, 7439–7443.
15. (a) Lipshutz, B. (2009) *Synlett*, 509–524; (b) Deutsch, C., Krause, N., and Lipshutz, B.H. (2008) *Chem. Rev.*, **108**, 2916–2927.
16. Mahoney, W.S., Brestensky, D.M., and Stryker, J.M. (1988) *J. Am. Chem. Soc.*, **110**, 291–293.
17. Rendler, S. and Oestreich, M. (2007) *Angew. Chem., Int. Ed.*, **46**, 498–504.
18. Lam, H.W. and Joensuu, P.M. (2005) *Org. Lett.*, **7**, 4225–4228.
19. Deschamp, J., Chuzel, O., Hannedouche, J., and Riant, O. (2006) *Angew. Chem., Int. Ed.*, **45**, 1292–1297.
20. Zhao, D., Oisaki, K., Kanai, M., and Shibasaki, M. (2006) *Tetrahedron Lett.*, **47**, 1403–1407.
21. Lipshutz, B.H., Amorelli, B., and Unger, J.B. (2008) *J. Am. Chem. Soc.*, **130**, 14378–14379.
22. Hayashi, T. and Yamasaki, K. (2003) *Chem. Rev.*, **103**, 2829–2844.
23. Taylor, S.J., Duffey, M.O., and Morken, J.P. (2000) *J. Am. Chem. Soc.*, **122**, 4528–4529.
24. Russell, A.E., Fuller, N.O., Taylor, S.J., Aurriset, P., and Morken, J.P. (2004) *Org. Lett.*, **6**, 2309–2312.
25. Nicolaou, K.C., Tang, W., Dagneau, P., and Faraoni, R. (2005) *Angew. Chem., Int. Ed.*, **44**, 3874–3879.
26. Cauble, D.F., Gipson, J.D., and Krische, M.J. (2003) *J. Am. Chem. Soc.*, **125**, 1110–1111.
27. Hayashi, T., Takahashi, M., Takaya, Y., and Ogasawara, M. (2002) *J. Am. Chem. Soc.*, **124**, 5052–5058.
28. Bocknack, B.M., Wang, L.-C., and Krische, M.J. (2004) *Proc. Natl Acad. Sci. USA*, **101**, 5421–5424.
29. Zhao, C.-X., Duffey, M.O., Taylor, S.J., and Morken, J.P. (2001) *Org. Lett.*, **3**, 1829–1831.
30. van Lingen, H.L., Zhuang, W., Hansen, T., Rutjes, F.P.J.T., and Jørgensen, K.A. (2003) *Org. Biomol. Chem.*, **1**, 1953–1958.
31. Lyle, M.P.A., Draper, N.D., and Wilson, P.D. (2005) *Org. Lett.*, **7**, 901–904.
32. (a) Naasz, R., Arnold, L.A., Pineschi, M., Keller, E., and Feringa, B.L. (1999) *J. Am. Chem. Soc.*, **121**, 1104–1105; (b) Sibi, M.P. and Chen, J. (2001) *J. Am. Chem. Soc.*, **123**, 9472–9473; (c) Hayashi, T., Tokunaga, N., Yoshida, K., and Han, J.W. (2002) *J. Am. Chem. Soc.*, **124**, 12102–12103; (d) Shintani, R., Tokunaga, N., Doi, H., and Hayashi, T. (2004) *J. Am. Chem. Soc.*, **126**, 6240–6241; (e) Rathgeb, X., March, S., and Alexakis, A. (2006) *J. Org. Chem.*, **71**, 5737–5742; (f) Šebesta, R., Pizzuti, M.G., Minnaard, A.J., and Feringa, B.L. (2007) *Adv. Synth. Catal.*, **349**, 1931–1937.
33. Dijk, E.W., Panella, L., Pinho, P., Naasz, R., Meetsma, A., Minnaard, A.J., and Feringa, B.L. (2004) *Tetrahedron*, **60**, 9687–9693.
34. Alexakis, A., Trevitt, G.P., and Bernardinelli, G. (2001) *J. Am. Chem. Soc.*, **123**, 4358–4359.
35. Brown, M.K. and Hoveyda, A.H. (2008) *J. Am. Chem. Soc.*, **130**, 12904–12906.
36. (a) Brown, M.K., May, T.L., Baxter, C.A., and Hoveyda, A.H. (2007) *Angew. Chem., Int. Ed.*, **46**, 1097–1100; (b) May, T.L.,

Brown, M.K., and Hoveyda, A.H. (2008) *Angew. Chem., Int. Ed.*, **47**, 7358–7362.
37. Knopff, O. and Alexakis, A. (2002) *Org. Lett.*, **4**, 3835–3837.
38. Alexakis, A. and March, S. (2002) *J. Org. Chem.*, **67**, 8753–8757.
39. Doyle, A.G. and Jacobsen, E.N. (2007) *Chem. Rev.*, **107**, 5713–5743.
40. (a) Mukherjee, S., Yang, J.W., Hoffmann, S., and List, B. (2007) *Chem. Rev.*, **107**, 5471–5569; (b) Erkkil, A., Majander, I., and Pihko, P.M. (2007) *Chem. Rev.*, **107**, 5471–5569.
41. Kaneko, S., Yoshino, T., Katoh, T., and Terashima, S. (1998) *Tetrahedron Lett.*, **54**, 5471–5484.
42. (a) Corn, R.S.E., Lovell, A.V., Karady, S., and Weinstock, L.M. (1986) *J. Org. Chem.*, **51**, 4710–4711; (b) Takagi, K., Katayama, H., and Yamada, H. (1988) *J. Org. Chem.*, **53**, 1157–1161.
43. Betancort, J.M., Sakthivel, K., Thayumanavan, R., and Barbas, C.F.III (2001) *Tetrahedron. Lett.*, **42**, 4441–4444.
44. Yamamoto, Y., Momiyama, N., and Yamamoto, H. (2004) *J. Am. Chem. Soc.*, **126**, 5962–5963.
45. Li, H., Wang, J., Xie, H., Zu, L., Jiang, W., Duesler, E.N., and Wang, W. (2007) *Org. Lett.*, **9**, 965–968.
46. Sundén, H., Ibrahem, I., Eriksson, L., and Córdova, A. (2005) *Angew. Chem., Int. Ed.*, **44**, 4877–4880.
47. Carlone, A., Marigo, M., North, C., Landa, A., and Jørgensen, K.A. (2006) *Chem. Commun.*, 4928–4930.
48. Zhou, J. and List, B. (2007) *J. Am. Chem. Soc.*, **129**, 7498–7499.
49. Wang, J., Yu, F., Zhang, X., and Ma, D. (2008) *Org. Lett.*, **10**, 2561–2564.
50. Penon, O., Carlone, A., Mazzanti, A., Locatelli, M., Sambri, L., Bartoli, G., and Melchiorre, P. (2008) *Chem. Eur. J.*, **14**, 4788–4791.
51. Valero, G., Schimer, J., Cisarova, I., Vesely, J., Moyano, A., and Rios, R. (2009) *Tetrahedron Lett.*, **50**, 1943–1946.
52. Halland, N., Aburel, P.S., and Jørgensen, K.A. (2004) *Angew. Chem., Int. Ed.*, **43**, 1272–1277.
53. (a) Eder, U., Sauer, G., and Wiechert, R. (1971) *Angew. Chem., Int. Ed. Engl.*, **10**, 496–497; (b) Hajos, Z.G. and Parrish, D.R. (1974) *J. Org. Chem.*, **39**, 1615–1621; see also: (c) Arai, T., Sasai, H., Aoe, K.-I., Okamura, K., Date, T., and Shibasaki, M. (1996) *Angew. Chem., Int. Ed. Engl.*, **35**, 104–106; (d) Bui, T. and Barbas, C.F. III (2000) *Tetrahedron Lett.*, **41**, 6951–6954.
54. (a) Marigo, M., Schulte, T., Franzén, J., and Jørgensen, K.A. (2005) *J. Am. Chem. Soc.*, **127**, 15710–15711; (b) Huang, Y., Walji, A.M., Larsen, C.H., and MacMillan, D.W.C. (2005) *J. Am. Chem. Soc.*, **127**, 15051–15053.
55. Simmons, B., Walji, A.B., and MacMillan, D.W.C. (2009) *Angew. Chem., Int. Ed.*, **48**, 4349–4353.
56. (a) Carle, J.S. and Chrisophersen, C. (1981) *J. Org. Chem.*, **46**, 3440–3443; (b) Carle, J.S. and Chrisophersen, C. (1980) *J. Org. Chem.*, **45**, 1586–1589; (c) Carle, J.S. and Chrisophersen, C. (1979) *J. Am. Chem. Soc.*, **101**, 4012–4013.
57. Austin, J.F., Kim, S.-G., Sinz, C.J., Xiao, W.-J., and MacMillan, D.W.C. (2004) *Proc. Natl Acad. Sci. USA*, **101**, 5482–5487.
58. Franzén, J. and Fisher, A. (2009) *Angew. Chem., Int. Ed.*, **48**, 787–791.
59. Franzén, H., Rios, R., Ibrahem, I., Zhao, G.-L., Eriksson, L., and Córdova, A. (2007) *Adv. Synth. Catal.*, **349**, 827–832.
60. Sundén, H., Ibrahem, I., Zhao, G.-L., Eriksson, L., and Córdova, A. (2007) *Chem. Eur. J.*, **13**, 574–581.
61. Rueping, M., Sugiono, E., and Merino, E. (2008) *Angew. Chem., Int. Ed.*, **47**, 3046–3049.
62. Rios, R., Sundén, H., Ibrahem, I., and Córdova, A. (2007) *Tetrahedron Lett.*, **48**, 2181–2184.
63. Rueping, M., Kuenkel, A., Tato, F., and Bats, J.W. (2009) *Angew. Chem., Int. Ed.*, **48**, 3699–3702.
64. Enders, D., Hüttl, M.R.M., Grondal, C., and Raabe, G. (2006) *Nature*, **441**, 861–863.
65. Ibrahem, I., Rios, R., Vesely, J., Zhao, G.-L., and Córdova, A. (2007) *Chem. Commun.*, **13**, 849–851.

66. Jiang, H., Elsner, P., Jensen, K.L., Falcicchio, A., Marcos, V., and Jørgensen, K.A. (2009) *Angew. Chem., Int. Ed.*, **48**, 6844–6848.
67. Lygo, B., Allbutt, B., and James, S.R. (2003) *Tetrahedron Lett.*, **44**, 5629–5632.
68. Guerin, D.J. and Miller, S.J. (2002) *J. Am. Chem. Soc.*, **124**, 2134–2136.
69. Guo, R., Morris, R.H., and Song, D. (2005) *J. Am. Chem. Soc.*, **127**, 516–517.
70. Noyori, R., Yamakawa, M., and Hashiguchi, S. (2001) *J. Org. Chem.*, **66**, 7931–7944.
71. Källström, S., Jagt, R.B.C., Sillanpää, R., Feringa, B.L., Minnaard, A.J., and Leino, R. (2006) *Eur. J. Org. Chem.*, 3826–3822.
72. Lee, S.J. and Beak, P. (2006) *J. Am. Chem. Soc.*, **128**, 2178–2179.
73. Matsuda, T., Shigeno, M., and Murakami, M. (2007) *J. Am. Chem. Soc.*, **129**, 12086–12087.

9
Asymmetric Epoxidations of α,β-Unsaturated Carbonyl Compounds
Alessandra Lattanzi

9.1
Introduction

The asymmetric epoxidation of alkenes is a fundamental process in organic synthesis, and unquestionably the most exploited approach to synthesize epoxides [1]. Alternative useful strategies to access these molecules concern the enantioselective alkylidenation of carbonyl compounds by ylides [2] and the Darzens condensation [3]. Enantioenriched epoxides are important targets of biological and pharmaceutical interest. Moreover, their high synthetic potential as intermediates is exemplified by the stereoselective ring opening with nucleophiles, which affords a variety of compounds with two contiguous chiral centers, the absolute configuration of which can be controlled [4].

During the past few decades, metal-based electrophilic asymmetric methodologies of epoxidation have been successfully developed and found widespread application in multistep synthesis; these include: the Sharpless' epoxidation of allylic alcohols mediated by the Ti/tartrate system [5]; and the Jacobsen's and Katsuki's epoxidation of unfunctionalized alkenes by using Mn- and Ti-/Salen complexes [6]. More recently, chiral Ru–Pt and Fe–Pt complexes have proved to be effective in the enantioselective epoxidation of disubstituted aromatic and challenging terminal unfunctionalized alkenes [7]. Impressive results have been achieved by the groups of Yang [8] and Shi [9] for the electrophilic organocatalyzed asymmetric epoxidation of a great variety of olefins such as trisubstituted, disubstituted *trans*-alkenes, terminal alkenes, dienes, and enynes (to cite the most relevant examples), and by the *in situ* generation of dioxiranes derived from chiral ketones. Within the context of asymmetric epoxidation catalyzed by organic promoters, notable achievements have also been attained by using chiral oxaziridines [10], oxaziridinium salts [11], amines [12], and peptide-based peracids [13].

The epoxidation of α,β-unsaturated carbonyl compounds leads to the formation of another valuable class of functionalized epoxides [14], the carbonyl and epoxide groups of which can be selectively manipulated to provide enantioenriched products such as allylic epoxy alcohols, α- or β-hydroxy ketones, and α,β-epoxy esters [15]. The general approach for the epoxidation of electron-poor

Catalytic Asymmetric Conjugate Reactions. Edited by Armando Córdova
Copyright © 2010 WILEY-VCH Verlag GmbH & Co. KGaA, Weinheim
ISBN: 978-3-527-32411-8

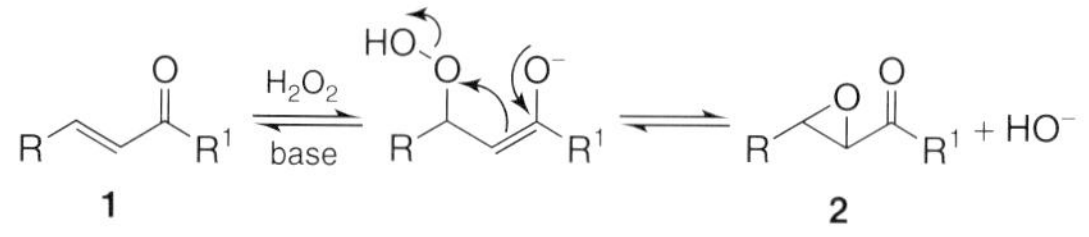

H2O2
base
HO
O
O−
R
R1
+ HO−
1
2

Scheme 9.1 The Weitz–Scheffer epoxidation.

alkenes requires the formation of nucleophilic oxidative species. The first example of the epoxidation of α,β-unsaturated carbonyl compounds was reported by Weitz and Scheffer in 1921 (Scheme 9.1) [16].

This two-step process first involves the conjugate addition to the α,β-unsaturated ketone **1** of the hydroperoxide anion generated after deprotonation of hydrogen peroxide in an alkaline medium. The anionic intermediate then gives rise to the epoxy ketone **2** by ring closure. Although control of the alkene geometry can, in principle, be problematic, highly diastereoselective and enantioselective epoxidation reactions of electron-poor alkenes have been developed. The metal-based and organocatalyzed methodologies for the asymmetric epoxidation of α,β-unsaturated carbonyl compounds reported over the past two decades will be described in this chapter. Illustrations of the scope and limitations, mechanistic aspects and – whenever possible – the applications of different systems, will be included in an attempt to provide an update for both academic and industrial chemists.

9.2 Metal-Catalyzed Epoxidations

9.2.1 Epoxidation of α,β-Unsaturated Ketones Mediated by Chirally Modified Zn- and Mg-Alkyl Peroxides

During the 1990s, a wide variety of methodologies for the enantioselective epoxidation of electron-poor alkenes employing chiral ligand–metal peroxide systems was reported. For example, Enders discovered a simple and powerful system for the epoxidation of *trans*-α,β-unsaturated ketones under stoichiometric amounts of Et_2Zn in the presence of molecular oxygen as a benign oxidant and easily available (1*R*,2*R*)-*N*-methylpseudoephedrine **3** as ligand at 0 °C in toluene (Scheme 9.2) [17].

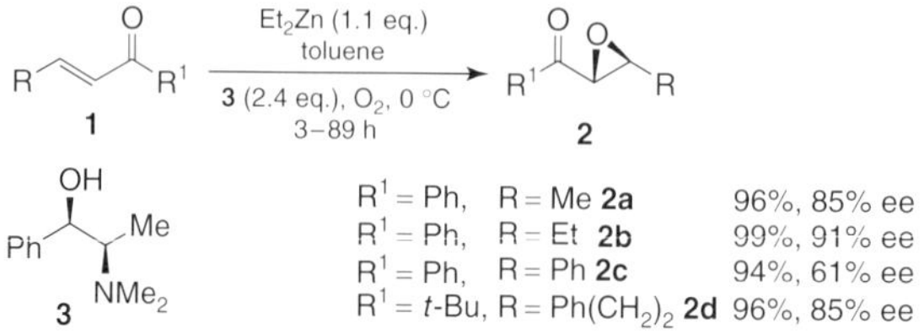

Et2Zn (1.1 eq.)
toluene
3 (2.4 eq.), O2, 0 °C
3–89 h
R
R1
1
2
OH
Me
Ph
NMe2
3
R1 = Ph, R = Me 2a 96%, 85% ee
R1 = Ph, R = Et 2b 99%, 91% ee
R1 = Ph, R = Ph 2c 94%, 61% ee
R1 = t-Bu, R = Ph(CH2)2 2d 96%, 85% ee

Scheme 9.2 Asymmetric epoxidation of *trans*-enones with O_2/Et_2Zn/ligand **3** system.

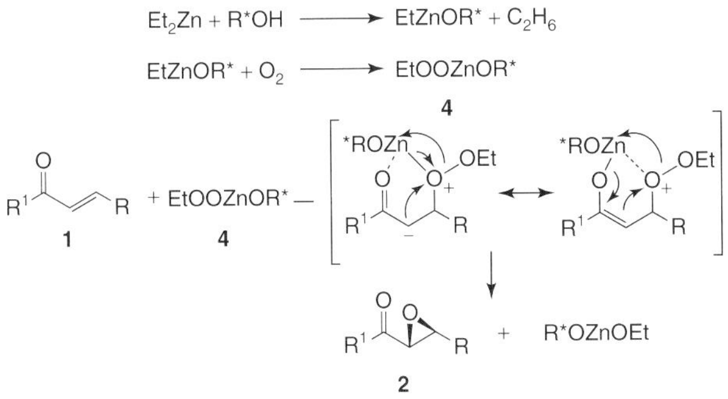
$Et_2Zn + R^*OH \longrightarrow EtZnOR^* + C_2H_6$
$EtZnOR^* + O_2 \longrightarrow EtOOZnOR^*$
4
*ROZn
OEt
O
R1
R
1
+ EtOOZnOR*
4
R*OZnOEt
2

Figure 9.1 Postulated mechanism for asymmetric epoxidation.

In this way, aryl alkyl, dialkyl ketones, and *trans*-chalcone could be almost quantitatively transformed in a highly diastereoselective fashion (de >99/1) to the *trans*-configured epoxy ketones **2** in good to high ee-values. The mechanism proposed for the epoxidation is illustrated in Figure 9.1. In this case, ethyl zinc alkoxide is formed from diethylzinc and the chiral alcohol with the evolution of ethane. The insertion of O_2 into the residual Zn-C bond then affords the chiral alkoxy(ethylperoxy)zinc **4**, which is the real epoxidizing species. This represents the first example of the generation of a chiral zinc alkyl peroxide, where the peroxide moiety is nucleophilic in nature. In the epoxidation process, the carbonyl oxygen coordinates the zinc atom in **4**, and a concurrent stereoselective 1,4-addition of the ethylperoxy anion occurs to form the intermediate enolate. Cyclization of the intermediate produces epoxide **2** and a zinc dialkoxide which, after hydrolysis, affords ligand **3** quantitatively. The same system has been successfully used for the epoxidation of conformationally fixed β-alkylidene-α-tetralones **5** (Scheme 9.3).

The enantioselectivity was increased according to size of the β-substituent. From these findings it has been proposed that, during the epoxidation, a high stereocontrol is assured if the α,β-unsaturated ketones can adopt an *s-cis* conformation, which was also supported by the unreactive nature of 2-cyclohexenones toward this epoxidizing system. Interestingly, *trans*-nitroalkenes could be epoxidized under the same conditions, affording the epoxides in moderate yields (47–64%) and up to 82% ee. It is worth noting here that this methodology represents the most efficient protocol ever reported for the asymmetric epoxidation of *trans*-nitroalkenes [18].

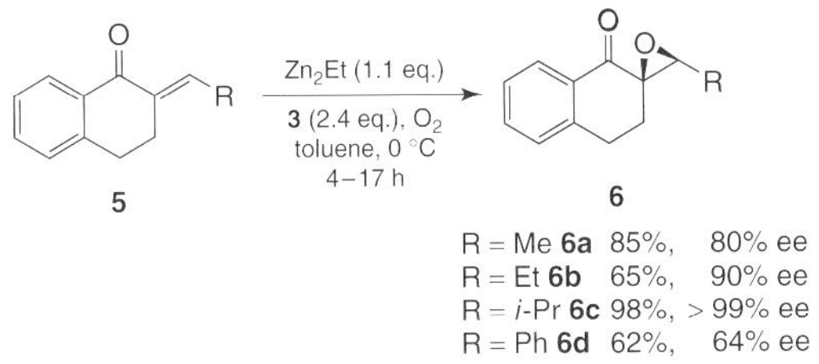
Zn_2Et (1.1 eq.)
3 (2.4 eq.), O_2
toluene, 0 °C
4–17 h
5
6
R = Me **6a** 85%, 80% ee
R = Et **6b** 65%, 90% ee
R = *i*-Pr **6c** 98%, > 99% ee
R = Ph **6d** 62%, 64% ee

Scheme 9.3 Asymmetric epoxidation of β-alkylidene-α-tetralones.

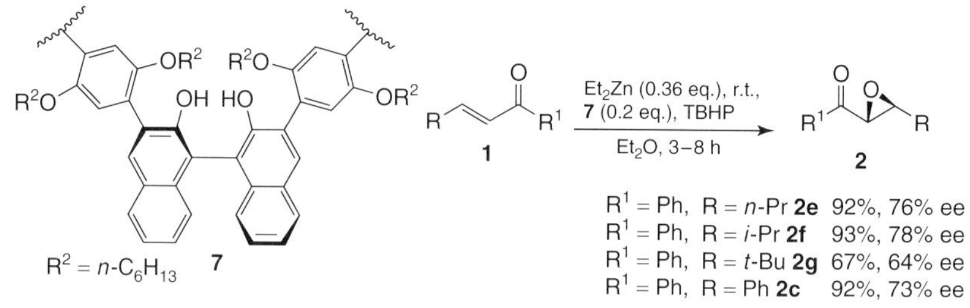

Scheme 9.4 Asymmetric epoxidation of *trans*-enones with TBHP/Et$_2$Zn/ligand **7**.

R^1 = Ph, R = n-Pr **2e** 92%, 76% ee
R^1 = Ph, R = i-Pr **2f** 93%, 78% ee
R^1 = Ph, R = t-Bu **2g** 67%, 64% ee
R^1 = Ph, R = Ph **2c** 92%, 73% ee

Pu developed a modified version of the Enders' system by using catalytic amounts of polymeric binaphthyl ligand **7** and Et$_2$Zn in the presence of *t*-butyl hydroperoxide (TBHP) as the oxygen source (Scheme 9.4) [19]. In this case, β-alkyl phenyl enones and chalcones were converted to the epoxides in high yield and moderate to good enantioselectivity. The mechanism postulated is illustrated in Figure 9.2, where the first-formed zinc alkoxide **8** reacts with TBHP to generate the active chiral epoxidizing species **9**. A conjugate addition to the enone of the peroxide anion occurs in analogy with Enders' proposal to give the epoxide. A ligand exchange between zinc alkoxide **10** and TBHP then regenerates the active peroxide species **9**.

In 1997, Jackson reported the enantioselective epoxidation of *trans*-chalcones using 10 mol% loadings of dibutylmagnesium, (+)-diethyl tartrate (DET) with TBHP as the oxidant in toluene/tetrahydrofuran (THF) mixture at room temperature (Scheme 9.5) [20]. In this case, the epoxides **2** were recovered in moderate yields and good to high ee-values. A modified version of this system was later developed for the enantioselective epoxidation of more challenging aliphatic enones, which are prone to enolization. The presence of powered activated molecular sieves was found to be beneficial for the *in situ* preparation of the catalytic system. Following the screening of different sterically demanding tartrate esters, commercially available (+)-di-*tert*-butyl tartrate (D*t*BT) was found to be the optimum ligand. A range of dialkyl enones afforded the epoxides in satisfactory yield and good to high ee-values (Scheme 9.6).

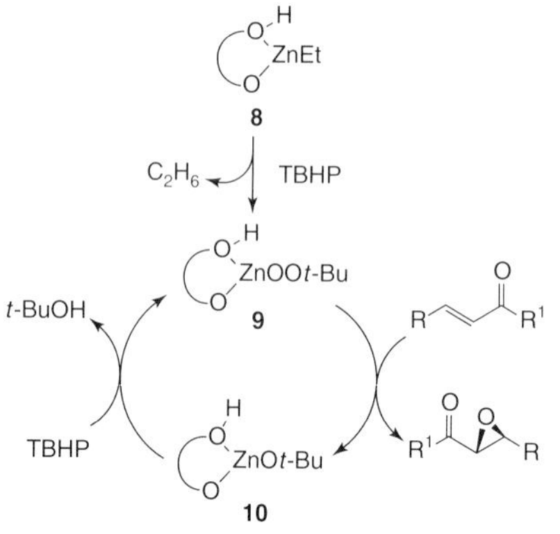

Figure 9.2 Proposed mechanism for asymmetric epoxidation.

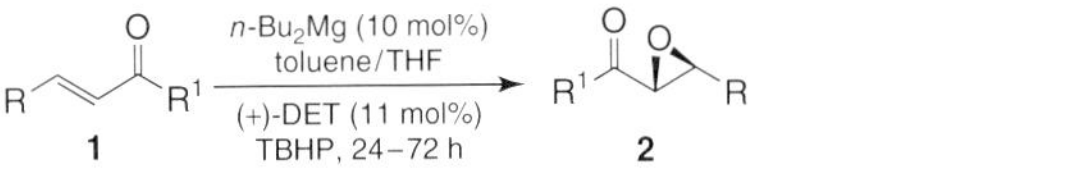

Scheme 9.5 Asymmetric epoxidation of trans-chalcones with TBHP/n-Bu$_2$Mg/(+)-DET system.

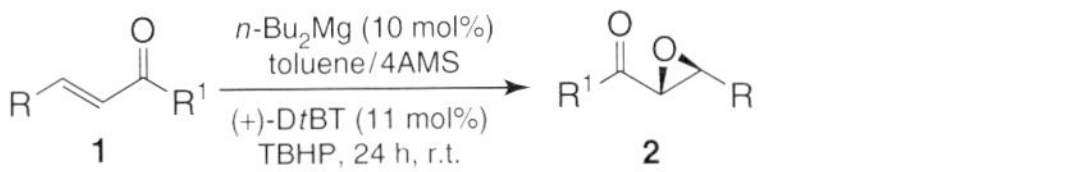

R^1 = Ph, R = Ph **2c** 61%, 94% ee
R^1 = Ph, R = p-ClC$_6$H$_4$ **2h** 54%, 81% ee
R^1 = Ph, R = p-MeC$_6$H$_4$ **2i** 36%, 84% ee
R^1 = β-naphthyl, R = Ph, **2j** 46%, 92% ee

R^1 = Me, R = n-C$_4$H$_9$ **2k** 53%, 91% ee
R^1 = Me, R = n-C$_6$H$_{13}$ **2l** 63%, 92% ee
R^1 = Et, R = Me **2m** 67%, 71% ee
R^1 = Me, R = 3,5-Br$_2$C$_6$H$_3$ **2n** 46%, 92% ee

Scheme 9.6 Asymmetric epoxidation of trans-dialkyl enones.

Bulkier tartrate esters led to a magnesium tartrate derivative which proved to be more active in the epoxidation, most likely because of their reduced attitude to be hydrolyzed and hence to inactivation [21]. Further studies by the same group showed the importance of additives to develop a more efficient protocol for the epoxidation. The presence of ethanol (0.48 equiv.)/molecular sieves helped to reduce the amount of either n-Bu$_2$Mg (0.06 equiv.) and DtBT (0.08 equiv.) while preserving the level of asymmetric induction (up to 96% ee). The addition of a well-established amount of water in the presence of a molecular sieve afforded results similar to those observed when using ethanol as additive [22]. From a mechanistic point of view, a linear dependence of the ee of the product on the ee of the tartrate ester was observed, which was consistent with a monomeric active species. The catalytic cycle postulated for the epoxidation is illustrated in Figure 9.3. It is interesting to note that this nucleophilic epoxidation reminds

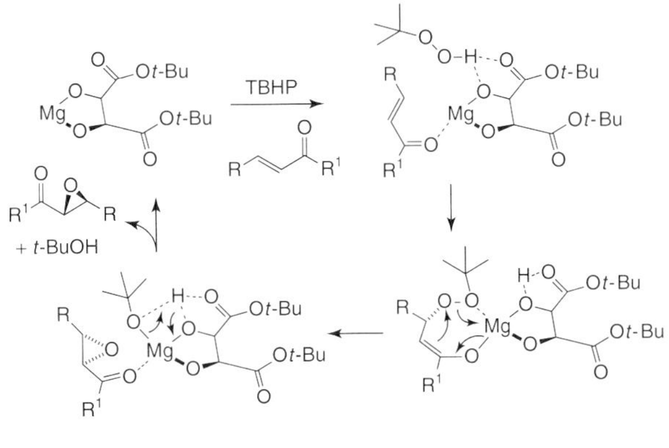

Figure 9.3 Proposed catalytic cycle for epoxidation.

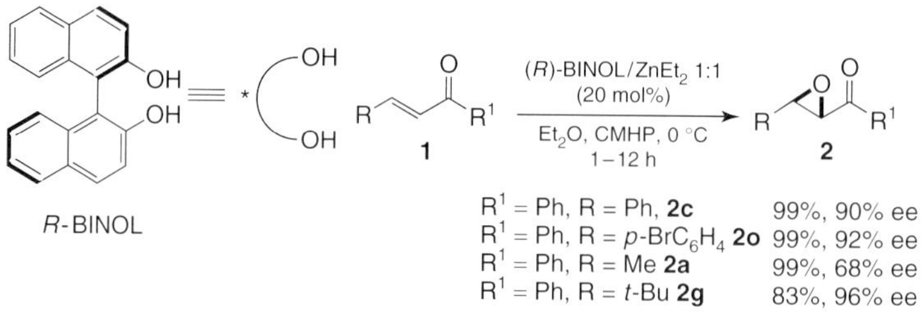

R-BINOL
(R)-BINOL/ZnEt2 1:1
(20 mol%)
Et2O, CMHP, 0 °C
1–12 h
1
2
R1 = Ph, R = Ph, 2c 99%, 90% ee
R1 = Ph, R = p-BrC6H4 2o 99%, 92% ee
R1 = Ph, R = Me 2a 99%, 68% ee
R1 = Ph, R = t-Bu 2g 83%, 96% ee

Scheme 9.7 Asymmetric epoxidation of *trans*-enones with CMHP/$ZnEt_2$/(*R*)-BINOL system.

and complements the Sharpless' system for the electrophilic epoxidation of allylic alcohols in the use of tartrate esters as ligands, TBHP as oxygen donor and activated molecular sieves as additives.

More recently, Dötz reported a simple system for the asymmetric epoxidation of *trans*-chalcones and β-alkyl phenyl enones. The use of Et_2Zn, 1,1′-bis(2-naphthol) [(*R*)-BINOL] in the presence of cumyl hydroperoxide (CMHP) as oxidant gave the *trans*-configured epoxides in excellent yield and up to 96% ee (Scheme 9.7) [23]. Nevertheless, the efficiency of this system was found to be heavily dependent on the substitution pattern and steric requirements of the enone. Chalcones with electron-donating groups in the β-phenyl ring were not epoxidized, while electron-poor substituents greatly enhanced their reactivity and the epoxides were obtained in excellent yield and high enantioselectivity. Moreover, the epoxidation of enones with β-alkyl substituents of increasing steric hindrance proceeded with a higher control of the asymmetric induction. The authors proposed a different mechanistic cycle for the epoxidation with respect to the systems of Enders and Pu. According to Dözt's proposal, an *in situ* zinc (*R*)-BINOLate complex would be formed, which would serve exclusively as a Lewis acid. This provided activation of the enone for the nucleophilic attack by complexation of the carbonyl functionality, while affording the *si*-face shielding of the enone (Figure 9.4). In this case, the alkyl hydroperoxide was supposed to give rise to an external attack at the β-position of the complexed enone.

Very recently, a heterogeneous catalyst for this epoxidizing system has been reported by Ding, based on a new strategy – that is, a "self-supporting" approach [24] for the immobilization of homogeneous catalysts through the assembly of chiral multitopic ligands and metal ions, without the use of any support. The *in situ*-prepared chiral self-supported BINOL–Zn catalyst with CMHP, furnished

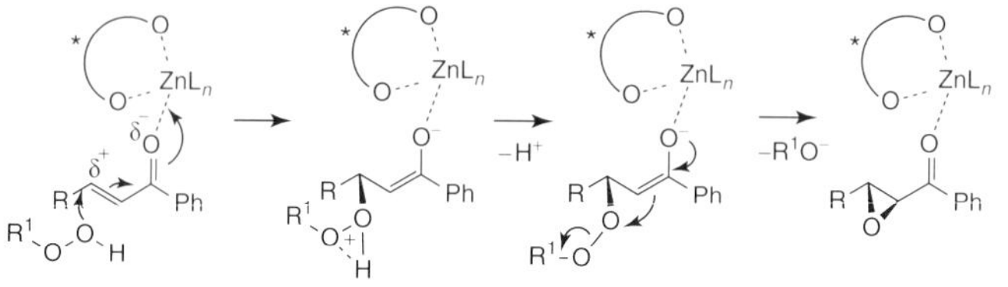

ZnLn
Ph
−H+
−R1O−

Figure 9.4 Postulated catalytic cycle for epoxidation.

9.2.2
Epoxidation of α,β-Unsaturated Ketones, Amides, and Esters Mediated by Lanthanide–BINOL Systems

During the past decade, Shibasaki has developed the most efficient catalytic systems for the highly enantioselective epoxidation of *trans-* and *cis-*enones, and for poorly reactive *trans*-unsaturated amides and esters. Based on their successful achievements [26] of two types of catalyst – namely the heterobimetallic complex $Na_3[La(binol)_3]$ (LSB)-**11** and the alkali-metal-free-lanthanum-BINOL complex **12** in the Michael addition of malonate esters to enone – Shibasaki *et al.* first checked these complexes in the epoxidation of *trans*-chalcone **1c**, using TBHP and CMHP as the oxidants [27]. The epoxidation promoted by 10 mol% of catalysts **11** and **12** afforded a better result when using catalyst **12**, prepared by mixing $La(i-PrO)_3$ and (*R*)-BINOL in the presence of activated 4 Å molecular sieves and CMHP at room temperature in THF (Scheme 9.8).

trans-Enones could be efficiently epoxidized with excellent enantioselectivity when using only 5 mol% of catalyst **12** and CMHP, or the alkali-metal-free-lanthanium complex **13** with ligand **14**, which improved the enantiocontrol of the process (Scheme 9.9).

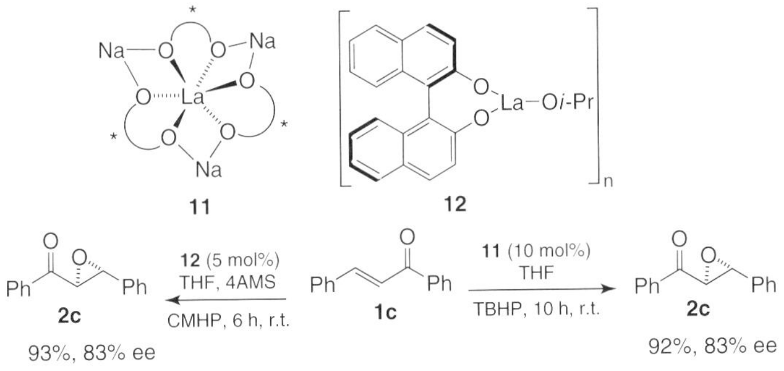

Scheme 9.8 Asymmetric epoxidation of *trans*-chalcone **1c** with lanthanoids-BINOL complexes/alkyl hydroperoxides.

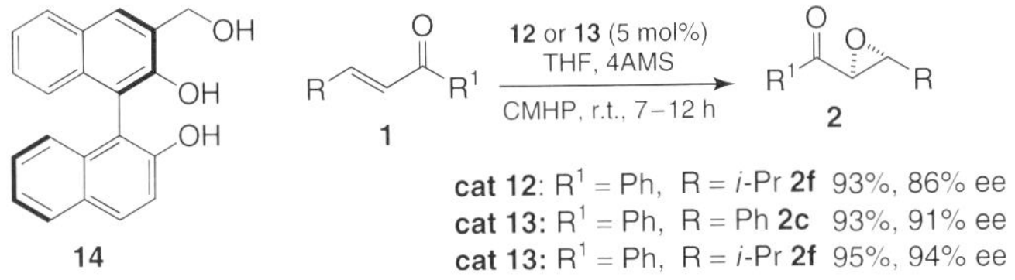

cat **12**: R¹ = Ph, R = *i*-Pr **2f** 93%, 86% ee
cat **13**: R¹ = Ph, R = Ph **2c** 93%, 91% ee
cat **13**: R¹ = Ph, R = *i*-Pr **2f** 95%, 94% ee

Scheme 9.9 Asymmetric epoxidation of *trans*-enones with lanthanoids-BINOL complexes/CMHP system.

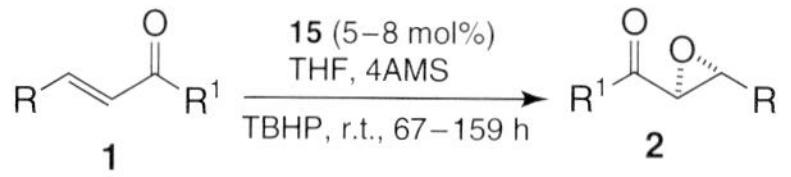

Scheme 9.10 Asymmetric epoxidation of aliphatic trans-enones with Yb-complex **15**/TBHP system.

R¹ = Me, R = Ph **2p** — 83%, 94% ee
R¹ = i-Pr, R = Ph **2q** — 55%, 88% ee
R¹ = Me, R = (CH$_2$)$_2$Ph **2r** — 91%, 88% ee
R¹ = Me, R = n-C$_5$H$_{11}$ **2s** — 71%, 91% ee

Conditions: **15** (5–8 mol%), THF, 4AMS, TBHP, r.t., 67–159 h; substrate **1** → **2**.

Ytterbium complex **15**, generated from Yb(Oi-Pr)$_3$ and ligand **14** in THF with TBHP and a 4 Å molecular sieve (4AMS), was found to be a superior catalyst for the epoxidation of aliphatic enones, affording the products in good yield and high enantioselectivity (Scheme 9.10). The differences in ionic radius and Lewis acidities between lanthanum and ytterbium were invoked to be responsible for the unconformity of the enantioselectivities observed with structurally different enones. Interestingly, the ytterbium complex **15**/TBHP system proved to be effective in the diastereo- and enantioselective epoxidation of cis-enones **16** to the corresponding cis-epoxy ketones **17**, which were formed in excellent enantiomeric excess (Scheme 9.11) [28].

A high control of the diastereoselectivity was observed, since trans-epoxy ketones were formed in negligible amounts for all of the substrates reported in Scheme 9.11, except for compound **17d**, the trans-epoxy ketone of which was additionally recovered in 19% yield and 58% ee. It should be noted that this system represents the only methodology reported to date for the asymmetric epoxidation of acyclic cis-enones.

Inanaga and coworkers improved the catalytic efficiency of this system by adding triphenylphosphine oxide in larger amounts with respect to the lanthanum metal, thus reducing the reaction time for the epoxidation [29]. Shibasaki showed that the addition of catalytic loadings of (Ph)$_3$AsO would enable a further improvement in the efficiency of the epoxidizing system, reducing the reaction time from hours to minutes [30]. The epoxidation of trans-chalcone was completed in only 3 min, affording the product in 95% yield and 97% ee when using 10 mol% of (Ph)$_3$AsO and 10 mol% of complex **12** at room temperature. A mechanistic cycle for the epoxidation under these conditions was proposed as illustrated in Figure 9.5.

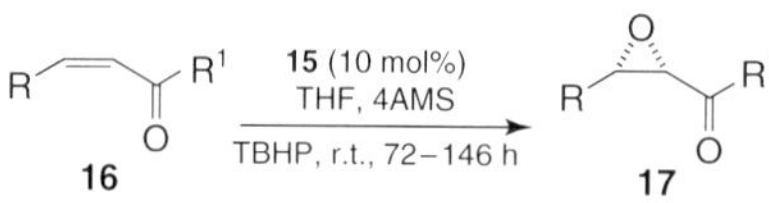

R¹ = Me, R = C$_5$H$_{11}$ **17a** — 74%, 94% ee
R¹ = (CH$_2$)$_2$Ph, R = C$_3$H$_7$ **17b** — 78%, 93% ee
R¹ = C$_3$H$_7$, R = C$_5$H$_{11}$ **17c** — 80%, 96% ee
R¹ = Ph, R = C$_3$H$_7$ **17d** — 51%, 88% ee

Conditions: **15** (10 mol%), THF, 4AMS, TBHP, r.t., 72–146 h; substrate **16** → **17**.

Scheme 9.11 Asymmetric epoxidation of cis-enones with Yb-complex **15**/TBHP system.

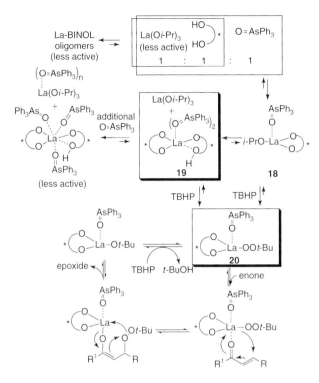

Figure 9.5 Proposed mechanism for epoxidation.

With the aid of NMR, mass spectrometry and X-ray analyses, the real catalytic species **18** was proposed to be the major component in the solution when using La(Oi-Pr)$_3$/(R)-BINOL/(Ph)$_3$AsO in 1:1:1 ratio.

The lanthanum alkoxide moiety in complex **18** functions as a Brønsted base to generate, *in situ*, a lanthanum-peroxide **20** which, at the same time, can activate the enone as a Lewis acid. A 1,4-addition of the La-peroxide to the enone then follows, according to the general mechanism of the Weitz–Scheffer epoxidation (see Scheme 9.1). An excess of La(Oi-Pr)$_3$ in solution would favor the equilibrium shift from complex **19** to the catalytic species **20**, thus accelerating the epoxidation.

Qian and de Vries studied the influence of various substituents in the BINOL ligand on the effectiveness of the epoxidizing system, and found that 6,6'-diphenyl-substituted (S)-BINOL with 5 mol% of Ga(Oi-Pr)$_3$ and CMHP system afforded chalcone epoxides in up to 95% ee at room temperature [31]. Sasai developed a modified system for the epoxidation using different polymeric supported-BINOL ligands with Yb(Oi-Pr)$_3$ and La(Oi-Pr)$_3$ metal complexes [32]. These heterogeneous catalysts provided the epoxides of *trans*-chalcone and benzalacetone in high yield, up to 89% ee, and could be recovered and reused for the epoxidation, while maintaining efficiency. Ding reported an effective heterogeneous version of the Shibasaki catalyst, based on the chiral self-supporting strategy. The epoxidation of different enones with the optimized La-based heterogeneous catalyst afforded the epoxides in excellent yields (91–99%) and high enantioselectivity (85–97% ee) [33]. Moreover, the catalyst could be recycled and reused for at least six cycles, without any significant loss of efficiency.

Shibasaki extended the use of the La(O*i*-Pr)$_3$/(*S*)-BINOL/(Ph)$_3$AsO complex **21** to the epoxidation of a variety of *trans*-α,β-unsaturated carboxylic acid imidazolides and acylpyrroles [34]. The reaction on the representative compound **22** gave the expected *trans*-configured epoxide **23** as a transient species that underwent facile transesterification to the corresponding perester **24**. This compound could then be easily converted into the epoxide methyl ester **25**, providing an indirect approach to the synthesis of enantioenriched epoxy esters (Scheme 9.12).

The protocol was extended to different *trans*-α,β-unsaturated carboxylic acid imidazolides **26**, affording the corresponding β-aryl and β-alkyl α,β-epoxyesters **27** in good yield and with high ee-values (Scheme 9.13).

The same group reported a modified version for the first successful example of the asymmetric epoxidation of *trans*-α,β-unsaturated amides **28** [35]. The Sm-(*S*)-BINOL–Ph$_3$AsO complex, generated from Sm(O*i*-Pr)$_3$, (*S*)-BINOL, and Ph$_3$AsO in a ratio of 1:1:1, was found to be a more effective catalyst for the epoxidation of compounds **28** (Scheme 9.14). In this way, a wide range of epoxides was recovered in excellent yield and ee-vales, and a useful one-pot tandem procedure was developed to access compounds **30**. The sequential process proved to be efficient, affording the corresponding α-hydroxy amides **30d,e** in excellent overall yield and ee-values, while excluding the occurrence of any racemization in

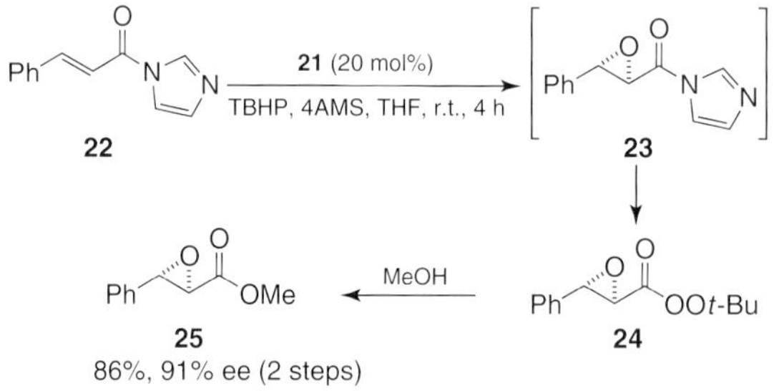

Scheme 9.12 Shibasaki's system for the epoxidation of cinnamic acid imidazolide **22**.

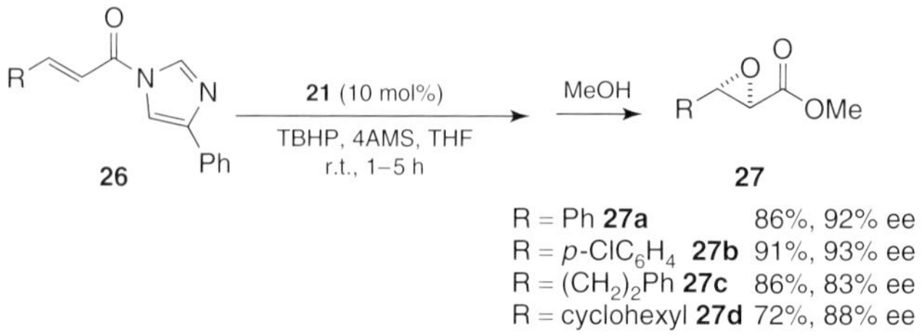

Scheme 9.13 Asymmetric epoxidation of *trans*-α,β-unsaturated carboxylic acid imidazolides **26** with complex **21**/TBHP system.

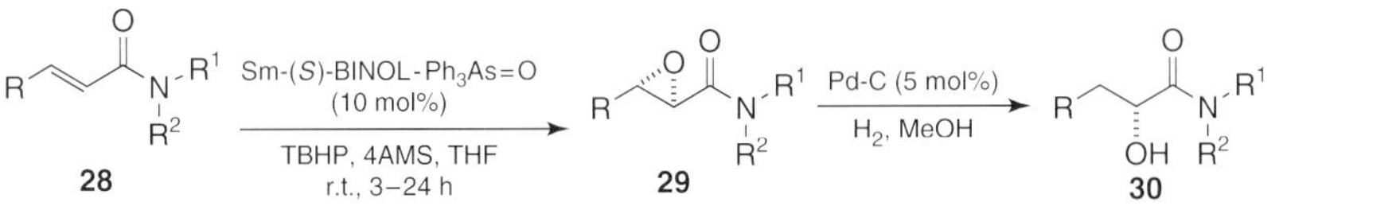

R = $(CH_2)_2$Ph, R^1 = Me, R^2 = H **29a** 99%, >99% ee
R = $(CH_2)_2$Ph, R^1 = Allyl, R^2 = H **29b** 95%, 98% ee
R = C_3H_7, R^1 = Bn, R^2 = H **29c** 94%, 94% ee
R = Ph, R^1 = R^2 = Me **29d** 96%, >99% ee

R = Ph, R^1 = R^2 = Me **30d** 90%, 99% ee
R = Ph, R^1 = Me, R^2 = H **30e** 97%, 97% ee

Scheme 9.14 Asymmetric epoxidation of *trans*-α,β-unsaturated amides **28** with Sm-BINOL-Ph_3AsO/TBHP system.

the tandem process. This one-pot process could be used to construct biologically active dipeptide fragments such as those found in β-aryllactyl-Leu sequences [36].

The lanthanum–BINOL complex/TBHP systems proved to be poorly active in the epoxidation of α,β-unsaturated esters, which are less electrophilic – and hence less reactive – compounds than ketones for the 1,4-addition type process. Indeed, very few examples, based on chiral dioxiranes [37] and manganese–salen [6b,38] complexes as catalysts have been reported for their asymmetric epoxidation.

Shibasaki developed an efficient system for the asymmetric epoxidation of β-aromatic and aliphatic α,β-unsaturated esters **31** by means of a complex generated from Y(O*i*-Pr)$_3$, $(Ph)_3$AsO and ligand **32** in 1 : 1 : 1 ratio, in the presence of a 4AMS and TBHP [39]. In this case, the correct combination of a less sterically demanding biphenyldiol **32** as well as yttrium, a metal of higher Lewis acidity, led to a catalytically active complex for the epoxidation of unsaturated esters (Scheme 9.15).

The *trans*-epoxy esters could be isolated in high yield and with excellent ee-values. Functionalized esters bearing different carbon–carbon double bonds and carbonyl groups could be selectively epoxidized, without observing any overoxidation. Subsequently, Shibasaki's systems have been used successfully as key steps for the synthesis of several natural compounds with biological activities, including (+)-decursin **34** [40], (−)-marmesin **35** [40], and strictifolione **36** [41] (Figure 9.6).

The synthesis of the antidepressant drug fluoxetin **39**, starting with the asymmetric epoxidation of α,β-unsaturated amide **28e** [34], is illustrated in Scheme 9.16.

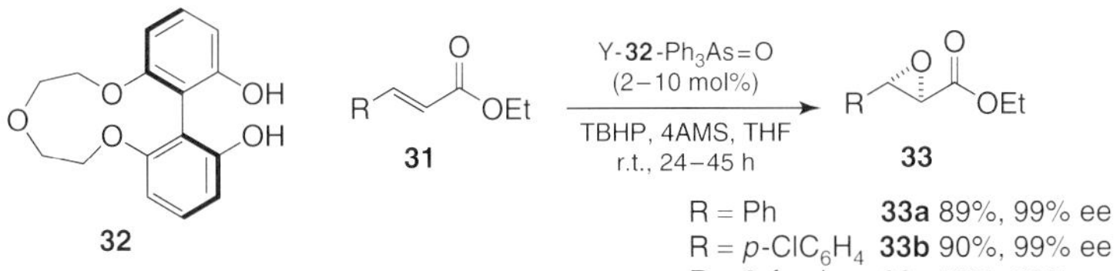

R = Ph **33a** 89%, 99% ee
R = *p*-ClC_6H_4 **33b** 90%, 99% ee
R = 3-furyl **33c** 78%, 92% ee
R = Ph$(CH_2)_2$ **33d** 86%, 91% ee

Scheme 9.15 Asymmetric epoxidation of *trans*-α,β-unsaturated esters **31** with Y-**32**-Ph_3AsO/TBHP system.

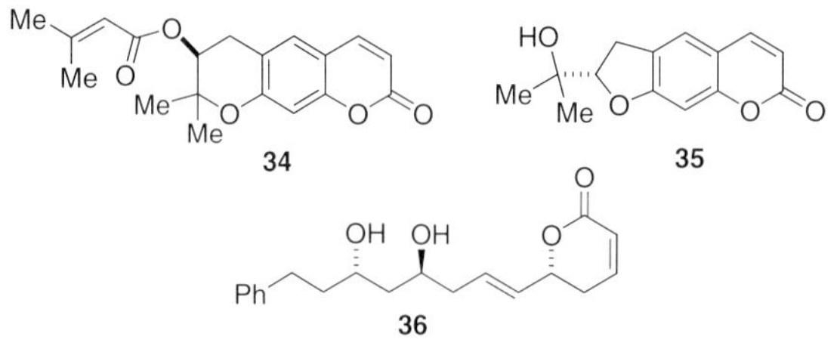

Figure 9.6 Natural products synthesized exploiting Shibasaki's epoxidizing systems as the key step.

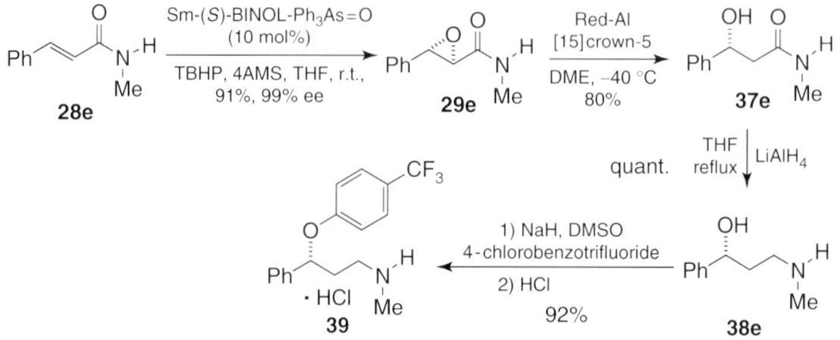

Scheme 9.16 Enantioselective epoxidation of amide **28e** as a key step for the synthesis of (R)-fluoxetine.

9.3
Organocatalyzed Epoxidations

9.3.1
Phase-Transfer Catalysts

The first examples of organocatalyzed methodologies for the enantioselective epoxidation of enones appeared during the 1970s, with the use of readily accessible organic molecules deriving from the chiral pool as cinchona alkaloid-based and poly-aminoacid-based catalysts. More recently, efficient and practical methodologies for their epoxidation have been elaborated by modifying the original catalysts and reaction conditions. It has been only during the past five years that small chiral molecules, which are easily available via low-cost amino acids such as L-proline and amines from cinchona alkaloids, have been disclosed as promoters for the asymmetric epoxidation of enones. Moreover, challenging compounds such as α,β-unsaturated aldehydes and cyclic enones have been epoxidized for the first time, in highly enantioselective manner.

In 1976, Wynberg reported the use of quaternary ammonium salts **40** and **41** derived from the cinchona alkaloids quinine and quinidine, respectively, as phase-transfer catalysts (PTCs), using hydrogen peroxide, alkyl hydroperoxides, or sodium hypochlorite as the oxidants in a biphasic medium for the epoxidation of *trans*-chalcones and naphthoquinones, achieving up to 54% ee (Figure 9.7) [42].

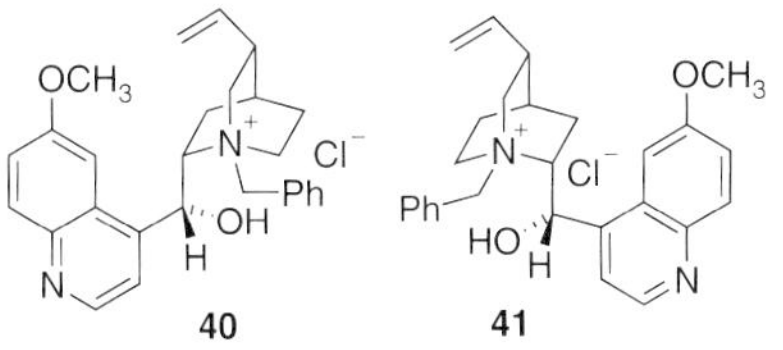

Figure 9.7 Wynberg's cinchona alkaloids-based quaternary ammonium salts.

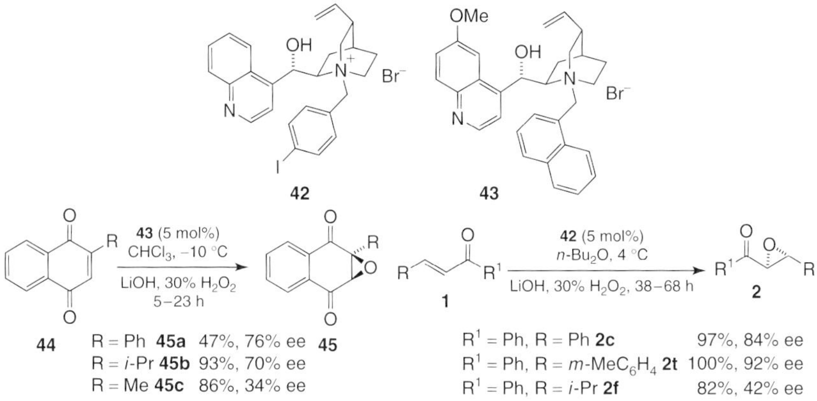

Scheme 9.17 Asymmetric epoxidation of *trans*-chalcones and naphthoquinones with PTC **42** and **43**/H$_2$O$_2$ system.

Arai and Shioiri reported the epoxidation of *trans*-chalcones and naphthoquinones promoted by improved PTC reagents derived from cinchonine or quinidine **42** and **43** in the presence of H$_2$O$_2$ (30% w/w in water) as a convenient oxidant (Scheme 9.17) [43]. The asymmetric induction was found to depend on the stereo and electronic effects of the substituents placed on the secondary alcohol and benzyl moieties of the catalysts. The substituents on chalcones and naphthoquinones also played a relevant role, as the ee-values ranged from modest to high when using optimized catalysts **42** and **43**.

The groups of Lygo [44] and Corey [45] reported, independently, the existence of modified quaternary ammonium salts via the alkylation of nitrogen with the (9-anthracenyl)methyl group, and subsequent benzylation of the secondary hydroxyl group to afford a highly effective catalyst **46**. The epoxidation of *trans*-enones, carried out in toluene using compound **46** [44] and an aqueous solution of NaOCl at 25 °C or an aqueous solution of KOCl at −40 °C [45], was greatly improved and the epoxides were isolated in excellent yield and ee (Scheme 9.18).

Complete control of the diastereoselectivity was detected in favor of *trans*-epoxides, while the highest level of enantioselectivity was achieved for the epoxidation of *trans*-chalcones, and good ee-values were obtained when a β-alkyl substituent was present in the enone. A somewhat limited applicability of this system was observed with less-reactive enones, such as those which bore an alkyl group at the carbonyl position and which required long reaction times so

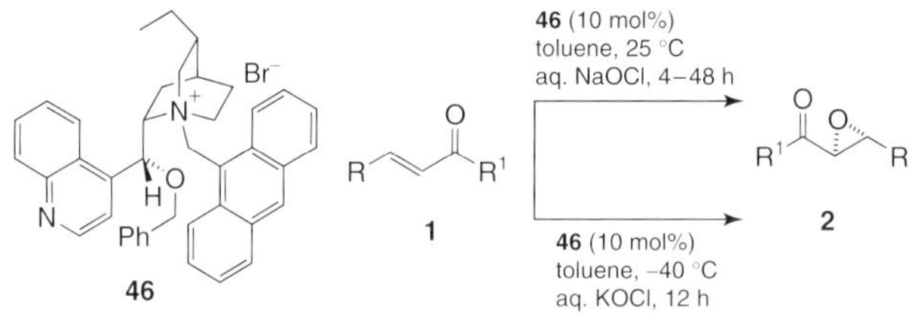

Scheme 9.18 Asymmetric epoxidation of *trans*-enones with PTC **46**/NaOCl or KOCl systems.

Lygo's system

R^1 = Ph, R = Ph **2c**	90%, 86% ee
R^1 = p-BrC$_6$H$_4$, R = Ph **2u**	99%, 88% ee
R^1 = Ph, R = p-MeOC$_6$H$_4$ **2v**	87%, 82% ee
R^1 = Ph, R = n-C$_6$H$_{13}$ **2w**	92%, 77% ee

Corey's system

R^1 = Ph, R = Ph **2c**	96%, 93% ee
R^1 = p-FC$_6$H$_4$, R = p-ClC$_6$H$_4$ **2x**	94%, 98.5% ee
R^1 = Ph, R = p-MeOC$_6$H$_4$ **2v**	70%, 95% ee
R^1 = Ph, R = c-C$_6$H$_{11}$ **2y**	85%, 94% ee

as to attain acceptable conversions. A detailed study on the reaction parameters led to the successful employment of catalyst **46** at only 1 mol% under Lygo's conditions [46]. Subsequently, trichloroisocyanuric acid (TCCA) was shown to be an alternative for use with catalyst **46**, as a safe and inexpensive oxygen source [47].

Corey suggested a model to account for the formation of the ($\alpha S, \beta R$)-enantiomer of the product in the epoxidation of *trans*-chalcones promoted by catalyst **46** (Figure 9.8). The hypochlorite ion could form a tight ion pair with the sole accessible face of the charged nitrogen with the Cl atom of ClO$^-$. In this way, the chalcone in the complex would be located so that the 4-fluorophenyl group would be wedged between the ethyl and quinoline substituents on the quinuclidine ring; simultaneously, the carbonyl oxygen would be located as close to the positively charged nitrogen as permitted by the van der Waals forces. The nucleophilic oxygen of ClO$^-$ would be close to the β-carbon of chalcone for the face-selective nucleophilic 1,4 addition. The negative charge, developed at the carbonyl oxygen in the transition state after the addition, would be stabilized by the proximate charged nitrogen of the catalyst. The cationic charge acceleration of the nucleophilic attack would also favor the formation of the complex hypothesized in Figure 9.8.

Interestingly, Lygo showed that also *cis*-enones such as compounds **47** and **49** could be epoxidized using catalyst **46** to provide *cis*-epoxides with complete diastereoselectivity, although with a lower level of asymmetric induction (Scheme 9.19) [48].

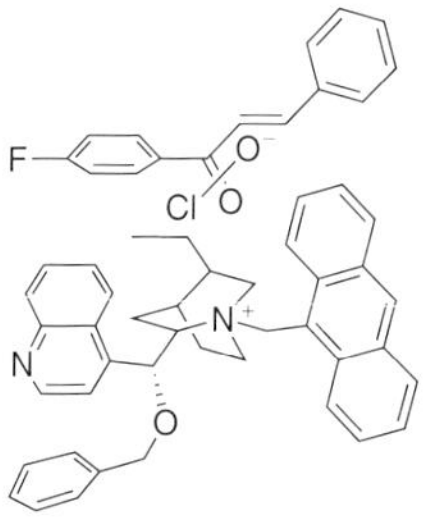

Figure 9.8 Proposed model (front view) of the ion pair involved in the epoxidation catalyzed by PTC **46**.

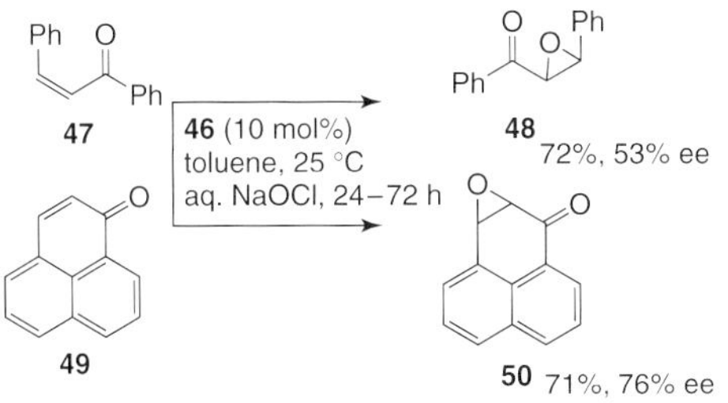

9.3 Organocatalyzed Epoxidations

Scheme 9.19 Asymmetric epoxidation of *cis*-enones with PTC **46**/NaOCl system.

Taking into account the two-step mechanism of the Weizt–Scheffer epoxidation, it has been suggested that the high control of diastereoselectivity observed might be ascribed to electrostatic interactions in the intermediate ion pair, which would slow down the rate of bond rotation relative to ring closure.

Isoflavones **51** can be efficiently epoxidized, as reported by Adam using PTC catalyst **53** and CMHP as the oxidant (Scheme 9.20) [49].

Berkessel has recently employed PTC **54** in the asymmetric epoxidation of vitamin K_3, a particularly challenging naphthoquinone (Scheme 9.21) [50].

The presence of a phenolic group in compound **54** was considered crucial to achieve a high level of asymmetric induction. This effect was ascribed to the additional site being able to provide a stabilizing hydrogen-bonding interaction with the carbonyl at C4 position of the substrate in the enolate–catalyst complex. When using only 2.5 mol% of catalyst and commercial bleach, the epoxide was obtained in good yield and 85% ee (notably, this was the best ee-value ever achieved for this biologically important epoxide).

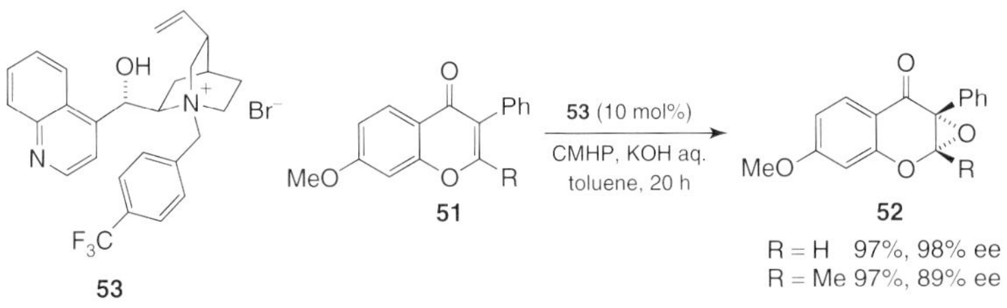

Scheme 9.20 Asymmetric epoxidation of isoflavones with PTC **53**/CMHP system.

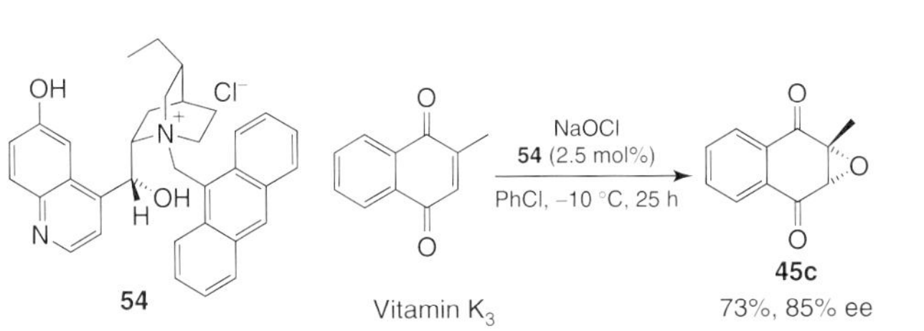

Scheme 9.21 Asymmetric epoxidation of vitamin K_3 with PTC **54**/bleach system.

9 Asymmetric Epoxidations of α,β-Unsaturated Carbonyl Compounds

Quaternary ammonium salts of dimeric cinchona alkaloids, such as compound **55**, proved to be highly efficient PTCs in the presence of the surfactant Span 20 for the epoxidation of *trans*-chalcones [51]. Surfactant use led to a dramatic increase in both the reaction rate and enantioselectivity; typically, with only 1 mol% of Span 20, 1 mol% of catalyst **55**, hydrogen peroxide (30% w/w in water) and aqueous KOH in diisopropyl ether, the epoxides were isolated with excellent yield and ee-value at room temperature (Scheme 9.22).

Dimeric cinchonidine and quinine have been anchored via nitrogen to long linear polyethylene glycol (PEG) chains to provide soluble supported chiral ammonium salts that are easily separated and potentially recyclable [52]. Nevertheless, these supported catalysts afforded moderate asymmetric induction (33–57% ee) in the epoxidation of *trans*-chalcones.

Maruoka developed a novel type of highly efficient quaternary spiro ammonium salt with axial chirality [53]. In this case, the most effective compound **56**, which incorporated two diaryl methanol groups, was used at 3 mol% loading in the presence of 13% NaOCl aqueous solution in toluene at 0 °C. This methodology proved to be fairly general, as it could be applied to a variety of *trans*-enones, the epoxides of which were recovered in excellent yield and ee (Scheme 9.23).

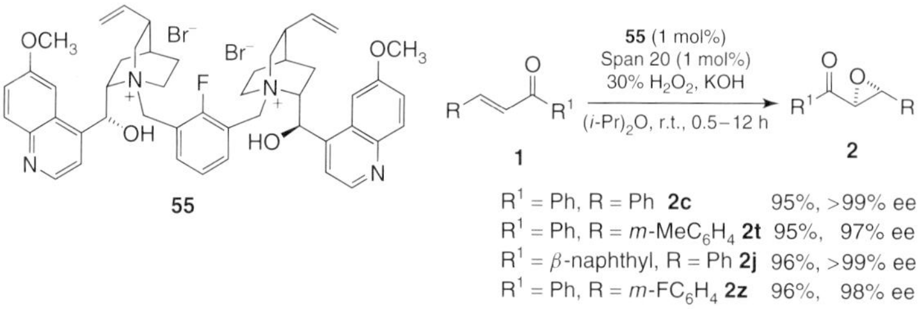

Scheme 9.22 Asymmetric epoxidation of *trans*-chalcones with PTC **55**/Span 20/H_2O_2 system.

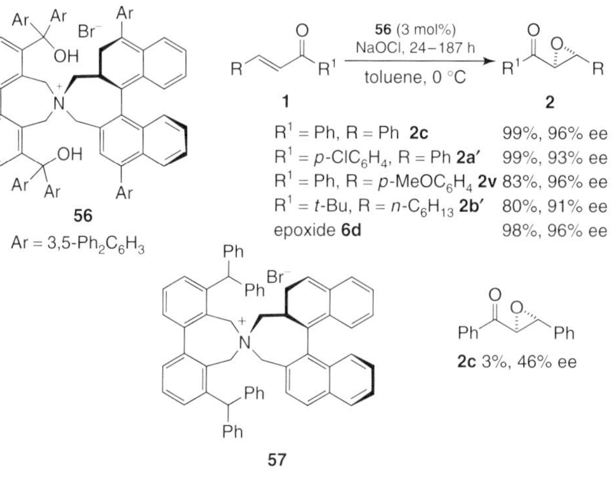

Scheme 9.23 Asymmetric epoxidation of *trans*-enones with PTC **56**/NaOCl system.

Interestingly, the catalyst OH groups showed a dramatic effect on the activity and enantioselectivity of the process. Typically, catalyst **57** afforded epoxide **2c** in negligible yield and with modest ee. However, on the basis of an X-ray analysis of catalyst **56**-PF_6, the biphenyl and binaphthyl subunits of the *N*-spiro structure were seen to be almost perpendicular, creating a chiral reaction cavity around the central nitrogen cation. The high asymmetric induction could be ascribed to hydrogen-bonding recognition of the enone by the OH groups, which brought the enone inside the chiral cavity and into close proximity with the hypochlorite ion for the conjugate addition step.

Bakó reported the use of chiral crown ethers, derived from D-mannose and D-glucose, as PTCs in the epoxidation of *trans*-chalcones [54]. Compound **58** catalyzed the reaction at 7 mol% loading, under basic conditions (20% NaOH aqueous solution) with TBHP as the oxygen donor in toluene at 5 °C. The epoxides were obtained with modest to good yield and ee-values (Scheme 9.24). The length of the chain bearing the hydroxyl group plays an important role in controlling the level of asymmetric induction. Moreover, the catalyst in which the hydroxyl group was protected afforded inferior results when used in the epoxidation. It has been suggested that the polar OH moiety might assure a better transport of the catalyst between the toluene–water phases, and would help in the formation of a transient complex with chalcone.

A PTC of type **59**, where the quaternary ammonium salt was embedded in the azacrown ether derived from BINOL, was designed (Figure 9.9) [55]. In this case, the epoxidation of *trans*-chalcones carried out using 10 mol% of catalyst **59**, and H_2O_2 (30% w/w in water) under basic conditions, in toluene at 0 °C, afforded the epoxy ketones in good yield and with variable levels of asymmetric induction (22–83% ee). The enantioselectivity was shown to depend on the length of the carbon chains at the quaternary nitrogen atom, as well as on the nature of the base used.

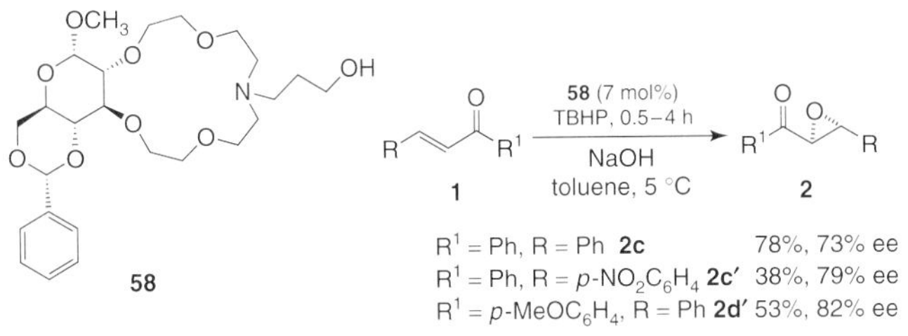

Scheme 9.24 Asymmetric epoxidation of *trans*-chalcones with PTC **58**/TBHP system.

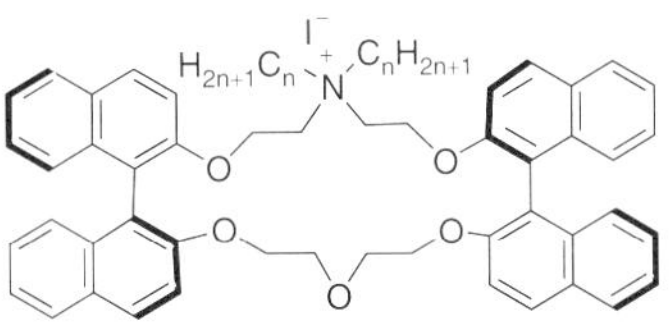

Figure 9.9 Mixed azacrown ether/BINOL quaternary ammonium salt **59**.

The first example of a C_2-symmetric guanidinium salt **60** was synthesized by Murphy [56], and used as a catalyst in the epoxidation of two representative enones at 5 mol% loading in toluene, with an aqueous solution of NaOCl at room temperature, affording the products with high ee-values (Scheme 9.25).

Very recently, Nagasawa reported a series of bifunctional C_2-symmetric guanidinium–urea catalysts, employing α-amino acids as chiral spacers [57]. The epoxidation of a variety of *trans*-chalcones proceeded efficiently and in a highly enantioselective manner, when using PTC **61** (5 mol%) with H_2O_2 (30% w/w in water) as the oxidant, and NaOH as base, in a mixture of toluene : water (19 : 1 at −10 °C (Scheme 9.26). Subsequently, catalyst **61** was quantitatively recovered from the reaction mixture and reused for five runs, with no loss of either catalytic activity or enantioselectivity.

It has been proposed that the cooperative interactions of the guanidinium and urea moieties with the peroxide anion and the carbonyl of the enone, respectively, would be involved in activating the reactive partners (Figure 9.10). Consequently, catalysts devoid of either the urea or the guanidinium moieties were shown to be inactive in the epoxidation.

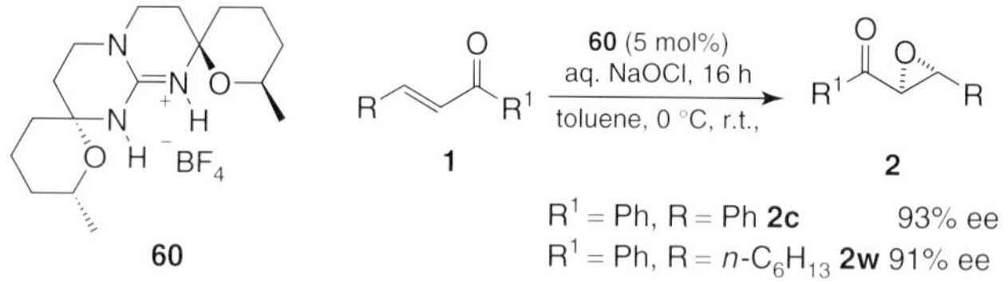
60
60 (5 mol%)
aq. NaOCl, 16 h
toluene, 0 °C, r.t.,
1
2
R¹ = Ph, R = Ph 2c 93% ee
R¹ = Ph, R = n-C6H13 2w 91% ee

Scheme 9.25 Asymmetric epoxidation with C_2-symmetric guanidium salt **60**/NaOCl system.

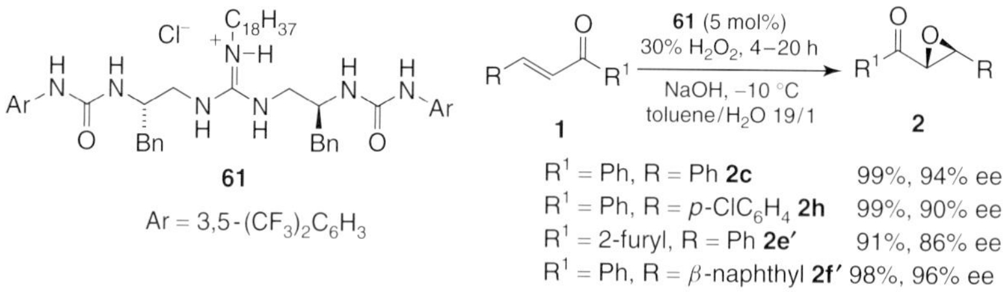
61
Ar = 3,5-(CF3)2C6H3
61 (5 mol%)
30% H2O2, 4–20 h
NaOH, −10 °C
toluene/H2O 19/1
1
2
R¹ = Ph, R = Ph 2c 99%, 94% ee
R¹ = Ph, R = p-ClC6H4 2h 99%, 90% ee
R¹ = 2-furyl, R = Ph 2e′ 91%, 86% ee
R¹ = Ph, R = β-naphthyl 2f′ 98%, 96% ee

Scheme 9.26 Asymmetric epoxidation of *trans*-chalcones with bifunctional PTC **61**/H_2O_2 system.

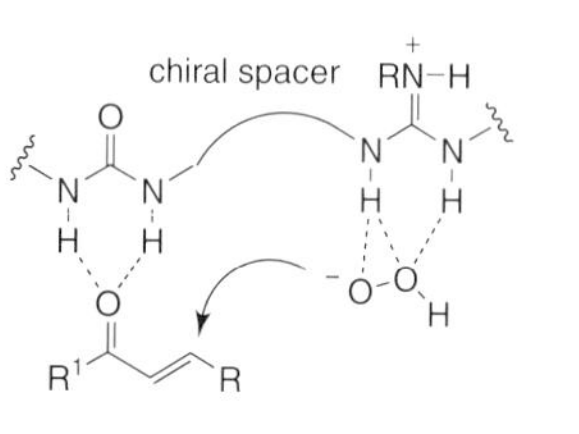
chiral spacer

Figure 9.10 Proposed cooperative interactions of guanidine-urea catalyst **61** with enone and hydrogen peroxide.

9.3.2
Polyamino Acids

The polyamino acid-catalyzed enantioselective epoxidation of enones was reported by Juliá and Colonna during the early 1980s [58]. The asymmetric epoxidation of *trans*-chalcones was efficiently mediated by homo-oligopeptides such as poly-L-alanine (PLA) or poly-L-leucine (PLL), in an aqueous solution of NaOH and H_2O_2 at room temperature. Good to high levels of enantioselectivity (up to 98% ee) were obtained when more than 10 amino acids were present in the oligopeptide, with an optimal number of 30 residues achieving the best level of asymmetric induction. Normally, the polyamino acid is prepared via the polymerization of an amino acid N-carboxy anhydride, induced by amines. The initial procedure comprised three phases, where *trans*-chalcones were first dissolved in toluene or hexane, the basic H_2O_2 constituted the second phase, and the homo-oligopeptide (which proved to be insoluble) formed the third phase. Under these conditions, *trans*-chalcones, dienones, and unsaturated ketoesters were epoxidized in good yield and with generally good to high ee-values (Scheme 9.27).

This reaction can be considered the first reliable system for the asymmetric epoxidation of different types of enone, based on the availability of the catalyst in both enantiomeric forms, and the simplicity of the reaction conditions. Although the enantioenriched epoxy ketones could be accessed according to the chirality of the poly-leucine catalyst used [59], the reaction times were often long, the enolizable enones proved to be problematic substrates, and the catalyst separation and recycling was difficult because the poly-aminoacid formed a gel during the reaction.

Nonetheless, several different research groups were able to improve the system by modifying the reaction parameters. For example, Roberts optimized the conditions and achieved a two-phase system that consisted of anhydrous THF as the solvent, a soluble urea–hydrogen peroxide complex (UHP) as the oxygen source, and 1,8-diazabicyclo[5.4.0]undec-7-ene (DBU) as a soluble base [60]. Alternatively, sodium percarbonate in a dimethylformamide (DMF)–water mixture could be successfully and conveniently used as a source of base and oxidant [61]. As a result of these modifications, in both cases a significant reduction in reaction time was observed, the catalyst loading could be reduced to 5 mol%, and the epoxidation of enolizable enones proved equally feasible as those enones with alkyl groups at the β-position. Under such two-phase conditions, the recovery of PLL was enhanced by adsorbing the catalyst onto silica for flash chromatography [62]. Those epoxidations which used the silica-adsorbed poly-L-leucine (PLL) produced better rates of conversion and higher ee-values, while the catalyst loading could be reduced to 2.5 mol% and the catalyst completely recycled. In order to solve problems caused

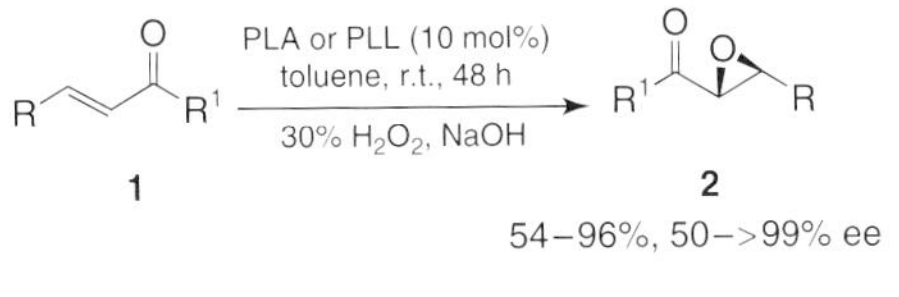

Scheme 9.27 Asymmetric epoxidation of *trans*-enones with triphasic system polyamino acids/H_2O_2.

by handling the gel catalyst, Tang and coworkers grafted the PLL to a silica gel that had been functionalized with primary 3-aminopropylsilyl (AMPSi) groups, so as to obtain a silica-grafted PLL [63]. Subsequent epoxidation with 8 mol% of PLL-AMPSi with sodium percarbonate in a DMF–water mixture at room temperature afforded the epoxides in good to high yield (50–94%) and ee (70–93%). The separation and recovery of the PLL-AMPSi catalyst was also improved, such that it could be reused for different runs with no significant loss of activity. In order to solubilize the poly-aminoacid catalyst and to obtain a monophase system, the incorporation of a catalyst onto a soluble polymer was also investigated. For example, when Roberts bonded PLL to PEG chains, a copolymer was obtained that formed an homogeneous monophasic system for the asymmetric epoxidation of *trans*-chalcone **1c**, using UHP as the oxidant and DBU as the base (Scheme 9.28) [64].

Remarkably, catalyst **62**, which contained smaller quantities of polypeptide, proved to be more efficient than compound **63**. This increased catalytic activity was related to the larger number of amino termini present in **62**, as the catalyst contained the same weights of poly-aminoacid in each experiment. The results of Fourier transform infra-red (FT-IR) studies showed the catalytically active components of the copolymers to have an α-helical structure, while a minimum of four leucine residues was sufficient to obtain a high conversion and enantioselectivity. Tsogoeva reported that the catalytic activity of the aforementioned polymer could be increased by enlarging the PEG chain [65]. Moreover, another catalyst, obtained via the polymerization of a styrene/aminomethyl copolymer and L-Leu N-carboxyanhydride, was synthesized. When used with UHP in THF for the continuously operated asymmetric epoxidation of *trans*-chalcones in a membrane reactor system, these soluble catalysts afforded the epoxide in excellent conversion after 15–30 min, with ee-values up to 97%.

Later, when Roberts investigated the role of the amino acid at the N-terminus of the catalyst [66], it was shown that those residues close to the N-terminus of the chain determined the sterochemical outcome of the oxidation. If the amino acids in the oligomers were changed and tested in the epoxidation of *trans*-chalcone **1c**, the penultimate and penultimate-plus-one residues were shown to play a crucial role. However, when glycine residues – which had no stereocenters – were inserted into these positions, the enantioselectivity fell dramatically.

Berkessel synthesized a series of solid-phase-bound leucine oligomers linked to TentaGel S NH$_2$ and inverse L-leucine oligomers bound to the support through the N-terminus [67]. Four to five amino acids were found to be sufficient to achieve a high asymmetric induction (96–98% ee), and the activity was increased with

catalyst	C(%)	ee(%)
62	80	98
63	63	95

H(L-Leu)$_{3.9}$NHCH$_2$CH$_2$(OCH$_2$CH$_2$)$_{71}$NH(L-Leu)$_{3.9}$H **62**
H(L-Leu)$_{11.6}$NHCH$_2$CH$_2$(OCH$_2$CH$_2$)$_{71}$NH(L-Leu)$_{11.6}$H **63**

Scheme 9.28 Asymmetric epoxidation of *trans*-chalcone **1c** with soluble PLL-PEG **62** and **63**/UHP system.

chain length. The results of both experimental studies and molecular modeling suggested that the catalytic active portion consisted of the N-terminal triad of an α-helical segment. A mechanistic model for the asymmetric epoxidation was proposed, where the carbonyl oxygen of chalcone was hydrogen-bonded to the NH of the amino acids at the N-terminus of position n-2, while the hydroperoxide anion was delivered face-selectively to the β-carbon atom of chalcone by the NH group of amino acid n-1. Consequently, the sense of asymmetric induction in the epoxidation would be determined by the helicity of the catalyst.

Both, experimental results and computational studies reported by Kelly and Roberts, let them to propose a model similar to that suggested by Berkessel, where the four N-terminal residues of helical PLL, that were not involved in intrachain hydrogen bonding, would be available to form an oxyanion hole capable of stabilizing the hydroperoxy enolate transition state by a pair of NH groups [68]. A third NH group would facilitate ejection of the hydroxyl group during ring closure.

An improvement of the original three-phase system of the Juliá–Colonna epoxidation was recently developed by Geller [69], whereby addition of the phase-transfer agent tetrabutylammonium bromide (TBAB) as co-catalyst in the asymmetric epoxidation of *trans*-chalcone caused an increase in the concentration of peroxide in the organic phase. In addition, the reaction rate was notably higher, while the epoxide was produced in >99% conversion and 94% ee in only 1.5 h. Likewise, the amount of oxidant and base could be reduced from 30 to 1.3 equiv. for H_2O_2, and from 4 to 1.3 equiv. for NaOH. Moreover, when PPL was synthesized by polymerization at higher temperatures, it proved to be more active than the "normal" PLL (which was used at 0.01 mol% in the presence of TBAB), and afforded the epoxide of *trans*-chalcone in 60% yield and 80% ee. These new triphasic/PTC conditions were successfully scaled up to the 100 g substrate level, which permitted differently substituted enones to be epoxidized under triphasic/PTC conditions with a high yield, a good to high ee-value, and short reaction times [70].

In mechanistic terms, the Weitz–Scheffer epoxidation can proceed as either: (i) a fast addition of the hydroperoxide anion to the enone, followed by a slow ring closure to the epoxide; or (ii) a slow addition of the hydroperoxide anion, followed by rapid ring closure (see Scheme 9.1). Following initial investigations of correctly deuterated enones under biphasic conditions by Roberts, the first of these hypotheses was suggested to be the most plausible [68a].

Kinetic studies of the Juliá–Colonna epoxidation under monophasic conditions, led to a more complex mechanistic proposal of a steady-state random bireactant system, suggestive of the enzyme-like catalysis provided by PLL (Figure 9.11) [71].

Subsequent reaction progress kinetic analyses showed that the epoxidation would proceed via a reversible addition of chalcone to PLL-bound hydroperoxide (PLL:HOO$^-$); this would lead to the formation of a transitory hydroperoxy enolate which, on acquiring the stereoelectronic requirements for ring-closure, would lead irreversibly to the epoxide [72].

Based on the valuable improvements achieved over the past three decades, the Juliá–Colonna epoxidation can be applied successfully to a wide variety of *trans*-enones (Scheme 9.29). Beside typical *trans*-chalcones, functionalized acyclic

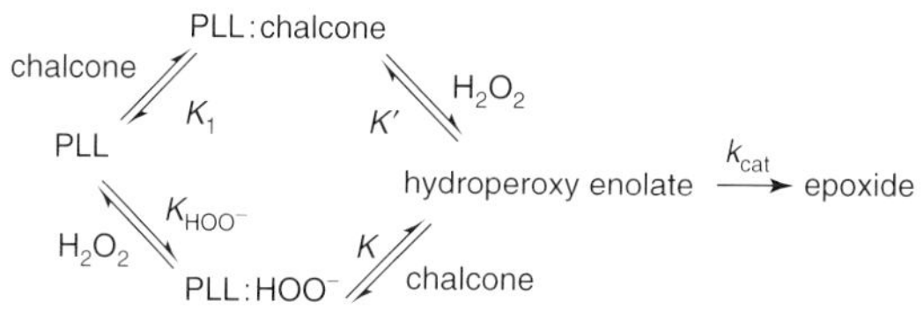

372 | *9 Asymmetric Epoxidations of α,β-Unsaturated Carbonyl Compounds*

Figure 9.11 Mechanistic proposal of consecutive-parallel reactions network for the epoxidation.

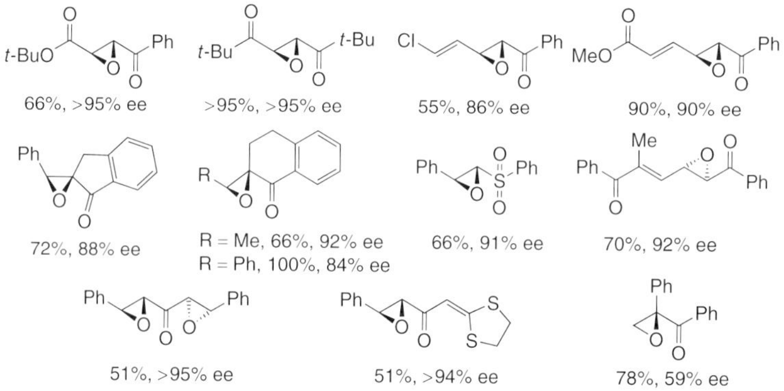

Scheme 9.29 Scope of the PLL-catalyzed asymmetric epoxidation.

enones, tetralones, *trans*-vinyl sulfones may be efficiently epoxidized using PLL as a catalyst; however, neither trisubstituted alkenones, *cis*-enones, nor less-reactive α,β-unsaturated esters would be suitable compounds for this system, and would continue to constitute a particularly challenging classes of enones for epoxidation in an enantioselective manner.

An important application of this methodology was reported by Roberts in relation to the preparation of the potent blood pressure-lowering agent, diltiazem **66** (Scheme 9.30) [60a]. Central to the approach was the Baeyer–Villiger reaction on the highly enantioenriched epoxide, to furnish the intermediate epoxy ester **64**.

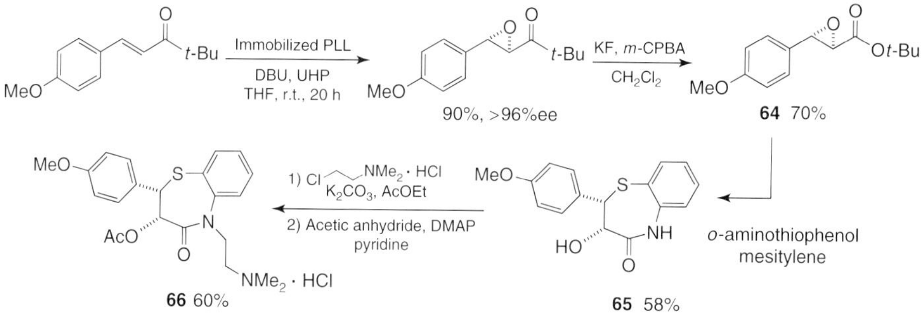

Scheme 9.30 Enantioselective epoxidation with PLL/UHP system as a key step for the synthesis of diltiazem.

9.3 Organocatalyzed Epoxidations

Analogously, the phenylisoserine side-chain of Taxol™ was synthesized by the same group by exploiting the PLL-catalyzed asymmetric epoxidation of enone **2g**, followed by the Baeyer–Villiger reaction to create the crucial epoxy ester intermediate [60a].

9.3.3
Optically Pure Alkyl Hydroperoxides or Ligands

In the Weitz–Scheffer epoxidation, a nucleophilic oxidative species can be generated by employing stoichiometric quantities of an optically pure alkyl hydroperoxide under basic conditions, so that a chiral peroxyalkyl anion is involved in the 1,4-addition to the electron-poor alkene. The restricted number of efficient approaches for the synthesis of enantiopure alkyl hydroperoxides, coupled to their general modest chemical and thermal stability, has hampered the development of this alternative route for the enantioselective epoxidation of enones.

Adam reported the use of a simple secondary optically pure alkyl hydroperoxide **67**, using KOH at −40 °C, in the epoxidation of different enones (Scheme 9.31) [73]. The *trans*-($\alpha S,\beta R$)-epoxides were recovered in high yield and with moderate to good ee-values, according to the reaction conditions and the substrate employed. In the presence of the potassium-chelating agent 18-crown-6, the enantioselectivity of the process was dramatically decreased, however. Accordingly, the authors rationalized the result as a "template effect" of the potassium cation, which coordinated both the lone pair of the carbonyl oxygen atom and the distal oxygen atom of the hydroperoxide anion. This simultaneous coordination most likely governs the π-facial differentiation.

Optically pure sugar-derived hydroperoxide **68** has been synthesized by Taylor, and used in the enantioselective epoxidation of naphthoquinones with DBU as a stoichiometric base (Scheme 9.32) [74]. A good conversion to the epoxides and moderate to good enantioselectivity were observed. Subsequently, Taylor reported a more in-depth investigation with different sugar-derived hydroperoxides in the DBU-promoted epoxidation of a precursor of the antibiotic alysamicin and naphthoquinones [75]. The epoxidation of vitamin K$_3$ afforded the epoxide in low yield, but in 78% ee. Under the same conditions, *trans*-chalcone proved to be a

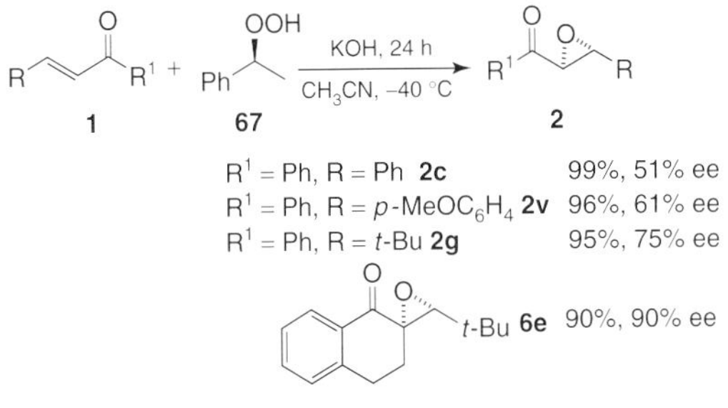

R^1 = Ph, R = Ph **2c**   99%, 51% ee
R^1 = Ph, R = *p*-MeOC$_6$H$_4$ **2v**   96%, 61% ee
R^1 = Ph, R = *t*-Bu **2g**   95%, 75% ee

6e 90%, 90% ee

Scheme 9.31 Asymmetric epoxidation of *trans*-enones with alkyl hydroperoxide **67**/KOH system.

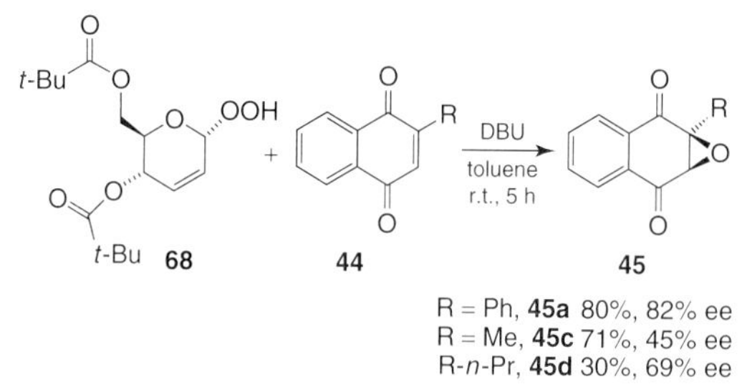
t-Bu
O
OOH
+
R
DBU
toluene
r.t., 5 h
R
t-Bu
68
44
45
R = Ph, 45a 80%, 82% ee
R = Me, 45c 71%, 45% ee
R-n-Pr, 45d 30%, 69% ee

Scheme 9.32 Asymmetric epoxidation of naphthoquinones with alkyl hydroperoxide **68**/DBU system.

poor substrate, as the asymmetric induction achieved was low (up to 13% ee). It was suggested that this intervention might be due to a "template effect," whereby the conjugated acid of the DBU had formed an ion pair with the hydroperoxide anion in apolar toluene. Certainly, this hypothesis was supported when molecular modeling studies of the reaction were conducted.

Recently, Chiemlewski synthesized a range of more effective alkyl hydroperoxides that had been derived from 2-deoxygalactose and monitored in the epoxidation of some *trans*-enones with different bases at room temperature [76]. The most efficient combination was found to be oxidant **69**, with NaOH as the base (Scheme 9.33), with a high conversion to epoxide being observed, as well as a good to high enantioselectivity. The s-*cis*-conformation of the enone was a fundamental prerequisite to achieve high stereocontrol, as the epoxidation of vitamin K_3 proceeded with only modest control of the asymmetric induction (40% ee). Again, the "template effect" of the sodium cation was found to decisively affect the enantiocontrol during epoxidation, with KOH, LiOH, and DBU each affording poorer results. The hemi-acetal **70** was recovered after epoxidation in 80% yield, and conveniently reoxidized to the hydroperoxide **69**.

The first tertiary enantiopure alkyl hydroperoxide was reported by Seebach [77]. In this case, compound **71**, which is easily obtained from commercially available (−)-TADDOL (2,2-dimethyl-α,α,α′,α′-tetraphenyldioxolane-4,5-dimethanol), was used in the epoxidation of *trans*-chalcone **1c** with *n*-BuLi as the base in THF

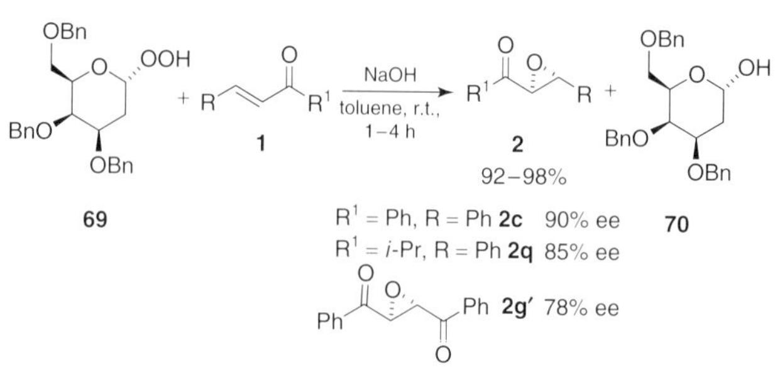
OBn
OOH
+ R
R1
NaOH
toluene, r.t.,
1–4 h
1
BnO
OBn
R1
R +
2
92–98%
OBn
OH
BnO
OBn
69
R1 = Ph, R = Ph 2c 90% ee
R1 = i-Pr, R = Ph 2q 85% ee
Ph
Ph 2g′ 78% ee
70

Scheme 9.33 Asymmetric epoxidation of *trans*-enones with alkyl hydroperoxide **69**/NaOH system.

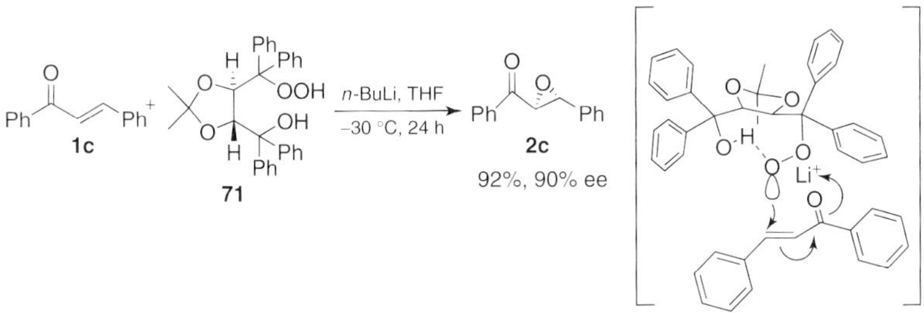

9.3 Organocatalyzed Epoxidations

Scheme 9.34 Asymmetric epoxidation of *trans*-chalcone **1c** with hydroperoxide **71**/*n*-BuLi system.

at −30 °C (Scheme 9.34). As a result, epoxide **2c** was isolated in high yield and high ee. When the different enones were reacted under similar conditions, this led to the products in high yield and in a variable range of enantioselectivities (10–82% ee). The diol (−)-TADDOL, which was a byproduct of the reaction, was isolated during the purification process and recycled for the synthesis of oxidant **71**.

Seebach suggested a model to account for the enantioselectivity observed in the epoxidation of *trans*-chalcone (Scheme 9.34). If the conjugate addition is assumed to be the rate- and stereoselectivity-determining step, then the approach of the lithium peroxide to the enone would lead to a preferential formation of the ($\alpha S, \beta R$)-configured epoxide.

Lattanzi developed an easy route to tertiary optically pure alkyl hydroperoxides based on the (*S*)-norcamphor skeleton [78]. Oxidant **72**, when synthesized in two steps and high yield from (*S*)-norcamphor, was employed in the asymmetric epoxidation of *trans*-chalcones and naphthoquinones. The optimal conditions for this required the use of *n*-BuLi as the base and THF as the solvent at −20 °C. As a consequence, the *trans*-chalcones were transformed to *trans*-epoxides in good yield and with up to 50% ee (Scheme 9.35). The epoxidation of vitamin K$_3$ afforded the epoxide **45c** in 80% yield and 51% ee. In addition, alcohol **73** could be recovered in high yields (>90%) during the purification process and conveniently reused for the one-step synthesis of oxidant **72**.

Tomioka developed a strategy for the epoxidation of *trans*-enones that relied on the double activation of an achiral alkyl hydroperoxide by lithiation and subsequent

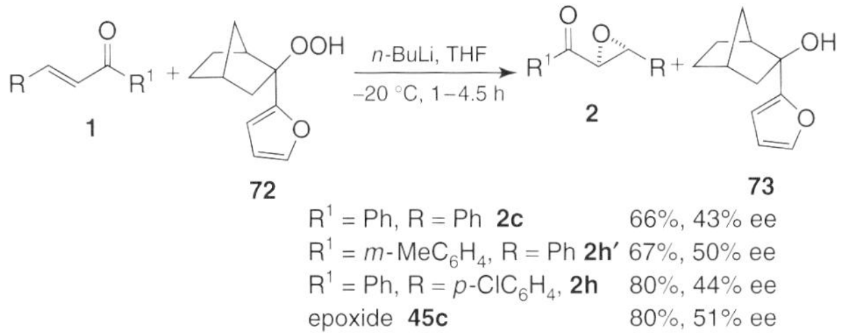

R^1 = Ph, R = Ph **2c** 66%, 43% ee
R^1 = *m*-MeC$_6$H$_4$, R = Ph **2h′** 67%, 50% ee
R^1 = Ph, R = *p*-ClC$_6$H$_4$, **2h** 80%, 44% ee
epoxide **45c** 80%, 51% ee

Scheme 9.35 Asymmetric epoxidation of *trans*-chalcones and vitamin K$_3$ with alkyl hydroperoxide **72**/*n*-BuLi system.

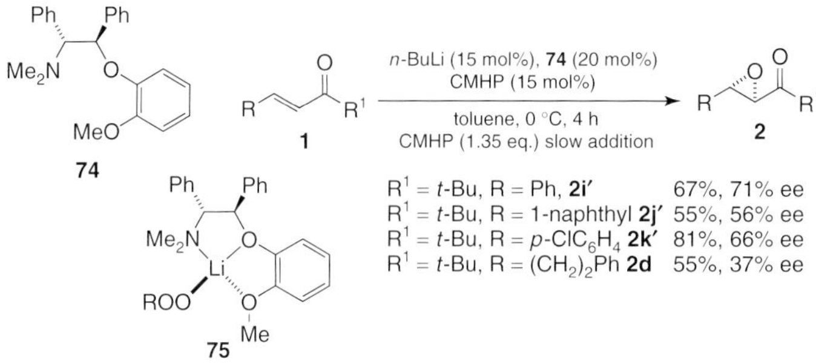

Scheme 9.36 Asymmetric epoxidation of *trans*-enones with CMHP/*n*-BuLi/**74** system.

chelate formation with a chiral ligand [79]. After checking different bidentate and tridentate ligands, as well as various alkyl hydroperoxides in model epoxidation processes, tridentate compound **74** and CMHP was shown to provide the best combination in toluene as the solvent for the epoxidation of compounds **1** (Scheme 9.36).

A tridentate ligand was necessary to generate a bicyclo[3.3.0] complex **75** by coordination to lithium, while the hemilabile methoxyphenyl side chain allowed a coordinating atom exchange with the carbonyl oxygen of the enone. Enones which bore a bulky substituent on the carbonyl, and were capable of directing the preferential coordination to the lithium of the carbonyl lone-pair at the side of the carbon–carbon double bond, afforded the epoxides with a greater control of the asymmetric induction. In addition, the ligand could be recovered in quantitative fashion after the reaction, and reused.

9.3.4
Guanidine-Based Catalysts

Alkyl hydroperoxides and organic bases such as DBU represent useful combinations to perform the Weitz–Scheffer epoxidation [80]. Taylor reported the first asymmetric version using enantiopure guanidine as bases [81]. Optically pure cyclic guanidines with a pK_a similar to that of DBU have been synthesized and used in stoichiometric amounts for the epoxidation of cyclohexenone **76** and *trans*-chalcone **1c** with TBHP in a toluene/5% isopropanol mixture (Scheme 9.37).

Although the conversions were moderate after seven days, these bases catalyzed the epoxidation with an encouraging level of asymmetric induction. The linear or cyclic guanidines were of comparable efficiency, while steric effects might be responsible for the decreased ee-values, probably due to a destabilization of the peroxide anion–guanidinium cation ion pair. It is interesting to note that the *O*-methylated C_2-symmetric compound **83** was less enantioselective than the corresponding hydroxyl substituted compound **82**. Monocyclic analogs of guanidine **79**, such as compounds **84** and **85**, proved to be more efficient catalysts in the epoxidation of cyclohexenone **76** under the same conditions, affording the epoxide in up to 60% ee (Figure 9.12).

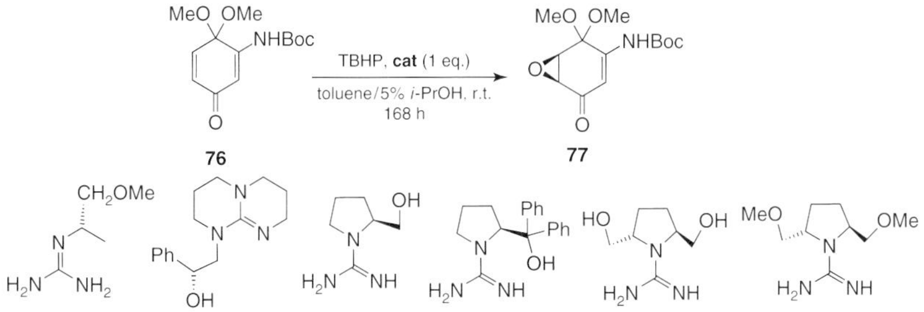

78 — 39%, 30% ee; 79 — 31%, 35% ee; 80 — 45%, 28% ee; 81 — 40%, 17% ee; 82 — 20%, 29% ee; 83 — 24%, 18% ee

Scheme 9.37 Asymmetric epoxidation of *trans*-enones with chiral guanidines/TBHP system.

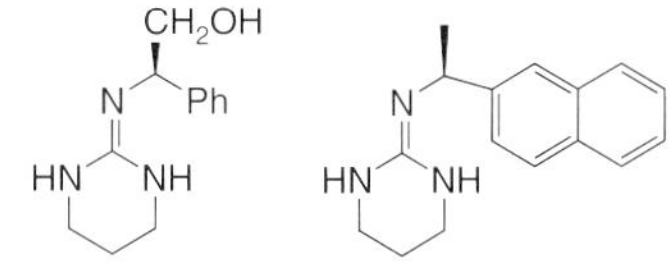

84 — 34%, 60% ee; 85 — 50%, 50% ee

Figure 9.12 Improved guanidines for the asymmetric epoxidation of cyclohexenone **76**.

The presence of the free hydroxyl group played a major role with respect to steric hindrance at the chiral center of guanidines **84** and **85** in the control of enantioselectivity.

9.3.5 Pyrrolidine-Based Catalysts

Lattanzi investigated the feasibility of the asymmetric epoxidation of *trans*-enones promoted by simple bifunctional molecules derived from low-cost and easily accessible L-proline [82]. Commercially available α,α-L-diphenyl prolinol **86** was shown to promote the enantioselective epoxidation of *trans*-chalcone **1c** at 30 mol% loading in hexane at room temperature, with TBHP as the oxygen donor (Scheme 9.38). The presence of the phenyl rings was fundamental for the control of the asymmetric induction as well as the free OH group, as both activity and enantioselectivity were markedly influenced when employing catalysts **87–89**. The enantioselective epoxidation of different *trans*-enones when using catalyst **86** and TBHP at room temperature was developed, and the products **2** were isolated in good yields and with

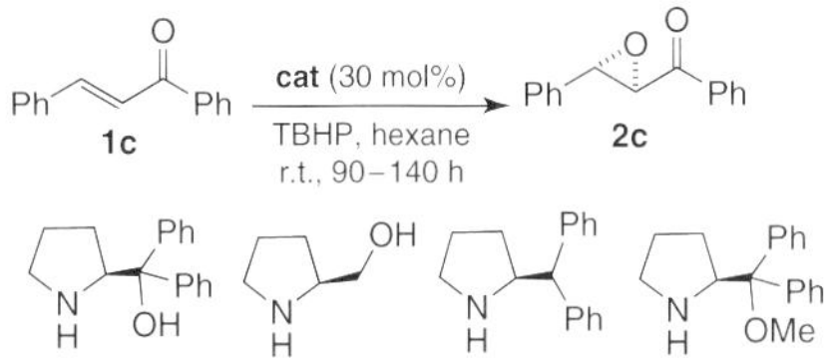

86 — 72%, 76% ee; 87 — 82%, rac; 88 — 22%, 23% ee; 89 — 14%, 63% ee

Scheme 9.38 Asymmetric epoxidation of *trans*-chalcone **1c** with different L-proline derivatives/TBHP systems.

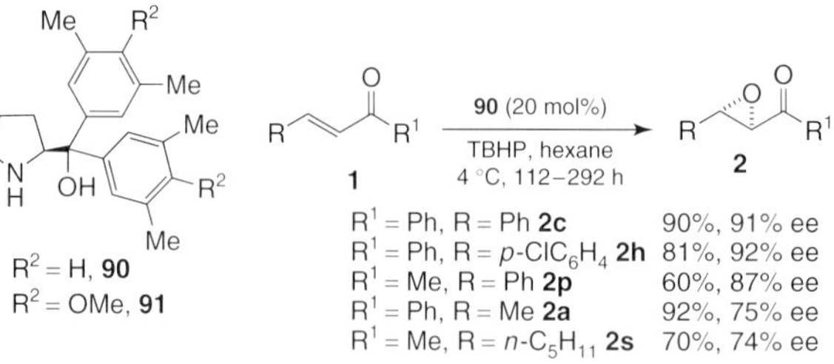

9 Asymmetric Epoxidations of α,β-Unsaturated Carbonyl Compounds

Scheme 9.39 Asymmetric epoxidation of *trans*-enones with the prolinol **90**/TBHP system.

R^1 = Ph, R = Ph **2c**	90%, 91% ee
R^1 = Ph, R = p-ClC$_6$H$_4$ **2h**	81%, 92% ee
R^1 = Me, R = Ph **2p**	60%, 87% ee
R^1 = Ph, R = Me **2a**	92%, 75% ee
R^1 = Me, R = n-C$_5$H$_{11}$ **2s**	70%, 74% ee

up to 80% ee. Investigations of the activity of phenyl-substituted prolinols, readily accessible from L-proline, helped to shed more light on those factors that affect the efficiency of this oxidative system [83]. As a general trend, electron-withdrawing groups on the phenyl rings depressed the activity, while electron-donating groups caused it to be enhanced. The fine tuning of the substituents allowed the discovery of the commercially available compound **90**, as a more effective catalyst (Scheme 9.39).

The epoxidation of a variety of *trans*-enones, performed with a 20 mol% loading of promoter **90** at 4 °C, allowed the isolation of *trans*-epoxides in higher yield and ee. Apart from chalcones, this system could be employed in the epoxidation of more challenging compounds such as alkyl, aryl, and dialkyl enones, achieving good levels of asymmetric induction. However, the main drawback of this system was the low reaction rates. The phenyl-trisubstituted catalyst **91** could be used at 10 mol% loading at room temperature so as to maintain an asymmetric induction (up to 90% ee) [84]. A catalytic cycle has been suggested, where the promoter **86** serves as a bifunctional organocatalyst, giving rise to the simultaneous activation of enone and TBHP by the hydroxyl and amino groups, respectively, and the formation of an ion pair as the catalytically active species (Figure 9.13).

The proposed mechanism, in which no covalent bonds were involved in the activation of both reagents by the catalyst, was also supported by the observation that polar, protic, and coordinating solvents such as THF and CH$_3$CN (which solvate the ion pair) caused drastic decreases in the efficiency and enantioselectivity of the reaction. Moreover, catalysts **88** and **89** were poorly active and less enantioselective (Scheme 9.38). Epoxidation carried out with promoter **90** in the presence of organic

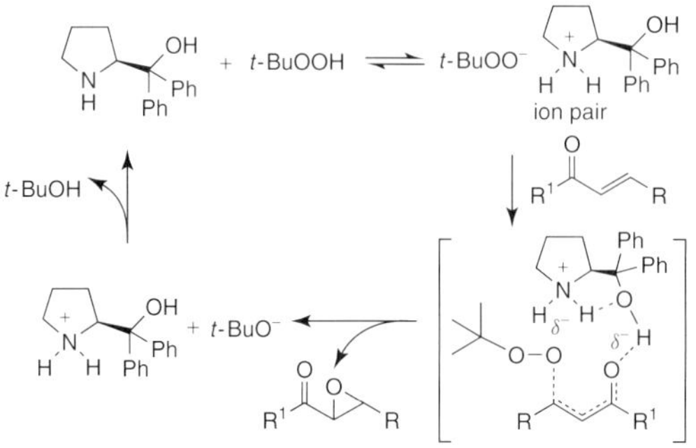

Figure 9.13 Proposed catalytic cycle for the epoxidation.

acids as co-catalysts was prevented from occurring, thus supporting the proposal that iminium ions are unlikely to be involved in the activation of enones with this type of catalyst. When nonlinear effects were investigated in the epoxidation of *trans*-chalcone **1c**, using compound **86** of different degrees of optical purity, the linearity observed strongly suggested that one molecule of the catalyst was involved in the enantiodifferentiating step.

Zhao synthesized derivative **92**, which is structurally similar to compound **90**, starting from *trans*-4-hydroxy-L-proline [85]. For this, a sterically hindered moiety was introduced into the pyrrolidine ring with a view to influence the asymmetric induction. Catalyst **92**, when employed at 30 mol% loading at room temperature with TBHP, afforded epoxides in satisfactory yield and with slightly improved ee-values with respect to catalysts **90** and **91** (Scheme 9.40).

The same group prepared the fluoro-derivative **93**, which was employed at 30 mol% loading in CCl_4 at room temperature, to afford the epoxides in moderate yield and with good ee-values [86]. Moreover, the catalyst could be recycled three times in the epoxidation, with very little loss of activity or asymmetric induction.

Recently, Loh synthesized compound **94**, starting from (R)-1-phenylethylamine, and used this as catalyst in the epoxidation of *trans*-enones; the products were isolated in good yield and enantioselectivity [87]. Lattanzi further disclosed that either tertiary β-amino alcohols such as *N*-alkylated prolinols or secondary acyclic β-amino alcohols were almost inactive in the epoxidation [88]. Interestingly, commercially available primary and flexible β-amino alcohols such as compound **95**, or catalysts readily obtained from α-amino acid esters, such as compounds **96** and **97**, could provide a moderate promotion of the epoxidation of *trans*-chalcone **1c** at a 30 mol% loading (Scheme 9.41). Moreover, their efficiency can be significantly tuned by introducing electron-donating substituents onto the phenyl rings.

Today, the catalytic asymmetric epoxidation of *trans*-α,β-unsaturated aldehydes represents a challenging transformation. In fact, a direct approach to this class of epoxides has become available only very recently, when Jørgensen reported the asymmetric epoxidation of enals, under environmentally friendly conditions, using H_2O_2 (35% w/w in water) and the *O*-trimethylsilyl-protected L-diaryl pyrrolidinemethanol **98** (10 mol%) in dichloromethane at room temperature [89]. Different oxygen donors, such as UHP, TBHP, and CMHP, proved to be

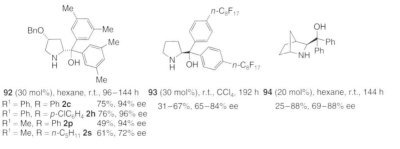

92 (30 mol%), hexane, r.t., 96–144 h   **93** (30 mol%), r.t., CCl_4, 192 h   **94** (20 mol%), hexane, r.t., 144 h
R^1 = Ph, R = Ph **2c**   75%, 94% ee   31–67%, 65–84% ee   25–88%, 69–88% ee
R^1 = Ph, R = *p*-ClC_6H_4 **2h** 76%, 96% ee
R^1 = Me, R = Ph **2p**   49%, 94% ee
R^1 = Me, R = *n*-C_5H_{11} **2s** 61%, 72% ee

Scheme 9.40 Asymmetric epoxidation of *trans*-enones with catalysts **92–94**/TBHP system.

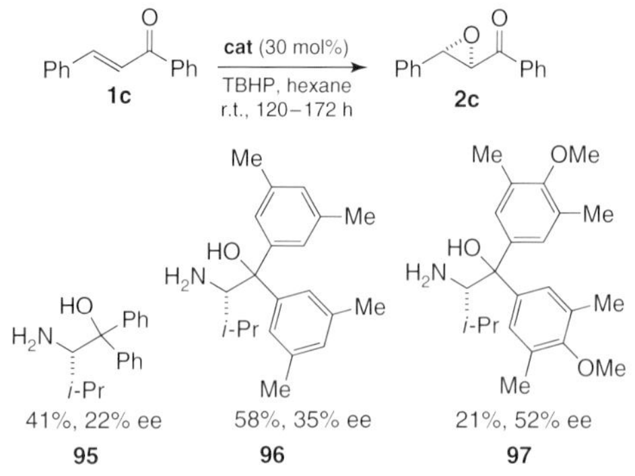

Scheme 9.41 Asymmetric epoxidation of *trans*-chalcone **1c** with the primary β-amino alcohols **95–97**/TBHP system.

of comparable efficiency in the epoxidation, while the reaction could be carried out in a wide range of nonpolar, protic, and halogenated solvents. A variety of epoxides were obtained in good to high yield, and with a high diastereoselectivity in favor of the *trans*-epoxide, which showed excellent ee-values (Scheme 9.42).

The formation of enantioenriched β-disubstituted epoxides is also feasible, as demonstrated by the epoxidation of citral **99f** to afford compound *trans*-**100f**, the sex pheromone from an acaric mite, in good yield and with fairly good control of the diastereoselectivity and enantioselectivity. The mechanism proposed for the epoxidation involved the generation of a chiral iminium ion [90] as the catalytically active species (Figure 9.14). In this case, the enantioselective step was the conjugate addition of H_2O_2 at the β-position of the less sterically hindered face of the iminium ion carbon–carbon double bond, leading to formation of the chiral enamine intermediate. The same group improved the methodology by using organocatalyst **98** and H_2O_2 in a benign solvent mixture (e.g., ethanol/water) at room temperature [91].

Córdova reported that the asymmetric epoxidation of *trans*-α,β-unsaturated aldehydes could be promoted by various L-proline derivatives with variable degrees of

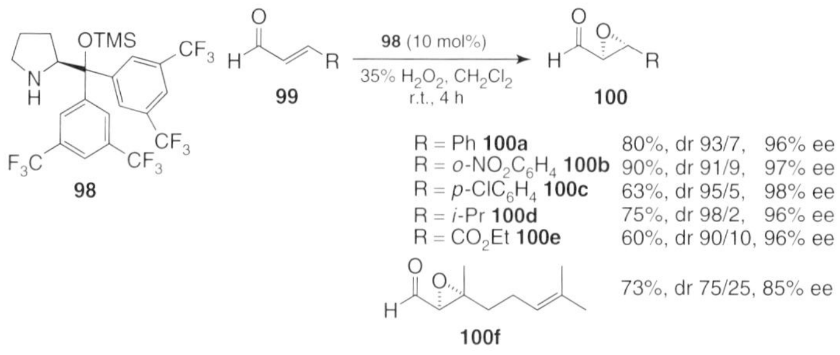

Scheme 9.42 Asymmetric epoxidation of *trans*-α,β-unsaturated aldehydes with the TMS-prolinol **98**/H_2O_2 system.

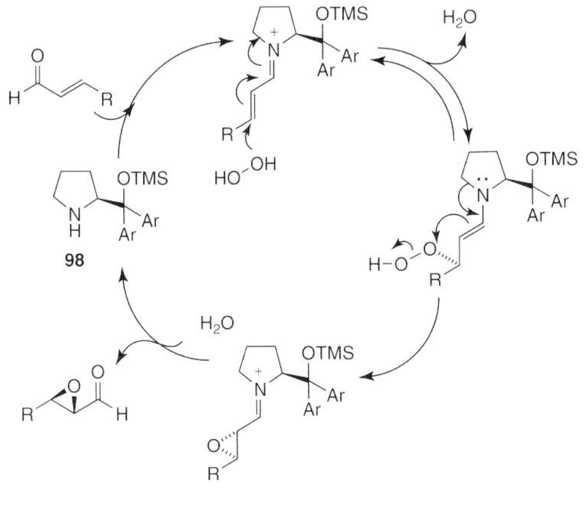
OTMS
H2O
Ar
Ar
R
HO OH
98
H-O
H2O

Figure 9.14 Proposed catalytic cycle for the asymmetric epoxidation of *trans*-enals.

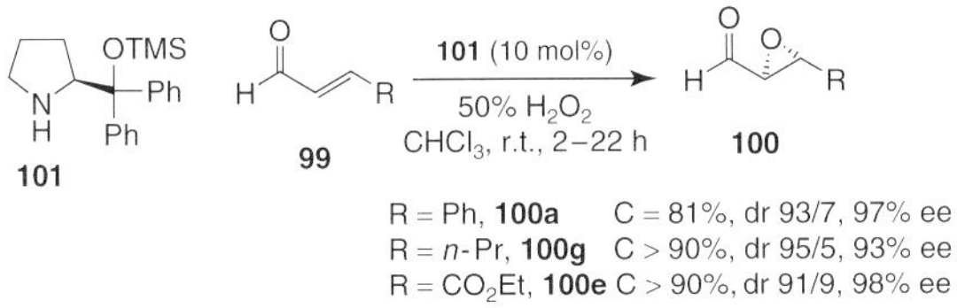
OTMS
Ph
Ph
101
99
101 (10 mol%)
50% H2O2
CHCl3, r.t., 2–22 h
100
R = Ph, 100a C = 81%, dr 93/7, 97% ee
R = n-Pr, 100g C > 90%, dr 95/5, 93% ee
R = CO2Et, 100e C > 90%, dr 91/9, 98% ee

Scheme 9.43 Asymmetric epoxidation of *trans*-α,β-unsaturated aldehydes with the **101**/H_2O_2 system.

activity and stereoselectivity [92]. In this case, *O*-protected prolinol **101** proved to be the most efficient when used at 10 mol% loading at room temperature in chloroform with H_2O_2 (50% w/w in water), affording the epoxides in high conversion, and excellent diastereoselectivity and enantioselectivity (Scheme 9.43).

The epoxidation of *trans*-α,β-unsaturated aldehydes can be also carried out by using 2-(fluorodiphenylmethyl)pyrrolidine **102** at 10 mol% loading in $CHCl_3$ with H_2O_2, as recently attested by Gilmour [93]. It was envisaged that the presence of fluorine beta to an electron-withdrawing group would lead to a preference for a gauche conformation for the iminium ion, resulting from condensation of the secondary β-fluoroamine with an aldehyde – an effect that would provide an extra degree of torsional rigidity and might assist the asymmetric induction. Indeed, this proved to be true as the epoxides were recovered in good to high yield, diastereoselectivity, and excellent enantiocontrol (Scheme 9.44).

9.3.6 Imidazolidinone Salt Catalysts

MacMillan reported imidazolidinone salt **103** as the catalyst in the epoxidation of *trans*-α,β-unsaturated aldehydes with [(nosylimino)iodo]benzene (NsNIPh) as the oxidant [94]. In a solvent mixture of dichloromethane or chloroform/acetic acid

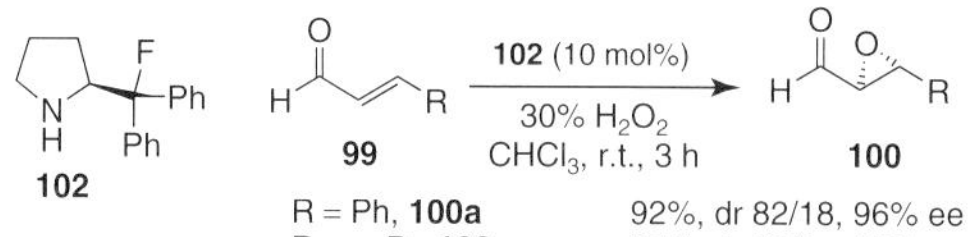

R = Ph, **100a** 92%, dr 82/18, 96% ee
R = *n*-Pr, **100g** 87%, dr 92/8, 95% ee
citral epoxide, **100f** 91%, dr 60/40, 81% ee
R = *p*-$NO_2C_6H_4$, **100h** 93%, dr 69/31, 96% ee

Scheme 9.44 Asymmetric epoxidation of *trans*-α,β-unsaturated aldehydes with the **102**/H_2O_2 system.

at −30 °C, and with 20 mol% of compound **103**, **the** different α,β-unsaturated aldehydes were converted into their corresponding epoxides in good to high yield, and with complete diastereocontrol for the *trans*-epoxides and high enantioselectivity (Scheme 9.45).

The mechanistic hypothesis proposed to justify these results was based on the formation of a covalent iminium species between the catalyst and the aldehyde, as shown previously for the proline-based catalysts **98**, **101**, **102**. The *t*-butyl and benzyl groups in compound **103** regulated the iminium geometry through nonbonding steric interactions, and the face selectivity for the nucleophilic attack at β-position by effective face-shielding, respectively (Figure 9.15).

9.3.7 Primary Amines

Among the different classes of alkenes, the cyclic enones are recognized as being extremely difficult to epoxidize in a highly enantioselective manner. The first powerful system to achieve asymmetric epoxidation of the cyclic enones, using H_2O_2 (50% w/w in water) and a 10 mol% loading of different chiral primary diamine salts, was developed by List [95]. It was proposed that these bifunctional catalysts could activate both the enone via iminium ion **I** formation and H_2O_2 via a general base catalysis (Scheme 9.46). Subsequently, 9-amino-9-deoxyepiquinine salt **104** proved to be the most effective promoter for the epoxidation of different

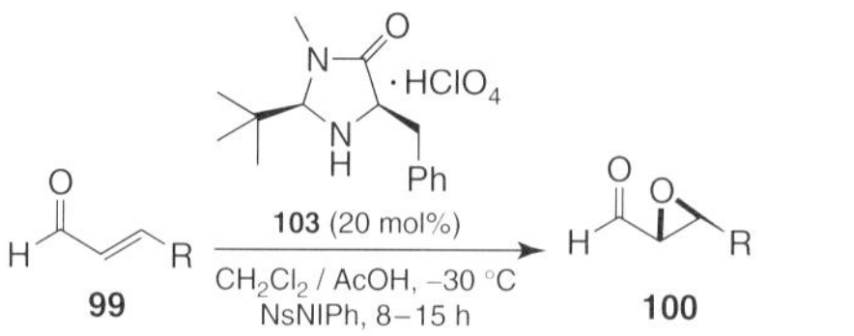

R = Ph, **100a** 92%, 92% ee
R = *n*-Pr, **100g** 72%, 88% ee
R = *p*-$NO_2C_6H_4$, **100h** 89%, 97% ee
R = CH_2OBn, **100i** 89%, 85% ee

Scheme 9.45 Asymmetric epoxidation of *trans*-α,β-unsaturated aldehydes with the **103**/NsNIPh system.

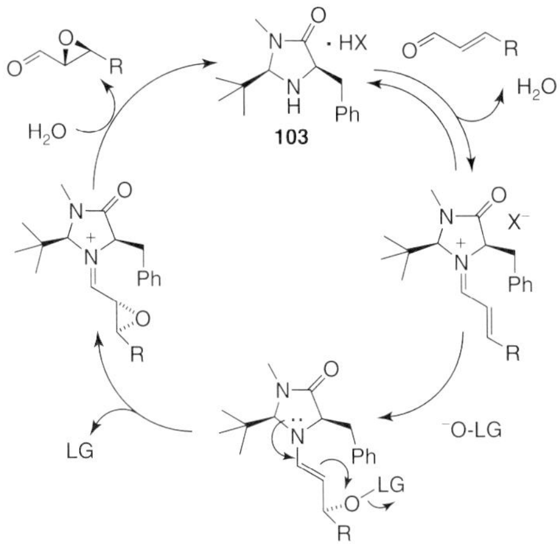

Figure 9.15 Proposed catalytic cycle for the asymmetric epoxidation of *trans*-enals mediated by **103**/NsNIPh system.

substituted cyclohexenones, cycloheptenones and cyclopentenone, with all products being isolated in good yield and with excellent ee-values.

The same activation strategy was independently applied by Deng [96] and List [97] in the asymmetric epoxidation of acyclic aliphatic *trans*-enones, mediated by organocatalyst **104** under different conditions (Scheme 9.47). In this case, Deng reported the epoxidation of compounds **1** using 10 mol% of catalyst **104** at 23 or 55 °C in toluene and with CMHP as the oxidant. The epoxides were isolated in moderate to high yield and excellent enantioselectivity. The presence of β-peroxide **107** was also observed, although the reaction conditions were optimized to reduce such formation. Under conditions previously optimized for cyclic enones, List observed the preferential formation of cyclic peroxyhemiketal intermediates **108**, and of the epoxides **2** as minor products. A basic treatment of the crude reaction mixture then led to a quantitative transformation of the peroxyhemiketals **108** to the epoxides, in generally good to high yields and with excellent enantioselectivity.

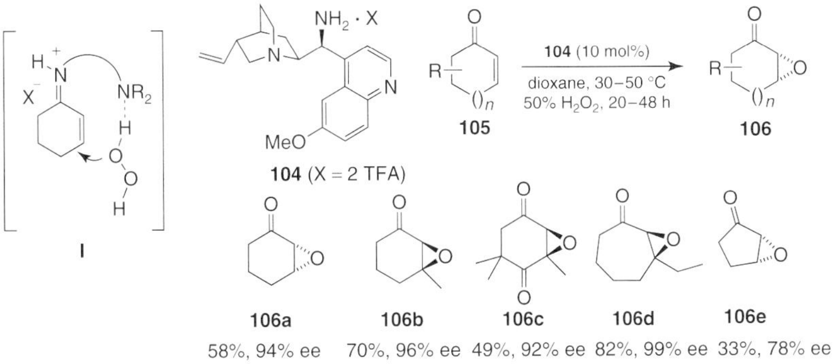

Scheme 9.46 Asymmetric epoxidation of cyclic enones with the **104**/H_2O_2 system.

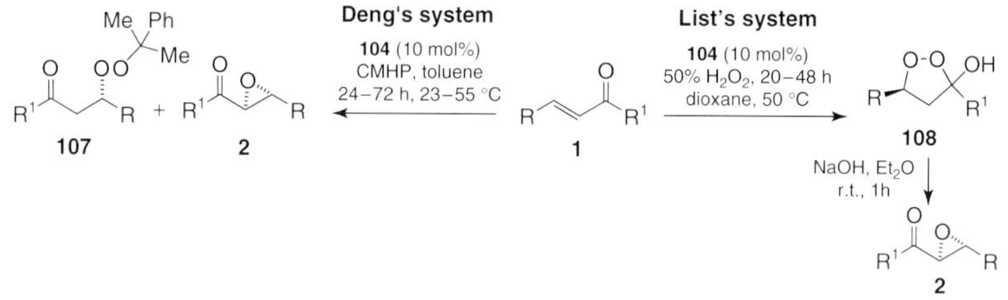

9 Asymmetric Epoxidations of α,β-Unsaturated Carbonyl Compounds

Scheme 9.47 Asymmetric epoxidation of aliphatic enones with the **104**/ROOH systems.

Deng's system: **104** (10 mol%), CMHP, toluene, 24–72 h, 23–55 °C

R^1 = Me, R = $(CH_2)_2$Ph **2r** 2/107 99/1, 88%, 97% ee
R^1 = Me, R = n-C_5H_{11} **2s** 2/107 99/1, 91%, 97% ee
R^1 = Me, R = Me **2l′** 2/107 68/32, 55%, 97% ee
R^1 = n-Bu, R = Me **2m′** 2/107 87/13, 71%, 97% ee

List's system: **104** (10 mol%), 50% H_2O_2, 20–48 h, dioxane, 50 °C; then NaOH, Et_2O, r.t., 1h

R^1 = Me, R = $(CH_2)_2$Ph **2r** 85%, 97% ee
R^1 = Me, R = n-C_6H_{13} **2l** 72%, 97% ee
R^1 = Et, R = Me **2m** 55%, 97% ee
R^1 = i-Bu, R = n-C_5H_{11} **2m′** 81%, 97% ee

Enones **1** with an aromatic residue at the double bond, or with a trisubstituted double bond, were shown to be unreactive under these conditions.

Recently, List has developed a new concept for enantioselective synthesis, termed "asymmetric counteranion-directed catalysis" (ACDC) [98]. The proposal is that, catalytic reactions which proceed via cationic intermediates can be performed in an enantioselective manner if a chiral counteranion is incorporated into the catalyst. A highly effective and general system for the epoxidation of trans-α,β-unsaturated aldehydes has been developed based on this strategy, using 10 mol% of the dibenzylammonium salt of 3,3′-bis-(2,4,6-triisopropyl-phenyl)-1,1′-binaphthyl-2,2′-diyl hydrogen phosphate (TRIP) **109** in dioxane or tert-butyl methyl ether with TBHP at 35 °C (Scheme 9.48) [99].

Both, aromatic and aliphatic 1,2-disubstituted enals gave the epoxides in good yield, with excellent diastereoselectivity and good to high enantiocontrol. Moreover, β,β-disubstituted α,β-unsaturated aldehydes **99lm** were, for the first time, obtained in very high enantioselectivity. Interestingly, the epoxidation of citral afforded a result for the trans-epoxide **100f** comparable to that achieved by using Jørgensen's

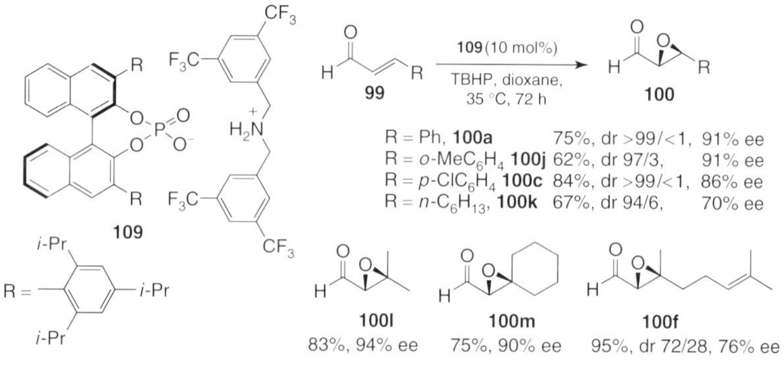

99 → **109** (10 mol%), TBHP, dioxane, 35 °C, 72 h → **100**

R = Ph, **100a** 75%, dr >99/<1, 91% ee
R = o-MeC$_6$H$_4$ **100j** 62%, dr 97/3, 91% ee
R = p-ClC$_6$H$_4$ **100c** 84%, dr >99/<1, 86% ee
R = n-C$_6$H$_{13}$, **100k** 67%, dr 94/6, 70% ee

100l 83%, 94% ee
100m 75%, 90% ee
100f 95%, dr 72/28, 76% ee

Scheme 9.48 Asymmetric epoxidation of trans-α,β-unsaturated aldehydes with the **109**/TBHP system.

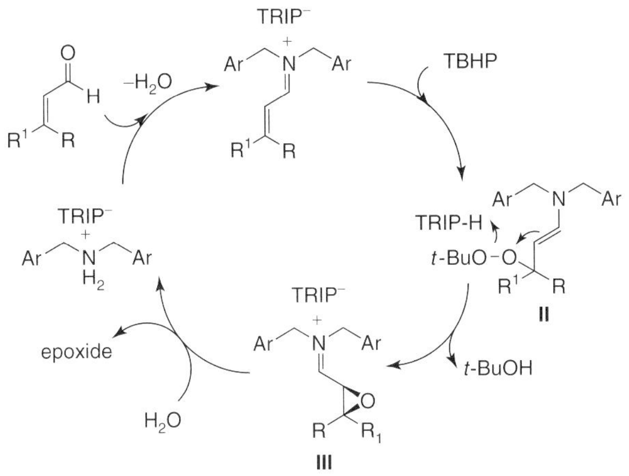

Figure 9.16 Proposed catalytic cycle for the asymmetric epoxidation of *trans*-enals.

system, and the *cis*-epoxide **100f** was obtained with 92% ee. From a mechanistic point of view, a catalytic cycle has been suggested where the most intriguing point is control of the asymmetric induction (Figure 9.16).

The stereogenic center in the product is not created in the conjugate addition step, as is usually observed, since the intermediate **II** is achiral. Rather, it is only in the cyclization to iminium ion **III** that the stereogenic center is created, which implies that the chiral TRIP anion is involved in the carbon–oxygen bond formation. This is an unusual example of an enantioselective Michael addition reaction organocatalyzed by an achiral amine with a chiral Brønsted acid, and contrasts with the commonly reported iminium activation that occurs via the use of a secondary or primary chiral amine with an achiral Brønsted acid [100].

9.4
Conclusions

As shown in this chapter, several metal-catalyzed and organocatalyzed oxidative systems have been developed during the past two decades for the enantioselective nucleophilic epoxidation of electron-poor alkenes. The level of diastereoselectivity is very high in the majority of protocols reported. Moreover, the control of enantioselectivity has reached useful levels (>90% ee) in the epoxidation of *trans*-α,β-unsaturated ketones such as chalcones, alkyl aryl ketones, and dialkyl ketones. The enantioselective epoxidation of α,β-unsaturated esters, amides, and derivatives thereof, can be performed only with lanthanide/BINOL-based systems. Hence, apart from the alternative use of chiral dioxiranes as epoxidizing systems for enoates, the development of nucleophilic protocols for their epoxidation would be highly desirable in view of their synthetic applications. Recent years have witnessed remarkable improvements in the area of organocatalyzed procedures, with the use of secondary and primary amines derived from α-amino acids or cinchona alkaloids as catalysts. Although currently employed at relatively high loadings (10–20 mol%), these systems have also solved the longstanding problem of the asymmetric epoxidation of α,β-unsaturated aldehydes and simple cyclic enones, achieving a high control of both diastereoselectivity and enantioselectivity. Besides the asymmetric epoxidation of acyclic *cis*-α,β-unsaturated carbonyl compounds, which is still at a poor stage of development, effort should be devoted to expand the substrate scope of the electron-poor alkenes that are susceptible to asymmetric epoxidation. In this context, the organocatalytic systems may play a major role, given the great variety of chiral promoters that is accessible (at least potentially) and the covalent and noncovalent types of activation that can be taken advantage of.

9.5
Experimental

Typical Procedure for Asymmetric Epoxidation of α,β-Unsaturated Esters [39]

The biphenyl diol was prepared in three steps from 2,2′,6,6′-tetrahydroxybiphenyl. To a stirred mixture of 4AMS (250 mg; 1000 mg mmol^{-1} of starting material; the 4AMS was dried for 3 h at 180 °C under reduced pressure) biphenyldiol **32** was added (7.2 mg, 0.025 mmol) and triphenylarsine oxide (8.1 mg, 0.025 mmol) as a THF solution (1.125 ml). Y(O-*i*-Pr)$_3$ (0.025 mmol, 0.2 M solution in THF) was then added to the reaction mixture at room temperature. After stirring for 45 min, TBHP (0.375 ml, 1.5 mmol, 4.0 M solution in toluene) was added. After stirring for a further 10 min, *trans*-α,β-unsaturated ester (220.3 mg, 1.25 mmol) was added and the mixture stirred at room temperature. Following consumption of the starting material, the reaction mixture was diluted with ethyl acetate and quenched with 2% citric acid (2.5 ml). The aqueous layer was extracted with ethyl acetate, and the combined organic layers were washed with brine and dried over sodium sulfate. After concentration under reduced pressure, the residue was purified by flash chromatography on silica gel (hexane:ethyl acetate mixtures, 100:1 to 50:1) to give the *trans*-epoxy ester (212.7 mg, 89%) in 99% ee. The ee-value of the product was determined using chiral stationary-phase HPLC analysis (DAICEL CHIRALPAK AD-H; *i*-PrOH:hexane, 2:98; flow rate 0.4 ml min^{-1}; t_R (2*S*,3*R*) = 31.5 min; t_R(2*R*,3*S*) = 38.0 min; detection at 254 nm).

Asymmetric Epoxidation of *trans*-Chalcone Catalyzed by Diaryl Prolinol 90/TBHP System [83]

TBHP (5–6 M decane solution, 33 µl, 0.18 mmol) was added to a stirred solution of commercially available catalyst **90** (9.2 mg, 0.03 mmol) and *trans*-chalcone (31.2 mg, 0.15 mmol) in hexane (0.215 ml) at 4 °C. Stirring was maintained until consumption of the starting material. The crude reaction mixture was directly purified by flash chromatography on silica gel (petroleum ether : diethyl ether, 99 : 1) to provide the *trans*-epoxy ketone (30.3 mg, 90%) in 91% ee. The ee-value of the product was determined using chiral stationary-phase HPLC analysis (DAICEL CHIRALCEL OD, 254 nm, *i*-PrOH : hexane, 2 : 98; flow rate 1.0 ml min^{-1}; t_R(2S,3R) = 17.3 min; t_R(2R,3S) = 18.5 min; detection at 254 nm).

References

1. For reviews on asymmetric epoxidation, see: (a) Bäckvall, J.E. (ed.) (2004) *Modern Oxidation Methods*, Wiley-VCH Verlag GmbH, Weinheim; (b) Xia, Q.H., Ge, H.Q., Ye, C.P., Liu, Z.M., and Su, K.X. (2005) *Chem. Rev.*, **105**, 1603–1662.
2. (a) Furukawa, N., Sugihara, Y., and Fujihara, H. (1989) *J. Org. Chem.*, **54**, 4222–4224; (b) Li, A.-H., Dai, L.-X., Hou, X.-L., Huang, Y.-Z., and Li, F.-W. (1996) *J. Org. Chem.*, **61**, 489–493; (c) Solladié-Cavallo, A. and Diep-Vohuule, A. (1995) *J. Org. Chem.*, **60**, 3494–3498; (d) Saito, T., Akiba, D., Sakairi, M., and Kanazawa, S. (2001) *Tetrahedron Lett.*, **42**, 57–59; (e) Julienne, K. and Metzner, P. (1998) *J. Org. Chem.*, **63**, 4532–4534; (f) Aggarwal, V.K., Ford, J.G., Thompson, A., Jones, R.V.H., and Standen, M. (1996) *J. Am. Chem. Soc.*, **118**, 7004–7005; (g) Aggarwal, V.K., Alonso, E., Hynd, G., Lydon, K.M., Palmer, M.J., Porcelloni, M., and Studley, J.R. (2001) *Angew. Chem., Int. Ed.*, **40**, 1430–1433.
3. (a) Colonna, S., Fornasier, R., and Pfeiffer, U. (1978) *J. Chem. Soc., Perkin Trans. 1*, 8–11; (b) Hummelen, J.C. and Wynberg, H. (1978) *Tetrahedron Lett.*, **19**, 1089–1092; (c) Bako, P., Szöllosy, A., Bombicz, P., and Töke, L. (1997) *Synlett*, 291–292; (d) Achard, T.J.R., Belokon, Y.N., Ilyin, M., Moskalenko, M., Northa, M., and Pizzato, F. (2007) *Tetrahedron Lett.*, **48**, 2965–2969.
4. Yudin, A.K. (ed.) (2006) *Aziridines and Epoxides in Organic Synthesis*, Chapters 7-9, Wiley-VCH Verlag GmbH, Weinheim.
5. (a) Katsuki, T. and Sharpless, K.B. (1980) *J. Am. Chem. Soc.*, **102**, 5974–5976; (b) Gao, Y., Hanson, R.M., Klunder, J.M., Ko, S.Y., Masamune, H., and Sharpless, K.B. (1987) *J. Am. Chem. Soc.*, **109**, 5765–5780; (c) Katsuki, T. (1999) in *Comprehensive Asymmetric Catalysis* (eds E.N. Jacobsen, A. Pfaltz, and H. Yamamoto), Springer, Berlin, pp. 621–648.
6. (a) Jacobsen, E.N. and Wu, M.H. (1999) in *Comprehensive Asymmetric*

Catalysis (eds E.N. Jacobsen, A. Pfaltz, and H. Yamamoto), Springer, Berlin, pp. 649–678; (b) Katsuki, T. (1995) *Coord. Chem. Rev.*, **140**, 189–214.

7. (a) Tse, M.K., Döbler, C., Bhor, S., Klawonn, M., Mägerlein, W., Hugl, H., and Beller, M. (2004) *Angew. Chem., Int. Ed.*, **43**, 5255–5260; (b) Gelalcha, F.G., Bitterlich, B., Anilkumar, G., Tse, M.K., and Beller, M. (2007) *Angew. Chem., Int. Ed.*, **46**, 7293–7296; (c) Colladon, M., Scarso, A., Sgarbossa, P., Michelin, R.A., and Strukul, G. (2007) *J. Am. Chem. Soc.*, **129**, 7680–7689.

8. (a) Yang, D., Yip, Y.-C., Tang, M.-W., Wong, M.-K., Zheng, J.-H., and Cheung, K.-K. (1996) *J. Am. Chem. Soc.*, **118**, 491–492; (b) Yang, D., Wong, M.-K., Yip, Y.-C., Wang, X.-C., Tang, M.-W., Zheng, J.-H., and Cheung, K.-K. (1998) *J. Am. Chem. Soc.*, **120**, 5943–5952; (c) Yang, D. (2004) *Acc. Chem. Res.*, **37**, 497–505.

9. (a) Tu, Y., Wang, Z.-X., and Shi, Y. (1996) *J. Am. Chem. Soc.*, **118**, 9806–9807; (b) Shi, Y. (2004) *Acc. Chem. Res.*, **37**, 488–496; (c) Goeddel, D., Shu, L., Yuan, Y., Wong, O.A., Wang, B., and Shi, Y. (2006) *J. Org. Chem.*, **71**, 1715–1717; (d) Wong, O.A. and Shi, Y. (2006) *J. Org. Chem.*, **71**, 3973–3976; (e) Burke, C.P. and Shi, Y. (2006) *Angew. Chem., Int. Ed.*, **45**, 4475–4478.

10. Davies, F.A., Harakal, M.E., and Awad, S.B. (1983) *J. Am. Chem. Soc.*, **105**, 3123–3126.

11. (a) Bohé, L., Hanquet, G., Lusinchi, M., and Lusinchi, X. (1993) *Tetrahedron Lett.*, **34**, 7271–7274; (b) Aggarwal, V.K. and Wang, F.M. (1996) *Chem. Commun.*, 191–192; (c) Page, P.C.B., Rassias, G.A., Bethell, D., and Schilling, M.B. (1998) *J. Org. Chem.*, **63**, 2774–2777; (d) Lacour, J., Monchaud, D., and Marsol, C. (2002) *Tetrahedron Lett.*, **43**, 8257–8260; (e) Armstrong, A., Ahmed, G., Garnett, I., and Goacolou, K. (1997) *Synlett*, 1075–1076.

12. (a) Adamo, M.F.A., Aggarwal, V.K., and Sage, M.A. (2000) *J. Am. Chem. Soc.*, **122**, 8317–8318; (b) Aggarwal, V.K., Lopin, C., and Sandrinelli, F. (2003) *J. Am. Chem. Soc.*, **125**, 7596–7601; (c) Ho, C.Y., Chen, Y.-C., Wong, M.-K., and Yang, D. (2005) *J. Org. Chem.*, **70**, 898–906.

13. Peris, G., Jacobsche, C.E., and Miller, S.J. (2007) *J. Am. Chem. Soc.*, **129**, 8710–8711.

14. For reviews, see: (a) Porter, M.J. and Skidmore, J. (2000) *Chem. Commun.*, 1215–1225; (b) Nemoto, T., Ohshima, T., and Shibasaki, M. (2002) *J. Synth. Org. Chem. Jpn.*, **60**, 94–105; (c) Lattanzi, A. (2008) *Curr. Org. Synth.*, **5**, 117–133; (d) Díez, D., Núñez, M.G., Antón, A.B., García, P., Moro, R.F., Garrido, N.M., Marcos, I.S., Basabe, P., and Urones, J.G. (2008) *Curr. Org. Synth.*, **5**, 186–216.

15. Lauret, C. (2001) *Tetrahedron: Asymmetry*, **12**, 2359–2383.

16. Weitz, E. and Scheffer, A. (1921) *Chem. Ber.*, **54**, 2327–2344.

17. (a) Enders, D., Zhu, J., and Raabe, G. (1996) *Angew. Chem., Int. Ed.*, **35**, 1725–1728; (b) Enders, D., Zhu, J., and Kramps, L. (1997) *Liebigs Ann. Recl.*, 1101–1113.

18. Enders, D., Kramps, L., and Zhu, J. (1998) *Tetrahedron: Asymmetry*, **9**, 3959–3962.

19. Yu, H.-B., Zheng, X.-F., Lin, Z.-M., Hu, Q.-S., Huang, W.-S., and Pu, L. (1999) *J. Org. Chem.*, **64**, 8149–8155.

20. Elston, C.L., Jackson, R.F.W., MacDonald, S.J.F., and Murray, P.J. (1997) *Angew. Chem., Int. Ed.*, **36**, 410–412.

21. Jacques, O., Richards, S.J., and Jackson, R.F.W. (2001) *Chem. Commun.*, 2712–2713.

22. Hinch, M., Jacques, O., Drago, C., Caggiano, L., Jackson, R.F.W., Dexter, C., Ansone, M.S., and Macdonald, S.J.F. (2006) *J. Mol. Catal. A: Chem.*, **251**, 123–128.

23. (a) Minatti, A. and Dötz, K.H. (2004) *Synlett*, 1634–1636; (b) Minatti, A. and Dötz, K.H. (2006) *Eur. J. Org. Chem.*, 268–276.

24. Ding, K., Wang, Z., and Shi, L. (2007) *Pure Appl. Chem.*, **79**, 1531–1540.

25. Wang, H., Wang, Z., and Ding, K. (2009) *Tetrahedron Lett.*, **50**, 2200–2203.
26. For a review, see: Shibasaki, M. and Yoshikawa, N. (2002) *Chem. Rev.*, **102**, 2187–2209.
27. Bougauchi, M., Watanabe, S., Arai, T., Sasai, H., and Shibasaki, M. (1997) *J. Am. Chem. Soc.*, **119**, 2329–2330.
28. Watanabe, S., Arai, T., Sasai, H., Bougauchi, M., and Shibasaki, M. (1998) *J. Org. Chem.*, **63**, 8090–8091.
29. Daikai, K., Kamaura, M., and Inanaga, J. (1998) *Tetrahedron Lett.*, **39**, 7321–7322.
30. Nemoto, T., Kakei, H., Ganadesikan, V., Tosaki, S.-Y., Ohshima, T., and Shibasaki, M. (2001) *J. Am. Chem. Soc.*, **123**, 2725–2732.
31. Chen, R., Qian, C., and de Vries, J.G. (2001) *Tetrahedron*, **57**, 9837–9842.
32. Jayaprakash, D., Kobayashi, Y., Arai, T., Hu, Q.-S., Zheng, X.-F., Pu, L., and Sasai, H. (2003) *J. Mol. Catal. A: Chem.*, **196**, 145–149.
33. Wang, X., Shi, L., Li, M., and Ding, K. (2005) *Angew. Chem., Int. Ed.*, **44**, 6362–6366.
34. (a) Nemoto, T., Ohshima, T., and Shibasaki, M. (2001) *J. Am. Chem. Soc.*, **123**, 9474–9475; (b) Kinoshita, T., Okada, S., Park, S.-R., Matsunaga, S., and Shibasaki, M. (2003) *Angew. Chem., Int. Ed.*, **42**, 4680–4684; (c) Kakei, H., Nemoto, T., Ohshima, T., and Shibasaki, M. (2004) *Angew. Chem., Int. Ed.*, **43**, 317–320.
35. Nemoto, T., Kakei, H., Gnanadesikan, V., Tosaki, S., Ohshima, T., and Shibasaki, M. (2002) *J. Am. Chem. Soc.*, **124**, 14544–14545.
36. Ishida, K., Okita, Y., Matsuda, H., Okino, T., and Murakami, M. (1999) *Tetrahedron*, **55**, 10971–10988.
37. (a) Wang, Z.-X. and Shi, Y. (1997) *J. Org. Chem.*, **62**, 8622–8623; (b) Wang, Z.X., Miller, S.M., Anderson, O.P., and Shi, Y. (1999) *J. Org. Chem.*, **64**, 6443–6458; (c) Wu, X.-Y., She, X., and Shi, Y. (2002) *J. Am. Chem. Soc.*, **124**, 8792–8793.
38. Jacobsen, E.N., Deng, L., Furukawa, Y., and Martinez, L.E. (1994) *Tetrahedron*, **50**, 4323–4334, and references therein cited.
39. Kakei, H., Tsuji, R., Ohshima, T., and Shibasaki, M. (2005) *J. Am. Chem. Soc.*, **127**, 8962–8963.
40. Nemoto, T., Ohshima, T., and Shibasaki, M. (2000) *Tetrahedron Lett.*, **41**, 9569–9574.
41. Tosaki, S., Nemoto, T., Ohshima, T., and Shibasaki, M. (2003) *Org. Lett.*, **5**, 495–498.
42. (a) Helder, R., Hummelen, J.C., Laane, R.W.P.M., Wiering, J.S., and Wynberg, H. (1976) *Tetrahedron Lett.*, **17**, 1831–1834; (b) Wynberg, H. and Greijdanus, B. (1978) *J. Chem. Soc. Chem Commun.*, 427–428; (c) Wynberg, H. and Marsman, B. (1980) *J. Org. Chem.*, **45**, 158–161; (d) Plium, H., and Wynberg, H. (1980) *J. Org. Chem.*, **45**, 2498–2502.
43. Arai, S., Tsuge, H., Oku, M., Miura, M., and Shioiri, T. (2002) *Tetrahedron*, **58**, 1623–1630.
44. (a) Lygo, B. and Wainwright, P.G. (1998) *Tetrahedron Lett.*, **39**, 1599–1602; (b) Lygo, B. and Wainwright, P.G. (1999) *Tetrahedron*, **55**, 6289–6300.
45. Corey, E.J. and Zhang, F.-Y. (1999) *Org. Lett.*, **1**, 1287–1290.
46. Lygo, B. and To, D.C.M. (2001) *Tetrahedron Lett.*, **42**, 1343–1346.
47. Ye, J., Wang, Y., Liu, R., Zhang, G., Zhang, Q., Chen, J., and Liang, X. (2003) *Chem. Commun.*, 2714–2715.
48. Lygo, B., Gardiner, S.D., McLeod, M.C., and To, D.C.M. (2007) *Org. Biomol. Chem.*, **5**, 2283–2290.
49. Adam, W., Rao, P.B., Degen, H.-G., Levai, A., Patonay, T., and Saha-Möller, C.R. (2002) *J. Org. Chem.*, **67**, 259–264.
50. Berkessel, A., Guixà, M., Schmidt, F., Neudörfl, J.M., and Lex, J. (2007) *Chem. Eur. J.*, **13**, 4483–4498.
51. Jew, S.-S., Lee, J.-H., Jeong, B.-S., Yoo, M.-S., Kim, M.-J., Lee, Y.-J., Lee, J., Choi, S.-H., Lee, K., Lah, M.-S., and Park, H.-G. (2005) *Angew. Chem., Int. Ed.*, **44**, 1383–1385.
52. Lu, J., Wang, X., Liu, J., Zhang, L., and Wang, Y. (2006) *Tetrahedron: Asymmetry*, **17**, 330–335.

53. Ooi, T., Ohara, D., Tamura, M., and Maruoka, K. (2004) *J. Am. Chem. Soc.*, **126**, 6844–6845.
54. Bakó, P., Bakó, T., Mészáros, A., Keglevich, G., Szöllösy, À., Bodor, S., Makó, A., and Töke, L. (2004) *Synlett*, 643–646.
55. Hori, K., Tamura, M., Tani, K., Nishiwaki, N., Ariga, M., and Tohda, Y. (2006) *Tetrahedron Lett.*, **47**, 3115–3118.
56. Allingham, M.T., Howard-Jones, A., Murphy, P.J., Thomas, D.A., and Caulkett, P.W.R. (2003) *Tetrahedron Lett.*, **44**, 8677–8680.
57. Tanaka, S. and Nagasawa, K. (2009) *Synlett*, 667–670.
58. (a) Juliá, S., Masana, J., and Vega, J.C. (1980) *Angew. Chem., Int. Ed.*, **19**, 929–931; (b) Juliá, S., Guixer, J., Masana, J., Rocas, J., Colonna, S., Annunziata, R., and Molinari, H. (1982) *J. Chem. Soc., Perkin Trans. 1*, 1317–1324.
59. Lasterra-Sanchéz, M.E. and Roberts, S.M. (1995) *J. Chem. Soc., Perkin Trans. 1*, 1467–1468.
60. (a) Adger, B.M., Barkley, J.V., Bergeron, S., Cappi, M.W., Flowerdew, B.E., Jackson, M.P., MacCague, R., Nugent, T.C., and Roberts, S.M. (1997) *J. Chem. Soc., Perkin Trans. 1*, 3501–3508; (b) Bentley, P.A., Bergeron, S., Cappi, M.W., Hibbs, D.E., Hurst-House, M.B., Nugent, T.C., Pulido, R., Roberts, S.M., and Wu, L.E. (1997) *J. Chem. Soc., Chem. Commun.*, 739–740.
61. Allen, J.V., Drauz, K.-H., Flood, R.W., Roberts, S.M., and Skidmore, J. (1999) *Tetrahedron Lett.*, **40**, 5417–5420.
62. Geller, T.P. and Roberts, S.M. (1999) *J. Chem. Soc., Perkin Trans. 1*, 1397–1398.
63. Yi, H., Zou, G., Li, Q., Chen, Q., Tang, J., and He, M.-Y. (2005) *Tetrahedron Lett.*, **46**, 5665–5668.
64. Flood, R.W., Geller, T.P., Petty, S.A., Roberts, S.M., Skidmore, J., and Volk, M. (2001) *Org. Lett.*, **3**, 683–686.
65. Tsogoeva, S.B., Wöltinger, J., Jost, C., Reichert, D., Kühnle, A., Krimmer, H.-P., and Drauz, K. (2002) *Synlett*, 707–710.
66. Bentley, P.A., Flood, R.W., Roberts, S.M., Skidmore, J., Smith, C.B., and Smith, J.A. (2001) *Chem. Commun.*, 1616–1617.
67. Berkessel, A., Gasch, N., Glaubitz, K., and Koch, C. (2001) *Org. Lett.*, **3**, 3839–3842.
68. (a) Kelly, D.R., Caroff, E., Flood, R.W., Heal, W., and Roberts, S.M. (2004) *Chem. Commun.*, 2016–2017; (b) Kelly, D.R. and Roberts, S.M. (2004) *Chem. Commun.*, 2018–2020; (c) Kelly, D.R. and Roberts, S.M. (2006) *Biopolymers*, **84**, 74–89.
69. Geller, T., Gerlach, A., Krüger, C.M., and Militzer, H.-C. (2004) *Tetrahedron Lett.*, **45**, 5065–5067.
70. Geller, T., Krüger, C.M., and Militzer, H.-C. (2004) *Tetrahedron Lett.*, **45**, 5069–5071.
71. Carrea, G., Colonna, S., Meek, A.D., Ottolina, G., and Roberts, S.M. (2004) *Tetrahedron: Asymmetry*, **15**, 2945–2949.
72. Mathew, S.P., Gunathilagan, S., Roberts, S.M., and Blackmond, D.G. (2005) *Org. Lett.*, **7**, 4847–4850.
73. (a) Adam, W., Rao, P.B., Degen, H.-G., and Saha-Möller, C.R. (2000) *J. Am. Chem. Soc.*, **122**, 5654–5655; (b) Adam, W., Rao, P.B., Degen, H.-G., and Saha-Möller, C.R. (2002) *Eur. J. Org. Chem.*, 630–639.
74. Dwyer, C.L., Gill, C.D., Ichihara, O., and Taylor, R.J.K. (2000) *Synlett*, 704–706.
75. Bundu, A., Berry, N.G., Gill, C.D., Dwyer, C.L., Stachulski, A.V., Taylor, R.J.K., and Whittall, J. (2005) *Tetrahedron: Asymmetry*, **16**, 283–293.
76. Kośnik, W., Bocian, W., Kozerski, L., Tvaroška, I., and Chmielewski, M. (2008) *Chem. Eur. J.*, **14**, 6087–6097.
77. Aoki, M. and Seebach, D. (2001) *Helv. Chim. Acta*, **84**, 187–207.
78. Lattanzi, A., Cocilova, M., Iannece, P., and Scettri, A. (2004) *Tetrahedron: Asymmetry*, **15**, 3751–3755.
79. Tanaka, Y., Nishimura, K., and Tomioka, K. (2003) *Tetrahedron*, **59**, 4549–4556.
80. (a) Schlessinger, R.H., Bebernitz, G.R., Lin, P., and Poss, A.J. (1985)

J. Am. Chem. Soc., **107**, 1777–1778; (b) Yadav, V.K. and Kapoor, K.K. (1995) *Tetrahedron*, **51**, 8573–8584.
81. (a) MacManus, J.C., Carey, J.S., and Taylor, R.J.K. (2003) *Synlett*, 365–368; (b) MacManus, J.C., Genski, T., Carey, J.S., and Taylor, R.J.K. (2003) *Synlett*, 369–371.
82. Lattanzi, A. (2005) *Org. Lett.*, **7**, 2579–2582.
83. Lattanzi, A. (2006) *Adv. Synth. Catal.*, **348**, 339–346.
84. Lattanzi, A. and Russo, A. (2006) *Tetrahedron*, **62**, 12264–12269.
85. Li, Y., Liu, X., Yang, Y., and Zhao, G. (2007) *J. Org. Chem.*, **72**, 288–291.
86. Cui, H., Li, Y., Zheng, C., Zhao, G., and Zhu, S. (2008) *J. Fluorine Chem.*, **129**, 45–50.
87. Lu, J., Xu, Y.-H., Liu, F., and Loh, T.-P. (2008) *Tetrahedron Lett.*, **49**, 6007–6008.
88. Russo, A. and Lattanzi, A. (2008) *Eur. J. Org. Chem.*, 2767–2773.
89. Marigo, M., Franzén, G., Poulsen, T.B., Zhuang, W., and Jørgensen, K.A. (2005) *J. Am. Chem. Soc.*, **127**, 6964–6965.
90. For a review, see: Erkkilae, A., Majander, I., and Pihko, P.M. (2007) *Chem. Rev.*, **107**, 5416–5470.
91. Zhuang, W., Marigo, M., and Jørgensen, K.A. (2005) *Org. Biomol. Chem.*, **3**, 3883–3885.
92. (a) Sundén, H., Ibrahem, I., and Córdova, A. (2006) *Tetrahedron Lett.*, **47**, 99–103; (b) Zhao, G.L., Ibrahem, I., Sundén, H., and Córdova, A. (2007) *Adv. Synth. Catal.*, **349**, 1210–1224.
93. Sparr, C., Schweizer, W.B., Senn, H.M., and Gilmour, R. (2009) *Angew. Chem., Int. Ed.*, **48**, 3065–3068.
94. Lee, S. and MacMillan, D.W.C. (2006) *Tetrahedron*, **62**, 11413–11424.
95. Wang, X., Reisinger, C.M., and List, B. (2008) *J. Am. Chem. Soc.*, **130**, 6070–6071.
96. Lu, X., Liu, Y., Sun, B., Cindric, B., and Deng, L. (2008) *J. Am. Chem. Soc.*, **130**, 8134–8135.
97. Reisinger, C.M., Wang, X., and List, B. (2008) *Angew. Chem., Int. Ed.*, **47**, 8112–8115.
98. Mayer, S. and List, B. (2006) *Angew. Chem., Int. Ed.*, **45**, 4193–4195.
99. Wang, X. and List, B. (2008) *Angew. Chem., Int. Ed.*, **47**, 1119–1122.
100. (a) Berkessel, A. and Gröger, H. (eds) (2005) *Asymmetric Organocatalysis: From Biomimetic Concepts to Applications in Asymmetric Synthesis*, Wiley-VCH Verlag GmbH, Weinheim; (b) Dalko, P.I. (ed.) (2007) *Enantioselective Organocatalysis: Reactions and Experimental Procedures*, Wiley-VCH Verlag GmbH, Weinheim.

10
Catalytic Asymmetric Baylis–Hillman Reactions and Surroundings
Gui-Ling Zhao

10.1
Introduction

The Baylis–Hillman reaction [1, 2], which is also known as Morita–Baylis–Hillman (MBH) reaction, was first described by Morita (as a phosphine-catalyzed reaction) in 1968 and subsequently, in 1972, in a German patent by Baylis and Hillman (as an amine-catalyzed reaction). As one of the most important tools for forming carbon–carbon single bonds, the Baylis–Hillman reaction can be used to convert simple starting materials into densely functionalized products in a catalytic and atom-economic [3] fashion, without generating byproducts or waste. The reaction can be broadly defined as a coupling reaction of an electrophile **1** with an activated alkene **2**, catalyzed by a tertiary amine or phosphine to afford an α-methylene-β-hydroxy derivative **3** (Scheme 10.1). As well as aldehydes, imines can also serve as the electrophile if they are activated to a sufficient degree, and in this case the reaction is normally referred to as the *aza–Morita–Baylis–Hillman reaction* [2g-i]. The merits of the Baylis–Hillman reaction are its superior mild conditions, its atom economy, and the great potential for further transformation. There are, however, certain drawbacks associated with the reaction, such as a low reaction rate, a low conversion, and a limited substrate scope. Fortunately, the extensive effort that has been made in recent years, by applying physical or chemical methods [4], has led to a partial resolution of these problems. Since the potential applications of the Baylis–Hillman reaction are in the synthesis of biologically important compounds and natural products [5], the asymmetric version of the reaction has attracted much attention over the past few years. The main focus of this chapter will be on the development of this asymmetric reaction, and its mechanism and applications in synthesis will be discussed.

10.2
The Reaction Mechanism

Although the first arrow-pushing mechanism was reported by Hoffman and Rabe in 1986 [6], Hill and Isaacs [7] later proposed the commonly accepted mechanism

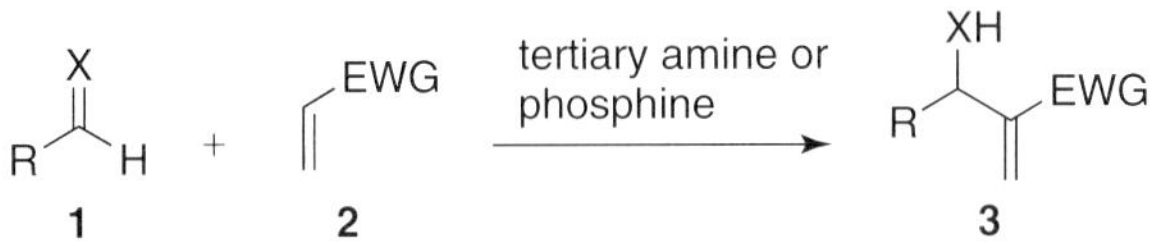

X = O or NTs, NCOOR,
$NP(O)Ph_2$, NSO_2Ph etc.
EWG = electron-withdrawing group

Scheme 10.1 The Morita–Baylis–Hillman (MBH) reaction.

for the Baylis–Hillman reaction, which was subsequently refined by others [8] (Scheme 10.2). The reaction involves first a reversible Michael addition of the nucleophilic catalyst onto the activated alkene **2** to produce an enolate **4** (step 1); this is followed by an aldolic reaction of **4** to aldehyde or imine **1** to result in a second zwitterionic intermediate **5** (step 2), an intramolecular proton transfer to form intermediate **6** (step 3), and an E2 or E1cb elimination to generate the product **3**, while the catalyst is released to complete the cycle (step 4). This mechanism was based on pressure dependence, rate, and kinetic isotope effect (KIE) data, and supported by the interception and structural characterization of each of the intermediates, using electrospray ionization mass spectrometry [9a].

Traditionally, the aldolic reaction between intermediate **4** and **1** (step 2) was the rate-determining step (RDS), since a KIE of 1.03 ± 0.1 was observed for the α-position of acrylonitrile (when the electron-withdrawing group (EWG) = CN), which means that no α-proton cleavage occurs in the RDS [7]. However, Aggarwal and coworkers [10] reported that at an early stage of the reaction (<20% conversion, in the absence of hydrogen-bond donors) the RDS is the proton transfer (step 3). These authors noted that the concentration of the product builds up as the reaction proceeds, and an autocatalysis of the proton transfer step occurs such that the RDS moves to step 2. Their proposal was supported by the observation that a substantial KIE ($k_H/k_D = 5 \pm 2$) in competition experiments between methyl

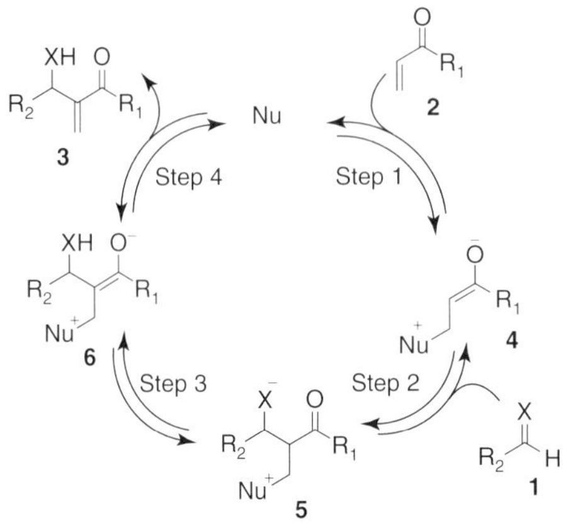

Scheme 10.2 Proposed mechanism for the MBH reaction.

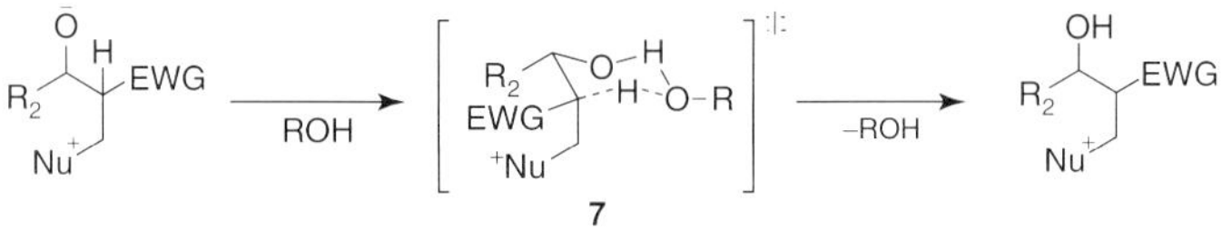

Scheme 10.3 Aggarwal's proposal for the alcohol-promoted proton-transfer mechanism.

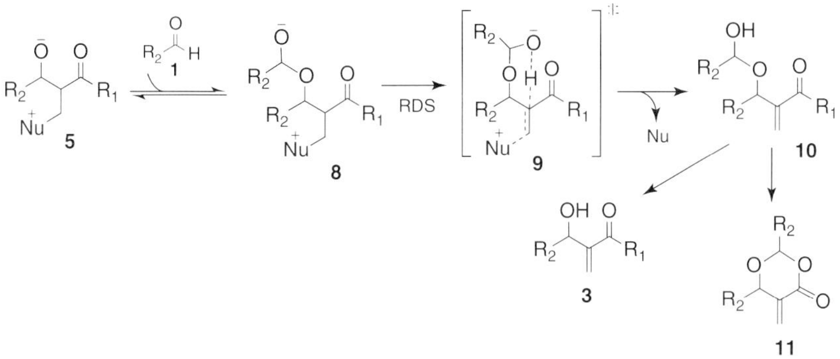

Scheme 10.4 McQuade's proposed mechanism of the MBH reaction.

acrylate and methyl α-^{2}H acrylate with PhCHO in the early stage of the reaction. The acceleration in the presence of protic solvents [11] also supported this proposal. An intermediate **7** for the alcohol-catalyzed (the product is an alcohol) proton transfer is shown in Scheme 10.3.

At the same time, McQuade and coworkers [12] proposed another new interpretation of the reaction mechanism, which is shown in Scheme 10.4. This mechanism was based on three observations:

- The RDS was second order in aldehyde and first order in DABCO and acrylate, by collecting the reaction rate data in aprotic solvents.
- That a primary KIE (2.2–5.2, depending on the solvent used) was received when methyl α-^{2}H acrylate was used.
- Large inverse-isotope effects (0.72–0.80) were shown when α-^{2}H-*p*-nitrobenzaldehyde was used in the reaction.

Based on these points, it was proposed that addition of the zwitterionic intermediate **5** to the second molecule of aldehyde **1** formed a hemiacetal intermediate **8**. This then underwent a rate-limiting proton transfer through a six-member transition state **9**, to afford **10**, which would break down to the BH product **3** in a series of post-rate-limiting steps. This proposal was also supported by the formation of a dioxane byproduct **11** in some cases [13] (Scheme 10.4).

Both intermediates **7** and **9** were successfully intercepted and structurally characterized by using electrospray ionization with mass spectrometry [9b]. Intermediates in the MBH reaction were successfully isolated by Krafft and Jacobsen [14a,b], in separate investigations. The results of an *ab initio* and density functional theory (DFT) study [15] showed that the activation barrier for the C–C bond formation was lower than for the proton-transfer step in the MBH reaction between methyl vinyl ketone (MVK) and benzaldehyde, which also supported the new explanation for the mechanism.

10.3
Asymmetric Intermolecular Baylis–Hillman Reaction

In principle, an asymmetric Baylis–Hillman reaction can be accomplished by any chiral source of the reaction component, including: (i) an enantiopure or enriched starting material **1** or **2**; (ii) a chiral catalyst; and (iii) a chiral solvent. In this respect, extensive studies have been conducted since the mid-1990s, with the earlier efforts being mainly devoted to the application of enantiopure starting material **1** or **2**, whereas the later studies concentrated on the various chiral catalysts that were available.

10.3.1
Diastereoselective Baylis–Hillman Reaction

In 1993, an enantiopure planar chiral benzaldehyde Cr(CO)$_3$ complex **12** was used successfully by Kündig and coworkers [16a] in the asymmetric Baylis–Hillman reaction with methyl acrylate or acrylonitrile (Scheme 10.5; X = O). The coordination of the aldehyde to the electrophilic Cr(CO)$_3$ group accelerated the reactions, and the corresponding products were obtained in good yields and with high diastereoselectivities. Moreover, it was found that such methodology could be extended equally well to the *o*-substituted planar chiral benzaldehyde imine complexes [16b] (Scheme 10.5; X = *N*-tosyl; NTs). Removal of the metal under mild conditions provided chiral allylic alcohols or amines.

The *N*-protected amino aldehydes **13** and **14** were explored in the reaction with methyl acrylate by Bussolari and coworkers [17] (Scheme 10.6). However, the reaction rate was very slow and the diastereoselectivities were poor.

Subsequently, enantiopure 4-oxoazetidine-2-carbaldehyde **15** was adopted by Alcaide and coworkers [18] in the reaction with MVK. In the presence of an equimolar amount of DABCO, Baylis–Hillman adducts were easily prepared in

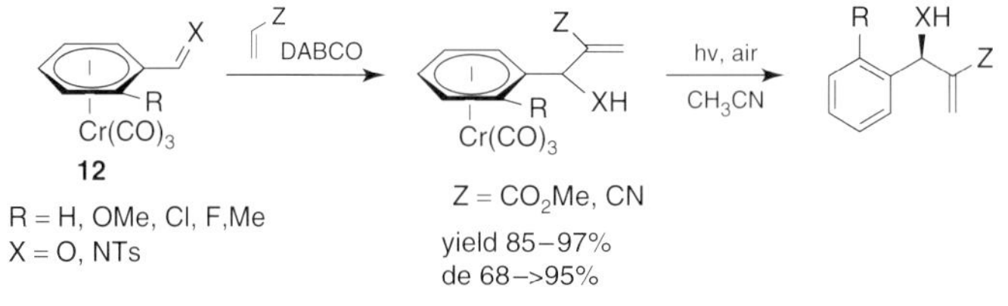

Scheme 10.5 MBH reactions of chiral benzaldehyde or imine complexes and methyl acrylate or acrylonitrile.

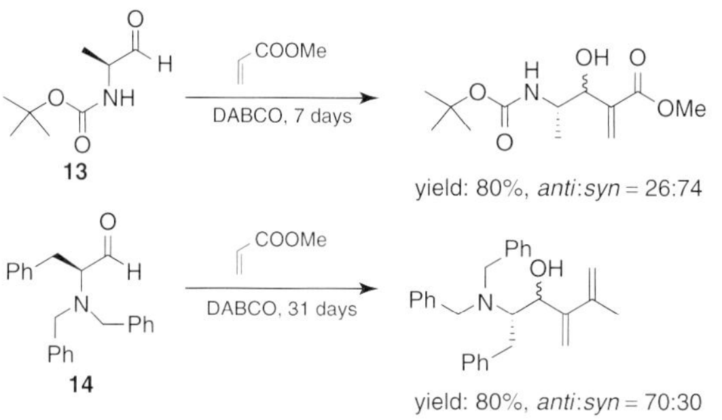

10.3 Asymmetric Intermolecular Baylis–Hillman Reaction

Scheme 10.6 MBH reactions of N-protected amino aldehydes and methyl acrylate.

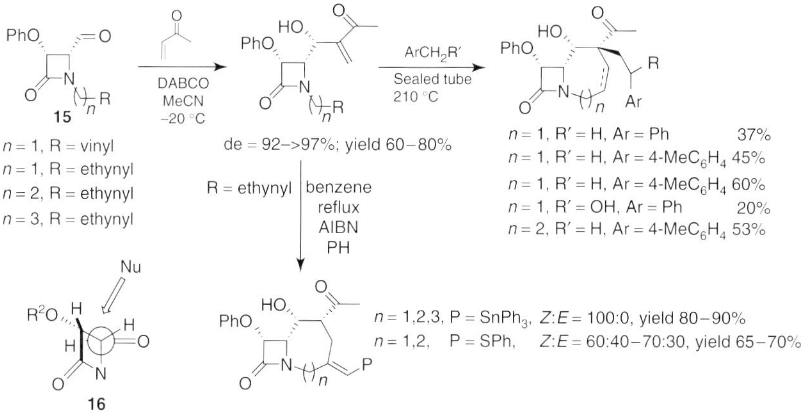

Scheme 10.7 MBH reactions of enantiopure aldehydes **15** and MVK, and their applications.

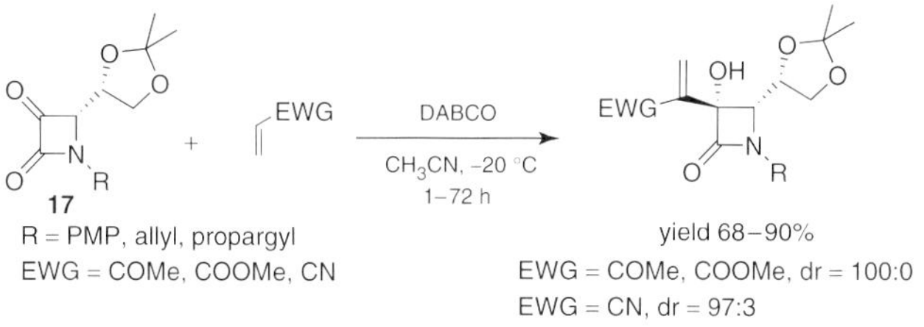

Scheme 10.8 MBH reactions of enantiopure aldehydes **17** with activated alkenes.

excellent yields and high diastereoselectivities (Scheme 10.7). The high diastereoselectivities can be interpreted by the Felkin–Anh model, through an *anti*-Felkin addition (Scheme 10.7; structure **16**). The products were also further transformed to the highly functionalized medium-sized ring-fused bicyclic β-lactams. Later, Alcaide and coworkers successfully developed enantiopure 3-oxo-2-azetidinones **17** for Baylis–Hillman coupling with activated alkenes to provide the corresponding products in high yields, and with excellent diastereoselectivities (Scheme 10.8).

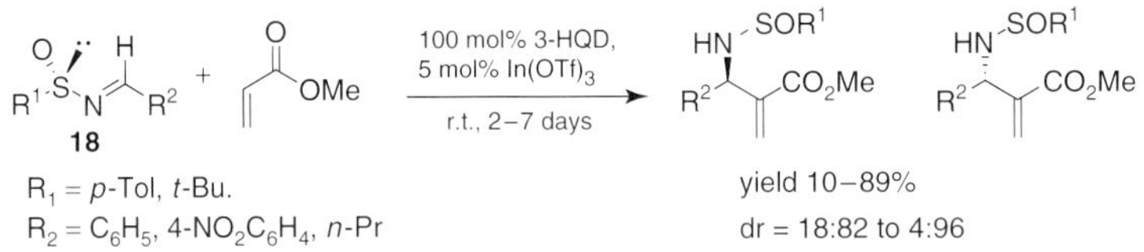

Scheme 10.9 MBH reactions of N-sulfinimines **18** with methyl acrylate.

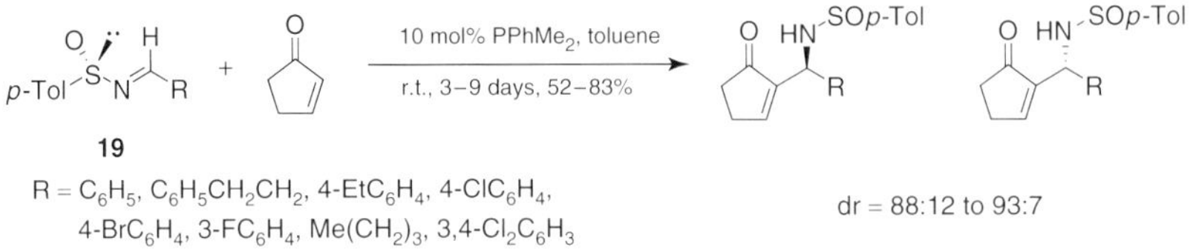

Scheme 10.10 MBH reactions of N-sulfinimines **19** with cyclopent-2-en-1-one.

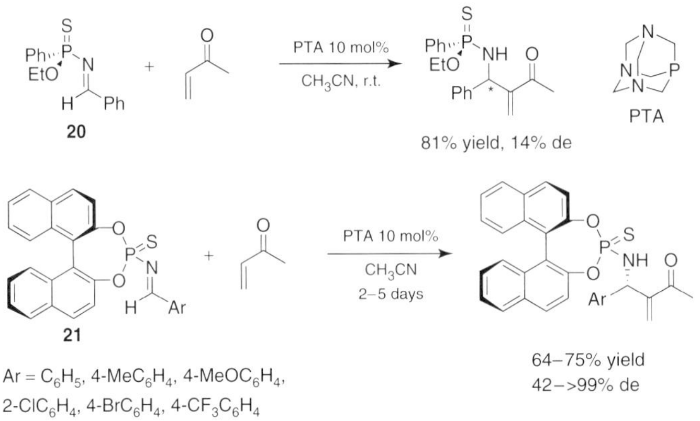

Scheme 10.11 MBH reactions of chiral imines **20** and **21** with MVK.

Shortly thereafter, Aggarwal and coworkers [19] reported the asymmetric Baylis–Hillman reaction of enantiopure N-sulfinimines **18** with methyl acrylate. In this case, 3-hydroxyquinuclidine (3-HQD) was used as the catalyst and In(OTf)$_3$ was shown to be the best Lewis acid additive (Scheme 10.9). The same group also investigated the N-tert-butanesulfinimine as the electrophile; this gave a higher diastereoselectivity but much poorer yields.

Enantiopure N-sulfinimine **19** was also used by Shi and coworkers [20] in reaction with cyclopent-2-en-1-one. The aza–Baylis–Hillman adducts were achieved in good yields and high diastereoselectivities (Scheme 10.10).

Recently, Tang and coworkers [21] synthesized two chiral imines **20** and **21** and applied them to the reaction with MVK in the presence of a Lewis base 1,3,5-triaza-7-phosphaadamantane (PTA). When imine **20** was used, only 14% diastereomeric excess (de) was obtained, but the reaction between **21** and MVK afforded the products in good yields and up to 99% de (Scheme 10.11).

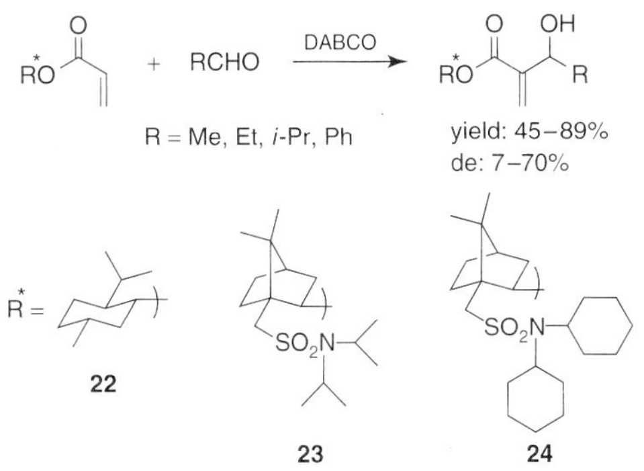

Scheme 10.12 MBH reactions of chiral acrylates and aldehydes.

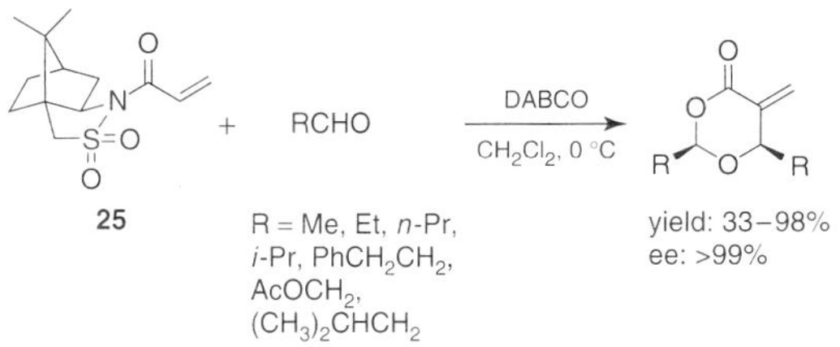

Scheme 10.13 MBH reactions of the Oppolzer camphor sultam and aldehydes.

Numerous attempts have been made by using chiral acrylate esters derived from chiral auxiliary. The easily accessible (−)-menthyl acrylate **22** was first used in the asymmetric reaction with propionaldehyde by Basavaiah and coworkers [22] in the presence of DABCO; the product was obtained in 16% de. These authors also investigated the reactions of chiral acrylates **23** and **24**, when the corresponding products were obtained in up to 70% de (Scheme 10.12). Later, Isaacs and coworkers [23] reported that, under high pressure (7.5 Kbar), the reaction between benzaldehyde and (−)-menthyl acrylate **22** resulted in high de (100%).

The most successful methodological approach in this direction has been developed by Leahy [13b,c], using the Oppolzer camphor sultam **25**. The reaction proceeded well to provide products with high yields and excellent enantioselectivities (Scheme 10.13). It should be noted that the sultam auxiliary is automatically removed by incorporation of a second molecule of aldehyde, and recovered in the purification of the adducts. The reaction product can be easily transformed to tulipalin B, the contact dermatitic agent in tulip bulbs (Scheme 10.14).

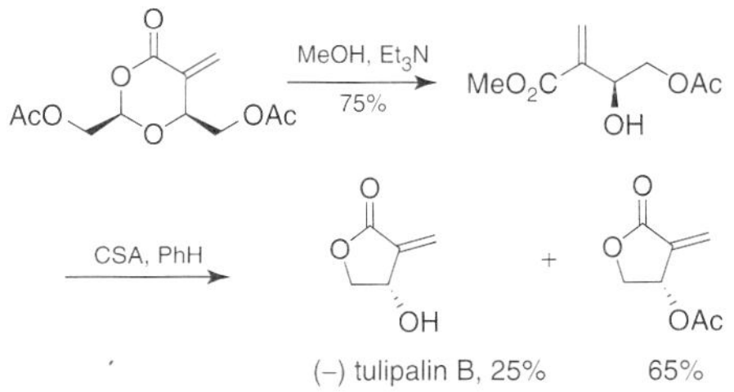

Scheme 10.14 Transformation of the MBH reaction product to tulipalin B.

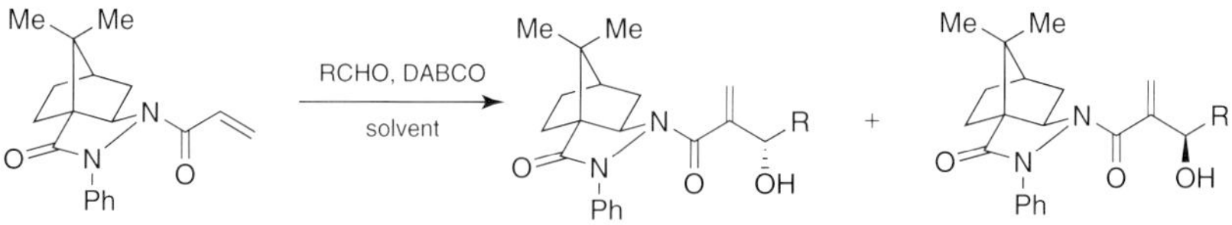

26 **A** **B**

R = Me, Et, $PhCH_2CH_2$, Me_2CHCH_2, Ph solvent = DMSO, **A**:**B** = 97:3–99:1, yield: 75–88%

R = Me, Et, $PhCH_2CH_2$, Me_2CHCH_2 solvent = THF/H_2O, **A**:**B** = 3:97–1:99, yield: 73–81%

Scheme 10.15 MBH reactions of compound **26** and aldehydes.

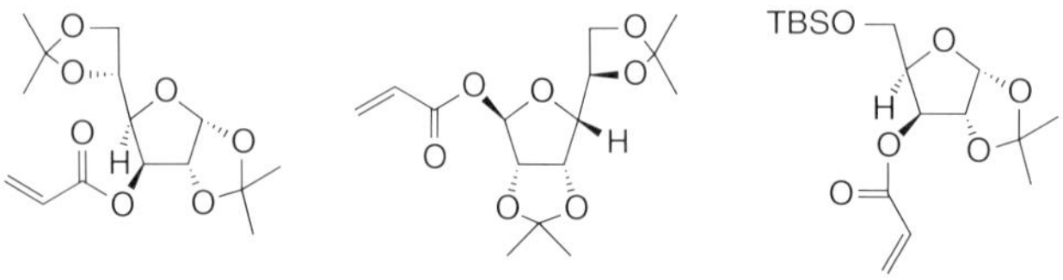

27 **28** **29**

Figure 10.1 Acrylates with chiral auxiliaries.

Subsequently, Chen and coworkers [24] prepared compound **26** having the same camphor scaffold. In the presence of DABCO, the Baylis–Hillman adducts β-hydroxy-α-methylene carbonyl derivatives were obtained in practical levels (up to 98% de). Moreover, both diastereomers of the product with high optical purity could be prepared from the same chiral auxiliary, by the correct choice of solvent (Scheme 10.15).

Several monosaccharides (Figure 10.1) have been developed as chiral auxiliaries by Krishna and coworkers [25] for the Baylis–Hillman reaction, with products having been achieved in high yields and with moderate de-values (up to 40%).

Bauer and coworkers [26] reported that chiral glyoxylates **30** can react with cyclic enones in the presence of Me_2S-$TiCl_4$, to give the corresponding Baylis–Hillman adducts in high diastereoselectivities (Scheme 10.16). Other activated olefins such as methyl acrylate or MVK cannot proceed via this reaction, and only the Me_2S-$TiCl_4$ catalytic system can promote this reaction.

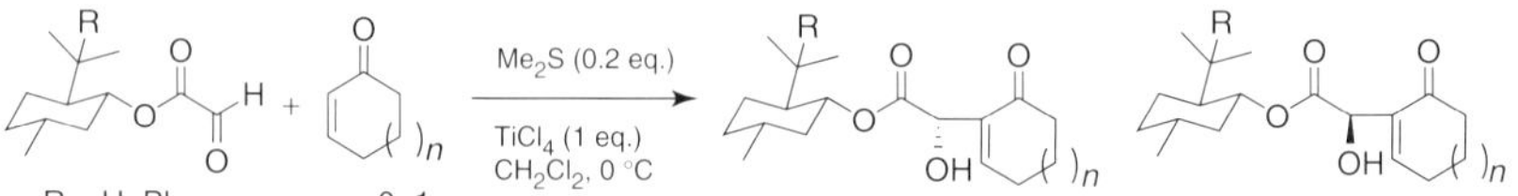

R = H, Ph *n* = 0, 1

30 A (major diastereoisomer) B (minor diastereoisomer)

R = H, *n* = 1, yield: 45%, de:8.7%
R = Ph, *n* = 1, yield: 78%,de >95%
R = Ph, *n* = 0, yield: 76%, de >95%

Scheme 10.16 MBH reactions of chiral glyoxylates **30** with cyclic enones.

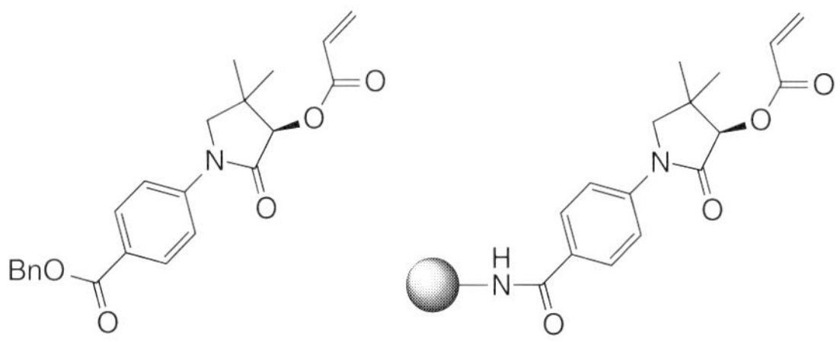

Figure 10.2 Acrylates with chiral auxiliaries and the supported acrylic ester.

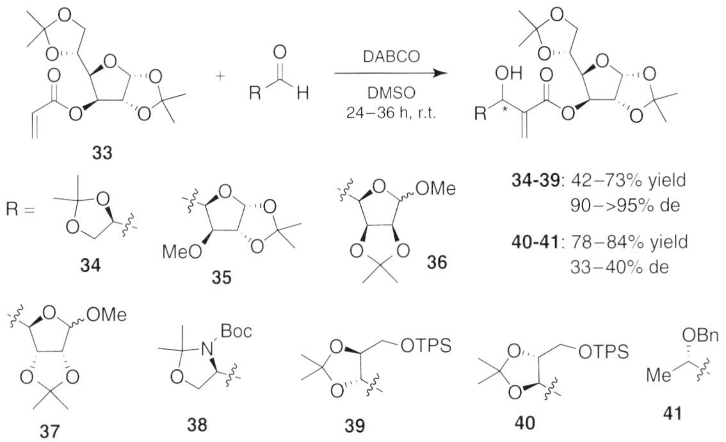

Scheme 10.17 MBH reactions of chiral acrylate **33** with chiral aldehydes **34–41**.

When Calmes and coworkers [27] applied **31** and **32** (Figure 10.2) to the asymmetric Baylis–Hillman reaction, both in solution and on solid support, the corresponding products were obtained in good yields and 34–40% enantiomeric excess (ee).

Krishna and coworkers [28] utilized both chiral acrylate **33** and chiral aldehydes **34–41** in Baylis–Hillman reactions, termed a *"double asymmetric induction"*. In the presence of DABCO, the corresponding adducts were obtained in high diastereoselectivities (in the case of aldehydes **34–39**). The poor selectivity of aldehydes **40** and **41** was due to the low steric bias of these aldehydes (Scheme 10.17).

Recently, Zhou and coworkers [29] found that in the medium of Me_3N/H_2O/low carbon alcohol or other polar solvent [e.g., tetrahydrofuran (THF), 1, 4-dioxane, acetonitrile], the rate of the Baylis–Hillman reaction can be accelerated. Thus, the asymmetric reactions of L-menthyl acrylate **42** and (+)-*N*-α-phenylethyl acrylamide **43** with aromatic aldehydes were realized in the Me_3N/H_2O/solvent homogeneous medium at 25 °C for seven days, which normally requires previously high-pressure treatment [22]. The corresponding adducts were obtained in good chemical yields and moderate to excellent diastereoselectivities (Scheme 10.18).

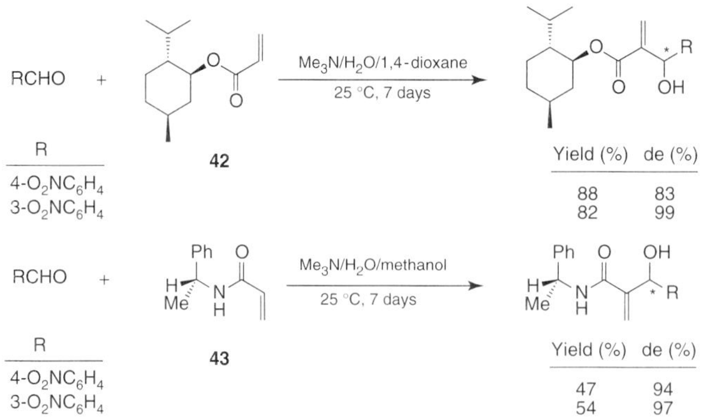

10 Catalytic Asymmetric Baylis–Hillman Reactions and Surroundings

Scheme 10.18 MBH reactions in the Me$_3$N/H$_2$O/solvent homogeneous medium.

10.3.2
Enantioselective Baylis–Hillman Reaction

During the past twenty years, important progress has been made in the development of the catalytic asymmetric version of the Baylis–Hillman reaction, using a variety of chiral catalysts. In particular, it was logical to exploit chiral tertiary amine or phosphine, as the reaction is efficiently catalyzed by these materials. Over the years, many different chiral tertiary amines and phosphines have been investigated, among which the most successful strategy has been to use a bifunctional catalyst [30] that involves a Lewis base group to initiate the reaction, and a Brønsted acid to stabilize the intermediate by the hydrogen-bonding interactions. Chiral chalcogenide, together with a Lewis acid, represents another alternative in some cases. Chiral Brønsted acids or Lewis acids, in addition to chiral or achiral tertiary amines or phosphines have been extensively studied for the asymmetric Baylis–Hillman reaction.

10.3.2.1 Chiral Tertiary Amine Catalysts
The first chiral tertiary amine catalyst to be developed for the asymmetric Baylis–Hillman reaction was quinidine [31] **44** (Figure 10.3), although this gave the Baylis–Hillman adduct only in 27% ee in the reaction between MVK and cyclohexyl carboxaldehyde, under high pressure.

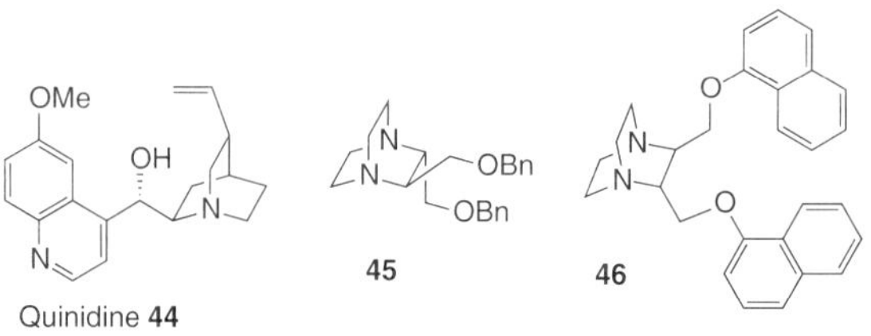

Figure 10.3 Chiral tertiary amine catalysts.

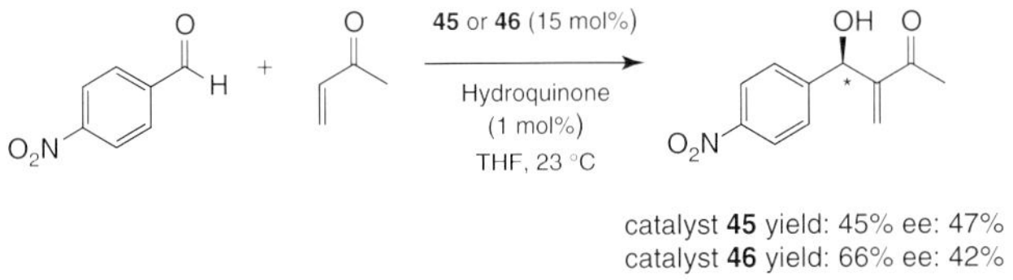
45 or 46 (15 mol%)
Hydroquinone
(1 mol%)
THF, 23 °C
O2N
OH O
catalyst 45 yield: 45% ee: 47%
catalyst 46 yield: 66% ee: 42%

Scheme 10.19 MBH reactions between 4-nitrobenzaldehyde and MVK, catalyzed by **45** and **46**.

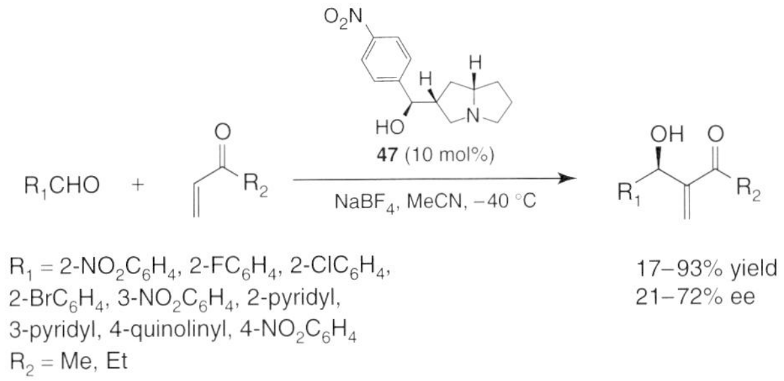
O2N
HO
47 (10 mol%)
R1CHO
R2
NaBF4, MeCN, −40 °C
OH O
R1
R2
R1 = 2-NO2C6H4, 2-FC6H4, 2-ClC6H4,
2-BrC6H4, 3-NO2C6H4, 2-pyridyl,
3-pyridyl, 4-quinolinyl, 4-NO2C6H4
R2 = Me, Et
17–93% yield
21–72% ee

Scheme 10.20 MBH reactions catalyzed by a chiral pyrrolizidine base **47**.

Subsequently, Hirama and coworkers [32] developed the chiral catalysts **45** and **46** (Figure 10.3), both of which were derived from DABCO, and showed that the reaction between 4-nitrobenzaldehyde and MVK could proceed in their presence under high pressure, with a 45–66% yield and an ee-value of 42–47% (Scheme 10.19).

When Barrett and coworkers [33a] explored a chiral pyrrolizidine base **47** derived from Boc-L-prolinol for the reaction between aromatic aldehydes and MVK or ethyl vinyl ketone (EVK), the desired products were provided with 21–72% ee in the presence of $NaBF_4$ (Scheme 10.20). These authors noted that a predominant formation of the intermediate **A**, which was less sterically hindered than the isomeric intermediate **B**, was the key factor for the enantiomeric excess (Figure 10.4). The presence of a hydroxyl group in the base was shown to be very important, both for good conversions and high enantioselectivities.

Some time later, the same group also developed a chiral catalyst [33b] **48** for the reaction between *p*-nitrobenzaldehyde and EVK. In this case, the reaction could be completed in 2 h, although the Baylis–Hillman reaction product was obtained only in 26% ee (Scheme 10.21).

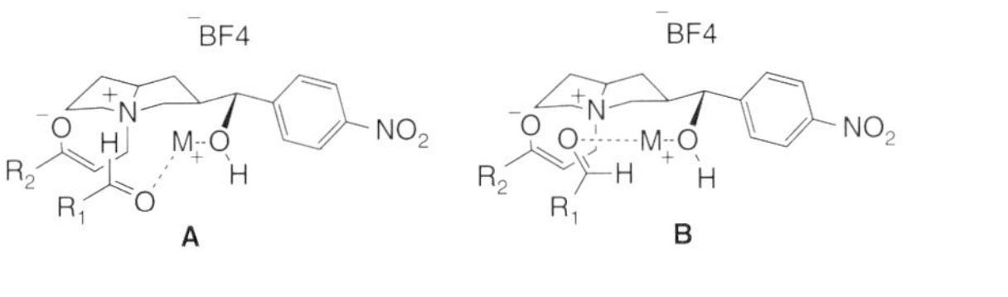
BF4
NO2
R2
R1
A
BF4
NO2
R2
R1
B

Figure 10.4 Intermediates for the reactions catalyzed by catalyst **47**.

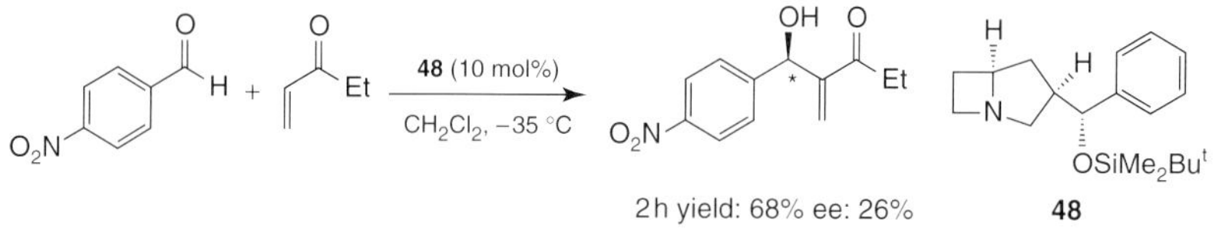

Scheme 10.21 MBH reactions between p-nitrobenzaldehyde and EVK, catalyzed by catalyst **48**.

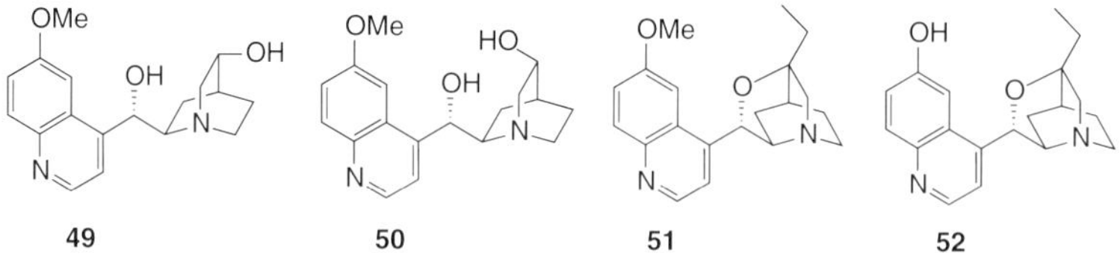

Figure 10.5 Chiral tertiary amines derived from cinchona alkaloids.

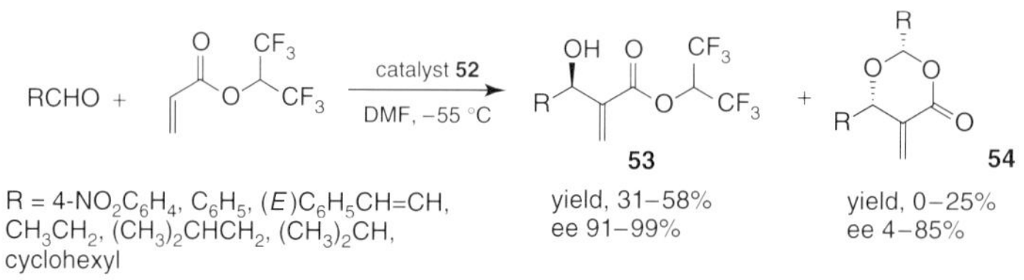

Scheme 10.22 MBH reactions between HFIPA and aldehydes, catalyzed by catalyst **52**.

Hatakayama and coworkers [34] found that very good enantioselectivities could be achieved by using chiral tertiary amines **52** in the reaction between 1,1,1,3,3,3-hexafluoroisopropyl acrylate (HFIPA) and various aldehydes. When the different chiral amines derived from cinchona alkaloids **49–52** were examined (Figure 10.5), the catalysts **51** and **52** produced good chemical yields, although only **52** showed excellent enantioselectivities (91–99%) (Scheme 10.22). The authors proposed that the rate enhancement observed for catalysts **51** and **52** had resulted from their increased nucleophilicity, due to a reduced steric hindrance around the nucleophilic nitrogen of the catalyst, which restrained the conformational freedom of the bulky aromatic moiety. A reasonable mechanism for the high enantioselectivity obtained with catalyst **52** was also proposed (Scheme 10.23), namely that the phenolic hydroxyl group of catalyst **52** plays a key role in stabilizing the intermediates **A** and **B**. Unfortunately, intermediate **B** suffers from severe steric interactions, as can be seen from the Newman projection **C**, and thus undergoes a reaction with a second molecule of aldehyde to form dioxanone (S)-**54**. However, intermediate **A** undergoes a facile elimination to produce (R)-ester **53**. The same reasoning was provided to explain why the absolute configurations of the products **53** and **54** were opposite.

The same methodology was applied successfully to synthesize the important biologically active molecules **56**, **57**, and **58** (Schemes 10.24 and 10.25) [35a,b].

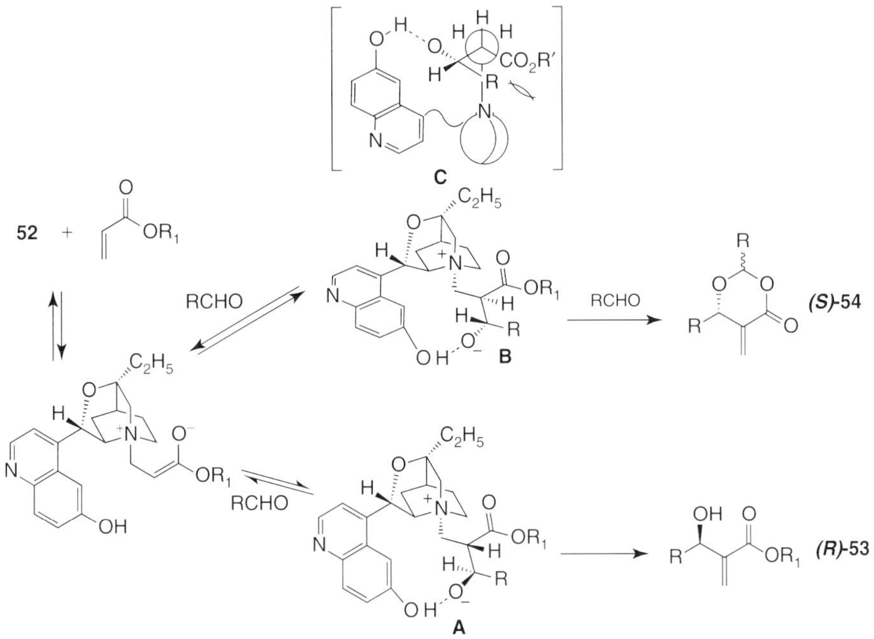

Scheme 10.23 Proposed mechanism for the enantioselective MBH reaction, catalyzed by **52**.

Later, Shi and coworkers [36] applied this efficient Lewis base catalyst **52** to the reaction between aldehydes and MVK or α-naphthyl acrylate (Scheme 10.26). However, only very poor enantioselectivities were observed, despite many different Lewis acids and Brønsted acids being used as additives. Ultimately, $LiClO_4$ was shown to be the best additive in the reaction between p-bromobenzaldehyde and MVK (Scheme 10.26).

Subsequently, Shi et al. applied catalyst **52** to the reactions between N-sulfonated imines and various activated alkenes such as alkyl vinyl ketone, acrolein, and acrylates [37a,b]. The corresponding adducts were obtained in moderate to good yields, and with good to excellent enantiomeric excess in most cases (Scheme 10.27). Various activated olefins such as MVK, EVK, acrolein, methyl acrylate, phenyl acrylate, or α-naphthyl acrylate may be used in this reaction. The authors proposed that the key factor for the high ee-values was the intramolecular hydrogen bonding between the phenolic OH and the nitrogen anion, stabilized by the sulfonyl group (Figure 10.6). Likewise, catalyst **52** was believed to act as a bifunctional chiral ligand promoter in this reaction, with the nitrogen atom in the quinuclidine moiety acting as a Lewis base to initiate the Baylis–Hillman reaction, while the phenolic OH group served as a Lewis acid (Brønsted acid as Lewis acid) through hydrogen bonding to stabilize the intermediate. It is hoped that this bifunctional chiral catalyst will inspire new avenues for the design of catalysts for Baylis–Hillman reactions.

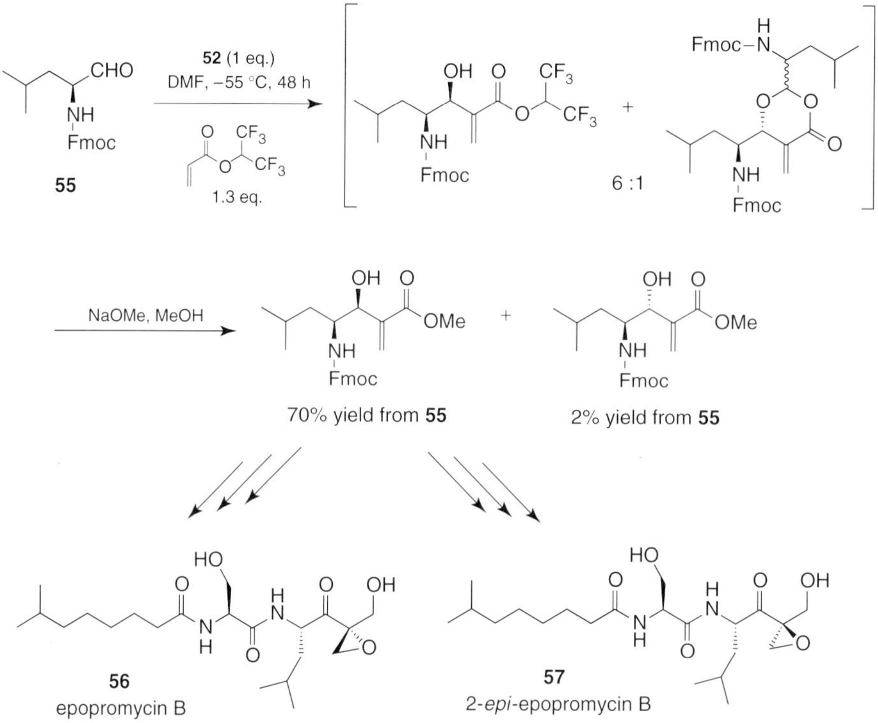

Scheme 10.24 Synthesis of the biologically active molecules **56** and **57** using MBH reaction as the key step.

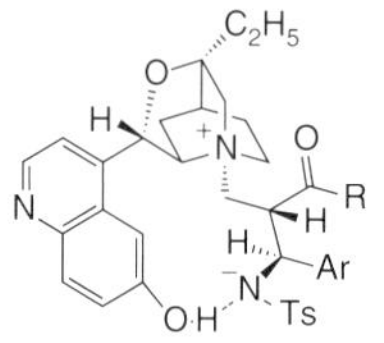

Figure 10.6 Proposed intermediate for the MBH reaction catalyzed by **52**.

Hatakayama and coworkers [38a,b] adopted catalyst **52** in the reaction between N-(arylmethylene)diphenylphosphineamides and HFIPA, giving the desired products in 42–97% yields with 54–73% ee (Scheme 10.28). Hatakayama's group synthesized an amine **59** (Figure 10.7) from quinine, and proved that it behaved as an enantiocomplementary catalyst to β-isocupreidine (**52**) in the asymmetric Baylis–Hillman reaction, affording the opposite enantiomer of the product.

Various congeners of β-isocupreidine (β-ICD) **60–69** (Figure 10.7) were systematically investigated by Hatakayama and coworkers [38c] using the reaction between p-nitrobenzaldehyde and HFIPA as a model. It was found that only those catalysts with both a cage-like tricyclic structure and a phenolic OH group (catalyst **65**, **66**, **69**) were capable of providing similar results to catalyst **52**; the remainder proved

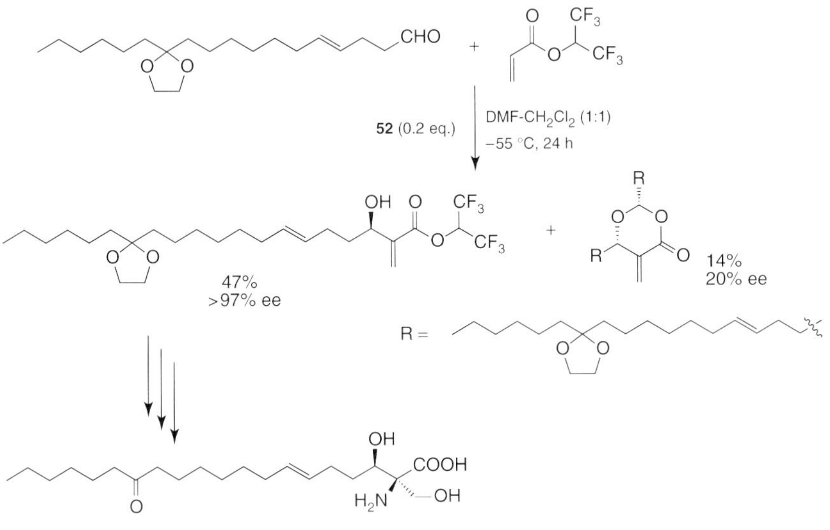

Scheme 10.25 Synthesis of the biologically active molecules **58** using MBH reaction as the key step.

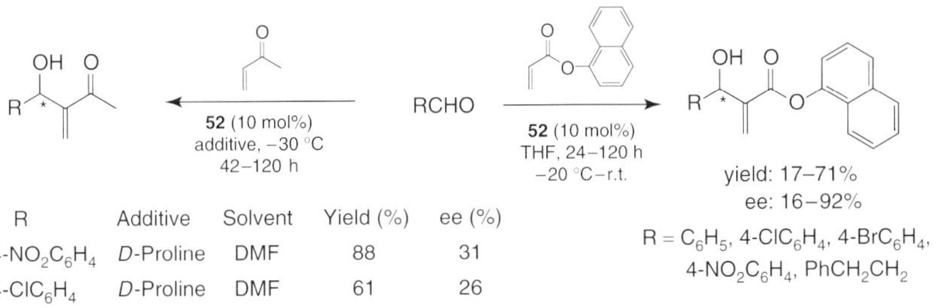

R	Additive	Solvent	Yield (%)	ee (%)
4-$NO_2C_6H_4$	D-Proline	DMF	88	31
4-ClC_6H_4	D-Proline	DMF	61	26
4-BrC_6H_4	$LiClO_4$	THF	43	49
$PhCH_2CH_2$	none	THF	69	49

R = C_6H_5, 4-ClC_6H_4, 4-BrC_6H_4, 4-$NO_2C_6H_4$, $PhCH_2CH_2$

Scheme 10.26 MBH reactions catalyzed by catalyst **52**.

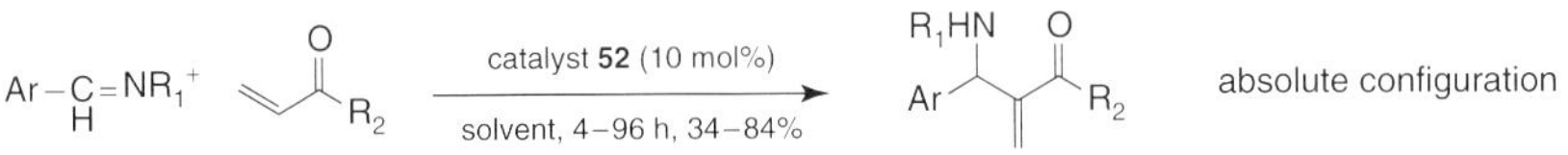

Ar = C_6H_5, 4-MeC_6H_4, 4-EtC_6H_4, 4-FC_6H_4, 4-ClC_6H_4, 4-BrC_6H_4, 4-$NO_2C_6H_4$, 3-FC_6H_4, 2,3-$Cl_2C_6H_3$, $C_6H_5CH=CH$, 2-furyl
R_1 = Ts, Ms, SES
R_2 = H, Me, Et, OMe, OPh, ONaph

46–99% ee (R_2 = Me, Et) R
43–89% ee (R_2 = H,   S
OMe, OPh, ONaph)

Scheme 10.27 Aza–MBH reactions catalyzed by catalyst **52**.

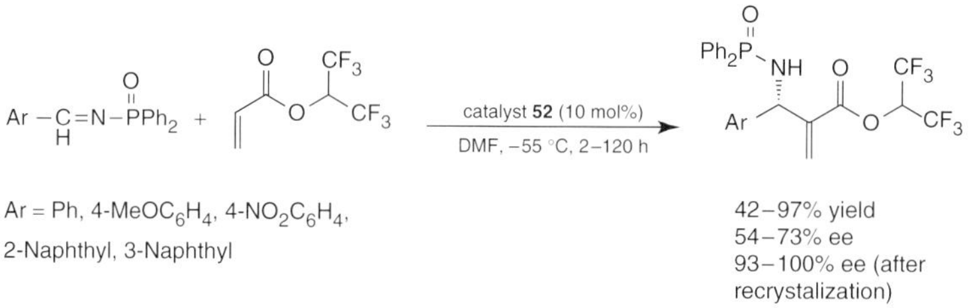

10 Catalytic Asymmetric Baylis–Hillman Reactions and Surroundings

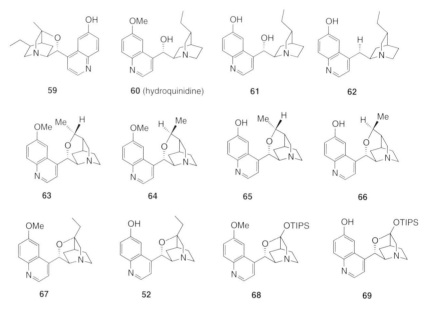

Scheme 10.28 Aza–MBH reactions catalyzed by catalyst **52**.

Figure 10.7 Various congeners of β-ICD used in the Baylis–Hillman reaction.

to be less-efficient catalysts (**60–64, 67, 68**). When different fluorine-containing acrylates were also examined in the presence of catalyst **52** (Scheme 10.29), only HFIPA with branched alkoxy groups were shown to have great effect on both the reaction rate and enantioselectivity. Neither methyl acrylate nor any other fluorine-containing acrylates with a linear alkoxy group were able to induce any appreciable enantioselectivity. Moreover, the azeotropically dried β-ICD had an enhanced catalytic ability.

When Shi and coworkers [39] tested catalyst **52** in the reactions between N-(arylmethylene)diphenylphosphineamides and MVK, acrolein, and acrylates, the aza–Baylis–Hillman adducts were obtained only in moderate yields and with very poor ee-values (Scheme 10.30).

Adolfsson and coworkers [40] applied catalyst **52** to the asymmetric one-pot, three component aza–Baylis–Hillman reaction (Scheme 10.31). In the presence

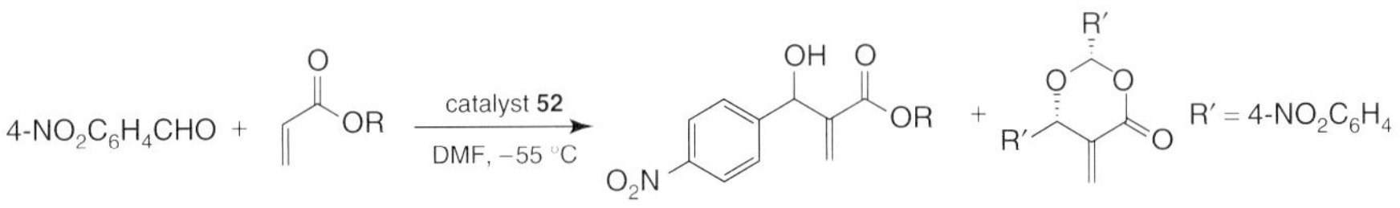

10.3 Asymmetric Intermolecular Baylis–Hillman Reaction

4-NO$_2$C$_6$H$_4$CHO + CH$_2$=CHCOOR →(catalyst 52, DMF, –55 °C) [4-NO$_2$C$_6$H$_4$-CH(OH)-C(=CH$_2$)-COOR] + [dioxanone byproduct, R' = 4-NO$_2$C$_6$H$_4$]

R	Yield (%) (ee %)	Yield (%) (ee %)
Me	69 (8)	0
CH$_2$CF$_3$	43 (3)	6 (6)
CH$_2$CF$_2$CF$_3$	50 (2)	8 (4)
CH$_2$CF$_2$CF$_2$CF$_3$	53 (0)	4 (4)
HFIPA	58 (98)	11 (4)

Scheme 10.29 MBH reactions of aldehydes with different fluorine-containing acrylates.

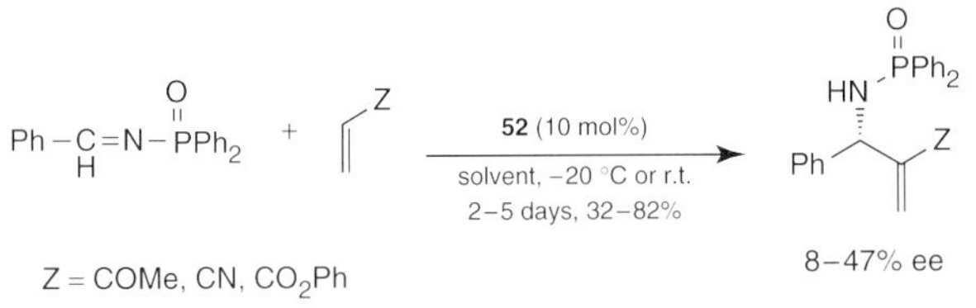

Ph–CH=N–P(=O)Ph$_2$ + CH$_2$=CH–Z →(52 (10 mol%), solvent, –20 °C or r.t., 2–5 days, 32–82%) Ph-CH(NHP(=O)Ph$_2$)-C(=CH$_2$)-Z

Z = COMe, CN, CO$_2$Ph

8–47% ee

Scheme 10.30 Aza–MBH reactions between N-(arylmethylene)diphenylphosphineamides and activated alkenes.

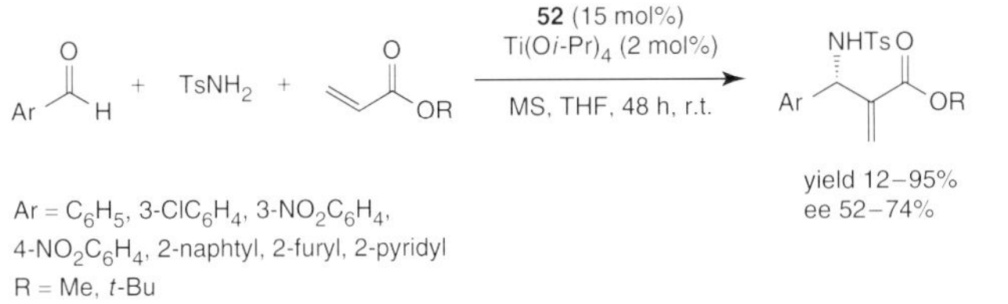

Ar-CHO + TsNH$_2$ + CH$_2$=CHCOOR →(52 (15 mol%), Ti(Oi-Pr)$_4$ (2 mol%), MS, THF, 48 h, r.t.) Ar-CH(NHTs)-C(=CH$_2$)-COOR

Ar = C$_6$H$_5$, 3-ClC$_6$H$_4$, 3-NO$_2$C$_6$H$_4$, 4-NO$_2$C$_6$H$_4$, 2-naphtyl, 2-furyl, 2-pyridyl
R = Me, t-Bu

yield 12–95%
ee 52–74%

Scheme 10.31 One-pot, three-component aza–Baylis–Hillman reaction, catalyzed by **52**.

of Ti(Oi-Pr)$_4$ and catalyst **52**, and with molecular sieves as additives, the aza–Baylis–Hillman adducts were obtained in good yields and with 49–74% ee.

In addition to quinidine-derivatives, a number of investigations of the Baylis–Hillman reaction have been conducted with (DHQD)$_2$AQN **70** [41], N-methylprolinol **71** [42], chiral diamine **72** [43], and chiral imidazolines **73–74** [44] (Figure 10.8). As an example, the Sharpless ligand **70**, together with EtCOOH, can catalyze the reaction between p-nitrobenzaldehyde and methyl acrylate to provide the corresponding product in 4% yield and 77% ee. Krishna and coworkers have reported that the small organic molecule **71** was capable of promoting the reaction between aromatic aldehydes and MVK or acrylates, giving good yields and moderate to good ee-values (Scheme 10.32). Hayashi and coworkers applied the chiral diamine **72** to the reaction between aromatic aldehyde and MVK, and provided the corresponding products with good yields and moderate ee-values (Scheme 10.33). The chiral imidazole derivative **73** was

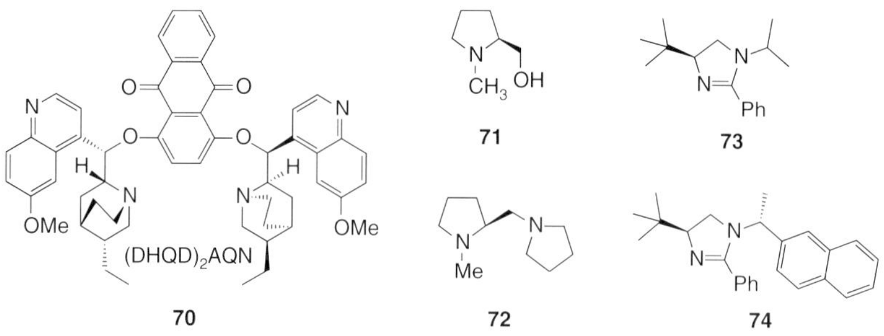

Figure 10.8 Other chiral tertiary amines used in the asymmetric MBH reaction.

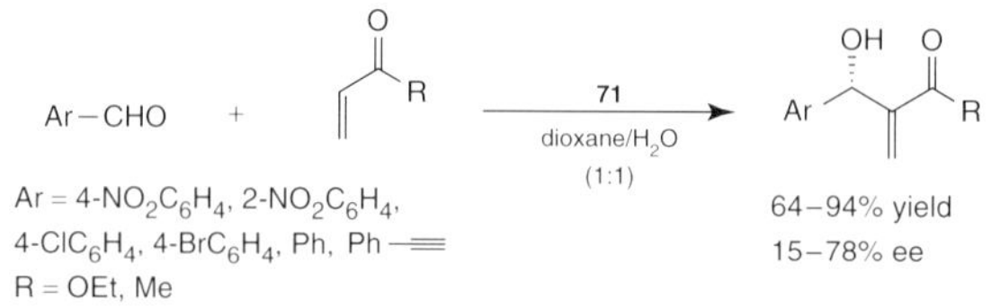

Scheme 10.32 MBH reaction between aromatic aldehydes and MVK or acrylates, catalyzed by **71**.

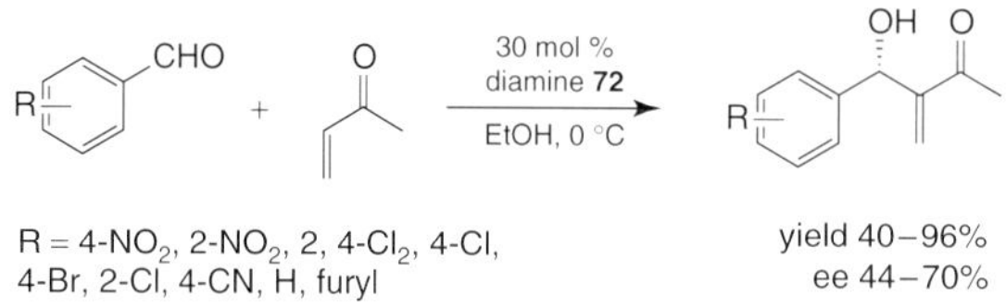

Scheme 10.33 MBH reaction between aromatic aldehydes and MVK, catalyzed by diamine **72**.

used in the reaction between aromatic aldehydes and unactivated acrylates, to generate the corresponding products with moderate ee-values (Scheme 10.34). Thus, by investigating a series of chiral imidazolines in the reaction between 4-nitrobenzaldehyde and methyl acrylate, **74** was found to be the most effective catalyst, giving the corresponding products in 84% yield and 60% ee, although the reaction time was very long (11 days). Notably, when catalyst **74** was used for the reaction between aromatic aldehydes and alkyl vinyl ketone, good yields and enantioselectivities were achieved (Scheme 10.35).

Following the development of the first successful bifunctional catalyst **52**, several such catalysts have been discovered. For example, Sasai and coworkers [45] identified a bifunctional organocatalyst **75** that was derived from BINOL for the aza–Baylis–Hillman reaction between N-tosyl imines and acrolein or alkyl vinyl ketone (Scheme 10.36), and which produced the corresponding adducts in excellent yields and 87–95% ee. In particular, when the i-Pr group of the catalyst was replaced with Me, H, Et, or t-Bu, a lower efficiency was observed and both the yields and enantioselectivities were decreased.

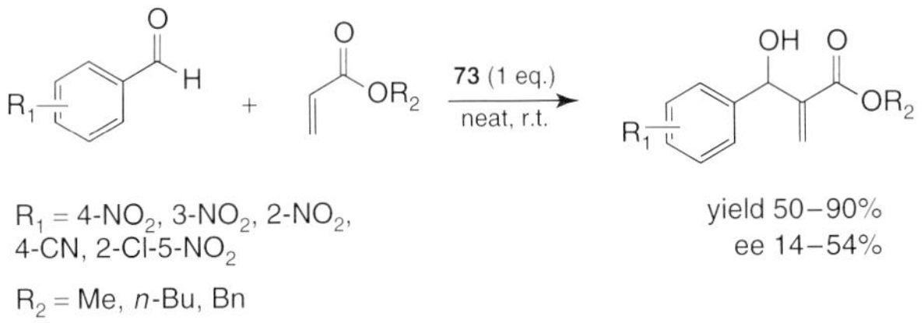
O
H
R1
+
O
OR2
73 (1 eq.)
neat, r.t.
OH O
OR2
R1
R1 = 4-NO2, 3-NO2, 2-NO2,
4-CN, 2-Cl-5-NO2
R2 = Me, n-Bu, Bn
yield 50–90%
ee 14–54%

Scheme 10.34 MBH reaction between aromatic aldehydes and unactivated acrylates, catalyzed by **73**.

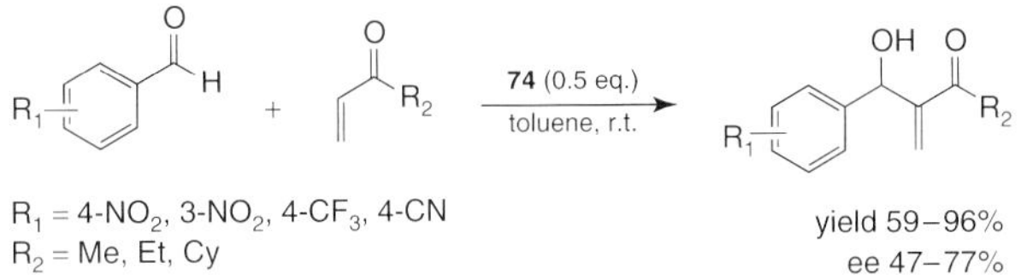
O
H
R1
+
O
R2
74 (0.5 eq.)
toluene, r.t.
OH O
R2
R1
R1 = 4-NO2, 3-NO2, 4-CF3, 4-CN
R2 = Me, Et, Cy
yield 59–96%
ee 47–77%

Scheme 10.35 MBH reaction between aromatic aldehydes and alkyl vinyl ketone, catalyzed by **74**.

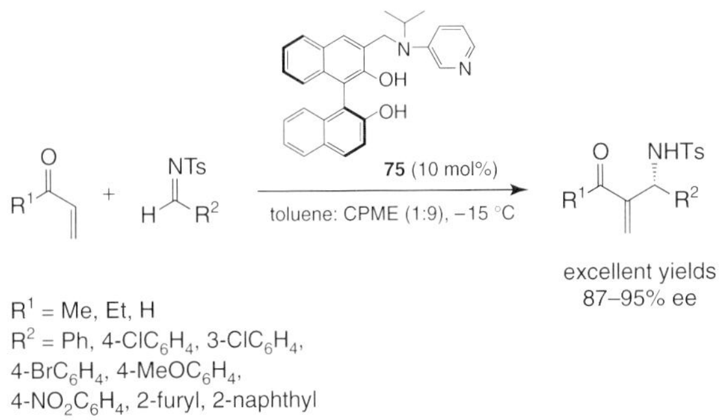
N
N
OH
OH
O
R1
+
NTs
H
R2
75 (10 mol%)
toluene: CPME (1:9), −15 °C
O NHTs
R1
R2
excellent yields
87–95% ee
R1 = Me, Et, H
R2 = Ph, 4-ClC6H4, 3-ClC6H4,
4-BrC6H4, 4-MeOC6H4,
4-NO2C6H4, 2-furyl, 2-naphthyl

Scheme 10.36 Aza–Baylis–Hillman reaction catalyzed by organocatalyst **75**.

Recently, Zhu and coworkers [46] developed a bifunctional β-ICD derivative **76** as the catalyst for the enantioselective aza–Baylis–Hillman reaction. By using 10 mol% of catalyst in combination with 10 mol% β-naphthol, the reaction products between *N*-(*p*-methoxybenzenesulfonyl)imine and β-naphthyl acrylates were observed in high yields and enantioselectivities. Besides aromatic imines, the aliphatic *N*-sulfinyl imines have been used successfully, and have led to the corresponding adduct with over 84% ee (Scheme 10.37). In this case, the H-bonds between *E*-enolate and β-naphthol and between the imine and amide NH were thought to be important, and a possible transition state was proposed (Scheme 10.38).

Shi and coworkers [47] reported a very interesting functional group-mediated reversal of asymmetric induction in the reaction of *N*-tosyl salicylaldehyde imines with MVK (Scheme 10.39). By introducing an *ortho*-phenol group to the imine substrate, the absolute configuration of the product was changed; the reason for such change was suggested to have resulted from the OH group in the substrate

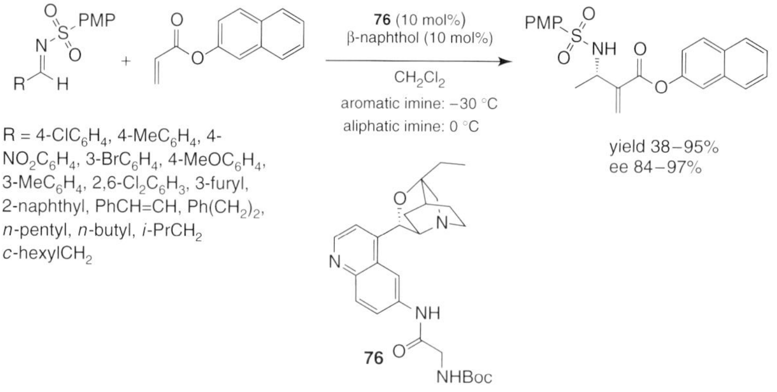

Scheme 10.37 Aza–Baylis–Hillman reaction catalyzed by a β-isocupreidine derivative **76**.

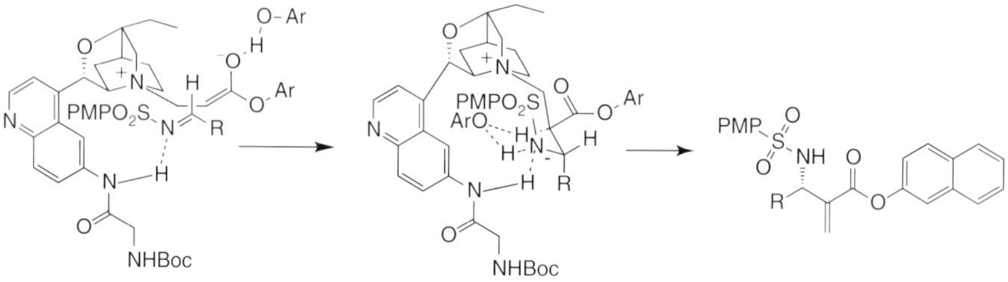

Scheme 10.38 The proposed transition state.

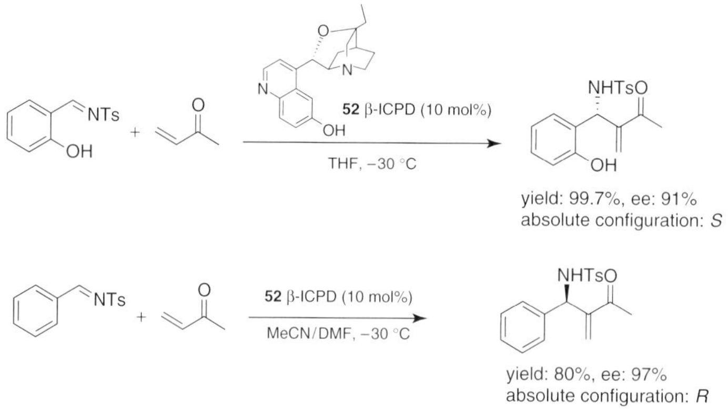

Scheme 10.39 Aza–MBH reaction between N-tosyl salicylaldehyde imines with MVK.

forming a stronger net type or branched hydrogen-bonding system in the reaction intermediate.

10.3.2.2 Chiral Tertiary Phosphines

The first chiral tertial phosphine catalyst to be developed for the Baylis–Hillman reaction was (S)-BINAP **77**, as reported by Soai and coworkers [48]. The reaction

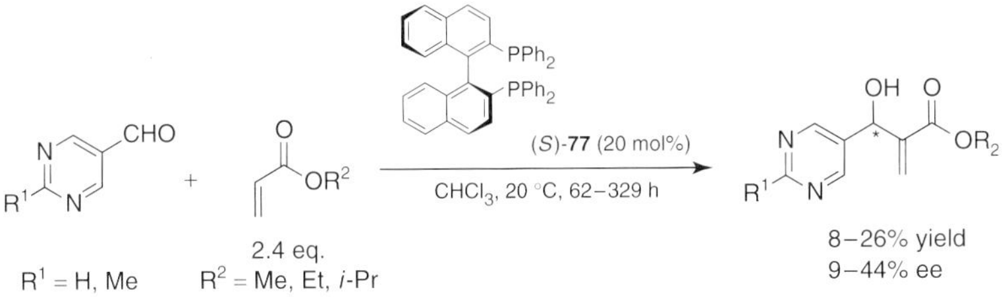

10.3 Asymmetric Intermolecular Baylis–Hillman Reaction

Scheme 10.40 MBH reaction catalyzed by (S)-BINAP.

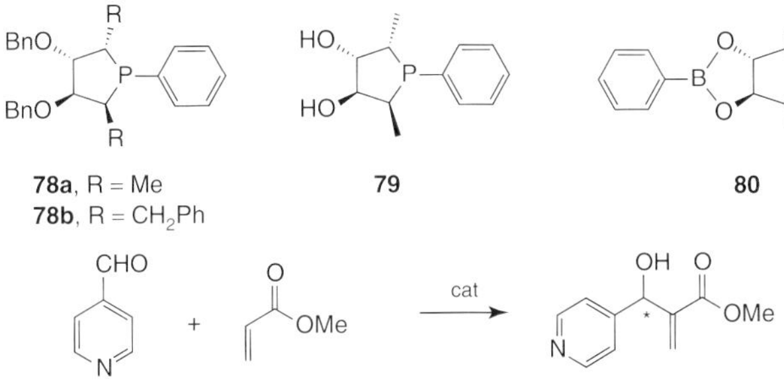

Scheme 10.41 MBH reactions catalyzed by chiral phosphines **78–80**.

Cat	Time (h)	Yield (%)	ee (%)
78a	70	29	19 (+)
78b	94	18	2 (−)
79	9	83	17 (+)
80	31	56	18 (+)

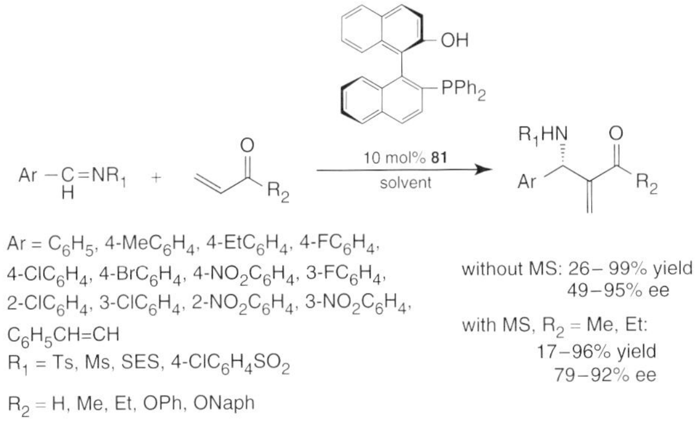

Scheme 10.42 Aza–Baylis–Hillman reactions catalyzed by chiral tertiary amine **81**.

Ar = C_6H_5, 4-MeC_6H_4, 4-EtC_6H_4, 4-FC_6H_4, 4-ClC_6H_4, 4-BrC_6H_4, 4-$NO_2C_6H_4$, 3-FC_6H_4, 2-ClC_6H_4, 3-ClC_6H_4, 2-$NO_2C_6H_4$, 3-$NO_2C_6H_4$, C_6H_5CH=CH
R_1 = Ts, Ms, SES, 4-Cl$C_6H_4SO_2$
R_2 = H, Me, Et, OPh, ONaph

without MS: 26–99% yield
49–95% ee

with MS, R_2 = Me, Et:
17–96% yield
79–92% ee

between various acrylates and pyrimidine carboxaldehyde catalyzed by (s)-BINAP gave the corresponding products in 9–44% enantiomeric excess (Scheme 10.40).

Zhang and coworkers [49] investigated the use of various chiral phosphines **78–80** as catalysts for the Baylis–Hillman reaction but, unfortunately, very poor enantiomeric excesses were achieved (Scheme 10.41).

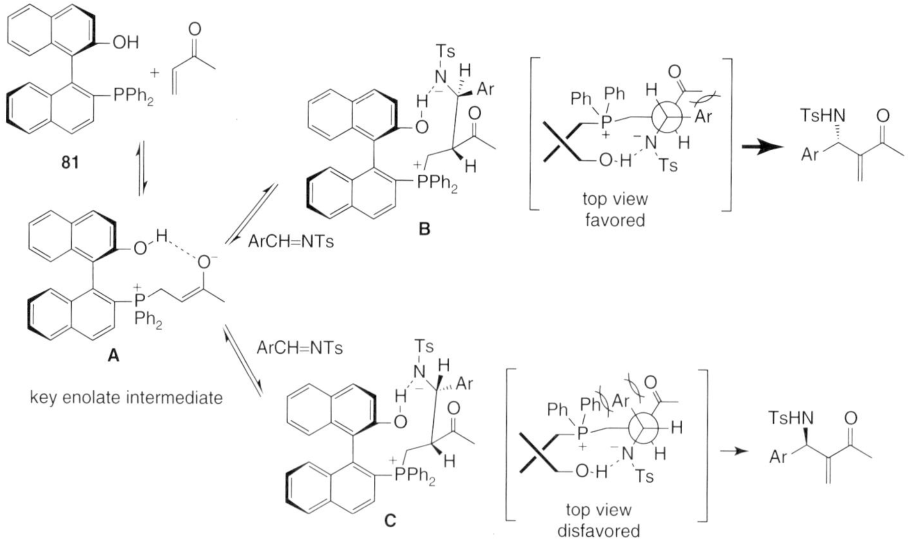

Scheme 10.43 The proposed mechanism.

Nonetheless, great progress has been made since Shi and coworkers [50a,b] applied the efficient chiral phosphine base, (R)-2'-diphenylphosphanyl-[1,1'] binaphthalenyl-2-ol (**81**) to the aza–Baylis–Hillman reaction. The asymmetric reactions between N-sulfonated imines and different activated alkenes, such as alkyl vinyl ketone, phenyl acrylate, and acrolein, gives the corresponding products in good yields and high ee-values (49–95%) (Scheme 10.42), although it was interesting to note that in the case of methyl acrylate, only 18% ee was obtained. Compared to the first efficient chiral tertiary amine catalyst **52**, this phosphine catalyst also has a phenolic hydroxyl group, which may explain the high enantioselectivity. The detailed mechanistic explanation (see Scheme 10.43) is that the phosphine atom acts as a Lewis base, which initiates the aza–Baylis–Hillman reaction, while the phenolic OH group serves as a Lewis acid which stabilizes the *in situ*-generated enolate intermediate **A**. The enolate zwitterionic intermediate **A** then undergoes a Mannich reaction with tosylimine to form the intermediates **B** and **C**, which are mostly stabilized by intramolecular hydrogen bonding. However, **C** suffers from steric repulsion, as shown in Scheme 10.43 from the Newman projection. Finally, the intermediate **B** undergoes a facile elimination to produce the (S)-enriched product.

In order to tune the reactivity and enantioselectivity of the transformation, several BINOL-derived bifunctional phosphines for the aza–Baylis–Hillman reaction were explored (Figure 10.9). Thus, by changing PPh_2 group to PMe_2 and PPhBu, catalysts of greater nucleophilicity were obtained (**82** and **90**). Catalyst **82** was used in the aza–Baylis–Hillman reaction between N-sulfonyl imines and cyclohex-2-en-1-one or cyclopent-2-en-1-one [51], and the corresponding products were isolated in good yields and with moderate ee-values (Scheme 10.44). Catalyst **90** [52a] was first

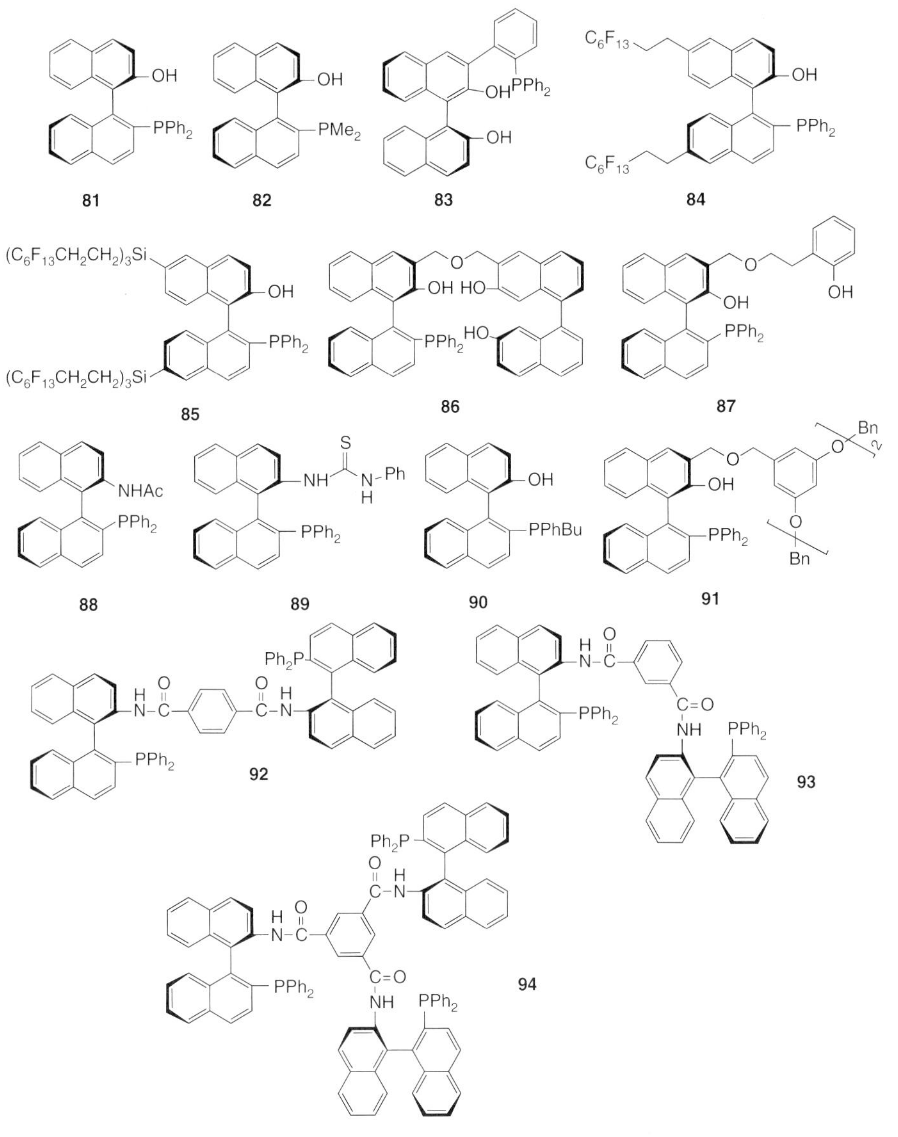

Figure 10.9 Additional BINOL-derived bifunctional phosphine catalysts.

investigated for the aza–Baylis–Hillman reaction between N-sulfonated imines and MVK. This catalyst is very efficient; the reactions can be completed within 1–5 h, while the catalyst loading can be reduced to 1 mol%. The corresponding adducts were obtained in excellent yields and moderate to good ee-values. By comparison, catalyst **81** was more stereoselective (Table 10.1, entry 9). When catalyst **90** was applied to the general Baylis–Hillman reaction between aldehydes and alkyl vinyl ketone [52b], the corresponding products were obtained in good yields

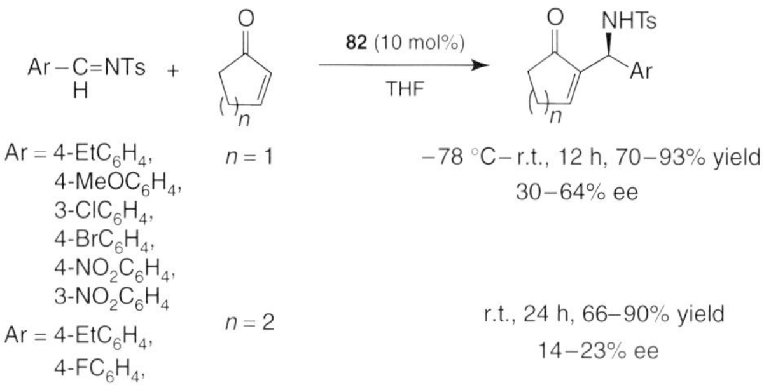
82 (10 mol%)
THF
NHTs
Ar
Ar–C=NTs +
H
Ar = 4-EtC6H4, 4-MeOC6H4, 3-ClC6H4, 4-BrC6H4, 4-NO2C6H4, 3-NO2C6H4
n = 1
−78 °C–r.t., 12 h, 70–93% yield
30–64% ee
Ar = 4-EtC6H4, 4-FC6H4,
n = 2
r.t., 24 h, 66–90% yield
14–23% ee

Scheme 10.44 Aza–Baylis–Hillman reaction catalyzed by catalyst **82**.

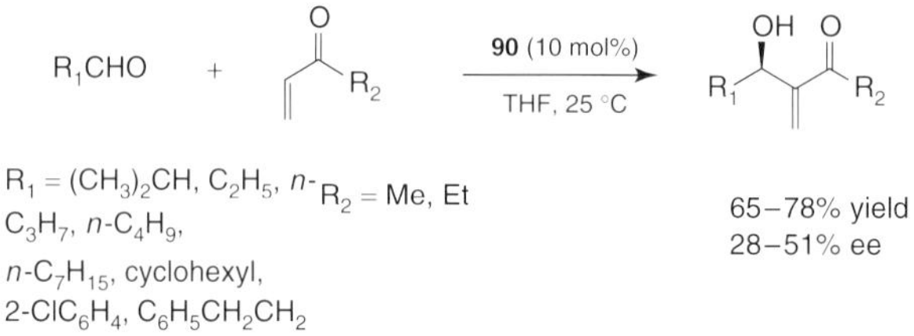
R1CHO +
90 (10 mol%)
THF, 25 °C
OH O
R1 R2
R1 = (CH3)2CH, C2H5, n-C3H7, n-C4H9, n-C7H15, cyclohexyl, 2-ClC6H4, C6H5CH2CH2
R2 = Me, Et
65–78% yield
28–51% ee

Scheme 10.45 MBH reactions catalyzed by catalyst **90**.

but with moderate ee-values (Scheme 10.45). When Sasai and coworkers [53] applied catalyst **83** to the aza–Baylis–Hillman reaction (Table 10.1, entry 2) it showed better results for some of the substrates compared to the reported organocatalyst **75** (Scheme 10.36). It is interesting to note that the absolute configuration of the major adduct obtained with **83** was opposite to that obtained with **75**, which contains a 3-pyridinylamino unit [45].

By introducing two perfluoroalkane long chains at the 6,6′-positions of the naphthalene framework, Shi and coworkers [54] were able to synthesize catalysts **84** and **85**, in the presence of which the aza–Baylis–Hillman reaction proceeded smoothly and the corresponding products were obtained in good yields with high ee-values (Table 10.1, entries 3 and 4). In addition, the chiral phosphine Lewis base **84** was more effective than the original Lewis base **81** (Table 10.1, entry 1).

Catalyst **86**, which bore multiple phenol groups, was applied to the same aza–Baylis–Hillman reaction [55] and gave excellent yields and ee-values (Table 10.1, entry 5). The same strategy was also employed by Ito and coworkers [56], who replaced the binaphthol unit with a simple phenol unit to produce the catalyst **87**. Excellent yields and high enantioselectivities were obtained when **87** was applied to the aza–Baylis–Hillman reaction between *N*-tosylimines and MVK (Table 10.1, entry 6).

The chiral Lewis base catalysts **81** and **86** were also applied to the general Baylis–Hillman reaction between arylaldehyde and MVK [57], but poor enantioselectivities were achieved (Scheme 10.46).

10.3 Asymmetric Intermolecular Baylis–Hillman Reaction

Table 10.1 The aza–Baylis–Hillman reaction catalyzed by different Lewis bases **81–94**.

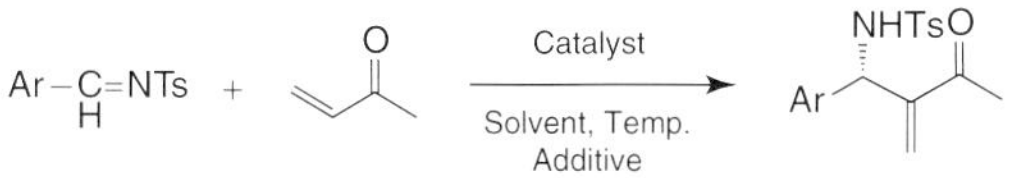

$$\text{Ar-C(H)=NTs} + \text{methyl vinyl ketone} \xrightarrow[\text{Solvent, Temp. Additive}]{\text{Catalyst}} \text{Ar-CH(NHTs)-C(=CH}_2\text{)-C(=O)-}$$

Entry	Catalyst	Solvent	Temperature (°C)	Additive	Time (h)	Yield (%)	ee (%)
1	81	THF	−30	Molecular sieves	12–36	82–96	79–92
2	83	t-BuOMe	−20	–	2–12	85–100	89–95
3	84	THF	−20	–	12–48	69–98	71–94
4	85	THF	15	–	24–48	53–88	52–82
5	86	THF	−20	–	24–48	70–97	90–96
6	87	THF	0	–	17–164	71–100	87–96
7	88	CH_2Cl_2	0	–	7–48	80–99	65–90
8	89	CH_2Cl_2	r.t.	PhCOOH	5–80	61–98	70–97
9	90	THF	r.t.	–	1–5	80–98	44–88
10	91	THF	−20	–	12–48	80–99	90–97
11	92	CH_2Cl_2	0	–	24–48	72–98	70–87
12	93	CH_2Cl_2	r.t.	–	48	60–83	64–82
13	94	CH_2Cl_2	r.t.	–	48	64–92	35–91

r.t. = room temperature.

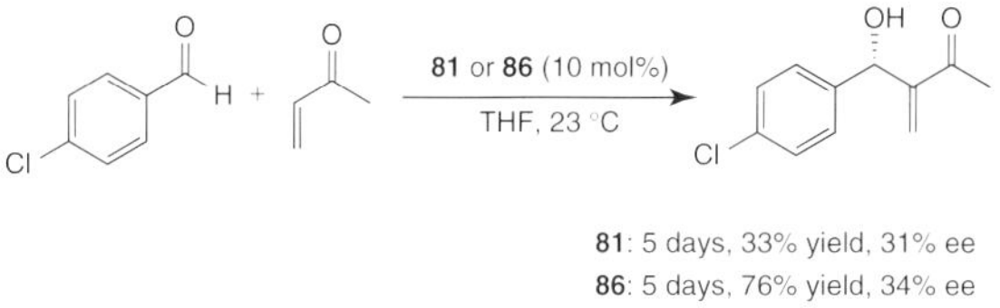

81: 5 days, 33% yield, 31% ee
86: 5 days, 76% yield, 34% ee

Scheme 10.46 MBH reactions catalyzed by catalysts **81** and **86**.

The chiral Lewis base catalyst **81** was fairly effective in the aza–Baylis–Hillman reaction of ethyl (arylimino)acetates with alkyl vinyl ketone, affording the corresponding adducts in good yields and excellent enantiomeric excess [58] (Scheme 10.47).

Both, amides and urea can act as hydrogen-bonding donors and stabilize the zwitterionic intermediates. Thus, catalysts **88** and **89** were prepared and investigated by Shi and coworkers for the aza–Baylis–Hillman reaction. The reactions between N-sulfonated imines and MVK, catalyzed by **88**, proceeded very well under mild and concise conditions to afford the corresponding adducts in good to excellent yields with high ee-values [59] (Table 10.1, entry 7); these results compared well to those achieved with catalyst **81** (Table 10.1, entry 1). The reason for this was that the amide proton had adequate acidity to stabilize the zwitterionic intermediates. The new chiral thiourea–phosphine catalyst **89** was thoroughly investigated for the reaction between N-sulfonated imines and alkyl vinyl ketone or acrolein [60]. In the presence of **89** and benzoic acid in dichloromethane at room temperature,

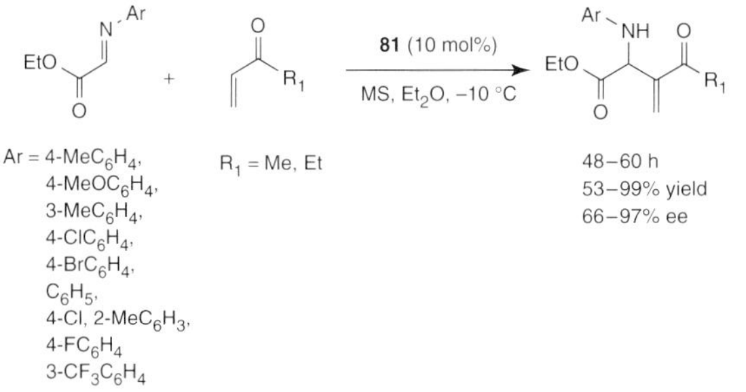

10 Catalytic Asymmetric Baylis–Hillman Reactions and Surroundings

Scheme 10.47 Aza–MBH reactions catalyzed by catalyst **81**.

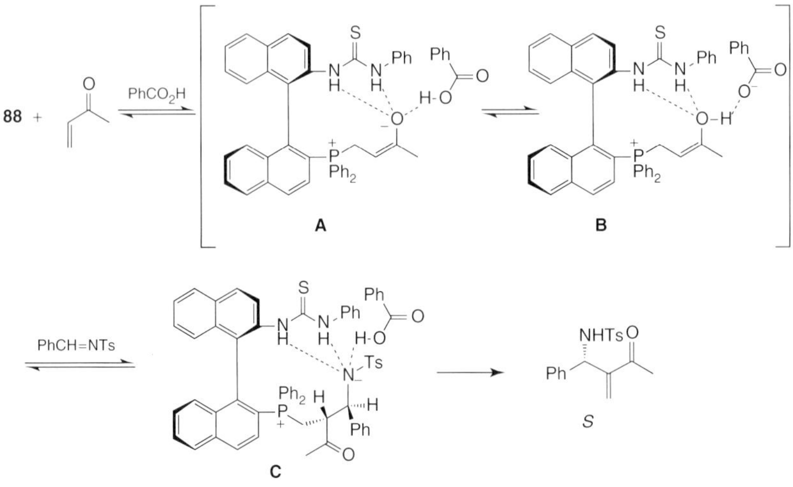

Ar = 4-MeC$_6$H$_4$, 4-MeOC$_6$H$_4$, 3-MeC$_6$H$_4$, 4-ClC$_6$H$_4$, 4-BrC$_6$H$_4$, C$_6$H$_5$, 4-Cl, 2-MeC$_6$H$_3$, 4-FC$_6$H$_4$, 3-CF$_3$C$_6$H$_4$

R$_1$ = Me, Et

81 (10 mol%), MS, Et$_2$O, –10 °C

48–60 h
53–99% yield
66–97% ee

Scheme 10.48 A rational mechanism.

the corresponding products were obtained in 61–98% yields with 70–97% ee (Table 10.1, entry 8). It is interesting to note that the use of benzoic acid caused dramatic increases in both the yield and enantioselectivity (without benzoic acid, only 12% yield and 16% ee). Consequently, when the role of benzoic acid in the reaction was investigated, a rational mechanism was proposed by the authors (Scheme 10.48). It was suggested that the intermediate **A** might be protonated by PhCOOH to form intermediate **B** [as confirmed with ^{31}P nuclear magnetic resonance (NMR) measurements] which further stabilized the intermediate. In addition, the benzoic acid may have accelerated the proton-transfer step, favoring product formation.

Shi and coworkers [61] synthesized the chiral phosphine Lewis base **91**, supported by a polyether dendrimer, and applied it to the aza–Baylis–Hillman reaction between N-sulfonated imines and alkyl vinyl ketone or acrolein. The corresponding adducts were obtained in good to excellent yields and excellent enantioselectivities (Table 10.1, entry 10). In addition, the dendrimer-supported chiral phosphine Lewis base was readily recovered and reused after the reaction. Subsequently, the same

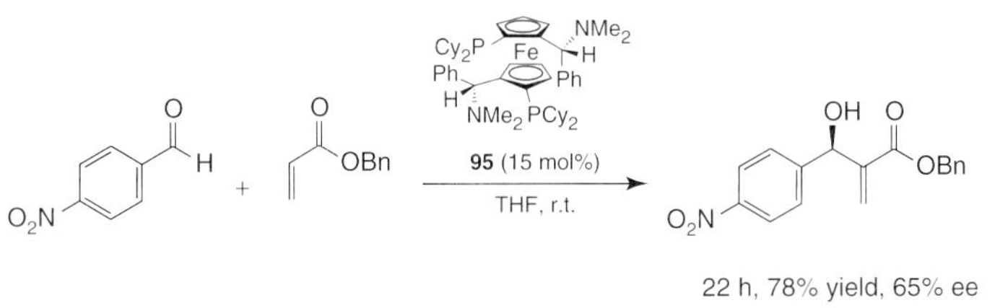

Scheme 10.49 MBH reactions catalyzed by chiral ferrocenylphosphines **95**.

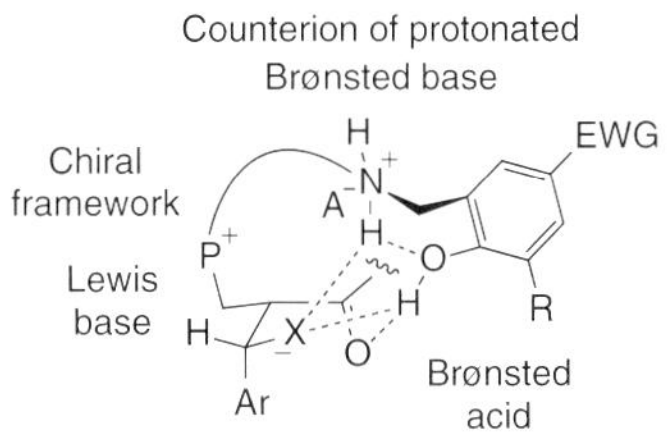

Figure 10.10 A trifunctional organocatalyst.

group reported the synthesis of novel chiral sterically congested phosphine-amide bifunctional organocatalysts **92**, **93**, and **94**, and applied them to the asymmetric aza–MBH reactions. Although similar results were obtained [62] (Table 10.1, entries 11–13) as were achieved with catalyst **81** (Table 10.1, entry 1), the catalysts **92–94** may have a more practical use as the reaction can be carried out at 0 °C or room temperature, rather than at −30 °C with catalyst **81**.

Chiral ferrocenylphosphines **95** were investigated by Pereira and coworkers [63] in the asymmetric Baylis–Hillman reaction between *p*-nitrobenzaldehyde and benzylacrylate. Chiral catalyst **95**, which combined both planar and central chirality, gave the Baylis–Hillman adducts in 78% yield and 65% ee (Scheme 10.49).

Recently, a trifunctional organocatalyst was designed and applied to the aza–Baylis–Hillman reactions by Liu and coworkers [64]. As well as a Lewis base nucleophile and a Brønsted acid for the H-bonding interactions used in the bifunctional catalyst above, this trifunctional system used another Brønsted base which could switch the activity via protonation by a strong Brønsted acid (Figure 10.10). In the presence of catalyst **96** and benzoic acid, the reactions between *N*-sulfonated imines and MVK afforded the corresponding products in high yields and with excellent ee-values (Scheme 10.50) in a shorter reaction time compared to the bifunctional catalyst **81**.

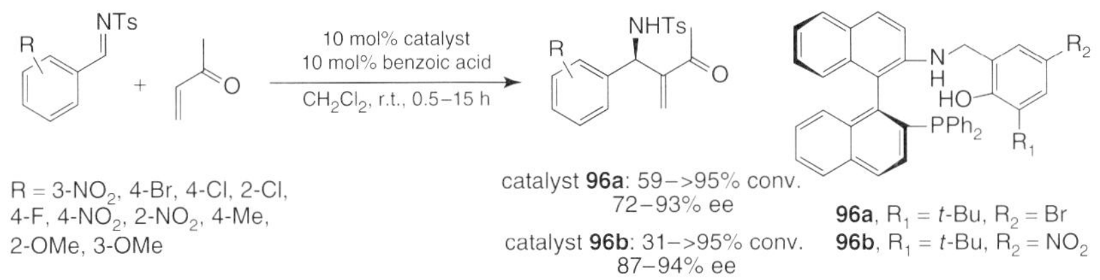

R = 3-NO$_2$, 4-Br, 4-Cl, 2-Cl, 4-F, 4-NO$_2$, 2-NO$_2$, 4-Me, 2-OMe, 3-OMe

catalyst **96a**: 59–>95% conv. 72–93% ee

catalyst **96b**: 31–>95% conv. 87–94% ee

96a, R$_1$ = *t*-Bu, R$_2$ = Br
96b, R$_1$ = *t*-Bu, R$_2$ = NO$_2$

Scheme 10.50 Aza–MBH reaction catalyzed by trifunctional catalysts **96**.

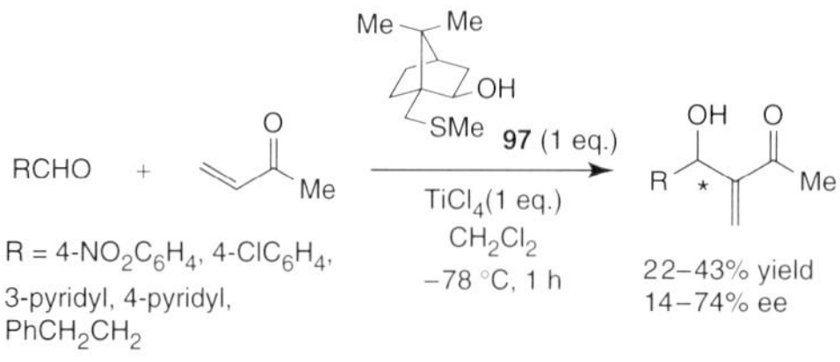

Scheme 10.51 MBH reactions catalyzed by chiral chalcogenide **97**.

10.3.2.3 Chiral Sulfides

In addition to tertiary amine and tertiary phosphine, chalcogenides such as sulfides or selinides may also promote the Baylis–Hillman reaction in the presence of a Lewis acid. Thus, the effect of a combination of chiral sulfide and Lewis acid on asymmetric Baylis–Hillman reactions has been investigated. The first enantioselective reaction was reported in 1999 by Kataoka and coworkers [65a], who used a chiral hydroxyl chalcogenide **97** together with $TiCl_4$ in the reaction of aldehydes and MVK. The corresponding adducts were achieved in moderate yields and 14–74% ee (Scheme 10.51). Moreover, a large number of diol ligands and bisoxazoline ligands, together with Me_2S and $TiCl_4$, were systematically investigated [65b]. The reaction proceeded smoothly, but gave very low ee-values of up to only 7%.

Goodman and coworkers [66] developed a sulfide-BF_3OEt_2 catalyst system for the reaction between various aldehydes and MVK. When the C_2-symmetric tricyclic sulfide **98** was used, the Baylis–Hillman reaction between aldehydes and MVK could proceed smoothly, giving the corresponding products in 41–60% yields and 14–46% ee (Scheme 10.52).

The most excellent result of the chalcogenide-mediated asymmetric Baylis–Hillman reaction was reported by Aggarwal and coworkers [67], who applied the chiral sulfide **99** together with Lewis acid TMSOTf to the reaction between cyclic enone and N-acry iminium ions, a new class of electrophile. The corresponding products were obtained in 80–98% ee (Scheme 10.53), and the origin of the enantioselectivity is shown in Scheme 10.54. The formation of intermediates **A** and **B** by the addition of catalyst **99** to cyclopentenone was confirmed using low-temperature NMR. The intermediate **A** may then approach the N-acry iminium ions in four possible ways (**C**, **D**, **E**, and **F**). Intermediate **C** suffers from the nonbonding interactions between the two rings, while intermediates **E** and **F** were excluded by the observation that no difference in ee-value occurred when changing the silyl moiety from trimethylsilyl to

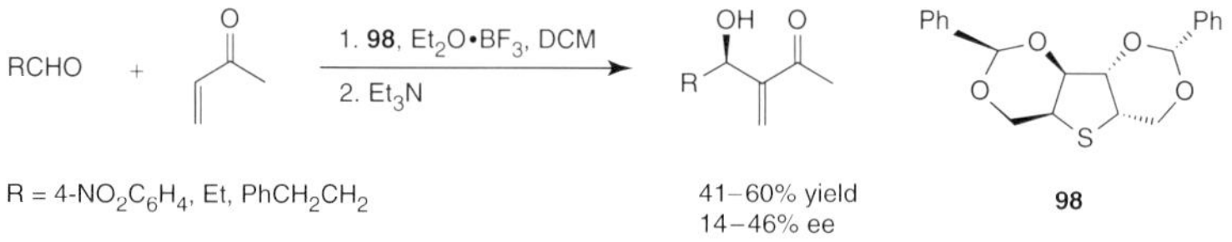

Scheme 10.52 MBH reactions catalyzed by sulfide **98**.

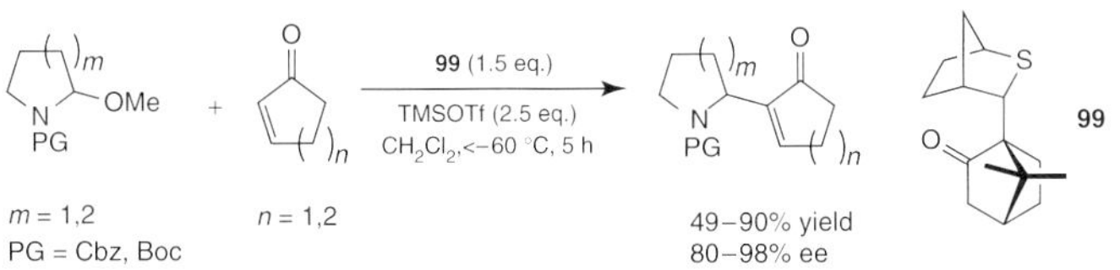

Scheme 10.53 MBH reactions catalyzed by the chiral sulfide **99**.

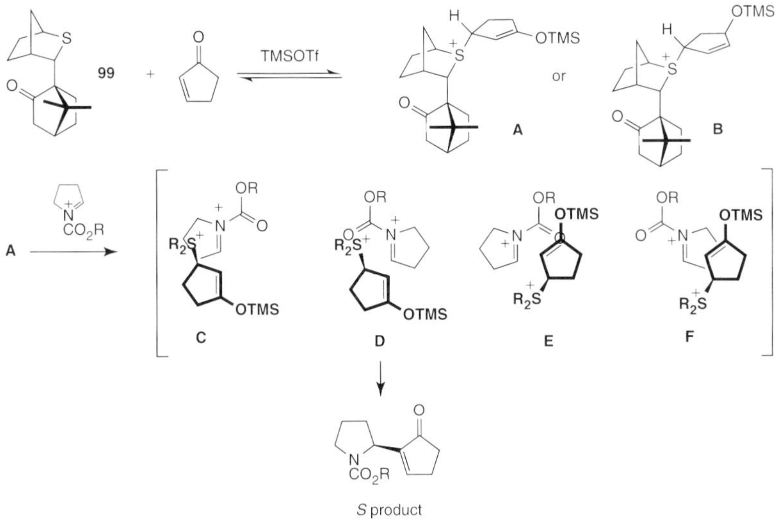

Scheme 10.54 The proposed mechanism for the origin of the enantioselectivity.

triisopropylsilyl. Consequently, the attack of intermediate **A** on the *Re* face of the imine ions in pathway **D** afforded the (s)-enriched product.

10.3.2.4 Chiral Acids

When investigating catalysts that might better activate reaction components, both Brønsted acids and Lewis acids have been shown to activate carbonyl groups involved in the Baylis–Hillman reaction. A coordination of the lone pairs of the carbonyl oxygen atom to the Lewis acid, or to the hydrogen bond between the carbonyl oxygen atom and the proton from the Brønsted acid, render the carbonyl group more favorable to nucleophilic attack by lowering the energy of the lowest unoccupied molecular orbital (LUMO) (Figure 10.11) [68]. Thus, the inclusion of chiral acid additives into the Baylis–Hillman reaction may represent a viable strategy for obtaining enantioselective reaction products.

L-Proline Combined with Achiral or Chiral Amines Shi and coworkers [69] were the first to show that L-proline and Lewis bases such as imidazole and triethylamine could promote the Baylis–Hillman reaction between arylaldehydes and MVK, although the products were obtained with very low optical activity.

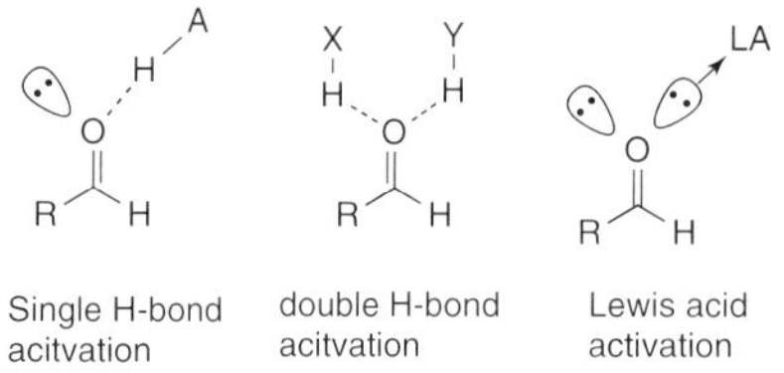

Figure 10.11 Alternative methods for Brønsted acid and Lewis acid activation of the carbonyl group.

Single H-bond acitvation

double H-bond acitvation

Lewis acid activation

Later, Miller and coworkers [70] suggested that a combination of the nucleophile-loaded peptides **100** and L-proline could promote the asymmetric Baylis–Hillman reaction between arylaldehydes and MVK, affording the corresponding products in good yields and enantioselectivities (Scheme 10.55). However, only highly activated aromatic aldehydes could proceed via this reaction.

When Tang and coworkers [71] adopted various chiral tertiary amine catalysts **101–104** (Figure 10.12), together with L-proline, in reactions between arylaldehydes and MVK (Scheme 10.56), moderate to good enantioselectivities were achieved for the electron-deficient aromatic aldehydes. However, a clear match/mismatch effect between the chiral amine and L-proline was observed when **103a** or **103b** and **104a** or **104b** were combined with L-proline. Under the same conditions, **103b** and **104b** gave much lower ee-values than were obtained with **103a** and **104a**, which suggested a mismatch with L-proline. The absolute configuration of the product was determined by the chirality of the L-proline; hence, an enamine intermediate was proposed for the reaction mechanism.

Usually, the MBH reactions of β-substituted α,β-unsaturated aldehydes or ketones are more problematic than those of the β-substituted substrates. However, Barbas and coworkers [72] reported that a combination of L-proline and imidazole could promote the reactions between β-substituted α,β-unsaturated aldehydes and α-imino esters protected with a p-methoxyphenyl (PMP) group. Although the corresponding Baylis–Hillman adducts were obtained in good yields

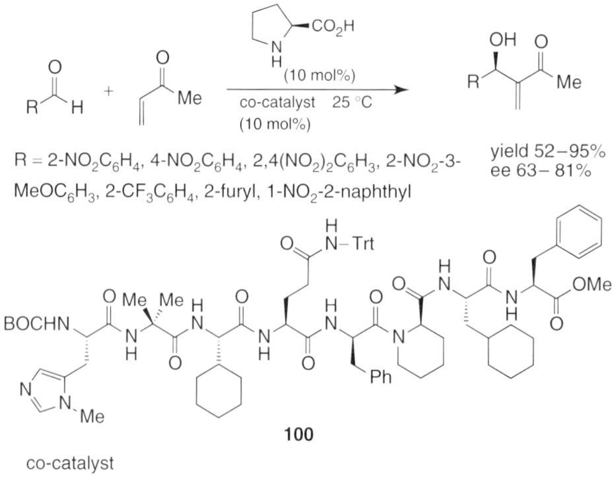

R = 2-$NO_2C_6H_4$, 4-$NO_2C_6H_4$, 2,4-$(NO_2)_2C_6H_3$, 2-NO_2-3-MeOC_6H_3, 2-$CF_3C_6H_4$, 2-furyl, 1-NO_2-2-naphthyl

yield 52–95%
ee 63–81%

100
co-catalyst

Scheme 10.55 MBH reactions catalyzed by peptides **100** and L-proline.

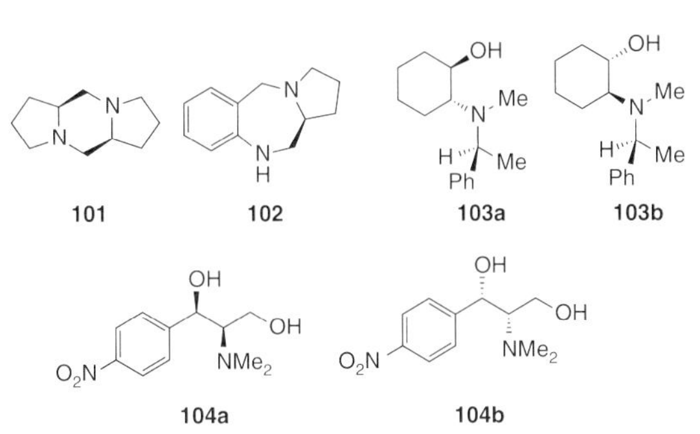
101
102
103a
103b
104a
104b

Figure 10.12 Chiral tertiary amine catalysts **101–104**.

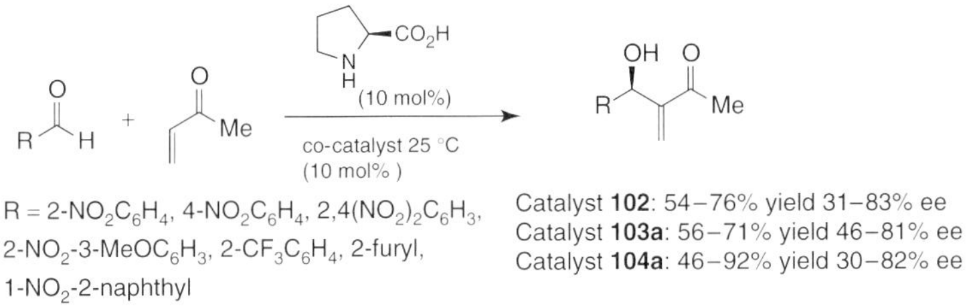
(10 mol%)
co-catalyst 25 °C
(10 mol%)
R = 2-NO2C6H4, 4-NO2C6H4, 2,4(NO2)2C6H3,
2-NO2-3-MeOC6H3, 2-CF3C6H4, 2-furyl,
1-NO2-2-naphthyl
Catalyst 102: 54–76% yield 31–83% ee
Catalyst 103a: 56–71% yield 46–81% ee
Catalyst 104a: 46–92% yield 30–82% ee

Scheme 10.56 MBH reactions catalyzed by tertiary amine catalysts and L-proline.

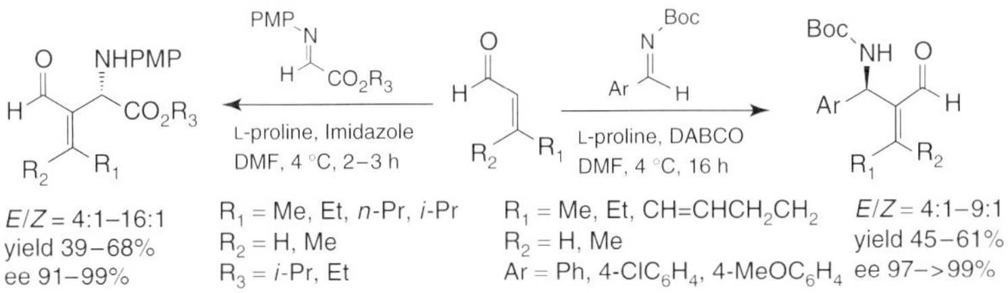
L-proline, Imidazole
DMF, 4 °C, 2–3 h
L-proline, DABCO
DMF, 4 °C, 16 h
E/Z = 4:1–16:1
yield 39–68%
ee 91–99%
R1 = Me, Et, n-Pr, i-Pr
R2 = H, Me
R3 = i-Pr, Et
R1 = Me, Et, CH=CHCH2CH2
R2 = H, Me
Ar = Ph, 4-ClC6H4, 4-MeOC6H4
E/Z = 4:1–9:1
yield 45–61%
ee 97–>99%

Scheme 10.57 The aza–Baylis–Hillman reactions co-catalyzed by L-proline and tertiary amine.

and excellent enantioselectivities (Scheme 10.57), the mechanism was considered to be a Mannich-type reaction, followed by isomerization of the double bond instead of the typical MBH reaction pathway. Córdova and coworkers [73] explored the combination of L-proline and DABCO for the reactions between unmodified aldehydes and *N*-Boc-protected imines. The reactions were shown to proceed with excellent chemoselectivity and enantioselectivity, while the Baylis–Hillman products were obtained in good yields with 97–99% ee (Scheme 10.57). This allowed highly enantiomerically enriched aza–Baylis–Hillman reaction products to be obtained with relative ease.

BINOL-Derivatives Chiral BINOL-derived Brønsted acid **105** was found by Schaus and coworkers [74] to catalyze the reaction between cyclohexenones and aldehydes,

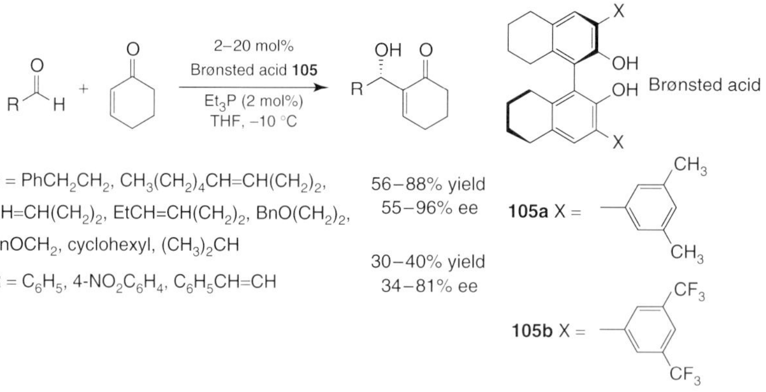

Scheme 10.58 MBH reactions catalyzed by chiral BINOL-derived Brønsted acid **105** and Et_3P.

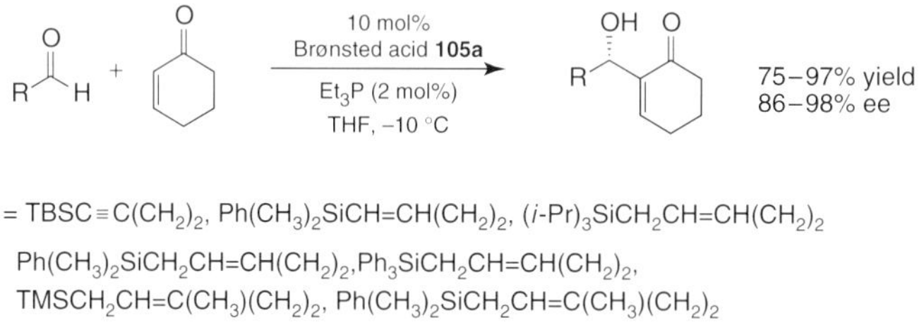

Scheme 10.59 MBH reactions of aldehydes containing the allyl silyl moiety.

together with Et_3P (Scheme 10.58). The corresponding Baylis–Hillman adducts were obtained in good yields and good enantioselectivities for the aliphatic aldehydes, whereas conjugated aldehydes such as benzaldehyde and cinnamaldehyde resulted in low yields and low enantioselectivities. Catalyst **105b** afforded the best results with the more hindered aldehydes. The same reaction was later applied to the synthesis of the core structure of the clerodane (Schemes 10.59 and 10.60). Initially, the Baylis–Hillman reaction between aldehydes containing an allyl silyl moiety (Scheme 10.59) with cyclohexenone in the presence of **105a** and Et_3P was explored. Except for the alkyne-containing substrate, all allyl silane-containing products were obtained with over 90% ee. The core structure of the clerodane was synthesized via a two-step asymmetric Baylis–Hillman reaction and Lewis acid BF_3OEt_2-promoted ring-annulation strategy (Scheme 10.60).

Thiourea Chiral ligands derived from thiourea have also recently been used successfully in the asymmetric Baylis–Hillman reaction.

For example, bis(thio)urea [75] **106** derived from chiral *trans*-1,2-diaminocyclohexane was shown to be a suitable organocatalyst for the asymmetric MBH reaction of various aldehydes and 2-cyclohexen-1-one, as reported by Nagasawa (Scheme 10.61). In the case of the aliphatic aldehydes, the corresponding products

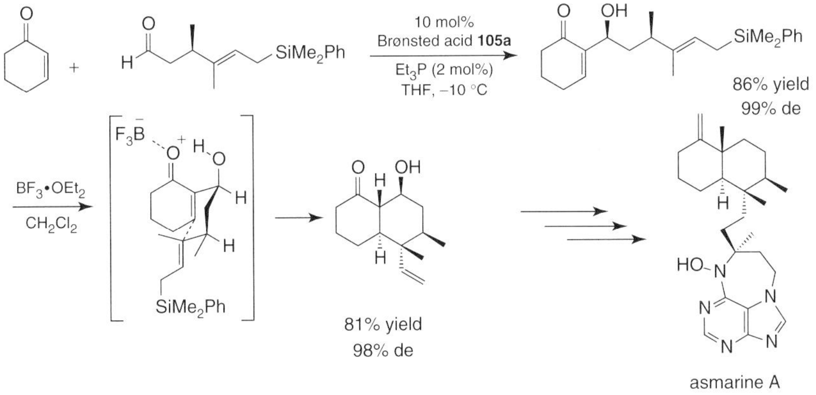

Scheme 10.60 Synthesis of the core structure of the clerodane.

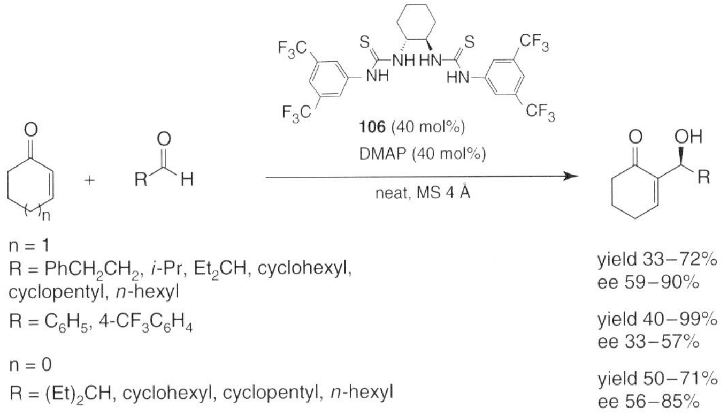

Scheme 10.61 MBH reactions catalyzed by bis(thio)urea **106** and DMAP.

were obtained in 59–90% ee, whereas for the aromatic aldehydes the ee-values of the products were much lower.

Jacobsen [14b] found that a chiral cocatalyst **107**, together with DABCO, could catalyze the aza–Baylis–Hillman reaction between aromatic nosylimine derivatives and methyl acrylate with high enantioselectivities, but with isolated yields that ranged between only 25% and 49% (Scheme 10.62). The authors suggested that the high ee-value had been obtained under conditions where a zwitterionic intermediate

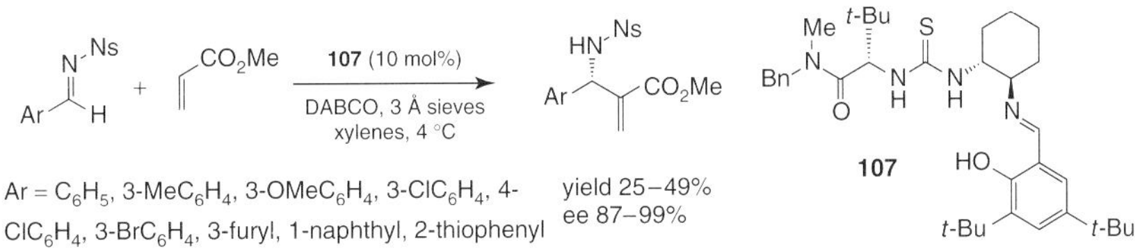

Scheme 10.62 Aza–MBH reactions catalyzed by catalyst **107** and DABCO.

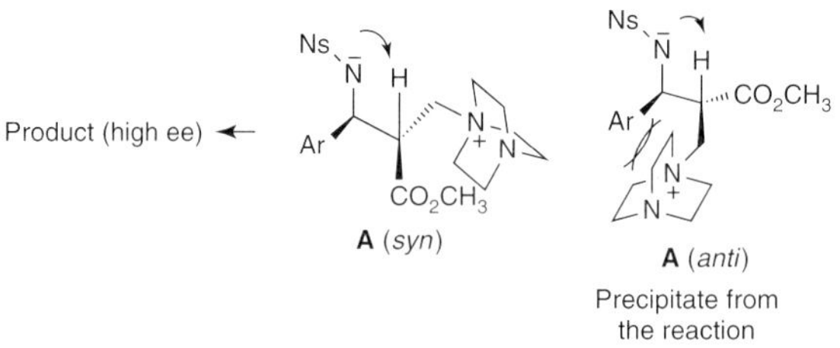

Figure 10.13 The reaction intermediate.

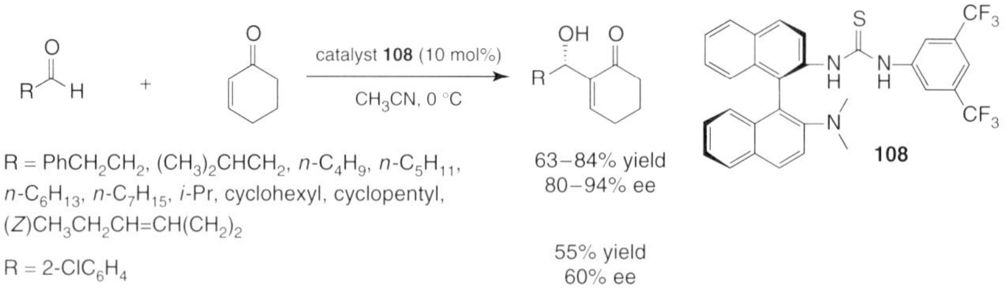

Scheme 10.63 MBH reaction catalyzed by the binaphthyl-derived amine thiourea catalyst **108**.

A (*anti*) (Figure 10.13) underwent precipitation from the reaction mixture, most likely because of its slow rate of elimination to form the product.

Wang and coworkers [76] developed a new bifunctional binaphthyl-derived amine thiourea catalyst **108** to promote the reaction between cyclohexenone and aldehydes (Scheme 10.63). The resultant products were obtained in high enantioselectivities for aliphatic aldehydes, while the aromatic aldehydes gave low ee-values.

Subsequently, Berkessel and coworkers [77] applied the bis(thio)urea catalyst **109**, derived from isophorone-diamine[3-(amino-methyl)-3,5,5-trimethylcyclohexylamine] (IPDA), together with DABCO to the same reaction (Scheme 10.64),

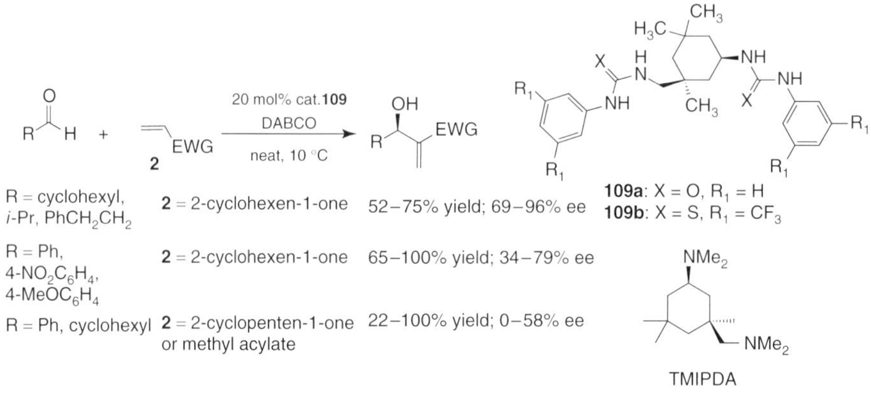

Scheme 10.64 MBH reactions catalyzed by the bis(thio)urea catalyst **109** and DABCO.

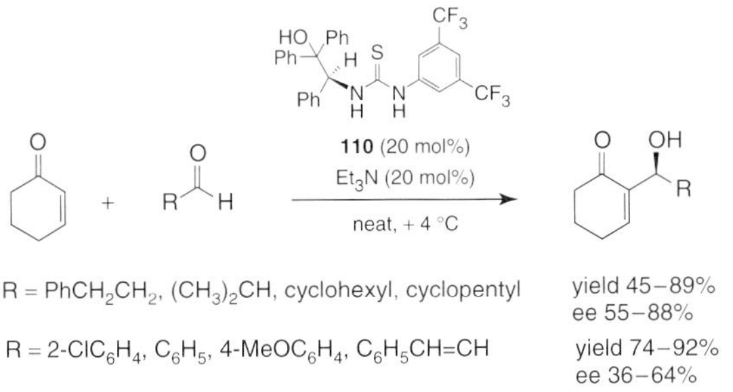

10.3 Asymmetric Intermolecular Baylis–Hillman Reaction

Scheme 10.65 MBH reactions catalyzed by the thiourea catalyst **110** and Et$_3$N.

R = PhCH$_2$CH$_2$, (CH$_3$)$_2$CH, cyclohexyl, cyclopentyl — yield 45–89%, ee 55–88%

R = 2-ClC$_6$H$_4$, C$_6$H$_5$, 4-MeOC$_6$H$_4$, C$_6$H$_5$CH=CH — yield 74–92%, ee 36–64%

and in some cases the Lewis base TMIPDA proved superior to DABCO. In this case, Michael-acceptors as well as 2-cyclohexen-1-one were investigated, but low enantioselectivities were observed.

Recently, thiourea [78] **110**, a simple amino alcohol derivative was prepared and tested in the Baylis–Hillman reaction, and similar results were obtained (Scheme 10.65).

Shi and coworkers [79] reported that an improved bis(thio)urea organocatalyst **111**, derived from axially chiral (R)-5,5′,6,6′,7,7′,8,8′-octahydro-1,1′-binaphthyl-2,2′-diamine (H$_8$-BINAM), was fairly effective in the MBH reaction between arylaldehydes and 2-cyclohexen-1-one. The corresponding adducts were obtained in good to excellent yields, and with up to 88% ee (Scheme 10.66).

Wu and coworkers [80] designed a chiral phosphinothiourea organocatalyst **112** for the asymmetric MBH reaction, and applied it to the general Baylis–Hillman reaction between aldehydes and MVK. The reactions proceeded well with aromatic aldehydes bearing electron-withdrawing groups, and the corresponding products were obtained rapidly, in good to excellent yields, and generally with an excellent enantiomeric excess (Scheme 10.67). A transition state was also proposed (Figure 10.14).

Lewis Acids In addition to chiral Brønsted acids, chiral Lewis acids have also been applied to the asymmetric Baylis–Hillman reaction. For example, Aggarwal and

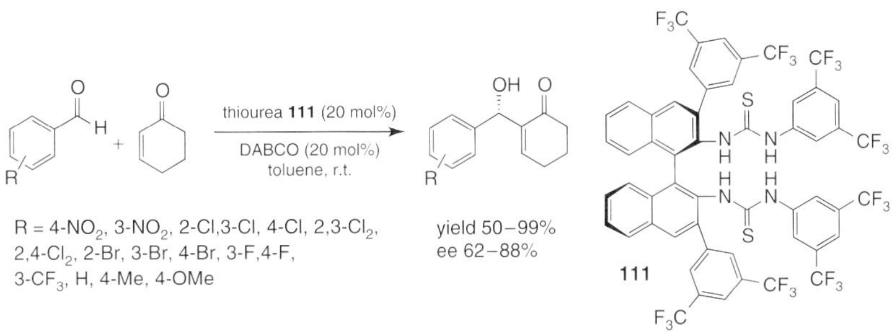

R = 4-NO$_2$, 3-NO$_2$, 2-Cl, 3-Cl, 4-Cl, 2,3-Cl$_2$, 2,4-Cl$_2$, 2-Br, 3-Br, 4-Br, 3-F, 4-F, 3-CF$_3$, H, 4-Me, 4-OMe — yield 50–99%, ee 62–88%

Scheme 10.66 MBH reactions catalyzed by the thiourea catalyst **111** and DABCO.

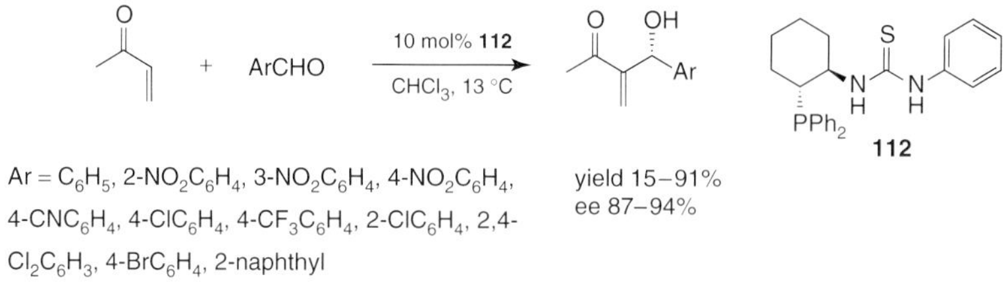
10 mol% 112
ArCHO
CHCl3, 13 °C
OH
Ar
PPh2
112
Ar = C6H5, 2-NO2C6H4, 3-NO2C6H4, 4-NO2C6H4, 4-CNC6H4, 4-ClC6H4, 4-CF3C6H4, 2-ClC6H4, 2,4-Cl2C6H3, 4-BrC6H4, 2-naphthyl
yield 15–91%
ee 87–94%

Scheme 10.67 MBH reactions catalyzed by the phosphinothiourea catalyst **112**.

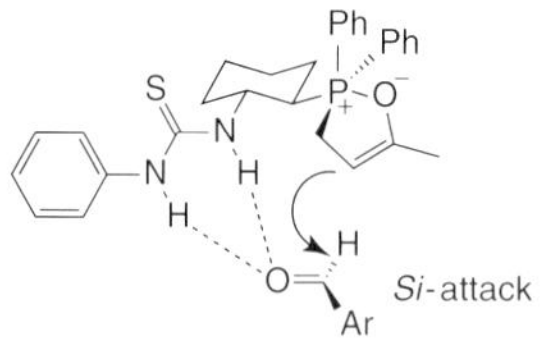
Si-attack

Figure 10.14 The proposed transition state.

coworkers [81] reported the use of lanthanide ions and different chiral ligands such as salen, aminodiols, aminotriols, bisoxazoline, and diisopropyltartrate in the reactions between aldehydes and *tert*-butyl acrylate. Unfortunately, however, the MBH adducts were obtained only in 5% ee, at best.

By using (*R*)-1,1′-bi-2-naphthol as co-catalyst together with tributylphosphine, the Baylis–Hillman reaction between 2-cyclopenten-1-one and 3-phenyl-1-propanal gave <10% ee [11h]. However, moderate ee-values (56%) could be achieved when an optically active calcium catalyst **113** was adopted (Scheme 10.68).

Later, Chen and coworkers [82] reported an effective chiral Lewis acid-catalyzed reaction between various aldehydes and acrylates (Scheme 10.69), whereby good to high enantioselectivities could be obtained using 3 mol% catalyst **114** and 3 mol% $La(OTf)_3$, together with DABCO. Likewise, when naphthyl acrylate was used as a Michael acceptor, the reaction was complete within 20 min, with reasonable yields and ee-values of 70–95%. A proposed intermediate which explains the enantiofacial bias is shown in Figure 10.15. In this case, the hexadentate ligand is first bound to the metal atom, after which the azaenolate and aldehyde are coordinated to the metal atom, from the less-hindered bottom side. The enolate then attacks the *Re* face of the aldehyde, leading to the (*s*)-product.

Sasai and coworkers [83] found that the combined use of a heterobimetallic asymmetric complex **115** and tributylphosphine could promote the Baylis–Hillman

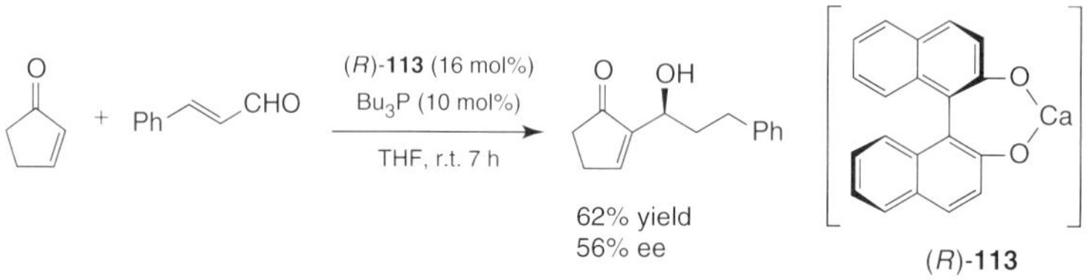
(R)-113 (16 mol%)
Bu3P (10 mol%)
THF, r.t. 7 h
CHO
Ph
OH
Ph
62% yield
56% ee
Ca
(R)-113

Scheme 10.68 MBH reaction catalyzed by the chiral calcium catalyst **113**.

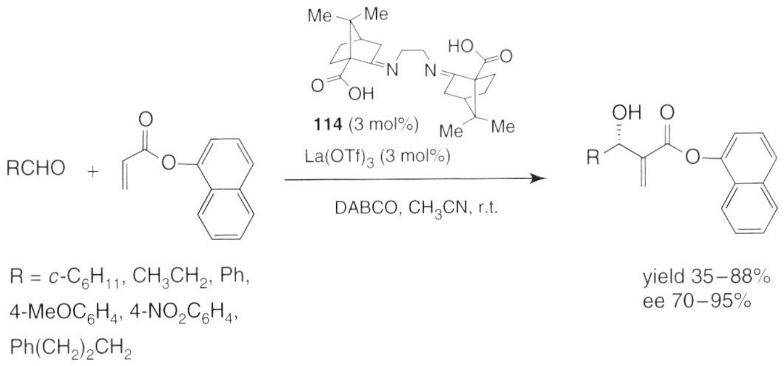

10.3 Asymmetric Intermolecular Baylis–Hillman Reaction

RCHO + [1-naphthyl acrylate] →(114 (3 mol%), La(OTf)$_3$ (3 mol%), DABCO, CH$_3$CN, r.t.) product

R = c-C$_6$H$_{11}$, CH$_3$CH$_2$, Ph, 4-MeOC$_6$H$_4$, 4-NO$_2$C$_6$H$_4$, Ph(CH$_2$)$_2$CH$_2$

yield 35–88%
ee 70–95%

Scheme 10.69 MBH reaction catalyzed by catalyst **114** and La(OTf)$_3$ and DABCO.

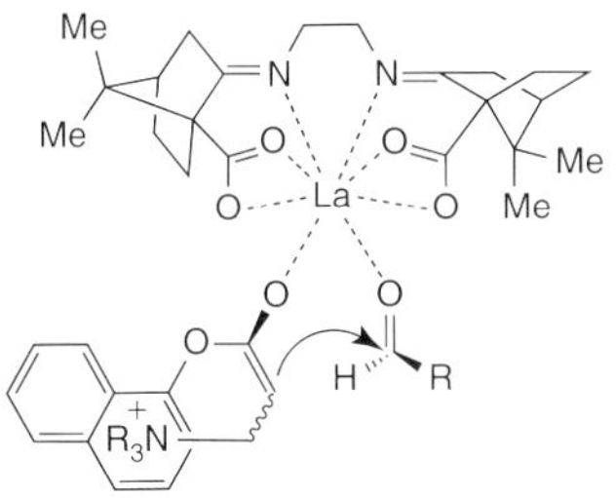

Figure 10.15 The proposed intermediate.

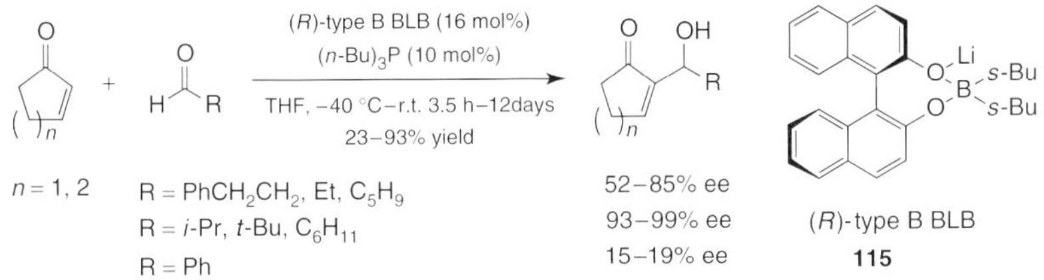

cyclopentenone/cyclohexenone + RCHO →((R)-type B BLB (16 mol%), (n-Bu)$_3$P (10 mol%), THF, −40 °C–r.t. 3.5 h–12days, 23–93% yield) product

n = 1, 2
R = PhCH$_2$CH$_2$, Et, C$_5$H$_9$ — 52–85% ee
R = i-Pr, t-Bu, C$_6$H$_{11}$ — 93–99% ee
R = Ph — 15–19% ee

(R)-type B BLB
115

Scheme 10.70 MBH reaction catalyzed by catalyst **115** and (n-Bu)$_3$P.

reactions between cyclohexenone or cyclopentenone and aldehydes (Scheme 10.70). The adducts were obtained in good yields and with moderate to high enantioselectivities. Although, in the case of aliphatic branched aldehydes, excellent ee-values were obtained, only low values were achieved when benzaldehyde was used.

10.3.3
Chiral Reaction Media

Rather than a chiral catalyst, it is possible to use a chiral reaction medium in the asymmetric Baylis–Hillman reaction. Vo-Thanh and coworkers [84] were the first to report the application of a chiral ionic liquid (IL) **116** in the reaction between arylaldehydes and methyl acrylate. The Baylis–Hillman reaction products were obtained in good yields and up to 44% ee (Scheme 10.71). Notably, the presence of

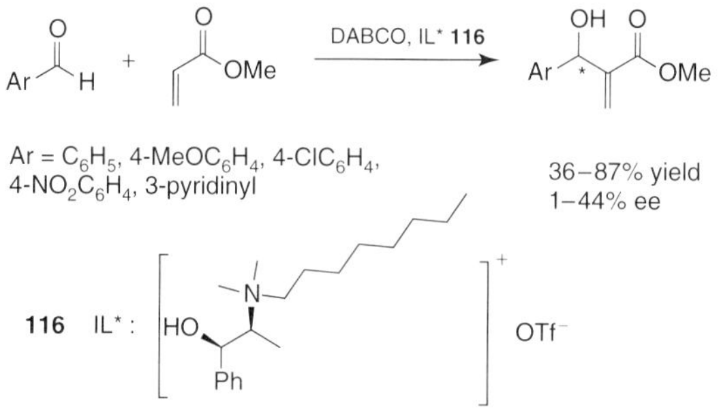

10 Catalytic Asymmetric Baylis–Hillman Reactions and Surroundings

Scheme 10.71 MBH reaction in chiral ionic liquids **116**.

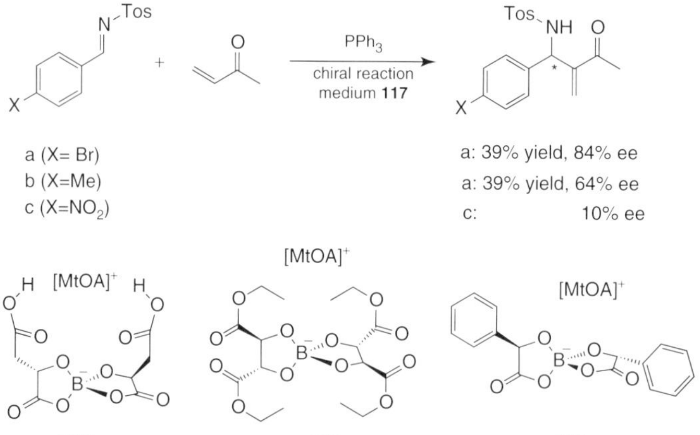

Scheme 10.72 Aza–MBH reaction in chiral ionic liquids **117**.

a hydroxyl function on the chiral IL was considered to play a key role in the transfer of the chirality.

When Leitner and coworkers [85] applied chiral ILs methyltrioctylammonium **117** to the aza–Baylis–Hillman reactions between N-sulfonated imines and MVK (Scheme 10.72), up to 84% ee was achieved in the reaction between N-(4-bromobenzylidene)-4-toluenesulfonamide and MVK. A Brønsted acidic moiety on the chiral ILs was considered necessary for both the reaction yield and asymmetric induction. For example when the structurally related chiral ILs **118** and **119** were used, the reaction gave very low conversions (15% and 14%, respectively) and racemic products.

Recently, the groups of both Ni and Headly [86] designed a new chiral IL (**120**) by incorporating chiral side chains on the C-2 position of the imidazolium cation ring. This IL was adopted as a reaction medium for reactions between aldehydes and acrylates in the presence of DABCO, and the corresponding adducts were obtained in good yields and 2–24% ee (Scheme 10.73). The proposed mechanism for the asymmetric discrimination is shown in Scheme 10.73 (structure **121**).

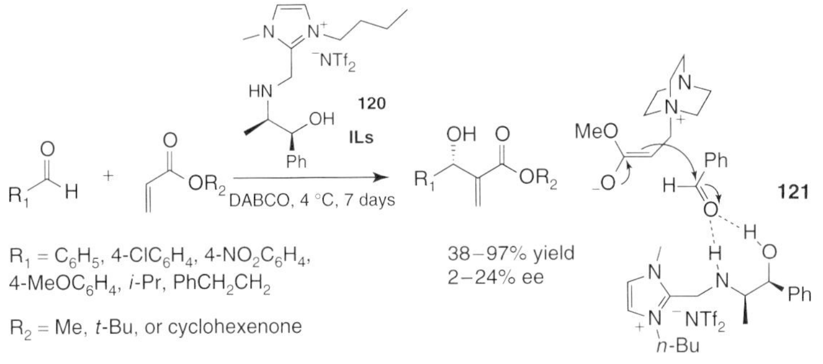

Scheme 10.73 MBH reaction in chiral ionic liquids **120**.

10.4
Asymmetric Intramolecular Morita–Baylis–Hillman Reaction

Unlike the intermolecular Baylis–Hillman reaction, very few examples of enantioselective intramolecular reactions have been reported. The first such example was developed by Frater and coworkers [87] (Scheme 10.74) who showed that, in the presence of chiral tertiary phosphine **122**, the α,β-unsaturated-ε-keto ester could be transformed to the corresponding Baylis–Hillman product in 40% yield and 14% ee after 10 days (Scheme 10.74). The authors considered the low ee-value was due to the reversibility of the ring-closure.

In a second example, Miller and coworkers [88] attempted to apply their co-catalyst system, which had afforded good enantioselectivities in intermolecular Baylis–Hillman reactions, to the intramolecular Baylis–Hillman reaction. Following an intensive screening of the amino acid and solvent, it was found that pipercolinic acid and N-methylimidazole could catalyze the intramolecular MBH reaction to give the corresponding products in moderate yields and good enantioselectivities (Scheme 10.75). Furthermore, by using acetic anhydride together with a peptide-based asymmetric acylation catalyst **123**, a kinetic resolution of the MBH reaction adduct occurred such that the final compound could be obtained in >98% ee.

Hong and coworkers [89] investigated the intramolecular Baylis–Hillman reaction of hept-2-enedial by using a dual catalyst system (Scheme 10.76). A good enantioselectivity was obtained when using L-proline as catalyst, the most interesting phenomenon being that the configuration of the product was changed

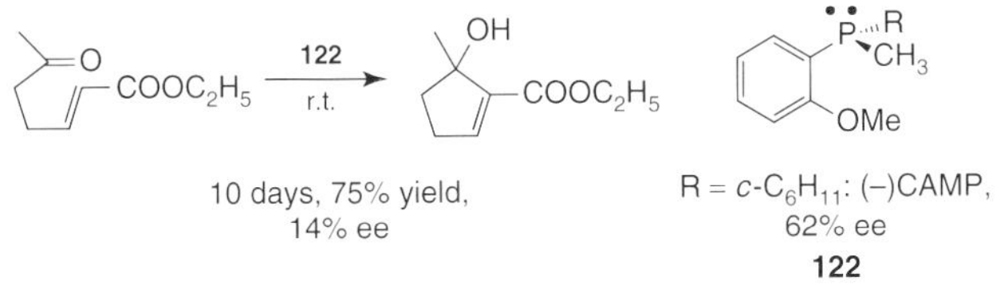

Scheme 10.74 Intramolecular MBH reaction catalyzed by chiral tertiary phosphine **122**.

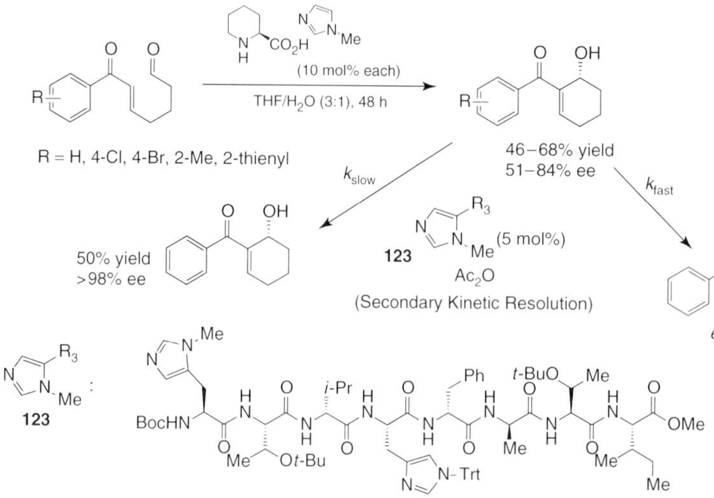

Scheme 10.75 Intramolecular MBH reaction catalyzed by pipercolinic acid and N-methylimidazole.

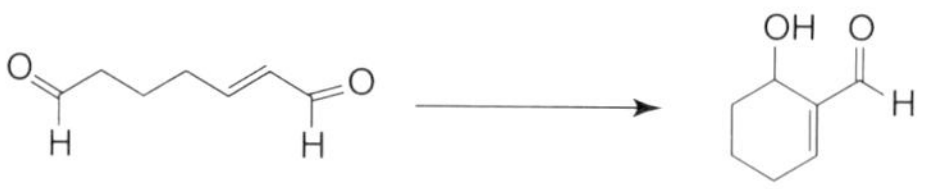

L-Pro (0.1 eq.) in DMF, 70 °C, 1 h   yield: 69%; ee 60%; configuration: S

L-Pro (1 eq.) +Imid (1 eq.) in CH$_3$CN, 0 °C, 7 h   yield: 72%; ee 93%; configuration: R

Scheme 10.76 Intramolecular MBH reaction catalyzed by L-proline and imidazole.

by the addition of imidazole to the reaction. The reason for this inversion of selectivity was considered to be the formation of a new reactive intermediate **T′** (see Scheme 10.77). The reaction was shown to start through an iminium activation, after which the enamine aldol reaction without imidazole proceeded favorably through a Zimmerman–Traxler transition state **A**, which was proposed to be more stable than **B**, so as to generate the (S)-product. In the presence of imidazole, the (s)-imidazolium **T′** was formed by an intramolecular hydrogen bonding. The following transition state **A′** suffered from an axial imidazonium group, which meant that **B′** would be more favorable and the (R)-product generated. This mechanism was confirmed by Gil Santos and coworkers, using DFT [14c].

A new chiral rhenium-containing phosphine catalyst family has been tested in the intramolecular MBH reactions of the dicarbonyl compounds. The corresponding products were obtained in excellent yields and with moderate enantiomeric excess [90] (Scheme 10.78). When the intramolecular Rauhut–Currier reactions catalyzed by this catalyst were subsequently explored, the products were obtained in moderate yields and ee-values (Scheme 10.79).

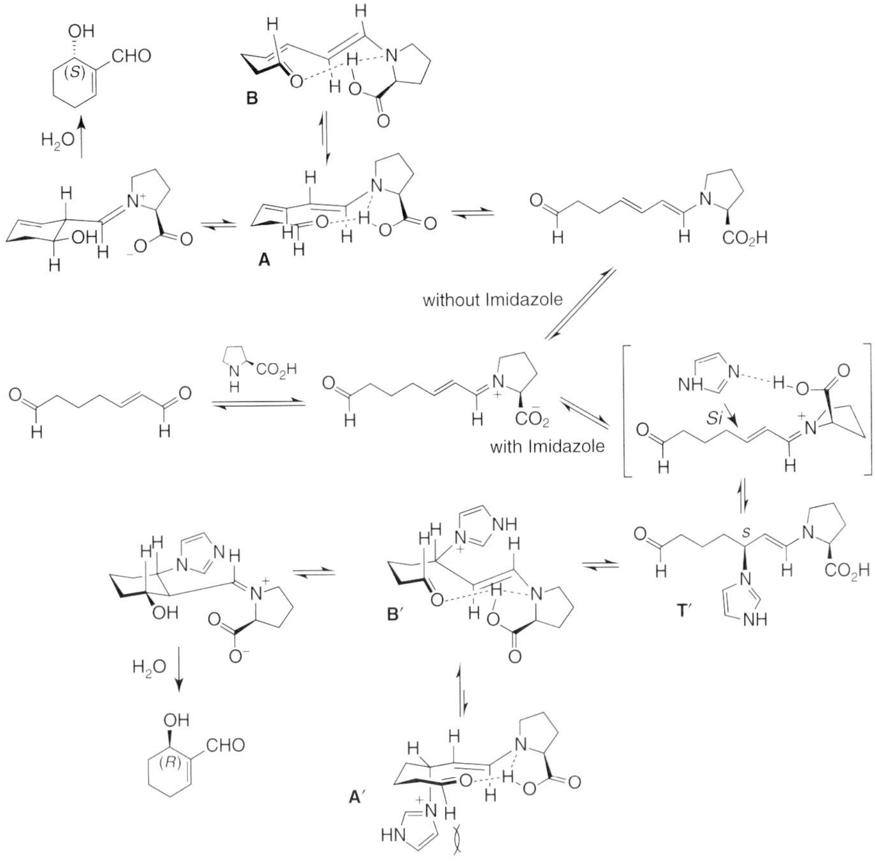

Scheme 10.77 Mechanism for the inversion of selectivity.

An asymmetric intramolecular Baylis–Hillman reaction of chiral substrates **125** and **126** afforded the corresponding products in good yields and with >95% de [91] (Scheme 10.80), thus offering an alternative route for the synthesis of α-methylene-β-hydroxylactones.

10.5
Conclusions

The asymmetric Baylis–Hillman reaction has undergone extensive examination during the past few years, and several excellent catalysts have been developed. However, there remains much room for improvement in terms of substrate generality, as all of the successful catalysts developed to date either are efficient only for imine substrates or are applicable only to aliphatic or activated aromatic aldehydes. Hence, more general catalysts that are suitable for a broader substrate range must

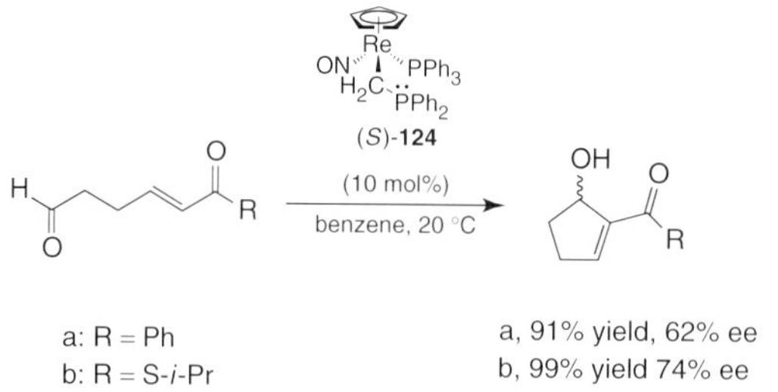
Re
ON
PPh3
H2C
PPh2
(S)-124
(10 mol%)
benzene, 20 °C
H
O
R
OH
a: R = Ph
b: R = S-i-Pr
a, 91% yield, 62% ee
b, 99% yield 74% ee

Scheme 10.78 Intramolecular MBH reaction catalyzed by chiral catalyst **124**.

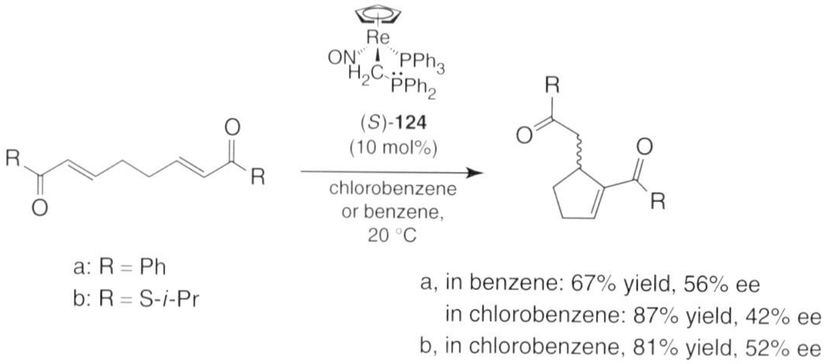
Re
ON
PPh3
H2C
PPh2
(S)-124
(10 mol%)
chlorobenzene
or benzene,
20 °C
R
O
a: R = Ph
b: R = S-i-Pr
a, in benzene: 67% yield, 56% ee
in chlorobenzene: 87% yield, 42% ee
b, in chlorobenzene, 81% yield, 52% ee

Scheme 10.79 Intramolecular Rauhut–Currier reactions catalyzed by catalyst **124**.

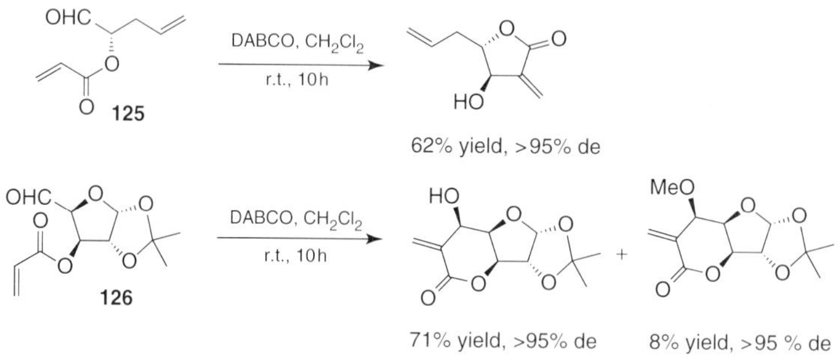
OHC
O
125
DABCO, CH2Cl2
r.t., 10h
HO
62% yield, >95% de
126
MeO
71% yield, >95% de
8% yield, >95 % de

Scheme 10.80 Asymmetric intramolecular MBH reaction of chiral substrates **125** and **126**.

be sought. Based on recent mechanistic developments, a good enantioselective catalyst for the Baylis–Hillman reaction must not only favor a major diastereomer in the Michael addition step (Scheme 10.2, step 2), but also promote a selective proton-transfer step (Scheme 10.2, step 3) of this major diastereomer. Likewise, the stereocontrol of both key steps (step 2, addition to aldehydes, and step 3, the proton transfer) must be of major concern in the design of a new catalyst. Whilst the results of the enantioselective Baylis–Hillman reaction are encouraging, there is a clear need for further development in this stimulating area of catalysis.

References

1. (a) Morita, K.-I., Suzuki, Z., and Hirose, H. (1968) *Bull. Chem. Soc. Jpn*, **41**, 2815–2819; (b) Baylis, A.B., Hillman, M.E.D. (1972) *Chem. Abstr.*, **77**, 34174q; Baylis, A.B., Hillman, M.E.D. (1972) Reaction of acrylic type compounds with aldehydes and certain ketones. Ger. Offen. 2155113; (c) Hillman, M.E.D., Baylis, A.B. (1973) US Patent 3,743,669; (d) Rauhut, M. and Currier, H. (1963) American Cyanamid Co., US patent 3074999; Rauhut, M. and Currier, H. (1963) *Chem. Abstr.*, **58**, 11224a.

2. For the reviews of Morita–Baylis–Hillman reaction: (a) Drewes, S.E. and Roos, G.H.P. (1988) *Tetrahedron*, **44**, 4653–4670; (b) Basavaiah, D., Rao, P.D., and Hyma, R.S. (1996) *Tetrahedron*, **52**, 8001–8062; (c) Ciganek, E. (1997) in *Organic Reactions*, vol. 51 (ed. L.A. Paquette), John Wiley & Sons, Inc., New York, pp. 201–359 (d) Basavaiah, D., Rao, A.J., and Satyanarayana, T. (2003) *Chem. Rev.*, **103**, 811–892; (e) Basavaiah, D., Rao, K.V., and Reddy, R.J. (2007) *Chem. Soc. Rev.*, **36**, 1581–1588; (f) Singh, V. and Batra, S. (2008) *Tetrahedron*, **64**, 4511; For reviews of aza-Baylis-Hillman reaction: (g) Shi, Y.-L. and Shi, M. (2007) *Eur. J. Org. Chem.*, 2905–2916; (h) Masson, G., Housseman, C., and Zhu, J. (2007) *Angew. Chem., Int. Ed.*, **46**, 4614–4628; (i) Declerck, V., Martinez, J., and Lamaty, F. (2009) *Chem. Rev.*, **109**, 1–48.

3. Trost, B.M. (1991) *Science*, **254**, 1471.

4. For selected recent examples, see: (a) Aggarwal, V.K., Dean, D.K., Mereu, A., and Williams, R. (2002) *J. Org. Chem.*, **67**, 510; (b) Aggarwal, V.K., Emme, I., and Fulford, S.Y. (2003) *J. Org. Chem.*, **68**, 692; (c) You, J., Xu, J., and Verkade, J.G. (2003) *Angew. Chem., Int. Ed.*, **42**, 5054–5056; (d) Kumar, A. and Pawar, S.S. (2003) *Tetrahedron*, **59**, 5019–5026; (e) Krishna, P.R., Manjuvani, A., Kannan, V., and Sharma, G.V.M. (2004) *Tetrahedron Lett.*, **45**, 1183–1185, and references therein.

5. For recent applications in total syntheses, see: (a) Lei, X. and Porco, J.A. (2006) *J. Am. Chem. Soc.*, **128**, 14790–114791; (b) Jogireddy, R. and Maier, M.E. (2006) *J. Org. Chem.*, **71**, 6999–7006; (c) Nag, S., Pathak, R., Kumar, M., Shukla, P.K., and Batra, S. (2006) *Bioorg. Med. Chem. Lett.*, **16**, 3824–3828; (d) Motozaki, T., Sawamura, K., Suzuki, A., Yoshida, K., Ueki, T., Ohara, A., Munakata, R., Takao, K.-I., and Tadano, K.-I. (2005) *Org. Lett.*, **7**, 2265–2267; (e) Trost, B.M., Machacek, M.R., and Tsui, H.C. (2005) *J. Am. Chem. Soc.*, **127**, 7014–7024; (f) Reddy, L.J., Fournier, L.F., Reddy, B.V.S., and Corey, E.J. (2005) *Org. Lett.*, **7**, 2699–2701; (g) Ho, A., Cyrus, K., and Kim, K.-B. (2005) *Eur. J. Org. Chem.*, 4829–4834; (h) Reddy, L.R., Saravanan, P., and Corey, E.J. (2004) *J. Am. Chem. Soc.*, **126**, 6230–6231; (i) Mergott, D.J., Frank, S.A., and Roush, W.R. (2004) *Proc. Natl Acad. Sci. USA*, **101**, 11955–11959.

6. Hoffmann, H.M.R. and Rabe, J. (1983) *Angew. Chem., Int. Ed.*, **22**, 796–797.

7. Hill, J.S. and Isaacs, N.S. (1990) *J. Phys. Org. Chem.*, **3**, 285–290.

8. (a) Kaye, P.T. and Bode, M.L. (1991) *Tetrahedron Lett.*, **32**, 5611–5614; (b) Fort, Y., Berthe, M.-C., and Caubère, P. (1992) *Tetrahedron*, **48**, 6371–6384.

9. (a) Santos, L.S., Pavam, C.H., Almeida, W.P., Coelho, F., and Eberlin, M.N. (2004) *Angew. Chem., Int. Ed.*, **43**, 4330–4333; (b) Amarante, G.W., Milagre, H.M.S., Vaz, B.G., Vilachã Ferreira, B.R., Eberlin, M.N., and Coelho, F. (2009) *J. Org. Chem.*, **74**, 3031–3037.

10. (a) Aggarwal, V.K., Fulford, S.Y., and Lloyd-Jones, G.C. (2005) *Angew. Chem., Int. Ed.*, **44**, 1706–1708; (b) Robiette, R., Aggarwal, V.K., and Harvey, J.N. (2007) *J. Am. Chem. Soc.*, **129**, 15513–15525.

11. (a) Ameer, F., Drewes, S.E., Freese, S., and Kaye, P.T. (1988) *Synth. Commun.*, **18**, 495–500; (b) Bailey, M., Markó, I.E., Ollis, D., and Rasmussen, P.R. (1990) *Tetrahedron Lett.*, **31**, 4509–4512; (c) Augé, J., Lubin, N., and Lubineau, A. (1994) *Tetrahedron Lett.*, **35**, 7947–7948;

(d) Rezgui, F. and Gaied, M.M.E. (1998) *Tetrahedron Lett.*, **39**, 5965–5968; (e) Basavaiah, D., Krishnamacharyulu, M., and Rao, A.J. (2000) *Synth. Commun.*, **30**, 2061–2069; (f) Yu, C., Liu, B., and Hu, L. (2001) *J. Org. Chem.*, **66**, 5413–5418; (g) Cai, J., Zhou, Z., Zhao, G., and Tang, C. (2002) *Org. Lett.*, **4**, 4723–4725; (h) Yamada, Y.M.A. and Ikegami, S. (2000) *Tetrahedron Lett.*, **41**, 2165–2169.

12. (a) Price, K.E., Broadwater, S.J., Jung, H.M., and McQuade, D.T. (2005) *Org. Lett.*, **7**, 147–150; (b) Price, K.E., Broadwater, S.J., Walker, B.J., and McQuade, D.T. (2005) *J. Org. Chem.*, **70**, 3980–3987.

13. (a) Perlmutter, P., Puniani, E., and Westman, G. (1996) *Tetrahedron Lett.*, **37**, 1715–1718; (b) Brzezinski, L.J., Rafel, S., and Leahy, J.W. (1997) *J. Am. Chem. Soc.*, **119**, 4317–4318; (c) Brzezinski, L.J., Rafel, S., and Leahy, J.W. (1997) *Tetrahedron*, **53**, 16423–16434.

14. (a) Krafft, M.E., Haxell, T.F.N., Seibert, K.A., and Abboud, K.A. (2006) *J. Am. Chem. Soc.*, **128**, 4174–4175; (b) Raheem, I.T. and Jacobsen, E.N. (2005) *Adv. Synth. Catal.*, **347**, 1701–1708.

15. (a) Xu, J.J. (2006) *Theochem*, **767**, 61–66; (b) Roy, D. and Sunoj, R.B. (2007) *Org. Lett.*, **9**, 4873–4876; (c) Duarte, F.J.S., Cabrita, E.J., Frenking, G., and Gil Santos, A. (2009) *Chem. Eur. J.*, **15**, 1734–1746.

16. (a) Kündig, E.P., Xu, L.H., Romanens, P., and Bernardinelli, G. (1993) *Tetrahedron Lett.*, **34**, 7049–7052; (b) Kündig, E.P., Xu, L.H., and Schnell, B. (1994) *Synlett*, 413–414.

17. Bussolari, J.C., Beers, K., Lalan, P., Murray, W.V., Gauthier, D., and McDonnell, P. (1998) *Chem. Lett.*, **27**, 787–788.

18. (a) Alcaide, B., Almendros, P., and Aragoncillo, C. (1999) *Chem. Commun.*, 1913–1914; (b) Alcaide, B., Almendros, P., and Aragoncillo, C. (2001) *J. Org. Chem.*, **66**, 1612–1620; (c) Alcaide, B., Almendros, P., and Aragoncillo, C. (1999) *Tetrahedron Lett.*, **40**, 7537–7540.

19. Aggarwal, V.K., Catsro, A.M.M., Mereu, A., and Adams, H. (2002) *Tetrahedron Lett.*, **43**, 1577–1581.

20. Shi, M. and Xu, Y.-M. (2002) *Tetrahedron: Asymmetry*, **13**, 1195–1200.

21. Lu, A., Xu, X., Gao, P., Zhou, Z., Song, H., and Tang, C. (2008) *Tetrahedron: Asymmetry*, **19**, 1886–1890.

22. Basavaiah, D., Gowriswari, V.V.L., Sarma, P.K.S., and Dharma Rao, P. (1990) *Tetrahedron Lett.*, **31**, 1621–1624.

23. Gilbert, A., Heritage, T.W., and Isaacs, N.S. (1991) *Tetrahedron: Asymmetry*, **10**, 969–972.

24. Yang, K.-S. and Chen, K. (2000) *Org. Lett.*, **2**, 729–731.

25. Krishna, P.R., Kannan, V., Ilangovan, A., and Sharma, G.V.M. (2001) *Tetrahedron: Asymmetry*, **12**, 829–837.

26. Bauer, T. and Tarasiuk, J. (2001) *Tetrahedron: Asymmetry*, **12**, 1741–1745.

27. Calmès, M., Akkari, R., Barthes, N., Escale, F., and Martinez, J. (2005) *Tetrahedron: Asymmetry*, **16**, 2179–2185.

28. Krishna, P.R., Sachwani, R., and Kannan, V. (2004) *Chem. Commun.*, 2580–2581.

29. He, K., Zhou, Z., Zhao, G., and Tang, C. (2006) *Heteroat. Chem.*, **17**, 317–321.

30. A study of the bifunctional activation and racemization in asymmetric aza–Baylis–Hillman reaction was reported: Buskens, P., Klankermayer, J., and Leitner, W. (2005) *J. Am. Chem. Soc.*, **127**, 16762–16763.

31. Drewes, S.E. and Roos, G.H.P. (1988) *Tetrahedron*, **44**, 4653–4670.

32. (a) Oishi, T. and Hirama, M. (1992) *Tetrahedron Lett.*, **33**, 639–642; (b) Oishi, T., Oguri, H., and Hirama, M. (1995) *Tetrahedron: Asymmetry*, **6**, 1241–1244.

33. (a) Barrett, A.G.M., Cook, A.S., and Kamimura, A. (1998) *Chem. Commun.*, 2533–2534; (b) Barrett, A.G.M., Dozzo, P., White, A.J.P., and Williams, D.J. (2002) *Tetrahedron*, **58**, 7303–7313.

34. Iwabuchi, Y., Nakatani, M., Yokoyama, N., and Hatakeyama, S. (1999) *J. Am. Chem. Soc.*, **121**, 10219–10220.

35. (a) Iwabuchi, Y., Sugihara, T., Esumi, T., and Hatakeyama, S. (2001) *Tetrahedron Lett.*, **42**, 7867–7871; (b) Iwabuchi, Y., Furukawa, M., Esumi, T., and

Hatakeyama, S. (2001) *Chem. Commun.*, 2030–2031.

36. Shi, M. and Jiang, J.-K. (2002) *Tetrahedron: Asymmetry*, **13**, 1941–1947.
37. (a) Shi, M. and Xu, Y.-M. (2002) *Angew. Chem., Int. Ed.*, **41**, 4507–4510; (b) Shi, M., Xu, Y.-M., and Shi, Y.-L. (2005) *Chem. Eur. J.*, **11**, 1794–1802.
38. (a) Kawahara, S., Nakano, A., Esumi, T., Iwabuchi, Y., and Hatakeyama, S. (2003) *Org. Lett.*, **5**, 3103–3105; (b) Nakano, A., Ushiyama, M., Iwabuchi, Y., and Hatakeyama, S. (2005) *Adv. Synth. Catal.*, **347**, 1790–1796; (c) Nakano, A., Kawahara, S., Akamatsu, S., Morokuma, K., Nakatani, M., Iwabuchi, Y., Takahashi, K., Ishihara, J., and Hatakeyama, S. (2006) *Tetrahedron*, **62**, 381–389.
39. Shi, M. and Zhao, G.-L. (2004) *Adv. Synth. Catal.*, **346**, 1205–1219.
40. Balan, D. and Adolfsson, H. (2003) *Tetrahedron Lett.*, **44**, 2521–2524.
41. Mocquet, C.M. and Warriner, S.L. (2004) *Synlett*, 356–358.
42. Krishna, P.R., Kannan, V., and Reddy, P.V.N. (2004) *Adv. Synth. Catal.*, **346**, 603–606.
43. Hayashi, Y., Tamura, T., and Shoji, M. (2004) *Adv. Synth. Catal.*, **346**, 1106–1110.
44. Xu, J., Guan, Y., Yang, S., Ng, Y., Peh, G., and Tan, C.-H. (2006) *Chem. Asian J.*, **1**, 724–729.
45. (a) Matsui, K., Takizawa, S., and Sasai, H. (2005) *J. Am. Chem. Soc.*, **127**, 3680–3681; (b) Matsui, K., Tanaka, K., Horii, A., Takizawa, S., and Sasai, H. (2006) *Tetrahedron: Asymmetry*, **17**, 578–583.
46. Abermil, N., Masson, G., and Zhu, J. (2008) *J. Am. Chem. Soc.*, **130**, 12596–12597.
47. Shi, M., Qi, M.-J., and Liu, X.-G. (2008) *Chem. Commun.*, 6025–6027.
48. Hayase, T., Shibata, T., Soai, K., and Wakatsuki, Y. (1998) *Chem. Commun.*, 1271–1272.
49. Li, W., Zhang, Z., Xiao, D., and Zhang, X. (2000) *J. Org. Chem.*, **65**, 3489–3496.
50. (a) Shi, M. and Chen, L.-H. (2003) *Chem. Commun.*, 1310–1311; (b) Shi, M., Chen, L.-H., and Li, C.-Q. (2005) *J. Am. Chem. Soc.*, **127**, 3790–3800.
51. Shi, M. and Li, C.-Q. (2005) *Tetrahedron: Asymmetry*, **16**, 1385–1397.
52. (a) Lei, Z.-Y., Ma, G.-N., and Shi, M. (2008) *Eur. J. Org. Chem.*, 3817–3820; (b) Lei, Z.-Y., Liu, X.-G., Shi, M., and Zhao, M. (2008) *Tetrahedron: Asymmetry*, **19**, 2058–2062.
53. Matsui, K., Takizawa, S., and Sasai, H. (2006) *Synlett*, 761–765.
54. Shi, M., Chen, L.-H., and Teng, W.-D. (2005) *Adv. Synth. Catal.*, **347**, 1781–1789.
55. Ito, K., Nishida, K., and Gotanda, T. (2007) *Tetrahedron Lett.*, **48**, 6147–6149.
56. Liu, Y.-H., Chen, L.-H., and Shi, M. (2006) *Adv. Synth. Catal.*, **348**, 973–979.
57. Shi, M., Liu, Y.-H., and Chen, L.-H. (2007) *Chirality*, **19**, 124–128.
58. Shi, M., Ma, G.-N., and Gao, J. (2007) *J. Org. Chem.*, **72**, 9779–9781.
59. Qi, M.-J., Ai, T., Shi, M., and Li, G. (2008) *Tetrahedron*, **64**, 1181–1186.
60. Shi, Y.-L. and Shi, M. (2007) *Adv. Synth. Catal.*, **349**, 2129–2135.
61. Liu, Y.-H. and Shi, M. (2008) *Adv. Synth. Catal.*, **350**, 122–128.
62. Guan, X.-Y., Jiang, Y.-Q., and Shi, M. (2008) *Eur. J. Org. Chem.*, 2150–2155.
63. Pereira, S.I., Adrio, J., Silva, A.M.S., and Carretero, J.C. (2005) *J. Org. Chem.*, **70**, 10175–10177.
64. Garnier, J.-M. and Liu, F. (2009) *Org. Biomol. Chem.*, **7**, 1272–1275.
65. (a) Kataoka, T., Iwama, T., Tsujiyama, S., Kanemastu, K., Iwamura, T., and Watanabe, S. (1999) *Chem. Lett.*, 257–258; (b) Iwama, T., Tsujiyama, S., Kinoshita, H., Kanemastu, K., Tsurukami, Y., Iwamura, T., Watanabe, S., and Kataoka, T. (1999) *Chem. Pharm. Bull.*, **47**, 956–961.
66. Walsh, L.M., Winn, C.L., and Goodman, J.M. (2002) *Tetrahedron Lett.*, **43**, 8219–8222.
67. Myers, E.L., de Vries, J.G., and Aggarwal, V.K. (2007) *Angew. Chem., Int. Ed.*, **46**, 1893–1896.
68. Pihko, P.M. (2004) *Angew. Chem., Int. Ed.*, **43**, 2062–2064.
69. Shi, M., Jiang, J.-K., and Li, C.-Q. (2002) *Tetrahedron Lett.*, **43**, 127–130.
70. (a) Imbriglio, J.E., Vasbinder, M.M., and Miller, S.J. (2003) *Org. Lett.*, **5**, 3741–3743; (b) Vasbinder, M.M.,

Imbriglio, J.E., and Miller, S.J. (2006) *Tetrahedron*, **62**, 11450–11459.
71. (a) Tang, H., Gao, P., Zhao, G., Zhou, Z., He, L., and Chu, C. (2007) *Catal. Commun.*, **8**, 1811–1814; (b) Tang, H., Zhao, G., Zhou, Z., Gao, P., He, L., and Chu, C. (2008) *Eur. J. Org. Chem.*, 126–135.
72. Utsumi, N., Zhang, H., Tanaka, F., and Barbas, C.F. III (2007) *Angew. Chem., Int. Ed.*, **46**, 1878–1880.
73. Vesely, J., Dziedzic, P., and Córdova, A. (2007) *Tetrahedron Lett.*, **48**, 6900–6904.
74. (a) McDougal, N.T. and Schaus, S.E. (2003) *J. Am. Chem. Soc.*, **125**, 12094–12095; (b) McDougal, N.T., Trevellini, W.L., Rodgen, S.A., Kliman, L.T., and Schaus, S.E. (2004) *Adv. Synth. Catal.*, **346**, 1231–1240; (c) Rodgen, S.A. and Schaus, S.E. (2006) *Angew. Chem., Int. Ed.*, **45**, 4929–4932.
75. (a) Sohtome, Y., Tanatani, A., Hashimoto, Y., and Nagasawa, K. (2004) *Tetrahedron Lett.*, **45**, 5589–5592; (b) Sohtome, Y., Takemura, N., Takagi, R., Hashimoto, Y., and Nagasawa, K. (2008) *Tetrahedron*, **64**, 9423–9429.
76. Wang, J., Li, H., Yu, X., Zu, L., and Wang, W. (2005) *Org. Lett.*, **7**, 4293–4296.
77. Berkessel, A., Roland, K., and Neudrfl, J.M. (2006) *Org. Lett.*, **8**, 4195–4198.
78. Lattanzi, A. (2007) *Synlett*, 2106–2110.
79. Shi, M. and Liu, X.-G. (2008) *Org. Lett.*, **10**, 1043–1046.
80. Yuan, K., Zhang, L., Song, H.-L., Hu, Y., and Wu, X.-Y. (2008) *Tetrahedron Lett.*, **49**, 6262–6264.
81. (a) Aggarwal, V.K., Tarver, G.J., and McCague, R. (1996) *Chem. Commun.*, 2713–2714; (b) Aggarwal, V.K., Mereu, A., Tarver, G.J., and McCague, R. (1998) *J. Org. Chem.*, **63**, 7183–7189.
82. Yang, K.-S., Lee, W.-D., Pan, J.-F., and Chen, K. (2003) *J. Org. Chem.*, **68**, 915–919.
83. Matsui, K., Takizawa, S., and Sasai, H. (2005) *Tetrahedron Lett.*, **46**, 1943–1946.
84. Pégot, B., Vo-Thanh, G., Gori, D., and Loupy, A. (2004) *Tetrahedron Lett.*, **45**, 6425–6428.
85. Gausepohl, R., Buskens, P., Kleinen, J., Bruckmann, A., Lehmann, C.W., Klankermayer, J., and Leitner, W. (2006) *Angew. Chem., Int. Ed.*, **45**, 3689–3692.
86. Garre, S., Parker, E., Ni, B., and Headley, A.D. (2008) *Org. Biomol. Chem.*, **6**, 3041–3043.
87. Roth, R., Gygax, P., and Fráter, G. (1992) *Tetrahedron Lett.*, **33**, 1045–1048.
88. Aroyan, C.E., Vasbinder, M.M., and Miller, S.J. (2005) *Org. Lett.*, **7**, 3849–3851.
89. Chen, S.-H., Hong, B.-C., Su, C.-F., and Sarshar, S. (2005) *Tetrahedron Lett.*, **46**, 8899–8903.
90. Seidel, F. and Gladysz, J.A. (2007) *Synlett*, 986–988.
91. Krishna, P.R., Kannan, V., and Sharma, G.V.M. (2004) *J. Org. Chem.*, **69**, 6467–6469.

Index

a

acidic carbonyl compounds as nucleophiles 235–238
addition/enantioselective protonation, Rh-catalyzed 34–37
– 1,4-addition/enantioselective protonation 34–37
– applications 36
– to diphenylphosphinylallenes 37
– unnatural peptides synthesis through site interconversion 37
– using guaiacol of 36
aldehyde conjugate additions 205–212
– to alkylidene malonates 212
– asymmetric organocatalytic triple cascade 212
– α-branched aldehyde additions to nitrostyrene 209
– diphenylprolinol silyl ether-catalyzed conjugate addition 207
– to enones 210
– intramolecular reaction 211
– to maleimides 211
– to nitroolefins 206–208
– tandem reaction 211
– to vinyl disulfones 209
– to vinyl sulfones 208–209
aldehydes as nucleophiles 240–241
alkenes, Rh-catalyzed ECA of 30–37
– 1,4-addition/enantioselective protonation 34–37
– alkenylphosphonates 31–32
– nitroalkenes 31–33
– sulfones 33–34
alkenylphosphonates, Rh-catalyzed ECA of 31–32
alkynes, Rh-catalyzed ECA of 44–47
– alkyne onto α,β-unsaturated ketones 46

– alkynylsilanols onto α,β-unsaturated ketones 46
– rac-alkynyl-alkenyl into chiral β-alkynyl ketones 45
aluminum–lithium complex (ALB) 177
amine-catalyzed enantioselective formal ene reaction 237
amine–thiourea catalysts 202–204
amino acid-based phosphine ligands 93–100
– in enantioselective conjugated additions 93–100
– – to enones 95
– – to N-acyloxazolidines 94
– – to nitroalkenes 96–97
– – phosphines 100–101
– – phosphites 101–107
– – to β-silyl-α,β-unsaturated ketones 97–98
– Hoveyda's 94
β-amino acids synthesis 136
amino alcohols 124–126
aminoamides 128
aminothiolates 124–127
asymmetric bifunctional catalysis 169–189
– using heterobimetallic and multimetallic systems in ECAs 169–189
asymmetric conjugated additions 132–137
asymmetric counteranion-directed catalysis (ACDC) 384
asymmetric epoxidations, 351–387, See also organocatalyzed epoxidations
– of β-alkylidene-α-tetralones 353
– of amides, mediated by Lanthanide–BINOL systems 357–362
– catalytic cycle for 355–356
– of esters, mediated by Lanthanide–BINOL systems 357–362
– postulated mechanism for 353
– of trans-enones 354–355

asymmetric epoxidations, 351–387, *See also* organocatalyzed epoxidations (*contd.*)
– of α,β-unsaturated carbonyl compounds, 386–387, *See also* metal-catalyzed epoxidations
– – Weitz–Scheffer epoxidation 352
asymmetric intermolecular Baylis–Hillman reaction 396–431
– diastereoselective 396–402
– enantioselective 402–429
asymmetric intramolecular Morita–Baylis–Hillman reaction 431–433
asymmetric organocatalysis (AO) 191
asymmetric organocatalytic triple cascade 212
axanes synthesis 137–138
aza–MBH reactions 407–412
– mechanism 414
aziridines 302

b

Baylis–Hillman reactions, 270, 393–434, *See also* asymmetric intermolecular Baylis–Hillman reaction
– reaction mechanism 393–396
– – Aggarwal's proposal for 395
– – McQuade's proposal 395
bidentate ligand rhodium complexes 20
bidentate phosphorus ligand, Rh complexes 11–13
BINOL-derivatives 423–424
Bis-sulfoxide, Rh complexes 18–19
boron enolates, trapping of 19–20
boronic acids 9–10
boron–zinc exchange, organozincs preparation by 76–77

c

carborhodation/conjugate addition, tandem Rh-catalyzed 50–53
cascade reactions 212–214, 259–287
– indolo[2,3*a*]quinolizidines synthesis 267
– initiated by conjugate Friedel–Crafts reaction 277–279
– initiated by hetero-Michael reactions 282–287
– intramolecular Morita–Baylis–Hillman process 261
– Knoevenagel/hydrogenation/Robinson annulation cascade process 261
– Michael/aldol cascade reactions 260–265

– Michael/Knoevenagel cascade reactions 266–267
– Michael/Michael cascade reactions 268–269
– Michael/Michael/aldol triple cascade reactions 270–274
– Michael/Morita–Baylis–Hilman cascade reactions 269–270
– Michael/*N*-acyliminium cyclization reaction 267–268
– Michael/α-alkylation cascade reactions 274–277
– Robinson annulations 261
– sulfa–Michael/aldol cascade for tetrahydrothiophenes synthesis 284
– sulfur-initiated thia–Michael 314–315
catalytic cycle, Pd 62–63
chiral acids 421–429
– binol-derivatives 423–424
– Lewis acids 427–429
– L-proline combined with 421–423
– thiourea 424–427
chiral cationic rhodium complexes 11
chiral reaction media 429–431
chiral sulfides 420–421
chiral tertiary amine catalysts 402–412
chiral tertiary phosphines 412–420
Clavukerins synthesis 132–133
Clavularin B synthesis 136–137
combinatorial transition-metal catalysis 15
1,4-conjugate additions
– followed by cycloaddition 344–347
– followed by hydrogenation 344–347
– followed by rearrangement 344–347
– Rh-catalyzed 345
– sparteine–allylamine complex reaction 346
1,6-conjugate additions, Rh-catalyzed 54–56
– Ir-catalyzed, competitive 55
– onto α,β,γ,δ-diunsaturated esters 54
– onto α,β,γ,δ-diunsaturated ketones 55
conjugate addition/tandem carborhodation 50–53
conjugate Friedel–Crafts reaction
– alkylations 241–245
– – LUMO-lowering effect 241
– cascade processes initiated by 277–279
– – hydrogen-transfer reaction 280–282
– – silyloxydienals and furans, reaction between 279
conjugate hydrogen-transfer reactions 245–249
– asymmetric counteranion-directed catalysis concept 247

– of enals 246
– of enones 246
conjugated additions 78–81
Cu-catalyzed conjugated additions 78–79
– enantioselective conjugated additions, ligands for, 81–123, See also amino acid-based phosphine ligands
– – [AcO(CH$_3$)$_4$]$_2$Zn addition to tert-butyl 2-methyl-6-oxocyclohex-1-enecarboxylate 123
– – alkenylalane to 3-methylcyclohexenone 92
– – alkenylalane to cycloenones 91
– – alkylaluminums to α,β-unsaturated lactams 91
– – alkylzincs N-acylpyrrolidinones 88
– – alkylzincs to ketimines addition 88
– – with biphenyl backbone 83
– – enantiopure cyclohexanones, synthesis of 86–87
– – (E)-(2-phenyl-1-propen-1-yl) dimethylaluminum addition to cyclohexenone 119
– – Et$_2$Zn addition to (E)-1-[(3-Phenyl)acryloyl]pyrrolidin-2-one 118
– – Et$_2$Zn addition to [3-[6-(tert-butyldimethylsilanyloxy) hex-2-enoyl]oxazolidin-2-one 120–121
– – Et$_2$Zn addition to 4,4-(dimethyl) cyclohexenone 121–122
– – Et$_2$Zn addition to cycloheptenone 119
– – Et$_2$Zn addition to N-[4-chlorophenyl (toluene-4-sulfonyl)methyl] formamide 117
– – i-Pr$_2$Zn addition to 1-acetylcyclopentene 120
– – Me$_3$Al addition to 5-methylpent-3-en-2-one 118
– – N-heterocyclic carbenes (NHC) compounds 107–109
– – Ph$_2$Zn addition to 3-methylcyclohexenone 122–123
– – phosphoramidites 81–93
– – with spirocyclic backbone 83
– – spirophosphoramidite ligands 89
– – taddol- and (R,R)-diaminocyclohexane-derived phosphoramidite ligand 83
– – taddol-based phosphoramidite ligands 90

– – trialkylaluminums to trisubstituted cycloenones 92
– ligands with mixed functionalities 109–113
– – peptide-based ligands 113
– – phosphino-amine ligands 111
– – sulfur-containing ligands 111–112
– – thioether ligands 112–113
– – thiophosphoramidite ligands 112–113
– of organozincs and organoaluminums to α,β-unsaturated carbonyl 71–113
– reaction mechanisms 72–73
Cu-complexes as ECAs catalysts 159–164
– 2-(trimethylsilyloxy)furans to oxazolidinone enoates 160
– aromatic C–H bonds to alkenes 162
– carbamates 163
– cyclic 1,3-dicarbonyl compounds 163
– enolsilanes to alkylidene malonates 160
– enolsilanes to fumaroyl oxazolidinones 161
– enolsilanes to β-enamidomalonates 161
– Friedel–Crafts alkylation
– – of heteroaromatic compounds 161
– – of indoles 163
– indoles to alkylidene malonate 162
(±)-β-Cuparenone, synthesis 129
cyanobisoxazolines 128–129
cyclohexene derivatives synthesis 215–216
cyclopropanation reactions 276–277

d
1,3-dicarbonyl compounds as nucleophiles 224–229
diamines 127–128
diastereoselective Baylis–Hillman reaction 396–402
diastereoselective conjugate addition 26
Diels–Alder reaction 213–214
dihydroxyacetone, conjugate addition of 202
2,6-dimethylphenol (DMP) 186
2,6-di-tert-butyl-4-methoxyphenyl (BHA) esters 146–147
dimethylsulfoxide (DMSO) 195
dinuclear Zn-complexes in catalytic ECAs 171–175
diphenylprolinol silyl ether-catalyzed conjugate addition 207
domino reactions involving ECAs 321–347
– metal-mediated domino Michael–aldol reactions 322–332
double asymmetric induction 401

e

enamides as nucleophiles 238–239
enamines in ECAs, 191–216, *See also* aldehyde conjugate additions; amine–thiourea catalysts; ketone conjugate additions
– mechanistic considerations 193–195
– – catalyst trapping 193
– – general mechanism 193
– – transition states 194
– Michael/Michael/aldol condensation sequence 213–214
– tandem or cascade reactions 212–214
enantiopure cyclohexanones, synthesis of 86–87
enantioselective Baylis–Hillman reaction 402–429
– BINOL-derived bifunctional phosphine catalysts 415
– chiral acids 421–429
– chiral sulfides 420–421
– chiral tertiary amine catalysts 402–412
– chiral tertiary phosphines 412–420
enantioselective conjugate additions (ECAs), 145–164, *See also under* Cu-complexes as ECAs catalysts; Grignard reagents; lithium reagents
erogorgiaene synthesis 134

f

Friedel–Crafts/chlorination cascade reaction, 278, *See also* conjugate Friedel–Crafts reaction
fumarates 26–29

g

gallium–sodium complex (GASB) 177
Grignard reagents, ECAs of 152–159
– Cu-catalyzed ACA using chiral ligands 153
– – for enantioselective conjugate additions 159–164
– – to trisubstituted enones 158
– – to α,β-unsaturated esters 156–158
– – to α,β-unsaturated sulfones 159
– – to α,β-unsaturated thioesters 156
– ligands for 152
guanacastepene intermediate synthesis 130–131
guanidine-based catalysts 376–377

h

1,1,1,3,3,3-hexafluoroisopropyl acrylate (HFIPA) 404

Hantzsch esters 245, 247–248, 280
heteroatoms to α,β-unsaturated carbonyl compounds 295–317
– intramolecular aza–Michael reactions 299–300
– organocatalytic ECAs, 295–317, *See also* intermolecular aza–Michael reactions
heterobimetallic systems in ECAs, asymmetric bifunctional catalysis using 169–189
– dinuclear Zn-complexes 171–175
– rare-earth–alkali metal-binol complexes 175–179
hetero-Michael reactions, cascade processes initiated by 282–287
– aminosulfenylation of α,β-unsaturated aldehydes 283
heteronucleophiles, conjugate additions of 249–259
– aza–Michael reaction of phenyltetrazole to enals 253
– aza–Michael reaction to enones 254
– intramolecular aza–Michael reaction 253
– nitrogen heterocycles 251
– N-nucleophiles 250–255
– O-nucleophiles 256–258
– phospha–Michael reaction 255
– P-nucleophiles 255–256
hydrogen-transfer reaction, cascade processes initiated by 280–282
hydroquinidine-2,5-diphenyl-4,6-pyrimidinediyl diether [(DHQD)$_2$PYR] 311

i

(+)-ibuprofen synthesis 133–134
imidazolidinone salt catalysts 381–382
iminium activation in ECAs, 219–287, *See also* cascade reactions; conjugate Friedel-Crafts alkylations; conjugate hydrogen-transfer reactions; Michael reactions
– Diels–Alder cycloadditions 221–222
– Lewis acid-mediated transformations 221
– mechanistic aspects 220–224
– primary versus secondary amine catalysts in 223
– stereochemical issues in 223
intermolecular aza–Michael reactions 295–299
– 1,2,3-benzotriazole to enone 299
– benzylhydroxylamine to α,β-unsaturated pyrazoles 297
– hydrazones to cyclic enones 297
– N-benzyloxyamine to enones 297

- N-silyloxycarbamates to enals 296
- pyrazoles to α, β-unsaturated carbonyls 299
- succinimides and N-heterocycles to enals 298
- TMSN$_3$ to α, β-unsaturated imides 296
intermolecular oxa–Michael reactions 304–306
- enals epoxidation 306
- enals reaction with alcohols 304
- enals reaction with oximes 305
- enones epoxidation 306
- enones reaction with oximes 305
intermolecular thia–Michael reactions 310–314
- 4-hydroxypyrrolidine derivative-catalyzed 311
- alkylthiols to α, β-unsaturated oxazolidinones 312
- arylthiols to α, β-unsaturated carbonyls 312
- hydroquinidine-2,5-diphenyl-4,6-pyrimidinediyl diether [(DHQD)$_2$PYR] 311
- sulfur-centered nucleophiles 312
- thioacetic acid addition to enones 312
- thiols to tert-butyl 2-phthalimidoacrylates 313
- thiols to α, β-unsaturated aldehydes 313
intramolecular aza–Michael reactions 299–300
- diphenylprolinol-catalyzed aza-[3+3] cycloadditions 300
- enantioselective organocatalytic 300
intramolecular Morita–Baylis–Hillman process 261
intramolecular oxa–Michael reactions 306–307
intramolecular reductive domino Michael–aldol reaction 327–328
iodine–zinc exchange reaction, organozincs preparation by 76

j
Juliá–Colonna epoxidation 371
Julia–Kociensky olefination 33

k
ketone conjugate additions 195–205
- of aldehydes 215
- to alkylidene malonates 205
- amine–thiourea catalysts 202–204
- of dihydroxyacetone 202
- of nitroalkenes 215
- to nitroolefins 195–202
- – asymmetric acyclic ketones, alkylation 197
- – Michael addition 199–200
- – pyrrolidine–pyridine-based catalyst 198
- – secondary amines 196–202
- – supramolecular assembled organocatalyst 200
- – syn:anti diastereoselectivity of 198
- – Zhao's self-assembled organocatalyst 199
- to nitrostyrenes 215
- primary amines 202
- to α,β-unsaturated carbonyl compounds 204
- to vinylphosphonates 205
- vinylphosphonates as acceptors 204
Knoevenagel/hydrogenation/Robinson annulation cascade process 261

l
lanthanide–BINOL Systems 357–362
- mediating amides epoxidation 357–362
- mediating esters epoxidation 357–362
- mediating α,β-Unsaturated Ketones epoxidation 357–362
leaf miner pheromones synthesis 135–136
Lewis acid/Lewis acid catalysis 181, 427–429
ligand systems, Rh complexes 11
lithium reagents, ECAs of 145–152
- aryllithium to naphthalene BHA-esters 146
- aryllithium to α,β-unsaturated tert-butyl esters 147
- to BHA-esters 147
- ligands for 146
- stannylation/lithiation sequence use 149
- to α,β-unsaturated aldimines 146
lithium trimethylarylborate 10
lithium–zinc transmetallation, organozincs preparation by 75–76
l-proline combined with achiral or chiral amines 421–423

m
MacMillan catalyst 243–244
- recyclable chiral catalyst 248
maleimides 26–29
Mannich reaction/tandem ECA 49–50
metal-catalyzed epoxidations 352–362
- of α,β-unsaturated ketones 352–357
- – mediated by chirally modified Zn- and Mg-alkyl peroxides 352–357

metal-mediated domino Michael–aldol reactions, 322–332, *See also* organocatalytic domino Michael reactions/electrophilic trapping
– catalytic cycles 326, 330
– electrophile trapping with noncarbonyl compounds 332–335
– – addition–silylation–cyclopropanation reaction 335
– – carbonyl analogs 334
– intramolecular reductive 327–328
– Ir-catalyzed reductive 331
– Mg-Box-catalyzed 331
– Rhodium-catalyzed 328–329
– stereochemical model 330
– Zimmerman–Traxler transition-state model 324
– Zimmerman–Traxler-type transition state 330
Michael cyclization/tandem conjugate addition 50
Michael reactions/additions 193, 224–241
– 1,3-dicarbonyl compounds as nucleophiles 224–229
– 2-hydroxy-1,4-naphtoquinones addition with enals 229
– acidic carbonyl compounds as nucleophiles 235–238
– aldehydes as nucleophiles 240–241
– amine-catalyzed enantioselective formal ene reaction 237
– diastereoselectivity problem 233
– facial selectivity in 225
– malonates addition to enals 228
– malonates addition to enones 226
– nitroalkanes as nucleophiles 230–234
– nitroalkanes to cyclic enones 230–231
– nitroethane addition to enals 232
– (S)-rolipram synthesis 234
– silyl enol ethers and enamides as nucleophiles 238–239
– stereocontrol in, factors influencing 220–224
– α,β-unsaturated aldehydes as Michael acceptors 231
– vinylogous Michael addition 235
Michael/aldol cascade reactions 260–265
– cyclohexenes synthesis 265
– cyclopentenes synthesis 265
– nitro-Michael/Henry reaction 264
– *tert*-butyl β-ketoesters 262
– α,β-unsaturated aldehydes 263
Michael/Henry reaction 212

Michael/Knoevenagel cascade reactions 266–267
– bicyclo[3.3.1]non-2-enes synthesis 266
– cyclohexanols synthesis 266
– cyclohexenones synthesis 266
Michael/Michael cascade reactions 268–269
– cyclopentanes synthesis 268
– vinylogous 269
Michael/Michael/aldol condensation sequence 213–214
Michael/Michael/aldol triple cascade reactions 270–274
– followed by one-pot intramolecular Diels–Alder reaction 272
Michael/Morita–Baylis–Hilman cascade reactions 269–270
Michael/N-acyliminium cyclization reaction 267–268
Michael/α-alkylation cascade reactions 274–277
– amine-catalyzed 275
– cyclopentanecarbaldehydes synthesis 276–277
– cyclopropanecarbaldehdyes synthesis 276
mixed donor ligands, Rh complexes 19
Mizoroki–Heck-type product 41
monodentate ligand, Rh complexes 14–16
Morita–Baylis–Hillman (MBH) reaction, *See* Baylis–Hillman reactions
Mukaiyama–Michael reaction 238–239
multimetallic systems in ECAs, asymmetric bifunctional catalysis using 169–189
muscone synthesis 133

n

'*N*' as nucleophiles 295–304
– intermolecular aza–Michael reactions, 295–299, *See also individual entry*
– nitrogen initiated aza–Michael cascade reactions, 300–304, *See also individual entry*
– pyrazoles as 298
N-heterocyclic carbenes (NHC) compounds 107–109
N-nucleophiles 250–255
natural compounds synthesis
– conjugated additions application in 129–137
– – β-amino acids synthesis 136
– – axanes 137–138
– – clavukerins synthesis 132–133
– – clavularin B synthesis 136–137
– – (±)-β-Cuparenone, synthesis 129
– – erogorgiaene synthesis 134

– – guanacastepene intermediate synthesis 130–131
– – (+)-ibuprofen synthesis 133–134
– – leaf miner pheromones synthesis 135–136
– – muscone synthesis 133
– – non-asymmetric 129–132
– – Phthiocerol synthesis 135
– – prostaglandins 131
– – (−)-Pumiliotoxin C synthesis 134–135
– – (±)-scopadulcic acid, synthesis 131–132
Ni-catalyzed conjugated additions 79–81
– arylaluminums to unsaturated steroids 80
– ligands for 124–129
– – amino alcohols 124–126
– – Aminoamides 128
– – aminothiolates 124–127
– – cyanobisoxazolines 128–129
– – diamines 127–128
– – pyridino-alcohols 127
– – Sulfoximines 128
– – thioethers 126–127
– of organozincs and organoaluminums to α, β-unsaturated carbonyl 71–113
– reaction mechanisms 73–74
– to steroid substrates 79
– vinylaluminum compounds to chalcone 80
nitroalkanes as nucleophiles 230–234
nitroalkenes, Rh-catalyzed ECA of 31–33
nitrogen heterocycles 251
nitrogen initiated aza–Michael cascade reactions 300–304
– (S)-proline-catalyzed 301
– organocatalytic enantioselective 301–302
– – aminoxylation/aza–Michael reactions 303
– – aza–Michael–hemiacetal reaction 302
– – aziridination of enals 302
– – enals reactions with N-(benzylthio)succinimide 303
– – with enones 303
nitro-Michael/Henry reaction 264
nitroolefins, aldehyde conjugate additions to 206–208
nitroolefins, ketone conjugate additions to 195–202
non-asymmetric conjugated additions application 129–132

O

'O' as nucleophiles 256–258, 304–310
– enantioselective hydroperoxidation of enones 258
– intermolecular oxa–Michael reactions 304–306
Oppolzer camphor 399
optically pure alkyl hydroperoxides or ligands 373–376
organoaluminum compounds 71–138
– conjugated additions to α, β-unsaturated carbonyl, Cu- and Ni-catalyzed 71–113
– preparation of 77–78
– properties of 72
organoboron reagents, Rh-catalyzed ECA of 1–37
– alkenes 30–37
– α,β-unsaturated aldehydes 20–22
– α,β-unsaturated ketones, 2–22, See also individual entry
organocatalytic domino Michael reactions/electrophilic trapping 335–344
– Diels–Alder products via 337
– α, β-unsaturated ketones reaction with β-ketoesters 339
organocatalytic ECAs 295–317
– of heteroatoms to α,β-unsaturated carbonyl compounds 295–317
organocatalyzed epoxidations 362–385
– guanidine-based catalysts 376–377
– imidazolidinone salt catalysts 381–382
– optically pure alkyl hydroperoxides or ligands 373–376
– phase-transfer catalysts 362–368
– polyamino acids 369–373
– – Juliá–Colonna epoxidation 371
– – with PLL/UHP system 372
– – trans-enones with triphasic system 369
– primary amines 382–385
– pyrrolidine-based catalysts 377–381
organosilicon reagents, Rh-catalyzed ECA of 41–44
– arylsilanol to acrylates 41
– onto α,β-unsaturated carbonyls 42
– with (pin)B-SiMe$_2$Ph 44
– tetraorganosilicon reagent to α,β-unsaturated carbonyls 43
– tetraorganosilicon reagents, regeneration 43
organotitanium reagents, Rh-catalyzed ECA 38–41
– phenyltitanium reagent onto 2-cyclohexenone 38

organozinc reagents, Rh-catalyzed ECA 38–41
– arylzinc onto piperidones 40
– arylzinc/ClSiMe$_3$ onto enals 40
– heteroaromatic zinc reagents to α,β-unsaturated ketones and esters 41
organozincs compounds 71–138
– commercial availability 77
– conjugated additions to α,β-unsaturated carbonyl, Cu- and Ni-catalyzed 71–113
– preparation 74–77
– – boron–zinc exchange 76–77
– – by direct insertion into carbon–halide bond 74–75
– – by iodine–zinc exchange reaction 76
– – by lithium–zinc transmetallation 75–76
– – by other metal–zinc exchange 77
– properties of 72
oxygen-initiated oxa–Michael cascade reactions 307–310
– boronic acids to enones 308
– cinchona alkaloids-catalyzed 310
– diarylpyrrolidinol silyl ether-catalyzed 309
– diphenylpyrrolidinol silyl ether-catalyzed 308
oxygenophilicity 72

p

'*P*' as nucleophiles 316–317
– hydrophosphination addition to enals 316
– trimethyl phosphite addition to enals 316
Palladium catalyzed asymmetric conjugate additions 56–63
– of arylboronic acids
– – onto α,β-unsaturated carbonyls 61
– – to β-arylenones 59
– catalytic cycle 62–63
– cationic Pd/dppe complex 58
– of organoboronic acids onto β-aryl enals 60
– onto α,β-unsaturated ketones 58
– organoboronic acids 57
– – conjugate addition 62
– – to α,β-unsaturated esters and amides 60
– tandem Pd-catalyzed ECA/aldol condensation 59
peptide-based ligands 113

phase-transfer catalysts 362–368
phospha–Michael reaction 255
phosphines 100–101
phosphites 101–107
phosphoramidites 81–93
phthiocerol synthesis 135
P-nucleophiles 255–256
polyamino acids, organocatalyzed epoxidations in 369–373
potassium aryltrifluoroborate salts 10
potassium vinyltrifluoroborate 10
precatalysts, Rh 10–11
primary amines
– ECAs in 202
– organocatalyzed epoxidation 382–385
prostaglandins (PGs), synthesis 130–132
(−)-Pumiliotoxin C synthesis 134–135
pyridino-alcohols 127
pyrrolidine-based catalysts 377–381

r

rare-earth–alkali metal-binol complexes in catalytic ECAs 175–179
– heteroatom nucleophiles 179–184
– – Shibasaki's epoxidation 183
Rauhut–Currier reactions 434
retro-aza–Michael reaction 270
Rh-catalyzed tandem processes, 47–53, See also tandem processes, Rh-catalyzed
Rhodium catalyzed asymmetric conjugate additions, 1–64, See also organoboron reagents, Rh-catalyzed ECA of
Rhodium-catalyzed reductive aldol reaction sequence 328–329
Robinson annulations 261

s

'*S*' as nucleophiles 310–315
– intermolecular thia–Michael reactions 310–314
(±)-scopadulcic acid, synthesis 131–132
secondary amines 196–202
Shibasaki's epoxidation 183
silyl enol ethers as nucleophiles 238–239
singly occupied molecular orbital (SOMO) catalysis 191–192
S-nucleophiles 258–259
spirophosphoramidite ligands 89
sulfa–Michael/aldol cascade for tetrahydrothiophenes synthesis 284
sulfones 33–34
sulfoximines 128
sulfur-initiated thia–Michael cascade reactions 314–315

– between enals and
  2-mercapto-1-phenylethanone 315
– hydrogen bond-mediated 315
synthetically useful acceptors for
  Rh-catalyzed ECAs 29–30

t

tandem processes, Rh-catalyzed 47–53
– conjugate addition/1,2-addition 48–49
– – onto an imine function 50
– intramolecular tandem ECA/aldol
  reaction 49
– tandem carborhodation/conjugate
  addition 50–53
– tandem conjugate addition/Michael
  cyclization 50
– tandem ECA/Mannich reaction 49–50
tandem reactions 211–214
– Pd-catalyzed ECA/aldol
  condensation 59
'template effect' 373
tetrahydrothiophenes synthesis,
  sulfa–Michael/aldol cascade for 284
thioethers 112–113, 126–127
thiophosphoramidite ligands 112–113
thiourea 424–427
transmetallation step in Rhodium catalyzed
  additions 6–7
tributylsilyl cyanide (TBSCN) 186
trichloroisocyanuric acid (TCCA) 364
trimethylsilyl cyanide (TMSCN) 185
triple Michael/Michael/aldol cascade
  273–274

u

α,β-unsaturated aldehydes 20–22
– arylboronic acids to enals Rh-catalyzed
  conjugate addition 21–22
α,β-unsaturated amides, Rh-catalyzed
  ECA on 22–30
– diastereoselective conjugate addition 26
α,β-unsaturated carbonyl compounds
– asymmetric epoxidations of, 351–387,
  See also under asymmetric
  epoxidations

– ketone conjugate additions to 204
α,β-unsaturated esters, Rh-catalyzed
  ECA on 22–30
– diastereoselective conjugate
  addition 26
– fumarates 26–29
– maleimides 26–29
– phenylboronic acid onto di-*tert*-butyl
  fumarate 27
– synthetically useful acceptors for
  Rh-catalyzed ECAs 29–30
α,β-unsaturated ketones 2–20
– bidentate phosphorus ligand 11–13
– Bis-sulfoxide 18–19
– boron enolates, trapping of 19–20
– catalytic cycle 4–7
– – catalyzed by a hydroxyrhodium
  complex 4–5
– – transmetallation step 6–7
– catalyzed by [Rh(acac)(C$_2$H$_4$)]/(*S*)-binap
  3–4
– chiral cationic rhodium complexes 11
– *cis* olefins, stereocontrol model for 8
– diene ligands 16–18
– enantioselection model 7–9
– ligand systems 11
– mixed donor ligands 19
– monodentate ligand 14–16
– organoboron sources other than boronic
  acids 9–10
– Rh precatalysts 10–11
– stereocontrol model 7
– *trans* olefins, stereocontrol model for 8

v

vinyl sulfones, aldehyde conjugate
  additions to 208–209
vinylogous Michael/Michael reaction 269

w

Weitz–Scheffer epoxidation 352, 359

z

Zimmerman–Traxler-type transition state
  324, 330